Quantum Theory of Magnetism

Wolfgang Nolting · Anupuru Ramakanth

Quantum Theory of Magnetism

Springer

Prof. Dr. Wolfgang Nolting
Humboldt-Universität zu Berlin
Institut für Physik
Newtonstr. 15
12489 Berlin
Germany
wolfgang.nolting@physik.hu-berlin.de

Prof. Dr. Anupuru Ramakanth
Kakatiya University
Department of Physics
Warangal-506009
India

ISBN 978-3-540-85415-9 e-ISBN 978-3-540-85416-6
DOI 10.1007/978-3-540-85416-6
Springer Heidelberg Dordrecht London New York

Library of Congress Control Number: 2009927326

Cover design: eStudio Calamar S.L.

Printed on acid-free paper

Springer is part of Springer Science+Business Media (www.springer.com)

Preface

In the literature, theory of magnetism is available in books at two levels. At the beginners level it is dealt as a chapter in books on solid state physics. There certain basic facts about magnetism are presented and some of the theoretical ideas are mentioned almost in passing. If one wants to persue the subject in depth then one has to refer to the books at the higher level. At this level there already do exist excellent books which have either a large canvas or limit the area but go to considerable depth. These books, however, are useful to those who already are practitioners of theoretical research in magnetism and are not exactly suitable for those who want to prepare themselves for research. In other words, we find that to our knowledge, there does not exist a book on the quantum theory of magnetism which serves as a text book and also one which helps and guides one in self-study. That is, a book where every step is worked out in detail and also contains a number of problems which help in self-assessment and also which supplement the material dealt with in the text of the book. It is precisely to fill this gap, we have attempted to write this book. This need was felt by one of the authors (WN) while giving special courses to advanced students almost two decades ago. As a result he published a book with exactly the same title but in German. The book was well received and many a reader has been asking since then if there is a revised edition. As a result, the present book is an english, thoroughly revised and updated version of the original German version. The book presumes the reader to have certain basic understanding of the concepts and techniques of quantum mechanics and statistical mechanics. Except for that, the book is, we hope, self-contained in the sense that every single step has been worked out both in the main text and also in the solutions to the problems. Some of the problems have rather long solutions. Then, these represent the results which most often are assumed to be well known but in fact they need to be worked out somewhere. Some of the problems are such that the solutions impart considerable training to one who wants to start out on research and helps him in learning certain "standard" tricks in order to understand some seminal papers and also in implementing his own research programme.

The first chapter introduces, starting from Maxwell's equations, certain basic facts about magnetism such as magnetic moment, magnetization and susceptibility and also contains a section on thermodynamics as applicable to magnetic systems. If one wants to understand magnetism of materials, one has to be first clear about the magnetism of individual atoms. This purpose is served by Chap. 2 (Atomic magnetism), in which all the important magnetic properties of atoms are discussed. We derive the electron spin and spin–orbit interaction from the relativistic Dirac equation and investigate the behaviour of an atomic electron in the field of the nucleus and also in the presence of an external magnetic field. The third chapter is devoted to "diamagnetism" which is a property possessed by *all* materials, which, however, is observable only when it is not overwhelmed by some other forms of magnetism (para-, ferro-, ferri- or antiferromagnetism). Diamagnetism in some measure can be explained as an induction effect, particularly because of the negative susceptibility. Chapter 4 deals with paramagnetism, which in contrast to diamagnetism, presumes the existence of permanent magnetic moments. These moments can either be localized stemming from partially filled electron shells of the ions in solid or they can be the moments of the quasi-freely moving band electrons. An external magnetic field tries to order them, whereas the thermal motion opposes the ordering tendency. The result is a positive susceptibility which in general is temperature dependent. Paramagnets are characterized by the fact that a direct interaction between the permanent moments, to a good approximation, can be neglected. In contrast to this, collective magetism (ferro, ferri and antiferromagnetism) is characterized by a spontaneous ordering of the magnetic moments below a critical temperature, and therefore a necessary precondition for this is the existence of a microscopic interaction between the moments. The so-called exchange interaction even though has its origin in pure electrostatic interaction cannot be understood from classical point of view. The general experience is that for a beginner this poses certain conceptual difficulties but, at the same time, it is the basis of understanding collective magnetism. Therefore exchange interaction has been discussed in considerable detail in Chap. 5. The so-called *direct* exchange interaction is determined via the overlap integrals of the wavefunctions of the participating magnetic ions. As a result it is of very short range and therefore is seldom realized in nature as compared to certain *indirect* exchange interactions which use the electrons in the conduction band (RKKY interaction) or the diamagnetic ions (superexchange, double exchange) as "catalysts" for an interaction between the localized moments. The coupling mechanisms are explained using simple cluster models and it is shown that ultimately all the interactions have the same operator form (Heisenberg model).

Having established the required conceptual basis, the last three chapters are devoted to the three important models of magnetism, namely, the Ising, the Heisenberg and the Hubbard model, respectively. In these chapters, an attempt has been made to present material such that the approach is pedagogic and at the same time gives the latest results available in literature. In doing this care is taken to derive all the results systematically and in every detail. Some of the important derivations are treated as problems whose complete solutions are given. While discussing quantum theory of magnetism it is imperative that one uses the techniques of many-body theory. In order to famialiarize the reader with these techniques, two appendices are added. The first one deals with the formalism of second quantization where all the results are worked out and to provide sufficient training to the self-learner, a set of problems is added. The second apendix is concerned with the many-body theory.

The topics in this appendix are so chosen that they are directly relevant to the theory of magnetism. Again in this appendix, too, problems are provided, some of which elucidate certain further results which are left out in the actual text.

The preperation of this book took about 3 years during which the authors were able to get together at either Berlin or Warangal for short periods, the financial support for which was provided by the Volkswagen Foundation and Kakatiya University. It is a pleasure to acknowledge the help of Dr. G. Gangadhar Reddy in various forms during the entire period of writing the book.

Berlin, Germany
Warangal, India
July 2009

W. Nolting
A. Ramakanth

Contents

Chapter 1
Basic Facts

Before we start the explanation and discussion of the characteristic features of the phenomenon "magnetism", we have to fix the notation and define the key terms of magnetism. We use the SI units and begin with a non-controversial definition of quantities such as the magnetic moment, magnetization and magnetic susceptibility, which are important for the theory of magnetism.

1.1 Macroscopic Maxwell Equations

Magnetism is a phenomenon observed in matter. Therefore we need the Maxwell's equations in matter. Matter is made up of charged or neutral particles, which are either bound or quasi-free. The charged particles respond to external fields in a complex manner which leads to induced multipoles and consequently to additional fields in matter.

Postulate: *Maxwell's equations of vacuum are universally valid microscopically!* Denoting the microscopic fields with small letters, using the customary notation, we have the well-known *microscopic Maxwell's equations*:

$$\begin{aligned} \nabla \times \mathbf{e} &= -\dot{\mathbf{b}} \qquad & \nabla \cdot \mathbf{b} &= 0 \\ \nabla \cdot \mathbf{e} &= \rho/\varepsilon_0 \qquad & \nabla \times \mathbf{b} &= \mu_0\, \mathbf{j} + \varepsilon_0\, \mu_0\, \dot{\mathbf{e}} \end{aligned} \tag{1.1}$$

$$\begin{aligned} \varepsilon_0 &= 8.854188 \times 10^{-12}\, \tfrac{As}{Vm} \\ \mu_0 &= 4\pi \times 10^{-7}\, \tfrac{Vs}{A \cdot m} \\ c^2 &= \tfrac{1}{\mu_0\, \varepsilon_0} \end{aligned} \tag{1.2}$$

here ε_0 is the permittivity, μ_0 the free-space permeability, and c is the velocity of light in vacuum. The problem becomes unsolvable because of the fact that, in matter, there are, on an average about 10^{23} molecular (atomic, subatomic) particles per cm^3 which are in constant motion (lattice vibrations, orbital motion of the atomic electrons, etc.) creating both in space and time rapidly oscillating

W. Nolting, A. Ramakanth, *Quantum Theory of Magnetism*,
DOI 10.1007/978-3-540-85416-6_1, © Springer-Verlag Berlin Heidelberg 2009

fields **e** and **b** whose determination appears to be a hopeless undertaking. On the other hand, a macroscopic measurement means a rough sampling of a microscopic area. That means, the measurement of a field quantity automatically implies an averaging over a finite space–time element which *smoothens* the rapid fluctuations. Typical variations in space are of the order of 1 Å(10^{-10} m) and typical variations in time are of the order of 10^{-17} s (nucleons) and of 10^{-13} s (atomic electrons).

Therefore, a theory is meaningful only for average quantities. A microscopically exact theory is impossible but, fortunately, it is also not necessary since it would contain superfluous information which is not accessible to experiment. Now the question is how does one theoretically describe the averaging process involved in the experiment?

Let us assume

$f(\mathbf{r}, t)$ is a microscopic field quantity,
$v(\mathbf{r})$ is a microscopically large and macroscopically small sphere with centre at $\mathbf{r}$, for example, a volume $10^{-6}\,\text{cm}^3$ which still has about 10^{17} particles.

Defintion: *Phenomenological average value:*

$$\overline{f(\mathbf{r}, t)} = \frac{1}{v(\mathbf{r})} \int_{v(\mathbf{r})} d^3r'\, f(\mathbf{r}', t) \tag{1.3}$$

In view of the large number of particles in the macroscopic volume $v(\mathbf{r})$, the average is simultaneously over space and time. Fast microscopic fluctuations are automatically smoothened by the space averaging. It should be noted that (1.3) is *not* the only possible way of averaging; fortunately this averaging does not require the specifying of the weight function (here it is $1/v$).

For the following discussion, we make use of the fundamental assumptions that the processes of differentiation and averaging are interchangeable:

$$\frac{\partial}{\partial t}\bar{f} = \overline{\frac{\partial f}{\partial t}} \quad ; \quad \nabla \bar{f} = \overline{\nabla f} \tag{1.4}$$

This is obviously satisfied by the averaging process (1.3).

We now define *macroscopic fields* by averaging the microscopic fields,

$$\mathbf{E}(\mathbf{r}, t) = \overline{\mathbf{e}(\mathbf{r}, t)} \quad ; \quad \mathbf{B}(\mathbf{r}, t) = \overline{\mathbf{b}(\mathbf{r}, t)} \tag{1.5}$$

so that we obtain the *macroscopic Maxwell's equations*

$$\begin{aligned} \nabla \times \mathbf{E} &= -\dot{\mathbf{B}} \qquad & \nabla \cdot \mathbf{B} &= 0 \\ \nabla \cdot \mathbf{E} &= \bar{\rho}/\varepsilon_0 \qquad & \nabla \times \mathbf{B} &= \mu_0\, \bar{\mathbf{j}} + \varepsilon_0\, \mu_0\, \dot{\mathbf{E}} \end{aligned} \tag{1.6}$$

If we use the *continuity equation*,

$$\nabla \cdot \bar{\mathbf{j}} + \dot{\bar{\rho}} = 0 \tag{1.7}$$

the quantity still to be determined is the average current density $\bar{\mathbf{j}}$. It is made up of two terms, namely, $\mathbf{j}_f$ originating from the free charge carriers (electrons) and $\mathbf{j}_{bound}$ originating from the localized ions:

$$\bar{\mathbf{j}} = \mathbf{j}_f + \mathbf{j}_{bound} \tag{1.8}$$

For the contribution of the free charges, we write

$$\mathbf{j}_f = \overline{\rho_f \cdot \mathbf{v}} \tag{1.9}$$

which we call *free current density*, where ρ_f is the respective charge density of the free charges.

The current density due to ions, which we call the *bound current density*, can be again split into two terms:

$$\mathbf{j}_{bound} = \mathbf{j}_p + \mathbf{j}_m \tag{1.10}$$

(1) $\mathbf{j}_p$ is the *current density of polarization charges* which results from the time-dependent dipole moments, charge displacements in ions, etc. In order to determine this, we start by recapitulating certain results of basic electrodynamics.

Potential of the jth ion:

$$4\pi\varepsilon_0\,\phi_j(\mathbf{r}) = \frac{q_j}{|\mathbf{r}-\mathbf{R}_j|} + \mathbf{P}_j\,\frac{\mathbf{r}-\mathbf{R}_j}{|\mathbf{r}-\mathbf{R}_j|^3} + \cdots. \tag{1.11}$$

where $\mathbf{R}_j$ is the position vector, q_j the charge and $\mathbf{P}_j$ the dipole moment of the jth ion (Fig. 1.1).

The contribution of (1.11) must be summed over all particles. Here, we are interested only in the second summand which we reformulate in terms of the *microscopic dipole density*,

$$\Pi(\mathbf{r}) = \sum_j \mathbf{P}_j \cdot \delta(\mathbf{r}-\mathbf{R}_j) \tag{1.12}$$

and obtain

$$4\pi\varepsilon_0\,\phi_p(\mathbf{r}) = \int d^3r'\,\Pi(\mathbf{r}')\cdot\frac{\mathbf{r}-\mathbf{r}'}{|\mathbf{r}-\mathbf{r}'|^3} = \int d^3r'\,\Pi(\mathbf{r}').\,\nabla_{\mathbf{r}'}\,\frac{1}{|\mathbf{r}-\mathbf{r}'|}.$$

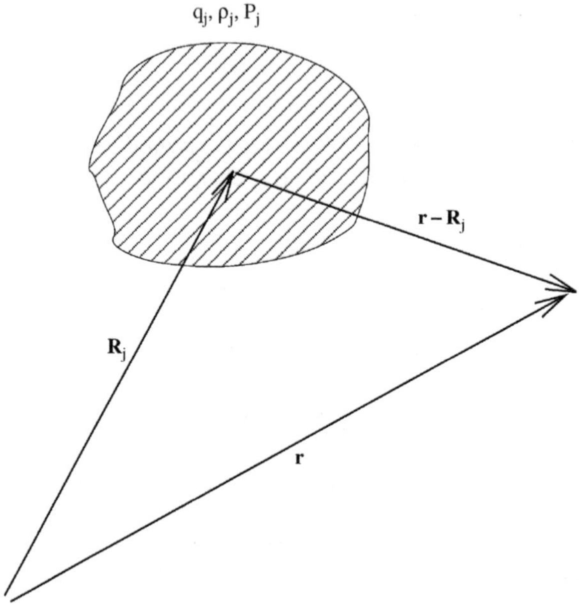

Fig. 1.1 Schematic plot for the definition of the scalar ion-potential.

From this it follows that

$$\begin{aligned} 4\pi\varepsilon_0 \cdot \overline{\phi_p}(\mathbf{r}) &= \frac{1}{v}\int_{|x|\le R} d^3\,x \int d^3r'\; \Pi(\mathbf{r}')\; \nabla_{\mathbf{r}'}\frac{1}{|\mathbf{r}+\mathbf{x}-\mathbf{r}'|} \\ &= \frac{1}{v}\int d^3\,x \int d^3\,r''\; \Pi(\mathbf{r}''+\mathbf{x})\; \nabla_{\mathbf{r}''}\frac{1}{|\mathbf{r}-\mathbf{r}''|} \\ &= \int d^3r''\; \overline{\Pi(\mathbf{r}'')}\; \nabla_{\mathbf{r}''}\frac{1}{|\mathbf{r}-\mathbf{r}''|}. \end{aligned} \tag{1.13}$$

where $\mathbf{r}'' = \mathbf{r}' - \mathbf{x}$. $v(\mathbf{r})$ is the averaging volume which is taken as a sphere with radius r.

We define the *macroscopic polarization* $\mathbf{P}(\mathbf{r})$ as

$$\mathbf{P}(\mathbf{r}) = \overline{\Pi(\mathbf{r})} = \frac{1}{v}\sum_{j\in v} \mathbf{P}_j \tag{1.14}$$

From (1.13), by taking gradient, we obtain the electric field $\mathbf{E}_p(\mathbf{r})$ created by the polarization:

$$4\pi\varepsilon_0\, \mathbf{E}_p(\mathbf{r}) = -\nabla_r \int d^3r''\mathbf{P}(\mathbf{r}'')\; \nabla_{\mathbf{r}''}\, \frac{1}{|\mathbf{r}-\mathbf{r}''|} \tag{1.15}$$

which can further be rewritten as

$$4\pi\varepsilon_0 \nabla \cdot \mathbf{E}_p(\mathbf{r}) = -\int d^3r''\mathbf{P}(\mathbf{r}'') \nabla_{\mathbf{r}''} \underbrace{\Delta_{\mathbf{r}} \frac{1}{|\mathbf{r}-\mathbf{r}''|}}_{-4\pi\delta(\mathbf{r}-\mathbf{r}'')}$$

$$= -4\pi \nabla_{\mathbf{r}} \int d^3r''\mathbf{P}(\mathbf{r}'')\,\delta(\mathbf{r}-\mathbf{r}'')$$

$$= -4\pi \nabla \cdot \mathbf{P}(\mathbf{r})$$

This helps us in defining the *polarization density* $\overline{\rho_p}$ as

$$\overline{\rho_p} = -\nabla \cdot \mathbf{P}(\mathbf{r}) \tag{1.16}$$

From the corresponding continuity equation, we get the current density

$$\overline{\mathbf{j}_p} = \frac{\partial}{\partial t}\mathbf{P}(\mathbf{r}) \tag{1.17}$$

(2) $\overline{\mathbf{j}_m}$ is the *magnetization current density*. It results from the internal motion of the atomic electrons in their stationary orbits. Here stationary means equal amount of charge flows into and out of a given volume:

$$\int_{\partial V} d\mathbf{f} \cdot \mathbf{j}_m = \int_V d^3r\, \nabla \cdot \mathbf{j}_m = 0. \tag{1.18}$$

where $\int_{\partial V}$ indicates integral over the surface that encloses the volume V. Since the volume V can be arbitrarily chosen, (1.18) is equivalent to

$$\nabla \cdot \mathbf{j}_m = 0 \tag{1.19}$$

For magnetic considerations, $\mathbf{j}_m$ is the primarily important quantity, since it defines the magnetic moment $\mathbf{m}_i$ of the ion localized at $\mathbf{R}_i$:

$$\mathbf{m}_i = \frac{1}{2}\int d^3r\, (\mathbf{r}-\mathbf{R}_i) \times \mathbf{j}_m^{(i)} \tag{1.20}$$

The index i indicates the lattice point.

It is instructive to evaluate (1.20) for a known example. Assuming that the magnetic moment is affected only by the electrons (assumed to be point charges) and not by the motion of the nucleus, we can write for the current density

$$\mathbf{j}_m = -e\sum_{j=1}^{p} \mathbf{v}_j\, \delta(\mathbf{r}-\mathbf{r}_j). \qquad (\mathbf{R}_i = 0) \tag{1.21}$$

The summation is over the p atomic electrons. Substituting in (1.20) we get the moment

$$\mathbf{m} = -\frac{e}{2}\sum_j \mathbf{r}_j \times \mathbf{v}_j = -\frac{e}{2m}\sum_j \mathbf{l}_j \tag{1.22}$$

where $\mathbf{l}_j$ is the angular momentum of the jth electron. This is the well-known relation between the angular momentum and the magnetic moment.

Eqs. (1.19) and (1.20) are fulfilled by the following generalized representation of the current density:

$$\mathbf{j}_m^{(i)} = -\mathbf{m}_i \times \nabla f(|\mathbf{r}-\mathbf{R}_i|) = \nabla \times (\mathbf{m}_i \cdot f(|\mathbf{r}-\mathbf{R}_i|)) \tag{1.23}$$

Note that $\mathbf{m}_i$ is a particle property and therefore not space dependent. Here the function $f(|\mathbf{r}-\mathbf{R}_i|)$ is almost arbitrary. It should, however, satisfy two conditions:

$$\begin{aligned} &(a)\ f \equiv 0 \text{ outside the ion at } \mathbf{R}_i \\ &(b) \int_{Ion} d^3r\ f(|\mathbf{r}-\mathbf{R}_i|) = 1 \end{aligned} \tag{1.24}$$

That $\mathbf{j}_m^{(i)}$ in the form of (1.23) indeed satisfies (1.19) and (1.20) can easily be checked by substitution (Problem 1.1).

We now define

$$\mathbf{j}_m(\mathbf{r}) = \nabla \times (\overline{\mathbf{m} \cdot f(|\mathbf{r}-\mathbf{R}|)}) = \nabla \times \mathbf{M}(\mathbf{r}) \tag{1.25}$$

and therewith

$$\mathbf{M}(\mathbf{r}) = \overline{\mathbf{m} \cdot f(|\mathbf{r}-\mathbf{R}|)} \tag{1.26}$$

where $\mathbf{M}(\mathbf{r})$ is called the *magnetization*.

With these definitions, we can now gather the macroscopic Maxwell equations:

$$\begin{aligned} \nabla \times \mathbf{B} &= \mu_0\,(\mathbf{j}_f + \bar{\mathbf{j}}_p + \bar{\mathbf{j}}_m) + \varepsilon_0\,\mu_0\,\dot{\mathbf{E}} \\ &= \mu_0\,\mathbf{j}_f + \mu_0\,\dot{\mathbf{P}} + \mu_0\,\nabla \times \mathbf{M} + \varepsilon_0\,\mu_0\,\dot{\mathbf{E}} \end{aligned}$$

This can be further rearranged as

$$\nabla \times (\mathbf{B} - \mu_0\,\mathbf{M}) = \mu_0\,\mathbf{j}_f + \mu_0\,(\varepsilon_0\,\dot{\mathbf{E}} + \dot{\mathbf{P}}) \tag{1.27}$$

Similarly, when we denote the macroscopic charge density by ρ, we find

$$\nabla \cdot \mathbf{E} = \frac{1}{\varepsilon_0}\,(\rho + \overline{\rho}_p) = \frac{1}{\varepsilon_0}\,(\rho - \nabla \cdot \mathbf{P}) \tag{1.28}$$

so that we have

$$\nabla \cdot (\varepsilon_0 \, \mathbf{E} + \mathbf{P}) = \rho \tag{1.29}$$

We can now summarize as follows:
Material equations:

$$\mathbf{B} = \mu_0 \, (\mathbf{H} + \mathbf{M}) \qquad ; \qquad \mathbf{D} = \varepsilon_0 \, \mathbf{E} + \mathbf{P} \tag{1.30}$$

Maxwell equations:

$$\begin{aligned} \nabla \times \mathbf{E} &= -\dot{\mathbf{B}} & \qquad \nabla \cdot \mathbf{B} &= 0 \\ \nabla \cdot \mathbf{D} &= \rho & \qquad \nabla \times \mathbf{H} &= \mathbf{j}_f + \dot{\mathbf{D}} \end{aligned} \tag{1.31}$$

We want to recall the fact that a charge q moving with a velocity $\mathbf{v}$, placed in an electric field $\mathbf{E}$ and a magnetic field $\mathbf{B}$ experiences a force, known as the *Lorentz force* given by

$$\mathbf{F} = q \cdot (\mathbf{E} + \mathbf{v} \times \mathbf{B}) \tag{1.32}$$

One should pay attention to the fact that a charged particle actually "sees" $\mathbf{E}$ and $\mathbf{B}$ since they are the actual physical fields. In contrast, $\mathbf{H}$ and $\mathbf{D}$ are only auxiliary quantities.

1.2 Magnetic Moment and Magnetization

The relationships for $\mathbf{m}$ and $\mathbf{M}$ derived so far are rather inconvenient to handle. Therefore, we will now try to express them in terms of the energy W of the magnetic system. W is defined through the magnetization current density $\mathbf{j}_m$ and *not* through the free current density, which produces a magnetic field in which the magnetic system finds itself. First we consider the magnetic energy of a single ion, starting with a classical treatment. How does the energy of an ion change when a magnetic field $\mathbf{B}_0$ is switched on? A change in energy appears due to the work done by the external field on the magnetization current density. This happens through the $\mathbf{E}$-field induced by $\mathbf{B}_0$:

$$\nabla \times \mathbf{E} = -\dot{\mathbf{B}}_0 \qquad \mathbf{B}_0 = \mu_0 \mathbf{H} \tag{1.33}$$

Denoting the change in energy of the ion in time dt by dW, we find

$$\begin{aligned} dW &= \int_{ion} \mathbf{j}_m \cdot \mathbf{E}\, d^3r dt \\ &= -dt \int_{ion} (\mathbf{m} \times \nabla f(|\mathbf{r}-\mathbf{R}|)) \cdot \mathbf{E}\, d^3r \\ &= -dt\, \mathbf{m} \cdot \int_{ion} (\nabla f \times \mathbf{E})\, d^3r \\ &= -dt\, \mathbf{m} \cdot \int_{ion} (\nabla \times (f\, \mathbf{E}) - f\, \nabla \times \mathbf{E})\, d^3r \\ &= -\mathbf{m} \cdot \int_{ion} d\mathbf{B}_0 \cdot f\, d^3r \end{aligned}$$

In the second step we have used (1.23), in the third step, we use the fact that $\mathbf{m}$ is a space-independent intrinsic particle property, and in the fourth step, the well-known vector formula

$$\nabla \times (\phi\, \mathbf{a}) = \phi\, \nabla \times \mathbf{a} - \mathbf{a} \times \nabla\phi$$

where $\mathbf{a}$ is a vector field and ϕ is a scalar field. In the last step we exploit the property (a) in (1.24) and the generalized Gauss theorem ($\mathbf{a}$: arbitrary vector field)

$$\int_V d^3r\, \nabla \times \mathbf{a} = \int_{\partial V} d\mathbf{S} \times \mathbf{a}$$

where $\int_{\partial V}$ indicates integral over the surface that encloses the volume V. Finally, the Maxwell equation (1.31) is inserted.

$d\mathbf{B}_0$ is certainly constant over the dimensions of an ion and therefore can be brought out of the integral. Exploiting the normalization of the function f (property (b) in (1.24)), the magnetic moment of the ion is then given by

$$\mathbf{m} = -\frac{\partial W}{\partial \mathbf{B}_0} = -\nabla_{\mathbf{B}_0} W \qquad (1.34)$$

So far we have calculated classically. The quantum mechanical generalization is quite straight forward.

Starting from the Schrödinger equation

$$(\widehat{H} - W)|\psi\rangle = 0 \quad ; \quad \langle\psi|\psi\rangle = 1 \qquad (1.35)$$

where $\widehat{H}$ is the Hamiltonian of the ion, by differentiating with respect to the external field $\mathbf{B}_0$, we obtain

$$\left(\frac{\partial \widehat{H}}{\partial \mathbf{B}_0} - \frac{\partial W}{\partial \mathbf{B}_0}\right)|\psi\rangle + (\widehat{H} - W)|\frac{\partial \psi}{\partial \mathbf{B}_0}\rangle = 0 \qquad (1.36)$$

By multiplying from the left with the bra $\langle\psi|$ we get

$$\langle\psi|\frac{\partial \widehat{H}}{\partial \mathbf{B}_0}|\psi\rangle = \frac{\partial W}{\partial \mathbf{B}_0} - \langle\psi|(\widehat{H} - W)|\frac{\partial \psi}{\partial \mathbf{B}_0}\rangle \tag{1.37}$$

The last term in the above equation vanishes since $\widehat{H}$ is Hermitean. Comparing this expression with (1.35) and exploiting the quantum mechanical correspondence principle we get for the *magnetic moment operator*

$$\widehat{\mathbf{m}} = -\frac{\partial}{\partial \mathbf{B}_0}\widehat{H} \tag{1.38}$$

As an example, we will now evaluate $\widehat{\mathbf{m}}$ for an ion in a homogeneous magnetic field. For this purpose, we need the Hamiltonian $\widehat{H}$ whose exact derivation is dealt with in the next section. Here we consider only the atomic electrons, treating the nucleus to be at rest which only provides the charge neutrality, and, in particular, does not influence the external magnetic field $\mathbf{B}_0 = \mu_0\mathbf{H}$ appreciably. The (complicated) interaction among the electrons, the interaction with the nucleus and the spin–orbit coupling effects will not be needed in detail for the moment. We are mainly interested in the terms that arise due to the external magnetic field. We choose the vector potential $\mathbf{A}$ such that

$$\mathbf{B}_0 = \nabla \times \mathbf{A} \quad ; \quad \nabla \cdot \mathbf{A} = 0 \quad (\textit{Coulomb gauge}) \tag{1.39}$$

This can be fulfilled by

$$\mathbf{A} = \frac{1}{2}\mathbf{B}_0 \times \mathbf{r} \tag{1.40}$$

The kinetic energy of the electrons *without* field is given by

$$T_0 = \sum_{i=1}^{p} \frac{\mathbf{p}_i^2}{2m} \tag{1.41}$$

where p is the number of atomic electrons. We denote the charge of the electron by $(-e)$ so that $e > 0$. In the presence of the field, we must distinguish the canonical momentum $\mathbf{p}_i$ from the mechanical momentum $m\mathbf{v}_i$:

$$\mathbf{p}_i = m\,\mathbf{v}_i - e\,\mathbf{A}(\mathbf{r}_i) \tag{1.42}$$

The kinetic energy in the presence of the field is then given by

$$T = \frac{1}{2m} \sum_{i=1}^{p} (\mathbf{p}_i + e\,\mathbf{A}(\mathbf{r}_i))^2$$
$$= \frac{1}{2m} \sum_{i=1}^{p} \left[\mathbf{p}_i^2 + e\,(\mathbf{p}_i \cdot \mathbf{A}(\mathbf{r}_i) + \mathbf{A}(\mathbf{r}_i) \cdot \mathbf{p}_i) + e^2\,\mathbf{A}^2(\mathbf{r}_i)\right]$$

One should pay attention that, in general, the operators $\mathbf{p}_i$ and $\mathbf{A}(\mathbf{r}_i)$ do not commute. In the Coulomb gauge, however, they do commute as can be seen from the following:

$$\mathbf{p}_i \cdot \mathbf{A}(\mathbf{r}_i) = \frac{\hbar}{i} \nabla_i \cdot \mathbf{A}(\mathbf{r}_i) = \frac{\hbar}{i} (\underbrace{\nabla \cdot \mathbf{A}}_{=0} + \mathbf{A} \cdot \nabla_i)$$
$$= \mathbf{A}(\mathbf{r}_i) \cdot \mathbf{p}_i \tag{1.43}$$

Therefore, we finally get the following expression for the kinetic energy:

$$T = T_0 + \frac{e}{m} \sum_{i=1}^{p} \mathbf{A}(\mathbf{r}_i) \cdot \mathbf{p}_i + \frac{e^2}{2m} \sum_{i=1}^{p} \mathbf{A}^2(\mathbf{r}_i) \tag{1.44}$$

Let the homogeneous magnetic field be oriented in the z-direction:

$$\mathbf{B}_0 = (0,\ 0,\ B_0) \tag{1.45}$$

Substituting in (1.40), the vector potential $\mathbf{A}$ is given by

$$\mathbf{A} = \frac{B_0}{2}(-y,\ x,\ 0) \tag{1.46}$$

The scalar product $\mathbf{A}_i \cdot \mathbf{p}_i$ appearing in (1.44) can then be expressed by the orbital angular momentum $\mathbf{l}_i$ of the ith electron:

$$\mathbf{A}_i \cdot \mathbf{p}_i = \frac{B_0}{2}(-y_i\ p_{ix} + x_i\ p_{iy}) = \frac{1}{2} B_0\, l_{iz}$$
$$= \frac{1}{2} \mathbf{B}_0 \cdot \mathbf{l}_i \tag{1.47}$$

Introducing the total orbital angular momentum $\mathbf{L}$ of the atomic electrons, which is given by

$$\mathbf{L} = \sum_{i=1}^{p} \mathbf{l}_i \tag{1.48}$$

Equation (1.44) eventually reduces to

$$T = T_0 + \frac{1}{\hbar}\,\mu_B\,\mathbf{L}\cdot\mathbf{B}_0 + \frac{e^2\,B_0^2}{8\,m}\sum_{i=1}^{p}(x_i^2 + y_i^2) \tag{1.49}$$

where μ_B the *Bohr magneton* is given by

$$\begin{aligned}\mu_B = \frac{e\hbar}{2m} &= 9.274 * 10^{-24}\,\frac{J}{T}\\ &= 0.579 * 10^{-4}\,\frac{eV}{T}\end{aligned} \tag{1.50}$$

The total magnetic moment of the atomic electrons due to their orbital angular momenta follows by inserting (1.49) in (1.38),

$$\mathbf{m}_L = -\frac{1}{\hbar}\,\mu_B\,\mathbf{L} \tag{1.51}$$

in agreement with (1.22). That means, the magnetic moment and angular momentum are always oriented antiparallel to each other. The reason for this is the negative charge of the electron. The energy is minimum when the moment and the field are parallel or when the field and the angular momentum are antiparallel.

So far, we have not taken the spin of the electron into account. The existence of spin is experimentally established, for example, by the Einstein–de Haas experiment. Its rigorous proof needs the Dirac theory (see Sect. 2.3).

A magnetic moment is also associated with the total spin S of the atomic electrons:

$$\mathbf{m}_S = -\frac{1}{\hbar}\,\mu_B\,g_e\,\mathbf{S} \tag{1.52}$$

where g_e is the Lande's g-factor given by

$$g_e = 2\left(1 + \frac{\alpha}{2\pi} + \mathcal{O}(\alpha^2)\right) \approx 2.0023, \tag{1.53}$$

and $\alpha \approx \frac{1}{137}$ is the Sommerfeld's fine structure constant. For our purpose, it is always enough to take $g_e \approx 2$. The interaction of $\mathbf{m}_S$ with the field gives one more term in the Hamiltonian:

$$H_s = g_e\,\frac{1}{\hbar}\,\mu_B\,\mathbf{S}\cdot\mathbf{B}_0 \tag{1.54}$$

Denoting by $\widehat{H}_0$ the Hamiltonian in the absence of the field, absorbing, e.g. the electron–electron and the electron–nucleus interaction, which we consider as field independent, the Hamiltonian of the ion in the presence of the field is given by

$$\widehat{H} = \widehat{H}_0 + \frac{\mu_B}{\hbar}(\mathbf{L} + 2\mathbf{S}) \cdot \mathbf{B}_0 + \frac{e^2 B_0^2}{8m} \sum_{i=1}^{p} \left(x_i^2 + y_i^2\right) \tag{1.55}$$

Using (1.38) we can now evaluate the magnetic moment:

$$\mathbf{m} = -\frac{\mu_B}{\hbar}(\mathbf{L} + 2\mathbf{S}) - \frac{e^2}{4m}\mathbf{B}_0 \sum_{i=1}^{p}(x_i^2 + y_i^2) \tag{1.56}$$

This result is valid only when the spin–orbit coupling is completely neglected. Also, only in this case can one uniquely define **L** and **S** (see Chap. 2). One recognizes that the second term in (1.56) corresponds to an *induced* magnetic moment, disappearing as soon as the field is switched off. On the contrary, the first term is a *permanent* moment that is present as long as $(\mathbf{L} + 2\mathbf{S})$ does not vanish. This requires an incompletely filled electron shell (Chap. 2).

We now come to the concept of *magnetization*. According to (1.26), to evaluate the magnetization, we have to average the magnetic current density over a volume $v(\mathbf{r})$ which is macroscopically small but still contains a large number of ions in it. This is *the* basic problem of every theory of magnetism. The motion of the charge carriers is strongly correlated and therefore it is a genuine many-body problem which can be handled only in certain limiting cases, namely, for

(a) *strongly localized moments* where the currents are restricted to particular lattice cells (*localized magnetism*);
(b) *itinerant moments* which are carried by quasi-free conduction electrons (*band magnetism*).

It is not always possible to classify magnetic materials into one or the other of these two classes. In addition, even these limiting cases are not completely understood, either.

For the localized moments (a), we have

$$\begin{aligned}\mathbf{M}(\mathbf{r}) &= \overline{\mathbf{m} \cdot f(|\mathbf{r} - \mathbf{R}|)} = \frac{1}{v(\mathbf{r})} \int_v d^3r' \mathbf{m} f \\ &= \frac{1}{v(\mathbf{r})} \sum_{i=1}^{N(v(\mathbf{r}))} \mathbf{m}_i \underbrace{\int_{v_i} d^3r' f(|\mathbf{r}' - \mathbf{R}_i|)}_{=1}\end{aligned}$$

where $N(v(\mathbf{r}))$ is the number of localized moments located in $v(\mathbf{r})$ and v_i denotes a volume containing only the ith ion:

$$\mathbf{M}(\mathbf{r}) = \frac{1}{v(\mathbf{r})} \sum_{i=1}^{N(v(\mathbf{r}))} \mathbf{m}_i \tag{1.57}$$

This expression further simplifies if we assume identical ions. Then the averaging over many identical ions in different states can be replaced by thermodynamic average over a single ion:

$$\mathbf{M}(\mathbf{r}) = n(\mathbf{r})\langle \mathbf{m} \rangle \tag{1.58}$$

here $n(\mathbf{r}) = N(v(\mathbf{r}))/v(\mathbf{r})$ is the ion density and $\langle \cdots \rangle$ means the thermodynamic average wherein lies the magnetic field and the temperature dependence. The evaluation of the magnetization $\mathbf{M}(\mathbf{r})$ under given assumptions is, therefore, actually a thermodynamic averaging over the individual ion moment.

1.3 Susceptibility

We introduce a further important quantity in the field of magnetism, the knowledge of which provides a lot of information. This is one of the so-called *response functions*. A response function describes the response of a system to external perturbations. In the case of magnetism, the perturbation is the external magnetic field $\mathbf{H}$ and the response is the magnetization $\mathbf{M}(\mathbf{r}, t)$. These two are related by susceptibility. In this section, we will restrict ourselves only to some definitions.

One distinguishes *linear media* where the response is directly proportional to the perturbation so that the susceptibility is independent of the field, and *non-linear media* where the higher powers of the perturbation are not negligible so that the susceptibility is field dependent.

Every magnetic material has a so-called *linear response region* for sufficiently small external perturbations, where the higher powers of the perturbation do not play any role. This is in general the case for "normal" experimentally realizable fields. One of the non-linear effects is, for example, the phenomenon of hysteresis.

Using the Fourier transforms of the magnetization and the field,

$$\mathbf{M}(\mathbf{r}, t) = \frac{1}{2\pi\ V} \sum_{\mathbf{q}} \int d\omega\, \mathbf{M}(\mathbf{q}, \omega)\, e^{i(\mathbf{q}\cdot\mathbf{r}-\omega t)} \tag{1.59}$$

$$\mathbf{H}(\mathbf{r}, t) = \frac{1}{2\pi\ V} \sum_{\mathbf{q}} \int d\omega\, \mathbf{H}(\mathbf{q}, \omega)\, e^{i(\mathbf{q}\cdot\mathbf{r}-\omega t)} \tag{1.60}$$

one defines the *generalized susceptibility* χ:

$$M_\alpha(\mathbf{q}, \omega) = \sum_{\mathbf{k}} \int d\bar{\omega} \sum_{\beta} \chi_{\alpha\beta}(\mathbf{q}, \mathbf{k}; \omega, \bar{\omega})\, H_\beta(\mathbf{k}, \bar{\omega}) \tag{1.61}$$

$$\alpha,\ \beta \in \{x,\ y,\ z\}$$

This is the most general case where χ is a field and temperature-dependent tensor. For the $(\mathbf{r}, t)$-dependent Fourier transforms, we have

$$\mathbf{M}(\mathbf{r}, t) = \int\int d^3r'dt'\underline{\chi}(\mathbf{r}, \mathbf{r}' \,;\, t, t')\, \mathbf{H}(\mathbf{r}', t') \tag{1.62}$$

$$\begin{aligned}\underline{\chi}(\mathbf{r}, \mathbf{r}' \,;\, t, t') = \frac{1}{2\pi V}\sum_{\mathbf{k},\mathbf{q}}\int\int d\omega\, d\bar{\omega}\; \underline{\chi}(\mathbf{q}, \mathbf{k}; \omega, \bar{\omega})\, e^{i\mathbf{q}\cdot(\mathbf{r}-\mathbf{r}')} *\\ *e^{-i\,\omega(t-t')}\, e^{i(\mathbf{q}-\mathbf{k})\cdot\mathbf{r}'}\, e^{-i(\omega-\bar{\omega})t'}\end{aligned} \tag{1.63}$$

Under certain preconditions, this expression takes a simpler form. For example, if the translational invariance and a stationary medium are imposed, then $\underline{\chi}$ can depend only on the differences $(\mathbf{r} - \mathbf{r}')$ and $(t - t')$. That means

$$\underline{\chi}(\mathbf{q}, \mathbf{k}; \omega, \bar{\omega}) = \underline{\chi}(\mathbf{q}, \omega)\, \delta_{\mathbf{kq}}\, \delta(\omega - \bar{\omega}) \tag{1.64}$$

This is actually the case, with which we will in general be concerned:

$$\mathbf{M}(\mathbf{r}, t) = \int\int d^3r'dt'\underline{\chi}(\mathbf{r} - \mathbf{r}', t - t')\, \mathbf{H}(\mathbf{r}', t') \tag{1.65}$$

$$\underline{\chi}(\mathbf{r} - \mathbf{r}', t - t') = \frac{1}{2\pi V}\sum_{\mathbf{q}}\int d\omega\; \underline{\chi}(\mathbf{q}, \omega)\, e^{i(\mathbf{q}(\mathbf{r}-\mathbf{r}')-\omega(t-t'))} \tag{1.66}$$

$\underline{\chi}(\mathbf{q}, \omega)$ is the *dynamic susceptibility* and $\underline{\chi}(\mathbf{q}, \omega = 0)$ is the *static susceptibility*.

If we further assume that the field is homogeneous and static, so that **H** is not dependent on **r** and t and that the magnetizable medium is homogeneous so that **M** is also not dependent on **r** and t, then we are left with the tensor

$$\underline{\chi} = \underline{\chi}(\mathbf{q} = 0, \omega = 0) \tag{1.67}$$

the elements of which are given by

$$\chi_{\alpha\beta} = \left(\frac{\partial M_\alpha}{\partial H_\beta}\right)_T \qquad \alpha,\ \beta\ \in\ \{x,\ y,\ z\}$$

Obviously, $\underline{\chi}$ may depend on T and **H**:

$$\chi_{\alpha\beta} = \chi_{\alpha\beta}(T, \mathbf{H}) \tag{1.68}$$

This quantity is eminently suitable to classify solid materials based on their magnetic properties. This will be demonstrated in the next section of this introductory chapter.

1.4 Classification of Magnetic Materials

One can classify the magnetic phenomena into three main groups.

1.4.1 Diamagnetism

It is defined by

$$\chi^{dia} < 0 \quad ; \quad \chi^{dia} = const. \tag{1.69}$$

It has to do basically with an induction effect. The external field induces magnetic dipoles, which, according to the Lenz's law, are oriented antiparallel to the field and therefore χ is negative (see the second term in (1.56)).

Diamagnetism is a property displayed by *all* materials. However, one speaks of diamagnetism only when no other, i.e. neither paramagnetism nor collective magnetism is present, because if any other magnetism is present, then diamgnetism, which is weak, is buried under the stronger effects.

Examples:

1. Almost all the organic substances
2. Metals like Bi, Zn and Hg
3. Nonmetals like S, J and Si
4. Superconductors for $T < T_C$ are perfect diamagnets, i.e. $\chi^{dia} = -1$ (*Meissner–Ochsenfeld effect*)

1.4.2 Paramagnetism

Typical for this class is

$$\chi^{para} > 0 \quad ; \quad \chi^{para} = \chi^{para}(T) \tag{1.70}$$

The essential requirement for paramagnetism is the existence of permanent magnetic dipoles, which are, more or less, oriented by the field **H** and this orientation is opposed by the thermal motion. Paramagnetism can be of two types depending on whether the permanent moments are localized or itinerant.

1.4.2.1 Localized Moments

The moments arise due to one of the inner partially filled electron shells.

Examples:

3d: transition metals
4f: rare earths
5f: actinides

Paramagnetism of these materials is the so-called *Langevin paramagnetism*. In this case, we have

$$\chi^{para} = \chi^{para}(T) \tag{1.71}$$

At high temperatures the well-known Curie's law is satisfied:

$$\chi(T) = \frac{C}{T} \tag{1.72}$$

1.4.2.2 Itinerant Moments

Quasi-free conduction electrons carry a permanent moment of one Bohr magneton ($1\mu_B$) each. In this case, we have *Pauli-paramagnetism*, where χ^{Pauli}, to a first approximation, is temperature independent (due to Pauli's principle). Schematically one can distinguish the two types (1.4.2.1) and (1.4.2.2) from Fig. 1.2 (in general, however, $\chi^{Pauli} \ll \chi^{Langevin}$).

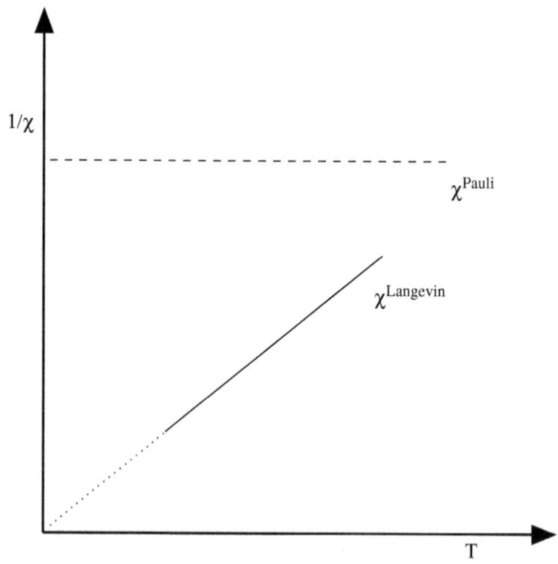

Fig. 1.2 Schematic plot of the inverse susceptibility as a function of temperature for a localized ($\chi^{Langevin}$) and for an itinerant (χ^{Pauli}) magnetic moment system (paramagnetism)

1.4.3 *Collective Magnetism*

The susceptibility in this case is in general a complicated function of the field, temperature, and also often of the *pre-treatment* of the sample:

$$\chi^C = \chi^C(T,\ H,\ history) \tag{1.73}$$

Collective magnetism arises due to a characteristic, only quantum mechanically explainable *exchange interaction* between the permanent magnetic dipoles. That leads to a critical temperature T^* below which there exists a *spontaneous magnetization*, i.e. a spontaneous, not forced externally, ordering of the dipoles. The permanent magnetic dipoles can again be either *localized* (Gd, EuO, Rb_2MnCl_4, etc.) or *itinerant* (Fe, Co, Ni, etc.). Collective magnetism can be further divided into three sub-classes.

1.4.3.1 Ferromagnetism

In this case the critical temperature is called the *Curie temperature*:

$$T^* = T_C \tag{1.74}$$

For temperatures $0 < T < T_C$ the permanent moments have a preferential orientation. At absolute zero $T = 0$, all the moments are oriented parallel to each other (Figs. 1.3 and 1.4).

Fig. 1.3 Ferromagnetic order at $0 < T < T_C$

Fig. 1.4 Ferromagnetic order at $T = 0$

1.4.3.2 Ferrimagnetism

In this case the lattice is divided into two sub-lattices A and B with different, normally antiparallel magnetizations $\mathbf{M}_A$ and $\mathbf{M}_B$ such that (Fig. 1.5)

$$\mathbf{M}_A \neq \mathbf{M}_B \tag{1.75}$$

and

$$\mathbf{M}_A + \mathbf{M}_B \neq 0 \qquad for\ T < T_C \tag{1.76}$$

Fig. 1.5 Ferrimagnetic order at $T = 0$

1.4.3.3 Antiferromagnetism

The critical temperature is called the *Neel temperature*:

$$T^* = T_N \tag{1.77}$$

It is a special case of ferrimagnetism (Fig. 1.6):

$$|\mathbf{M}_A| = |\mathbf{M}_B| \neq 0 \quad ; \quad \mathbf{M}_A = -\mathbf{M}_B \quad for\ T < T_N. \tag{1.78}$$

Fig. 1.6 Antiferromagnetic order at $T = 0$

The total magnetization $\mathbf{M} = \mathbf{M}_A + \mathbf{M}_B$ is therefore always zero.

Above the critical temperature T^*, collective magnetism goes into paramagnetism with the characteristic behaviour of the inverse susceptibility sketched in Fig. 1.7. θ is the so-called *paramagnetic Curie temperature.*

The different features, listed above, will all be derived in the course of this book.

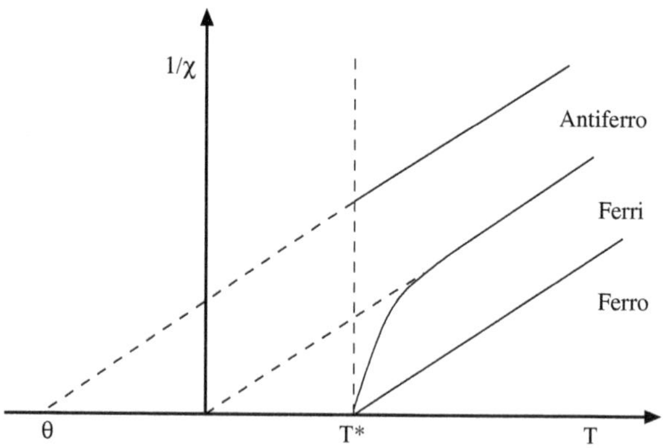

Fig. 1.7 Characteristic temperature behaviour of the inverse susceptibility for antiferromagnetic, ferrimagnetic and ferromagnetic localized moment systems. θ is the paramagnetic Curie temperature

1.5 Elements of Thermodynamics

We presume that the reader is familiar with the basic principles of Thermodynamics and Statistical Mechanics. We therefore refer here only to some peculiarities of the magnetic system which are vital for the consistency of later derivations.

For a magnetic system the *first law of thermodynamics* can either be written as

$$dU = TdS - VMdB_0 + \mu \, dN \tag{1.79}$$

or

$$dU = TdS + VB_0dM + \mu \, dN \tag{1.80}$$

where U is the internal energy, defined as the statistical expectation value of the Hamiltonian $\widehat{H}$ of the system,

$$U = \langle \widehat{H} \rangle \tag{1.81}$$

T, S, B_0, M , V, μ and N are, respectively, temperature, entropy, magnetic induction, magnetization, volume, chemical potential and particle number.

The existence of two different Eqs. (1.79) and (1.80) for the internal energy means that the magnetic analogue to the volume work ($-pdV$, p : pressure) of the fluid system is not unique. The difference lies in the energy stored in the magnetic field. Shall this contribution be counted or not? Both versions are correct and do not lead to any contradiction. The version (1.80) has the advantage that according to

$p \longleftrightarrow B_0$ (intensive)
$V \longleftrightarrow -m = -VM$ (extensive)

intensive and extensive thermodynamic variables correspond to each other.

However, the *total* internal energy contains the pure field energy, too. That is accounted for by the version (1.79). The change of the corresponding free energy,

$$F = U - TS,$$

reads in this case

$$dF = -SdT - VMdB_0 + \mu \, dN \tag{1.82}$$

That means the magnetic induction B_0 is one of the *natural variables* of the thermodynamic potential F :

$$F = F(T, B_0, N) \tag{1.83}$$

We will use this version rather exclusively in the following chapters due to the following two reasons. First, the free energy then depends on the variables (T, B_0, N) which are most convenient for magnetic systems since they are related to the usual experimental situation. Second, it fits the definition (1.81) of the internal energy U as the expectation value of the system Hamiltonian. This can be seen as follows.

On switching on a magnetic field at $T = 0$, the internal energy changes by

$$\Delta U = \langle \widehat{H}(B_0) - \widehat{H}(0) \rangle \tag{1.84}$$

According to our definition (1.38) of the magnetic moment this means

$$\begin{aligned} dU &= \langle \nabla_{B_0} \widehat{H} \rangle dB_0 \\ &= -\langle \widehat{m} \rangle dB_0 \\ &= -VMdB_0 \end{aligned} \tag{1.85}$$

For $T = 0$ and $N = const.$ this is just (1.79)

For the free energy $F = F(T, B_0)$, we can derive a fundamental relationship to the canonical partition function :

$$Z_N = Z_N(T, B_0) = Tre^{-\beta \widehat{H}} \tag{1.86}$$

We inspect for a system with $N = const.$

$$\begin{aligned} d(-\beta F) &= \frac{F}{k_B T^2} dT + \frac{1}{k_B T}(SdT + VMdB_0) \\ &= \beta \left(\frac{U}{T} dT + VMdB_0 \right) \\ &= \beta \left(\frac{1}{T} \langle \widehat{H} \rangle dT - \langle \nabla_{B_0} \widehat{H} \rangle dB_0 \right) \end{aligned}$$

From the definition (1.86) of the partition function one gets immediately

$$\begin{aligned} \langle \widehat{H} \rangle &= \frac{1}{Z_N} Tr \left(e^{-\beta \widehat{H}} \widehat{H} \right) \\ &= -\frac{\partial}{\partial \beta} \ln Z_N(T, B_0) \\ &= k_B T^2 \frac{\partial}{\partial T} \ln Z_N(T, B_0) \end{aligned} \tag{1.87}$$

$$\begin{aligned} \langle \nabla_{B_0} \widehat{H} \rangle &= \frac{1}{Z_N} Tr \left(e^{-\beta \widehat{H}} \nabla_{B_0} \widehat{H} \right) \\ &= -\frac{1}{\beta} \frac{\partial}{\partial B_0} \ln Z_N(T, B_0) \end{aligned} \tag{1.88}$$

It follows

$$d(-\beta F) = \left(\frac{\partial}{\partial T} \ln Z_N\right)_{B_0} dT + \left(\frac{\partial}{\partial B_0} \ln Z_N\right)_T dB_0$$

and therewith the fundamental relation between free energy and canonical partition function:

$$F(T, B_0) = -k_B T \ln Z_N(T, B_0) \tag{1.89}$$

The magnetization $M(T, B_0)$ can then be derived with (1.82)

$$\begin{aligned} M(T, B_0) &= -\frac{1}{V}\left(\frac{\partial F}{\partial B_0}\right)_{T,N} \\ &= \frac{k_B T}{V}\left(\frac{\partial}{\partial B_0} \ln Z_N(T, B_0)\right)_T \end{aligned} \tag{1.90}$$

Let us repeat the above considerations for the grand canonical ensemble. Central thermodynamic potential in this case is the grand canonical potential

$$\Omega = F - G \tag{1.91}$$

where G is the Gibbs enthalpy. Using the Gibbs–Duhem relation, $G = \mu N$, one gets with (1.82)

$$\begin{aligned} d\Omega &= dF - \mu dN - N d\mu \\ &= -SdT - VMdB_0 - Nd\mu \end{aligned} \tag{1.92}$$

That means

$$\Omega = \Omega(T, B_0, \mu)$$

Since the particle number is not fixed, N is to be read as $N = \langle \widehat{N} \rangle$, where $\widehat{N}$ is the particle number operator and $\langle \cdots \rangle$ is now the grand canonical average

$$\langle \widehat{N} \rangle = \frac{1}{\Xi} Tr\left(e^{-\beta(\widehat{H} - \mu N)} \widehat{N}\right) \tag{1.93}$$

where Ξ is the grand partition function,

$$\begin{aligned} \Xi &= Tr\left(e^{-\beta(\widehat{H} - \mu N)}\right) \\ &= \Xi_\mu(T, B_0) \end{aligned} \tag{1.94}$$

One recognizes

$$\langle \widehat{N} \rangle = \frac{1}{\beta} \frac{\partial}{\partial \mu} \ln \Xi_\mu(T, B_0) \tag{1.95}$$

$$\langle \nabla_{B_0} \widehat{H} \rangle = -\frac{1}{\beta} \frac{\partial}{\partial B_0} \ln \Xi_\mu(T, B_0) \tag{1.96}$$

$$U = \langle \widehat{H} \rangle = -\frac{\partial}{\partial \beta} \ln \Xi_\mu(T, B_0) + \mu \langle \widehat{N} \rangle \tag{1.97}$$

We inspect now

$$\begin{aligned}
d(\beta\Omega) &= \\
&= -\frac{\Omega}{k_B T^2} dT - \frac{1}{k_B T} \left(S dT + V M d B_0 + \langle \widehat{N} \rangle d\mu \right) \\
&= -\frac{U - \mu \langle \widehat{N} \rangle}{k_B T^2} dT - \frac{1}{k_B T} (V M d B_0 + \langle \widehat{N} \rangle d\mu) \\
&= -\left(\frac{\partial}{\partial T} \ln \Xi_\mu(T, B_0) \right)_{\mu, B_0} dT - \left(\frac{\partial}{\partial B_0} \ln \Xi_\mu(T, B_0) \right)_{T, \mu} d B_0 \\
&\quad - \left(\frac{\partial}{\partial \mu} \ln \Xi(T, B_0) \right)_{T, B_0} d\mu
\end{aligned}$$

Except for an unimportant constant we therefore get

$$\Omega(T, B_0, \mu) = -k_B T \ln \Xi_\mu(T, B_0) \tag{1.98}$$

The magnetization now reads

$$\begin{aligned}
M(T, B_0) &= -\frac{1}{V} \left(\frac{\partial \Omega}{\partial B_0} \right)_{T, \mu} \\
&= \frac{k_B T}{V} \left(\frac{\partial}{\partial B_0} \ln \Xi_\mu(T, B_0) \right)_{T, \mu}
\end{aligned} \tag{1.99}$$

Equations (1.98) and (1.99), derived in the grand canonical ensemble, are fully equivalent to the “canonical” expressions (1.89) and (1.90). All the four will be frequently used in the later chapters.

1.6 Problems

Problem 1.1 According to classical electrodynamics, the magnetization current $\mathbf{j}_m^{(i)}$ inside a particle (e.g. electrons in atomic shells) produces a magnetic moment

$$\mathbf{m}_i = \frac{1}{2} \int d^3r \, (\mathbf{r} - \mathbf{R}_i) \times \mathbf{j}_m^{(i)}$$

Here $\mathbf{R}_i$ is the position vector of the centre of the particle. Show that this relation is equivalent to

$$\mathbf{j}_m^{(i)} = \nabla \times (\mathbf{m}_i f(|\mathbf{r} - \mathbf{R}_i|))$$

where the function f, which is differentiable sufficient number of times, can be chosen "almost arbitrarily" so long as it satisfies the following conditions:

1. $f \equiv 0$ outside the i-th particle,
2. $\int_{particle} f(|\mathbf{r} - \mathbf{R}_i|) d^3r \equiv 1$.

Problem 1.2 Let $\widehat{H}$ be the Hamiltonian of a magnetic system which is placed in a homogeneous magnetic field B_0. The operator for the magnetic moment $\widehat{m}$ is defined by

$$\widehat{m} = -\frac{d}{dB_0} \widehat{H}$$

$\widehat{m}$ refers to a permanent moment and diamagnetic components are neglected. That means

$$\frac{d}{dB_0} \widehat{m} = 0$$

The magnetization $\mathbf{M}$ and susceptibility χ_T are mainly determined by the statistical average of the magnetic moment:

$$M = \frac{1}{V} \langle \widehat{m} \rangle \; ; \; \chi_T = \mu_0 \left(\frac{\partial M}{\partial B_0} \right)_T$$

where μ_0 is the permeability of vacuum. Verify the following relation between the susceptibility and the fluctuations of the magnetic moment (*fluctuation–dissipation theorem*):

$$\chi_T = \frac{1}{k_B T} \frac{\mu_0}{V} \langle (\widehat{m} - \langle \widehat{m} \rangle)^2 \rangle$$

Problem 1.3 A small magnet of dipole moment $\mathbf{m}$ is suspended at $\mathbf{r}_0 = (x_0, 0, 0)$ such that it can freely rotate in the x–y plane. A homogeneous magnetic field $\mathbf{B}_0 = B_0 \mathbf{e}_x$ acts on it. In addition to that, let there be a magnetic field $\mathbf{B}_I$ due to a line of current I flowing in the z-direction. Calculate the angle α the dipole makes with the x-axis.

Problem 1.4 The ideal paramagnet is characterized by the Curie law (1.72):

$$\frac{M}{H} = \frac{C}{T} \qquad C : \textit{Curie constant}$$

Calculate the work done on an ideal paramagnet when the applied magnetic field H is isothermally increased from H_1 to $H_2 > H_1$.

Problem 1.5 For not very low temperatures, the Curie law (1.72) represents the equation of state of an ideal paramagnet.

1. Show that for the heat capacities

$$c_{M,H} = T \left(\frac{\partial S}{\partial T} \right)_{M,H}$$

holds

$$c_M = \left(\frac{\partial U}{\partial T} \right)_M \; ; \; c_H = \left(\frac{\partial U}{\partial T} \right)_H + \mu_0 \frac{V}{C} M^2$$

M, magnetization; H, magnetic field; $V = const.$, volume; S, entropy; C, Curie constant; μ_0, permeability of vacuum.

2. For adiabatic change of states, derive the following relation:

$$\left(\frac{\partial M}{\partial H} \right)_S = \frac{c_M}{c_H} \cdot \frac{\mu_0 V M - \left(\frac{\partial U}{\partial H} \right)_T}{\mu_0 V H - \left(\frac{\partial U}{\partial M} \right)_T}$$

Problem 1.6 Assume that a paramagnetic substance with known heat capacity c_H satisfies the Curie law

$$\frac{M}{H} = \frac{C}{T} \qquad C : \textit{Curie constant}$$

For a reversible adiabatic change of state calculate

$$\left(\frac{\partial T}{\partial H} \right)_S$$

Chapter 2
Atomic Magnetism

2.1 Hund's Rules

We consider a single atom or an ion, which, in addition to completely filled electronic shells, has exactly *one* partially filled shell.

We denote by l the orbital angular momentum quantum number of the partially filled shell and by p the number of electrons in this shell ($p < 2(2l + 1)$).

When there are no interactions, the degeneracy of the ground state will be equal to the degeneracy corresponding to the possible distributions of the p-electrons into $2(2l + 1)$ levels. This degeneracy will be partially lifted by the Coulomb interaction H_C and the spin–orbit coupling H_{SO}.

Under certain conditions for the relative magnitudes of these interactions, namely, if

$$H_C \gg H_{SO} \quad ; \quad (\text{“}\textit{light}\text{” } \textit{nuclei}) \tag{2.1}$$

very simple rules are valid in order to find energetically lowest energy terms. First we will simply list them up. A more detailed discussion of the cases that do not satisfy (2.1) will be done in later sections.

2.1.1 Russell–Saunders (LS-) Coupling

This is realized, when, to a first approximation, the spin–orbit interaction is negligible. Then the Hamiltonian can be written to a good approximation as

$$H \approx \sum_{i=1}^{p} \left(\frac{p^2}{2m} + V(r_i) \right) + \frac{1}{2} \sum_{i,j}^{i \neq j} \frac{e^2}{4\pi \varepsilon_0 \, r_{ij}} \quad ; \quad r_{ij} = |\mathbf{r}_i - \mathbf{r}_j| \tag{2.2}$$

In this case, the total orbital angular momentum $\mathbf{L}$ and the total spin angular momentum $\mathbf{S}$ which are given by

W. Nolting, A. Ramakanth, *Quantum Theory of Magnetism*,
DOI 10.1007/978-3-540-85416-6_2, © Springer-Verlag Berlin Heidelberg 2009

$$\mathbf{L} = \sum_{i=1}^{p} \mathbf{l}_i \qquad ; \qquad \mathbf{S} = \sum_{i=1}^{p} \mathbf{s}_i \tag{2.3}$$

commute with H:

$$[\mathbf{S}\,,\ H]_- = 0 \qquad ; \qquad [\mathbf{L}\,,\ H]_- = 0 \tag{2.4}$$

Equation (2.4) is valid because, firstly, H does not contain $\mathbf{S}$ at all and then H is rotationally invariant. Further it is always valid that

$$[\mathbf{J}\,,\ H]_- = 0 \tag{2.5}$$

where $\mathbf{J}$ is the total angular momentum of the shell:

$$\mathbf{J} = \mathbf{L} + \mathbf{S} \tag{2.6}$$

Physically it means that

$$L = \sum_{i=1}^{p} m_l^{(i)}\,,\ S = \sum_{i=1}^{p} m_s^{(i)} \ and\ J \tag{2.7}$$

are *good quantum numbers*. $m_l^{(i)}$ and $m_s^{(i)}$ are the magnetic quantum numbers of the individual electrons. In other words, there exists a simultaneous set of eigenstates for the operators

$$H,\ J^2,\ J_z,\ L^2,\ L_z,\ S^2,\ S_z$$

and the states can be labelled by the corresponding quantum numbers:

$$|\cdots\rangle = |J, M_J, L, M_L, S, M_S\rangle \tag{2.8}$$

That means for example,

$$\begin{array}{ll} J^2|\cdots\rangle = \hbar^2 J(J+1)|\cdots\rangle & J_z|\cdots\rangle = \hbar M_J|\cdots\rangle \\ J = |L-S|, \cdots, L+S & -J \le M_J \le +J \end{array} \tag{2.9}$$

The other angular momentum operators also operate in exactly similar fashion. For the Hamiltonian we have

$$H|\cdots\rangle = E_{JLS}^{(0)}\,|\cdots\rangle \tag{2.10}$$

The energy eigenvalues will depend on J, L, S, but, in the absence of a magnetic field, they will degenerate with respect to M_J, M_L and M_S.

2.1.2 Hund's Rules for LS Coupling

For given L and S all the possible levels build the so-called *LS multiplet*. For the energetically lowest of them, the following rules are valid:

1. *Maximum S*, as far as being consistent with Pauli's principle.

$$S = \frac{1}{2}\left[(2l+1) - |2l+1-p|\right] \tag{2.11}$$

2. *Maximum L*, as far as being consistent with both the Pauli's principle and with (1):

$$L = S\,|2l+1-p| \tag{2.12}$$

For magnetic problems, only the multiplet built according to (1) and (2) is meaningful. All the other multiplets, i.e. those which correspond to different (L, S)-values, lie, in general, energetically much higher ($> 1\,\mathrm{eV}$, compare with $k_B T_R \approx 1/40\,\mathrm{eV}$ (T_R: room temperature), $\mu_B B \leq 10^{-2}\,\mathrm{eV}$).
Notation:

$$\begin{array}{l} L = 0\ \ 1\ \ 2\ \ 3 \cdots \\ X = S\ \ P\ \ D\ \ F \cdots \end{array}$$

$$\text{"}Term\text{"}:\ {}^{2S+1}X_J \tag{2.13}$$

Every multiplet constructed according to (1) and (2) further contains

$$\sum_{J=|L-S|}^{L+S} (2J+1) = (2S+1)(2L+1) \tag{2.14}$$

distinct states. Among these, again, often, though not always, only those states are important which satisfy the third Hund's rule:

3. $\mathbf{J} = |\mathbf{L} - \mathbf{S}|$ in case the shell is less than half-filled ($p \leq (2l+1)$)
$\mathbf{J} = \mathbf{L} + \mathbf{S}$ in case the shell is more than half-filled ($p \geq (2l+1)$)
which means

$$J = S\,|2l - p| \tag{2.15}$$

A special case is represented by the completely filled shell with $p = 2(2l+1)$. This is because, in this case, all the angular momenta vanish. A consequence of this is, on the other hand, that the total angular momentum of the whole atom is *identically equal* to the angular momentum of the partially filled shell.

Attention!

Let us conclude this section with an important remark! From (1.52) and (1.57) we know that as a consequence of the negative electron charge, angular momentum and magnetic moment are oriented antiparallel to each other. In a magnetic field, e.g. the energetically favoured orientation of the magnetic moment is parallel and that of the corresponding angular momentum is antiparallel to the field. Nevertheless, in almost all textbooks, the different sign of moment and momentum is disregarded. Although being, strictly speaking, wrong, it does not change any relevant physical statement. In order not to confuse the reader, when referring to other textbooks, we will follow the trend and disregard the different sign of moment and momentum. As an example, we will assume that a *spin up* electron (spin projection $+\frac{1}{2}$) has a lower energy in the magnetic field than a *spin down* electron (spin projection $-\frac{1}{2}$). This has already been used in (2.7) and (2.11) as well as in Table 2.1.

Table 2.1 Atomic term scheme for an f-shell occupied by p electrons (e.g. 4f-shell of a rare earth atom). Here $\uparrow$ represents spin projection $1/2$ and $\downarrow$ spin projection $-1/2$

p	m_l										
	3	2	1	0	−1	−2	−3	S	L	J	Term
1	↑							1/2	3	5/2	$^2F_{5/2}$
2	↑	↑						1	5	4	$^3H_4$
3	↑	↑	↑					3/2	6	9/2	$^4I_{9/2}$
4	↑	↑	↑	↑				2	6	4	$^5I_4$
5	↑	↑	↑	↑	↑			5/2	5	5/2	$^6H_{5/2}$
6	↑	↑	↑	↑	↑	↑		3	3	0	$^7F_0$
7	↑	↑	↑	↑	↑	↑	↑	7/2	0	7/2	$^8S_{7/2}$
8	↓↑	↑	↑	↑	↑	↑	↑	3	3	6	$^7F_6$
9	↓↑	↓↑	↑	↑	↑	↑	↑	5/2	5	15/2	$^6H_{15/2}$
10	↓↑	↓↑	↓↑	↑	↑	↑	↑	2	6	8	$^5I_8$
11	↓↑	↓↑	↓↑	↓↑	↑	↑	↑	3/2	6	15/2	$^4I_{15/2}$
12	↓↑	↓↑	↓↑	↓↑	↓↑	↑	↑	1	5	6	$^3H_6$
13	↓↑	↓↑	↓↑	↓↑	↓↑	↓↑	↑	1/2	3	7/2	$^2F_{7/2}$
14	↓↑	↓↑	↓↑	↓↑	↓↑	↓↑	↓↑	0	0	0	$^1S_0$

2.2 Dirac Equation

We now want to discuss, more quantitatively, the ideas that were presented rather qualitatively so far. In particular, we are concerned with establishing a rigorous basis for the electron spin. The starting point for this is the Dirac equation which is a result of the linearization of the relativistic generalization of the Schrödinger equation.

The Dirac equation describes

1. Relativistic electrons
2. Spin 1/2 particles
3. Magnetic moment of spin
4. Spin–orbit coupling

Thus, even the spin–orbit coupling is basically a relativistic effect. We will consider, in the following, *a single electron.*

The energy equation for a free classical relativistic particle is given by

$$E^2 = c^2\mathbf{p}^2 + m^2c^4 \tag{2.16}$$

where m is the mass of the electron, $\mathbf{p} = \gamma\,(v_x,\ v_y,\ v_z)$ is the relativistic mechanical momentum with $\gamma = (1 - v^2/c^2)^{1/2}$ and c is the velocity of light. From (2.16) one can obtain the wave equation using the *Schrödinger's correspondence rule*

$$E \rightarrow i\hbar\,\frac{\partial}{\partial t} \qquad ; \qquad \mathbf{p} \rightarrow \frac{\hbar}{i}\,\nabla. \tag{2.17}$$

This can be written compactly in relativistically covariant four-dimensional form:

$$p_\mu \rightarrow i\hbar\,\partial^\mu \tag{2.18}$$

where

$$p_\mu = (E/c,\ \gamma\ m\ v_x,\ \gamma\ m\ v_y,\ \gamma\ m\ v_z) = (E/c,\ \gamma\ m\ \mathbf{v}) \tag{2.19}$$

is the four-momentum and

$$\partial^\mu = \left(\frac{1}{c}\frac{\partial}{\partial t},\ -\nabla\right) \tag{2.20}$$

is the contravariant four-gradient.

With this we obtain the so-called *Klein-Gordon equation* which is the relativistic generalization of the Schrödinger equation:

$$\left(\Delta - \frac{1}{c^2}\frac{\partial^2}{\partial t^2} - \frac{m^2c^2}{\hbar^2}\right)\psi = 0 \tag{2.21}$$

There is a problem with (2.21). It is a differential equation of second order in t. Therefore, its solution requires two initial conditions, namely, for ψ and $\dot{\psi}$. In contrast, the non-relativistic Schrödinger equation is obviously linear in t. Therefore one should question, whether inclusion of the relativistic effects should lead to such drastic changes in the requirement on the initial information.

Dirac's idea was to *linearize* the wave equation (2.21):

$$\left(E - c\sum_i \alpha_i\, p_i - \beta\, m\, c^2\right)\left(E + c\sum_j \alpha_j\, p_j + \beta\, m\, c^2\right) = 0 \qquad i,\ j \in \{x,\ y,\ z\} \tag{2.22}$$

This equation is identical to (2.16) provided the operators (!) α_i and β are so chosen to satisfy the following relations:

$$\alpha_i\, \alpha_j + \alpha_j\, \alpha_i = 2\, \delta ij \cdot 1 \tag{2.23}$$

$$\alpha_i\, \beta + \beta\, \alpha_i = 0 \tag{2.24}$$

$$\beta^2 = 1 \tag{2.25}$$

Further, α's and β's must commute with $\mathbf{p}$. Every solution of the linearized equations

$$\left(E \pm c \sum_i \alpha_i\, p_i \pm \beta\, m\, c^2\right) \psi = 0 \tag{2.26}$$

is also a solution of the Klein–Gordon equation; the converse is, however, not necessarily true. We thus have the *Dirac equation for a free particle*:

$$\left(i\hbar\, \frac{\partial}{\partial t} - c\, \alpha \cdot \mathbf{p} - \beta\, m\, c^2\right) \psi = 0. \tag{2.27}$$

This form of the wave equation is satisfying from the relativistic view point, too. The special theory of relativity is typically known to treat the space and time components on equal footing. Therefore, a relativistic wave equation is expected to possess the symmetry between the space coordinates and time, which means, it should be of first order in the space coordinates, too. Obviously, that is the case here!

The above considerations can easily be extended to the case, where the particle is in the presence of an electromagnetic field, yielding the *Dirac equation for an electron in an electromagnetic field.*

$$\left\{i\hbar\, \frac{\partial}{\partial t} - c\, \alpha \cdot (\mathbf{p} + e\mathbf{A}) - \beta\, m\, c^2 + e\, \phi\right\} \psi = 0 \tag{2.28}$$

It is obtained by the usual substitutions:

$$\mathbf{p} \rightarrow \mathbf{p} + e\, \mathbf{A} \tag{2.29}$$

$$E \to E + e\,\phi \tag{2.30}$$

They can be expressed in the covariant form by

$$p_\mu \to p_\mu + e\,A_\mu \tag{2.31}$$

where A_μ is the four-potential

$$A_\mu \equiv (\mathbf{A}, \frac{1}{c}\,\phi) \tag{2.32}$$

The conditions (2.23), (2.24), and (2.25) on α's and β's cannot be satisfied by normal c-numbers. Rather similar relations are obeyed by the Pauli's spin matrices:

$$\sigma = (\sigma_x,\ \sigma_y,\ \sigma_z) \tag{2.33}$$

$$\sigma_x = \begin{pmatrix} 0 & 1 \\ 1 & 0 \end{pmatrix} \quad \sigma_y = \begin{pmatrix} 0 & -i \\ i & 0 \end{pmatrix} \quad \sigma_z = \begin{pmatrix} 1 & 0 \\ 0 & -1 \end{pmatrix} \tag{2.34}$$

Eqs. (2.23), (2.24), and (2.25) are satisfied by the following 4×4 matrices (see Problem 2.2):

$$\widehat{\alpha} = \begin{pmatrix} 0 & \sigma \\ \sigma & 0 \end{pmatrix}; \quad \widehat{\beta} = \begin{pmatrix} \mathbb{1}_2 & 0 \\ 0 & -\mathbb{1}_2 \end{pmatrix} \tag{2.35}$$

Here $\mathbb{1}_2$ indicates a 2×2 unit matrix. The wavefunction also naturally will be a four-component object:

$$\psi = \begin{pmatrix} \psi_1 \\ \psi_2 \\ \psi_3 \\ \psi_4 \end{pmatrix} \tag{2.36}$$

According to (2.27), the Dirac's Hamiltonian $H_D^{(0)}$ for the free electron, which is defined by

$$i\hbar\,\frac{\partial}{\partial t}\,\psi = H_D^{(0)}\,\psi \tag{2.37}$$

satisfies the equation

$$H_D^{(0)} = c\,\widehat{\alpha}\,\mathbf{p} + \widehat{\beta}\,m\,c^2 \tag{2.38}$$

$H_D^{(0)}$ is given in the matrix form by

$$H_D^{(0)} = \begin{pmatrix} mc^2 & 0 & cp_z & c(p_x - ip_y) \\ 0 & mc^2 & c(p_x + ip_y) & -cp_z \\ cp_z & c(p_x - ip_y) & -mc^2 & 0 \\ c(p_x + ip_y) & -cp_z & 0 & -mc^2 \end{pmatrix} \tag{2.39}$$

Obviously,

$$\left[H_D^{(0)}, \mathbf{p}\right]_- = 0 \tag{2.40}$$

which means we can choose the eigenfunctions of $H_D^{(0)}$ as momentum eigenfunctions. These are the plane waves whose direction of propagation may be chosen as the z-axis ($p_x = p_y = 0, \;\; p_z = p$):

$$\psi \equiv \widehat{a}\, e^{\frac{i}{\hbar}(pz - Et)} \tag{2.41}$$

Here $\widehat{a}$ is a spinor with four position-independent components a_i:

$$\widehat{a} = \begin{pmatrix} a_1 \\ a_2 \\ a_3 \\ a_4 \end{pmatrix} \tag{2.42}$$

Now we have to solve the eigenvalue equation

$$\begin{pmatrix} mc^2 - E & 0 & cp & 0 \\ 0 & mc^2 - E & 0 & -cp \\ cp & 0 & -mc^2 - E & 0 \\ 0 & -cp & 0 & -mc^2 - E \end{pmatrix} \begin{pmatrix} a_1 \\ a_2 \\ a_3 \\ a_4 \end{pmatrix} = 0 \tag{2.43}$$

where p is no longer an operator but a c-number. The condition that the secular determinant should vanish gives the energy eigenvalues

$$E = \varepsilon E_p \; ; \quad \varepsilon = \pm 1 \; ; \quad E_p = \sqrt{c^2\, p^2 + m^2\, c^4} \tag{2.44}$$

Each eigenvalue is doubly degenerate, i.e. for every eigenvalue there exist two linearly independent eigensolutions. As a reasonable result, we retrieve the relativistic energy law which is to be expected.

For the un-normalized spinors, we can easily find
$\varepsilon = +1$

$$\widehat{a}_1^{(+)} \equiv \begin{pmatrix} 1 \\ 0 \\ \frac{c\,p}{E_p+mc^2} \\ 0 \end{pmatrix} \; ; \; \widehat{a}_2^{(+)} \equiv \begin{pmatrix} 0 \\ 1 \\ 0 \\ -\frac{c\,p}{E_p+mc^2} \end{pmatrix} \tag{2.45}$$

$\varepsilon = -1$

$$\widehat{a}_1^{(-)} \equiv \begin{pmatrix} -\frac{c\,p}{E_p+mc^2} \\ 0 \\ 1 \\ 0 \end{pmatrix} \; ; \; \widehat{a}_2^{(-)} \equiv \begin{pmatrix} 0 \\ \frac{c\,p}{E_p+mc^2} \\ 0 \\ 1 \end{pmatrix} \tag{2.46}$$

The *general solution* will be a linear combination

$$\psi^{(\pm)} = \left(A_{\pm}\,\widehat{a}_1^{(\pm)} + B_{\pm}\,\widehat{a}_2^{(\pm)}\right) exp[\frac{i}{\hbar}\,(p\,z - E\,t)] \tag{2.47}$$

We now define the *Dirac spin operator*

$$\widehat{\mathbf{s}} = \frac{\hbar}{2}\,\widehat{\sigma} \qquad ; \qquad \widehat{\sigma} = \begin{pmatrix} \sigma & 0 \\ 0 & \sigma \end{pmatrix} \tag{2.48}$$

where $\widehat{\sigma}$ is the four-component relativistic generalization of the Pauli spin operator $\sigma = (\sigma_x,\ \sigma_y,\ \sigma_z)$. In particular, it holds

$$\widehat{\sigma}_z = \begin{pmatrix} 1 & 0 & 0 & 0 \\ 0 & -1 & 0 & 0 \\ 0 & 0 & 1 & 0 \\ 0 & 0 & 0 & -1 \end{pmatrix} \tag{2.49}$$

Then it directly follows that

$$\widehat{s}_z\,\widehat{a}_1^{(\pm)} = +\frac{\hbar}{2}\,\widehat{a}_1^{(\pm)} \tag{2.50}$$

$$\widehat{s}_z\,\widehat{a}_2^{(\pm)} = -\frac{\hbar}{2}\,\widehat{a}_2^{(\pm)} \tag{2.51}$$

The four-component spinors distinguish themselves through a "new" quantum number, which we call from now onwards as *spin*:

$$\begin{aligned}\widehat{a}_1^{(+)} &\Leftrightarrow \left(+E_p,\ +\tfrac{\hbar}{2}\right)\\ \widehat{a}_2^{(+)} &\Leftrightarrow \left(+E_p,\ -\tfrac{\hbar}{2}\right)\\ \widehat{a}_1^{(-)} &\Leftrightarrow \left(-E_p,\ +\tfrac{\hbar}{2}\right)\\ \widehat{a}_2^{(-)} &\Leftrightarrow \left(-E_p,\ -\tfrac{\hbar}{2}\right)\end{aligned} \tag{2.52}$$

It should be noted that the general solution $\psi^{(\pm)}$ is *not* an eigenfunction of $\widehat{s}_z$. This is obvious because $\widehat{s}_z$ does not commute with $H_D^{(0)}$ (see Problem 2.5).

2.3 Electron Spin

In the last section, from the Dirac equation, we have obtained an indication about a new quantum number, namely the spin. We now want to show, at first more or less qualitatively, that this new degree of freedom is something like an angular momentum.

We extend the Dirac's Hamiltonian operator $H_D^{(0)}$ (2.38) by including a central potential $V(r)$.

$$H_D^{(V)} = c\,\widehat{\alpha}\cdot\mathbf{p} + \widehat{\beta}\,m\,c^2 + V(r) \tag{2.53}$$

Non-relativistically, we expect that in a central potential, the orbital angular momentum

$$\mathbf{l} = \mathbf{r}\times\mathbf{p} \tag{2.54}$$

of the electron is a constant of motion. However, one finds (Problem 2.5) that

$$\left[\mathbf{l}\,,\ H_D^{(V)}\right]_- \neq 0 \tag{2.55}$$

As an example, we explicitly calculate the commutator (2.55) for the x-component $l_x = yp_z - zp_y$. The other components behave analogously.

$$\begin{aligned}\left[l_x\,,\ H_D^{(V)}\right]_- &= c\sum_{i=1}^{3}[l_x\,,\ \alpha_i p_i]_- + mc^2\,[l_x\,,\ \beta]_- + [l_x\,,\ V(r)]_-\\ &= c\sum_{i=1}^{3}\alpha_i\left([yp_z\,,\ p_i]_- - \left[zp_y\,,\ p_i\right]_-\right)\\ &= c\left(\alpha_y\left[y\,,\ p_y\right]_- p_z - \alpha_z\,[z\,,\ p_z]_-\,p_y\right)\end{aligned}$$

This finally results in

$$\left[l_x\,,\;H_D^{(V)}\right]_- = i\hbar c\left(\alpha_y p_z - \alpha_z p_y\right) \tag{2.56}$$

For the other components we analogously get (see Problem 2.5)

$$\left[l_y\,,\;H_D^{(V)}\right]_- = i\hbar c\left(\alpha_z p_x - \alpha_x p_z\right) \tag{2.57}$$

$$\left[l_z\,,\;H_D^{(V)}\right]_- = i\hbar c\left(\alpha_x p_y - \alpha_y p_x\right) \tag{2.58}$$

A theory is certainly not satisfactory, if it violates the principle of conservation of angular momentum, when the force is a central force. The reason probably is that the orbital angular momentum is not the total angular momentum. Therefore, let us examine the Dirac's spin operator (2.48),

$$\widehat{\mathbf{s}} = \frac{\hbar}{2}\,\widehat{\sigma} \tag{2.59}$$

and consider again the x-component:

$$\widehat{s}_x = \frac{\hbar}{2}\begin{pmatrix}\sigma_x & 0\\ 0 & \sigma_x\end{pmatrix} = \frac{\hbar}{2}\begin{pmatrix}0\,1\,0\,0\\ 1\,0\,0\,0\\ 0\,0\,0\,1\\ 0\,0\,1\,0\end{pmatrix} \tag{2.60}$$

In calculating the commutators of $\widehat{s}_x$ with the Dirac's Hamiltonian operator $H_D^{(V)}$, we exploit the fact that $\widehat{s}_x$ commutes with p_i. Then we have

$$\left[\widehat{s}_x\,,\;H_D^{(V)}\right]_- = c\sum_{i=1}^{3}\left[\widehat{s}_x\,,\;\alpha_i\right]_-\,p_i + mc^2\left[\widehat{s}_x\,,\;\beta\right]_- \tag{2.61}$$

Using the matrix representation (2.35) for β, we find that

$$\widehat{s}_x\,\beta = \frac{\hbar}{2}\begin{pmatrix}0 & 1 & 0 & 0\\ 1 & 0 & 0 & 0\\ 0 & 0 & 0 & -1\\ 0 & 0 & -1 & 0\end{pmatrix} = \beta\,\widehat{s}_x \tag{2.62}$$

Therefore, the second summand in (2.61) vanishes:

$$\left[\widehat{s}_x\,,\;\beta\right]_- = 0 \tag{2.63}$$

Using the definition of α_x (2.35), we get

$$\widehat{s}_x\,\alpha_x = \frac{\hbar}{2}\begin{pmatrix} 0&0&1&0\\ 0&0&0&1\\ 1&0&0&0\\ 0&1&0&0 \end{pmatrix} = \alpha_x\,\widehat{s}_x \tag{2.64}$$

Therefore, we have

$$[\widehat{s}_x\ ,\ \alpha_x]_- = 0 \tag{2.65}$$

Only the commutators of $\widehat{s}_x$ with α_y and α_z give non-zero contributions. We note that

$$\widehat{s}_x\,\alpha_y = \frac{\hbar}{2}\begin{pmatrix} 0&0&i&0\\ 0&0&0&-i\\ i&0&0&0\\ 0&-i&0&0 \end{pmatrix} = -\alpha_y\,\widehat{s}_x = +\,i\,\frac{\hbar}{2}\,\alpha_z \tag{2.66}$$

Therefore, we have

$$\left[\widehat{s}_x\ ,\ \alpha_y\right]_- = +\,i\,\hbar\,\alpha_z \tag{2.67}$$

Analogously we also get from

$$\widehat{s}_x\,\alpha_z = \frac{\hbar}{2}\begin{pmatrix} 0&0&0&-1\\ 0&0&1&0\\ 0&-1&0&0\\ 1&0&0&0 \end{pmatrix} = -\alpha_z\,\widehat{s}_x = -\,i\,\frac{\hbar}{2}\,\alpha_y \tag{2.68}$$

the third commutator:

$$[\widehat{s}_x\ ,\ \alpha_z]_- = -i\,\hbar\,\alpha_y \tag{2.69}$$

Thus, we finally have

$$\left[\widehat{s}_x\ ,\ H_D^{(V)}\right]_- = i\,\hbar\,c\left(\alpha_z\,p_y - \alpha_y\,p_z\right) \tag{2.70}$$

We can calculate analogously for the other components:

$$\left[\widehat{s}_y\ ,\ H_D^{(V)}\right]_- = i\,\hbar\,c\,(\alpha_x\,p_z - \alpha_z\,p_x) \tag{2.71}$$

$$\left[\widehat{s}_z\ ,\ H_D^{(V)}\right]_- = i\,\hbar\,c\left(\alpha_y\,p_x - \alpha_x\,p_y\right) \tag{2.72}$$

Neither $\widehat{\mathbf{s}}$ nor $\mathbf{l}$ commutes with $H_D^{(V)}$ but we have

$$\left[\mathbf{l}\,,\; H_D^{(V)}\right]_- = -\left[\widehat{\mathbf{s}}\,,\; H_D^{(V)}\right]_- \tag{2.73}$$

so that the *sum* of spin and orbital angular momentum does commute with $H_D^{(V)}$. This leads to the following interpretation:

$$\begin{aligned} \mathbf{l}+\widehat{\mathbf{s}} = \mathbf{l} + \frac{\hbar}{2}\widehat{\sigma} \quad &- \quad \textit{Total angular momentum operator} \\ \widehat{\mathbf{s}} = \frac{\hbar}{2}\widehat{\sigma} \quad &- \quad \textit{Spin operator} \end{aligned} \tag{2.74}$$

Therefore, the spin, which follows quite naturally from Dirac's theory can obviously be interpreted as angular momentum. The total angular momentum $\mathbf{l}+\widehat{\mathbf{s}}$ is, in a central field, a conserved quantity.

We now return to the problem of *Dirac particle in an electromagnetic field.* According to (2.28) the Hamiltonian is

$$H_D = c\widehat{\alpha}\cdot(\mathbf{p}+e\mathbf{A}) + \widehat{\beta}mc^2 - e\phi \tag{2.75}$$

The field could be due to the nuclear charge. From now onwards, we will consider only the positive energy solutions (electrons).

It is convenient to split the four-component Dirac spinor into two components.

$$\psi = \begin{pmatrix} \psi_1 \\ \psi_2 \\ \psi_3 \\ \psi_4 \end{pmatrix} = \begin{pmatrix} \psi_+ \\ 0 \end{pmatrix} + \begin{pmatrix} 0 \\ \psi_- \end{pmatrix} \tag{2.76}$$

with

$$\psi_+ = \begin{pmatrix} \psi_1 \\ \psi_2 \end{pmatrix} \quad ; \quad \psi_- = \begin{pmatrix} \psi_3 \\ \psi_4 \end{pmatrix} \tag{2.77}$$

At the moment, it is only another way of writing, where both the summands are the eigenstates of the operator β with eigenvalues ± 1, respectively:

$$\widehat{\beta}\begin{pmatrix} \psi_+ \\ 0 \end{pmatrix} = \begin{pmatrix} \psi_+ \\ 0 \end{pmatrix} \quad ; \quad \widehat{\beta}\begin{pmatrix} 0 \\ \psi_- \end{pmatrix} = -\begin{pmatrix} 0 \\ \psi_- \end{pmatrix} \tag{2.78}$$

The motivation for this splitting is the following. We are ultimately interested in the non-relativistic limit $v \ll c$ of the Dirac's theory. In this limit, the energy eigenvalue of the *free particle* differ from the rest energy mc^2 only by a small amount. The difference

$$\begin{aligned} T = E_p - mc^2 &= \sqrt{c^2p^2 + m^2c^4} - mc^2 \\ &= mc^2\left(1 + \frac{p^2}{m^2c^2}\right)^{1/2} - mc^2 \\ &\approx \frac{p^2}{2m} \end{aligned} \tag{2.79}$$

is just the same as the well-known expression for the kinetic energy. We can estimate the relative magnitudes of the individual components $\widehat{a}_{1,2}^{(+)}$ (2.45) of the solution for the free Dirac particle:

$$\begin{aligned} \left(\frac{a_{13}^{(+)}}{a_{11}^{(+)}}\right)^2 = \left(\frac{a_{24}^{(+)}}{a_{22}^{(+)}}\right)^2 &= \frac{cp^2}{(E_p + mc^2)^2} \\ &= \frac{E_p^2 - (mc^2)^2}{(E_p + mc^2)^2} \\ &= \frac{E_p - mc^2}{E_p + mc^2} \\ &= \frac{T}{T + 2mc^2} = \mathcal{O}\left(\frac{v^2}{c^2}\right) \ll 1 \end{aligned} \tag{2.80}$$

In the non-relativistic limit, the two components $a_{13}^{(+)}$ and $a_{24}^{(+)}$ are negligibly small. Even when a "normal" electromagnetic field is switched on, these order of magnitude estimates will not change. On the other hand, $a_{13}^{(+)}$ and $a_{24}^{(+)}$ determine the corresponding components ψ_-. Thus, in the limit $v \ll c$, between ψ_- and ψ_+, there will be an order of magnitude difference. This is actually the motivation for the above-performed splitting, so that, in the non-relativistic limit, the Dirac's theory reduces to a two-component theory *(Pauli's theory).*

Using (2.75) for H_D and (2.35) for $\widehat{\alpha}$ and $\widehat{\beta}$, we can write the Dirac equation in the following form:

$$\begin{aligned} H_D\begin{pmatrix}\psi_+\\ \psi_-\end{pmatrix} &= c\,(\mathbf{p} + e\mathbf{A})\begin{pmatrix}\sigma\,\psi_-\\ \sigma\,\psi_+\end{pmatrix} + mc^2\begin{pmatrix}\psi_+\\ -\psi_-\end{pmatrix} \\ &- e\phi\begin{pmatrix}\psi_+\\ \psi_-\end{pmatrix} \doteq E\begin{pmatrix}\psi_+\\ \psi_-\end{pmatrix} \end{aligned} \tag{2.81}$$

This matrix equation means that we have to solve the following system of equations:

$$(E - mc^2 + e\phi)\psi_+ = c\,(\mathbf{p} + e\mathbf{A})\cdot\sigma\;\psi_- \tag{2.82}$$

$$(E + mc^2 + e\phi)\psi_- = c\,(\mathbf{p} + e\mathbf{A})\cdot\sigma\;\psi_+ \tag{2.83}$$

From the second equation, it follows that

$$\psi_- = (E + mc^2 + e\phi)^{-1} c\,(\mathbf{p} + e\mathbf{A}) \cdot \sigma\ \psi_+ \tag{2.84}$$

Once again from this, the orders of magnitude are clear.

$$\underbrace{\psi_-}_{small} \approx \frac{\text{“}v\text{''}}{c} \underbrace{\psi_+}_{large} \tag{2.85}$$

Within the framework of the *Pauli's theory*, which represents the non-relativistic limit of Dirac's theory, that is, if

$$(E - mc^2),\ e\phi,\ \frac{1}{2m}(\mathbf{p} + e\mathbf{A})^2 \ll mc^2 \tag{2.86}$$

we can write up to an error of order of magnitude v^2/c^2

$$\psi_- \approx \frac{1}{2mc}\,(\mathbf{p} + e\mathbf{A}) \cdot \sigma\ \psi_+ \tag{2.87}$$

The system of Eqs. (2.82) and (2.83) can then be written within the corresponding error as eigenvalue equation for ψ_+ by inserting (2.87) into (2.82):

$$H_p\ \psi_+ = (E - mc^2)\ \psi_+ \tag{2.88}$$

with

$$H_p = \frac{1}{2m}\,[(\mathbf{p} + e\mathbf{A}) \cdot \sigma][(\mathbf{p} + e\mathbf{A}) \cdot \sigma] - e\phi \tag{2.89}$$

The *Pauli's Hamiltonian* is then a 2×2 matrix. Using the identity (proof as Problem 2.6)

$$(\mathbf{a} \cdot \sigma)(\mathbf{b} \cdot \sigma) = (\mathbf{a} \cdot \mathbf{b})1 + i(\mathbf{a} \times \mathbf{b}) \cdot \sigma \tag{2.90}$$

we can further rewrite (2.89). For that we have to evaluate

$$\begin{aligned}(\mathbf{p} + e\mathbf{A}) \times (\mathbf{p} + e\mathbf{A})\psi &= (\mathbf{p} \times \mathbf{p} + e^2\mathbf{A} \times \mathbf{A})\ \psi \\ &\quad + e\ (\mathbf{A} \times \mathbf{p} + \mathbf{p} \times \mathbf{A})\ \psi \\ &= e\ \frac{\hbar}{i}\ (\mathbf{A} \times \nabla\psi + \nabla \times (\mathbf{A}\psi)) \\ &= e\ \frac{\hbar}{i}\ (\nabla \times \mathbf{A})\ \psi\end{aligned}$$

With $\nabla \times \mathbf{A} = \mathbf{B}_0$ and $\mathbf{s} = \frac{\hbar}{2}\,\sigma$, we then have

$$H_p = \frac{1}{2m}\,(\mathbf{p} + e\mathbf{A})^2 + 2\,\frac{\mu_B}{\hbar}\,(\mathbf{s}\cdot\mathbf{B}_0) - e\phi \tag{2.91}$$

This is the Hamiltonian of a particle of mass m, the charge e and magnetic eigenmoment

$$\widehat{\mathbf{m}}_s = -2\,\frac{\mu_B}{\hbar}\,\mathbf{s} \tag{2.92}$$

We now summarize the important results of this section:

1. $\mathbf{s} = \frac{\hbar}{2}\,\sigma$ is classified as angular momentum with the name, "spin" with eigenvalues $\pm\,\frac{\hbar}{2}$.
2. Magnetic eigenmoment of the electron is established with a Lande's g-factor $g_e = 2$.

2.4 Spin–Orbit Coupling

Spin–orbit coupling is a relativistic effect and therefore it must be derived from the Dirac's theory, which we will do in the following. However, the essence of the argument can already be recognized based on a simple-minded physical estimate. Therefore, we will first present this estimate before undertaking the exact derivation. The electron moves in the rest system of the charged nucleus which creates an electrostatic field

$$\mathbf{E} = -\nabla\phi \tag{2.93}$$

Since the electron moves with respect to the nucleus, it "sees" a magnetic field $\mathbf{B}$ which according to relativistic electrodynamics is given by

$$\bar{\mathbf{B}} = \gamma\left(\mathbf{B} - \frac{1}{c}\,\beta\times\mathbf{E}\right) - \frac{\gamma^2}{\gamma+1}\,\beta\,(\beta\cdot\mathbf{B}) \qquad (\beta = \frac{1}{c}\,\mathbf{v}) \tag{2.94}$$

Here $\mathbf{E}$ and $\mathbf{B}$ shall be the fields in the rest frame of the nucleus ($\mathbf{B} = 0$) and $\bar{\mathbf{E}}$ and $\bar{\mathbf{B}}$ are the fields in the rest frame of the electron. Since $v \ll c$, $\gamma \approx 1$, we are left with

$$\bar{\mathbf{B}} \approx -\frac{1}{c^2}\,\mathbf{v}\times\mathbf{E} \tag{2.95}$$

If we postulate the existence of a spin moment, then, due to this field, there is an additional term in the Hamiltonian (2.91) of the electron:

$$\bar{H}_{SO} = 2\,\frac{\mu_B}{\hbar}\,\bar{\mathbf{B}}\cdot\mathbf{s} = \frac{e}{mc^2}\,(\mathbf{E}\times\mathbf{v})\cdot\mathbf{s} \tag{2.96}$$

If we assume a spherically symmetric nuclear potential,

$$\mathbf{E} = -\frac{\mathbf{r}}{r}\frac{d\phi}{dr} \tag{2.97}$$

then we get using $\mathbf{l} = \mathbf{r} \times \mathbf{p} = m\,\mathbf{r} \times \mathbf{v}$,

$$\bar{H}_{SO} = -\frac{e}{m^2c^2}\left(\frac{1}{r}\frac{d\phi}{dr}\right)(\mathbf{l} \cdot \mathbf{s}) \tag{2.98}$$

This extra term clearly describes the coupling of the electron spin with the orbital motion in the field of the nucleus. However, it comes out that, H_{SO} is too large by a factor of 2.

For the rigorous derivation of spin–orbit interaction, we make use of the *non-relativistic limit of the Dirac's theory*. Now, however, we have to extend the approximation a step further as compared to what we have done in the last section. In discussing the spin magnetic moment, we have neglected the terms of the order of (v^2/c^2). It was done, so that, in the limit $v \ll c$, the "small" component ψ_- can be eliminated and the theory can be reduced from a four-component theory to a two-component theory, which is the normal situation in non-relativistic physics. However, neglecting ψ_- might possibly lead to an error in the normalization of the wavefunction. This we examine more carefully now.

$$\psi = \begin{pmatrix} \psi_+ \\ \psi_- \end{pmatrix} \Rightarrow \psi^*\psi = \psi_+^*\psi_+ + \psi_-^*\psi_-. \tag{2.99}$$

When we go over to a two-component theory, the normalization should be conserved. Therefore, we start with the following ansatz:

$$\psi_+ = N\chi \tag{2.100}$$

The normalization constant will be fixed by the condition

$$\psi^*\,\psi \stackrel{!}{=} \chi^*\,\chi = N^2\,\chi^*\,\chi + \psi_-^*\,\psi_- \tag{2.101}$$

χ is now the "new" wavefunction of the two-component theory. Let the electron move in the **E**-field ($\mathbf{B}_0 = 0$) created by the nucleus. Then we have according to (2.84)

$$\psi_- = \frac{c}{E + mc^2 + e\phi}\,(\mathbf{p} \cdot \sigma)\,\psi_+ \tag{2.102}$$

The quantity, which is "small" in the non-relativistic limit is not E but

$$T = E - mc^2 \tag{2.103}$$

Therefore, we substitute appropriately in (2.102) and expand in powers of v/c:

$$\begin{aligned}\psi_- &= \frac{1}{2mc}\left(1+\frac{T+e\phi}{2mc^2}\right)^{-1}(\mathbf{p}\cdot\sigma)\,\psi_+ \\ &= \frac{1}{2mc}\left(1-\frac{T+e\phi}{2mc^2}+\mathcal{O}(v^4/c^4)\right)(\mathbf{p}\cdot\sigma)N\,\chi\end{aligned} \tag{2.104}$$

Substituting this in the normalization condition (2.101),

$$\chi^*\chi = N^2\chi^*\chi\left(1+\frac{1}{4m^2c^2}(\mathbf{p}\cdot\sigma)^2+\mathcal{O}(v^4/c^4)\right) \tag{2.105}$$

and using the vector identity (2.90) to evaluate the scalar product,

$$(\mathbf{p}\cdot\sigma)^2 = \mathbf{p}^2 + i(\mathbf{p}\times\mathbf{p})\cdot\sigma = \mathbf{p}^2 \tag{2.106}$$

finally yields for the normalization constant N

$$N = \left(1+\frac{p^2}{4m^2c^2}+\mathcal{O}(v^4/c^4)\right)^{-1/2} = 1-\frac{p^2}{8m^2c^2}+\mathcal{O}(v^4/c^4) \tag{2.107}$$

With this, we have the following intermediate result for ψ_+:

$$\psi_+ \approx \left(1-\frac{p^2}{8m^2c^2}\right)\chi \tag{2.108}$$

Substituting this in (2.104) we get for ψ_-

$$\psi_- \approx \frac{1}{2mc}\left(1-\frac{T+e\phi}{2mc^2}\right)(\mathbf{p}\cdot\sigma)\left(1-\frac{p^2}{8m^2c^2}\right)\chi$$

and finally

$$\psi_- \approx \frac{1}{2mc}\left((\mathbf{p}\cdot\sigma)-(\mathbf{p}\cdot\sigma)\frac{p^2}{8m^2c^2}-\frac{T+e\phi}{2mc^2}(\mathbf{p}\cdot\sigma)\right)\chi \tag{2.109}$$

Compared to the result in Sect. 2.3, the last two terms in the above equation are extra. We will substitute this expression for ψ_- in (2.82) which is still exact and which is rewritten as

$$(T+e\phi)\,\psi_+ = c\,(\mathbf{p}\cdot\sigma)\,\psi_- \tag{2.110}$$

By doing this, we have transformed a four-component theory into a two-component one. With (2.108) and (2.109), we then obtain an eigenvalue equation for χ:

$$(T+e\phi)\left(1-\frac{p^2}{8m^2c^2}\right)\chi \approx \frac{1}{2m}\left[(\mathbf{p}\cdot\sigma)^2\left(1-\frac{p^2}{8m^2c^2}\right) - (\mathbf{p}\cdot\sigma)\frac{T+e\phi}{2mc^2}(\mathbf{p}\cdot\sigma)\right]\chi \tag{2.111}$$

We will expand $T+e\phi$ also in powers of v/c:

$$\begin{aligned} T+e\phi &= \sqrt{c^2p^2+m^2c^4}-mc^2 \\ &= mc^2\left[\left(1+\frac{p^2}{m^2c^2}\right)^{1/2}-1\right] \\ &= mc^2\left[1+\frac{p^2}{2m^2c^2}-\frac{1}{8}\frac{p^4}{m^4c^4}-1+\mathcal{O}(v^6/c^6)\right] \\ &\approx \frac{p^2}{2m}-\frac{p^4}{8m^3c^2} \end{aligned} \tag{2.112}$$

This simplifies our eigenvalue equation for χ:

$$(T+e\phi)\chi \approx \left[\frac{p^2}{2m}-(\mathbf{p}\cdot\sigma)\,\frac{T+e\phi}{4m^2c^2}\,(\mathbf{p}\cdot\sigma)\right]\chi \tag{2.113}$$

When proceeding further, we have to take into account the fact that ϕ, which is a function of r, and $\mathbf{p}$ do not commute with each other. In contrast, T is treated as the energy eigenvalue and therefore as a c-number (see (2.115)). Then we have

$$\begin{aligned} (\mathbf{p}\cdot\sigma)\frac{T+e\phi}{4m^2c^2}(\mathbf{p}\cdot\sigma) &= \frac{T+e\phi}{4m^2c^2}p^2+\frac{\hbar}{i}\frac{e}{4m^2c^2}(\nabla\phi\cdot\sigma)(\mathbf{p}\cdot\sigma) \\ &\approx \frac{p^4}{8m^3c^2}+\frac{\hbar}{i}\frac{e}{4m^2c^2}[(\nabla\phi\cdot\mathbf{p})+i(\nabla\phi\times\mathbf{p})\cdot\sigma] \end{aligned}$$

The last step is obtained by again using the vector identity (2.90). We define an operator, which, we from now on, will interpret as spin–orbit interaction, which is given by ($\mathbf{s}=\frac{\hbar}{2}\,\sigma$)

$$H_{SO}=\frac{-e}{2m^2c^2}\,[(\nabla\phi\times\mathbf{p})\cdot\mathbf{s}] \tag{2.114}$$

Then, we have the following eigenvalue equation for χ:

$$\tilde{H}\chi = T\chi \tag{2.115}$$

where

$$\tilde{H}=\frac{p^2}{2m}-\frac{p^4}{8m^3c^2}-e\phi+\frac{i\hbar e}{4m^2c^2}\,(\nabla\phi\cdot\mathbf{p})+H_{SO} \tag{2.116}$$

The individual terms have the following meaning:

$\frac{p^2}{2m}$	Non-relativistic kinetic energy of the electron
$-\frac{p^4}{8m^3c^2}$	First relativistic correction to kinetic energy
$-e\phi$	Potential energy of the electron in the Coulomb field of the nucleus, e.g. $4\pi\varepsilon_0\phi(r) = \frac{Z^*e}{r}$ Z^* = Effective nuclear charge
$\frac{i\hbar e}{4m^2c^2}(\nabla\phi \cdot \mathbf{p})$	*Darwin term*, relativistic correction to the potential energy of the electron (which has no classical analogue)
H_{SO}	Spin–orbit interaction

H_{SO} is different from the "phenomenological" result $\bar{H}_{SO}$ (2.98) exactly by a factor of 1/2:

$$H_{SO} = \frac{1}{2}\bar{H}_{SO} \tag{2.117}$$

For a spherically symmetric potential $\phi(r)$, i.e. $\nabla\phi(r) = \frac{\mathbf{r}}{r}\frac{d\phi}{dr}$, H_{SO} can be rewritten as

$$H_{SO} = -\frac{e}{2m^2c^2}\left(\frac{1}{r}\frac{d\phi}{dr}\right)(\mathbf{l}\cdot\mathbf{s}) \tag{2.118}$$

or using the simple expression

$$4\pi\ \varepsilon_0\ \phi(r) = \frac{Z^*e}{r}$$

it can be written as

$$H_{SO} = \lambda\,(\mathbf{l}\cdot\mathbf{s}) \quad ; \quad \lambda = \frac{1}{8\pi\,\varepsilon_0 m^2c^2}\cdot\frac{Z^*e^2}{r^3} \tag{2.119}$$

H_{SO} decreases rapidly with increasing nuclear distance r.

We discuss a few *consequences of spin–orbit coupling:*

1. H_{SO} is the reason why even in the absence of an external magnetic field ($\mathbf{B}_0 = 0$) $\mathbf{l}$ and $\mathbf{s}$ do not commute with H_D (see Sect. 2.3). One can easily show that (Problem 2.8)

$$[\mathbf{l}\cdot\mathbf{s}\ ,\ \mathbf{l}]_- = i\hbar\,(\mathbf{l}\times\mathbf{s}) = -\,[\mathbf{l}\cdot\mathbf{s}\ ,\ \mathbf{s}]_- \tag{2.120}$$

On the other hand,

$$[\mathbf{l}\cdot\mathbf{s}\ ,\ \mathbf{j}]_- = 0 \quad ; \quad \mathbf{j} = \mathbf{l}+\mathbf{s} \tag{2.121}$$

and further we have

$$[\mathbf{l} \cdot \mathbf{s}\, ,\, \mathbf{j}^2]_- = [\mathbf{l} \cdot \mathbf{s}\, ,\, \mathbf{l}^2]_- = [\mathbf{l} \cdot \mathbf{s}\, ,\, \mathbf{s}^2]_- = 0 \tag{2.122}$$

It means that the energy eigenstates can be classified by j, m_j, l and s (*good* quantum numbers) but not by m_l and m_s. H_{SO} couples (hybridizes) the states with different m_l and m_s.

2. H_{SO} partially lifts the degeneracy of the LS-multiplets (here the doublet, since our treatment is valid only for one electron: $j = l \pm 1/2$). Because of

$$\mathbf{j} = \mathbf{l} + \mathbf{s} \Rightarrow 2(\mathbf{l} \cdot \mathbf{s}) = \mathbf{j}^2 - \mathbf{l}^2 - \mathbf{s}^2$$

H_{SO} produces a *fine structure of the energy terms*:

$$E_{nlj}^{(0)} = E_{nl}^{(0)} + \lambda_{nl}\, \hbar^2\, [j\,(j+1) - l\,(l+1) - s\,(s+1)] \tag{2.123}$$

Here n is the principal quantum number and $E_{nl}^{(0)}$ is the energy in the absence of the spin–orbit coupling. The spin–orbit coupling constant is given according to (2.119) by

$$\lambda_{nl} = -\frac{e}{2m^2c^2}\, \langle nls | \frac{1}{r}\frac{d\phi}{dr} | nls \rangle \tag{2.124}$$

Because of the spin–orbit coupling, the terms $j = l \pm 1/2$ have different energies for $l \neq 0$. However, the $2j+1$-fold degeneracy due to m_j still remains.

2.5 Wigner–Eckart Theorem

Using Dirac equation, we were able to explain the existence of electron spin and the appearance of the spin–orbit coupling. We are now interested in the behaviour of the electron in the presence of the nuclear field and an external magnetic field. That is, we want to find out which of the energy levels are available, in the presence of these two fields, to the electron. This is certainly a non-trivial problem. For this purpose, we first discuss the so-called *Wigner–Eckart theorem*, which, at first glance, appears to be highly specialized but turns out to be very useful, particularly, in the context of the magnetic problems with which we are here concerned. In order to formulate and later apply the theorem, we first present a few necessary pre-considerations.

2.5.1 Rotation

Let Σ and $\bar{\Sigma}$ be two coordinate systems which have a common origin and are rotated with respect to each other. Let their axes be defined through the unit vectors

$$\mathbf{e}_i,\ \bar{\mathbf{e}}_i \quad ; \quad i = 1,\ 2,\ 3$$

The $\bar{\mathbf{e}}_i$ can be written as a linear combination of $\mathbf{e}_i$:

$$\bar{\mathbf{e}}_i = \sum_{j=1}^{3} R_{ji}\, \mathbf{e}_j \tag{2.125}$$

The coefficients R_{ji} which are given by

$$R_{ji} = (\mathbf{e}_j \cdot \bar{\mathbf{e}}_i) \tag{2.126}$$

are uniquely fixed by the angle and the axis of rotation. As a result of rotation, a vector $\mathbf{r}$ in Σ transforms itself as follows:

$$\mathbf{r} = (x_1,\ x_2,\ x_3) \rightarrow \bar{\mathbf{r}} = (\bar{x}_1,\ \bar{x}_2,\ \bar{x}_3) \tag{2.127}$$

In the co-rotated coordinate system $\bar{\Sigma}$, we naturally have

$$\bar{\mathbf{r}} = (x_1,\ x_2,\ x_3) \tag{2.128}$$

That means

$$\bar{\mathbf{r}} = R\,[\mathbf{r}] = \sum_j x_j\, \bar{\mathbf{e}}_j = \sum_i \bar{x}_i\, \mathbf{e}_i$$

and therefore,

$$\bar{x}_i = \sum_j R_{ij}\, x_j \tag{2.129}$$

R_{ij} are the elements of the 3×3 rotation matrix with the well-known *properties:*

1. $R_{ij} = R^*_{ij}$
2. $\tilde{R} = R^{-1}$ ($\tilde{R}$: transposed matrix)
3. $\det R = 1$

Example: Rotation through an angle ϕ about the z-axis:

$$R_z(\phi) = \begin{pmatrix} \cos\phi & -\sin\phi & 0 \\ \sin\phi & \cos\phi & 0 \\ 0 & 0 & 1 \end{pmatrix} \tag{2.130}$$

That means

$$\begin{aligned} \bar{x}_1 &= x_1 \cos\phi - x_2 \sin\phi \\ \bar{x}_2 &= x_1 \sin\phi + x_2 \cos\phi \\ \bar{x}_3 &= x_3 \end{aligned} \tag{2.131}$$

2.5.2 Rotation Operator

We now want to consider, quantum mechanically, the rotation of a physical system. Let $|\psi\rangle$ and $|\bar{\psi}\rangle$ denote the state of the system before and after the rotation. They are related by

$$|\bar{\psi}\rangle = R\,|\psi\rangle \tag{2.132}$$

where R is the rotation operator. The states $|\psi\rangle$ and $|\bar{\psi}\rangle$ obey the condition that the results of a measurement in the state $|\bar{\psi}\rangle$ follow from the rotation of the results of the same measurement done in the state $|\psi\rangle$. As an example, let us consider the measurement of position. Suppose in the state $|\psi\rangle$, the result of measurement being $\mathbf{r}_1$ has the probability density $|\psi(\mathbf{r}_1)|^2$ and the result of the same measurement in the state $|\bar{\psi}\rangle$ being $\mathbf{r}$ has the probability density $|\bar{\psi}(\mathbf{r})|^2$. If (2.132) is valid, it is required that

$$\mathbf{r} = R\,[\mathbf{r}_1] \quad ; \quad |\bar{\psi}(\mathbf{r})|^2 = |\psi(\mathbf{r}_1)|^2 \tag{2.133}$$

In order to satisfy this, it is sufficient to have

$$\bar{\psi}(\mathbf{r}) = R\,\psi(\mathbf{r}) = \psi(R^{-1}\,\mathbf{r}) \tag{2.134}$$

Since normalization should not be affected by rotation, the rotation operator must necessarily be unitary:

$$R^\dagger\,R = R\,R^\dagger = \mathbb{1} \Leftrightarrow R^\dagger = R^{-1} \tag{2.135}$$

With this, the behaviour of the states under rotation is clear. Now the question is, How do the observables behave? Rotation of an observable means nothing but the rotation of the measuring instruments. Logically, the measurement of A in the state $|\psi\rangle$ must mean the same as the measurement of $\bar{A}$ in the state $|\bar{\psi}\rangle$. That is,

$$\langle\psi|A|\psi\rangle \stackrel{!}{=} \langle\bar{\psi}|\bar{A}|\bar{\psi}\rangle = \langle\psi|R^\dagger\,A\,R|\psi\rangle \tag{2.136}$$

which means

$$\begin{aligned} A &= R^{-1}\,\bar{A}\,R \\ \bar{A} &= R\,A\,R^{-1} \end{aligned} \tag{2.137}$$

We conclude that the observables under rotation have the same transformation properties as the states.

There are two types of operators which are of special interest, namely, the *scalar operator* S which remains uninfluenced by rotation, i.e.

$$\bar{S} \stackrel{!}{=} S = R\,S\,R^{-1} \;\Leftrightarrow\; [R\,,\;S]_- = 0 \tag{2.138}$$

and the *vector operator* $\mathbf{V} = (V_1,\; V_2,\; V_3)$ with the components

$$V_i = \mathbf{V} \cdot \mathbf{e}_i \tag{2.139}$$

which transforms under rotation as follows:

$$\bar{V}_i = \mathbf{V} \cdot \bar{\mathbf{e}}_i = \sum_j R_{ji}\,\mathbf{V} \cdot \mathbf{e}_j = \sum_j R_{ji}\,V_j \tag{2.140}$$

Comparing this with (2.129) we see that the components of a vector *operator* behave under rotation R as the components of a vector under R^{-1}.

2.5.3 Angular Momentum

We now consider the special rotation (2.130):

$$\begin{aligned} R_z(\phi)\,\psi(x,\;y,\;z) &= \psi\,(R_z^{-1}(\phi)\,\mathbf{r}) \\ &= \psi(x\cos\phi + y\sin\phi,\;-x\sin\phi + y\cos\phi,\;z) \end{aligned} \tag{2.141}$$

If ϕ is an infinitesimally small angle, $\phi \to \varepsilon = 0^+$, we can replace $\cos\phi$ by 1 and $\sin\phi$ by $\phi = \varepsilon$ and perform a Taylor expansion:

$$\begin{aligned} R_z(\varepsilon)\,\psi(x,\;y,\;z) &= \psi(x + \varepsilon\,y,\;-\varepsilon\,x + y,\;z) \\ &= \psi(x,\;y,\;z) + \varepsilon\left(\frac{\partial\psi}{\partial x}\,y - \frac{\partial\psi}{\partial y}\,x\right) + \cdots \\ &= \left(1 - \frac{i}{\hbar}\,\varepsilon\,l_z\right)\psi(x,\;y,\;z) \end{aligned}$$

This leads to the following important relation:

$$R_z(\varepsilon) = 1 - \frac{i}{\hbar}\,\varepsilon\,l_z \tag{2.142}$$

where l_z is the z-component of the orbital angular momentum operator

$$\mathbf{l} = \mathbf{r} \times \mathbf{p} \tag{2.143}$$

A more general form of the relation (2.142) is, when $\mathbf{n}$ denotes the unit vector in the direction of the axis of rotation,

$$R_{\mathbf{n}}(\varepsilon) = 1 - \frac{i}{\hbar}\,\varepsilon\,(\mathbf{n} \cdot \mathbf{l}) \tag{2.144}$$

A complete generalization of these results to an arbitrary system defines the total angular momentum **j**.

$$R_{\mathbf{n}}(\varepsilon) = 1 - \frac{i}{\hbar}\, \varepsilon\, (\mathbf{n} \cdot \mathbf{j}) \tag{2.145}$$

The above definition of **j** is meaningful because, it is supported by the fact that the well-known commutation relations for **j** can be derived from it. That can be shown as follows:

Let us consider the infinitesimal rotation of the vector **V** about the axis **n**

$$\bar{\mathbf{V}} = \mathbf{V} + d\mathbf{V}$$

From Fig. 2.1, we can see that $d\mathbf{V}$ can be written as

$$d\mathbf{V} = \varepsilon\, \mathbf{n} \times \mathbf{V}$$

Therefore, we have

$$\bar{\mathbf{V}} = \mathbf{V} + \varepsilon\, \mathbf{n} \times \mathbf{V} \tag{2.146}$$

Now let

$$\mathbf{K} = \sum_{i=1}^{3} K_i\, \mathbf{e}_i \tag{2.147}$$

be an arbitrary vector operator. Using (2.137) and (2.145), we find that the components K_i transform as

$$\bar{K}_i = R_{\mathbf{n}}(\varepsilon)\, K_i\, R_{\mathbf{n}}^{-1}(\varepsilon) \approx K_i + \frac{i}{\hbar}\, \varepsilon\, [K_i\ ,\ \mathbf{n} \cdot \mathbf{j}]_- \tag{2.148}$$

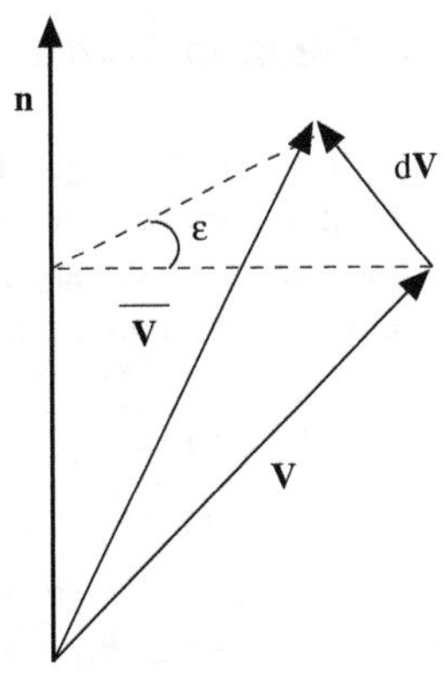

Fig. 2.1 Rotation of a vector **V** by a small angle ε about the axis **n**

where we could neglect the terms of the order of ε^2. On the other hand, using (2.146) we also have

$$\begin{aligned} \bar{K}_i = \mathbf{K} \cdot \bar{\mathbf{e}}_i &= \mathbf{K} \cdot (\mathbf{e}_i + \varepsilon\, \mathbf{n} \times \mathbf{e}_i) \\ &= K_i + \varepsilon\, \mathbf{K} \cdot (\mathbf{n} \times \mathbf{e}_i) \end{aligned} \tag{2.149}$$

Comparing with (2.148), we get the important relation

$$[(\mathbf{n} \cdot \mathbf{j})\,,\; (\mathbf{e}_i \cdot \mathbf{K})]_- = i\,\hbar\,(\mathbf{n} \times \mathbf{e}_i) \cdot \mathbf{K} \tag{2.150}$$

Substituting $\mathbf{K} = \mathbf{j}$ we get, as stated above, the fundamental commutation relations for the angular momentum $\mathbf{j}$.

From (2.145) it follows that the angular momentum of a system is defined by its transformation behaviour under an infinitesimal rotation of the coordinate system. This is, of course, also valid for *finite rotations*. In order to show that, we exploit the relation

$$\begin{aligned} R_{\mathbf{n}}(\phi + d\phi) &= R_{\mathbf{n}}(d\phi)\, R_{\mathbf{n}}(\phi) \\ &= \left(1 - \frac{i}{\hbar} d\phi(\mathbf{n} \cdot \mathbf{j})\right) R_{\mathbf{n}}(\phi) \end{aligned}$$

which can also be written as a differential equation

$$\frac{d}{d\phi} R_{\mathbf{n}}(\phi) = -\frac{i}{\hbar}\,(\mathbf{n} \cdot \mathbf{j})\, R_{\mathbf{n}}(\phi)$$

which can be easily integrated to give

$$R_{\mathbf{n}}(\phi) = exp\left(-\frac{i}{\hbar}(\mathbf{n} \cdot \mathbf{j})\,\phi\right) \tag{2.151}$$

2.5.4 Rotation Matrices

Let $|p;\; j\; m\rangle$ be the eigenstates of the angular momentum operators j^2 and j_z. p is a set of some quantum numbers that are necessary to specify the states but are not concerned with the angular momentum. The following relations are well known from elementary quantum mechanics:

$$j^2\,|p;\; j\; m\rangle = \hbar^2 j(j+1)\,|p;\; j\; m\rangle \tag{2.152}$$

$$j_z\,|p;\; j\; m\rangle = \hbar\, m\,|p;\; j\; m\rangle \tag{2.153}$$

$$j_\pm = j_x \pm i j_y \tag{2.154}$$

$$j_\pm |p;\; j\; m\rangle = \hbar\sqrt{j(j+1) - m(m \pm 1)}\,|p;\; j\; m \pm 1\rangle \tag{2.155}$$

We denote by $\mathcal{H}^{(j)}$ the space spanned by the $2j+1$ states $|p;\ j\ m\rangle$ corresponding to a given j. Then, in the space $\mathcal{H}^{(j)}$, we have

$$\sum_m |p;\ j\ m\rangle\langle p;\ j\ m| = \mathbb{1} \tag{2.156}$$

From the special form of the rotation operators, it is obvious that the space $\mathcal{H}^{(j)}$ is invariant under rotations. An application of the operator $R_n(\phi)$ affects only the magnetic quantum number m. The following equation:

$$\begin{aligned} R\,|p;\ j\ m\rangle &= \sum_{m'} |p;\ j\ m'\rangle\langle p;\ j\ m'|R|p;\ j\ m\rangle \\ &= \sum_{m'} |p;\ j\ m'\rangle\ R^{(j)}_{m'm} \end{aligned} \tag{2.157}$$

defines the $(2j+1)$-dimensional rotation matrix

$$R^{(j)}_{m'm} = \langle p;\ j\ m'|\ R\ |p;\ j\ m\rangle \tag{2.158}$$

$\mathcal{H}^{(j)}$ is called an *irreducible space*. What it means is that, all the vectors $R\ |\psi\rangle$, which result from the application of the rotation operator R on an arbitrarily chosen vector $|\psi\rangle$ in this space, are sufficient to span this space. Even if there exists only one vector $|\phi\rangle$, such that $\{R\ |\phi\rangle\}$ spans the space only partially, then the space is reducible under rotations. In $\mathcal{H}^{(j)}$ such a $|\phi\rangle$ does not exist.

As examples, we consider three special rotation matrices:

Let ε again be an infinitesimal rotation angle.

(a) $R_z(\varepsilon) = 1 - \frac{i}{\hbar}\ \varepsilon\ j_z$
Using (2.153), it directly follows that

$$(R_z(\varepsilon))^{(j)}_{m'm} = \delta_{m'm}\ (1 - i\varepsilon\ m) \tag{2.159}$$

(b) $R_x(\varepsilon) = 1 - \frac{i}{\hbar}\ \varepsilon\ j_x$:
Since $j_x = \frac{1}{2}\ (j_+ + j_-)$ and using (2.155), it follows that

$$\begin{aligned} (R_x(\varepsilon))^{(j)}_{m'm} &= \delta_{m'm} - \frac{i\varepsilon}{2}\sqrt{j(j+1) - m(m+1)}\ \delta_{m',m+1} \\ &\quad - \frac{i\varepsilon}{2}\sqrt{j(j+1) - m(m-1)}\ \delta_{m',m-1} \end{aligned} \tag{2.160}$$

(c) $R_y(\varepsilon) = 1 - \frac{i}{\hbar}\ \varepsilon\ j_y$
With $j_y = \frac{1}{2i}(j_+ - j_-)$ and again using (2.155), it follows that

$$\left(R_y(\varepsilon)\right)^{(j)}_{m'm} = \delta_{m'm} - \frac{\varepsilon}{2}\sqrt{j(j+1) - m(m+1)}\,\delta_{m',m+1} + \frac{\varepsilon}{2}\sqrt{j(j+1) - m(m-1)}\,\delta_{m',m-1} \quad (2.161)$$

2.5.5 Tensor Operators

A tensor is defined by its transformation properties under a coordinate transformation. A *tensor of rank k in an n-dimensional space* consists of a group of n^k numbers which transform linearly under rotation, according to specific laws:

$k = 0$	:	Scalar $\bar{x} = x$
$k = 1$	:	Vector of n-components x_i with $\bar{x}_i = \sum_j R_{ij}\, x_j$
$k = 2$	:	n^2-components F_{ij} with $\bar{F}_{ij} = \sum_{l,m} R_{il}\, R_{jm}\, F_{lm}$

The transition from a tensor to a tensor operator is exactly analogous to the one from vector to vector operator.

Tensor operator: A group of operators (components) which transform linearly within among themselves under a rotation.

For our purpose, in the following, only the *irreducible tensor operators* are interesting. These are the tensor operators which operate in an irreducible space. One can now show that an *irreducible tensor operator of rank k* consists of a set of $(2k+1)$ operators (*standard components*)

$$T_q^{(k)}, \quad q = -k,\ -k+1,\ \cdots,\ +k$$

which under rotation transform as

$$R\, T_q^{(k)}\, R^{-1} = \sum_{q'=-k}^{+k} R_{q'q}^{(k)}\, T_{q'}^{(k)} \quad (2.162)$$

We notice that they follow the same transformation laws as the states $|j,\ m\rangle$ in $\mathcal{H}^{(j)}$ (2.158). All this appears to be a very special case, but actually, it is of wide applicability, i.e. there exist many physically relevant operators which belong to this class.

The relation (2.162) is valid if and only if it is valid for every infinitesimal rotation. If we substitute R for such a rotation and use the rotation matrices that were evaluated as examples (2.159), (2.160), and (2.161), then we recognize that the above definition is strictly equivalent to the following commutation relations (Problem 2.9):

$$\begin{aligned}\left[j_z\,,\,T_q^{(k)}\right]_- &= \hbar\, q\, T_q^{(k)}\\ \left[j_\pm\,,\,T_q^{(k)}\right]_- &= \hbar\,\sqrt{k(k+1)-q(q\pm 1)}\;T_{q\pm 1}^{(k)}\end{aligned}$$

Therefore, one can treat these "handy" commutation relations also as definitions.

Examples:

1. $\widehat{T}^{(0)}$: *Scalar operators* are irreducible tensor operators of rank zero. They commute with the angular momentum j.
2. $\widehat{T}^{(1)}$: *Every vector operator* K is an irreducible tensor operator of rank one with the standard components

$$T_0^{(1)} = K_z \tag{2.163}$$

$$T_{\pm 1}^{(1)} = \mp\frac{1}{\sqrt{2}}\,(K_x \pm i\, K_y) \tag{2.164}$$

With the help of the general relation (2.150), one can easily prove and verify the commutation relations (Problem 2.10). The angular momentum j itself is a special example:

$$j_0 = j_z \quad ; \quad j_{\pm 1} = \mp\frac{1}{\sqrt{2}}\, j_\pm \tag{2.165}$$

3. $\widehat{T}^{(l)}$: One can formally treat the spherical harmonics $Y_{lm}(\theta, \phi)$ to be operators. Then they fulfil the commutation relations for the irreducible tensor operators of rank l.

$$\begin{aligned}\left[j_\pm\,,\,Y_{lm}\right]_- &= \left[l_\pm\,,\,Y_{lm}\right]_-\\ &= l_\pm\, Y_{lm} - Y_{lm}\, l_\pm\\ &= (l_\pm\, Y_{lm}) + Y_{lm}\, l_\pm - Y_{lm}\, l_\pm\\ &= \hbar\sqrt{l(l+1)-m(m\pm 1)}\; Y_{l\; m\pm 1}\end{aligned} \tag{2.166}$$

Analogously we find

$$[j_z\,,\,Y_{lm}]_- = \hbar\, m\, Y_{lm} \tag{2.167}$$

Therefore, $T_m^{(l)} = Y_{lm}$ is an irreducible tensor operator of rank l.

2.5.6 Wigner–Eckart Theorem

Tensors have their own algebra with a number of theorems. One of the most useful among them is, surely, the Wigner–Eckart theorem which we present here without proof. It is concerned with the matrix elements of the standard components of irreducible tensor operators in the space $\mathcal{H}^{(j)}$:

$$\langle p\,;\, j\, m_j|T_q^{(k)}|p'\,;\, j'\, m_{j'}\rangle = T_{red}^{(k)}\,(p\; j; p'\; j')\langle j'\; k\; m_{j'}q|j\; m_j\rangle \tag{2.168}$$

$T_{red}^{(k)}\,(p\; j; p'\; j')$ is called the *reduced matrix element* of the tensor $\widehat{T}^{(k)}$. It is *independent* of the quantum numbers $m_{j'}$, m_j and q. $\langle j'\; k\; m_{j'}q|j\; m_j\rangle$ is a Clebsch–Gordon coefficient known in connection with the addition of two angular momenta:

$$|j_1\; j_2, j\; m_j\rangle = \sum_{m_1, m_2} |j_1\; m_1\rangle|j_2\; m_2\rangle\langle j_1\; j_2\; m_1\; m_2|j\; m_j\rangle \tag{2.169}$$

Here we have

$$m_j = m_1 + m_2 \quad ; \quad |j_1 - j_2| \le j \le j_1 + j_2 \tag{2.170}$$

That means the matrix elements (2.168) is different from zero only if

$$q = m_j - m_{j'} \quad ; \quad \left|j - j'\right| \le k \le j + j' \tag{2.171}$$

The actual meaning of the theorem lies in the factorization that is achieved by (2.168). The Clebsch–Gordon coefficient is independent of T and $T_{red}^{(k)}$ is independent of m_j, $m_{j'}$ and q. That means the *matrix elements of the standard components of tensor operators with the same rank are proportional to each other*. In the following discussion, we will exploit this fact very often .

We will be specially interested in tensor operators of rank one, i.e. in vector operators. Therefore, we want to discuss these operators in more detail. For the *angular momentum* j, and for its standard components j_q, in view of (2.165), the Wigner–Eckart theorem becomes

$$\langle p;\; j\; m_j|j_q|p';\; j\; m_{j'}\rangle = j_{red}(p\; j;\; p'\; j')\delta_{pp'}\;\delta_{jj'}\;\langle j'\; 1\; m_{j'}\; q|j\; m_j\rangle \tag{2.172}$$

We can, as a result from the beginning, restrict ourselves to the matrix elements which are diagonal in p and j. Then, for an arbitrary tensor operator of rank one, we have

$$\langle p;\; j\; m_j|T_q^{(1)}|p;\; j\; m_{j'}\rangle = \frac{T_{red}^{(1)}(p\; j)}{j_{red}(p\; j)}\langle p;\; j\; m_j|j_q|p;\; j\; m_{j'}\rangle \tag{2.173}$$

The forefactor is determined from the scalar product

$$T^{(1)} \cdot \mathbf{j} = \sum_q^{0,\,1,\,-1} T_q^{(1)}\; j_q^{\dagger} \tag{2.174}$$

The scalar product $T^{(1)} \cdot \mathbf{j}$ must be direction independent. Therefore, for the purpose of the calculation, we can assume any arbitrary m_j:

$$\begin{aligned}
\langle T^{(1)} \cdot \mathbf{j} \rangle &= \langle p\ j\ m_j | T^{(1)} \cdot \mathbf{j} | p\ j\ m_j \rangle \\
&= \sum_q \sum_{p' j' m_{j'}} \langle p j m_j | T_q^{(1)} | p' j' m_{j'} \rangle \underbrace{\langle p' j' m_{j'} | j_q^\dagger | p j m_j \rangle}_{\sim \delta_{pp'}\ \delta_{jj'}} \\
&= \frac{T_{red}^{(1)}(p\ j)}{j_{red}(p\ j)} \sum_{q, m_{j'}} \langle p j m_j | j_q | p j m_{j'} \rangle \langle p j m_{j'} | j_q^\dagger | p j m_j \rangle \\
&= \frac{T_{red}^{(1)}(p\ j)}{j_{red}(p\ j)} \underbrace{\langle p\ j\ m_j | \mathbf{j}^2 | p\ j\ m_j \rangle}_{\hbar^2\ j(j+1)}
\end{aligned}$$

Here we have exploited (2.165): $j^2 = \sum_q j_q \cdot j_q^\dagger$. With this, we get the following important relation:

$$\langle p\ j\ m_j | T_q^{(1)} | p\ j\ m_{j'} \rangle = \frac{\langle T^{(1)} \cdot \mathbf{j} \rangle}{\hbar^2 j(j+1)} \langle p\ j\ m_j | j_q | p\ j\ m_{j'} \rangle \tag{2.175}$$

which holds of course only in the subspace of fixed p and j.

2.5.7 Examples of Application

1. $\widehat{T}^{(1)} = \mathbf{l} + 2\mathbf{s} = \mathbf{j} + \mathbf{s}$
   This operator is important in connection with the Zeeman effect, which will be discussed towards the end of the chapter. We are interested in the $q = 0$ component:

   $$T_0^{(1)} = l_z + 2s_z = j_z + s_z \tag{2.176}$$

   In (2.175) we require the expression

   $$\begin{aligned}
   \langle (T^{(1)} \cdot \mathbf{j}) \rangle &= \langle j^2 + \mathbf{s} \cdot \mathbf{j} \rangle = \langle j^2 + s^2 + \mathbf{s} \cdot \mathbf{l} \rangle \\
   &= \langle j^2 + s^2 + \frac{1}{2}(j^2 - l^2 - s^2) \rangle \\
   &= \langle j^2 + \frac{1}{2}(j^2 - l^2 + s^2) \rangle
   \end{aligned}$$

   If we restrict ourselves to the space in which, simultaneously, j^2, l^2 and s^2 are diagonal, then, we have

$$\langle (T^{(1)} \cdot \mathbf{j} \rangle = \hbar^2 \left\{ j(j+1) + \frac{1}{2} \left[j(j+1) - l(l+1) + s(s+1) \right] \right\} \quad (2.177)$$

At this point, we define the *Lande-factor*

$$g_j(l, s) = 1 + \frac{j(j+1) - l(l+1) + s(s+1)}{2j(j+1)} \quad (2.178)$$

With this, we finally get

$$\langle p\ j\ m_j | (l_z + 2s_z) | p\ j\ m_{j'} \rangle = g_j(l, s)\, \hbar\, m_j\, \delta_{m_j m_{j'}} \quad (2.179)$$

When $q = 0$, then, because of (2.171), naturally $m_j = m_{j'}$.

2. $\widehat{T}^{(1)} = \mathbf{s}$
Here also, we will be interested only in the $q = 0$ component

$$T_0^{(1)} = s_z \quad (2.180)$$

We again require the scalar product

$$T^{(1)} \cdot \mathbf{j} = \mathbf{s} \cdot \mathbf{j} = s^2 + \frac{1}{2} \left(j^2 - l^2 - s^2 \right) = \frac{1}{2} \left(j^2 - l^2 + s^2 \right)$$

Substituting this in (2.175) we get under the same conditions as in (1),

$$\langle p\ j\ m_j | s_z | p\ j\ m_{j'} \rangle = (g_j(l, s) - 1)\, \hbar\, m_j\, \delta_{m_j m_{j'}} \quad (2.181)$$

This result agrees of course with (2.179) in example (1).
In the space of the states $|ls,\ j\ m_j\rangle$ with fixed $l,\ s,\ j$ obviously

$$s_z = (g_j(l, s) - 1)\, j_z \quad (2.182)$$

is an operator identity. That is the reason, why, in many models of magnetism, only "interacting spins" are considered (see the Heisenberg model which appears later) even though, in general, the total angular momentum is involved.

2.6 Electron in an External Magnetic Field

After the background prepared in the earlier sections, we already have at least a rough picture of how the Hamiltonian should look for an electron moving in

(a) a nuclear field $\mathbf{E} = -\nabla \phi$,
(b) a homogeneous external magnetic field $\mathbf{B}_0 = (0,\ 0,\ B_0)$.

We will assume, for the moment, that the nuclear field is spherically symmetric (the nucleus is treated as approximately a positive point charge). The deviations from this simplifying assumption will be discussed in later sections. The Hamiltonian without the relativistic corrections, in this case, should have the following form:

$$H = \frac{p^2}{2m} + V(r) + H_{dia} + H_{SO} + \frac{\mu_B}{\hbar}\,(\mathbf{l} + 2\mathbf{s}) \cdot \mathbf{B}_0 \tag{2.183}$$

The individual terms have the following meaning:

(a) $H_0 = \frac{p^2}{2m} + V(r)$: This describes the motion of a spinless particle of mass m and charge $-e$ in the field of the positively charged nucleus ($V(r) = -e\,\phi(r)$). This is considered to be solved (hydrogen atom problem in basic quantum mechanics). That means, we presume the *energy levels* $E_{nl}^{(0)}$ as known. Here n is the principal quantum number and l the orbital angular momentum quantum number.
(b) $H_{dia} = \frac{e^2 B_0^2}{8m}\,(x^2 + y^2)$: This is the diamagnetic part, which has been derived in Sect. 1.2. It results from the substitution $\mathbf{p} \to \mathbf{p} + e\mathbf{A}$ which is necessary in the presence of an electromagnetic field. This term is usually very small and is meaningful only when the other terms do not play any role. Therefore, in this section we will not take into account H_{dia}. We will discuss diamagnetic effects in considerable detail in Chap. 3.
(c) $H_{SO} = \lambda\,(\mathbf{l} \cdot \mathbf{s})$: The spin–orbit coupling was already discussed in Sect. 2.4. It produces a *fine structure of the terms*, i.e. the energy levels $E_{nl}^{(0)}$ are split into an (l, s)-multiplet for $j \neq 0$:

$$E_{nlj}^{(0)} = E_{nl}^{(0)} + \frac{1}{2}\,\lambda_{nl}\,\hbar^2\{j(j+1) - l(l+1) - s(s+1)\} \tag{2.184}$$

which, in our one-electron system, is naturally a doublet.

$$j = l \pm \frac{1}{2} \tag{2.185}$$

That means, the spin–orbit interaction lifts the j-degeneracy. The following commutators, which have already been evaluated, are important:

$$\begin{aligned}
\left[H_{SO}\,,\, j^2\right]_- &= 0\\
\left[H_{SO}\,,\, l^2\right]_- &= 0\\
\left[H_{SO}\,,\, s^2\right]_- &= 0\\
\left[H_{SO}\,,\, j_z\right]_- &= 0\\
\left[H_{SO}\,,\, l_z\right]_- &\neq 0\\
\left[H_{SO}\,,\, s_z\right]_- &\neq 0
\end{aligned} \tag{2.186}$$

These commutation relations show that only j, m_j, l and s are "good" quantum numbers and not m_l and m_s.

(d) $H_z = -\frac{\mu_B}{\hbar}\,(\mathbf{l}+2\mathbf{s})\cdot\mathbf{B}_0 = -\frac{\mu_B}{\hbar}\,(l_z+2s_z)\,B_0$: This term will be called the *Zeeman term*. With respect to the sign before the scalar product we refer to the remark at the end of Sect. 2.1.One can easily show that j^2 and s_z do not commute.

$$\left[j^2\,,\;s_z\right]_- = -2\,i\,\hbar\,(l_x s_y - l_y s_x) \neq 0 \tag{2.187}$$

From this it follows that

$$\left[j^2\,,\;H_z\right]_- = \frac{\mu_B}{\hbar}\,B_0\left[j^2\,,\;s_z\right]_- \neq 0 \tag{2.188}$$

Therefore, when the external field is switched on, j is also not a good quantum number. But it is still valid that

$$[j_z\,,\;H_z]_- = 0 \tag{2.189}$$

and, therefore, m_j still remains a good quantum number. As a result, the energy eigenstates can still be classified, as before, by $(n,\;m_j,\;l,\;s)$ though not by $j,\;m_l,\;m_s$. Physically, that means that the external field forces transitions between states with different j but with the same m_j. The energy eigenstates will therefore be corresponding linear combinations:

$$|\psi\rangle = \alpha_+\left|j = l + \frac{1}{2}\;m_j\right\rangle + \alpha_-\left|j = l - \frac{1}{2}\;m_j\right\rangle \tag{2.190}$$

We use the following shorthand notation:

$$|+\rangle = \left|j = l + \frac{1}{2}\;m_j\right\rangle \quad ; \quad |-\rangle = \left|j = l - \frac{1}{2}\;m_j\right\rangle \tag{2.191}$$

Then we have to solve the following Schrödinger equation:

$$H|\psi\rangle = \alpha_+\;H|+\rangle + \alpha_-\;H|-\rangle = E|\psi\rangle \tag{2.192}$$

Since

$$\langle +|+\rangle = \langle -|-\rangle = 1 \quad ; \quad \langle +|-\rangle = \langle -|+\rangle = 0$$

from (2.192) it follows that

$$\begin{aligned} E\,\alpha_+ &= \alpha_+\langle +|H|+\rangle + \alpha_-\langle +|H|-\rangle \\ E\,\alpha_- &= \alpha_+\langle -|H|+\rangle + \alpha_-\langle -|H|-\rangle \end{aligned} \tag{2.193}$$

With the notation

$$E_{\pm} = \langle \pm | H | \pm \rangle \quad ; \quad \eta = \langle - | H | + \rangle \tag{2.194}$$

the solvability condition for the system of homogeneous equations

$$(E - E_+)\,\alpha_+ - \eta^*\,\alpha_- = 0$$
$$-\eta\,\alpha_+ + (E - E_-)\,\alpha_- = 0$$

gives the required energy eigenvalues:

$$E_{1,\,2}(l\ s\ m_j) = \frac{1}{2}\left[(E_+ - E_-) \pm \sqrt{(E_+ - E_-)^2 + 4\,\eta^2}\right] \tag{2.195}$$

We calculate the matrix elements E_+, E_- and η with the help of the Wigner–Eckart theorem:

$$\begin{aligned} E_{\pm} = &\left\langle j = l \pm \frac{1}{2},\ m_j \,|(H_0 + H_{SO})|\, j = l \pm \frac{1}{2},\ m_j \right\rangle - \\ &- \frac{\mu_B}{\hbar} B_0 \left\langle j = l \pm \frac{1}{2},\ m_j \,|(l_z + 2 s_z)|\, j = l \pm \frac{1}{2},\ m_j \right\rangle \end{aligned} \tag{2.196}$$

The first summand is exactly $E^{(0)}_{nlj=l\pm\frac{1}{2}}$ and is therefore known from (2.184). The second summand was evaluated as an application example of Wigner–Eckart theorem. The result is given in (2.179). So, we have

$$E_{\pm} = E^{(0)}_{nlj=l\pm\frac{1}{2}} - \mu_B\, m_j\, B_0\, g_{j=l\pm\frac{1}{2}}\,(l,\ s) \tag{2.197}$$

The Lande-factor, defined in (2.178), becomes in this case

$$g_{j=l\pm\frac{1}{2}} = 1 \pm \frac{1}{2l+1} \tag{2.198}$$

Further, we find with (2.184),

$$E^{(0)}_{nlj=l+\frac{1}{2}} = E^{(0)}_{nl} + \frac{1}{2}\,\lambda_{nl}\,\hbar^2\,l \tag{2.199}$$

$$E^{(0)}_{nlj=l-\frac{1}{2}} = E^{(0)}_{nl} - \frac{1}{2}\,\lambda_{nl}\,\hbar^2\,(l+1) \tag{2.200}$$

With this, we have determined the matrix elements E_+ and E_-:

$$E_+ = E_{nl}^{(0)} + \frac{1}{2} \lambda_{nl} \hbar^2 l - \mu_B B_0 m_j \frac{2l+2}{2l+1} \tag{2.201}$$

$$E_- = E_{nl}^{(0)} - \frac{1}{2} \lambda_{nl} \hbar^2 (l+1) - \mu_B B_0 m_j \frac{2l}{2l+1} \tag{2.202}$$

The "mixed" matrix element η can also be easily evaluated using Wigner–Eckart theorem. From the definition (2.194) we have

$$\eta = \langle j = l - \frac{1}{2},\ m_j | (H_0 + H_{SO}) | j = l + \frac{1}{2},\ m_j \rangle - \frac{\mu_B}{\hbar} B_0 \times \\ \times \langle j = l - \frac{1}{2},\ m_j | (l_z + 2s_z) | j = l + \frac{1}{2},\ m_j \rangle \tag{2.203}$$

The first summand vanishes because for $(H_0 + H_{SO})$, j is a good quantum number, i.e. $(H_0 + H_{SO})$ is diagonal in the $|j, m_j\rangle$ space. In the second summand, if $(l_z + 2s_z)$ is replaced by $(j_z + s_z)$, then the matrix element also vanishes for the term with j_z. Therefore, what remains is

$$\eta = -\frac{\mu_B}{\hbar} B_0 \langle j = l - \frac{1}{2},\ m_j | s_z | j = l + \frac{1}{2},\ m_j \rangle \tag{2.204}$$

This is, in general, unequal to zero since s_z induces transitions. According to the definition of s_z given in (2.34) it holds

$$s_z^2 = \frac{\hbar^2}{4} \mathbb{1} \tag{2.205}$$

This can be exploited as follows:

$$\frac{\hbar^2}{4} = \langle j\ m_j | s_z^2 | j\ m_j \rangle = \sum_{j', m_{j'}} \langle j\ m_j | s_z | j'\ m_{j'} \rangle \langle j'\ m_{j'} | s_z | j\ m_j \rangle$$

Since the commutator $[s_z,\ j_z]_- = 0$, both the factors contribute only if $m_j = m_{j'}$. So what remains is

$$\frac{\hbar^2}{4} = \sum_{j'} \langle j\ m_j | s_z | j'\ m_j \rangle \langle j'\ m_j | s_z | j\ m_j \rangle \tag{2.206}$$

For $j = l + \frac{1}{2}$ and $j' = l \pm \frac{1}{2}$, we have, in particular,

$$\frac{\hbar^2}{4} = \left| \langle l + \frac{1}{2},\ m_j | s_z | l + \frac{1}{2} \rangle \right|^2 + \left| \langle l - \frac{1}{2},\ m_j | s_z | l + \frac{1}{2},\ m_j \rangle \right|^2 \tag{2.207}$$

The second summand is identical with η (2.204) except for the factors. Therefore, we get

$$\frac{\hbar^2}{\mu_B^2\, B_0^2}\,|\eta|^2 = \frac{\hbar^2}{4} - \left|\langle j = l + \frac{1}{2}\, m_j|s_z|j = l + \frac{1}{2}\, m_j\rangle\right|^2 \qquad (2.208)$$

The remaining matrix element has been evaluated as an application example for Wigner–Eckart theorem. From (2.181) and (2.198) we get for η

$$|\eta|^2 = \mu_B^2\, B_0^2 \left(\frac{1}{4} - \frac{m_j^2}{(2l+1)^2}\right) \qquad (2.209)$$

Substituting (2.209) along with (2.201) and (2.202) in (2.195), we obtain, for the given (n, l), a somewhat complicated expression for the $2(2l + 1)$ possible levels, which are available for an electron in a magnetic field:

$$E_{1,2}(lsm_j) = \left(E_{nl}^{(0)} - \frac{1}{4}\,\hbar^2\,\lambda_{nl} - \mu_B\, B_0\, m_j\right) \pm \frac{1}{2}\sqrt{\lambda_{nl}^2\,\hbar^4\left(l+\frac{1}{2}\right)^2 - 2\, m_j\, \lambda_{nl}\hbar^2\mu_B\, B_0 + \mu_B^2 B_0^2} \qquad (2.210)$$

We will now discuss certain limiting cases:

1. *Weak fields:* $\mu_B\, B_0 \ll \lambda_{nl}$
   In this case, one can neglect the terms containing B_0^2 and therefore also $|\eta|^2$. That means (2.195), $E_\pm$ are already the solutions. In this approximation, j^2 commutes with H and, therefore, j remains a good quantum number.

$$E_{nljm_j} = E_{nlj}^{(0)} - g_j(l, s)\, m_j\, \mu_B\, B_0 \qquad (2.211)$$

   This result is known as *anomalous Zeeman effect* which is characterized by the linear dependence of the energy levels on the field.
2. *Strong fields:* $\mu_B\, B_0 \gg \lambda_{nl}$
   Now, to a first approximation, we can neglect the spin–orbit coupling so that m_l and m_s are still good quantum numbers.

$$m_j = m_{l\pm\frac{1}{2}} E_{nlm_lm_s} = E_{nl}^{(0)} \qquad - (m_l + 2m_s)\, \mu_B\, B_0 \qquad (2.212)$$

   This result is known as the *normal Zeeman effect* which also has linear field dependence.
3. *Intermediate fields:* $\mu_B\, B_0 \approx \lambda_{nl}$
   Now we must apply the full expressions for $E_{1,2}(l,\ m_j)$. Then the field dependence is not linear anymore. A *special case* is worth mentioning, namely

$$\left|m_j\right| = l + \frac{1}{2} \Leftrightarrow |\eta|^2 = 0 \qquad (2.213)$$

   In this case, (2.211) is valid independent of the strength of the magnetic field.

2.7 Nuclear Quadrupole Field

So far we have treated the influence of the nucleus on the electron motion in the simplest form, namely, we considered the nucleus to be a positive point charge. The potential energy of the electron in the field of the nucleus is then given by

$$V(r) = -e\phi(r) = -\frac{1}{4\pi\varepsilon_0}\frac{Z^* e^2}{r} \tag{2.214}$$

Strictly speaking, that is too simple because, we are neglecting the higher multipole moments of the nucleus. We will now treat this more rigorously. However, we will continue to assume that the electron remains outside the region where nuclear charge and nuclear currents are appreciable. This assumption is naturally a little problematic for s-electrons which have a finite probability of existence within the nucleus. We denote by $\mathbf{R}(R, \Theta, \Phi)$, the position vector within the nucleus and by $\mathbf{r}(r, \theta, \phi)$, the position vector outside the nucleus (Fig. 2.2). Then the general solution for the nuclear potential $\phi(\mathbf{r})$ is given by

$$4\pi\varepsilon_0\,\phi(\mathbf{r}) = \int_{(nucl)} d^3R\,\frac{\rho(\mathbf{R})}{|\mathbf{r}-\mathbf{R}|} \tag{2.215}$$

We have assumed here that the nuclear charge is confined to a finite volume. From electrodynamics, we know the corresponding *multipole expansion*:

$$4\pi\varepsilon_0\,\phi(\mathbf{r}) = \sum_{l=0}^{\infty}\frac{4\pi}{2l+1}\frac{1}{r^{l+1}}\sum_{m=-l}^{+l} Q_{lm}\,Y_{lm}(\theta,\phi) \tag{2.216}$$

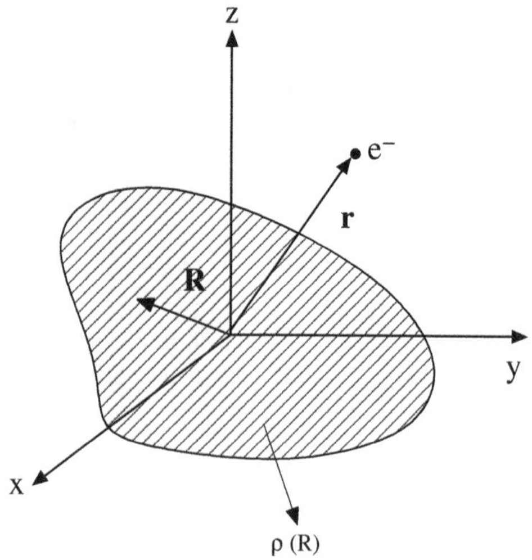

Fig. 2.2 Schematic plot of the nuclear charge density seen by the atomic electron e^-

The nuclear coordinates are present exclusively in the *spherical multipole moments*:

$$Q_{lm} = \int d^3R\, Y^*_{lm}(\Theta, \Phi)\, \rho(R)\, R^l \tag{2.217}$$

Since we can assume that $r \gg R$, we can restrict ourselves to the first summands of the expansion. Using

$$Y_{00}(\Theta, \Phi) = Y^*_{00}(\Theta, \Phi) = \frac{1}{\sqrt{4\pi}} \tag{2.218}$$

and

$$\int d^3R\, \rho(\mathbf{R}) = Ze \tag{2.219}$$

we get for the *monopole moment* ($l = 0$)

$$4\pi\varepsilon_0\, \phi^{(0)}(\mathbf{r}) = \frac{Ze}{r} \tag{2.220}$$

This is nothing but the potential of a point charge at the origin. The simplest way to evaluate the *dipole moment ($l = 1$)* is by using the addition theorem for the spherical harmonics, which is given by

$$\sum_{m=-l}^{+l} Y^*_{lm}(\Theta, \Phi)\, Y_{lm}(\theta, \phi) = \frac{2l+1}{4\pi}\, P_l(\cos\gamma) \stackrel{l=1}{\longrightarrow} \frac{3}{4\pi}\cos\gamma \tag{2.221}$$

γ is the angle between the space directions (Θ, Φ) and (θ, ϕ) and therefore is given by

$$\cos\gamma = \frac{\mathbf{r}\cdot\mathbf{R}}{rR} \tag{2.222}$$

Using the definition of the dipole moment $\mathbf{P}$

$$\mathbf{P} = \int d^3R\, \rho(\mathbf{R})\, \mathbf{R} \tag{2.223}$$

the ($l = 1$) summand of the multipole expansion (2.216) can be written as

$$4\pi\varepsilon_0\, \phi^{(1)}(\mathbf{r}) = \frac{\mathbf{r}\cdot\mathbf{P}}{r^3} \tag{2.224}$$

One knows that the electric dipole moment of a nucleus is in general zero. Therefore, we must consider the multipole expansion one step further, i.e. to the *quadrupole moment* ($l = 2$):

$$
\begin{aligned}
4\pi\,\varepsilon_0\,\phi^{(2)}(\mathbf{r}) = \frac{4\pi}{5}\,\frac{1}{r^3}\,(Q_{20}Y_{20} + Q_{21}Y_{21} + Q_{2-1}Y_{2-1} + \\
+ Q_{22}Y_{22} + Q_{2-2}Y_{2-2}) \qquad (2.225)
\end{aligned}
$$

With the spherical harmonics

$$
\begin{aligned}
Y_{20}(\Theta,\,\Phi) &= \frac{1}{2}\sqrt{\frac{5}{\pi}}\,\frac{1}{R^2}\,(3Z^2 - R^2) \\
Y_{2\pm 1}(\Theta,\,\Phi) &= \mp\sqrt{\frac{15}{8\pi}}\,\frac{Z}{R^2}\,(X \pm iY) \qquad (2.226) \\
Y_{2\pm 2}(\Theta,\,\Phi) &= \frac{1}{4}\sqrt{\frac{15}{2\pi}}\,\frac{1}{R^2}\,(X \pm iY)^2
\end{aligned}
$$

we get the following expressions for the *components of the electric quadrupole moment* ($\mathbf{R} = (X,\ Y,\ Z)$):

$$
\begin{aligned}
Q_{20} &= \frac{1}{2}\sqrt{\frac{5}{4\pi}}\int d^3R\,\rho(\mathbf{R})\,(3Z^2 - R^2) \\
Q_{2\pm 1} &= \mp\sqrt{\frac{15}{8\pi}}\int d^3R\,\rho(\mathbf{R})\,Z(X \pm iY) \qquad (2.227) \\
Q_{2\pm 2} &= \frac{1}{4}\sqrt{\frac{5}{2\pi}}\int d^3R\,\rho(\mathbf{R})\,(X \pm iY)^2
\end{aligned}
$$

We can now summarize the quadrupole part of the potential in the familiar form

$$
\begin{aligned}
4\pi\,\varepsilon_0\,\phi^{(2)}(r) &= \frac{1}{2}\sum_{i,j} q_{ij}\,\frac{x_i x_j}{r^5} \\
& \qquad\qquad (2.228) \\
q_{ij} &= \int d^3R\,\rho(\mathbf{R})\,(3X_iX_j - \delta_{ij}R^2)
\end{aligned}
$$

With this, we have actually completed the discussion. The nuclear quadrupole field appears in the Hamiltonian in the form $H_Q = -e\,\phi^{(2)}(\mathbf{r})$. It is, of course, obvious that it will not be very simple to evaluate the eigenvalues, eigenstates and matrix elements for such an operator.

We will therefore, with the help of the Wigner–Eckart theorem, formulate an equivalent Hamiltonian in terms of the angular momentum operators **I** and **j**.

I: *Operator for the total nuclear spin*
j: *Operator for the total angular momentum of the electron*

The procedure adopted here is based on the following idea: Since $Q_{2m} \sim Y^*_{2m}$, $\widehat{Q}_2$ is an irreducible tensor operator of rank two with the standard components Q_{2m}

(see Example (3) with (2.162)). We now construct a special tensor operator of rank two out of a combination of nuclear spin operators. In the space, in which I^2 and I_z are diagonal, their matrix elements are, according to the Wigner–Eckart theorem, proportional to the matrix elements of the quadrupole operator.

How do we find the above-mentioned combinations? According to (2.162), they must obey the following relations:

$$\left[I_\pm \,,\, T_q^{(2)}\right]_- = \sqrt{6 - q(q \pm 1)}\,\hbar\, T_{q\pm 1}^{(2)} \tag{2.229}$$

$$\left[I_z \,,\, T_q^{(2)}\right]_- = \hbar\, q\, T_q^{(2)} \tag{2.230}$$

The structure of Q_{20} (2.227) suggests the following ansatz:

$$T_0^{(2)} = I^2 - 3I_z^2 \tag{2.231}$$

We can immediately see that

$$\left[I_z \,,\, T_0^{(2)}\right]_- = 0 \tag{2.232}$$

and therefore (2.230) is satisfied. Using

$$I_\pm = I_x \pm i\, I y \tag{2.233}$$

we also have

$$T_0^{(2)} = \frac{1}{2}\left(I_- I_+ + I_+ I_- - 4I_z^2\right) \tag{2.234}$$

so that we get

$$\left[I_\pm \,,\, T_0^{(2)}\right]_- = \pm\, 3\,\hbar\,(I_z I_\pm + I_\pm I_z) \doteq \sqrt{6}\,\hbar\, T_{\pm 1}^{(2)}$$

The last step is required by (2.229). Therefore, we have to choose

$$T_{\pm 1}^{(2)} = \pm\frac{1}{2}\sqrt{6}\,(I_z I_\pm + I_\pm I_z) \tag{2.235}$$

We then evaluate

$$\left[I_\pm \,,\, T_{\pm 1}^{(2)}\right]_- = -\sqrt{6}\,\hbar\, I_\pm^2 \doteq \sqrt{6 - 2}\,\hbar\, T_{\pm 2}^{(2)}$$

and define

$$T_{\pm 2}^{(2)} = -\frac{1}{2}\sqrt{6}\,(I_\pm)^2 \tag{2.236}$$

By verifying the commutation relations (2.229) and (2.230), we finally show that, $T_0^{(2)}$, $T_{\pm 1}^{(2)}$ and $T_{\pm 2}^{(2)}$ are actually the standard components of an irreducible tensor of rank two (see Problem 2.11). One should notice that these operators have a structure similar to that of Q_{2m} (2.227).

We now exploit the Wigner–Eckart theorem whose main statement is about the proportionality of the matrix elements of different tensor operators of the same rank. Let $|I, M\rangle$ be the eigenstates of the nuclear operators I^2 and I_z. Then, we have

$$\langle IM|Q_{2m}^{+}|IM'\rangle = \alpha_k \langle IM|T_m^{(2)}|IM'\rangle \tag{2.237}$$

Here the proportionality constant α_k is independent of M, M' and m! One normally defines the *quadrupole moment of the nucleus* by

$$+e\, Q = \alpha_k \langle II|T_0^{(2)}|II\rangle \tag{2.238}$$

As far as we are concerned, here, Q is an experimentally determined parameter. α_k is fixed by this parameter:

$$+e\, Q = \alpha_k\ \langle II|(I^2 - 3I_z^2)|II\rangle = \alpha_k\ \hbar^2\ \{I(I+1) - 3I^2\} \tag{2.239}$$

For $I = 1/2$, we have $Q = 0$. For $I \neq 1/2$, one can solve this equation for α_k:

$$\alpha_k = \frac{-eQ}{\hbar^2 I(2I-1)} \tag{2.240}$$

In the space, in which I is a good quantum number, one can treat (2.237) as an operator identity:

$$Q_{2m} = \frac{-eQ}{\hbar^2 I(2I-1)} \left(T_m^{(2)}\right)^{\dagger} \tag{2.241}$$

With this, we have fixed the nuclear part of the quadrupole term in the Hamiltonian:

$$\begin{aligned} 4\pi\varepsilon_0\, \phi^{(2)}(\mathbf{r}) &= \frac{4\pi}{5}\frac{1}{r^3}\sum_{m=-2}^{+2} Q_{2m}\, Y_{2m}(\theta,\phi) \\ &= \sum_{m=-2}^{+2} \underbrace{\left[\frac{4\pi}{5}\frac{1}{r^3}\, Y_{2m}(\theta,\phi)\right]}_{ele.\ contribution} \underbrace{\left[\alpha_k \left(T_m^{(2)}\right)^{\dagger}\right]}_{nucl.\ contribution} \end{aligned} \tag{2.242}$$

Since the electronic part is proportional to $Y_{2m}(\theta, \phi)$, it can also be considered as an irreducible tensor operator of rank two, and we can repeat the same procedure for the electronic part as we have done for the nuclear part. The only difference is that, now, we have to replace I and M by j and m_j. With

$$
\begin{aligned}
\widehat{t}^{(2)} : t_{\pm 2}^{(2)} &= -\frac{1}{2}\sqrt{6}\, j_{\pm}^2 \\
t_{\pm 1}^{(2)} &= \pm\frac{1}{2}\sqrt{6}\,(j_z j_{\pm} + j_{\pm} j_z) \\
t_0^{(2)} &= j^2 - 3 j_z^2
\end{aligned}
\tag{2.243}
$$

we obtain the following quadrupole term:

$$
4\pi\varepsilon_0\, \phi^{(2)}(\mathbf{r}) = \alpha_e\, \alpha_k \sum_{m=-2}^{+2} t_m^{(2)} \left(T_m^{(2)}\right)^{\dagger} \tag{2.244}
$$

Here α_e is proportional to the reduced matrix element of the electronic tensor $\widehat{t}^{(2)}$.

Finally, by substituting and rearranging the expressions for $t_m^{(2)}$ and $T_m^{(2)}$, we obtain the quadrupole contribution to the Hamiltonian as

$$
\begin{aligned}
H_Q &= -e\, \phi^{(2)}(\mathbf{r}) \\
&= \frac{\alpha_e}{4\pi\varepsilon_0}\, \frac{e^2 Q}{\hbar^2 I(2I-1)} \left\{6(\mathbf{j}\cdot\mathbf{I})^2 + 3\hbar^2(\mathbf{j}\cdot\mathbf{I}) - 2\mathbf{I}^2\cdot\mathbf{j}^2\right\}
\end{aligned}
\tag{2.245}
$$

The quadrupole field is, in general, small compared to the other fields which act on the electron.

From the derivation it is clear, that, if the nuclear quadrupole interacts not with a single electron but with an arbitrarily complicated charge, a completely analogous expression will be obtained. In that case only the forefactor which, in any case is unknown, will be different.

2.8 Hyperfine Field

We have to take into account a second influence of the nucleus on the movement of the electron, which originates because of the motion of the nuclear charges, i.e. because of the *nuclear current density* $\mathbf{j}(\mathbf{R})$. This creates a vector potential $\mathbf{A}_N(\mathbf{r})$, which, according to electrodynamics, is given by

$$
\mathbf{A}_N(\mathbf{r}) = \frac{\mu_0}{4\pi} \int_{nucl} d^3R\, \frac{\mathbf{j}(\mathbf{R})}{|\mathbf{r}-\mathbf{R}|} \tag{2.246}
$$

Since $r \gg R$ we can approximate

$$
\frac{1}{|\mathbf{r}-\mathbf{R}|} \approx \frac{1}{r} + \frac{\mathbf{r}\cdot\mathbf{R}}{r^3} \tag{2.247}
$$

The first summand does not contribute in the case of a localized current density $\mathbf{j}(\mathbf{R})$ (Problem 2.13). The first non-vanishing term then has the form

$$\mathbf{A}_N(\mathbf{r}) \approx \frac{\mu_0}{4\pi} \frac{\mathbf{m}_N \times \mathbf{r}}{r^3} = \frac{\mu_0}{4\pi} \nabla \times \left(\frac{1}{r} \mathbf{m}_N \right) \tag{2.248}$$

Here $\mathbf{m}_N$ is the nuclear moment (see (1.20)):

$$\mathbf{m}_N = \frac{1}{2} \int d^3R \, (\mathbf{R} \times \mathbf{j}(\mathbf{R})) \tag{2.249}$$

The vector potential originating from the nuclear moment has naturally the same action as the one originating from an external magnetic field. That means, it appears in the *kinetic energy of the electron*

$$\begin{aligned}\frac{p^2}{2m} &\to \frac{1}{2m}(\mathbf{p} + e\mathbf{A}_N)^2 \\ &= \frac{p^2}{2m} + \frac{e}{m}\mathbf{p} \cdot \mathbf{A}_N + \mathcal{O}(A_N^2) \;\; (\textit{Coulomb gauge!})\end{aligned}$$

and in the *Zeeman energy* through

$$2 \frac{\mu_B}{\hbar} \mathbf{S} \cdot (\nabla \times \mathbf{A}_N) \tag{2.250}$$

This gives two extra terms in the Hamiltonian

$$H_{HF} = H_{HF}^{(1)} + H_{HF}^{(2)} \tag{2.251}$$

The individual terms have to be now evaluated. $H_{HF}^{(1)}$ acts on the orbital motion of the electron,

$$\begin{aligned}H_{HF}^{(1)} = \frac{e}{m}\mathbf{p} \cdot \mathbf{A}_N &= \frac{\mu_0}{4\pi} \frac{e}{m} \frac{1}{r^3} \mathbf{p} \cdot (\mathbf{m}_N \times \mathbf{r}) \\ &= \frac{\mu_0}{4\pi} \frac{e}{m} \frac{1}{r^3} \mathbf{m}_N \cdot (\mathbf{r} \times \mathbf{p})\end{aligned} \tag{2.252}$$

and gives the so-called *orbital hyperfine interaction*:

$$H_{HF}^{(1)} = 2 \frac{\mu_B}{\hbar} \frac{1}{r^3} \frac{\mu_0}{4\pi} (\mathbf{m}_N \cdot \mathbf{l}) \tag{2.253}$$

This part, obviously, vanishes when the electron is in a pure s-state.

Now, we will consider the second part of the hyperfine interaction. For that let us first evaluate (Δ: Laplace operator)

$$\begin{aligned}\nabla \times \mathbf{A}_N &= \frac{\mu_0}{4\pi} \nabla \times \nabla \times \left(\frac{1}{r}\, \mathbf{m}_N \right) \\ &= \frac{\mu_0}{4\pi} \left[\nabla \left(\nabla \cdot \left(\frac{1}{r} \mathbf{m}_N \right) \right) - \Delta \frac{1}{r} \mathbf{m}_N \right] \\ &= \frac{\mu_0}{4\pi} \left[\nabla \left(\mathbf{m}_N \cdot \nabla \frac{1}{r} \right) - \mathbf{m}_N \Delta \frac{1}{r} \right]\end{aligned}$$

We can neglect the last term because

$$\Delta \frac{1}{r} = -4\pi\ \delta(\mathbf{r})$$

and we always have $r \gg R$. Then what remains is

$$\nabla \times \mathbf{A}_N = -\frac{\mu_0}{4\pi}\ \nabla \left(\frac{1}{r^3} \mathbf{r} \cdot \mathbf{m}_N \right) \tag{2.254}$$

We calculate the x-component:

$$\frac{d}{dx} \left(\frac{\mathbf{r} \cdot \mathbf{m}_N}{r^3} \right) = \frac{(\mathbf{m}_N)_x}{r^3} - \frac{\mathbf{r} \cdot \mathbf{m}_N}{r^4}\, 3\, \frac{x}{r}$$

The other components can be evaluated analogously so that we get

$$\nabla \times \mathbf{A}_N = \frac{\mu_0}{4\pi}\, \frac{3\mathbf{r}(\mathbf{m}_N \cdot \mathbf{r}) - \mathbf{m}_N\, r^2}{r^5} \tag{2.255}$$

This is of course not a surprising result. It is nothing but the dipole field created by the nuclear moment $\mathbf{m}_N$. The magnetic spin moment of the electron

$$\mathbf{m}_s = -2\, \frac{\mu_B}{\hbar}\, \mathbf{s}$$

interacts with this field and gives the *dipolar hyperfine interaction*

$$H_{dip} = 2\, \frac{\mu_B}{\hbar}\, \frac{\mu_0}{4\pi}\, \mathbf{s} \cdot \left(\frac{3\, \mathbf{r}\, (\mathbf{m}_N \cdot \mathbf{r})}{r^5} - \frac{\mathbf{m}_N}{r^3} \right) \tag{2.256}$$

The expressions for $H_{HF}^{(1)}$ and H_{dip} were derived under the assumption that the electron is always outside the nucleus ($r \gg R$). This assumption is, of course, not realistic for s-electrons, since it is well known that there is a finite probability for the s-electrons to exist inside the nucleus. This plays no role in the case of $H_{HF}^{(1)}$ because the matrix elements any way vanish for $l = 0$. For H_{dip}, however, it means a restriction. That is why we would like to investigate this term once more in some more detail, so that the corresponding term in the Hamiltonian appears, at least, plausible.

Let $\psi(\mathbf{r})$ be the wavefunction of the electron (without the spin part). We leave out the unimportant pre-factors and evaluate the expectation value of $\mathbf{s} \cdot (\nabla \times \mathbf{A}_N)$ in the state $\psi(\mathbf{r})$. Let the nucleus be inside a fictitious, small sphere of radius R. Then we can split the expectation value, which we want to evaluate, into two parts:

$$
\begin{aligned}
&\int d^3r\ \psi^*(\mathbf{r})[\mathbf{s} \cdot (\nabla \times \mathbf{A}_N)]\psi(\mathbf{r}) \\
&= \int_{r<R} d^3r\ \psi^*(\mathbf{r})[\mathbf{s} \cdot (\nabla \times \mathbf{A}_N)]\psi(\mathbf{r}) + \\
&+ \int_{r>R} d^3r\ \psi^*(\mathbf{r})[\mathbf{s} \cdot (\nabla \times \mathbf{A}_N)]\psi(\mathbf{r})
\end{aligned}
\tag{2.257}
$$

The second term gives the dipolar hyperfine interaction derived above. The new thing is therefore only the first term. Using

$$
\nabla \cdot (\mathbf{A}_N \times \mathbf{s}) = \mathbf{s} \cdot (\nabla \times \mathbf{A}_N) - \mathbf{A}_N \cdot (\nabla \times \mathbf{s}) = \mathbf{s} \cdot (\nabla \times \mathbf{A}_N)
$$

and assuming that the modulus square of the wavefunction $|\psi(\mathbf{r})|^2$ changes only a little within the volume of the nucleus, we can write

$$
\begin{aligned}
&\int_{r<R} d^3r\ \psi^*(\mathbf{r})[\mathbf{s} \cdot (\nabla \times \mathbf{A}_N)]\psi(\mathbf{r}) \approx \\
&\approx \int_{r<R} d^3r \nabla \cdot ((\mathbf{A}_N \times \mathbf{s})|\psi(\mathbf{r})|^2) \\
&= \int_{S_R} d\mathbf{f} \cdot (\mathbf{A}_N \times \mathbf{s})|\psi(\mathbf{R}|^2 \\
&= \mathbf{s} \cdot \int_{S_R} d\mathbf{f} \times \mathbf{A}_N |\psi(\mathbf{R})|^2
\end{aligned}
$$

S_R means the surface of the sphere of radius R. Let the sphere of radius R, on the one hand, be so small that one can assume the electron wavefunction $|\psi(\mathbf{R})|^2 \approx |\psi(0)|^2$ and, on the other hand, be so large that we can use the expression (2.248) for $\mathbf{A}_N(\mathbf{R})$. Then, if the direction of the nuclear moment $\mathbf{m}_N$ defines the z-axis, we further have

$$\int_{r<R} d^3r\psi^*(\mathbf{r})[\mathbf{s}\cdot(\nabla\times\mathbf{A}_N)]\psi(\mathbf{r})$$
$$= \mathbf{s}\cdot\int_{S_R} R^2\, d\Omega \frac{\mathbf{R}}{R}\times\left(\frac{\mu_0}{4\pi}\frac{\mathbf{m}_N\times\mathbf{R}}{R^3}\right)|\psi(0)|^2$$
$$= \frac{\mu_0}{4\pi}|\psi(0)|^2\frac{1}{R^2}\,\mathbf{s}\cdot\int\int d\cos\Theta\, d\Phi(\mathbf{m}_N\cdot R^2 - \mathbf{R}(\mathbf{R}\cdot\mathbf{m}_N))$$
$$= \frac{\mu_0}{4\pi}|\psi(0)|^2\,\mathbf{s}\cdot\int\int d\Phi\, d\cos\Theta\, m_N *$$
$$* \{(0,\ 0,\ 1) - \cos\Theta(\sin\Theta\cos\Phi,\ \sin\Theta\sin\Phi,\ \cos\Theta)\}$$
$$= \frac{2}{3}\mu_0\,(\mathbf{m}_N\cdot\mathbf{s})\,|\psi(0)|^2$$

In operator form, this leads to the following contribution to the hyperfine interaction:

$$H_{cont} = \frac{4}{3}\mu_0\frac{\mu_B}{\hbar}(\mathbf{s}\cdot\mathbf{m}_N)\,\delta(\mathbf{r}) \tag{2.258}$$

This is called the *contact hyperfine interaction* which comes into play, obviously, only for s-electrons. The interaction of the electron spin $\mathbf{s}$ with the magnetic field created by the nuclear currents is then given by

$$H_{HF}^{(2)} = H_{dip} + H_{cont} \tag{2.259}$$

Expressing the nuclear moment $\mathbf{m}_N$ in terms of the nuclear spin $\mathbf{I}$

$$\mathbf{m}_N = g_N\frac{\mu_N}{\hbar}\mathbf{I} \tag{2.260}$$

where μ_N is the nuclear magneton and g_N is the nuclear g-factor, the total contribution of the nucleus to the Hamiltonian of the electron can be written as

$$H_{HF} = \frac{\mu_0}{2\pi}\frac{\mu_B}{\hbar^2}g_N\,\mu_N\left[\frac{1}{r^3}(\mathbf{I}\cdot\mathbf{l}) + \left(\frac{3(\mathbf{s}\cdot\mathbf{r})\,(\mathbf{I}\cdot\mathbf{r})}{r^5} - \frac{\mathbf{s}\cdot\mathbf{I}}{r^3}\right)\right.$$
$$\left. + \frac{8\pi}{3}(\mathbf{s}\cdot\mathbf{I})\delta(\mathbf{r})\right] \tag{2.261}$$

This is known as the *hyperfine interaction*. The interaction consists of the contribution due to the orbital angular momentum $\mathbf{l}$ as well as the spin $\mathbf{s}$ of the electron with the nuclear spin $\mathbf{I}$. H_{HF} is responsible for the hyperfine structure of the electron terms. Since $\mu_N \approx 10^{-3}\mu_B$ ($\mu_N = \frac{e\hbar}{2m_p}$; m_p is the mass of the proton), H_{HF} is hardly important for static magnetism, even though, it is important for frequency shifts in resonance experiments.

2.9 Magnetic Hamiltonian of the Atomic Electron

For a better overview, we want to summarize, in the form of a list, all the influences that we have discussed till now, which "probe" an electron.

(a) **Spinless particle:**
In a spherically symmetric central field and in a homogeneous magnetic field $\mathbf{B}_0 = (0,\ 0,\ B_0)$, the Hamiltonian operator consists of three parts:

$$H_a = H_0 + H_{dia} + H_{orb} \tag{2.262}$$

H_0 consists of the kinetic energy of the electron and its potential energy in the central field:

$$H_0 = \frac{p^2}{2m} + V(r) \tag{2.263}$$

The diamagnetic part

$$H_{dia} = \frac{e^2 B_0^2}{8m} (x^2 + y^2) \tag{2.264}$$

and the interaction of the field with the orbital moment of the electron

$$H_{orb} = -\frac{\mu_B}{\hbar} (\mathbf{l} \cdot \mathbf{B}_0) \tag{2.265}$$

appear because of the substitution $\mathbf{p} \rightarrow \mathbf{p} + e\mathbf{A}$, if the vector potential is chosen such that

$$\mathbf{A} = \frac{1}{2} B_0 (-y,\ x,\ 0) \tag{2.266}$$

and the Coulomb gauge $\nabla \cdot \mathbf{A} = 0$ is satisfied. Because of the

(b) **spin of the electron**
there appear two additional terms

$$H_b = H_{SO} + H_z \tag{2.267}$$

namely, the spin–orbit interaction

$$H_{SO} = -\frac{e}{2m^2c^2} \left(\frac{1}{r}\frac{d\phi}{dr}\right) (\mathbf{l} \cdot \mathbf{s}) = \lambda\, (\mathbf{l} \cdot \mathbf{s}) \tag{2.268}$$

which causes a fine structure of the terms, and the Zeeman term

$$H_z = -2\, \frac{\mu_B}{\hbar}\, \mathbf{s} \cdot \mathbf{B}_0 \tag{2.269}$$

which results from the interaction of the spin moment with the field, and induces transitions between states with different j but the same m_j. Note that we have agreed to assume the parallel orientation of the spin relative to the field as the energetically favoured orientation (see the remarks at the end of Sect. 2.1)

A full relativistic calculation adds mainly two more terms as

(c) **relativistic corrections:**

$$H_c = H_{0,rel} + H_D \tag{2.270}$$

$H_{0,rel}$ represents a relativistic correction to the kinetic energy

$$H_{0,rel} = -\frac{p^4}{8m^3c^2} \tag{2.271}$$

whereas the so-called *Darwin term* does not permit a simple-minded explanation and is given by

$$H_D = \frac{e\hbar^2}{4m^2c^2}\,(\nabla\phi \cdot \nabla) \tag{2.272}$$

In case the assumption of a spherically symmetric, electrostatic nuclear potential, as it appears in (2.263), is not acceptable, then four more terms have to be taken in to account, which are classified as

(d) **nuclear effects:**

$$H_d = H_Q + H_{orb} + H_{dip} + H_{cont} \tag{2.273}$$

H_Q originates because of the nuclear quadrupole field:

$$H_Q = \frac{\alpha}{4\pi\varepsilon_0}\left(6(\mathbf{j}\cdot\mathbf{I})^2 + 3\hbar^2(\mathbf{j}\cdot\mathbf{I}) - 2I^2j^2\right) \tag{2.274}$$

$\mathbf{I}$ is the nuclear spin and α is a constant. The existence of the nuclear spin still leads to an *orbital hyperfine interaction*

$$H_{orb} = \mu_0\, g_N\, \frac{\mu_B}{2\pi}\, \frac{\mu_N}{\hbar^2}\, \frac{(\mathbf{I}\cdot\mathbf{l})}{r^3} \tag{2.275}$$

and to the *dipolar hyperfine interaction*

$$H_{dip} = \mu_0\, g_N\, \frac{\mu_B}{2\pi}\, \frac{\mu_N}{\hbar^2}\left[\frac{3(\mathbf{s}\cdot\mathbf{r})(\mathbf{I}\cdot\mathbf{r})}{r^5} - \frac{\mathbf{s}\cdot\mathbf{I}}{r^3}\right] \tag{2.276}$$

For s-electrons, which have a finite probability of existing inside the nucleus, finally, we have to take into account one more interaction, namely, the *contact hyperfine interaction*:

$$H_{cont} = \mu_0\, g_N\, \frac{4\mu_B\,\mu_N}{3\hbar^2}\,(\mathbf{s}\cdot\mathbf{I})\,\delta(\mathbf{r}) \tag{2.277}$$

2.10 Many-Electron Systems

Till today, an exact theory for atoms with $Z > 1$ electrons has not been possible. Therefore, one has to depend on more or less drastic approximations whose only justification is derived from a comparison with experiments. Completely filled electron shells pose no difficulties, since they lead to spherically symmetric potentials, which do not shift the individual energy terms *against each other*. Since only the energy *differences* are measurable, a rigid shift of the complete energy spectrum by a spherically symmetric potential is not interesting.

We consider an atom with Z electrons out of which p are in a partially filled shell. What are the problems in treating such a system?

1. There is no Dirac's theory for many-electron systems. Therefore, one has problems in treating the spin–orbit interaction correctly.
2. The Coulomb interaction among the electrons has to be taken into account.

For the moment, we restrict ourselves to a more or less qualitative discussion of the relevant interacting systems.

2.10.1 Coulomb Interaction

This consists of two parts, namely the interaction H_N of the electrons with the positively charged nucleus and the interaction H_e of electrons among themselves.

$$H_C = H_N + H_e \tag{2.278}$$

If we neglect higher order corrections (nuclear quadrupole field and nuclear currents), then we have

$$H_N = \frac{-Ze^2}{4\pi\varepsilon_0}\sum_{i=1}^{p}\frac{1}{r_i} \tag{2.279}$$

$$H_e = \frac{e^2}{4\pi\varepsilon_0}\,\frac{1}{2}\sum_{i,j=1(i\neq j)}^{p}\frac{1}{|\mathbf{r}_i - \mathbf{r}_j|} \tag{2.280}$$

H_e makes the problem in general unsolvable. The theory must therefore accept approximations. Often one takes recourse to the so-called *central field approximation*, where the p-electrons move *independent of each other* in a central potential (*Hartree-potential*).

$$H_{cf} = -e\,\phi_{cf}(\mathbf{r}) = -e\,\phi_{cf}(r) \tag{2.281}$$

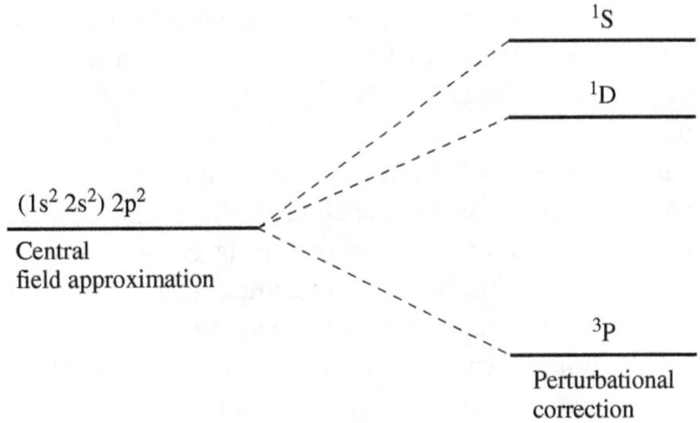

Fig. 2.3 Schematic of the ground state term scheme of carbon $(1s^2 2s^2 2p^1)$

This is a sum of the usual nuclear potential H_N and an average repulsive potential. The skill in the choice of the repulsive potential determines the quality of the approximation. Predominantly, the presence of the other electrons has the effect of screening the Coulomb field of the nucleus from the electron under consideration.

In the next stage of the approximation, one writes instead of (2.278),

$$H_C = H_{cf} + H_1 \tag{2.282}$$

where

$$H_1 = (H_N + H_e) - H_{cf} \tag{2.283}$$

and obtains the spectrum of H_C perturbatively from H_{cf}. In general, the eigenvalues of H_{cf} are highly degenerate and the degeneracy is either fully or partially lifted by H_1. One diagonalizes H_1 in the eigenspace spanned by the eigenstates $|E_{0,\alpha}\rangle$ belonging to the degenerate ground state energy E_0. The condition on the secular determinant

$$det\left(\langle E_{0,\beta}|H_1|E_{0,\alpha}\rangle - E\,\delta_{\alpha\beta}\right) \doteq 0 \tag{2.284}$$

yields an improved ground state and the first few excited states. On the whole, H_C determines the rough structure of the terms. Such a calculation, e.g. done for the ground state of carbon $(1s^2\,2s^2\,2p^2)$ gives the schematic result of (Fig. 2.3).

The $4s$-electrons occupy the closed shell. Therefore, it is necessary only to take care of the two p-electrons. We will present a more complete picture later.

2.10.2 Spin–Orbit Coupling

There appear two types of spin–orbit couplings, namely between the

1. spin and orbit of the same electron $\sim (\mathbf{l}_i \cdot \mathbf{s}_i)$, and between the
2. spin and orbit of different electrons $\sim ((\mathbf{l}_i \cdot \mathbf{s}_j)$.

One can easily see that the two couplings may be of different weightage. Seen from the rest system of the electron, in the case of "normal" spin–orbit coupling, the positively charged nucleus moves around the electron and builds a strong ring current, which creates a magnetic field, to which the magnetic spin moment couples. This interaction will have the same sign at all times and, therefore, will be of considerable magnitude. On the contrary, the interaction between two electrons moving around in different orbits, may, during one period, change sign many times. As a result, the contribution of (2) in comparison to (1) will be small.

That is why, one neglects (2) and compensates the error to a certain degree by replacing the potential ϕ by an effective potential which results in a modification of the forefactor for the spin–orbit coupling (1).

$$H_{SO} = \tilde{\lambda}\,(\mathbf{l} \cdot \mathbf{S}) \tag{2.285}$$

$$\tilde{\lambda} = -\frac{e}{2m^2c^2}\left(\frac{1}{r}\frac{d\tilde{\phi}}{dr}\right) \tag{2.286}$$

This approximation, at least for not so light atoms, has proved to be useful.

2.10.3 Further Couplings

We have discussed in Sect. 2.8, the coupling between the electron spin and the nuclear spin. Exactly the same arguments are valid for the coupling between the spin of the ith and the jth electron.

$$\underbrace{3\frac{(\mathbf{s}_i \cdot \mathbf{r}_{ij})(\mathbf{s}_j \cdot \mathbf{r}_{ij})}{r_{ij}^5} - \frac{\mathbf{s}_i \cdot \mathbf{s}_j}{r_{ij}^3}}_{dipolar\ interaction} + \underbrace{\frac{8\pi}{3}(\mathbf{s}_i \cdot \mathbf{s}_j)\delta(\mathbf{r}_{ij})}_{contact\ term}$$

$$\mathbf{r}_{ij} = \mathbf{r}_i - \mathbf{r}_j$$

Since here also we must average over the orbits of different electrons, this term also will be dominated by the other contributions in the Hamiltonian and, therefore, turns out to be relatively uninteresting. The same is also valid for orbit–orbit coupling, which, in the simplest form, can be assumed to be $\sim (\mathbf{l}_i \cdot \mathbf{l}_j)$.

In the lowest approximation, nuclear effects can be included additively from the ones derived in Sect. 2.7 and 2.8 for a single electron. Since $\mu_N \approx 10^{-3}\mu_B$, they are small and, therefore, will be taken into account only when "necessary".

Furthermore, relativistic corrections to the kinetic energy (2.271) or the Darwin term (2.272) are, for our purposes, unimportant.

So, finally, *the Hamiltonian, for p-electrons in a partially filled shell* reads as

$$
\begin{aligned}
H^{(0)} &= \sum_{i=1}^{p} \left(\frac{p_i^2}{2m} - e\,\phi_N(\mathbf{r}_i) + \tilde{\lambda}(r_i)\,(\mathbf{l}_i \cdot \mathbf{s}_i) \right) \\
&+ \frac{1}{2} \sum_{i,j=1}^{p} \frac{e^2}{4\pi\varepsilon_0\,|\mathbf{r}_i - \mathbf{r}_j|} \\
&= H_0 + H_{SO} + H_e
\end{aligned}
\tag{2.287}
$$

The first three summations can be separated according to the individual electrons so that, in the absence of the fourth summation, we simply would have to sum over the results obtained in the last section. But, even then, the problem at hand here, is more complicated. This is because, with

$$
\begin{aligned}
\mathbf{L} &= \sum_{i=1}^{p} \mathbf{l}_i \qquad \textit{total angular momentum} \\
\mathbf{S} &= \sum_{i=1}^{p} \mathbf{s}_i \qquad \textit{total spin}
\end{aligned}
\tag{2.288}
$$

we are no more interested in the quantum numbers l and s of the individual electrons but in the quantum numbers L and S of the incompletely filled shell. One immediately realizes that, because of the spin–orbit interaction, L^2 and S^2 do *not* commute with $H^{(0)}$. This is different from the case of a single electron (2.121). The reason for this is, e.g. L^2 is not simply equal to $\sum_{i=1}^{p} \mathbf{l}_i^2$. That means, H_{SO} induces transitions between the LS-multiplets. The question is, how strong are these "state mixtures" (hybridizations), that is, how probable are such transitions. This in turn depends on the relative strength of the individual terms in the Hamiltonian. One distinguishes two limiting cases:

$$
H_e \gg H_{SB}: \begin{cases} \mathbf{L} = \sum_i \mathbf{l}_i\,,\ \mathbf{S} = \sum_i \mathbf{s}_i \\ \mathbf{J} = \mathbf{L} + \mathbf{S} \end{cases} \qquad LS - \textit{coupling}
$$

$$
H_{SB} \gg H_e: \begin{cases} \mathbf{j}_i = \mathbf{l}_i + \mathbf{s}_i \\ \mathbf{J} = \sum_i \mathbf{j}_i \end{cases} \qquad jj - \textit{coupling}
$$

According to estimates, the details of which we will not go into here, H_e goes as $\sim Z^{1/2}$, whereas H_{SO} goes as Z^2. The LS-coupling is, therefore, an excellent approximation for light and medium heavy nuclei. In case of heavy nuclei, jj-coupling dominates. From Pb onwards, H_e and H_{SO} are of the same order of magnitude. Let us now consider the

(a) *LS* Coupling (Russell–Saunders Coupling)

in somewhat more detail. Because $H_e \gg H_{SO}$, L and S are still good quantum numbers. The energy differences between the LS-multiplets are about one or two orders of magnitude larger than the internal multiplet splittings. Therefore, one can, to a first approximation, neglect the transitions into the higher multiplets. The possible basis states can be labelled by the quantum numbers γ, L, S, M_L and M_S:

$$|\gamma LSM_LM_S\rangle$$

Here γ is the set of quantum numbers that is additionally necessary to specify the configuration. We will discuss the spin–orbit interaction in the space of a *fixed* LS pair. In this case again, the Wigner–Eckart theorem plays a decisive role. Using the above basis we can write

$$\begin{aligned}&\sum_{i=1}^{p}\langle\gamma LSM_LM_S|\tilde{\lambda}_i(\mathbf{l_i}\cdot\mathbf{s}_i)|\gamma LSM_L'M_S'\rangle\\&=\sum_{i=1}^{p}\sum_{q}^{0,\pm1}\sum_{M_L''M_S''}\langle\gamma LSM_LM_S|l_{qi}|\gamma LSM_L''M_S''\rangle\;*\\&\qquad *\;\langle\gamma LSM_L''M_S''|\tilde{\lambda}_i(s_{qi})^\dagger|\gamma LSM_L'M_S'\rangle\end{aligned}\tag{2.289}$$

The sum over q runs over the three standard components ((2.163) and (2.164)) of the vector operators $\mathbf{l}$ and $\mathbf{s}$. For both the matrix elements in the above equation, we make use of the Wigner–Eckart theorem. The matrix element

$$\langle\gamma LSM_LM_S|l_{qi}|\gamma LSM_L''M_S''\rangle=\langle\bar{\gamma}LM_L|l_{qi}|\bar{\gamma}LM_L''\rangle\tag{2.290}$$

is certainly unequal to zero only for $M_S = M_S''$ because $\mathbf{l}_i$ commutes with $\mathbf{S}=\sum_i\mathbf{s}_i$. As a result, $\bar{\gamma}$ does not, in principle, contain M_S'' at all. Because of the Wigner–Eckart theorem, we can use the proportionality of the matrix elements (2.290) to the respective matrix element of the total angular momentum:

$$\langle\bar{\gamma}LM_L|l_{qi}|\bar{\gamma}LM_L''\rangle=\alpha(\bar{\gamma},\,L)\langle\bar{\gamma}LM_L|L_q|\bar{\gamma}LM_L''\rangle\tag{2.291}$$

The coefficient α does *not* depend on M_L'' and M_S''.

The same procedure can be applied for the second matrix element of (2.289).

$$\langle\gamma LSM_L''M_S''|(s_{qi})^\dagger|\gamma LSM_L'M_S'\rangle=\langle\bar{\bar{\gamma}}SM_S''|(s_{qi})^\dagger|\bar{\bar{\gamma}}SM_S'\rangle\tag{2.292}$$

Now $\bar{\bar{\gamma}}$ does not contain M_L''. We can now use the Wigner–Eckart theorem to bring the total spin $\mathbf{S}$ into play.

$$\langle\bar{\bar{\gamma}}SM_S''|\tilde{\lambda}(s_{qi})^\dagger|\bar{\bar{\gamma}}SM_S'\rangle=\beta(\bar{\bar{\gamma}},S)\langle\bar{\bar{\gamma}}SM_S''|S_q^\dagger|\bar{\bar{\gamma}}SM_S'\rangle\tag{2.293}$$

Substituting (2.291) and (2.293) in (2.289), we can pull the coefficients α and β out of the summation into the front. That means finally

$$\langle \gamma LSM_L M_S | H_{SO} | \gamma LSM'_L M'_S \rangle \sim \langle \gamma LSM_L M_S | (\mathbf{L} \cdot \mathbf{S}) | \gamma LSM'_L M'_S \rangle \qquad (2.294)$$

From this we can, in the space of a given (LS)-multiplet, read off an operator identity:

$$H_{SO} = \Lambda(\gamma, LS)(\mathbf{L} \cdot \mathbf{S}) \qquad (2.295)$$

This form of the spin–orbit coupling is valid only if the transitions to other (LS)-multiplets can be neglected.

The consequences of the spin–orbit coupling in this case are completely analogous to those of one-electron system:

$$\mathbf{J}^2 = (\mathbf{L} + \mathbf{S})^2 \Rightarrow \mathbf{L} \cdot \mathbf{S} = \frac{1}{2}(J^2 - L^2 - S^2) \qquad (2.296)$$

One can easily evaluate the following commutators:

$$\left[(\mathbf{L} \cdot \mathbf{S}) \, , \, J^2\right] = 0 \Rightarrow \left[H^{(0)} \, , \, J^2\right] = 0 \qquad (2.297)$$

$$\left[(\mathbf{L} \cdot \mathbf{S}) \, , \, L_z\right] = -\left[(\mathbf{L} \cdot \mathbf{S}) \, , \, S_z\right] \neq 0 \qquad (2.298)$$

That means J, M_J, L and S are good quantum numbers and M_L and M_S are not good quantum numbers. Therefore, H_{SO} is not diagonal in the $|\gamma LSM_L M_S\rangle$ representation but it is diagonal in $|\gamma LSJM_J\rangle$ representation. H_{SO} splits the degenerate LS multiplet depending on the value of J,

$$|L - S| \leq J \leq L + S \qquad (2.299)$$

and causes a *fine structure of the terms*:

$$E^{(0)}_{\gamma LSJ} = E^{(0)}_{\gamma LS} + \frac{1}{2}\hbar^2 \Lambda(\gamma, LS)\{J(J+1) - L(L+1) - S(S+1)\} \qquad (2.300)$$

We can now complete the solution for the ground state of carbon sketched in Fig. 2.3 (Fig. 2.4). According to (2.301), in case L and S are not equal to zero, H_{SO} introduces an additional splitting:

In (2.300), one recognizes the famous *Lande's interval rule*, which makes a statement about the energy difference between two neighbouring terms in an LS multiplet:

$$\Delta E = E^{(0)}_{\gamma LSJ} - E^{(0)}_{\gamma LSJ-1} = \hbar^2 \, \Lambda(\gamma, LS) \, J \qquad (2.301)$$

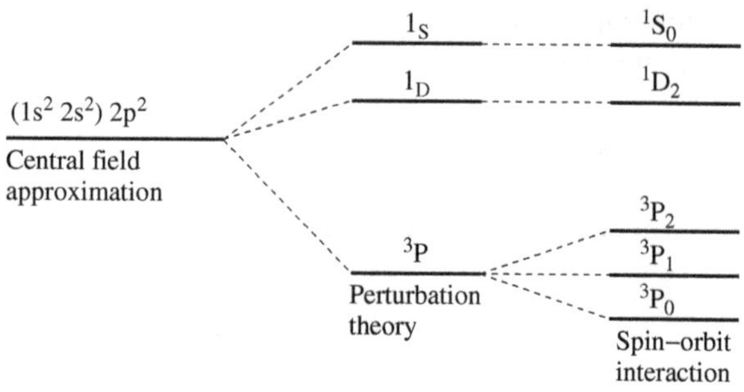

Fig. 2.4 Extended term scheme of carbon ($1s^2 2s^2 2p^2$)

This relation can be used both as an experimental criterion for LS coupling as well as for determining Λ.

Since

$$-J \le M_J \le +J$$

every $E^{(0)}_{\gamma LSJ}$ is still $(2J+1)$-fold directionally degenerate. This remaining degeneracy can be either fully or partially lifted by an external field.

The effect of an external magnetic field can be easily built into our Hamiltonian.

$$H = H^{(0)} - \frac{\mu_B}{\hbar}\,(L_z + 2S_z)\,B_0 \tag{2.302}$$

The same complications appear now as were discussed in Sect. 2.6 for a single-electron system.

$$\left[H\,,\,J^2\right]_- \neq 0 \Rightarrow J \text{ is no more a good quantum number}$$
$$\left[H\,,\,J_z\right]_- = 0 \Rightarrow M_j \text{ remains a good quantum number}$$

In the region of validity of the Russell–Saunders coupling, the magnetic field induces transitions between the states with different J but with the same M_J. The energy levels can be calculated according to the same scheme as was done in Sect. 2.6 for the single-electron system. We only have to replace the small letters by the corresponding capital letters. Naturally the limiting cases are also same.

1. *Weak fields:* $\mu_B B_0 \ll \Lambda(\gamma, LS)$
   J can still be treated as a good quantum number: The non-diagonal elements of the Hamiltonian in J are negligible:

$$E_{\gamma LSJM_J} = E^{(0)}_{\gamma LSJ} + g_J(L,S)\,M_J\,\mu_B\,B_0 \tag{2.303}$$

$$g_J(L,S) = 1 + \frac{J(J+1) - L(L+1) + S(S+1)}{2J(J+1)} \tag{2.304}$$

   This situation is known as the *anomalous Zeeman effect*.

2. *Strong fields:* $\mu_B B_0 \gg \Lambda(\gamma, LS)$
   Now M_L and M_S are still good quantum numbers.

$$E_{\gamma LSJM_J} = E^{(0)}_{\gamma LSJ} + (M_L + 2M_S)\,\mu_B\,B_0 \tag{2.305}$$

   This situation is known as the *normal Zeeman effect.*
3. *Transition region:* $\mu_B B_0 \approx \Lambda(\gamma, LS)$
   In this case, the field dependence is complicated, which, in principle can be evaluated as in Sect. 2.6. For $S > 1/2$, however, more non-diagonal elements appear as compared to the doublet case of Sect. 2.6.

(b) jj-Coupling

jj-coupling is realized in the "heavy" atoms which lie in the right bottom corner of the periodic table:

$$H_e \ll H_{SO} \tag{2.306}$$

To a first approximation, there is no coupling between different electron. Every electron is "a system by itself" with the total angular momentum

$$\mathbf{j}_i = \mathbf{l}_i + \mathbf{s}_i \tag{2.307}$$

H_{SO} lifts the degeneracy of the single-particle states, every level with $l \neq 0$ is split into two levels with $j = l \pm 1/2$ and the corresponding states are labelled by the quantum numbers $(\gamma\ l\ j\ m_j)$.

Because of H_e, strictly speaking, only J remains a good quantum number, j_i only approximately. The Hund's rules discussed in Sect. 2.1 are valid only for LS coupling and are not applicable for the jj-coupling.

(c) Intermediate Coupling

For the intermediate coupling,

$$H_e \approx H_{SO}, \tag{2.308}$$

which is realized in the middle of the periodic table, there exists practically no applicable theory. This case poses mathematically the most complicated problem.

2.11 Problems

Problem 2.1 How many states belong to a given (L, S)-multiplet? Check your answer with Eq. (2.14).

Problem 2.2 Show that the Pauli's spin matrices σ_x, σ_y, σ_z have the following properties:

1. $\sigma_x^2 = \sigma_y^2 = \sigma_z^2 = \mathbb{1}_2$

2. $\left[\sigma_x, \sigma_y\right]_+ = \left[\sigma_y, \sigma_z\right]_+ = [\sigma_z, \sigma_x]_+ = 0$

3. $\left[\sigma_x, \sigma_y\right]_- = 2i\sigma_z \; ; \; \left[\sigma_y, \sigma_z\right]_- = 2i\sigma_x \; ; \; [\sigma_z, \sigma_x]_- = 2i\sigma_y$

4. $\sigma_x\sigma_y\sigma_z = i\,\mathbb{1}_2$

Problem 2.3 Show that the Dirac matrices

$$\widehat{\alpha} = \begin{pmatrix} 0 & \sigma \\ \sigma & 0 \end{pmatrix} \quad ; \quad \widehat{\beta} = \begin{pmatrix} \mathbb{1}_2 & 0 \\ 0 & -\mathbb{1}_2 \end{pmatrix}$$

satisfy the conditions (2.23), (2.24) and (2.25).

Problem 2.4 Show that the cartesian components of the Dirac's spin operator (2.59),

$$\widehat{s}_{x,y,z} = \frac{\hbar}{2}\begin{pmatrix} \sigma_{x,y,z} & 0 \\ 0 & \sigma_{x,y,z} \end{pmatrix} \qquad \sigma_{x,y,z} : \text{ Pauli's spin matrices,}$$

satisfy the fundamental commutation relations of angular momentum.

Problem 2.5 For a free Dirac electron, calculate the commutators

$$\left[\widehat{s}, H_D^{(0)}\right]_- \quad ; \quad \left[\mathbf{l}, H_D^{(0)}\right]_- ,$$

where $\widehat{s}$ is the Dirac's spin operator and $\mathbf{l}$ is the orbital angular momentum operator.

Problem 2.6 Let **a** and **b** be two vector operators which commute with all the three Pauli's spin matrices. Prove the following relation:

$$(\sigma \cdot \mathbf{a})(\sigma \cdot \mathbf{b}) = \mathbf{a} \cdot \mathbf{b}\, \mathbb{1}_2 + i\sigma \cdot (\mathbf{a} \times \mathbf{b})$$

Problem 2.7 For a (relativistic) electron in an external electromagnetic field (vector potential $\mathbf{A}(\mathbf{r}, t)$, scalar potential $\varphi(\mathbf{r}, t)$) derive the equations of motion of the observables *position* $\mathbf{r}$ and *mechanical momentum* $\mathbf{p}_m = \mathbf{p} + e\mathbf{A}$ (Heisenberg picture). Show that the time-dependent operators satisfy the classical Lorentz force law

$$\frac{\mathrm{d}}{\mathrm{d}t}\mathbf{p}_m = -e(\mathbf{E} + \dot{\mathbf{r}} \times \mathbf{B})$$

(Ehrenfest theorem!).

Problem 2.8 For the operator of the spin–orbit interaction

$$H_{SB} = \lambda(\mathbf{L} \cdot \mathbf{S})$$

calculate the following commutators:

1. $[H_{SB}, \mathbf{L}]_-$
2. $[H_{SB}, \mathbf{S}]_-$
3. $\left[H_{SB}, \mathbf{L}^2\right]_-$
4. $\left[H_{SB}, \mathbf{S}^2\right]_-$
5. $\left[H_{SB}, \mathbf{J}^2\right]_-$ $(\mathbf{J} = \mathbf{L} + \mathbf{S})$

Problem 2.9 Let $\mathbf{J} = (J_x, J_y, J_z)$ be an angular momentum operator and $\hat{T}^{(k)}$ be an irreducible tensor operator with $(2k+1)$ standard components $T_q^{(k)}$, $q = -k, -k+1, \ldots, +k$. Show that the following commutation relations must hold:

1. $\left[J_z, T_q^{(k)}\right]_- = \hbar q T_q^{(k)}$
2. $\left[J_\pm, T_q^{(k)}\right]_- = \hbar\sqrt{k(k+1) - q(q \pm 1)}T_{q\pm 1}^{(k)}$

$$(J_\pm = J_x \pm i J_y)$$

Problem 2.10 Show that every vector operator $\mathbf{K} = (K_x, K_y, K_z)$ is an irreducible tensor operator of rank one $\hat{T}^{(1)}$ with the standard components.

$$T_0^{(1)} = K_z$$
$$T_{\pm 1}^{(1)} = \mp\frac{1}{\sqrt{2}}(K_x \pm i K_y)$$

Problem 2.11 Let $\mathbf{I} = (I_x, I_y, I_z)$ be the nuclear spin operator and I^2 its square. Show that the combinations

$$T_0^{(2)} = I^2 - 3I_z^2$$
$$T_{\pm 1}^{(2)} = \pm\frac{1}{2}\sqrt{6}(I_z I_\pm + I_\pm I_z)$$
$$T_{\pm 2}^{(2)} = -\frac{1}{2}\sqrt{6}(I_\pm)^2$$

with $I_\pm = I_x \pm i I_y$ are the standard components of an irreducible tensor of rank two. In doing this verify the commutation relations:

$$\left[I_\pm, T_q^{(2)}\right]_- = \hbar\sqrt{6 - q(q \pm 1)}T_{q\pm 1}^{(2)}$$
$$\left[I_z, T_q^{(2)}\right]_- = \hbar q T_q^{(2)}$$

Problem 2.12 Let $T_q^{(2)}$ be the standard components of the tensor $\hat{T}^{(2)}$ of rank two related to the nuclear spin **I** from Problem 2.11. The standard components $t_q^{(2)}$ of the tensor $\hat{t}^{(2)}$ are defined exactly analogously, where only in place of the nuclear spin components, the components of the electron angular momentum $\mathbf{J} = (J_x, J_y, J_z)$ appear:

$$t_0^{(2)} = J^2 - 3J_z^2$$
$$t_{\pm 1}^{(2)} = \pm\frac{1}{2}\sqrt{6}(J_z J_\pm + J_\pm J_z)$$
$$t_{\pm 2}^{(2)} = -\frac{1}{2}\sqrt{6}J_\pm^{(2)}$$

Express the scalar product

$$A = \sum_{q=-2}^{+2} t_q^{(2)} \cdot \left(T_q^{(2)}\right)^\dagger$$

in terms of I^2, J^2 and $(\mathbf{J} \cdot \mathbf{I})$.

Problem 2.13 Let the current density **j** be bounded in space and be free of sources (div $\mathbf{j} = 0$; magnetostatics). Let $f(\mathbf{r})$ and $g(\mathbf{r})$ be differentiable but otherwise arbitrary scalar fields.

1. Show that

$$D = \int d^3r (f(\mathbf{r})\mathbf{j} \cdot \nabla g(\mathbf{r}) + g(\mathbf{r})\mathbf{j} \cdot \nabla f(\mathbf{r})) = 0$$

2. Show that

$$\int d^3r\, \mathbf{j}(\mathbf{r}) = 0$$

3. Let **a** be an arbitrary vector. Verify the following vector identity:

$$\int d^3r\, (\mathbf{a} \cdot \mathbf{r})\mathbf{j}(\mathbf{r}) = \frac{1}{2}\left\{\mathbf{a} \times \int (\mathbf{r} \times \mathbf{j}(\mathbf{r}))d^3r\right\}$$

Chapter 3
Diamagnetism

3.1 Bohr–van Leeuwen Theorem

Usually diamagnetism is explained as an induction effect (see Sect. 1.4), as the magnetism of moving charges treated within a classical, easily picturizable, vector atom model. The Larmor precession of the orbital angular momentum about the direction of the magnetic field induces an extra moment, which according to the Lenz's law, is in the direction opposite to that of the applied field. In reality, there are inconsistencies hidden in this derivation, i.e. certain "quantum mechanical elements" are buried underneath the arguments. For there exists a theorem known as the Bohr–van Leeuwen theorem, which can be proved rigorously. The theorem states

Magnetism is a quantum mechanical effect. Strictly classically, there cannot be either dia-, para- or collective magnetism.

Proof Let the solid consist of identical ions and possess translational symmetry. Then, according to (1.58), the magnetization is given by

$$\mathbf{M} = \frac{N}{V}\langle \mathbf{m} \rangle \tag{3.1}$$

where $\mathbf{m}$ is the magnetic moment of the individual ion. N is the number of ions in volume V. Then according to (1.34), we have

$$\mathbf{m} = -\frac{\partial W}{\partial \mathbf{B}_0} = -\frac{\partial H}{\partial \mathbf{B}_0} \tag{3.2}$$

Here H is the classical Hamiltonian function of a single ion. The classical average is built in the following way:

$$\langle \mathbf{m} \rangle = \frac{1}{Z^*} \int \cdots \int dx_1 \cdots dx_{3N_e}\, dp_1 \cdots dp_{3N_e}\, \mathbf{m}\, e^{-\beta H} \tag{3.3}$$

W. Nolting, A. Ramakanth, *Quantum Theory of Magnetism*,
DOI 10.1007/978-3-540-85416-6_3,

$Z = \frac{1}{N_e! h^{3N_e}} Z^*$ is the classical partition function given by

$$Z^* = \int \cdots \int dx_1 \cdots dx_{3N_e}\, dp_1 \cdots dp_{3N_e} e^{-\beta H} \tag{3.4}$$

N_e is the number of electrons per ion. Using (3.1) and (3.2), we get

$$\langle \mathbf{m} \rangle = \frac{1}{\beta Z} \frac{\partial Z}{\partial \mathbf{B}_0} \qquad (\beta = \frac{1}{k_B T}) \quad (k_B = 1.3805\ 10^{-23} J/K) \tag{3.5}$$

The Bohr–van Leeuwen theorem is proved, if we can show that even when an external field is switched on, Z does not depend on this field.

In the presence of a magnetic field $\mathbf{B}_0 = \nabla \times \mathbf{A}$, H has the general form

$$H = \frac{1}{2m} \sum_{i=1}^{3N_e} (\mathbf{p}_i + e\mathbf{A}_i)^2 + H_1(x_1, \cdots x_{3N_e}) \tag{3.6}$$

where H_1 is the term representing the electron interactions. The partition function (3.4) can then be written as

$$Z^* = \int \cdots \int_V dx_1 \cdots dx_{3N_e} e^{-\beta H_1(x_1, \cdots x_{3N_e})} * \\ * \int_{-\infty}^{+\infty} \cdots \int_{-\infty}^{+\infty} dp_1 \cdots dp_{3N_e} exp\left(-\frac{\beta}{2m} \sum_{i=1}^{3N_e} (\mathbf{p}_i + e\mathbf{A}_i)^2\right) \tag{3.7}$$

Decisive is the fact that the momentum integration runs from $-\infty$ to $+\infty$. We can then substitute

$$u_i = p_i + eA_i \tag{3.8}$$

without changing the limits of integration.

$$Z^* = \int \cdots \int_V dx_1 \cdots dx_{3N_e} e^{-\beta H_1(x_1, \cdots x_{3N_e})} * \\ * \int_{-\infty}^{+\infty} \cdots \int_{-\infty}^{+\infty} du_1 \cdots du_{3N_e} exp\left(-\frac{\beta}{2m} \sum_{i=1}^{3N_e} u_i^2\right) \tag{3.9}$$

This means, obviously, that, Z is in fact, field independent.

$$Z \neq Z(B_0) \tag{3.10}$$

As a result, according to (3.5), the average magnetic moment vanishes in all cases.

$$\langle \mathbf{m} \rangle \equiv 0 \tag{3.11}$$

Thus, rigorously, classically, there is no magnetism. We will therefore, right from the beginning, argue quantum mechanically and refrain from using any semiclassical models.

Matter consists of charged particles in motion and so respond to an external magnetic field $\mathbf{B}_0$. We have to distinguish between two situations. (a) The system already contains permanent magnetic moments. Then, they will order themselves in a field. This gives rise to collective phenomena, namely *para-, ferro-, antiferro- and ferrimagnetism*, which will be discussed later. (b) The field itself induces magnetic moments. This phenomenon, called *diamagnetism* is a characteristic of *all* materials, being, however, observable only when (a) is not present. We will now be concerned with this phenomenon. There are qualitative differences between the diamagnetism of insulators (*Larmor diamagnetism*) and the diamagnetism of metals (*Landau diamagnetism*). Therefore, we will discuss them separately.

3.2 Larmor Diamagnetism (Insulators)

Diamagnetism is a property of all systems, but it is observable only if the given system does not exhibit, in addition to diamagnetism, any one of the para-, ferro-, ferri- or antiferromagnetism. Let us, therefore, consider a solid which is made up of ions with completely filled electron shells. Then, for the ground state, we have

$$\mathbf{J}|0\rangle = \mathbf{L}|0\rangle = \mathbf{S}|0\rangle = 0 \tag{3.12}$$

The condition that diamagnetism is observable only if the electron shells are completely filled is classically not understandable. Here already, one recognizes the advantage of the correct quantum mechanical description when compared to the "picturizable" semiclassical description.

We now switch on an external magnetic field on a diamagnet. Let the magnetic field be given by

$$\mathbf{B}_0 = \mu_0\,\mathbf{H} = (0,\ 0,\ B_0) \tag{3.13}$$

We look for the response of the system to this field, which means, the field-induced magnetic moment or in other words, the magnetization.

Let the solid we are considering be an insulator, i.e. all the electrons are strictly localized. Then we can write

$$\mathbf{M}(B_0) = \frac{N}{V}\langle 0|\mathbf{m}|0\rangle \tag{3.14}$$

N is the number of ions in volume V. Magnetic energies ($\approx \mu_B\, B_0$) are in general so small that the system remains in the ground state even when the field is switched on.

That is why we can, while performing the average, restrict ourselves to the ground state $|0\rangle$.

There are two terms in the Hamiltonian, which contain the magnetic field (see Sect. 2.10).

$$H_z = -\frac{\mu_B}{\hbar}\,(L_z + 2S_z)\,B_0 \tag{3.15}$$

$$H_{dia} = \frac{e^2 B_0^2}{8m}\sum_{j=1}^{N_e}(x_j^2 + y_j^2) \tag{3.16}$$

N_e is the number of electrons in the ion or the atom. In order to obtain **m**, we must differentiate with respect to B_0. Because of the completely filled shells, in view of (3.12),

$$\langle 0|\frac{\partial H_z}{\partial B_0}|0\rangle = 0 \tag{3.17}$$

So, what remains is

$$\mathbf{M}(B_0) = -\frac{N}{V}\langle 0|\frac{\partial H_{dia}}{\partial \mathbf{B}_0}|0\rangle \tag{3.18}$$

Because of the spherical symmetry of the individual ion (noble gas configuration), we have

$$\sum_{j=1}^{N_e}\langle 0|x_j^2|0\rangle = \sum_{j=1}^{N_e}\langle 0|y_j^2|0\rangle = \sum_{j=1}^{N_e}\langle 0|z_j^2|0\rangle = \frac{1}{3}\sum_{j=1}^{N_e}\langle 0|r_j^2|0\rangle \tag{3.19}$$

This gives for the z-component of the magnetization (x- component and y-component vanish!)

$$M(B_0) = -\frac{Ne^2}{6mV}\,B_0\sum_{j=1}^{N_e}\langle r_j^2\rangle \tag{3.20}$$

By differentiating once more with respect to B_0, we finally get for the *diamagnetic susceptibility*

$$\chi^{dia} = \mu_0\left(\frac{\partial M}{\partial B_0}\right)_T = -\frac{Ne^2\mu_0}{6mV}\sum_{j=1}^{N_e}\langle 0|r_j^2|0\rangle \tag{3.21}$$

The negative susceptibility is typical for diamagnets (see Sect. 1.4). A diamagnet is always repulsed by a magnetic pole. The negative sign can be understood as the expression of the Lenz's law. The external field induces a moment whose field is directed opposite to that of the applied field. Let us estimate the *order of magnitude* of χ^{dia}. In literature, usually, the numbers are given for the *molar susceptibility*:

$$\chi_m^{dia} = \chi^{dia} \frac{N_A}{N/V} \left(\frac{cm^3}{mol}\right) \tag{3.22}$$

where $N_A = 6.022 \times 10^{23}/mol$ is the Avogadro number. For the estimation, we introduce the average ion radius

$$\langle r^2 \rangle = \frac{1}{N_e} \sum_{j=1}^{N_e} \langle 0 | r_j^2 | 0 \rangle \tag{3.23}$$

and express it in units of the Bohr radius:

$$a_B = \frac{4\pi \varepsilon_0 \, \hbar^2}{me^2} = 0.529 \text{ Å} \tag{3.24}$$

Then we get for the molar susceptibility

$$\chi_m^{dia} = -0.995 \times 10^{-5} N_e \langle r^2/a_B^2 \rangle \left(\frac{cm^3}{mol}\right) \tag{3.25}$$

The average value $\langle r^2/a_B^2 \rangle$ is of the order of 1. Therefore, χ_m^{dia} is very small. That is why, diamagnetism is observable only when it is not buried under either the paramagnetism or collective magnetism.

Examples for diamagnets are

(a) Noble gases
(b) Simple ionic crystals (alkali halides)

In the latter case, to a first approximation, the contributions from both the ions add up.

Table 3.1 Examples for the diamagnetic molar susceptibility χ_m^{dia} in 10^{-6}(cm^3/mol)

	He: −1.9	Li^+: −0.7
F^- : −9.4	Ne : −7.2	Na^+ : −6.1
Cl^- : −24.2	Ar : −19.4	K^+ : −14.6
Br^- : −34.5	K : −28.0	Rb^+ : −22.0
J^- : −50.6	Xe : −43.0	Cs^+ : −35.1

From Table 3.1, we can note the following trends:

(a) In the table, in each column, the electron number increases from top to bottom and so does $|\chi|$ in conformity with (3.21).

(b) In a given row, the electron number is the same but the nuclear charge number Z increases from left to right. That means an increasing attractive force on the electron shell which results in decreasing the size of the ion from left to right. $\langle r^2 \rangle$ is a measure of the ion size and therefore again the trend is in conformity with (3.21).

The evaluation of the average ionic radius $\langle r^2 \rangle$ is a non-trivial quantum mechanical problem. For this purpose, one often uses the following semi-empirical method: Let us consider an electron in a state with the principal quantum number n and orbital quantum number l. Then, using the exactly known eigenfunctions of hydrogen atom [1], we get

$$\langle r_{nl}^2 \rangle = \frac{a_B^2\, n^2}{2Z^2}(5n^2 + 1 - 3l\,(l+1)) \tag{3.26}$$

Due to the spherical symmetry of the electron shells, the magnetic quantum number does not play any role. In the case of many electrons, one has to take into account the screening of the nuclear charge. That is done to a first approximation by defining an effective nuclear charge seen by the electron to be

$$Z^* = Z - \sigma_{nl} \tag{3.27}$$

where σ_{nl} is to be treated as a parameter which has to be determined by other independent measurements. In addition, if we take into account the $2(2l+1)$-fold degeneracy of the levels we get

$$\chi^{dia} = -\frac{Ne^2\mu_0}{6mV}\, a_B^2 \sum_n \sum_l \frac{(2l+1)\, n^2\, (5n^2 + 1 - 3l\,(l+1))}{(Z - \sigma_{nl})^2} \tag{3.28}$$

The summations run over all the occupied shells.

3.3 The Sommerfeld Model of a Metal

Metallic solids have two "sources of diamagnetism", namely

1. the completely filled shells of the ion core ($\sim$ *Larmor*),
2. the freely moving conduction electrons ($\sim$ *Landau*).

The contribution of (1) was discussed in Sect. 3.2. We will refer to the contribution of (2) in Sect. 3.4.

Since the conduction electrons have a spin and therefore a permanent magnetic moment, when a magnetic field is applied, there appear both para- and also diamagnetic effects. The coupling of the field to the spin leads to paramagnetism and the coupling of the field to the orbital motion leads to diamagnetism. The two couplings cannot, however, be separated. There exist interference terms, which, depending on the strength of the field, show either paramagnetic or diamagnetic behaviour. Therefore, we find that the isothermal magnetic susceptibility χ_T of conduction electrons is made up of three additive terms:

$$\chi_T = \chi_{Pauli} + \chi_{Landau} + \chi_{osc} \tag{3.29}$$

The *Pauli susceptibility* χ_{Pauli} is clearly due to the electron spin and is positive. χ_{Landau} is negative (diamagnetism!) and is a consequence of the orbital motion of the electrons. χ_{osc} is an oscillatory function of the applied field and leads to the *de Haas-von Alphen effect* (see Sect. 3.5)

In order to calculate the susceptibility (3.29), we will describe the conduction electrons within the framework of the so-called *Sommerfeld model.* As the model has extensive applications even in later chapters, we will at first present a detailed discussion of the most important aspects of this model.

3.3.1 Properties of the Model

The Sommerfeld model, to a good approximation, explains the properties of the so-called "simple metals" such as Na, K, Mg, Cu, Ag, Au. The model is defined by the following assumptions:

1. Ideal Fermi gas in volume $V = L^3$
2. Periodic boundary conditions on the volume V
3. Lattice potential $V(\mathbf{r}) \equiv const.$

We will list out the most important properties of the model.

3.3.1.1 Eigenfunctions and Eigenenergies

Because of (1) and (3), the plane waves

$$\psi_{\mathbf{k}\sigma}(\mathbf{r}) = \frac{1}{\sqrt{V}} e^{i\mathbf{k}\cdot\mathbf{r}} \chi_\sigma \tag{3.30}$$

where

$$\chi_\uparrow = \begin{pmatrix} 1 \\ 0 \end{pmatrix} \quad ; \quad \chi_\downarrow = \begin{pmatrix} 0 \\ 1 \end{pmatrix} \tag{3.31}$$

are, in the volume V, the eigenfunctions of the Hamiltonian

$$H = H_0 = -\frac{\hbar^2}{2m}\nabla^2 \tag{3.32}$$

The eigenenergies are then given by

$$\varepsilon(\mathbf{k}) = \frac{\hbar^2 k^2}{2m} \tag{3.33}$$

For the wavefunction, the assumption of periodic boundary conditions means

$$\begin{aligned}\psi_{\mathbf{k}\sigma}(x,\ y,\ z) &= \psi_{\mathbf{k}\sigma}(x+L,\ y,\ z) = \psi_{\mathbf{k}\sigma}(x,\ y+L,\ z)\\ &= \psi_{\mathbf{k}\sigma}(x,\ y,\ z+L)\end{aligned} \tag{3.34}$$

Because of (3.30), this is satisfied only for certain wave vectors

$$k_{x,y,z} = n_{x,y,z}\frac{2\pi}{L}, \qquad n_x, n_y,\ n_z \in \mathbb{N} \tag{3.35}$$

The density of allowed **k**-vectors in **k**-space is not arbitrarily large any more. The volume occupied by each allowed **k**-vector is

$$\Delta\mathbf{k} = \frac{(2\pi)^3}{V} \tag{3.36}$$

Of course, there will be two spin-degenerate states for each of these allowed **k**-vectors. As a result, the energy levels are discrete:

$$\varepsilon(\mathbf{k}) = \frac{\hbar^2}{2m}\left(k_x^2 + k_y^2 + k_z^2\right) = \frac{2\pi^2\,\hbar^2}{m\,L^2}\left(n_x^2 + n_y^2 + n_z^2\right) \tag{3.37}$$

The sum in the bracket is a non-negative integer.

3.3.1.2 Ground State of the N_e-Electron System

In the ground state ($T = 0$) the electrons occupy all the states with energy

$$\varepsilon(\mathbf{k}) \le \varepsilon_F = \frac{\hbar^2\,k_F^2}{2\,m} \tag{3.38}$$

ε_F is called the *Fermi energy*, which is the energy of the highest occupied energy level at $T = 0$. All the levels below and including this level are each occupied by two electrons with oppositely oriented spins. All the levels above this are unoccupied. In view of the isotropic energy dispersion, $\varepsilon(\mathbf{k}) = \varepsilon(k)$, the electrons in the ground state occupy all the states within the *Fermi sphere* (Fig. 3.1). The radius of the Fermi sphere defines the *Fermi wavevector* k_F. k_F is fixed by the number of electrons N_e:

$$N_e = 2\sum_{\mathbf{k}}^{k\le k_F} 1 = \frac{2}{\Delta\mathbf{k}}\int_{k\le k_F} d^3k = \frac{2V}{8\pi^3}\frac{4\pi}{3}k_F^3 \tag{3.39}$$

This gives us the important relations

$$k_F = \left(\frac{3\pi^2 N_e}{V}\right)^{1/3} \tag{3.40}$$

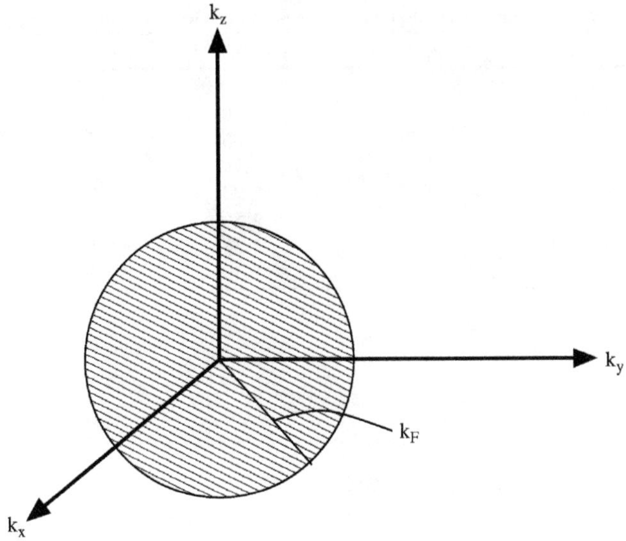

Fig. 3.1 Fermi sphere in **k**-space

$$\varepsilon_F = \frac{\hbar^2}{2m}\left(\frac{3\pi^2 N_e}{V}\right)^{2/3} \tag{3.41}$$

One can easily calculate (Problem 3.2) the average energy per electron to be

$$\bar{\varepsilon} = \frac{2}{N_e}\sum_{\mathbf{k}}^{k\le k_F}\frac{\hbar^2 k^2}{2m} = \frac{3}{5}\varepsilon_F \tag{3.42}$$

A few typical values for simple metals are given in Table 3.2. The last column is the so-called *Fermi temperature*:

$$T_F = \frac{\varepsilon_F}{k_B} \tag{3.43}$$

3.3.1.3 Density of States

Density of states is a fundamental quantity of solid state physics, which is defined as

$$\rho(E)dE = number\ of\ states\ in\ the\ energy\ interval\ [E,\ E + dE].$$

Table 3.2 Typical data, calculated within the Sommerfeld model, for simple metals.

	$N_e/V[\text{cm}^{-3}]$	$k_F[\text{cm}^{-1}]$	$v_F[\text{cms}^{-1}]$	$\varepsilon_F[\text{eV}]$	$T_F[\text{K}]$
Li	4.6×10^{22}	1.10×10^8	1.30×10^8	4.7	5.5×10^4
Na	2.50	0.90	1.10	3.1	3.7
K	1.34	0.73	0.85	2.1	2.4
Rb	1.08	0.68	0.79	1.8	2.1
Cs	0.86	0.63	0.73	1.5	1.8
Cu	8.50	1.35	1.56	7.0	8.2
Ag	5.76	1.19	1.38	5.5	6.4
Au	5.90	1.20	1.39	5.5	6.4

That means

$$\rho(E)dE = 2\frac{1}{\Delta\mathbf{k}} \int_{shell(E,E+dE)} d^3k \tag{3.44}$$

The integration is to be performed in **k**-space, over a shell which contains the **k**-vectors, all of which belong to energies between E and $E + dE$. In terms of the *phase volume*

$$\phi(E) = \int_{\varepsilon(\mathbf{k}) \le E} d^3k \tag{3.45}$$

we can also write

$$\rho(E)dE = \frac{2V}{(2\pi)^3} \left(\frac{d}{dE} \phi(E) \right) dE \tag{3.46}$$

For the Sommerfeld model, $\phi(E)$ can easily be calculated.

$$\phi_0(E) = \frac{4\pi}{3} k^3 \bigg|_{\varepsilon(k)=E} = \frac{4\pi}{3} \left(\frac{2mE}{\hbar^2} \right)^{3/2} \tag{3.47}$$

Differentiating this expression with respect to E,

$$\frac{d\phi_0}{dE} = 2\pi \left(\frac{2m}{\hbar^2} \right)^{3/2} \sqrt{E}$$

and substituting in (3.46), we get

$$\rho_0(E) = \begin{cases} d\ E^{1/2}, & if\ E \ge 0 \\ 0 & otherwise \end{cases} \tag{3.48}$$

with

$$d = \frac{V}{2\pi^2}\left(\frac{2m}{\hbar^2}\right)^{3/2} = \frac{3N_e}{2\varepsilon_F^{3/2}} \tag{3.49}$$

The $\sqrt{E}$-dependence of the density of states is typical for the Sommerfeld model (Problem 3.4). Since

$$\rho_0(\varepsilon_F) = \frac{3N_e}{2\varepsilon_F} \tag{3.50}$$

we can also write

$$d = \rho_0(\varepsilon_F)\,\frac{1}{\sqrt{\varepsilon_F}} \tag{3.51}$$

3.3.1.4 Occupation Probability

In Statistical Mechanics, the grand canonical partition function is defined by

$$\Xi(T, V, \mu) = Tr\left(e^{-\beta(H-\mu\widehat{N})}\right) \tag{3.52}$$

μ is the chemical potential and $\widehat{N}$ is the particle number *operator*. The simplest way to evaluate the trace is in the energy representation.

$$\Xi(T, V, \mu) = \sum_{N=0}^{\infty}\sum_{n} e^{-\beta(E_n^{(N)}-\mu N)} \tag{3.53}$$

N is now the particle *number* and $E_n^{(N)}$ is the nth eigenenergy of the N-particle system. For a non-interacting system, these quantities can be expressed in terms of the single-particle energies ε_i and their occupation numbers n_i.

$$N = \sum_i n_i \quad ; \quad E_n^{(N)} = \sum_i n_i\varepsilon_i \tag{3.54}$$

The index i runs over all the single-particle states. With the help of (3.54), we can reformulate the partition function.

$$\Xi(T, V, \mu) = \sum_{N=0}^{\infty}\sum_{\substack{\{n_i\}\\ \sum n_i=N}} exp\left(-\beta\sum_i n_i(\varepsilon_i-\mu)\right) \tag{3.55}$$

The second summation runs over all the conceivable distributions of the N particles into the available single-particle levels. With

$$exp\left(-\beta\sum_i n_i(\varepsilon_i-\mu)\right)=\Pi_i exp\left(-\beta n_i(\varepsilon_i-\mu)\right)$$

and

$$\sum_{N=0}^{\infty}\sum_{\substack{\{n_i\}\\ \sum n_i=N}}\rightarrow\sum_{n_1}\sum_{n_2}\sum_{n_3}\cdots\sum_{n_r}\cdots$$

we can write Ξ as

$$\Xi(T,V,\mu)=\prod_r\left(\sum_{n_r}exp\left(-\beta n_r(\varepsilon_r-\mu)\right)\right) \tag{3.56}$$

In our case, the particles are electrons, i.e. Fermions for which the allowed occupation numbers are either 0 or 1. So we get

$$\Xi(T,V,\mu)=\prod_r(1+exp(-\beta(\varepsilon_r-\mu))) \tag{3.57}$$

It is well-known that, in Statistical Mechanics, the expectation value of an observable is defined as

$$\langle\widehat{A}\rangle=\frac{1}{\Xi}Tr\left(\widehat{A}\,e^{-\beta(H-\mu N)}\right) \tag{3.58}$$

The occupation probability of a particular level of a Fermi system is the same as the expectation value of occupation number operator $\widehat{n}_r$:

$$\begin{aligned}\langle\widehat{n}_r\rangle&=\frac{1}{\Xi}Tr\left(\widehat{n_r}\,e^{-\beta(H-\mu N})\right)\\&=\frac{1}{\Xi}\sum_{N=0}^{\infty}\sum_{\substack{\{n_i\}\\ \sum n_i=N}}n_r exp\left(-\beta\sum_j n_j(\varepsilon_j-\mu)\right)\end{aligned} \tag{3.59}$$

Comparing this with (3.55) one gets the simple relation

$$\langle\widehat{n}_r\rangle=-\frac{1}{\beta}\frac{\partial}{\partial\varepsilon_r}\ln\Xi \tag{3.60}$$

For the explicit calculation, we use (3.57) and obtain

$$\langle\widehat{n}_r\rangle=\{1+exp(\beta(\varepsilon_r-\mu))\}^{-1} \tag{3.61}$$

This is the well-known *Fermi function* $f_-(\varepsilon_r)$, which gives the probability for the occupation ($0 \leq \langle \widehat{n}_r \rangle \leq 1$) of the level ε_r at temperature T. $1 - f_-(\varepsilon_r)$ is then the probability that the state is unoccupied. At $T = 0$, $f_-(\varepsilon_r)$ is a step function

$$f_-(E;\, T = 0) = \theta(\varepsilon_F - E) \tag{3.62}$$

which is in conformity with the fact that, at $T = 0$ all the states below the Fermi energy $\varepsilon_F = \mu(T = 0)$ are occupied and all the states above ε_F are empty. For $T > 0$, the Fermi function smoothens out around the Fermi edge. For all temperatures it holds

$$f_-(E = \mu) = \frac{1}{2} \tag{3.63}$$

The smoothening out takes place symmetrically, i.e. the state with energy $\mu + \Delta E$ has the same probability for occupation as the state with energy $E - \Delta E$ has the probability, for being empty (Fig. 3.2).

$$f_-(\mu + \Delta E) = 1 - f_-(\mu - \Delta E) \tag{3.64}$$

With (3.63) and

$$\frac{df_-}{dE} \overset{E \to \mu}{\longrightarrow} -\frac{1}{4k_B T}, \tag{3.65}$$

we can estimate the width of the smoothened *Fermi layer* to be about $4k_B T$. From Table 3.2, one recognizes that even at room temperature, the layer constitutes, at the most, 1% of the total distribution.

The high-energy tail of the distribution reproduces the classical Maxwell–Boltzmann distribution:

$$f_-(E) \approx exp(-\beta(E - \mu)) \quad if\ E - \mu \gg k_B T \tag{3.66}$$

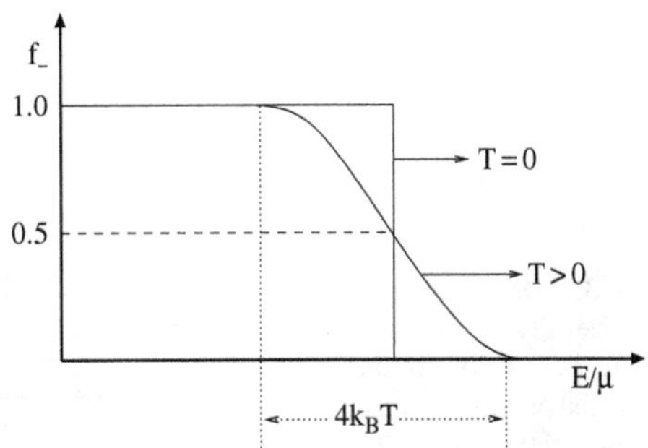

Fig. 3.2 Fermi distribution as a function of reduced energy E/μ. μ is the chemical potential

3.3.1.5 Thermodynamics of the Sommerfeld Model

The product of the density of states and the Fermi function gives the density of the occupied states (Fig. 3.3). Now, if an integral of this product over energy is taken, we get the *total electron number* N_e:

$$N_e = \int_{-\infty}^{+\infty} dE \; f_-(E)\,\rho_0(E) \tag{3.67}$$

The relation between the *internal energy* and the density of the occupied states lies on hand:

$$U = \int_{-\infty}^{+\infty} dE \; E \; f_-(E)\,\rho_0(E) \tag{3.68}$$

With this result, we have in principle, the complete thermodynamics of the Sommerfeld model is fixed. For example, the *specific heat* is given by

$$c_V = \left(\frac{\partial U}{\partial T}\right)_V \tag{3.69}$$

and the *free energy* (Problem B.10) by

$$F(T,V) = U(0) - T\int_0^T dT' \; \frac{U(T') - U(0)}{T'^2} \tag{3.70}$$

or the *entropy* by

$$\begin{aligned} S = -\left(\frac{\partial F}{\partial T}\right)_V &= \frac{U(T) - U(0)}{T} + \int_0^T dT' \; \frac{U(T') - U(0)}{T'^2} \\ &= \frac{1}{T}\,(U - F) \end{aligned} \tag{3.71}$$

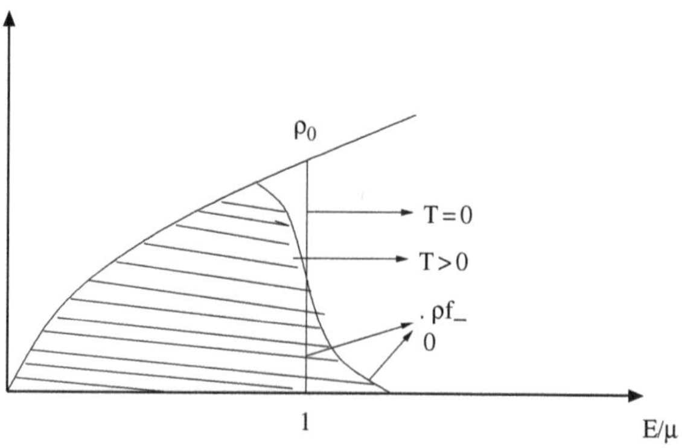

Fig. 3.3 Density of states ρ_0 and density of occupied states $\rho_0 f_-$ as functions of the reduced energy E/μ in the Sommerfeld model

The integrals that are required to be evaluated in order to determine U or N_e are of the type

$$I(T) = \int_{-\infty}^{+\infty} dE\ g(E)\ f_{-}(E) \tag{3.72}$$

where $f_{-}(E)$ is the Fermi function. This integral is different from its $T = 0$ value

$$I(T = 0) = \int_{-\infty}^{\varepsilon_F} dE\ g(E) \tag{3.73}$$

by an expression which practically is determined exclusively by the behaviour of the function $g(E)$ in the "Fermi layer". As this is small, power series expansions should be very promising.

3.3.2 *Sommerfeld Expansion*

We will now discuss as an insert, the Sommerfeld expansion, which is extremely useful in calculating integrals like (3.72), which are typical for solid state physics.

Let the function $g(E)$ satisfy three conditions:

1. $g(E)$ is non-singular in the Fermi layer
2. $g(E) \to 0$ as $E \to -\infty$
3. For $E \to \infty$, $g(E)$ diverges at the most like a power of E

Then we have (proof is given below)

$$\begin{aligned} I(T) &= \int_{-\infty}^{+\infty} dE\ g(E) f_{-}(E) \\ &= \int_{-\infty}^{\mu} dE\ g(E) + \sum_{n=1}^{\infty} \alpha_n (k_B T)^{2n}\ g^{(2n-1)}\big|_{E=\mu}\,. \end{aligned} \tag{3.74}$$

The coefficients α_n are given by

$$\alpha_n = 2\left(1 - \frac{1}{2^{2n-1}}\right)\zeta(2n) \tag{3.75}$$

where $\zeta(n)$ is the Riemann's ζ-function given by

$$\zeta(n) = \sum_{p=1}^{\infty} \frac{1}{p^n} \tag{3.76}$$

A few special values of this function are

$$\zeta(2) = \frac{\pi^2}{6}; \quad \zeta(4) = \frac{\pi^4}{90}; \quad \zeta(6) = \frac{\pi^6}{945} \cdots$$

The usefulness of the expansion is particularly clear, if the function $g(E)$ has the behaviour

$$g^{(n)}(E)\big|_{E=\mu} \approx \frac{g(\mu)}{\mu^n} \tag{3.77}$$

which is the case, for example, in the integrals for N_e and $U(T)$. Then, the series converges very fast, because the ratio of the successive terms is of the order of $\left(\frac{k_B T}{\mu}\right)^2 \ll 1$. In such a case, it is sufficient to restrict the sum to the first few terms:

$$I(T) = \int_{-\infty}^{\mu} dE\; g(E) + \frac{\pi^2}{6}(k_B T)^2 g'(\mu) + \frac{7\pi^4}{360}(k_B T)^4 g'''(\mu) + \mathcal{O}\left(\left(\frac{k_B T}{\mu}\right)^6\right) \tag{3.78}$$

Before deriving the Sommerfeld expansion, we want to discuss a few of its applications.

3.3.2.1 Chemical Potential μ

We can find the temperature dependence of the chemical potential easily using the Sommerfeld expansion in the expression for N_e (3.67). The density of states $\rho_0(E)$ of the Sommerfeld model (3.48) satisfies all the conditions of the Sommerfeld expansion. We substitute in (3.74) or (3.78) $g(E) = \rho_0(E)$ and obtain

$$N_e = \int_{-\infty}^{\mu} dE\; \rho_0(E) + \frac{\pi^2}{6}\,(k_B T)^2\,\rho_0'(\mu) \tag{3.79}$$

Using (3.48) and (3.49) one gets immediately

$$N_e \approx N_e \left(\frac{\mu}{\varepsilon_F}\right)^{3/2} + \frac{\pi^2}{6}(k_B T)^2\,\frac{3}{4}\,N_e \frac{1}{\varepsilon_F^{3/2}\,\mu^{1/2}}$$

The particle number gets cancelled and so we have

$$1 \approx \left(\frac{\mu}{\varepsilon_F}\right)^{3/2}\left[1 + \frac{\pi^2}{8}\left(\frac{k_B T}{\mu}\right)^2\right] \tag{3.80}$$

The second summand in the bracket is typically $<10^{-4}$ and therefore we can expand to get

$$\frac{\mu}{\varepsilon_F} \approx 1 - \frac{2}{3}\frac{\pi^2}{8}\left(\frac{k_B T}{\mu}\right)^2 \tag{3.81}$$

which finally leads to the well-known result of Statistical Mechanics:

$$\mu = \varepsilon_F \left[1 - \frac{\pi^2}{12}\left(\frac{k_B T}{\varepsilon_F}\right)^2\right] \tag{3.82}$$

Under normal conditions, in the case of a "degenerate" electron system, the temperature dependence of the chemical potential is practically negligible.

3.3.2.2 Internal Energy *U*

The function $E\ \rho_0(E)$ in (3.68) satisfies the preconditions for the Sommerfeld expansion

$$\begin{aligned} U(T) &\approx \int_0^{\mu} dE\ E\ \rho_0(E) + \frac{\pi^2}{6}(k_B T)^2\ [\mu\ \rho_0'(\mu) + \rho_0(\mu)] \\ &= \frac{2}{5}\ \mu^2\ \rho_0(\mu) + \frac{\pi^2}{4}\ (k_B T)^2\ \rho_0(\mu) \\ &= d\ \frac{2}{5}\ \varepsilon_F^{5/2}\left[\left(\frac{\mu}{\varepsilon}\right)^{5/2} + \frac{5\pi^2}{8}\left(\frac{k_B T}{\varepsilon_F}\right)^2\left(\frac{\mu}{\varepsilon_F}\right)^{1/2}\right] \end{aligned}$$

In view of the estimated orders of magnitude, we can simplify

$$\left(\frac{\mu}{\varepsilon_F}\right)^n \approx 1 - n\ \frac{\pi^2}{12}\left(\frac{k_B T}{\varepsilon_F}\right)^2 \tag{3.83}$$

With $U(0) = \frac{3}{5} N_e \varepsilon_F$, it follows that

$$\begin{aligned} U(T) = U(0)\Bigg[1 - \frac{5\pi^2}{24}\left(\frac{k_B T}{\varepsilon_F}\right)^2 + \frac{5\pi^2}{8}\left(\frac{k_B T}{\varepsilon_F}\right)^2 \\ + \mathcal{O}\left(\left[\frac{k_B T}{\varepsilon_F}\right]^4\right)\Bigg] \end{aligned}$$

Finally, we get

$$U(T) = U(0)\left[1 + \frac{5\pi^2}{12}\left(\frac{k_B T}{\varepsilon_F}\right)^2 + \mathcal{O}\left(\left[\frac{k_B T}{\varepsilon_F}\right]^4\right)\right] \tag{3.84}$$

3.3.2.3 Specific Heat c_V

$$c_V = \left(\frac{\partial U}{\partial T}\right)_V = U(0)\,\frac{5\pi^2}{6}\left(\frac{k_B}{\varepsilon_F}\right)^2 T \tag{3.85}$$

The characteristic feature is the linear temperature dependence, which is uniquely experimentally confirmed:

$$c_V = \gamma\, T \quad with\; \gamma = \frac{a}{\varepsilon_F} = b\,\rho_0(\varepsilon_F) \tag{3.86}$$

$$a = \frac{1}{2}N_e\,\pi^2\,k_B^2 \quad ; \quad b = \frac{1}{3}\,\pi^2\,k_B^2 \tag{3.87}$$

3.3.2.4 Proof of the Sommerfeld Expansion

We want to calculate

$$I(T) = \int_{-\infty}^{+\infty} dE\; g(E)\; f_-(E) \tag{3.88}$$

For that, we define

$$p(E) = \int_{-\infty}^{E} d\eta\; g(\eta) \;\Leftrightarrow\; g(E) = \frac{dp(E)}{dE} \tag{3.89}$$

Using this in (3.88) and integrating by parts

$$I(T) = \; p(E)f_-(E)|_{-\infty}^{+\infty} - \int_{-\infty}^{+\infty} dE\; p(E)f_-'(E) \tag{3.90}$$

The first term vanishes at the upper limit due to $f_-(E)$ and the precondition (3) and at the lower limit due to $p(E)$.

We now use the Taylor expansion of $p(E)$ around $E = \mu$, which is allowed because of the precondition (1):

$$p(E) = p(\mu) + \sum_{n=1}^{\infty} \frac{(E-\mu)^n}{n!}\; p^{(n)}(\mu) \tag{3.91}$$

The first summand gives the following contribution:

$$I_0(T) = -p(\mu)\int_{-\infty}^{+\infty} dE\; f_-'(E) = -p(\mu)\;\; f_-(E)|_{-\infty}^{+\infty} = +p(\mu) \tag{3.92}$$

In the sum in (3.91), only the terms with even powers of $(E-\mu)$ contribute to the integral $I(T)$, because

$$f'_{-}(E) = -\frac{\beta}{4\cosh^2\left(\frac{1}{2}\,\beta\,(E-\mu)\right)} \tag{3.93}$$

is an even function of $(E-\mu)$. Then we have the intermediate result

$$I(T) = I_0(T) + \beta \sum_{n=1}^{\infty} \frac{1}{(2n)!}\; g^{(2n-1)}(\mu)\; I_{2n}(T) \tag{3.94}$$

where we have abbreviated,

$$\begin{aligned}
I_{2n}(T) &= \int_{-\infty}^{+\infty} dE\,(E-\mu)^{2n} \frac{e^{\beta(E-\mu)}}{\left(e^{\beta(E-\mu)}+1\right)^2} \\
&= \beta^{-(2n+1)} \int_{-\infty}^{+\infty} dx\; x^{2n}\, \frac{e^x}{(e^x+1)^2} \\
&= -2\beta^{-(2n+1)} \left[\frac{d}{d\lambda} \int_0^{\infty} dx\, \frac{x^{2n-1}}{e^{\lambda x}+1}\right]_{\lambda=1} \\
&= -2\beta^{-(2n+1)} \left[\frac{d}{d\lambda} \lambda^{-2n} \int_0^{\infty} du\, \frac{u^{2n-1}}{e^{u}+1}\right]_{\lambda=1} \\
&= 4\,n\;\beta^{-(2n+1)} \int_0^{\infty} du\, \frac{u^{2n-1}}{e^{u}+1}
\end{aligned} \tag{3.95}$$

We now recognize the Riemann ζ-function which is defined by

$$\zeta(n) = \sum_{p=1}^{\infty} \frac{1}{p^n} = \frac{1}{(1-2^{1-n})\Gamma(n)} \int_0^{\infty} \frac{x^{n-1}}{e^x+1}\, dx \tag{3.96}$$

With this, the integral $I_{2n}(T)$ can be written as

$$\begin{aligned}
I_{2n}(T) &= 4\,n\;\beta^{-(2n+1)} \left(1-2^{1-2n}\right) \Gamma(2n)\,\zeta(2n) \\
&= 2\left(1-\frac{1}{2^{2n-1}}\right)(2n)!\;\zeta(2n)\;\beta^{-(2n+1)}
\end{aligned} \tag{3.97}$$

After substituting this relation in (3.94), the proof of Sommerfeld expansion is complete:

$$I(T) = p(\mu) + 2\sum_{n=1}^{\infty}\left(1-\frac{1}{2^{2n-1}}\right)\zeta(2n)\,(k_BT)^{2n}\; g^{(2n-1)}(\mu)$$

3.4 Landau Diamagnetism (Metals)

After this excursion into the Sommerfeld model, we come back to our actual problem of the diamagnetism of solid metals. We will calculate the susceptibility of the conduction electrons within the Sommerfeld model, using a procedure that was suggested by Landau [2]. This was later generalized by Peierls [3] by taking into account the periodic lattice potential. However, we will not discuss this part here, as the Landau theory itself contains all the important features.

We first want to solve the Schrödinger equation for the free electrons in the presence of a homogeneous static magnetic field. From the results obtained for the eigenstates and eigenenergies, we will calculate the grand canonical partition function $\Xi(T, B_0, \mu)$,

$$\Xi(T, B_0, \mu) = Tr\left(e^{-\beta(H-\mu\widehat{N})}\right)$$

From the partition function, we find the grand canonical potential $\Omega(T, B_0, \mu)$

$$\Omega(T, B_0, \mu) = -k_B T \ \ln \Xi(T, B_0, \mu) \tag{3.98}$$

$$d\Omega = -SdT - m\, dB_0 - N\, d\mu \tag{3.99}$$

and from $\Omega(T, B_0, \mu)$, we get the susceptibility

$$\chi_T = -\frac{\mu_0}{V}\left(\frac{\partial^2\Omega}{\partial B_0^2}\right)_T \tag{3.100}$$

This is the programme for the next sections.

3.4.1 Free Electrons in Magnetic Field (Landau Levels)

In the presence of an external magnetic field specified by the vector potential $\mathbf{A}(\mathbf{r})$, the Hamiltonian of a system of N_e non-interacting electrons is given by

$$H = \frac{1}{2m^*}\sum_{i=1}^{N_e}(\mathbf{p}_i + e\mathbf{A}(\mathbf{r}_i))^2 \tag{3.101}$$

The effective mass m^* should, in first approximation, take care of the lattice potential which will otherwise be neglected. However, it should be noted that m^* appears only in the orbital motion of the electrons and not in the spin interactions. When considering the latter, we have to use only the bare mass m. Since the electrons are not correlated with each other, we can limit our considerations to a single electron. We choose the vector potential such that

$$\nabla \times \mathbf{A} = \mathbf{B}_0 = B_0\, \mathbf{e}_z \tag{3.102}$$

$$\nabla \cdot \mathbf{A} = 0 \quad (Coulomb - gauge) \tag{3.103}$$

These conditions are fulfilled if we choose

$$\mathbf{A}(\mathbf{r}) = (0,\; B_0\, x,\; 0) \tag{3.104}$$

Substituting (3.104) in (3.101) and using (3.103), the Hamiltonian H_0 of a single electron is

$$\begin{aligned} H_0 &= \frac{1}{2m^*}(\mathbf{p} + e\mathbf{A})^2 = \frac{1}{2m^*}(p^2 + 2e\mathbf{A}\cdot\mathbf{p} + e^2\mathbf{A}^2) \\ &= \frac{1}{2m^*}\left(p_x^2 + p_z^2 + p_y^2 + 2e\; B_0\, x\; p_y + e^2\; B_0^2\, x^2\right) \end{aligned}$$

This can be further rewritten as

$$H_0 = \frac{1}{2m^*}\left(p_x^2 + p_z^2 + (p_y + e\; B_0\, x)^2\right) \tag{3.105}$$

For solving the Schrödinger equation

$$H_0\; \psi(x, y, z) = E\; \psi(x, y, z) \tag{3.106}$$

we choose the ansatz

$$\psi(x, y, z) = e^{ik_z z} e^{ik_y y}\; U(x) \tag{3.107}$$

This gives the eigenvalue equation

$$\left[-\frac{\hbar^2}{2m^*}\frac{d^2}{dx^2} + \frac{1}{2m^*}\left(\hbar\, k_y + e\; B_0\, x\right)^2\right] U(x) = \left(E - \frac{\hbar^2\, k_z^2}{2m^*}\right) U(x) \tag{3.108}$$

We will now introduce the *cyclotron frequency*

$$\omega_c^* = \frac{eB_0}{m^*} \qquad (\hbar\omega_c^* = 2\mu_B^* B_0) \tag{3.109}$$

and change the variable from x to ρ given by

$$\rho = x + \frac{1}{\omega_c^*}\frac{\hbar\, k_y}{m^*} \tag{3.110}$$

Then the eigenvalue equation to be solved becomes

$$\left[-\frac{\hbar^2}{2m^*}\frac{d^2}{d\rho^2} + \frac{1}{2}m^*\omega_c^{*2}\rho^2\right] U(\rho) = \left(E - \frac{\hbar^2\, k_z^2}{2m^*}\right) U(\rho) \tag{3.111}$$

We recognize that, this is the eigenvalue equation for a harmonic oscillator with a frequency ω_c^*. The solution is well known. The eigenfunctions are the Hermite polynomials with the eigenvalues:

$$E_n(k_z) = \hbar\,\omega_c^* \left(n + \frac{1}{2}\right) + \frac{\hbar^2\,k_z^2}{2m^*} \qquad n = 1,\ 2,\ \cdots \tag{3.112}$$

These energies are called the *Landau levels*. They describe

(a) the quantized motion of the electron in the plane perpendicular to the field (*orbital quantization*) and
(b) the unperturbed motion in the direction of the field.

In order to take into account the spin, we have to add one more term, namely the Zeeman term:

$$E_n^\sigma(k_z) = E_n(k_z) - z_\sigma\,\mu_B\,B_0 \tag{3.113}$$

z_σ is a sign factor:

$$z_\sigma = \delta_{\sigma\uparrow} - \delta_{\sigma\downarrow} \tag{3.114}$$

$\mu_B = \frac{e\hbar}{2m}$ is the "bare" Bohr magneton. Equations (3.112) and (3.113) can be summarized in the *term scheme* of Fig. 3.4:

Let the electron be enclosed in a rectangular box of the edge lengths $L_x,\ L_y,\ L_z$ so that $V = L_x L_y L_z$ and let us assume that the periodic boundary conditions are fulfilled. Then the allowed wave vectors are given by

$$k_y = n_y\,\frac{2\pi}{L_y} \quad ; \quad k_z = n_z\,\frac{2\pi}{L_z} \quad ; \quad n_y,\ n_z\ \in \mathbb{N} \tag{3.115}$$

An important point for the following discussion is the *degeneracy* of the Landau levels. The energies $E_n(k_z)$ are not dependent on k_y and are therefore degenerate

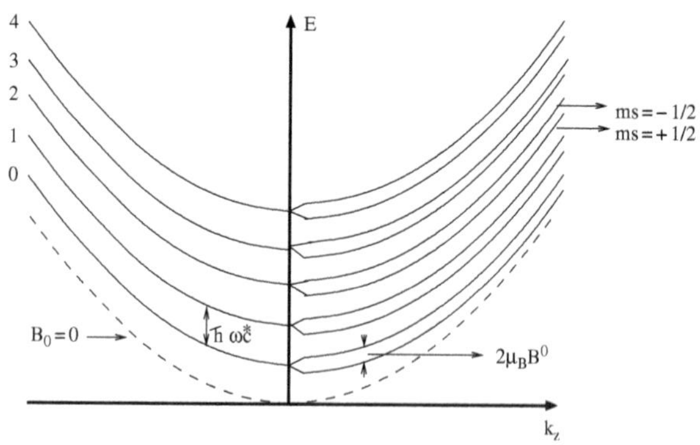

Fig. 3.4 Term scheme for a free electron in a homogeneous magnetic field (Landau levels). The *left* part disregards the electron spin, the *right* part shows additional Zeeman splitting due to the electron spin

with respect to the possible values of k_y. The number of the possible k_y-values is exactly the *degree of degeneracy* g_y of the Landau level. g_y is obtained by dividing the difference between the largest and the smallest values of the allowed k_y-values by the difference between adjacent k_y-values:

$$g_y = \frac{k_y^{max} - k_y^{min}}{2\pi/L_y} \tag{3.116}$$

k_y^{max} and k_y^{min} are determined as follows: The particle finds itself in a box whose side along the x-direction has a length L_x. Therefore, we have $-\frac{L_x}{2} \leq x \leq +\frac{L_x}{2}$ and as a result, in view of (3.110),

$$\frac{L_x}{2} + \rho \geq \frac{\hbar k_y}{eB_0} \geq \rho - \frac{L_x}{2}$$

what gives k_y^{max} and k_y^{min} as

$$k_y^{max} = \frac{eB_0}{\hbar}\left(\frac{L_x}{2} + \rho\right) \quad ; \quad k_y^{min} = \frac{eB_0}{\hbar}\left(-\frac{L_x}{2} + \rho\right)$$

Therefore, the degree of degeneracy reads

$$g_y(B_0) = \frac{eL_xL_y}{2\pi\hbar}B_0 \tag{3.117}$$

Each of the Landau levels is $g_y(B_0)$-fold degenerate, where the degree of degeneracy is, interestingly, proportional to the magnetic field B_0. For a better interpretation, let us, for the moment, consider a *two-dimensional electron gas*. At a very large field, all the electrons occupy the lowest Landau level ($n = 0$). A further increase of the field makes, because of ω_c^*, the total energy W increase linearly with the field while the magnetization remains constant. On the other hand, when the field is decreased, starting from a critical field $B_0^{(0)}$ which is decided by the condition

$$N_e = 2g_y(B_0^{(0)}) \tag{3.118}$$

electrons have to shift to the $n = 1$-Landau level. The factor two stands for the spin degeneracy. Consequently, the energy of the system will first increase a little bit. For $B_0 < \frac{1}{2}B_0^{(0)}$, the $n = 2$-Landau level will be populated and so on (Fig. 3.5). The critical field for which the nth Landau level is filled is given by

$$B_0^{(n)} = \frac{1}{n+1}B_0^{(0)}$$

If the external field is between two critical fields, $B_0^{(n-1)} \geq B_0 \geq B_0^{(n)}$ then the energy is given by

Fig. 3.5 Schematic plot of the energy of the two-dimensional electron gas as a function of the magnetic field. $B_0^{(0)}$: Critical field above which only the $n = 0$ Landau level is occupied

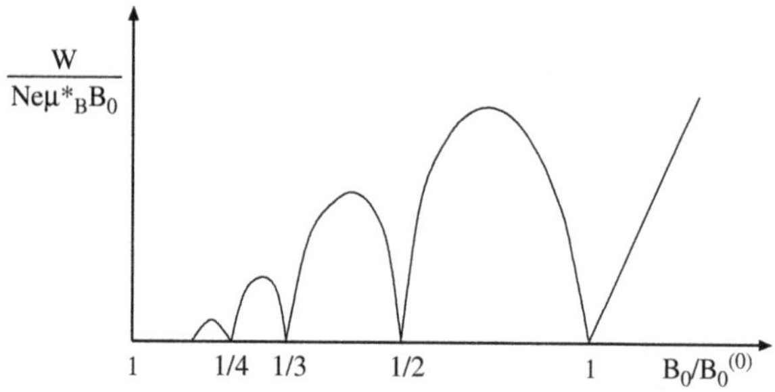

$$E(B_0) = N_e \mu_B^* B_0 (2n + 1 - n(n+1) B_0 / B_0^{(0)})$$

(see Problem 3.6). This yields the field dependence, schematically plotted in Fig. 3.5.

Since $\mathbf{m} = -\frac{\partial W}{\partial \mathbf{B}_0}$, these oscillations of the energy must manifest themselves as oscillations in the magnetization. This will be investigated in more detail in Sect. 3.4.3.

How does the quantization (3.112) appear in **k**-space? Since the kinetic energy of a non-interacting electron gas does not change in a magnetic field, the energy remains purely kinetic even if a magnetic field is switched on and therefore, the following correspondence must be valid:

$$\frac{\hbar^2}{2m^*}\left(k_x^2 + k_y^2\right) \Leftrightarrow \hbar\omega_c^* \left(n + \frac{1}{2}\right) \tag{3.119}$$

The originally regularly spaced k-values, in the field, condense onto the surfaces of cylinders whose axes coincide with the field direction (Figs. 3.6 and 3.7). The radius of the cylinder increases proportional to $\sqrt{B_0}$, since the cross-sectional area of the cylinder is given according to (3.119) by

$$S(E) = \pi\left(k^2 - k_z^2\right) = \pi\left(\frac{2m^* E_n}{\hbar^2} - k_z^2\right) = 2\pi\left(n + \frac{1}{2}\right)\frac{eB_0}{\hbar} \tag{3.120}$$

The number of states will naturally not change by the application of the magnetic field.

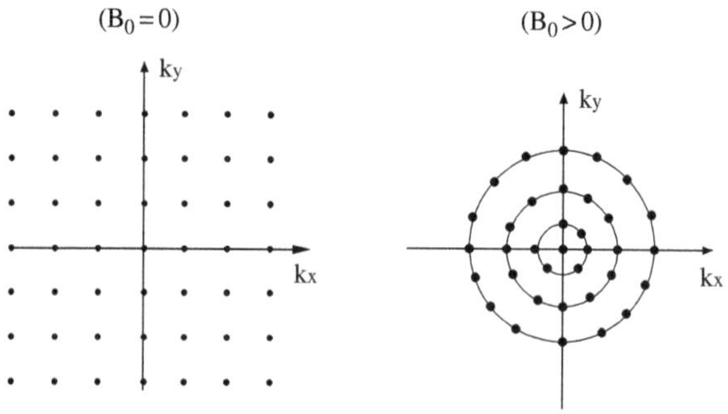

Fig. 3.6 **k**-states of a two-dimensional electron gas with and without magnetic field. Spin splitting is neglected

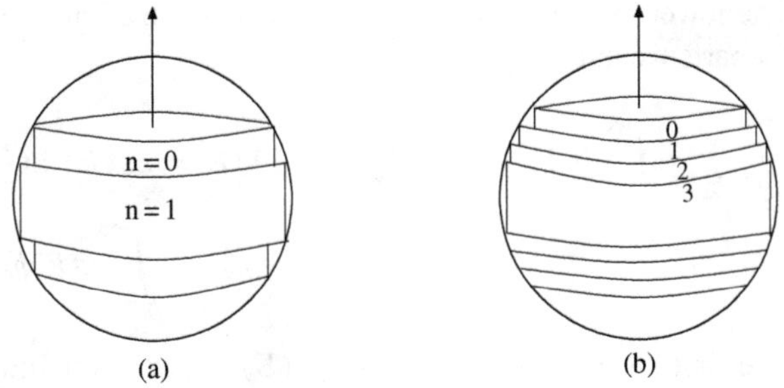

Fig. 3.7 Landau cylinders within the Fermi sphere for, respectively, strong (**a**) and weak (**b**) magnetic fields. Spin splitting is not shown

3.4.2 Grand Canonical Potential of the Conduction Electrons

We now know the allowed energy eigenvalues and the corresponding degeneracies so that we can calculate the grand canonical potential of the conduction electrons:

$$\Omega(T, B_0, \mu) = -k_B T \ln \Xi(T, B_0, \mu) \tag{3.121}$$

where μ is the chemical potential and Ξ is the grand canonical partition function. In the case of an ideal Fermi gas (e.g. non-interacting electrons), we have already seen (3.57) that

$$\Xi(T, B_0, \mu) = \prod_i (1 + exp(-\beta(\varepsilon_i - \mu))) \tag{3.122}$$

The index i runs over *all* states, that means, an energy ε_i is counted as many times as its degree of degeneracy.

So, the starting point is the following expression for the grand canonical potential of the non-interacting electrons:

$$\Omega(T, B_0, \mu) = -k_B T \sum_i \ln\left(1 + e^{-\beta(\varepsilon_i - \mu)}\right) \tag{3.123}$$

Here, ε_i are the Landau levels $E_n^\sigma(k_z)$ (3.113). Because of the terms $\hbar^2\, k_z^2/2m^*$, at least in the thermodynamic limit, these levels lie arbitrarily close to each other. Therefore, we replace the summation by an integration. We denote by

$\phi_\sigma(E) =$ the number of σ-states with $E_n^\sigma \le E$
$\rho_\sigma(E)dE =$ the number of σ-states in the energy interval $[E,\ E + dE]$
$\rho_\sigma(E)$ is the spin-polarized density of states with

$$\rho_\sigma(E) = \frac{d}{dE}\phi_\sigma(E)$$

Therewith, we can write instead of (3.123):

$$\Omega(T, B_0, \mu) = -k_B T \sum_\sigma \int_{\dots}^{\infty} \ln\left(1 + e^{-\beta(E-\mu)}\right) \rho_\sigma(E)\, dE \tag{3.124}$$

The lower limit is given by $\phi_\sigma(E) = 0$ and need not be specified here. Integrating by parts we have

$$\int_{\dots}^{\infty} \ln\left(1 + e^{-\beta(E-\mu)}\right) \rho_\sigma(E) dE = \phi_\sigma(E) \ln\left(1 + e^{-\beta(E-\mu)}\right)\Big|_{\dots}^{\infty} - \int_{\dots}^{\infty} dE\, \phi_\sigma(E) \frac{-\beta e^{-\beta(E-\mu)}}{1 + e^{-\beta(E-\mu)}}$$

The first term vanishes because $\phi_\sigma(E)$ at the lower limit and the logarithm at the upper limit are equal to zero. In the integrand of the second term, we recognize the Fermi function:

$$\Omega(T, B_0, \mu) = -\sum_\sigma \int_{\dots}^{\infty} \phi_\sigma(E) f_-(E) dE \tag{3.125}$$

The main problem now is to find out $\phi_\sigma(E)$. For a given n how many eigenvalues are there such that $E_n^\sigma(k_z) \le E$? Using (3.112), we can write this inequality as follows

$$k_z^2 \le \frac{2m^*}{\hbar^2} \left\{ E - \hbar\omega_c^* \left(n + \frac{1}{2}\right) - z_\sigma \mu_B B_0 \right\}$$

That means, there exists a maximum and a minimum k_z. The difference between the two, divided by the spacing $2\pi/L_z$ gives the number of the "appropriate" eigenvalues:

$$\frac{k_z^{max} - k_z^{min}}{2\pi/L_z} = \frac{\sqrt{2m^*}}{\hbar} \frac{L_z}{\pi} \sqrt{\{\cdots\}}$$

We still have to take into account the degree of degeneracy $g_y(B_0)$ and sum over the "appropriate" n:

$$\phi_\sigma(E) = \frac{\sqrt{2m^*}\, V}{2\pi^2\, \hbar^2} e\, B_0 \sum_{n=0}^{n_{max}} \sqrt{E - \hbar\, \omega_c^*\, (n + 1/2) - z_\sigma\, \mu_B\, B_0} \tag{3.126}$$

n_{max} is determined from the condition that the quantity under the square root must remain positive. The "usual" density of states $\rho_\sigma(E)$ (see Sect. 3.3.1) is then obtained as

$$\rho_\sigma(E) = \frac{d\phi_\sigma(E)}{dE}. \tag{3.127}$$

We will, however, use $\phi_\sigma(E)$ further. For a shorthand notation, we introduce the following reduced quantities:

$$\varepsilon = \frac{E}{\hbar\, \omega_c^*} \quad ; \quad \tilde{\mu} = \frac{\mu}{\hbar\, \omega_c^*} \quad ; \quad \tilde{\beta} = \frac{\hbar\, \omega_c^*}{k_B T} = \beta\, \hbar\, \omega_c^* \tag{3.128}$$

$$\tilde{f}_-(\varepsilon) = \{1 + exp(\tilde{\beta}(\varepsilon - \tilde{\mu}))\}^{-1} \tag{3.129}$$

Defining

$$\alpha = \frac{8}{3} V \frac{(\mu_B\ m\ B_0)^{5/2}}{m^*\ \pi^2\ \hbar^3} \tag{3.130}$$

and using (3.126) in (3.125), we get for the grand canonical potential

$$\Omega(T, B_0, \mu) = -\frac{3}{2}\alpha \sum_\sigma \int_{\dots}^{\infty} d\varepsilon \tilde{f}_-(\varepsilon) \sum_{n=0}^{n_{max}} \sqrt{\varepsilon - n - \frac{1}{2} - z_\sigma \frac{m^*}{2m}} \tag{3.131}$$

The integral can be exactly evaluated. It can be integrated by parts, where the integrated part vanishes because at the upper limit $\varepsilon = \infty$, the Fermi function is zero and at the lower limit, the sum is zero ($\phi_\sigma(E) = 0$). With

$$\eta = \varepsilon - z_\sigma \frac{m^*}{2m} \tag{3.132}$$

it then follows that

$$\Omega(T, B_0, \mu) = \alpha \sum_\sigma \int_{-\infty}^{+\infty} d\eta \tilde{f}'_- \left(\eta + z_\sigma \frac{m^*}{2m}\right) \sum_{n=0}^{n_{max}} \left(\eta - n - \frac{1}{2}\right)^{3/2} \tag{3.133}$$

$\tilde{f}'_-$ has the character of a δ-function at $\varepsilon = \tilde{\mu}$ ($\approx$ Fermi edge) which certainly lies above the lower limit of the integration. Therefore, we can use $-\infty$ as the lower limit of integration.

For the sum in the integrand we can write

$$\Sigma(\eta) = \sum_{n=0}^{n_{max}} \left(\eta - n - \frac{1}{2}\right)^{3/2} = \int_0^{\infty} dx(\eta - x)^{3/2} \sum_{n=0}^{n_{max}} \delta(x - (n + \frac{1}{2})) \tag{3.134}$$

n_{max} has to be chosen so that, $\eta \geq n + \frac{1}{2}$ for $n \leq n_{max}$. Therefore, we have

$$\Sigma(\eta) = \int_0^{\eta} dx(\eta - x)^{3/2} \sum_{n=-\infty}^{+\infty} \delta(x - (n + \frac{1}{2})) \tag{3.135}$$

Through the special choice of the limits of integration, we can let the summation to run from $-\infty$ to $+\infty$. The sum on the right-hand side can be written as a Fourier series (Problem 3.8):

$$\sum_{n=-\infty}^{+\infty} \delta(x - (n + \frac{1}{2})) = \sum_{p=-\infty}^{+\infty} e^{2\pi i p(x - \frac{1}{2})} = \sum_{p=-\infty}^{+\infty} (-1)^p e^{2\pi i p x}$$

Then we get

$$\Sigma(\eta) = \sum_{p=-\infty}^{+\infty} (-1)^p \int_0^{\eta} dx(\eta - x)^{3/2} e^{2\pi i p x} = \sum_{p=-\infty}^{+\infty} (-1)^p I_p(\eta) \tag{3.136}$$

For $p = 0$, it is easy to evaluate $I_p(\eta)$:

$$I_0(\eta) = \frac{2}{5}\,\eta^{5/2} \tag{3.137}$$

For $p \neq 0$, however, the calculation is more involved. It is convenient to substitute $u = \sqrt{\eta - x}$, and after integrating by parts twice, we get

$$I_{p\neq 0}(\eta) = -\frac{\eta^{3/2}}{2\pi i p} + \frac{3\eta^{1/2}}{8\pi^2 p^2} - \frac{3e^{2\pi i p \eta}}{8\pi^2 p^2} \int_0^{\sqrt{\eta}} due^{-2\pi i p u^2} \tag{3.138}$$

When we substitute this in $\Sigma(\eta)$, the first term vanishes in the sum over p. For the second summand, we use

$$\sum_{p=-\infty}^{+\infty} \frac{(-1)^p}{p^2} = -\frac{\pi^2}{6} \tag{3.139}$$

Then we get the intermediate result

$$\Omega(T, B_0, \mu) = \alpha \sum_{\sigma} \int_{-\infty}^{+\infty} d\eta \tilde{f}'_{-}\left(\eta + z_\sigma \frac{m^*}{2m}\right) \Sigma(\eta) \tag{3.140}$$

with

$$\Sigma(n) = \frac{2}{5}\eta^{5/2} - \frac{1}{16}\eta^{1/2} - \frac{3}{4\pi^2} \sum_{p=1}^{\infty} \frac{(-1)^p}{p^2} Re\left[e^{2\pi i p \eta} \int_0^{\sqrt{\eta}} due^{-2\pi i p u^2}\right] \tag{3.141}$$

The first two summands are relatively harmless to evaluate. Using (3.132) for η, we have to evaluate the integral

$$\Omega_0 = \alpha \sum_{\sigma} \int_{-\infty}^{+\infty} d\varepsilon \tilde{f}'_{-}(\varepsilon) \left\{\frac{2}{5}\left(\varepsilon - z_\sigma \frac{m^*}{2m}\right)^{5/2} - \frac{1}{16}\left(\varepsilon - z_\sigma \frac{m^*}{2m}\right)^{1/2}\right\} \tag{3.142}$$

We use the series expansion

$$\left(\varepsilon - z_\sigma \frac{m^*}{2m}\right)^{5/2} = \varepsilon^{5/2}\left\{1 - \frac{5}{2} z_\sigma \frac{m^*}{2m}\frac{1}{\varepsilon} + \frac{5\cdot 3}{2\cdot 4}\frac{m^{*2}}{4m^2}\frac{1}{\varepsilon^2} + \cdots\right\}$$

and

$$\left(\varepsilon - z_\sigma \frac{m^*}{2m}\right)^{1/2} = \varepsilon^{1/2}\left\{1 - \frac{1}{2} z_\sigma \frac{m^*}{2m}\frac{1}{\varepsilon} + \cdots\right\}$$

Because of the sign factor z_σ, the linear terms in z_σ will disappear after the summation over σ. The decisive factor is the fact that in the integrand of Ω_0, $\tilde{f}'_-(\varepsilon)$ has the character of a δ-function

$$\tilde{f}'_-(\varepsilon) \approx -\delta(\varepsilon - \varepsilon_0) \tag{3.143}$$

Further, in the case of normal metallic electron densities (μ: a few eV) and normal effective masses ($\hbar\omega_c^* = 2\mu_B^* B_0$: a few 10^{-3} eV)

$$\varepsilon \approx \tilde{\mu} = \frac{\mu}{\hbar\omega_c^*} \gg 1$$

Therefore, we can terminate the series expansion after a few terms:

$$\Omega_0 = -\alpha\tilde{\mu}^{\;5/2}\left(\frac{4}{5} + \frac{3}{8}\frac{m^{*2}}{m^2\varepsilon^2} + \cdots\right) + \frac{1}{8}\varepsilon^{1/2}(1 - \cdots)$$

The first three terms of the grand canonical potential are then given by

$$\Omega_0 \approx -\alpha\left\{\frac{4}{5}\tilde{\mu}^{\;5/2} - \frac{1}{8}\tilde{\mu}^{\;1/2}\left(1 - 3\left(\frac{m^*}{m}\right)^2\right)\right\} \tag{3.144}$$

This part of the grand canonical potential leads to *Landau diamagnetism* and *Pauli paramagnetism*, whereas the remaining part, which is oscillating and has to be still evaluated, is responsible for the *de Haas-van Alphen effect*.

The integral, that has to be still evaluated, has the form of an error integral

$$\begin{aligned} A_p(\eta) &= \int_0^{\sqrt{\eta}} du e^{-2\pi i p u^2} = \frac{1}{2\sqrt{2ip}}\frac{2}{\sqrt{\pi}}\int_0^{\sqrt{2\pi i p\eta}} dx\, e^{-x^2} \\ &= \frac{1}{2\sqrt{2ip}}\, erf\left(\sqrt{2\pi i p\eta}\right) \end{aligned} \tag{3.145}$$

which can be expanded into a fast converging series. This is because it appears in the integrand for the grand canonical potential (3.140), wherein, again, the δ-function character of $\tilde{f}'$ sees to it that we can assume $\eta \approx \tilde{\mu} \gg 1$. We can show that

$$erf(z) = 1 - erfc(z) = 1 - \frac{2}{\sqrt{\pi}} \int_z^{\infty} dt e^{-t^2} \tag{3.146}$$

the asymptotic representation (see [4]: formulae 7.1.1, 7.1.2, 7.1.14)

$$erf(z) = 1 - \frac{e^{-z^2}}{z\sqrt{\pi}} \left(1 - \frac{1}{2z^2} + \frac{3}{4z^4} + \cdots \right) \tag{3.147}$$

Using this, we get for the integral A,

$$A_p(\eta) = \frac{1}{2^{3/2}\sqrt{ip}} \left(1 - \frac{e^{-2\pi i p \eta}}{\sqrt{2\pi^2 i p \eta}} \left(1 - \frac{1}{4\pi i p \eta} + \frac{3}{4(2\pi i p \eta)^2} + \cdots \right)\right)$$

With

$$\frac{1}{\sqrt{i}} = e^{-i\frac{\pi}{4}}$$

we can finally estimate

$$A_p(\eta) = \frac{e^{-i\pi/4}}{2\sqrt{2p}} + \mathcal{O}(\eta^{-1/2}/p) \quad ; \quad (p > 0) \tag{3.148}$$

For the oscillatory part of $\Sigma(\eta)$ (3.141), we then have the intermediate result

$$\Sigma_{osc}(\eta) \approx -\frac{3}{4\pi^2\, 2^{3/2}} \sum_{p=1}^{\infty} \frac{(-1)^p}{p^{5/2}} \cos(2\pi p \eta - \pi/4) \tag{3.149}$$

That means, for the grand canonical potential, according to (3.140), if we remove the substitution (3.132) and perform the spin summation:

$$\Omega_{osc}(T, B_0, \mu) = -\frac{3\alpha}{4\pi^2\sqrt{2}} \sum_{p=1}^{\infty} \frac{(-1)^p}{p^{5/2}} \cos\left(p\pi \frac{m^*}{m}\right) * \\ * \int_{-\infty}^{+\infty} d\varepsilon\, \tilde{f}_-{}'(\varepsilon)\, \cos(2\pi\varepsilon - \pi/4) \tag{3.150}$$

The still remaining integral can be further worked out. However, we cannot now replace $\tilde{f}'(\varepsilon)$ by simply a δ-function, since the integrand strongly oscillates in the interesting region. But, the integral can be exactly evaluated using the residue theorem. That we will do as an auxiliary calculation.

According to (3.93), it holds for the derivative of the Fermi function:

$$\tilde{f}'_{-}(\varepsilon) = -\frac{\tilde{\beta}}{4\cosh^2\left(\frac{1}{2}\tilde{\beta}(\varepsilon - \tilde{\mu})\right)} \tag{3.151}$$

We substitute $\rho = \tilde{\beta}(\varepsilon - \tilde{\mu})$ and then we have to evaluate the real part of

$$\int_{-\infty}^{+\infty} d\varepsilon\, \tilde{f}'_{-}(\varepsilon) e^{2\pi i p \varepsilon - i\pi/4} = -e^{2\pi i p \tilde{\mu} - i\pi/4} \int_{-\infty}^{+\infty} d\rho \frac{exp\left(\frac{1}{\tilde{\beta}}\, 2\pi i p \rho\right)}{4cosh^2(\frac{1}{2}\,\rho)} \tag{3.152}$$

Let the integral be denoted by $J(p)$. We solve it with the help of the residue theorem, where, since $p > 0$, we close the path of integration in the upper half-plane. Since $\cosh(\frac{1}{2}\rho) = \cos(i\frac{1}{2}\rho)$, the integrand has poles at

$$\rho_n = i\,(2n+1)\,\pi \tag{3.153}$$

Only the poles with $n \geq 0$ lie inside the region of integration. Further, we have

$$\begin{aligned}\cosh\left(\frac{1}{2}\rho\right) &= i\ \sinh\left(\frac{1}{2}(\rho - \rho_n)\right)(-1)^n \\ &= \frac{i}{2}(\rho - \rho_n)\left(1 + \frac{1}{24}(\rho - \rho_n)^2 + \cdots\right)(-1)^n\end{aligned}$$

It follows as Taylor expansion

$$\frac{1}{\cosh^2\left(\frac{1}{2}\rho\right)} = \frac{-4}{(\rho - \rho_n)^2}\left(1 - \frac{1}{12}(\rho - \rho_n)^2 + \cdots\right) \tag{3.154}$$

Thus the integrand of $J(p)$ has, at ρ_n, a pole of second order with the residue

$$\begin{aligned}Res_{\rho_n} &= \lim_{\rho \to \rho_n} \frac{d}{d\rho}\left[(\rho - \rho_n)^2 \frac{e^{\frac{2\pi i p \rho}{\tilde{\beta}}}}{4\cosh^2\left(\frac{1}{2}\rho\right)}\right] \\ &= -\lim_{\rho \to \rho_n} \frac{d}{d\rho}\left[exp\left(\frac{2\pi i p}{\tilde{\beta}}\rho\right)\left(1 - \frac{1}{12}(\rho - \rho_n)^2 + \cdots\right)\right] \\ &= -\tilde{\beta}^{-1} 2\pi i p\, e^{-\frac{2\pi^2}{\tilde{\beta}}(2n+1)p}\end{aligned} \tag{3.155}$$

From the residue theorem, we eventually get

$$
\begin{aligned}
J(p) &= +4\pi^2\, \tilde{\beta}^{-1}\, p \sum_{n=0}^{\infty} exp\left(-\frac{2\pi^2}{\tilde{\beta}}(2n+1)\, p\right) \\
&= 4\pi^2\, p\, \tilde{\beta}^{-1} exp\left(-\frac{2\pi^2}{\tilde{\beta}}\, p\right) \sum_{n=0}^{\infty} exp\left(-\frac{4\pi^2}{\tilde{\beta}}\, n\, p\right) \\
&= 4\pi^2\, p\, \tilde{\beta}^{-1} exp\left(-\frac{2\pi^2}{\tilde{\beta}}\, p\right) \left(1 - exp\left(-\frac{4\pi^2}{\tilde{\beta}}\, p\right)\right)^{-1}
\end{aligned}
$$

That means

$$
J(p) = \frac{2\pi^2\, p}{\tilde{\beta} \sinh\left(2\pi^2 \frac{p}{\tilde{\beta}}\right)} \tag{3.156}
$$

This we use in (3.152)

$$
\int_{-\infty}^{+\infty} d\varepsilon\, \tilde{f}'(\varepsilon)\, \cos(2\pi p \varepsilon - \frac{\pi}{4}) = -\frac{2\pi^2 p}{\tilde{\beta}}\, \frac{\cos(\frac{\pi}{4} - 2\pi p \tilde{\mu})}{\sinh(2\pi^2 \frac{p}{\tilde{\beta}})} \tag{3.157}
$$

By substituting (3.157) in (3.150), the oscillatory part of the grand canonical potential is completely determined.

$$
\Omega_{osc}(T, B_0, \mu) = \frac{3\alpha}{2^{3/2}\, \tilde{\beta}} \sum_{p=1}^{\infty} \frac{(-1)^p}{p^{3/2}} \cos\left(p\pi \frac{m^*}{m}\right) \frac{\cos\left(\frac{\pi}{4} - 2\pi p \tilde{\mu}\right)}{\sinh\left(2\pi^2 \frac{p}{\tilde{\beta}}\right)} \tag{3.158}
$$

Going back to the original notation, finally, we get the grand canonical potential of the conduction electrons as

$$
\Omega(T, B_0, \mu) = \Omega_0(T, B_0, \mu) + \Omega_{osc}(T, B_0, \mu) \tag{3.159}
$$

with

$$
\begin{aligned}
\Omega_0(T, B_0, \mu) = -\, N \left(\frac{\mu}{\varepsilon_F}\right)^{3/2} \Bigg\{ \frac{2}{5}\mu + \\
+ \frac{1}{4\mu} \left(\mu_B^* B_0\right)^2 \left(3\left(\frac{m^*}{m}\right)^2 - 1\right) \Bigg\}
\end{aligned} \tag{3.160}
$$

$$\Omega_{osc}(T, B_0, \mu) = \frac{3}{2}\, k_B T \frac{N}{\varepsilon_F^{3/2}} \left(\mu_B^* B_0\right)^{3/2} *$$

$$* \sum_{p=1}^{\infty} \frac{(-1)^p}{p^{3/2}} \cos\left(p\pi \frac{m^*}{m}\right) \frac{\cos\left(\frac{\pi}{4} - p\,\frac{\pi\mu}{\mu_B^* B_0}\right)}{\sinh\left(p\,\frac{\pi^2 k_B T}{\mu_B^* B_0}\right)} \tag{3.161}$$

In deriving this, we have used

$$\mu_B^* = \frac{e\hbar}{2m^*} = \frac{m}{m^*}\mu_B$$

and

$$\varepsilon_F = \frac{\hbar^2}{2m^*}\left(3\pi^2 \frac{N_e}{V}\right)^{2/3}$$

We have thus determined the grand canonical potential of the conduction electrons completely as a function of T and B_0. From this, now, the magnetization and the susceptibility can be derived.

3.4.3 Susceptibility of the Conduction Electrons

We obtain the magnetization M from the relation (3.99)

$$M(T, B_0) = -\frac{1}{V}\left(\frac{\partial \Omega}{\partial B_0}\right)_{T,\mu} \tag{3.162}$$

Thereby, we have to express the chemical potential μ by T, B_0 and the (fixed) particle number. For this purpose, we exploit (3.99)

$$N = -\left(\frac{\partial \Omega}{\partial \mu}\right)_{T,B_0} \tag{3.163}$$

The partial differentiation is easily performed on (3.159):

$$N = \frac{N}{\varepsilon_F^{3/2}} \left\{ \mu^{3/2} + \frac{(\mu_B^* B_0)^2}{8\mu^{1/2}} \left(3 \left(\frac{m^*}{m} \right)^2 - 1 \right) \right\} -$$
$$- \frac{3}{2} k_B T \frac{N}{\varepsilon_F^{3/2}} \left(\mu_B^* B_0 \right)^{3/2} \sum_{p=1}^{\infty} \frac{(-1)^p}{p^{1/2}} \cos \left(p\pi \frac{m^*}{m} \right) *$$
$$* \, p \, \frac{\pi}{\mu_B^* B_0} \frac{\sin\left(\frac{\pi}{4} - p \frac{\pi\mu}{\mu_B^* B_0} \right)}{\sinh\left(p \frac{\pi^2 k_B T}{\mu_B^* B_0} \right)} \tag{3.164}$$

Rearranging leads to

$$\left(\frac{\mu}{\varepsilon_F} \right)^{3/2} = 1 - \gamma_1(B_0) \left(3 \left(\frac{m^*}{m} \right)^2 - 1 \right)$$
$$+ \frac{3}{2} \gamma_2(T, B_0) \sum_{p=1}^{\infty} \frac{(-1)^p \cos\left(p\pi \frac{m^*}{m} \right)}{p^{1/2}} \frac{\sin\left(\frac{\pi}{4} - p \frac{\pi\mu}{\mu_B^* B_0} \right)}{\sinh\left(p \frac{\pi^2 k_B T}{\mu_B^* B_0} \right)} \tag{3.165}$$

The coefficients γ_1 and γ_2

$$\gamma_1(B_0) = \frac{(\mu_B^* \, B_0)^2}{8 \, \mu^{1/2} \, \varepsilon_F^{3/2}} \tag{3.166}$$

$$\gamma_2(T, B_0) = \pi \left(\frac{k_B T}{\varepsilon_F} \right) \left(\frac{\mu_B^* \, B_0}{\varepsilon_F} \right)^{1/2} \tag{3.167}$$

are both very small compared to 1, as can be seen from the following: For a *degenerate electron gas* we can assume

$$1 \, eV \le \varepsilon_F \le 10 \, eV \tag{3.168}$$

Using furthermore,

$$\mu_B = 0.579 \times 10^{-4} \, \frac{eV}{T} \quad ; \quad k_B = 0.862 \times 10^{-4} \, \frac{eV}{K} \tag{3.169}$$

we can estimate

$$\gamma_1 \ll 1 \quad ; \quad \gamma_2 \ll 1$$

That means that, $\mu \approx \varepsilon_F$. Therefore, in (3.165), we can, without much error, replace μ by ε_F on the right-hand side. Then what remains is an expression of the form

$$\frac{\mu}{\varepsilon_F} = (1-x)^{2/3} \text{ with } x \ll 1$$

which we can further approximate by

$$\frac{\mu}{\varepsilon_F} \approx 1 - \frac{2}{3}x$$

That leads to the following result for the field dependence of the chemical potential:

$$\mu = \varepsilon_F \left[1 - \frac{2}{3}\gamma_1(B_0)\left(3\left(\frac{m^*}{m}\right)^2 - 1\right) + \right.$$
$$\left. + \gamma_2(T, B_0) \sum_{p=1}^{\infty} \frac{(-1)^p}{p^{1/2}} \cos\left(p\pi\frac{m^*}{m}\right) \frac{\sin\left(\frac{\pi}{4} - p\frac{\pi\varepsilon_F}{\mu_B^* B_0}\right)}{\sinh\left(p\frac{\pi^2 k_B T}{\mu_B^* B_0}\right)} \right] \tag{3.170}$$

If we want to compare this expression with the well-known result of Statistical Mechanics for the temperature dependence of μ (3.82)

$$\mu(T) \approx \varepsilon_F \left(1 - \frac{\pi^2}{12}\left(\frac{k_B T}{\varepsilon_F}\right)^2\right)$$

we have to remember that, in the present calculation, the correction term cannot appear, since we have, in several places, replaced the derivative f'_- by the δ-function. The finite width of f'_- around μ, however, produces the temperature effect. For the oscillating third summand in (3.170), however, this simplification has not been used.

The result (3.170) for μ along with the estimates for γ_1 and γ_2 makes it clear that, for our purpose, we can take, with sufficient accuracy,

$$\mu \approx \varepsilon_F \tag{3.171}$$

Using this, we will now calculate the magnetization resulting from the non-oscillating part of the grand canonical potential (3.144). The oscillating part will be considered especially in Section 3.5.

$$M_0(T, B_0) = -\frac{1}{V}\left(\frac{\partial \Omega_0}{\partial B_0}\right)_{T,\mu=\varepsilon_F} = \frac{N}{2V}\frac{\mu_B^{*2}}{\varepsilon_F}\left(3\left(\frac{m^*}{m}\right)^2 - 1\right) B_0 \tag{3.172}$$

From this we directly get the susceptibility of the conduction electrons that we are looking for:

$$\chi_0 = \mu_0 \left(\frac{\partial M_0}{\partial B_0}\right)_T = \frac{3}{2}\frac{N}{V}\mu_0\frac{\mu_B^2}{\varepsilon_F}\left(1 - \frac{1}{3}\left(\frac{m^*}{m}\right)^2\right) \tag{3.173}$$

Here we have used $\mu_B/\mu_B^* = m^*/m$. We see that χ_0 contains diamagnetic and paramagnetic components:

$$\chi_0 = \chi_{Pauli} + \chi_{Landau} \tag{3.174}$$

$$\chi_{Pauli} = \frac{3}{2}\frac{N}{V}\mu_0\frac{\mu_B^2}{\varepsilon_F} > 0 \tag{3.175}$$

$$\chi_{Landau} = -\frac{1}{2}\frac{N}{V}\mu_0\frac{\mu_B^{*2}}{\varepsilon_F} < 0 \tag{3.176}$$

The term χ_{Pauli} describes the so-called *Pauli spin paramagnetism.* It originates because of the permanent magnetic spin moment $-2\mu_B/\hbar\mathbf{s}$ of the conduction electrons. We will discuss this contribution in more physical detail in Chap. 4.

χ_{Landau} is a diamagnetic component. It is known as the *Landau–Peierls diamagnetism* which results from the quantization of the orbital moments in the presence of an external magnetic field. We want to add a few remarks on the result (3.174):

1. For really free electrons, naturally $m^* = m$, so that, with (3.175) and (3.176) we have

$$\chi_{Landau}^{(0)} = -\frac{1}{3}\chi_{Pauli}^{(0)} \tag{3.177}$$

   For many metals, however, m^* is distinctly different from m, so that the diamagnetic component can outweigh the paramagnetic component (e.g. Bi). In general, the diamagnetic and the paramagnetic components are of the same order of magnitude.
2. The concept of effective isotropic mass m^* is quite problematic (relatively good for the alkali metals). The effective mass is, in general, an (anisotropic) tensor:

$$\left(\frac{1}{m^*}\right)_{ij} = \frac{1}{\hbar^2}\frac{\partial^2\varepsilon_n(\mathbf{k})}{\partial k_i \partial k_j} \quad ;i,\ j \in \{x,\ y,\ z\} \tag{3.178}$$

   Therefore, in the expressions in this section, m^* is always a quantity, which is some kind of an *average* performed over the "Fermi layer".
3. The Coulomb interaction of the band electrons, their scattering by phonons and imperfections and also other temperature effects have been neglected. There are only few improvements that exist in literature in this direction. Often, one takes recourse to using phenomenological damping terms.
4. Measurements always give χ_{total}, which is a sum of χ_{Landau}, χ_{Pauli}, χ_{Larmor} and χ_{osc}, so that separate measurements of χ_{Landau} are not so simple. Calculations show that χ_{Landau} is very small (Table 3.3).

Table 3.3 Diamagnetic contributions to the susceptibility of the alkali metals. In the first column are given the effective masses calculated by Ham [5]. The second column gives the Landau susceptibility calculated for $m^* = m$ and the third column the value obtained by making the correction using the first column

m^*/m	$-\chi^{(0)}_{Landau}.10^6$	$-\chi_{Landau}.10^6$
Li 1.66	3.41	2.05
Na 1.00	4.99	4.99
K 1.09	7.62	6.99
Rb 1.21	8.71	7.20
Cs 1.76	10.21	5.80

5. The fact that χ_{Pauli} as well as χ_{Landau} are, to a first approximation, *independent* of temperature as well as the magnetic field is confirmed by experiment.
6. The derivation restricted itself to the s-electrons. In other situations, the orbital motion can also lead to paramagnetic effects [6].

3.5 The de Haas–Van Alphen Effect

The oscillations of the magnetic susceptibility χ as a function of the external magnetic field B_0 (more precisely $1/B_0$) is the *de Haas–van Alphen effect*. Similar oscillations are also observed in many other physical quantities (in particular, the transport quantities), e.g. in thermal and electrical conductivity, magnetostriction, Hall effect. Here we discuss only the oscillations of the magnetic susceptibility.

3.5.1 Oscillations in the Magnetic Susceptibility

These oscillations naturally arise from the not yet evaluated oscillatory part of the grand canonical potential:

$$\chi_{osc} = -\frac{\mu_0}{V}\left(\frac{\partial^2 \Omega_{osc}}{\partial B_0^2}\right)_{T,\mu=\varepsilon_F} \tag{3.179}$$

The calculation is simple but laborious. With the notation

$$\alpha(T,\ B_0) = \frac{3}{2}\frac{k_B T}{\varepsilon_F}\left(\frac{\mu_B^* B_0}{\varepsilon_F}\right)^{1/2} \tag{3.180}$$

$$\beta(T,\ B_0) = \frac{\pi^2 k_B T}{\mu_B^* B_0} \tag{3.181}$$

$$\gamma(B_0) = \frac{\varepsilon_F \pi}{\mu_B^* B_0} \tag{3.182}$$

$$D_r(p) = \frac{N}{V}(-1)^p\ p^r \cos\left(p\pi\frac{m^*}{m}\right)\mu_B^* \tag{3.183}$$

we first write the magnetization as

$$M_{osc}(T, B_0) = -\frac{1}{V}\left(\frac{\partial \Omega_{osc}}{\partial B_0}\right)_{T,\mu=\varepsilon_F} = M_1 + M_2 + M_3 \tag{3.184}$$

The three summands are defined as follows:

$$M_1 = -\frac{3}{2}\,\alpha \sum_{p=1}^{\infty} D_{-3/2}(p)\,\frac{\cos(\pi/4 - p\gamma)}{\sinh(p\beta)} \tag{3.185}$$

$$M_2 = \gamma\,\alpha \sum_{p=1}^{\infty} D_{-1/2}(p)\,\frac{\sin(\pi/4 - p\gamma)}{\sinh(p\beta)} \tag{3.186}$$

$$M_3 = -\beta\,\alpha \sum_{p=1}^{\infty} D_{-1/2}(p)\,\frac{\cos(\pi/4 - p\gamma)}{\sinh(p\beta)}\coth(p\beta) \tag{3.187}$$

Differentiating once again gives the susceptibility:

$$\chi_{osc} = \mu_0\left(\frac{\partial M_{osc}}{\partial B_0}\right)_T = \chi_1 + \chi_2 + \chi_3 \tag{3.188}$$

$$\begin{aligned}\chi_1 = \mu_0 \Bigg[& -\frac{3\alpha}{4B_0}\sum_{p=1}^{\infty} D_{-3/2}\,\frac{\cos(\pi/4 - p\gamma)}{\sinh(p\beta)} \\ & + \frac{3\alpha\gamma}{2B_0}\sum_{p=1}^{\infty} D_{-1/2}\frac{\sin(\pi/4 - p\gamma)}{\sinh(p\beta)} \\ & - \frac{3\alpha\beta}{2B_0}\sum_{p=1}^{\infty} D_{-1/2}\frac{\cos(\pi/4 - p\gamma)}{\sinh(p\beta)}\coth(p\beta)\Bigg]\end{aligned} \tag{3.189}$$

$$\begin{aligned}\chi_2 = \mu_0 \Bigg[& -\frac{\alpha\gamma}{2B_0}\sum_{p=1}^{\infty} D_{-1/2}\frac{\sin(\pi/4 - p\gamma)}{\sinh(p\beta)} \\ & + \frac{\alpha\gamma^2}{B_0}\sum_{p=1}^{\infty} D_{1/2}\frac{\cos(\pi/4 - p\gamma)}{\sinh(p\beta)} \\ & + \frac{\alpha\beta\gamma}{B_0}\sum_{p=1}^{\infty} D_{1/2}\frac{\sin(\pi/4 - p\gamma)}{\sinh(p\beta)}\coth(p\beta)\Bigg]\end{aligned} \tag{3.190}$$

$$\chi_3 = \mu_0 \left[\frac{\alpha\beta}{2B_0} \sum_{p=1}^{\infty} D_{-1/2} \frac{\cos(\pi/4 - p\gamma)}{\sinh(p\beta)} \coth(p\beta) \right.$$
$$+ \frac{\alpha\beta\gamma}{B_0} \sum_{p=1}^{\infty} D_{1/2} \frac{\sin(\pi/4 - p\gamma)}{\sinh(p\beta)} \coth(p\beta)$$
$$\left. - \frac{\alpha\beta^2}{B_0} \sum_{p=1}^{\infty} D_{1/2} \frac{\cos(\pi/4 - p\gamma)}{\sinh^3(p\beta)} (1 + \cosh^2(p\beta)) \right] \tag{3.191}$$

Inspite of the extremely simple model (*Sommerfeld model*), the response function χ turns out to be a rather complicated expression. If we assume that all the summations that appear in the expressions are of the same order of magnitude, then the pre-factors provide the real meaning of the terms. In general, however, these pre-factors have different orders of magnitude.

With the values given in (3.169) for μ_B and k_B and the value for ε_F, for normal fields and not too high temperatures, we get

$$\gamma \gg \beta \gg \alpha \tag{3.192}$$

Therefore, it is sufficient to consider only the term proportional to γ^2:

$$\chi_{osc} \approx \mu_0 \frac{\alpha\gamma^2}{B_0} \sum_{p=1}^{\infty} D_{1/2}(p) \frac{\cos(\pi/4 - p\gamma)}{\sinh(p\beta)} \tag{3.193}$$

We want to discuss this expression a little further.

1. $D_{1/2}(p)$ contains $\cos(p\pi \frac{m^*}{m})$. This term can be traced back to the electron spin. The other parts are associated with the orbital motion. That means the orbital and spin susceptibilities are not simply additive. Therefore, they cannot be handled separately.
2. Because of $\sinh(p\ \beta)$, the series converges very fast. As a result, often it is sufficient to consider only the $p = 1$ summand:

$$\chi_{osc} \approx -\mu_0 \frac{3}{2} N k_B T \frac{1}{V} \frac{1}{B_0^2} \pi^2 \left(\frac{\varepsilon_F}{\mu_B^* B_0} \right)^{1/2} \cos\left(\pi \frac{m^*}{m} \right) \times$$
$$\times \frac{\cos(\pi/4 - \pi\varepsilon_F/\mu_B^* B_0)}{\sinh(\pi^2 k_B T/\mu_B^* B_0)} \tag{3.194}$$

   Caution should be exercised in case $m^* \ll m$, which means, $\mu_B^* \gg \mu_B$. Then, it might be necessary to take into account more summands.
3. The characteristic of the de Haas–Van Alphen effect is the χ-oscillations with the period $\Delta\gamma$

$$p\,\Delta\gamma \stackrel{!}{=} 2\pi = p\,\frac{\pi\,\varepsilon_F}{\mu_B^*}\,\Delta\left(\frac{1}{B_0}\right) \tag{3.195}$$

Thus, χ oscillates as a function of $1/B_0$ with the period

$$\Delta(1/B_0) = p^{-1}\,\frac{2\mu_B^*}{\varepsilon_F} \tag{3.196}$$

$\Delta(1/B_0)$ does *not* depend on temperature. $p = 1$ is the ground oscillation.

4. The oscillations are naturally easier to observe if the period Δ is larger. This is the case if the Fermi energy ε_F is as small as possible, that is, if $\varepsilon_F = \frac{\hbar^2}{2m^*}(3\pi^2 n_e)^{3/2}$ is as small as possible. That means the electron density $n_e = N_e/V$ should be small.
5. The amplitude of the oscillations decreases for small fields as

$$exp\left(-p\,\pi^2\frac{k_B T}{\mu_B^* B_0}\right)$$

The above-mentioned facts have been clearly confirmed by experiment.

3.5.2 Electron Orbits in Magnetic Field

We want to understand the physical origin of the oscillations and discuss the possibility of applications. In order to do this it is necessary to first understand a few related things. First, we will consider the motion in **k**-space.

3.5.2.1 Motion in k-Space

The equation of motion of an electron state of wave number **k** under the influence of a magnetic field fulfils, at least approximately, the classical equation:

$$\hbar\dot{\mathbf{k}} = -e\,\mathbf{v}\times\mathbf{B}_0 \tag{3.197}$$

Here $\mathbf{v} = \hbar^{-1}\nabla_k\varepsilon(\mathbf{k})$ is the group velocity of the wave packet built from the Bloch functions. Equation (3.197) says that, $d\mathbf{k}$ is perpendicular to $\mathbf{B}_0$, $\mathbf{v}$ and $\nabla_{\mathbf{k}}\varepsilon(\mathbf{k})$ (Fig. 3.8). Further, $\nabla_{\mathbf{k}}\varepsilon(\mathbf{k})$ is perpendicular to the surface $\varepsilon =$ const. Therefore, the end of the **k**-vector moves along the curve given by the intersection of the plane $\varepsilon =$ const. with a plane perpendicular to $\mathbf{B}_0$. If the constant energy surface is a simply connected plane, then the electron moves in the **k**-space along a closed curve. The integrals of motion are thereby

(a) energy
(b) $k_z =$ component of **k**, parallel to the field

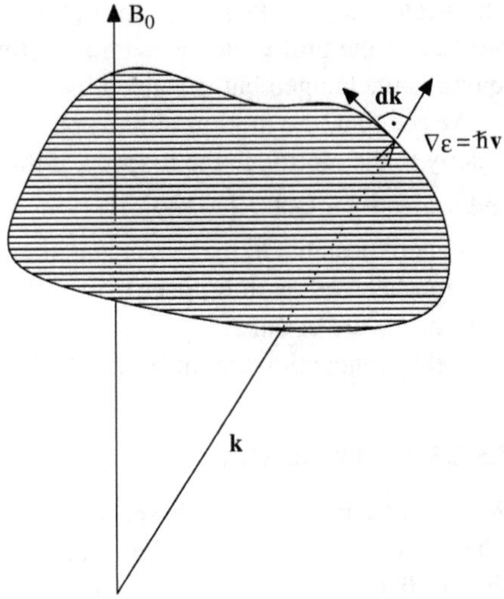

Fig. 3.8 Illustration for the connection of real space and **k**-space for the movement of an electron in a homogeneous magnetic field

In the reduced-zone scheme, the **k**-vector folds back as soon as it reaches the zone boundary (*open orbit*). When the zone boundary is not reached, then it is called *closed orbit.* In the periodic zone scheme, the "umklappen" of **k** is identical with crossing the boundary of the Brillouin zone. In such a situation, both closed and also open orbits may appear. The direction of a closed orbit depends on whether the energy increases outward or inward.

The connection between $\mathbf{v} = \dot{\mathbf{r}}$ and $\dot{\mathbf{k}}$ in (3.197) expresses a close relationship between the **k**-space motion and the motion in the real space.

3.5.2.2 Motion in the Real Space

Let $\mathbf{e}_z$ be the unit vector in the direction of the homogeneous field. Then,

$$\mathbf{r}_\perp = \mathbf{r} - (\mathbf{r} \cdot \mathbf{e}_z)\, \mathbf{e}_z \tag{3.198}$$

is the projection of the position vector on the plane perpendicular to $\mathbf{B}_0$. We will use this fact in the following reformulation:

$$\begin{aligned} \mathbf{e}_z \times \dot{\mathbf{k}} &= -\frac{e}{\hbar}\mathbf{e}_z \times \dot{\mathbf{r}} \times \mathbf{B}_0 = -\frac{e}{\hbar}\left(\dot{\mathbf{r}}\,(\mathbf{e}_z \cdot \mathbf{B}_0) - \mathbf{B}_0\,(\mathbf{e}_z \cdot \dot{\mathbf{r}})\right) \\ &= -\frac{eB_0}{\hbar}\left(\dot{\mathbf{r}} - \mathbf{e}_z\,(\mathbf{e}_z \cdot \dot{\mathbf{r}})\right) = -\frac{eB_0}{\hbar}\,\dot{\mathbf{r}}_\perp \end{aligned}$$

On integrating we get

$$\mathbf{r}_\perp(t) - \mathbf{r}_\perp(0) = -\frac{\hbar}{eB_0}\,\mathbf{e}_z \times (\mathbf{k}(t) - \mathbf{k}(0)) \tag{3.199}$$

The vector $(\mathbf{k}(t) - \mathbf{k}(0))$ lies in a plane which is perpendicular to $\mathbf{B}_0$. The vector product of the unit vector $\mathbf{e}_z$ with a vector perpendicular to it gives a vector whose length is unchanged but is rotated by $\pi/2$ about the direction of the field.

We see that, the motion in real space, projected onto the x–y plane, corresponds exactly to the motion in the $\mathbf{k}$-space, if this is rotated by $\pi/2$ about the field direction and scaled by a factor $\left(-\frac{\hbar}{eB_0}\right)$.

It is not possible to make any statements regarding the z-direction. The orbits (in position- as well as in $\mathbf{k}$-space) are called *cyclotron orbits*. Only for free electrons these orbits are circular.

In this connection we introduce the important concept of cyclotron mass.

3.5.2.3 Cyclotron Mass

We consider two orbits in $\mathbf{k}$-space of constant energy, namely for E and $E + \Delta E$ where $\Delta\mathbf{k}$ is the perpendicular vector between these two surfaces (Fig. 3.9). We should distinguish this from $(\Delta\mathbf{k})_\perp$, which is the perpendicular distance in the xy plane. Then the energy difference ΔE can be expressed in terms of $(\Delta\mathbf{k})_\perp$:

$$\Delta E = (\nabla_{\mathbf{k}}\varepsilon(\mathbf{k})) \cdot (\Delta\mathbf{k})_\perp = (\nabla\varepsilon)_\perp \cdot \Delta k_\perp \tag{3.200}$$

With this, we now calculate the time the electron needs to travel on the orbit from $\mathbf{k}_1$ to $\mathbf{k}_2$:

$$\begin{aligned} t_2 - t_1 &= \int_{t_1}^{t_2} dt = \int_1^2 \frac{dk}{|\dot{\mathbf{k}}|} = \frac{\hbar^2}{e}\int_1^2 \frac{dk}{|\nabla_{\mathbf{k}}\varepsilon(\mathbf{k}) \times \mathbf{B}_0|} \\ &= \frac{\hbar^2}{e}\int_1^2 \frac{dk}{(\nabla\varepsilon)_\perp B_0} = \frac{\hbar^2}{eB_0}\frac{1}{\Delta E}\int_{k_1}^{k_2} dk \cdot \Delta k_\perp \end{aligned}$$

The integral represents the area $\Delta S_{1,2}$ from $\mathbf{k}_1$ to $\mathbf{k}_2$ between the two orbits in the plane of the $\mathbf{k}$-space which is perpendicular to $\mathbf{B}_0$. In the limit $\Delta E \to 0$, it follows that

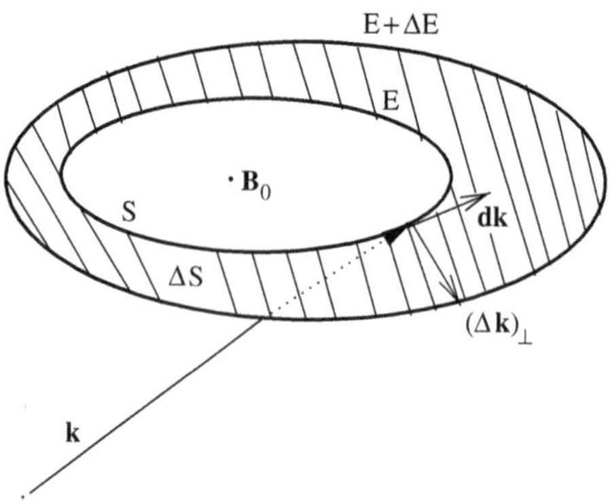

Fig. 3.9 Two electron orbits in $\mathbf{k}$-space perpendicular to the magnetic field $\mathbf{B}_0$ with constant energies E and $E + \Delta E$, respectively

$$t_2 - t_1 = \frac{\hbar^2}{eB_0}\frac{\partial S_{1,2}}{\partial E} \tag{3.201}$$

In the cases of closed orbits, the period of revolution τ is given by

$$\tau = \frac{\hbar^2}{eB_0}\frac{\partial S(E, k_z)}{\partial E} = \frac{2\pi}{\omega_c} \tag{3.202}$$

ω_c is called the *cyclotron frequency*. As a simple example, let us consider the case of free electrons. In this case, we have

$$\varepsilon(\mathbf{k}) = \frac{\hbar^2 k^2}{2m}$$

That is, the surfaces of constant energy are the surfaces of spheres. The orbits of constant energy in the x–y plane are then circles of the radius $\sqrt{k^2 - k_z^2}$. Then, we immediately have

$$S_0(E, k_z) = \pi(k^2 - k_z^2) = \pi\left(\frac{2m}{\hbar^2}E - k_z^2\right)$$

and

$$\frac{\partial S_0}{\partial E} = \frac{2m\pi}{\hbar^2} \qquad \tau_0 = \frac{2m\pi}{eB_0} = \frac{2\pi}{\omega_c^{(0)}} \tag{3.203}$$

In analogy with this special case, using (3.203), one defines

$$m_c(E, k_z) = \frac{\hbar^2}{2\pi}\frac{\partial S(E, k_z)}{\partial E} \tag{3.204}$$

as the *cyclotron mass*. Therefore, in general, we have

$$\tau = \frac{2\pi}{eB_0}m_c \Leftrightarrow \omega_c = \frac{eB_0}{m_c} \tag{3.205}$$

Normally, the cyclotron mass m_c is different from the effective mass m^*. Only for an isotropic dispersion $\left(\varepsilon(\mathbf{k}) = \frac{\hbar^2 k^2}{2m^*}\right)$, it holds $m^* = m_c$.

3.5.2.4 Landau Cylinder

We found that the electron motion in a magnetic field in the plane perpendicular to the field is quantized. For the cross-sectional area of the nth Landau cylinder, we have according to (3.113)

$$S_n^\sigma(k_z) = \pi\left(k_x^2 + k_y^2\right) = \pi\left(\frac{2m^*}{\hbar^2}E_n^\sigma(k_z) - k_z^2\right)$$
$$= \pi\left((2n+1) - z_\sigma\frac{m^*}{m}\right)\frac{eB_0}{\hbar} \tag{3.206}$$

We can now calculate the annular area between two adjacent cylinders

$$\Delta S^{\sigma\sigma'} = S_{n+1}^\sigma(k_z) - S_n^{\sigma'}(k_z) \tag{3.207}$$

as

$$\Delta S^{\sigma\sigma'} = \frac{2\pi e}{\hbar}B_0\left(1 - (z_\sigma - z_{\sigma'})\frac{m^*}{m}\right) \tag{3.208}$$

The annular area is thus independent of the Landau quantum number n. It increases linearly with the field.

If there were no field,

$$2\frac{\Delta S^{\sigma\sigma}}{4\pi^2/(L_xL_y)} = 2\frac{L_xL_y}{2\pi\hbar}eB_0 = 2g_y(B_0)$$

states would have been in $\Delta S^{\sigma\sigma}$. The factor 2 takes care of the spin degeneracy. Therefore, on a "Landau circle", in the x–y plane, there are exactly the same number of states as there are without field in the corresponding annular area. The annular area increases with the field by the same amount as the degree of degeneracy g_y of the Landau levels. According to (3.5.2.1) and (3.5.2.2), the k-values on a Landau cylinder do not remain constant but rotate with the frequency

$$\omega_c = 2\pi\frac{eB_0}{\hbar^2}\left(\frac{\partial S}{\partial E}\right)^{-1} = 2\pi\frac{eB_0}{\hbar^2}\frac{\hbar^2}{2\pi m^*} = \frac{eB_0}{m^*} = \omega_c^* \tag{3.209}$$

Now it is clear, what happens inside the Fermi sea, when the magnetic field $\mathbf{B}_0$ is switched on. The regularly ordered points in the $\mathbf{k}$-space in the absence of the field (one point per grid volume $\Delta\mathbf{k} = \frac{2\pi^3}{V}$) arrange themselves on cylinders and rotate on these with the cyclotron frequency. The number of the states, however, does not change.

3.5.3 Physical Origin of the Oscillations

For the cross-sectional area of the nth Landau cylinder, we can according to (3.206) write

$$S_n^\sigma = 2\pi(n + \phi_\sigma)\frac{eB_0}{\hbar} \tag{3.210}$$

ϕ_σ is an undetermined constant. In the next section, we will show that, S_n^σ in this form is valid not only for the Sommerfeld model but is valid in general.

The oscillations in the grand canonical potential and the susceptibility are related to the successive emptying of the Landau cylinders. This will now be discussed in a little more detail. If by changing of the field, every electron would stay put in its Landau level, then, since $\hbar\omega_c^* \sim B_0$, the energy of the system would increase proportional to B_0. By increasing field (at $T = 0$), what actually happens is the following: Since the degree of degeneracy on the cylinder increases,

1. the electrons may shift from outer to inner cylinder
2. the electrons on a cylinder may drop from larger to smaller $|k_z|$ (Fig. 3.10).

Because of (1) and (2), the total energy at $T = 0$ will remain at its lowest possible value. Therefore, on the whole, mainly, a rearrangement of the occupation of the states inside the Fermi body takes place. When a Landau cylinder goes out of the Fermi body, it will be emptied. Let A_0 be the maximum cross-sectional area of the Fermi body perpendicular to the field $\mathbf{B}_0$. The nth cylinder empties itself at the field strength $B_0^{(n)}$, exactly when the frontal area of the Landau cylinder equals A_0:

$$S_n^\sigma\left(B_0^{(n)}\right) = A_0 \tag{3.211}$$

In view of (3.210), it means

$$\frac{1}{B_0^{(n)}} = 2\pi(n+\phi_\sigma)\frac{e}{\hbar}\frac{1}{A_0} \tag{3.212}$$

Then, the $(n-1)$th cylinder empties itself when

$$S_{n-1}^\sigma\left(B_0^{(n-1)}\right) = A_0 \tag{3.213}$$

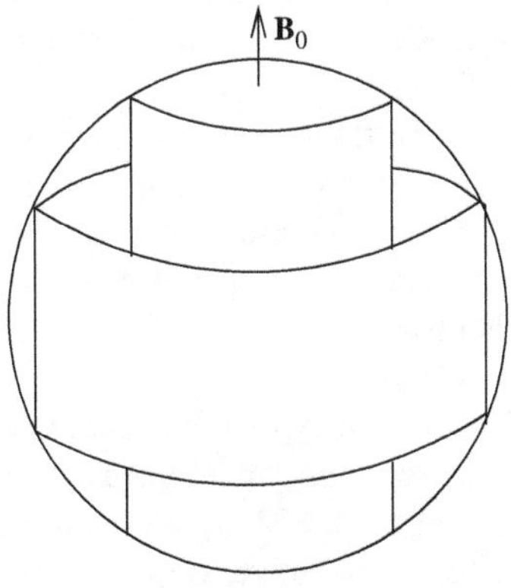

Fig. 3.10 Landau cylinders within the Fermi sphere

That is, when

$$\frac{1}{B_0^{(n-1)}} = 2\pi(n - 1 + \phi_\sigma)\frac{e}{\hbar}\frac{1}{A_0} \tag{3.214}$$

This gives a period, which is independent of the Landau quantum number:

$$\Delta\left(\frac{1}{B_0}\right) = \frac{1}{B_0^{(n)}} - \frac{1}{B_0^{(n-1)}} = \frac{e}{\hbar}\frac{2\pi}{A_0} \tag{3.215}$$

It is determined directly from the cross-sectional area A_0 of the Fermi sea. In the case of the Sommerfeld model, which has been used so far, it is simple to obtain:

$$A_0 = \pi k_F^2 = \frac{2m^*\pi}{\hbar^2}\varepsilon_F = \pi\frac{\varepsilon_F}{\mu_B^*}\frac{e}{\hbar} \tag{3.216}$$

From this we get the period

$$\Delta(1/B_0) = \frac{2\mu_B^*}{\varepsilon_F} \tag{3.217}$$

which agrees exactly with the oscillations in χ (3.196). This result is also valid even if the Fermi sea is not a simple sphere. Measurement of the period of oscillations in χ automatically provides the extremal cross-sectional area perpendicular to the field. By varying the field *direction*, one can obtain valuable information about the shape of the Fermi surface. This is actually, the real practical significance of the de Haas–Van Alphen effect.

Why the extremal (also the minimal) cross-sectional area of the Fermi body determines the period can be qualitatively understood in the following way:

The derivation of the grand canonical potential in Sect. 3.4.2 clearly shows that, the oscillations are essentially related to level density $\rho_\sigma(E)$, that is, the number of states in the energy interval between E and $E + \Delta E$. In the various summations and integrations, the derivative of the Fermi function $f'_- \sim \delta(E - \varepsilon_F)$ plays a special role, whose δ-function character sees to it that the level density is important mainly at $E = \varepsilon_F$.

Since the density of states in the z-direction is constant, the contribution of a Landau cylinder is proportional to the shaded area in the Fig. 3.11.

This area is quite obviously maximal when the cross section of the Landau cylinder equals the maximal cross-sectional area of the Fermi body (Fig. 3.12). This is also true in the case of minimal cross-sectional areas as one can easily see from Figs. 3.13 and 3.14.

The density of states is always then maximal, when S_n^σ is equal to the *extremal* cross-sectional area A_0. That explains why the oscillations and their period are related to A_0. Many electronic properties depend on the density of states $\rho_\sigma(E)$ at the Fermi edge ε_F. All these properties show the above-discussed oscillatory behaviour as a function of $1/B_0$.

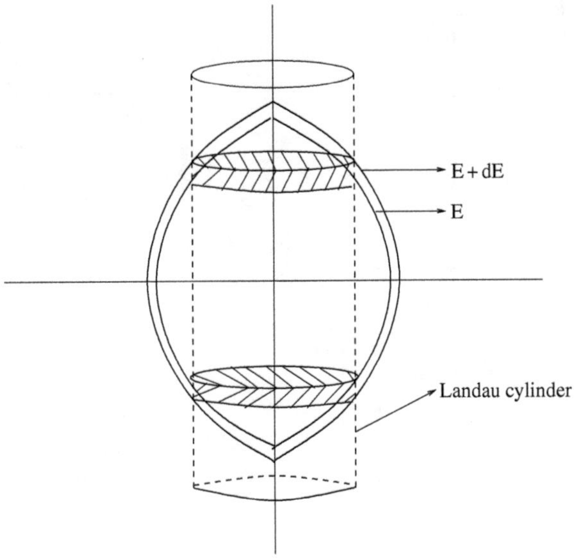

Fig. 3.11 Graphic representation of the states with energies in between E and $E + \Delta E$ which simultaneously are to be found on a Landau cylinder

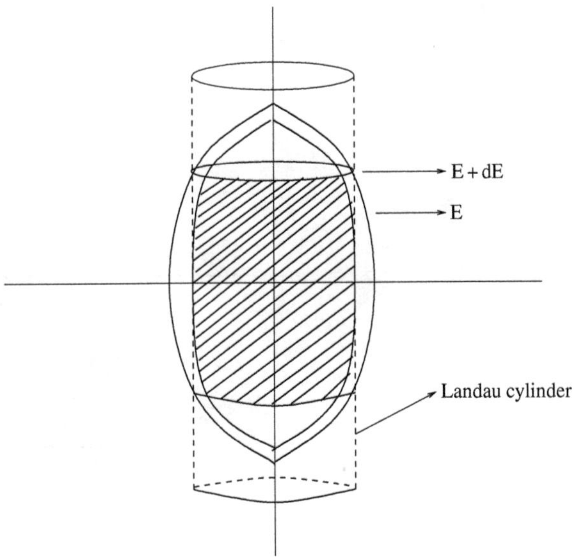

Fig. 3.12 The same as in Fig. 3.11 but now for a maximum of states on the Landau cylinder with energies in between E and $E + \Delta E$

3.5.4 Onsager Argument

The Landau theory, discussed so far, is based on the Sommerfeld model for quasi-free electrons, characterized by spherically shaped Fermi surfaces. The generalization to arbitrarily shaped simply connected Fermi surfaces is done by Onsager [7].

Fig. 3.13 The same as in Fig. 3.11, but now for a different shape of the E = const. area

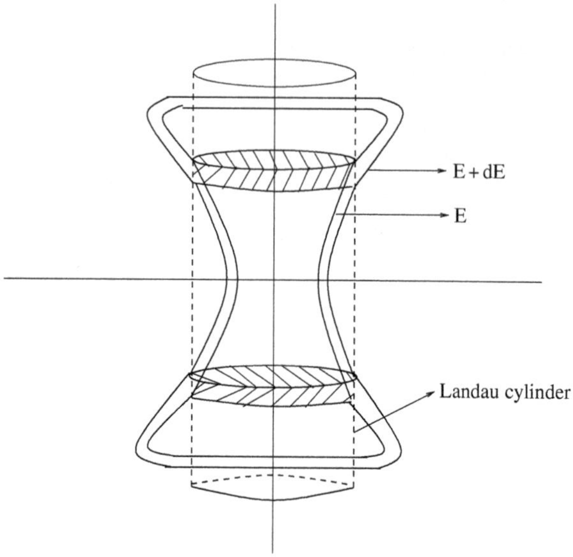

Fig. 3.14 The same as in Fig. 3.12, but now for a different shape of the E = const. area

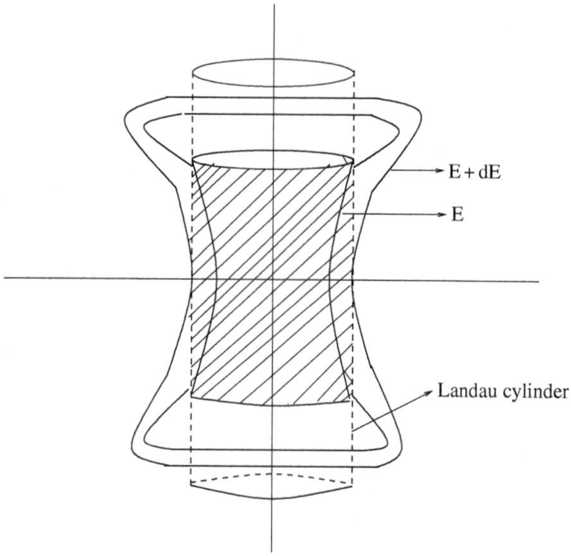

In Sect. 3.5.2, we have seen that the electron motion in real space follows a closed orbit in the plane perpendicular to the direction of the field. This motion can be quantized according to the *Bohr–Sommerfeld condition*:

$$\oint \mathbf{p} \cdot d\mathbf{r} = \oint (\hbar\, \mathbf{k} - e\, \mathbf{A}(\mathbf{r})) \cdot d\mathbf{r} = (n + \phi)\, h \tag{3.218}$$

This rule is valid for very large quantum numbers n, i.e. in the neighbourhood of the classical limit. This limit is ensured here because, with

$$n\,\hbar\,\omega_c^* = 2\mu_B^*\,B_0\,n \overset{!}{\leq} \varepsilon_F$$
$$(\mu_B \approx 0.579 \times 10^{-4}\,eV/T\ ;\ \varepsilon_F = 1\cdots 10\,eV) \qquad (3.219)$$

in normal fields, about 10^3–10^4 Landau levels are involved. ϕ is an undetermined constant between 0 and 1, and $n \in \mathbb{N}$. We will calculate the two summands of the integrand in (3.218) separately. From the Stoke's theorem, it follows that

$$e\oint_{\partial F} \mathbf{A}\cdot d\mathbf{r} = e\int_F \nabla\times\mathbf{A}\cdot d\mathbf{f} = e\mathbf{B}_0\cdot\int d\mathbf{f} = e\,B_0\,F_\perp \qquad (3.220)$$

$F_\perp$ is the projection of the plane of motion of the electron onto the x–y plane.

With the appropriate choice of the reference point in position- and momentum-space, according to (3.199), we have

$$\mathbf{r}_\perp(t) = -\frac{\hbar}{e\,B_0}\,\mathbf{e}_z \times \mathbf{k}(t) \qquad (3.221)$$

Here, $\mathbf{k}(t)$ must lie in the x–y plane so that we can solve for $\mathbf{k}(t)$.

$$\mathbf{k}(t) = -\frac{e\,B_0}{\hbar}\,(\mathbf{r}_\perp \times \mathbf{e}_z) = -\frac{e\,B_0}{\hbar}\,(\mathbf{r}\times\mathbf{e}_z) \qquad (3.222)$$

With this, we evaluate

$$\oint \hbar\,\mathbf{k}\cdot d\mathbf{r} = -e\,B_0\oint(\mathbf{r}\times\mathbf{e}_z)\cdot d\mathbf{r} = -e\,B_0\,\mathbf{e}_z\cdot\oint(d\mathbf{r}\times\mathbf{r})$$
$$= +2e\,B_0\,F_\perp \qquad (3.223)$$

The Bohr–Sommerfeld condition now reads

$$(n+\phi)h = 2e\,B_0\,F_\perp - e\,B_0\,F_\perp$$

from which finally we get

$$F_\perp = \frac{(n+\phi)\,h}{e\,B_0} \qquad (3.224)$$

This is the area in real space, which has still to be scaled by $\left(-\frac{eB_0}{\hbar}\right)^2$ in order to get the area in the $\mathbf{k}$-space:

$$S_n = \frac{(n+\phi)h}{e\,B_0}\,\frac{e^2 B_0^2}{\hbar^2} = 2\pi(n+\phi)\frac{e\,B_0}{\hbar} \qquad (3.225)$$

This is exactly the expression (3.210), which we have derived for the frontal area of the Landau cylinder within the Landau theory, which we have now derived without assuming free electrons. So, the expression for the χ-period, $\Delta(1/B_0)$ (3.217),

$$\Delta(1/B_0) = \frac{e}{\hbar}\frac{2\pi}{A_0} \tag{3.226}$$

is in general valid.

Up to now we have not taken into account the influence of the phonons ($T \neq 0$), impurities, etc. The scattering processes related to them can make the electrons move in orbits which are not closed. If τ is the average collision time, then the uncertainity in the energy is $\Delta E \sim \hbar/\tau$.

When ΔE is larger than the separation between the Landau levels, then the oscillatory behaviour is washed out. In order that the de Haas–Van Alphen effect is observable, we must demand that $\Delta E \ll \hbar\omega_c^*$. That means pure metals, low temperatures and large fields.

3.6 Problems

Problem 3.1 For the Larmor diamagnetism of insulators determine the magnetic moment induced by a homogeneous field $\mathbf{B}_0 = B_0\mathbf{e}z$ and the corresponding diamagnetic susceptibility χ^{dia}. To do this, use the classical picture that the electrons of the concerned atom move *classically* in stable orbits. The orbital angular momentum $\mathbf{l}$ associated with the orbital motion executes a Larmor precession about the direction of the field with a frequency $\omega_L = \frac{eB_0}{2m}$ ($-e$: electron charge, m: electron mass). Compare the result with the quantum mechanically correct expression (3.21). Is there a contradiction to Bohr–van Leeuwen theorem?

Problem 3.2 Verify that in the Sommerfeld model, the average energy of an electron at $T = 0$ is given by

$$\bar{\varepsilon} = \frac{3}{5}\varepsilon_F \qquad (\varepsilon_F\text{: Fermi energy})$$

gilt.

Problem 3.3 For the Sommerfeld model of simple metals, calculate

1. the grand canonical partition function $\Xi_\mu(T, V)$
2. the average occupation number of the single-particle level $\varepsilon(\mathbf{k})$:

$$\langle \hat{n}_{\mathbf{k}\sigma} \rangle = \langle a^+_{\mathbf{k}\sigma} a_{\mathbf{k}\sigma} \rangle \qquad (\sigma = \uparrow, \downarrow\text{: spin projection})$$

3. the entropy

$$S = \frac{\partial}{\partial T}\left(k_B T \ln \Xi_\mu(T, V)\right)$$

Verify the third law!

Problem 3.4 Within the Sommerfeld model, calculate the density of states of a

1. one dimensional
2. two dimensional
3. d-dimensional

metal.

Problem 3.5 Consider the Sommerfeld model for extreme relativistic, non-interacting Fermions in volume V. One-particle energies are

$$\sqrt{c^2p^2 + m^2c^4} \rightarrow cp = c\hbar k = \varepsilon(k)$$

1. Calculate the density of states $\rho_0(E)$!
2. What is the temperature dependence of the chemical potential μ?
3. How does the heat capacity depend on temperature?

Compare with the results of the non-relativistic case.

Problem 3.6 Consider two-dimensional electron gas in the presence of a perpendicular field $\mathbf{B}_0 = B_0\mathbf{e}_z$. According to Sect. 3.4.1, in the ground state, the N_e electrons occupy the Landau levels

$$E_n = \hbar\omega_c^*\left(n + \frac{1}{2}\right); \qquad n = 0, 1, 2, \ldots$$

$$\hbar\omega_c^* = 2\mu_B^* B_0$$

The spin splitting is neglected here.

1. What is the smallest field $B_0 = B_0^{(0)}$ at which all the electrons are placed in the $n = 0$ level?
2. What is the field $B_0 = B_0^{(n_0)} \leq B_0^{(0)}$ at which the N_e electrons are uniformly distributed in the Landau levels up to the quantum number n_0?
3. If the field B_0 lies between the two critical fields $B_0^{(n_0)}$ and $B_0^{(n_0-1)}$

$$B_0^{(n_0-1)} \geq B_0 \geq B_0^{(n_0)}$$

   Calculate the total energy $E(B_0)$ of the N_e-electron system!
4. What do you get for the special case $E\left(B_0^{(n_0)}\right)$?

Problem 3.7 Consider a system of N spinless electrons which are interacting with each other placed in a homogeneous field $\mathbf{B}_0 = B_0\mathbf{e}_z$.

1. Calculate the canonical partition function Z_1 of a single electron.

2. The temperature be so high that Boltzmann statistics can be applied to the N-electron system to a good approximation. This means, in particular, for the canonical partition function:

$$Z_N = \frac{Z_1^N}{N!}$$

Calculate the average magnetic moment m.

Problem 3.8 Show that

$$f(x) = \sum_{n=-\infty}^{+\infty} \delta\left[x - \left(n + \frac{1}{2}\right)\right]$$

can be written as a Fourier series as follows:

$$f(x) = \sum_{p=-\infty}^{+\infty} (-1)^p \, e^{i2\pi px}$$

References

1. H. A. Bethe, E. E. Salpeter, Handbuch der Physik XXXV, Springer, Berlin, 1957
2. L. D. Landau, Z. Phys. **64**, 629 (1930)
3. R. Peierls, Z. Phys. **81**, 186 (1933)
4. M. Abramowitz, T. A. Stegun, "Handbook of Mathematical Functions", Dover, New York, 1972
5. F.S. Ham, Phys. Rev. B **128**, 252 (1962)
6. R. Kubo, Y. J. Ohata, J. Phys. Soc. Japan **11**, 547 (1956)
7. L. Onsager, Phys. Mag. **43**, 1006 (1952)

Chapter 4
Paramagnetism

The subject of "paramagnetism" deals with the reaction of *permanent magnetic moments* to an external magnetic field. The permanent magnetic moments can be formed by

(a) the moments of partially filled atomic electron shells as, e.g. in
  3d: transition metals
  4f: rare earths
  5f: actinides
  In such a case, the moments are *localized* at definite lattice points (insulators).
(b) The moments can also be due to itinerant conduction electrons in metallic solids, in which case,

$$\mathbf{m}_s = -2\,\frac{\mu_B}{\hbar}\,\mathbf{s}$$

Let us remind the reader to what we have agreed upon after Table 2.1. In the following we will disregard the different sign of magnetic moment and spin.

In this chapter, we assume that there is no interaction worth mentioning among the moments, so that, in the absence of an external field, the total magnetization vanishes. In the presence of an external magnetic field $\mathbf{B}_0 = \mu_0\mathbf{H}$, the moments try to orient themselves parallel to the field in order to minimize the internal energy U of the system. Against this works the temperature, which by creating the maximum possible disorder, tries to maximize the entropy S. The requirement

$$F = U - TS \stackrel{!}{=} Minimum$$

where F is the free energy, finally determines the total magnetization. Therefore, we expect the susceptibility of a paramagnet to be

$$\chi > 0 \quad ; \quad \chi = \chi(T)$$

In paramagnetism also, there are qualitative differences between insulators and metals. That is why, they will be considered separately.

W. Nolting, A. Ramakanth, *Quantum Theory of Magnetism*,
DOI 10.1007/978-3-540-85416-6_4,

4.1 Pauli Spin Paramagnetism

We begin with the paramagnetism of the conduction electrons. The susceptibility of the conduction electrons was discussed in detail in Sects. 3.4 and 3.5 and is given by

$$\chi = \chi_{Landau} + \chi_{Pauli} + \chi_{osc} \tag{4.1}$$

χ_{Landau} is negative and so is a diamagnetic component, which arises due to the orbital motion of the conduction electrons. χ_{Pauli} is positive and so is a paramagnetic component, which is ascribed to the spin of the electrons. χ_{osc} oscillates with the field between positive and negative values and is connected with both the orbital motion and the spin.

We are here interested in χ_{Pauli} for which we have already found that (3.175)

$$\chi_{Pauli} = \frac{3}{2}\frac{N}{V}\mu_0\frac{\mu_B^2}{\varepsilon_F} \tag{4.2}$$

This expression is normally derived in a considerably simpler way than was done in Sect. 3.4, which is, on the other hand, physically more insightful. This simpler method will, therefore, be followed now. As mentioned at the beginning of this chapter, the susceptibility should be temperature dependent. However, we see from the above expression that χ_{Pauli} is *independent* of temperature. This has to be justified. In order to do that, we calculate the maximal corrections which give the temperature dependence. The starting point is the *Sommerfeld model* which was introduced in Sect. 3.3.

4.1.1 "Primitive" Theory of the Pauli Spin Paramagnetism

We split the density of states $\rho(E)$ of the conduction electrons into two parts

$$\rho(E) = \rho_\uparrow(E) + \rho_\downarrow(E) \tag{4.3}$$

$\rho_\uparrow$ is the density of states for the electrons with spin parallel to the field ($m_s = +\frac{1}{2}$), and $\rho_\downarrow$ for the electrons with spin antiparallel ($m_s = -\frac{1}{2}$). If the field is switched off, then both the densities of states are equal

$$\rho_\uparrow(E) = \rho_\downarrow(E) = \frac{1}{2}\rho_0(E) \tag{4.4}$$

That means the system contains equal number of $\uparrow$-electrons as $\downarrow$-electrons. That is why, the total magnetization is zero (Fig. 4.1).

When the field is switched on ($B_0 \neq 0$), the band states are shifted from their original position by the energy (Fig. 4.2)

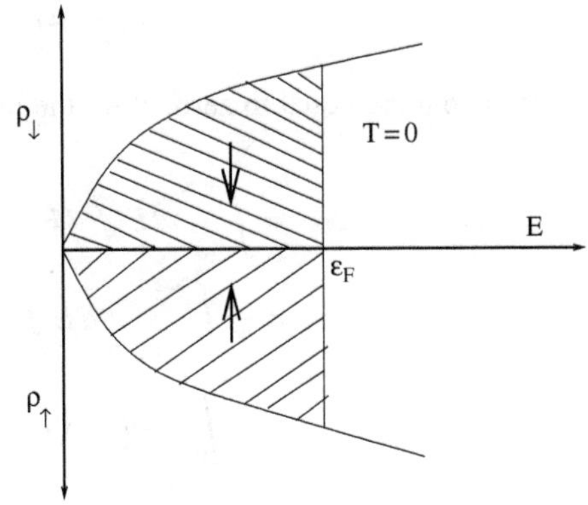

Fig. 4.1 Schematic plot of the spin-polarized densities of states, $\rho_\uparrow$ and $\rho_\downarrow$ of the Sommerfeld model as functions of the energy, at $T = 0$ and a vanishing external magnetic field $\mathbf{B}_0$

$$\Delta E_\sigma = -z_\sigma \, \mu_B \, B_0 \tag{4.5}$$

$$z_\sigma = \delta_{\sigma\uparrow} - \delta_{\sigma\downarrow} \tag{4.6}$$

In the field, the $\uparrow$-electrons have lower energy than the $\downarrow$-electrons (see remark after Table 2.1). The density of states $\rho_\uparrow$ and $\rho_\downarrow$ are shifted rigidly with respect to each other so that we have

$$\rho_\sigma(E) = \frac{1}{2}\,\rho_0(E + z_\sigma \mu_B B_0) \tag{4.7}$$

For the moment, we leave out the orbital quantization (Landau levels, see Sect. 3.4.1) from consideration. We will consider only the shift in energy, for which the electron spin is responsible. Because of this shift, the $\downarrow$-electrons will flow out into the $\uparrow$-states so that a new common Fermi edge is formed. That results in, for $B_0 \neq 0$, $N_\uparrow > N_\downarrow$, where N_σ is the number of electrons with spin projection σ. Then, the total moment is given by

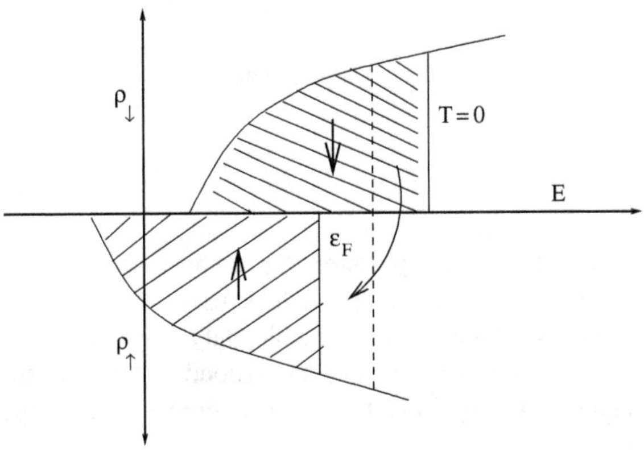

Fig. 4.2 The same as in Fig. 4.1 but now with a finite external field ($\mathbf{B}_0 \neq 0$)

$$m_{tot} = \mu_B(N_\uparrow - N_\downarrow) \tag{4.8}$$

which is no more equal to zero. Now, the job is to find the electron numbers $N_{\uparrow\downarrow}$:

$$\begin{aligned} N_\sigma &= \int_{-\infty}^{+\infty} dE\ f_-(E)\,\rho_\sigma(E) \\ &= \frac{1}{2}\int_{-z_\sigma\mu_B B_0}^{\infty} dE\ f_-(E)\,\rho_0(E + z_\sigma\mu_B B_0) \\ &= \frac{1}{2}\int_0^\infty d\eta\ f_-(\eta - z_\sigma\mu_B B_0)\,\rho_0(\eta) \end{aligned} \tag{4.9}$$

In the region, where the Fermi function (f_-) distinctly deviates from either one or zero, $\mu_B B_0$ is, in general, very small compared to η. Therefore, a Taylor expansion of the Fermi function can be terminated after the linear term:

$$N_\sigma \approx \frac{1}{2}\int_0^\infty d\eta\ \left\{f_-(\eta) - z_\sigma\ \mu_B\ B_0\ \frac{\partial f}{\partial\eta}\right\}\rho_0(\eta) \tag{4.10}$$

With this, we get for the magnetization

$$M = \frac{\mu_B}{V}\,(N_\uparrow - N_\downarrow) = -\frac{\mu_B^2}{V}\ B_0\int_0^\infty d\eta\ \frac{\partial f_-}{\partial\eta}\ \rho_0(\eta) \tag{4.11}$$

The susceptibility then has the form

$$\chi_{Pauli} = \mu_0\left(\frac{\partial M}{\partial B_0}\right)_T = -\frac{1}{V}\ \mu_0\ \mu_B^2\int_0^\infty dE\ f_-'(E)\ \rho_0(E) \tag{4.12}$$

To a good first approximation, according to (3.65), we can write

$$f_-'(E) \approx -\delta(E - \varepsilon_F) \tag{4.13}$$

At $T = 0$, this relation is exact. Then we have for χ_{Pauli},

$$\chi_{Pauli} = \frac{1}{V}\ \mu_0\ \mu_B^2\ \rho_0(\varepsilon_F) = \frac{3}{2}\ \frac{N_e}{V}\ \mu_0\ \frac{\mu_B^2}{\varepsilon_F} \tag{4.14}$$

This is exactly the expression that we have derived in another way as Eq. (3.175) in Sect. 3.4.3. In the last step, we have used $\rho_0(\varepsilon_F) = \frac{3}{2}N_e/\varepsilon_F$ (3.50).

Because of the Pauli's principle, only a very small fraction of the electrons, which are in the thin *Fermi layer* can respond to the field $\mathbf{B}_0$. This explains the order of magnitude ($\sim 10^{-6}$) and the almost temperature independence of the Pauli susceptibility.

4.1.2 Temperature Corrections

The Pauli susceptibility derived above is independent of temperature. This is essentially due to the fact that in (4.13), we have replaced the derivative of the Fermi function by a δ-function. Now, with the help of the Sommerfeld expansion, we will improve this result. The starting point is Eq. (4.12), which we first integrate by parts, and note that the integrated part vanishes. Then what remains is

$$\chi_{Pauli}(T) = \mu_0 \mu_B^2 \frac{1}{V} \int_0^\infty dE f_-(E) \rho_0'(E) \tag{4.15}$$

The integrand satisfies the preconditions for the Sommerfeld expansion (3.74). Therefore, we can use the result of this expansion:

$$\begin{aligned} \chi_{Pauli} &\approx \mu_0\, \mu_B^2\, \frac{1}{V} \left[\int_0^\mu dE\, \rho_0'(E) + \frac{\pi^2}{6}\, (k_B T)^2\, \rho_0''(\mu) \right] \\ &\approx \mu_0\, \mu_B^2\, \frac{1}{V} \left[\rho_0(\mu) + \frac{\pi^2}{6}\, (k_B T)^2\, \rho_0''(\mu) \right] \\ &\approx \mu_0\, \mu_B^2\, \frac{1}{V} \left[\rho_0(\varepsilon_F) + (\mu - \varepsilon_F)\, \rho_0'(\varepsilon_F) + \cdots \right. \\ &\quad \left. + \frac{\pi^2}{6}\, (k_B T)^2\, \rho_0''(\varepsilon_F) \right] \end{aligned} \tag{4.16}$$

The term $(\mu - \varepsilon_F)$ which stems from a Taylor expansion of $\rho_0(\mu)$ around $\mu = \varepsilon_F$, must be carefully determined, because it is of the same order of magnitude as the third summand. In the third term, we can replace $\rho_0''(\mu)$ by $\rho_0''(\varepsilon_F)$ because of the already very small pre-factor. We now exploit the fact that the electron number N_e is, of course, temperature independent.

$$N_e(T = 0) \stackrel{!}{=} N_e(T \neq 0) \tag{4.17}$$

At $T = 0$ we have

$$N_e(T = 0) = \int_{-\infty}^{\varepsilon_F} dE\, \rho_0(E) \tag{4.18}$$

and at $T \neq 0$,

$$\begin{aligned}N_e(T \neq 0) &= \int_{-\infty}^{+\infty} dE\ \rho_0(E) f_-(E) \\ &= \int_{-\infty}^{\mu} dE\ \rho_0(E) + \frac{\pi^2}{6}(k_B T)^2 \rho'(\mu) + \cdots \\ &= \int_{-\infty}^{\varepsilon_F} dE\ \rho_0(E) + (\mu - \varepsilon_F)\, \rho_0(\varepsilon_F) + \cdots \\ &\quad \cdots + \frac{\pi^2}{6}(k_B T)^2 \rho'(\varepsilon_F) + \cdots \end{aligned} \tag{4.19}$$

In view of (4.17) and (4.18) we get from (4.19)

$$\mu - \varepsilon_F \approx -\frac{\pi^2}{6}(k_B T)^2 \frac{\rho_0'(\varepsilon_F)}{\rho_0(\varepsilon_F)} \tag{4.20}$$

Substituting this in (4.16), we get

$$\chi_{Pauli} = \chi_{Pauli}^{(0)} \left[1 + \frac{\pi^2}{6}(k_B T)^2 \left\{ \frac{\rho_0''(\varepsilon_F)}{\rho_0(\varepsilon_F)} - \left(\frac{\rho_0'(\varepsilon_F)}{\rho_0(\varepsilon_F)} \right)^2 \right\} \right] \tag{4.21}$$

The above relation holds rather generally. Using the density of states of the Sommerfeld model (3.48) it simplifies to

$$\chi_{Pauli} = \chi_{Pauli}^{(0)} \left[1 - \frac{\pi^2}{12} \left(\frac{k_B T}{\varepsilon_F} \right)^2 \right] \tag{4.22}$$

Thus, the temperature corrections in the susceptibility in normal metals are very small and normally can be neglected.

4.1.3 Exchange Corrections

In this section, we want to begin to take into account, at least in a simple approximation, the Coulomb interaction among the conduction electrons, which we have so far neglected. In the process, we will know about the so-called *exchange interaction*, which will play a central role in the collective magnetism to be discussed later. The discussion will be limited to the $T = 0$ case and as we have seen in the last section, the temperature corrections are not expected to be important.

The solution of the problem consists of three steps:

1. The total energy E (at $T = 0$, it is the same as the free energy) is calculated as a function of the electron numbers $N_\uparrow$ and $N_\downarrow$.
2. E is varied with respect to $N_{\uparrow\downarrow}$. The minimum fixes the "true" $N_\uparrow$ and $N_\downarrow$.
3. E is expressed in terms of the "true" $N_\uparrow$ and $N_\downarrow$ and using this the susceptibility

$$\chi = -\frac{\mu_0}{V}\left(\frac{\partial^2 E}{\partial B_0^2}\right)_{T=0} \tag{4.23}$$

is calculated.

We demonstrate the procedure first using the Sommerfeld model.

4.1.3.1 Sommerfeld Model

On the application of an external field B_0, the density of states becomes spin dependent (4.7). To calculate the internal energy, we must add both the spin parts:

$$E^{(0)} = \frac{1}{2}\sum_{\sigma}\int_{-\infty}^{+\infty} dE\ Ef_{-}(E)\ \rho_0(E + z_\sigma \mu_B B_0) \tag{4.24}$$

Making simple rearrangements and exploiting (4.9) for the electron number, we get

$$E^{(0)} = -\sum_{\sigma} z_\sigma \mu_B B_0 N_\sigma + \frac{1}{2}\int_0^{\infty} d\eta\ \eta\rho_0(\eta)\sum_{\sigma} f_{-}(\eta - z_\sigma \mu_B B_0) \tag{4.25}$$

At $T = 0$, the Fermi function is a step function, therefore

$$E^{(0)} = -\sum_{\sigma} z_\sigma \mu_B B_0 N_\sigma + \frac{1}{2}\sum_{\sigma}\int_0^{\varepsilon_F^\sigma} d\eta\ \eta\rho_0(\eta) \tag{4.26}$$

Substituting (3.48) for ρ_0, we finally get

$$E^{(0)} = -\sum_{\sigma} z_\sigma \mu_B B_0 N_\sigma + \frac{d}{2}\frac{2}{5}\sum_{\sigma}\left(\varepsilon_F^\sigma\right)^{5/2} \tag{4.27}$$

ε_F^σ is determined exactly as ε_F was done in (3.41), when one fills a sphere in k-space with N_σ electrons:

$$\varepsilon_F^\sigma = \frac{\hbar^2}{2m}\left(k_F^\sigma\right)^2 \quad ; \quad k_F^\sigma = \left(6\pi^2\frac{N_\sigma}{V}\right)^{1/3} \tag{4.28}$$

The constant d is defined in (3.49):

$$\frac{1}{5}d\left(\varepsilon_F^\sigma\right)^{5/2} = \frac{3\hbar^2}{10m}\left(\frac{6\pi^2}{V}\right)^{2/3} N_\sigma^{5/3}$$

With this, the total energy is given by

$$E^{(0)} = \frac{3\hbar^2}{10m}\left(\frac{6\pi^2}{V}\right)^{2/3}\left(N_\uparrow^{5/3} + N_\downarrow^{5/3}\right) - \mu_B\, B_0(N_\uparrow - N_\downarrow) \tag{4.29}$$

In this, $N_\uparrow$ and $N_\downarrow$ are still unknown. The influence of the field is certainly very small. Therefore, we write

$$N_\sigma = \frac{N_e}{2} + z_\sigma x \tag{4.30}$$

Without the field, naturally, we have $N_\sigma = \frac{1}{2}N_e$. With the field, however, x will be unequal to zero, even though, it will be very small compared to N_e. So, we can use a series expansion

$$\begin{aligned} N_\sigma^{5/3} &= \left(\frac{N_e}{2}\right)^{5/3}\left(1 + \frac{2z_\sigma}{N_e}\,x\right)^{5/3} \\ &= \left(\frac{N_e}{2}\right)^{5/3}\left(1 + \frac{5}{3}\,\frac{2z_\sigma}{N_e}\,x + \frac{5}{9}\,\frac{4z_\sigma^2}{N_e^2}\,x^2 + \cdots\right) \end{aligned}$$

In (4.29), we require

$$\sum_\sigma N_\sigma^{5/3} \approx \left(\frac{1}{2}N_e\right)^{5/3} 2\left(1 + \frac{20}{9}\frac{x^2}{N_e^2} + \cdots\right) \tag{4.31}$$

Without the field, we have

$$E_0^{(0)} = \frac{3\hbar^2}{10m}\left(\frac{6\pi^2}{V}\right)^{2/3} 2\left(\frac{N_e}{2}\right)^{5/3} = \frac{3}{5}N_e\varepsilon_F$$

and *with* the field

$$E_x^{(0)} = E_0^{(0)} + \frac{4}{3}\varepsilon_F\frac{x^2}{N_e} - 2\mu_B B_0 x \tag{4.32}$$

In equilibrium, x will take such a value that $E^{(0)}$ is minimal. From the condition

$$0 \stackrel{!}{=} \left.\frac{\partial E_x^{(0)}}{\partial x}\right|_{x=x_0}$$

we get

$$x_0 = \frac{3}{4}N_e\,\frac{\mu_B\, B_0}{\varepsilon_F} \tag{4.33}$$

Substituting this in (4.32) gives

$$E^{(0)} = E^{(0)}_{x=x_0} = E^{(0)}_0 - \frac{3}{4} N_e \frac{(\mu_B B_0)^2}{\varepsilon_F} \tag{4.34}$$

From this we can now calculate the *susceptibility*:

$$\chi_{Pauli} = -\frac{\mu_0}{V} \left(\frac{\partial^2 E^{(0)}}{\partial B_0^2} \right)_{T=0} = \frac{3}{2} \mu_0 \frac{N_e}{V} \frac{\mu_B^2}{\varepsilon_F} \tag{4.35}$$

which is again our old result (3.175) for the Pauli susceptibility. We will now follow the same procedure, in order to obtain information about the interacting electron system. For this we use the so-called Jellium model.

4.1.3.2 Jellium Model

This model of a metallic solid is defined by the following assumptions:

1. There are N_e electrons in volume $V = L^3$ which interact among themselves via the Coulomb interaction

$$H_c = \frac{1}{2} \sum_{i,j}^{i \neq j} \frac{e^2}{4\pi \varepsilon_0 |\mathbf{r}_i - \mathbf{r}_j|} \tag{4.36}$$

2. There are singly charged positive ions

$$N_e = N_i \tag{4.37}$$

3. The charge of the ions is "uniformly smeared out" to give

    (1) charge neutrality
    (2) lattice potential $V(\mathbf{r}) = const.$

4. Periodic boundary conditions on the volume V.

This model is not exactly solvable. The following treatment has the character of a first-order perturbation theory for the ground state. The starting point is the following *Hamiltonian*:

$$H = H_e + H_+ + H_{e+} \tag{4.38}$$

H_e is the electronic part, which will be discussed below in more detail . H_+ describes the uniformly smeared out charge of the ions. "Uniformly smeared out" means that the ion density $n(\mathbf{r})$ is independent of the position:

$$n(\mathbf{r}) \to \frac{N_i}{V} \tag{4.39}$$

Then, it is easy to evaluate H_+:

$$\begin{aligned} H_+ &= \frac{1}{2}\frac{e^2}{4\pi\varepsilon_0}\int\int d^3r\, d^3r' \frac{n(\mathbf{r})n(\mathbf{r}')}{|\mathbf{r}-\mathbf{r}'|} e^{-\alpha|\mathbf{r}-\mathbf{r}'|} \\ &\to \frac{1}{2}\frac{e^2}{4\pi\varepsilon_0}\left(\frac{N_i}{V}\right)^2 \int\int d^3r\, d^3r' \frac{e^{-\alpha|\mathbf{r}-\mathbf{r}'|}}{|\mathbf{r}-\mathbf{r}'|} \end{aligned}$$

$e^{-\alpha|\mathbf{r}-\mathbf{r}'|}$ is a factor that ensures convergence. We finally get (the integral is solved as Problem 4.3),

$$H_+ = \frac{1}{2}\frac{e^2}{4\pi\varepsilon_0}\frac{N_i^2}{V}\frac{4\pi}{\alpha^2} \tag{4.40}$$

Because of the assumption (4), we need the thermodynamic limit ($N_i \to \infty$, $V \to \infty$, $N_i/V = const.$). In such a case the Coulomb integrals diverge. That is the reason for introducing the convergence factor.

The procedure is, first, all the integrals are evaluated for $\alpha > 0$ and finite N_i and finite V. Then the thermodynamic limit is taken. Only at the end the limit $\alpha \to 0$ is applied. In the limit $\alpha \to 0$, H_+ as given by (4.40), diverges, but it will be compensated by other terms which are still to be calculated.

The term H_{e+} in (4.38) describes the interaction of the electrons with the uniform positive background.

$$\begin{aligned} H_{e+} &= -\frac{e^2}{4\pi\varepsilon_0}\sum_{j=1}^{N_e}\int d^3r \frac{n(\mathbf{r})}{|\mathbf{r}-\mathbf{r}_j|} e^{-\alpha|\mathbf{r}-\mathbf{r}_j|} \\ &\to -\frac{e^2}{4\pi\varepsilon_0}\frac{N_i}{V}\sum_{j=1}^{N_e}\int d^3r \frac{e^{-\alpha|\mathbf{r}-\mathbf{r}_j|}}{|\mathbf{r}-\mathbf{r}_j|} \end{aligned}$$

Here $\mathbf{r}_j$ is treated as a classical variable. The integral can be easily evaluated to give (Problem 4.3)

$$H_{e+} = -\frac{e^2}{4\pi\varepsilon_0}\frac{N_i}{V}\sum_{j=1}^{N_e}\frac{4\pi}{\alpha^2} = -\frac{e^2}{4\pi\varepsilon_0}\frac{4\pi}{\alpha^2}\frac{N_i N_e}{V} \tag{4.41}$$

Since $N_i = N_e$, we have altogether

$$H = H_e - \frac{1}{2}\frac{e^2}{\varepsilon_0\alpha^2}\frac{N_i^2}{V} \tag{4.42}$$

The second summand still diverges in the limit $\alpha \to 0$, but, finally, we will see that it will be exactly compensated by a part of H_e.

We will now discuss the really interesting term H_e,

$$H_e = \sum_{j=1}^{N_e} \frac{p_j^2}{2m} + H_c \tag{4.43}$$

which consists of the kinetic energy of the electrons plus the Coulomb interaction among them. We want to perform the first-order perturbation calculation, which means, calculating the expectation value of H_c in the ground state of the non-interacting, indistinguishable electrons. As a consequence of the Pauli's principle, we have to deal with the totally antisymmetrized product of N_e single-particle states. Let $|\psi_{\alpha_r}^{(\mu)}\rangle$ be the orthonormal single-particle state, where α_r is a set of quantum numbers and μ is a particle number. Then, we can write the ground state as (see Appendix A)

$$|\psi_0\rangle = |\psi_{\alpha_1} \cdots \psi_{\alpha_N}\rangle^{(-)} = \frac{1}{\sqrt{N!}} \sum_{\mathcal{P}} (-1)^p \, \mathcal{P} \left\{ |\psi_{\alpha_1}^{(1)}\rangle \cdots |\psi_{\alpha_N}^{(N)}\rangle \right\} \tag{4.44}$$

The summation is over all the permutations of the particle indices given as superscripts. p is the number of transpositions in the permutation $\mathcal{P}$.

Because of the assumption (3) of the Jellium model, the single-particle states for this model are simply the plane waves multiplied by $s = 1/2$ spinors. We write this symbolically as follows:

$$|\psi_{\alpha_r}^{(\mu)}\rangle \equiv |\mathbf{k}^\mu, \sigma^\mu\rangle = |\mathbf{k}^\mu\rangle |\sigma^\mu\rangle = |k^\mu\rangle \tag{4.45}$$

Since the Coulomb interaction H_c is a two-particle interaction, the matrix element E_c, constructed with N_e-particle states, can be expressed exclusively in terms of two-particle states:

$$E_c = \frac{1}{2} \frac{e^2}{4\pi\varepsilon_0} \sum_{k,k'}{}' \, {}^{(-)}\langle k k'| \frac{1}{|\widehat{\mathbf{r}}^{(1)} - \widehat{\mathbf{r}}^{(2)}|} |k k'\rangle^{(-)} \tag{4.46}$$

$\widehat{\mathbf{r}}^{(i)}$ is the position operator in the Hilbert's space of the ith particle. In (4.46), the summation is over all the *occupied* states, indicated by the prime on the sum. The antisymmetrized two-particle state is given by

$$|k, k'\rangle^{(-)} = \frac{1}{\sqrt{2}} (|k_1\rangle |k_2'\rangle - |k_2\rangle |k_1'\rangle) \tag{4.47}$$

where now the lower indices 1, 2 number the particles. The matrix element consists of four summands out of which two each give the same contribution:

$$E_c = E_{dir} + E_{ex} \tag{4.48}$$

We will discuss the two terms in the above equation separately. The first summand is called the direct coulomb interaction.

Direct Coulomb Interaction

This term is characterized by the fact that the particle indices in the bra- and the ket-states of the matrix element (4.46) are the same:

$$E_{dir} = \frac{1}{2}\frac{e^2}{4\pi\varepsilon_0}\sum_{k,k'}{}' \langle k_1|\langle k_2'|\frac{1}{|\widehat{\mathbf{r}}^{(1)} - \widehat{\mathbf{r}}^{(2)}|}|k_1\rangle|k_2'\rangle \tag{4.49}$$

The operator $(|\widehat{\mathbf{r}}^{(1)} - \widehat{\mathbf{r}}^{(2)}|)^{-1}$ acts only on the space part and not on the spin part. The spin parts are orthonormal.

$$\begin{aligned} E_{dir} = &\frac{e^2}{8\pi\varepsilon_0}\sum_{\sigma,\sigma'}\sum_{\mathbf{k},\mathbf{k}'}\int\int d^3r_1\, d^3r_2 \times \\ &\times \langle \mathbf{k}^{(1)}|\langle \mathbf{k}'^{(2)}|\frac{1}{|\widehat{\mathbf{r}}^{(1)} - \widehat{\mathbf{r}}^{(2)}|}|\mathbf{r}_1\rangle|\mathbf{r}_2\rangle\langle \mathbf{r}_1|\mathbf{k}^{(1)}\rangle\langle \mathbf{r}_2|\mathbf{k}'^{(2)}\rangle \end{aligned}$$

Here, we have inserted a complete set of position eigenstates. This makes the application of the operator $(|\widehat{\mathbf{r}}^{(1)} - \widehat{\mathbf{r}}^{(2)}|)^{-1}$ trivial:

$$\begin{aligned} E_{dir} = &\frac{e^2}{8\pi\varepsilon_0}\sum_{\sigma,\sigma'}{}'\sum_{\mathbf{k},\mathbf{k}'}{}'\int\int d^3r_1\, d^3r_2 * \\ &* \frac{1}{|\mathbf{r}_1 - \mathbf{r}_2|}\langle \mathbf{k}|\mathbf{r}_1\rangle\langle \mathbf{k}'|\mathbf{r}_2\rangle\langle \mathbf{r}_1|\mathbf{k}\rangle\langle \mathbf{r}_2|\mathbf{k}'\rangle \\ = &\frac{e^2}{8\pi\varepsilon_0}\frac{1}{V^2}\sum_{\sigma,\sigma'}{}'\sum_{\mathbf{k},\mathbf{k}'}{}'\int\int d^3r_1\, d^3r_2 \frac{1}{|\mathbf{r}_1 - \mathbf{r}_2|} \end{aligned}$$

Here we have used the plane-wave representation:

$$\langle \mathbf{r}|\mathbf{k}\rangle = \frac{1}{\sqrt{V}}\, e^{i\mathbf{k}\cdot\mathbf{r}}$$

Just as in obtaining (4.40), the double integral must be solved with a convergence factor. Then, we finally have

$$E_{dir} = \frac{1}{2}\frac{e^2}{\varepsilon_0\alpha^2}\frac{N_e^2}{V} \tag{4.50}$$

Since $N_i = N_e$, the "direct" interaction exactly compensates the still remaining divergent part ($\alpha \to 0$) (4.42) of the positive background due to the ions.

The second summand in (4.48) is called the exchange interaction.

Exchange Interaction

The notation is due to the fact that the particle indices in the bra- and ket-states of the matrix element are interchanged:

$$E_{ex} = -\frac{1}{2}\frac{e^2}{4\pi\varepsilon_0}\sum_{k,k'}{}' \langle k_1|\langle k_2'|\frac{1}{|\widehat{r_1} - \widehat{r_2}|}|k_2\rangle|k_1'\rangle \tag{4.51}$$

In classical physics, there is no concept of indistinguishable particles. Therefore, there is no classical analogue to this matrix element. In view of the orthogonality of the spin states, we must have $\sigma = \sigma'$. Then

$$\begin{aligned} E_{ex} &= -\frac{1}{2}\frac{e^2}{4\pi\varepsilon_0}\sum_{\mathbf{k},\mathbf{k}',\sigma}^{k,k'\le k_F^\sigma}\int\int d^3r_1 d^3r_2 \, * \\ &* \, \langle \mathbf{k}^{(1)}|\langle \mathbf{k}'^{(2)}|\frac{1}{|\widehat{r_1} - \widehat{r_2}|}|\mathbf{r}_1\rangle|\mathbf{r}_2\rangle\langle \mathbf{r}_2|\mathbf{k}^{(2)}\rangle\langle \mathbf{r}_1|\mathbf{k}'^{(1)}\rangle \\ &= -\frac{1}{2}\frac{e^2}{4\pi\varepsilon_0}\frac{1}{V^2}\sum_{\mathbf{k},\mathbf{k}',\sigma}^{k,k'\le k_F^\sigma}\int\int d^3r_1 d^3r_2 \frac{1}{|r_1 - r_2|}e^{-i(\mathbf{k}-\mathbf{k}')(\mathbf{r}_1-\mathbf{r}_2)} \end{aligned}$$

k_F^σ is defined in (4.28). As is usually done, we convert the sum into an integral,

$$\sum_{\mathbf{k}} \to \frac{V}{(2\pi)^3}\int d^3k \tag{4.52}$$

and get as an intermediate result

$$\begin{aligned} E_{ex} &= \frac{-e^2}{8\pi\varepsilon_0(2\pi)^6}\int\int d^3r_1 d^3r_2 \sum_\sigma * \\ &* \int_{k\le k_F^\sigma}\int_{k'\le k_F^\sigma} d^3k d^3k' \frac{e^{-i(\mathbf{k}-\mathbf{k}')\cdot(\mathbf{r}_1-\mathbf{r}_2)}}{|\mathbf{r}_1 - \mathbf{r}_2|} \end{aligned} \tag{4.53}$$

In terms of the relative and centre of mass coordinates

$$\mathbf{r} = \mathbf{r}_1 - \mathbf{r}_2 \quad ; \quad \mathbf{R} = \frac{1}{2}(\mathbf{r}_1 + \mathbf{r}_2) \tag{4.54}$$

Equation (4.53) becomes

$$E_{ex} = -\frac{e^2 V}{8\pi\varepsilon_0(2\pi)^6}\int d^3r \sum_\sigma \int\int_{k,k'\le k_F^\sigma} d^3k d^3k' \frac{e^{-i(\mathbf{k}-\mathbf{k}')\cdot\mathbf{r}}}{r} \tag{4.55}$$

The k-integrations can be performed without any difficulty (Problem 4.4).

$$\int_{k\le k_F^\sigma} d^3k e^{i\mathbf{k}\cdot\mathbf{r}} = \frac{4\pi}{r}\int_0^{k_F^\sigma} dk\, k \sin kr = -4\pi\, \frac{k_F^\sigma\, r\cos k_F^\sigma r - \sin k_F^\sigma r}{r^3} \tag{4.56}$$

Then, we have for the exchange energy

$$E_{ex} = -\frac{e^2\, V}{32\pi^5\, \varepsilon_0}\sum_\sigma \int d^3r\, \frac{1}{r^7}\left(k_F^\sigma\, r\cos k_F^\sigma\, r - \sin k_F^\sigma\, r\right)^2$$

or after the substitution $x = rk_F^\sigma$:

$$E_{ex} = -\frac{e^2 V}{8\pi^4\varepsilon_0}\left\{\int_0^\infty dx \frac{1}{x^5}(x\cos x - \sin x)^2\right\}\sum_\sigma \left(k_F^\sigma\right)^4 \tag{4.57}$$

The integral inside the braces can be solved elementarily (Problem 4.4).

$$\int_0^\infty dx \frac{1}{x^5}(x\cos x - \sin x)^2 = \frac{1}{4} \tag{4.58}$$

With this, the exchange energy has the simple form

$$E_{ex} = -\frac{e^2 V}{32\pi^4\varepsilon_0}\sum_\sigma (k_F^\sigma)^4 \tag{4.59}$$

We substitute (4.28) for k_F^σ and get

$$E_{ex} = -\frac{3e^2}{16\pi^2\varepsilon_0}\left(6\pi^2\frac{1}{V}\right)^{1/3}\sum_\sigma N_\sigma^{4/3} \tag{4.60}$$

We again make the ansatz,

$$N_\sigma = \frac{N_e}{2} + z_\sigma x$$

and expand up to the quadratic term in x:

$$N_\sigma^{4/3} \approx \left(\frac{N_e}{2}\right)^{4/3}\left(1 + \frac{8}{3}\frac{z_\sigma x}{N_e} + \frac{2}{9}\frac{4x^2}{N_e^2} + \cdots\right)$$

In the summation over σ, the linear term in x vanishes so that

$$\sum_\sigma N_\sigma^{4/3} \approx 2\left(\frac{N_e}{2}\right)^{4/3}\left(1 + \frac{2}{9}\frac{4x^2}{N_e^2}\right). \tag{4.61}$$

We have, thus, determined the exchange energy as a function of x:

$$E_{ex}(x) = -\frac{3e^2}{8\pi\varepsilon_0} N_e \left(\frac{3N_e}{8\pi V}\right)^{1/3} \left(1 + \frac{8x^2}{9N_e}\right) \tag{4.62}$$

This is also, at the same time, the total contribution of the Coulomb interaction to the energy of the ground state of the Jellium model within the first-order perturbation theory. The contribution of the kinetic energy has already been calculated in connection with the Sommerfeld model (4.32). Adding them together, we get the energy of the ground state as a function of x:

$$\begin{aligned} E_0(x) &= E_x^{(0)} + E_{ex}(x) \\ &= E_0^{(0)} - \frac{3e^2}{8\pi\varepsilon_0} N_e \left(\frac{3N_e}{8\pi V}\right)^{1/3} \\ &+ \frac{4}{3}\varepsilon_F \frac{x^2}{N_e} - 2\mu_B B_0 x - \frac{e^2}{3\pi\varepsilon_0}\left(\frac{3N_e}{8\pi V}\right)^{1/3} \frac{x^2}{N_e} \end{aligned} \tag{4.63}$$

According to this, the ground state energy without the field ($x = 0$) is given by

$$E_{00} = E_0(x = 0) = E_0^{(0)} - \frac{3e^2}{8\pi\varepsilon_0} N_e \left(\frac{3N_e}{8\pi V}\right)^{1/3} \tag{4.64}$$

Before proceeding further, we introduce some standard notation:

$$\begin{aligned} n_e &= N_e/V \quad \textit{average electron density} \\ v_e &= \tfrac{1}{n_e} \quad \textit{average volume per electron} \end{aligned} \tag{4.65}$$

One defines a dimensionless parameter called the *density parameter* r_s via

$$v_e = \frac{4\pi}{3}(a_B r_s)^3 \tag{4.66}$$

where

$$a_B = \frac{\hbar^2 4\pi\varepsilon_0}{me^2} = 0.529\ \text{Å}$$

is the *Bohr radius*. The smaller the r_s is, the larger is the electron density. Typical values of r_s for metals lie between 1 and 6. In a similar way, we introduce a parameter for energy (Rydberg):

$$1\ ryd = \frac{1}{4\pi\varepsilon_0}\frac{e^2}{2a_B} = 13.605\ eV \tag{4.67}$$

With the abbreviation

$$\alpha = \left(\frac{9\pi}{4}\right)^{1/3} = 1.92$$

one gets the following expression for the Fermi energy of the Sommerfeld model:

$$\varepsilon_F = \left(\frac{e^2}{8\pi\,\varepsilon_0 a_B}\right)\frac{\alpha^2}{r_s^2} = \frac{\alpha^2}{r_s^2}\,[ryd] \tag{4.68}$$

According to (3.42), $\frac{3}{5}\varepsilon_F$ is the average kinetic energy per electron:

$$E_0^{(0)} = \frac{2.21}{r_s^2} N_e\,[ryd] \tag{4.69}$$

One can further, easily show that

$$\frac{3e^2}{8\pi\,\varepsilon_0} N_e \left(\frac{3N_e}{8\pi\,V}\right)^{1/3} = \frac{e^2}{8\pi\,\varepsilon_0 a_B} N_e \frac{3\alpha}{2\pi r_s}$$

With this, one gets for the total *ground state energy of the Jellium model without field*

$$E_{00}/N_e = \left(\frac{3}{5}\frac{\alpha^2}{r_s^2} - \frac{3\alpha}{2\pi r_s}\right)\left(\frac{e^2}{8\pi\,\varepsilon_0 a_B}\right) = \left(\frac{2.21}{r_s^2} - \frac{0.916}{r_s}\right)\,[ryd] \tag{4.70}$$

This is, of course, the result in the first-order perturbation theory, only. Let us discuss this expression a little further (Fig. 4.3).

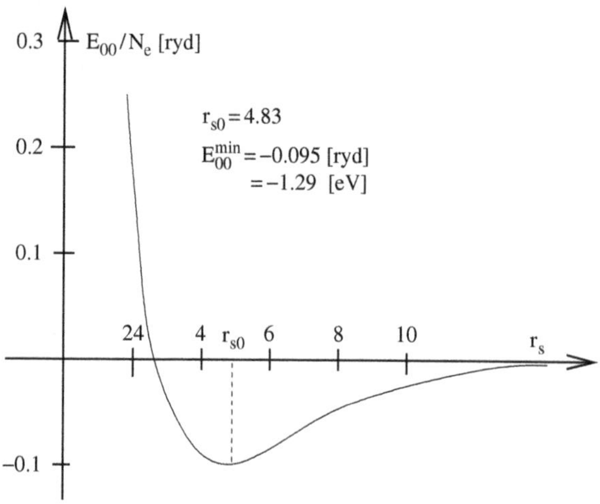

Fig. 4.3 Ground state energy of the Jellium model as a function of the density parameter r_s

The minimum in energy which is found at $r_s = 4.83$ with $E_{00}^{min} = -0.095[ryd] = -1.29$ eV indicates an *optimal* density, which actually means an optimal inter-ionic distance, and with that, one explains, e.g. the phenomenon of *metallic bonding*.

Since this is a result of a perturbation theory approximation, it is usually generalized in the following manner:

$$\frac{E_{00}}{N_e} = \frac{2.21}{r_s^2} - \frac{0.916}{r_s} + \varepsilon_{corr}\ [ryd] \tag{4.71}$$

The first summand is the kinetic energy per electron and the second summand is the *exchange energy*. As it is made clear in the derivation, it is a consequence of the indistinguishability of the particles and so it is a consequence of the Pauli's principle. This takes care of the fact that electrons with parallel spins do not come too close to each other and therefore automatically causes a reduction in the Coulomb interaction energy of particles of like charge. It explains the negative sign. The last summand is called the *correlation energy*. It is the deviation of the perturbation theory result from the exact result and therefore is naturally unknown. The conventional methods of perturbation theory fail to determine ε_{corr}. The modern methods of many-body theory have led to the following expression [1]:

$$\varepsilon_{corr} = \frac{2}{\pi^2}(1 - \ln 2)\ln r_s - 0.094 + \mathcal{O}(r_s \ln r_s) \tag{4.72}$$

We will now return to our actual problem. In terms of the new variables that we have introduced, the ground state energy of the Jellium model in the presence of the field (4.63) is

$$E_0(x) = E_{00} - 2\mu_B B_0 x + \frac{e^2}{8\pi\varepsilon_0 a_B}\frac{4\alpha}{3\pi r_s}\left(\frac{\pi\alpha}{r_s} - 1\right)\frac{x^2}{N_e} \tag{4.73}$$

The third and fourth summands are the modifications to the kinetic energy and exchange energy, respectively, caused by the field. This expression is varied with respect to x.

$$\frac{\partial E_0(x)}{\partial x} = -2\mu_B B_0 + \frac{e^2}{4\pi\varepsilon_0 a_B}\frac{4\alpha}{3\pi r_s}\left(\frac{\pi\alpha}{r_s} - 1\right)\frac{x}{N_e}$$

We get the extremum value at $x = x_0$ where

$$x_0 = \frac{3\pi r_s}{4\alpha} N_e \frac{\mu_B B_0}{\frac{e^2}{8\pi\varepsilon_0 a_B}\left(\frac{\pi\alpha}{r_s} - 1\right)} \tag{4.74}$$

With this, we get for the ground state of the Jellium model in the presence of an external field as

$$E_0(x = x_0) \equiv E_0 = E_{00} - 2\mu_B B_0\, x_0 + \mu_B B_0\, x_0$$

$$\begin{aligned} E_0 &= E_{00} - \frac{3}{4} N_e \frac{(\mu_B B_0)^2}{\frac{e^2}{8\pi\varepsilon_0 a_B}\frac{\alpha^2}{r_s^2}\left(1 - \frac{r_s}{\pi\alpha}\right)} \\ &= E_{00} - \frac{3}{4}\frac{N_e}{\varepsilon_F}\frac{(\mu_B B_0)^2}{1 - \frac{r_s}{\pi\alpha}} \end{aligned} \tag{4.75}$$

In the last step, we used the expression for ε_F given in (4.68). Now it is simple to calculate the *exchange corrected susceptibility*:

$$\chi_{ex} = -\frac{\mu_0}{V}\frac{\partial^2 E_0}{\partial B_0^2} = \frac{3}{2}\,\mu_0\,\mu_B^2\,\frac{N_e}{V}\frac{1}{\varepsilon_F}\frac{1}{1 - \frac{r_s}{\pi\alpha}} \tag{4.76}$$

If we substitute the earlier result (4.35) for the non-interacting electron system, we finally get

$$\chi_{ex} = \chi_{Pauli}\,\frac{1}{1 - \frac{r_s}{\pi\alpha}} \tag{4.77}$$

The smaller the value of r_s, the larger is the electron density and the more exact are the results of the simple Sommerfeld model. At first glance, this appears to be a surprising fact, but it is a consequence of the screening effects in an interacting electron gas.

In view of the facts, $\pi\alpha \approx 6.03$ and $1 < r_s < 6$, the influence of the exchange interaction, which actually means the Coulomb interaction, can be quite considerable for typical metals. One observes, however, that the drastic changes in χ caused by E_{ex} are, to a large extent, again annulled by the correlation energy $E_{corr} = N_e \varepsilon_{corr}$ (Table 4.1).

The main effect of E_{corr} is felt while taking into account the correlation between the electrons of *antiparallel* spins, because E_{ex} is calculated using the states of the Sommerfeld model, which, due to the Pauli's principle, includes only the correlations between parallel spins. The calculation of E_{corr} is extremely difficult. One knows, however, a series expansion (4.72) in r_s.

Table 4.1 Pauli susceptibility of conduction electrons described by the Jellium model. The fifth column contains the results of Silverstein [2] which were obtained by approximately taking into account the correlation energy. The sixth column of the table gives the experimental values of Schumacher and Vehse [3]

$10^6 \times$	χ_{Pauli}	$\chi_{Pauli}(m^*)$	χ_{ex}	χ_{corr}	χ_{exp}
Li	10	17	150	27.8	26.6 ± 1.3
Na	15	15	43	19.6	25.8 ± 2.6
k	23	25	195	31.7	

4.2 Paramagnetism of the Localized Electrons

We now discuss the paramagnetism of insulators. The electrons which are responsible for the paramagnetism are strictly localized on fixed lattice points and produce a permanent magnetic moment there. Rare earths and their compounds are almost ideal examples of this picture, and they are normally called the "$4f$-systems". The electron configuration of the *neutral* rare earth atom corresponds to the stable noble gas configuration of xenon, plus additional $4f$- and $6s$-parts:

$$[Xe](4f)^p(6s)^2 \tag{4.78}$$

In the periodic table, the rare earths start with the element La and from then on, successively add electrons in the $4f$-shell. So, they differ from each other by the number p of electrons in the $4f$-shell of the rare earth element. In the compounds, in general, the rare earth atom gives away three electrons, namely the two $6s$-electrons and one of the $4f$-electrons:

$$RE \rightarrow (RE)^{3+} + \{(6s)^2 + (4f)^1\} \tag{4.79}$$

These three electrons, in the case of insulators, participate in the bonding and in metals, they become the quasi-free conduction electrons.

The partially filled $4f$-shell is situated inside the xenon core and is strongly screened by the completely filled $(5s)^2$- and $(5p)^6$-shells, which lie outside the xenon core. For this reason, the $4f$-wavefunctions of the neighbouring rare earth ions practically do not overlap. Therefore, the $4f$-shells and the magnetic moments produced by them are strongly localized at the respective lattice points. Such a system can be described by the following extremely simple model: We assume that there are N identical, independent ions (atoms) in a volume V, and we are interested only in the magnetic moment produced by these ions (atoms). In view of the strong intra-atomic correlations, one can assume that the localized magnetic moment is determined by the Hund's rules of the atomic physics (see Sect. 2.1). Thus, the calculation of the temperature and field dependence of the magnetization M and the susceptibility is essentially an *atomic problem* (Fig. 4.4).

In addition to the $4f$-systems, this model is also relevant to certain $3d$- and $5f$-systems. Another decisive presumption shall be the *LS-coupling*, which was

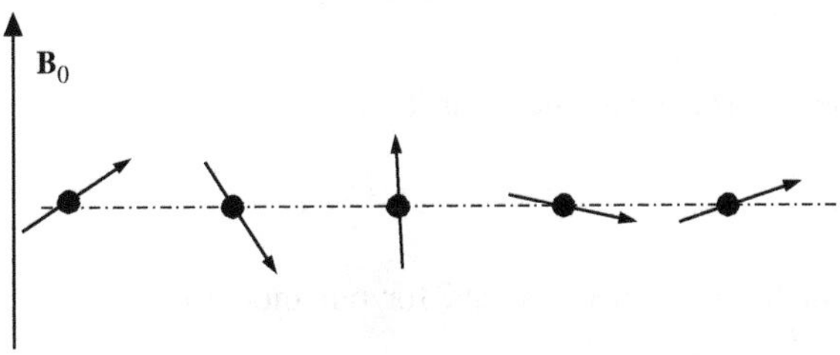

Fig. 4.4 Paramagnetic local moment system in an external magnetic field

discussed in Sect. 2.10. The distances between different (L,S)-multiplets are so large that the transitions among them are improbable and therefore we assume that the quantum numbers L and S of the $4f$-shell belonging to the squares L^2 and S^2 of total orbital angular momentum and total spin, respectively, are good quantum numbers. The magnetic moment localized at the lattice site R_j is given by

$$\mathbf{m}_j = -\frac{\mu_B}{\hbar}(\mathbf{L}_j + 2\mathbf{S}_j) = -\frac{\mu_B}{\hbar}(\mathbf{J}_j + \mathbf{S}_j) \tag{4.80}$$

According to the remark made in Sect. 2.1 (after Table 2.1), we disregard in the following the minus sign in the definition (4.80) of the magnetic moment.

We start with the model Hamiltonian

$$H = \sum_{j=1}^{N} \left(H_0^{(j)} + H_{SO}^{(j)} - \mathbf{m}_j \cdot \mathbf{B}_0 \right) = \sum_{j=1}^{N} H_1^{(j)} \tag{4.81}$$

$H_0^{(j)}$ determines the term scheme of the jth ion based on the Coulomb interaction of the electrons with the nucleus and also among themselves and therefore, in some sense, fixes the *coarse structure* of the terms. Since we want to restrict ourselves to a single (LS)-multiplet, $H_0^{(j)}$ by itself is not important for us. $H_{SO}^{(j)}$ is the spin–orbit coupling in the jth ion, which determines the *fine structure* of the terms. The last summand in parenthesis in (4.81) represents the Zeeman energy. The relative strength of the last two interactions is decisive in the calculation of the *magnetization*.

$$\mathbf{M} = n\langle \mathbf{m} \rangle \qquad n = \frac{N}{V} \tag{4.82}$$

The angular brackets $\langle \cdots \rangle$ means a quantum statistical average, which actually involves two averaging processes, namely

(a) the quantum mechanical expectation value of the operator in the given state of the atom
and

(b) the thermal average over all the states of the atom.

In general, the average value of an observable A is given by

$$\langle A \rangle = \frac{1}{Z} Tr \left(A e^{-\beta H} \right) \tag{4.83}$$

Here Z is the canonical partition function

$$Z = Tr \left(e^{-\beta H} \right) = Z_1^N \tag{4.84}$$

which has been factorized for our model (4.81) into a product of single-particle partition functions

$$Z_1 = Tr\left(e^{-\beta H_1}\right) \tag{4.85}$$

because there does not exist any interaction among the moments.

Let there be a homogeneous magnetic field $\mathbf{B}_0$ applied in the z-direction. Then naturally the x and y components of magnetization of the paramagnet vanish and we have to calculate only $M_z = M$:

$$M = n\,\frac{1}{Z_1}\,Tr\left(me^{-\beta H_1}\right) = k_B T\, n\,\frac{\partial}{\partial B_0}\ln Z_1 \tag{4.86}$$

That means, once we determine the single-particle partition function Z_1, the problem is solved. The calculation of the trace cannot, however, be done for a general situation; therefore, we will do it for particular situations and consider the limiting cases. Three factors influence M:

1. *Thermal energy* $k_B T$: This is of course obvious since it appears explicitly in the formulae.
2. *Spin–orbit interaction*: According to (2.295), we have

$$H_{SO} = \Lambda(\gamma, LS)\,(\mathbf{L}\cdot\mathbf{S}) \tag{4.87}$$

H_{SO} splits the (LS)-multiplet according to J (2.300)

$$E^{(0)}_{\gamma LSJ} = E^{(0)}_{\gamma LS} + \frac{1}{2}\hbar^2\Lambda(\gamma, LS)\{J(J+1) - L(L+1) - S(S+1)\} \tag{4.88}$$

The coupling constant Λ determines the separation between the individual terms of a multiplet.

3. *Magnetic field*:

$$H_z = -\frac{\mu_B}{\hbar}(J_z + S_z)B_0 \tag{4.89}$$

We have shown in Sect. 2.10 that H_z does not commute with J^2,

$$\left[H_z\,,\; J^2\right]_- \neq 0$$

so that, after the magnetic field is switched on, J is no more a good quantum number. That is, the magnetic field induces transitions among the individual terms of the (LS)-multiplets.

We can determine the partition function Z_1 in only those cases, where there is orders of magnitude difference among (1), (2) and (3).

4.2.1 Weak Spin–Orbit Interaction

Weak spin–orbit interaction means

$$\hbar^2\,\Lambda(\gamma,\ LS) \ll k_B T \tag{4.90}$$

in the just interesting region of temperature. We have to further differentiate the situations with respect to the field. If the field energy also is very much larger than the spin–orbit coupling,

4.2.1.1 $\hbar^2\,\Lambda(\gamma,\ LS) \ll k_B T,\ \mu_B B_0$

then we can assume two things:

1. The preconditions of the *normal* Zeeman effect are fulfilled (2.305)
2. All the terms of the (LS)-multiplet are occupied with almost equal probability.

In this case, M_L and M_S are "still good" quantum numbers, i.e. the eigenstates and eigenenergies can be classified according to these quantum numbers. In contrast, J is *not* a good quantum number.

$$E_{\gamma LSM_L M_S} = E^{(0)}_{\gamma LS} - (M_L + 2M_S)\mu_B B_0 \tag{4.91}$$

$E^{(0)}_{\gamma LS}$ are the energies *without* the field and *without* the spin–orbit coupling, which means, $E^{(0)}_{\gamma LS}$ are the eigenenergies of H_0. In order to calculate the trace that appears in the partition function Z_1, we naturally choose the energy representation:

$$Z_1 = exp\left(-\beta E^{(0)}_{\gamma LS}\right) \sum_{M_L=-L}^{+L} \sum_{M_S=s}^{+S} e^{\beta\mu_B B_0 (M_L + 2M_S)} \tag{4.92}$$

The pre-factor will be very small for the higher (LS)-multiplets, and therefore, as agreed upon earlier, we restrict ourselves only to the lowest of them. With the notation

$$b = \beta\mu_B\ B_0 > 0 \tag{4.93}$$

we calculate

$$\begin{aligned}\sum_{M_L=-L}^{+L} e^{bM_L} &= e^{bL}\sum_{n=0}^{2L}(e^{-b})^n = e^{bL}\,\frac{1-e^{-b(2L+1)}}{1-e^{-b}}\\ &= \frac{e^{b(L+1/2)} - e^{-b(L+1/2)}}{e^{1/2b} - e^{-1/2b}}\end{aligned} \tag{4.94}$$

The summation over the orbital magnetic quantum number therefore gives

$$\sum_{M_L=-L}^{+L} exp\,(-\beta\ \mu_B\ B_0 M_L) = \frac{\sinh(\beta\mu_B B_0(L+\frac{1}{2}))}{\sinh(\frac{1}{2}\beta\mu_B B_0)} \tag{4.95}$$

Exactly analogously we also get

$$\sum_{M_S=-S}^{+S} exp\,(2\beta\mu_B B_0 M_S) = \frac{\sinh(\beta\mu_B B_0(2S+1))}{\sinh(\beta\mu_B B_0)} \tag{4.96}$$

so that the partition function reads

$$Z_1 = e^{-\beta E^{(0)}_{\gamma LS}} \frac{\sinh(\beta\mu_B B_0(L+\frac{1}{2}))\ \sinh(\beta\mu_B B_0(2S+1))}{\sinh(\frac{1}{2}\beta\mu_B B_0)\ \sinh(\beta\mu_B B_0)} \tag{4.97}$$

By differentiating the logarithm of the partition function with respect to the field, we obtain the magnetization of a paramagnet (4.86):

$$\frac{\partial}{\partial B_0}\ln Z_1 = \frac{1}{Z_1^{(L)}}\frac{\partial Z_1^{(L)}}{\partial B_0} + \frac{1}{Z_1^{(S)}}\frac{\partial Z_1^{(S)}}{\partial B_0}$$

We will explicitly evaluate the first summand:

$$\begin{aligned}
&\frac{1}{Z_1^{(L)}}\frac{\partial Z_1^{(L)}}{\partial B_0}\\
&= \frac{\sinh(\frac{1}{2}b)}{\sinh(b(L+\frac{1}{2}))}\left\{\frac{\sinh(\frac{1}{2}b)\beta\mu_B(L+\frac{1}{2})\cosh(b(L+\frac{1}{2}))}{\sinh^2(\frac{1}{2}b)}\right.\\
&\left.- \frac{\frac{1}{2}\beta\mu_B\cosh(\frac{1}{2}b)\sinh(b(L+\frac{1}{2}))}{\sinh^2(\frac{1}{2}b)}\right\}\\
&= \beta\mu_B\left(L+\frac{1}{2}\right)\coth\left(b\left(L+\frac{1}{2}\right)\right) - \frac{1}{2}\beta\mu_B\coth\left(\frac{1}{2}b\right)
\end{aligned} \tag{4.98}$$

We now introduce a function, the so-called *Brillouin function*, which is central to the theory of magnetism:

$$B_D(x) = \frac{2D+1}{2D}\coth\left(\frac{2D+1}{2D}x\right) - \frac{1}{2D}\coth\left(\frac{x}{2D}\right) \tag{4.99}$$

Through this function, the *magnetization* given by (4.86) can be written in the following form:

$$M(T, B_0) = n\mu_B\{L B_L(\beta\mu_B B_0 L) + 2S B_S(2\beta\mu_B B_0 S)\} \tag{4.100}$$

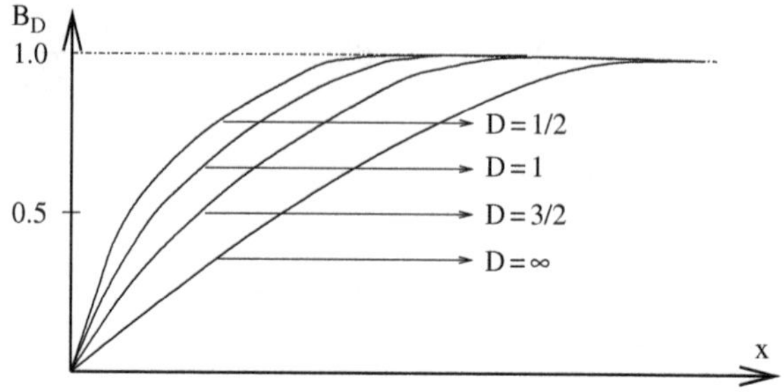

Fig. 4.5 Brillouin function $B_D(x)$ as a function of x for various parameters D

We will first discuss a few general properties of the Brillouin function (Fig. 4.5):

1. $D = \frac{1}{2}$: In this special case,

$$B_{1/2}(x) = \tanh x \tag{4.101}$$

2. $D \to \infty$: In this limit, the Brillouin function reduces to the *Langevin function* $L(x)$,

$$B_{D\to\infty}(x) \equiv L(x) = \coth x - \frac{1}{x} \tag{4.102}$$

which appears in the classical treatment of paramagnetism, where the space quantization of the orbital angular momentum is ignored.
3. *small* x: From the expansion of $\coth x$, one obtains,

$$B_D(x) = \frac{D+1}{3D}x - \frac{D+1}{3D}\,\frac{2D^2+2D+1}{30D^2}x^3 + \cdots \tag{4.103}$$

which means, in particular,

$$B_D(0) = 0 \tag{4.104}$$

Because of this property, according to (4.100), the magnetization of a paramagnet is zero, if either $B_0 = 0$ or $T = \infty$. Physically that means, a paramagnet does not possess *spontaneous magnetization.*
4. $B_D(-x) = -B_D(x)$: This means that, when the direction of the magnetic field is reversed, the direction of the magnetization is also reversed, which is physically obvious.
5. $\lim_{x\to\infty} B_D(x) = 1$: The magnetization shows saturation for $B_0 \to \infty$ or for $T \to 0$. Physically that means that all the moments are oriented parallel to the field.

$$M \to M_0 = n\mu_B(L+2S) \tag{4.105}$$

The high-temperature behaviour of the magnetization (4.100) is interesting. With the starting precondition, it means

$$\Lambda(\gamma, LS) \ll \mu_B B_0 \ll k_B T$$

or

$$\beta \mu_B B_0 \ll 1$$

In this case, the argument of the Brillouin function is small, and therefore we can terminate the series expansion (4.103) after the linear term:

$$M \approx \frac{n\mu_B^2}{3k_B T} B_0\{L(L+1) + 4S(S+1)\} \tag{4.106}$$

The susceptibility

$$\chi = \mu_0 \left(\frac{\partial M}{\partial B_0}\right)_T$$

shows a characteristic $1/T$ behaviour, which is called the *Curie law*.

$$\chi(T) = \frac{C_1}{T} \tag{4.107}$$

C_1 is the so-called "*Curie constant*", which is given by

$$C_1 = n\frac{\mu_0 \mu_B^2}{3k_B}(L(L+1) + 4S(S+1)). \tag{4.108}$$

A purely classical calculation (see (4.102)) would have given a similar high-temperature behaviour

$$\chi_{cl} = \frac{C_{cl}}{T} \quad ; \quad C_{cl} = n\,\frac{\mu_0 \mu^2}{3k_B} \tag{4.109}$$

where μ is the magnetic moment. In analogy, one therefore defines

$$\mu_{eff} = \mu_B p_{eff} \qquad p_{eff} = \sqrt{L(L+1) + 4S(S+1)} \tag{4.110}$$

where p_{eff} is called the *effective number of magnetons*.

Till now, we have assumed that, in addition to the thermal energy, the field energy is also large compared to the spin–orbit interaction. Now we will consider the situation.

4.2.1.2 $\hbar^2\Lambda(\gamma, LS)$, $\mu_B B_0 \ll k_B T$

Let the spin–orbit coupling be as before small but of the same order of magnitude as the magnetic energy. Then, we are no more in the region of the "normal" Zeeman effect. The spin–orbit coupling cannot be neglected any more.

In the partition function, H_z and H_{SO} appear, respectively, in the form βH_z and βH_{SO} so that $e^{\beta H}$ can be expanded up to the linear terms in βH_z and βH_{SO}. Since H_0 commutes with $(H_z + H_{SO})$, we can write $exp(-\beta(H_0 + H_z + H_{SO})) = exp(-\beta H_0)exp(-\beta(H_{SO} + H_z)$ and expand the second exponential up to the linear terms:

$$M \approx n \frac{\mu_B}{\hbar} \frac{Tr\left\{(J_z + S_z)(1 - \beta H_z - \beta H_{SO})e^{-\beta H_0}\right\}}{Tr\left\{(1 - \beta H_z - \beta H_{SO})e^{-\beta H_0}\right\}} \tag{4.111}$$

To calculate the trace, we choose the eigenstates of H_0, J^2 and J_z. The contribution from $e^{-\beta H_0}$ then cancels out and therefore need not be considered any more. Only *one* (LS)-multiplet will be considered, i.e. the operator $e^{-\beta H_0}$ gives only one eigenvalue. We want to evaluate the individual terms in (4.111) separately. First, because of the Wigner–Eckart theorem, we have

$$\begin{aligned} \langle \gamma LS; JM_J|(J_z + S_z)|\gamma LS; JM_J\rangle &\sim \langle \gamma LS; JM_J|J_z|\gamma LS; JM_J\rangle \\ &\sim M_J \end{aligned} \tag{4.112}$$

That means

$$Tr(J_z + S_z) \sim \sum_J C_J \sum_{M_J=-J}^{+J} M_J = 0 \tag{4.113}$$

Therefore, it follows also

$$Tr(H_z) = 0 \tag{4.114}$$

The fine structure of the terms caused by H_{SO} leads to

$$\begin{aligned} Tr(H_{SO}) &= \frac{\hbar^2}{2}\Lambda \sum_J \sum_{M_J} (J(J+1) - L(L+1) - S(S+1)) \\ &= \frac{\hbar^2}{2}\Lambda \sum_J (2J+1)(J(J+1) - L(L+1) - S(S+1)) \end{aligned}$$

Without any loss of generality, we can assume that $L > S$; then J runs over the values $J = L - S,\ L - S + 1,\ \cdots .\ L + S$. That means the trace is given by

$$Tr(H_{SO}) \sim \sum_{n=-S}^{+S} (2L+2n+1)((L+n)(L+n+1)$$
$$-L(L+1) - S(S+1))$$

Using the formula (Problem 4.5)

$$\sum_{n=-S}^{+S} n^2 = \frac{1}{3} S(S+1)(2S+1) \qquad (4.115)$$

it follows that

$$Tr(H_{SO}) = 0 \qquad (4.116)$$

This means that the spin–orbit interaction, which is responsible for the fine structure splitting, does not, however, shift the centre of gravity of the multiplet!

In (4.111), we still need the term

$$Tr\left((J_z + S_z)H_{SO}\right) \sim \frac{\hbar^2}{2} \Lambda \sum_J [J(J+1) - L(L+1)$$
$$-S(S+1)] C_J \sum_{M_j=-J}^{+J} M_J = 0 \qquad (4.117)$$

Here we have exploited (4.112) and the fact that the trace is built with the eigenstates of H_{SO}. The trace of the unit matrix gives the dimension of the Hilbert space under consideration:

$$Tr(1) = \sum_{J=|L-S|}^{L+S} (2J+1) = (2L+1)(2S+1) \qquad (4.118)$$

This is, of course, valid only because we are in the space of a particular (LS)-multiplet in which, H_0 has only one eigenvalue.

Then, according to (4.111) what remains for the magnetization is

$$M = -n\frac{\mu_B}{\hbar}\beta \frac{Tr((J_z + S_z)H_z)}{(2L+1)(2S+1)} = n\frac{\mu_B^2}{\hbar^2}\beta B_0 \frac{Tr(J_z + S_z)^2}{(2L+1)(2S+1)} \qquad (4.119)$$

The trace is independent of the basis. We choose as an appropriate set of states

$$|\gamma LS, M_L M_S\rangle$$

which represents a complete basis. These are, however, not the eigenstates of the Hamiltonian:

$$M = n \frac{\mu_B^2 B_0 \beta}{\hbar^2(2L+1)(2S+1)} \hbar^2 \sum_{M_s=-S}^{+S} \sum_{M_L=-L}^{+L} (M_L + 2M_S)^2 \tag{4.120}$$

Using (4.115), the double sum can be easily evaluated:

$$\begin{aligned}
&\sum_{M_S=-S}^{+S} \sum_{M_L=-L}^{+L} (M_L + 2M_S)^2 \\
&= (2S+1) \sum_{M_L} M_L^2 + 4(2L+1) \sum_{M_S} M_S^2 \\
&= \frac{1}{3}(2S+1)(2L+1)(L(L+1) + 4S(S+1))
\end{aligned} \tag{4.121}$$

Substituting this in (4.120), we finally get for the magnetization

$$M = n \frac{\mu_B^2}{3k_B T} B_0 \{L(L+1) + 4S(S+1)\} \tag{4.122}$$

For the susceptibility, for high temperatures, one gets exactly the same result as in case (a) (4.106):

$$\chi = n \mu_0 \frac{\mu_B^2}{3k_B T} \{L(L+1) + 4S(S+1)\} \tag{4.123}$$

Therefore, up to the linear term in $\beta = \frac{1}{k_B T}$, if the spin–orbit coupling is weak, there is no change in the regions of $\mu_B B_0 \gg \hbar^2 \Lambda$ (normal Zeeman effect) and $\mu_B B_0 \approx \hbar^2 \Lambda$. The change appears only for higher powers of β.

4.2.2 Strong Spin–Orbit Coupling

We now demand that

$$\hbar^2 \Lambda(\gamma, LS) \gg k_B T, \ \mu_B B_0$$

This is the case, which is discussed normally as the *Langevin paramagnetism*, which is realized for the $4f$-systems in normal fields. One does not have a uniform distribution over the fine structure terms of the (LS)-multiplets any more but, only the lowest term is actually occupied to a certain degree. Further, we are in the region of anomalous Zeeman effect, the non-diagonal terms of S^z play only a marginal role (see Sect. 2.6) and J is still a "good" quantum number.

For the energies in question, we have according to (2.303),

$$E_{\gamma LSJM_J} = E^{(0)}_{\gamma LSJ} + g_J(L,S)\, M_J\, \mu_B\, B_0 \tag{4.124}$$

With this, we get for the partition function

$$Z_1 = e^{-\beta E^{(0)}_{\gamma LSJ}} \sum_{M_J=-J}^{+J} e^{-\beta g_J M_J \mu_B B_0} \tag{4.125}$$

Because of the assumption made above, it is necessary to take into account only one J, namely the one which is energetically most favourable according to the third Hund's rule. The partition function Z_1 is then calculated in the same manner as demonstrated for (4.97):

$$Z_1 = e^{-\beta E^{(0)}_{\gamma LSJ}} \frac{\sinh(\beta g_J \mu_B B_0(J+\frac{1}{2}))}{\sinh(\frac{1}{2}\beta g_J \mu_B B_0)} \tag{4.126}$$

This results in the *magnetization*

$$M = M_0 B_J(\beta g_J \; J \; \mu_B \; B_0) \tag{4.127}$$

$$M_0 = n \; J \; g_J \; \mu_B \tag{4.128}$$

M_0 is the *saturation magnetization.*

The susceptibility is obtained, as usual, by differentiating with respect to the field B_0. Again, what is interesting is the high-temperature behaviour, which, just as in (4.107), gives the *Curie law* ($\beta \mu_B B_0 \ll 1$)

$$\chi = \frac{C}{T} \tag{4.129}$$

$$C = n \; \mu_0 \frac{p^2_{eff}}{3k_B} \mu^2_B \quad (\mathit{Curie\ constant}) \tag{4.130}$$

The interpretation of the effective magneton number is of course now different:

$$p_{eff} = g_J \sqrt{J(J+1)} \tag{4.131}$$

The Curie law is experimentally uniquely confirmed. It is interesting to compare the order of magnitudes of the Pauli paramagnetism which was discussed in Sect. 4.1 and the Langevin paramagnetism. From (4.14), (4.129), and (4.130) we have

$$\frac{\chi_{Pauli}}{\chi_{Langevin}} = \frac{9}{2} \frac{1}{g^2_J J(J+1)} \frac{k_B T}{\varepsilon_F} \tag{4.132}$$

Therefore, in general, $\chi_{Langevin} \gg \chi_{Pauli}$.

4.2.3 Van Vleck Paramagnetism

We want to close this chapter by discussing another special, however, somewhat subtle issue of paramagnetism. This is about the *van Vleck paramagnetism*, which leads, in the first approximation, to a temperature-independent susceptibility. This is observed in systems, where the localized magnetic moment arises due to an electron shell, which is one electron short of being half-filled. That means, the ground state term

$$J = S|2l - p| = 0 \tag{4.133}$$

is non-magnetic ($p = 2(2l + 1)$ means completely filled shell). This is the case, e.g. for Eu^{3+} in Eu_2O_3, where, there are six $4f$-electrons. Van Vleck's paramagnetism can be understood, if the preconditions of the last section are, to some extent, relaxed, i.e. when the fine structure splitting is not very large compared to k_BT any more. In this situation, one has to note that

(a) the higher terms of the multiplets cannot be neglected any more;
(b) the operator S_z (in H_z) induces transitions among the fine structure terms.

Since we can, as before, restrict ourselves to a single multiplet, the partition function is given by

$$Z_1 = e^{-\beta E^{(0)}_{\gamma LS}}\, Tr\left(e^{-\beta(H_{SO}+H_z)}\right) \tag{4.134}$$

The fine structure splitting shall be, even though not large compared to k_BT, but still large compared to $\mu_B B_0$. We, therefore, expand the exponential in the trace up to the linear term in B_0, that is, up to the linear term in H_z. We denote by $E_J(\Lambda)$ the eigenvalues of H_{SO},

$$E_J(\Lambda) = \frac{\hbar^2}{2}\Lambda(\gamma, LS)(J(J+1) - L(L+1) - S(S+1)) \tag{4.135}$$

so that we get

$$\begin{aligned} Z_1 &= e^{-\beta E^{(0)}_{\gamma LS}} \sum_{n=0}^{\infty} \frac{(-\beta)^n}{n!} \sum_{J,M_J} \langle J M_J | (H_{SO} + H_z)^n | J M_J \rangle \\ &\approx \cdots \langle J M_J | (H^n_{SO} + \sum_{r=0}^{n-1} H^{n-1-r}_{SO} H_z H^r_{SO}) | J M_J \rangle \\ &= e^{-\beta E^{(0)}_{\gamma LS}} \sum_{n=0}^{\infty} \frac{(-\beta)^n}{n!} * \\ &* \sum_J \left\{ (2J+1) E^n_J(\Lambda) + n E^{n-1}_J(\Lambda) \sum_{M_J=-J}^{+J} \langle J M_J | H_z | J M_J \rangle \right\} \end{aligned}$$

As shown in (4.113), the last term vanishes. Therefore, what remains for the partition function is

$$Z_1 \approx e^{-\beta E^{(0)}_{\gamma LS}} \sum_J (2J+1) e^{-\beta E_J(\Lambda)} \tag{4.136}$$

In order to calculate the magnetization,

$$M = n \frac{1}{Z_1} Tr\left(m e^{-\beta H_1}\right)$$

we now need another trace of the form

$$Tr\left(H_z e^{-\beta(H_{SO}+H_z)}\right) \approx \sum_{n=0}^{\infty} \frac{(-\beta)^n}{n!} \sum_{j,M_J} \langle J M_J | (H_z H^n_{SO} + H_z \sum_{r=0}^{n-1} H^{n-1-r}_{SO} H_z H^r_{SO}) | J M_J \rangle \tag{4.137}$$

The action of H_{SO} on the state $|JM_J\rangle$ is known. On the other hand, the action of H_z is not known because of the presence of S_z.

$$\sum_{J,M_J} \langle J M_J | H_z H^n_{SO} | J M_J \rangle = \sum_J E^n_J(\Lambda) \sum_{M_J} \langle J M_J | H_z | J M_J \rangle = 0 \tag{4.138}$$

With this, it follows that

$$Tr\left(H_z e^{-\beta(H_{SO}+H_z)}\right) \approx \sum_{n=0}^{\infty} \frac{(-\beta)^n}{n!} *$$

$$* \sum_{\substack{JM_J \\ J'M_{J'}}} \langle J M_J | H_z | J' M_{J'} \rangle \langle J' M_{J'} | H_z | J M_J \rangle \sum_{r=0}^{n-1} E^{n-1-r}_{J'}(\Lambda) E^r_J(\Lambda)$$

The first matrix element, since H_z and J_z commute, is unequal to zero only for $M_J = M_{J'}$. The rest reduces to

$$Tr\left(H_z e^{-\beta(H_{SO}+H_z)}\right) = \sum_{\substack{J,J' \\ M_J}} |\langle J M_J | H_z | J' M_J \rangle|^2 \, \frac{e^{-\beta E_{J'}} - e^{-\beta E_J}}{E_{J'} - E_J} \tag{4.139}$$

The operator H_z induces transitions between the terms with J and $J' = J \pm 1$. With the help of Wigner–Eckart theorem, one can see that no other quantum numbers come into the picture. S_z is an irreducible operator of rank one ($k = 1$). So its matrix elements are unequal to zero, according to (2.172), only when

$$|J - J'| \leq k \leq J + J'$$

is satisfied. Therefore, only $J = J'$ and $J' = J \pm 1$ come into question.

(a) $J' = J$: Because of the quotients, this must be understood as a limit

$$\lim_{J' \to J} \frac{e^{-\beta E_{J'}} - e^{-\beta E_J}}{E_{J'} - E_J} = -\beta e^{-\beta E_J} \tag{4.140}$$

The remaining matrix element has been calculated, with the help of Wigner–Eckart theorem, in (2.179) :

$$|\langle J M_J | H_z | J M_J \rangle|^2 = (g_J \mu_B M_J)^2 \ B_0^2 \tag{4.141}$$

Therefore, the total contribution of the summand with $J' = J$ in (4.138) is

$$-\sum_{JM_J} e^{-\beta E_J} \beta (g_J \mu_B M_J)^2 B_0^2$$

The non-diagonal elements
(b) $J' = J \pm 1$: are more difficult to evaluate:

$$\begin{aligned}
&\sum_{JM_J} \langle J M_J | H_z | J \pm 1 M_J \rangle|^2 \frac{e^{-\beta E_{J\pm1}} - e^{-\beta E_J}}{E_{J\pm1} - E_J} \\
&= \sum_{\bar{J} M_{\bar{J}}} |\langle \bar{J} \mp 1 M_{\bar{J}} | H_z | \bar{J} M_{\bar{J}} \rangle|^2 \frac{e^{-\beta E_{\bar{J}}}}{E_{\bar{J}} - E_{\bar{J}\mp1}} \\
&- \sum_{JM_J} |\langle J M_J | H_z | J \pm 1 M_J \rangle|^2 \frac{e^{-\beta E_J}}{E_{J\pm1} - E_J}
\end{aligned}$$

Substituting this in (4.139), we get

$$\begin{aligned}
&\sum_{JM_J} \sum_{J'=J\pm1} |\langle J M_J | H_z | J' M_J \rangle|^2 \ \frac{e^{-\beta E_{J'}} - e^{-\beta E_J}}{E_{J'} - E_J} \\
&= -2 \sum_{JM_J} e^{-\beta E_J} \left\{ \frac{|\langle J + 1 M_J | H_z | J M_J \rangle|^2}{E_{J+1} - E_J} \right. \\
&\quad \left. + \frac{|\langle J M_J | H_z | J - 1 M_J \rangle|^2}{E_{J-1} - E_J} \right\}
\end{aligned} \tag{4.142}$$

The non-diagonal matrix elements of H_z are not as easy to calculate as the diagonal matrix elements in (a). The procedure is the same as was used in Sect. 2.6. We give here only the result [4]:

$$
\begin{aligned}
|\langle JM_J|H_z|J-1M_J\rangle|^2 &= \frac{\mu_B^2}{\hbar^2}B_0^2\,|\langle JM_J|S_z|J-1M_J\rangle|^2 \\
&= \mu_B^2 B_0^2 (J^2-M_J^2)\,\frac{(J^2-(L-S)^2)((L+S+1)^2-J^2)}{4J^2(2J+1)(2J-1)}
\end{aligned}
\tag{4.143}
$$

If we substitute $J = l+\frac{1}{2}$, $S=\frac{1}{2}$, $L=1$ and $M_J = m$, we reproduce our earlier result (2.209). The other non-diagonal matrix element is obtained from (4.142) by replacing J by $J+1$.

We now have everything in order to find the magnetization of the paramagnet. The summations over M_J can be performed easily with the help of (4.115):

$$\sum_{M_J=-J}^{+J} M_J^2 = \frac{1}{3}J(J+1)(2J+1) \tag{4.144}$$

$$\sum_{M_J=-J}^{+J} (J^2-M_J^2) = \frac{1}{3}J(2J+1)(2J-1) \tag{4.145}$$

$$\sum_{M_J=-J}^{+J} ((J+1)^2-M_J^2) = \frac{1}{3}(2J+1)(J+1)(2J+3) \tag{4.146}$$

Using the following shorthand notation,

$$
\begin{aligned}
V(J) \equiv \mu_B^2 &\left\{ \frac{(J^2-(L-S)^2)((L+S+1)^2-J^2)}{6J(2J+1)(E_{J-1}-E_J)} \right. \\
&+ \left. \frac{((J+1)^2-(L-S)^2)((L+S+1)^2-(J+1)^2)}{6(J+1)(2J+1)(E_{J+1}-E_J)} \right\}
\end{aligned}
$$

we now finally have the magnetization

$$M = nB_0 \frac{\sum_{J=|L-S|}^{L+S}(2J+1)e^{-\beta E_J}\left\{\frac{J(J+1)}{3k_BT}(g_J\mu_B)^2 + V(J)\right\}}{\sum_{J=|L-S|}^{L+S}(2J+1)e^{-\beta E_J}} \tag{4.147}$$

This result was derived under the precondition

$$\hbar^2\Lambda(\gamma, LS),\ k_BT \gg \mu_B B_0$$

but not necessarily,

$$\hbar^2\Lambda \gg k_BT$$

The susceptibility here is simply $\mu_0 M / B_0$. The temperature-independent correction $V(J)$ is actually responsible for Van Vleck paramagnetism. Let us, for simplicity, assume that the multiplet splitting given by Λ is so large, that, only the lowest term is appreciably occupied. Then, we can restrict ourselves to the first term in the sum. This term, however, because of the non-diagonality of S_z, already contains an admixture of the higher J-levels which are hidden in $V(J)$:

$$\chi \approx n\mu_0 \left\{ \frac{J(J+1)}{3k_B T}(g_J \mu_B)^2 + V(J) \right\} \tag{4.148}$$

The first summand is exactly the result obtained in the last section (Curie law (4.129)) wherein, we have, from the start, neglected the non-diagonal elements of S_z. J is the quantum number of the ground state.

According to the *Lande's interval rule* (2.302), we have

$$E_{J+1} - E_J = \hbar^2 \Lambda(\gamma,\ LS)(J+1) \quad ; \quad E_{J-1} - E_J = \hbar^2 \Lambda(\gamma,\ LS)J \tag{4.149}$$

That means that, the Van Vleck term $V(J)$ is smaller by a factor

$$\frac{\hbar^2\ \Lambda(\gamma,\ LS)}{k_B T}$$

compared to the first term in the bracket of the above expression, which is, the *Langevin paramagnetism* discussed in the preceding section. This factor is, in general, very much smaller than 1 and therefore can be neglected. That is why, Van Vleck paramagnetism is clearly observable, only, if the ground state has $J = 0$, because, in that case, the first summand vanishes. This is always the case, if the electron shell falls short by one electron from being half-filled. Figure 4.6 shows the example of Eu^{3+} in Eu_2O_3.

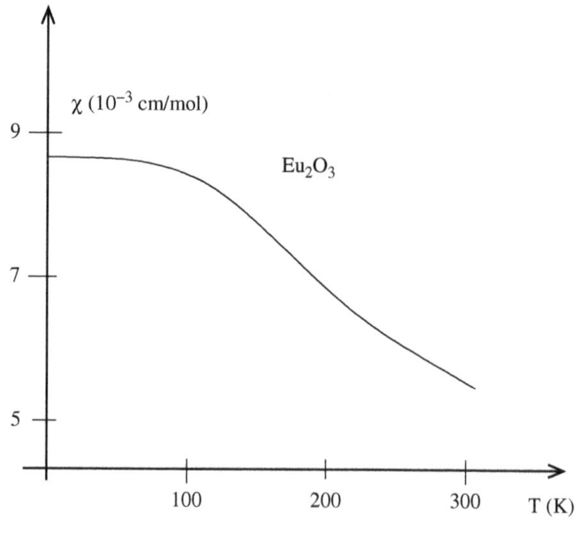

Fig. 4.6 Susceptibility of the Van Vleck paramagnet Eu_2O_3 as a function of temperature (after [5])

At high temperatures, χ is calculated using the complicated formula (4.147), which takes into account, the excited states also. At low temperatures, only the ground state has a role to play and the temperature-*independent* Van Vleck paramagnetism becomes noticeable.

4.3 Problems

Problem 4.1 As a result of the periodic boundary conditions on the volume

$$V = L_x L_y L_z$$

the wavenumbers **k** become discrete:

$$q_{x,y,z} = \frac{2\pi}{L_{x,y,z}} n_{x,y,z} \qquad \text{with } n_{x,y,z} \in \mathbb{Z}$$

Deduce the orthogonality relation

$$\frac{1}{V} \int_V d^3r \, e^{i(\mathbf{q}-\mathbf{q}')\cdot\mathbf{r}} = \delta_{\mathbf{q},\mathbf{q}'}$$

Problem 4.2 Consider a crystalline solid, i.e. the surface effects do not play a role. Let $\mathbf{a}_1$, $\mathbf{a}_2$ and $\mathbf{a}_3$ be the primitive translation vectors of the Bravais lattice from which the lattice vectors

$$\mathbf{R}^{\mathbf{n}} = \sum_{i=1}^{3} n_i \mathbf{a}_i$$

are constructed, where $\mathbf{n} = (n_1, n_2, n_3)$ holds and n_i are integers. Due to the assumed translational symmetry, we can imagine the solid to be a parallelpiped spanned by the three vectors $\mathbf{A}_i = N_i \mathbf{a}_i$ with $N_i \gg 1$ and even. That means for the lattice vectors: $n_i = 0, 1, 2, \cdots, N_i - 1$.

1. The solutions (Bloch functions) for the non-interacting electrons should satisfy the periodic boundary conditions of the crystal on the surface. What does it mean for the allowed wavevectors **k**? Which wavevectors lie within the first Brillouin zone?
2. Deduce the orthogonality relation

$$\frac{1}{N} \sum_{\mathbf{n}} e^{i(\mathbf{k}-\mathbf{k}')\cdot\mathbf{R}^{\mathbf{n}}} = \delta_{\mathbf{k},\mathbf{k}'}$$

3. Deduce the orthonormality relation

$$\frac{1}{N}\sum_{\mathbf{k}}^{1.BZ} e^{i\mathbf{k}\cdot(\mathbf{R^n}-\mathbf{R^m})} = \delta_{\mathbf{n,m}}$$

Problem 4.3 Calculate the following frequently used integrals:

1.

$$I_1 = \int_V d^3r \int_V d^3r' \frac{e^{-\alpha|\mathbf{r}-\mathbf{r}'|}}{|\mathbf{r}-\mathbf{r}'|} \quad ; \quad \alpha > 0$$

2.

$$I_2 = \int_V d^3r \int_V d^3r' \frac{e^{i(\mathbf{q}\cdot\mathbf{r}+\mathbf{q}'\cdot\mathbf{r}')}}{|\mathbf{r}-\mathbf{r}'|}$$

In both the cases let V be finite with $V \to \infty$.

Problem 4.4 In the ground state, N_e electrons occupy the states within a Fermi sphere of radius k_F (Sommerfeld model). For this, calculate the following integral:

$$I = \int d^3r \int_{FS} d^3k \int_{FS} d^3k' \frac{e^{i(\mathbf{k}-\mathbf{k}')\cdot\mathbf{r}}}{r}$$

Here the position integral should be taken over the entire space and the wavevector integral over the occupied states of the Fermi sphere (FS).

Problem 4.5 Let S be the spin of a partially filled electron shell so that it is either an integer or half-integer. Then show that the following holds:

$$\sum_{n=-S}^{+S} n^2 = \frac{1}{3} S(S+1)(2S+1)$$

Problem 4.6 A classical thermodynamic system consists of N atoms in volume V. Each of the atoms carries a magnetic moment μ_i ($|\ \mu_i\ | = \mu \ \ \forall\, i$). The Hamiltonian function is composed of two parts,

$$H(\mathbf{q},\mathbf{p}) = H_0(\mathbf{q},\mathbf{p}) + H_1(\mathbf{q},\mathbf{p})$$

out of which $H_0(\mathbf{q},\mathbf{p})$ describes the system in the absence of a magnetic field, while $H_1(\mathbf{q},\mathbf{p})$ describes the influence of a homogeneous magnetic field $\mathbf{B} = B\mathbf{e_z}$.

1. How does the field term H_1 read?
2. Calculate the canonical partition function.

3. Determine the temperature and field dependences of the average total magnetic moment:

$$\mathbf{m} = \langle \sum_{i=1}^{N} \mu_i \rangle$$

4. Discuss the total magnetic moment in the limits $\beta\mu B \gg 1$ and $\beta\mu B \ll 1$. (*Classical Langevin paramagnetism*).

Problem 4.7 Assume that a paramagnetic substance has the susceptibility $\chi_T = \frac{M}{H} = \chi_T(T)$.

1. Calculate magnetization dependence of the free energy.
2. Derive from that the corresponding dependencies of the internal energy and entropy.
3. Show that the Curie law is in conflict with the third law of thermodynamics, i.e. it can*not* be valid for arbitrarily low temperatures.

Problem 4.8 Let a magnetic moment system be specified by the variables temperature T, magnetic field H and magnetization M. (The pressure p and the volume V be constant and irrelevant for the following).

1. Assume that the internal energy $U = U(T, M)$ and the equation of state of the form $M = f(T, H)$ are known. Formulate the difference of the heat capacities $c_H - c_M$.
2. What is the result specially for the ideal paramagnet?

$$M = \frac{C}{T} H \ ; \ \left(\frac{\partial U}{\partial M}\right)_T = 0 \qquad C : \text{Curie constant}$$

3. Prove the following relations:

(a)

$$\left(\frac{\partial S}{\partial M}\right)_T = -\mu_0 V \left(\frac{\partial H}{\partial T}\right)_M$$

(b)

$$\left(\frac{\partial S}{\partial H}\right)_T = \mu_0 V \left(\frac{\partial M}{\partial T}\right)_H$$

(c)

$$\left(\frac{\partial S}{\partial M}\right)_T = \frac{1}{T}\left[\left(\frac{\partial U}{\partial M}\right)_T - \mu_0 V H\right]$$

4. With a part of the above results, verify the relation

$$c_H - c_M = -\mu_0 V T \left(\frac{\partial H}{\partial T}\right)_M \left(\frac{\partial M}{\partial T}\right)_H$$

5. For calculating $c_H - c_M$ use the equation of state

$$H = \frac{1}{C}(T - T_C)\, M + b\, M^3$$

   C, T_C and b are positive quantities!
6. Show that for such an equation of state c_M can *not* depend on M!
7. From the equation of state given in 5., calculate the free energy $F = F(T, M)$ and the entropy $S = S(T, M)$.
8. Show that the equation of state given in 5., in a certain temperature region, for $H \to 0$ in addition to the obvious solution $M = 0$, also has a non-trivial solution $M = M_S \neq 0$. Discuss the stability of the two solutions by comparing the free energies!
9. Calculate the temperature dependence of the isothermal susceptibility χ_T and the difference $c_H - c_M$ of the heat capacities in the limit $H \to 0$!

Problem 4.9 For the ideal paramagnet ($M = \frac{C}{T} H$, $c_M(T, M = 0) = \gamma T$) discuss *adiabatic demagnetization*

1. The paramagnet is placed in a heat bath $B(T_1)$. What is the heat released when the magnetic field is increased from zero to $H \neq 0$?
2. The system is decoupled from the heat bath and the field is switched off adiabatically reversibly. Calculate the final temperature T_f.

References

1. A. L. Fetter, J. D. Walecka, "Quantum Theory of Many-Particle Systems", McGraw-Hill, New York, 1971, p. 21
2. S. D. Silverstein, Phys. Rev. **130**, 1703 (1963)
3. R. T. Schumacher, W. E. Vehse, J. Phys. Chem. Solids **24**, 297 (1963)
4. E. U. Condon, G. H. Shortley, "The Theory of Atomic Spectra", Cambridge University Press, London, 1959
5. A. S. Borovik-Romanov, N. M. Kreines, Sov. Phys. JETP **2**, 657 (1956)

Chapter 5
Exchange Interaction

We could understand the concepts of *diamagnetism and paramagnetism* without assuming any explicit interaction among the magnetic moments (or among the electrons that build these moments). Even though the interactions may lead to drastic corrections (see Sect. 4.1.3), the basic phenomena themselves are not affected by it. It is completely different in the case of

Ferromagnetism
Ferrimagnetism
Antiferromagnetism

The signature of these forms of magnetism is the existence of *spontaneous* ordering of the magnetic moments in a solid at a temperature below a critical temperature T^*. The critical temperature is given different names as follows.

For both ferro- and ferrimagnetism, $T^* = T_c$ is called the *Curie temperature* and for antiferromagnetism, $T^* = T_N$ is called the *Neel temperature*.

Above T^*, the spontaneous ordering vanishes and the material behaves like a normal paramagnet. These magnetic phenomena cannot be explained without interactions. That is why one speaks of *cooperative* or *collective phenomena*.

The interactions which are decisive for the collective magnetism are called *exchange interactions*.

As we have already learnt in Sect. 4.1.3, the interaction is essentially the electrostatic Coulomb interaction. The matrix elements of this interaction, constructed with completely antisymmetrized wavefunctions, contain terms which are classically not understandable and correspond to an *exchange* of the indices of the identical particles (Fermions). Such *exchange phenomena* were discovered independently in 1926 by *Dirac* and *Heisenberg* to be decisive for the collective magnetism. This chapter is focussed on these fundamental phenomena.

Till today, there exists no closed theory of magnetism which can describe the totality of all the phenomena in a unified manner.

Model building is unavoidable and the models are, in general, relevant only for particular special features of magnetism.

To get a better overview, we begin with a kind of *classification*.

W. Nolting, A. Ramakanth, *Quantum Theory of Magnetism*,
DOI 10.1007/978-3-540-85416-6_5, © Springer-Verlag Berlin Heidelberg 2009

At the outset, there is a necessary precondition for collective magnetism in solids (liquid (anti) ferromagnets are not considered here)

There must exist permanent magnetic moments.

Just like in the cases of dia- and paramagnets, we distinguish between the magnetism of insulators and metals.

Insulators

The magnetism is produced by *localized* magnetic moments of a partially filled electron shell, e.g 3d-, 4d-, 4f- and 5f-shells.

Examples:

Ferro: $CrBr_3$, K_2CuF_4, EuO, EuS, $CdCr_2Se_4$, Rb_2CrCl_4, etc.

Antiferro: $EuTe$, MnO, $RbMnF_3$, Rb_2MnCl_4, etc.

Ferri: $EuSe$, etc.

These substances are very well described by the *Heisenberg model*:

$$H_f = -\sum_{i,j} J_{ij}\, \mathbf{S}_i \cdot \mathbf{S}_j$$

where J_{ij} is the exchange integral. The Heisenberg model will be studied in detail in Chap. 7.

The physical justification of the coupling constants J_{ij} will be the central theme of this chapter.

The Heisenberg Hamiltonian should be understood as an *effective* operator. The spin–spin interaction $\mathbf{S}_i \cdot \mathbf{S}_j$ acting on the corresponding spin states, simulates the contribution of the exchange matrix elements of the Coulomb interaction, which are believed to be responsible for the spontaneous magnetization (Fig. 5.1).

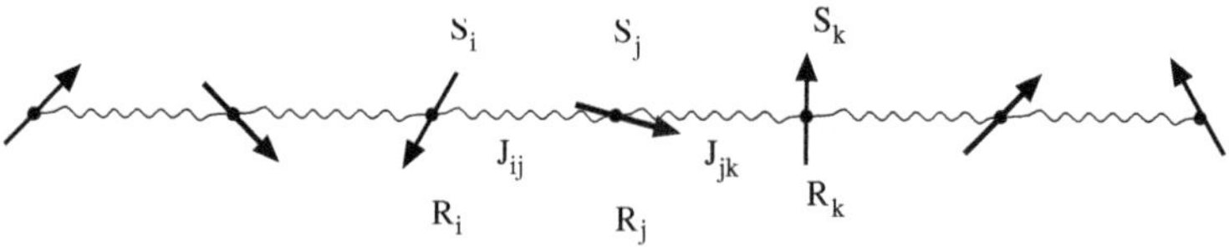

Fig. 5.1 Schematic plot of a linear Heisenberg chain of localized spins, coupled by exchange interaction J_{ij}. $\mathbf{S}_i$: spin operators, $\mathbf{R}_i$: lattice sites

Metals

Band Magnetism

Typical for this class is the fact that the same electrons are responsible for the electrical conduction and also for the magnetism. Examples are Fe, Co, Ni and their alloys.

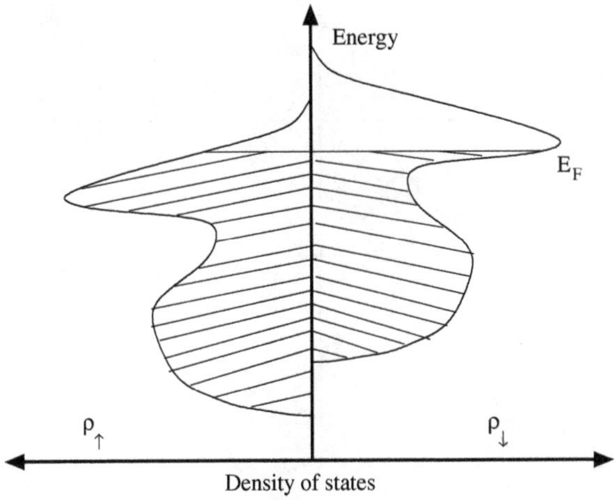

Fig. 5.2 Spin-polarized densities of states of a band-ferromagnet for a temperature below T_c

Here the exchange interaction produces a spin-dependent band shift for $T < T_c$ so that one particular spin orientation becomes preferred for the band electrons.

The simplest, even though, not exactly solvable model for this class is the *Hubbard model*

$$H_s = \sum_{ij\sigma} T_{ij}\, c_{i\sigma}^{\dagger}\, c_{j\sigma} + \frac{1}{2}\, U \sum_{i\sigma} n_{i\sigma}\, n_{i-\sigma}$$

where $c_{i\sigma}^{\dagger}$ is the creation operator for an electron of spin σ at the lattice site $\mathbf{R}_i$, $c_{j\sigma}$ the annihilation operator for an electron of spin σ at the lattice site $\mathbf{R}_j$ and $n_{i\sigma} = c_{i\sigma}^{\dagger}\, c_{i\sigma}$ is the electron occupation number operator at the lattice site $\mathbf{R}_i$.

In this model, the Coulomb interaction is taken into account in a simplified, intra-atomic version because a strong and strongly screened Coulomb interaction is considered as being responsible for the spin-dependent splitting of the relevant energy band (see Fig. 5.2). The influence of the lattice potential is present in the so-called "hopping integrals" T_{ij}. The Hubbard model is investigated in detail in Chap. 8.

Localized Magnetism

If magnetism and electrical conduction are carried by *different* groups of electrons, then one speaks of "localized" or "local-moment" magnetism (Fig. 5.3). Example for this type is the metallic 4f-system Gd. An appropriate model for this situation is the *s-f model*, sometimes also called *Kondo-lattice model*

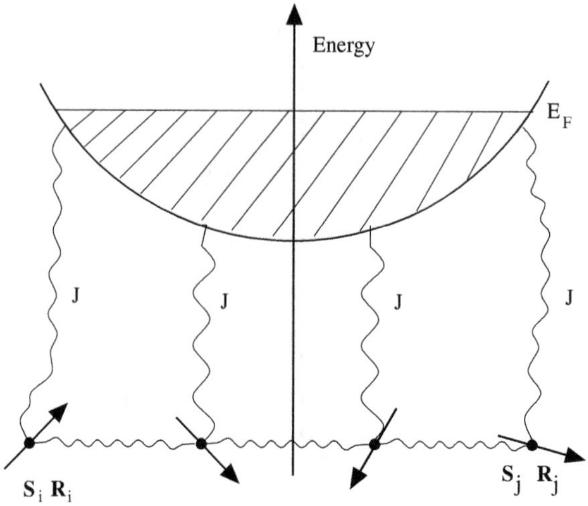

Fig. 5.3 Schematic plot of a local-moment system exchang-coupled to itinerant band electrons

$$H = H_s + H_f - J \sum_i \sigma_i \cdot \mathbf{S}_i$$

where σ_i is the spin operator for the conduction electron at the position $\mathbf{R}_i$ and $\mathbf{S}_i$ is the localized spin. J is the corresponding coupling constant, which can be positive as well as negative.

5.1 Phenomenological Theories

5.1.1 The Exchange Field

We want to understand the physical origin of the interaction, which forces a ferromagnet into an ordered state below T_c. In the case of paramagnets, such an ordering is caused by an external magnetic field. Therefore, it is reasonable to assume that, inside a ferromagnet, an *internal* magnetic field is produced, which, from now on, we call the *exchange field*, which orients the existing permanent magnetic moments. Before we inquire into the origin of this field, we want to make an estimate of its magnitude.

The temperature T_c, at which the phase transition ferromagnetism $\Leftrightarrow$ paramagnetism takes place, is derived from the general thermodynamic condition on the free energy:

$$F = U - TS \stackrel{!}{=} minimum \tag{5.1}$$

Two influences are decisive

(a) exchange field $B_{ex} \rightarrow$ order $\rightarrow U$ minimal
(b) temperature $\rightarrow$ disorder $\rightarrow S$ maximal

so that we can assume by plausibility:

$$T \lesseqgtr T_C \Leftrightarrow (a) \gtrless (b)$$

At T_C, the thermal energy $k_B T_C$ and the field energy $\mu_B B_{ex}$ obviously balance each other. Thus, T_C is a measure of the strength of the ferromagnetic coupling, whose order of magnitude can be found out with a simple consideration:

$$E_{ex} \approx \mu_B \, B_{ex} \approx k_B \, T_C \tag{5.2}$$

With

$$\mu_B = 0.579 * 10^{-4} \, \frac{eV}{T} \quad ; \quad k_B T = 0.862 * 10^{-4} \, \frac{eV}{K} \tag{5.3}$$

one can estimate B_{ex} with the help of the experimentally known values of T_C (see Table 5.1).

It should be noted that the largest fields created in the laboratory are $10\ T < B_0 < 20\ T$. Thus, the exchange fields are enormous. So, what is the origin of the exchange field? Since the system contains permanent magnetic moments, the first possibility is that it is a result of the classical dipole–dipole interaction. A simple estimate makes it immediately clear that this classical interaction *cannot* be the cause of ferromagnetism.

The dipole field at $\mathbf{R}_i$ due to a moment $\mathbf{m}_j$ localized at $\mathbf{R}_j$ is given by

$$\frac{\mu_0}{4\pi}\left[\frac{3(\mathbf{m}_j \cdot \mathbf{r}_{ij})\mathbf{r}_{ij} - \mathbf{r}_{ij}^2 \mathbf{m}_j}{r_{ij}^5}\right] \quad ; \quad \mathbf{r}_{ij} = \mathbf{R}_i - \mathbf{R}_j$$

Then, the moment $\mathbf{m}_i$ localized at $\mathbf{R}_i$, in the field of all the other moments, has the energy

Table 5.1 Estimation of the exchange field B_{ex} from the Curie temperature of some prominent ferromagnets. The numbers for B_{ex} follow from (5.2)

	$T_c(K)$	$k_B T_c(meV)$	$B_{ex}(T)$
Fe	1043	89.907	1552.79
Co	1393	120.077	2073.86
Ni	631	54.392	939.42
Gd	290	24.998	431.74
EuO	69.33	5.976	103.22

$$E_D^{(i)} = -\mathbf{m}_i \cdot \mathbf{B}_D(\mathbf{R}_i) = \frac{\mu_0}{4\pi} \sum_j{}' \frac{r_{ij}^2(\mathbf{m}_i \cdot \mathbf{m}_j) - 3(\mathbf{m}_i \cdot \mathbf{r}_{ij})(\mathbf{m}_j \cdot \mathbf{r}_{ij})}{r_{ij}^5} \tag{5.4}$$

The lattice sum that is appearing in the above equation can be exactly performed for simple lattices. We restrict ourselves here to a simple estimate

$$\left| E_D^{(i)} \right| \approx \frac{\mu_0}{4\pi} \frac{(p_{eff}\, \mu_B)^2}{r^3} z = 0.5371 \times 10^{-4} \frac{z\, p_{eff}^2}{r^3}\, eV\, \text{Å}^3 \tag{5.5}$$

where z is the number of the nearest neighbours, p_{eff} is the effective number of magnetons, $r \approx 2$Å is the typical distance between nearest neighbours. With this, we get

$$\left| E_D^{(i)} \right| \approx 10^{-4} eV\; (\widehat{=} 1.16K,\; \widehat{=} 1.73T) \tag{5.6}$$

From this estimate, it is clear that dipole interaction *cannot* be the origin of ferromagnetism. It can only be a correction to the actual exchange interaction and represents an anisotropic effect which has importance in itself for many practical applications. However, it has nothing to do with the *origin* of ferromagnetism. The concept of the "exchange field" obviously fails. Nevertheless, we will see in the next section that it can help in a good qualitative *description* of ferromagnetism, however, not in an *explanation* of it.

5.1.2 Weiss Ferromagnet

The first phenomenological theory of a ferromagnet, which is able to qualitatively explain the *phase transition "ferromagnetism ⇔ paramagnetism"*, is due to P. Weiss [1]. Without knowing about the exchange interaction in detail, or even wanting to investigate it, its existence is simply postulated in the form of an exchange field $\mathbf{B}_{ex}$.

5.1.2.1 Weiss Model

The postulates of the Weiss model are as follows:

1. There exists an exchange field $\mathbf{B}_{ex} = \mu_0 \mathbf{H}_{ex}$, which is proportional to the macroscopic magnetization $\mathbf{M}$

$$\mathbf{B}_{ex} = \lambda \mu_0 \mathbf{M} \tag{5.7}$$

2. The permanent magnetic moments of the ferromagnet respond to $\mathbf{B}_{ex}$ just as the moments in a Langevin paramagnet (see Sect. 4.2.2) respond to an external field.

Point (2) means according to Eq. (4.127),

$$M(T) = M_0 B_J(\beta J g_J \mu_B B_{ex}) = M_0 B_J(\beta J g_J \mu_B \mu_0 \lambda M(T)) \tag{5.8}$$

$$M_0 = \frac{N}{V} J g_J \mu_B \quad \text{“}saturation\ magnetization\text{”} \tag{5.9}$$

The assumption of the Brillouin function B_j is of course “historically incorrect” because P. Weiss had to use the classical (!) Langevin function L (4.102). Equation (5.9) is an implicite equation to determine $M(T)$, which we will now investigate in detail.

(a) $M(T) = 0$ is always a solution since $B_J(0) = 0$. Do other solutions exist?
(b) *Graphical solution*: Since $\lim_{M\to\infty} B_J \to 1$, a solution $M \neq 0$ exists only if the initial slope of the $M_0 B_J$ curve is greater than 1 (Fig. 5.4). That provides a criterion for *spontaneous magnetization* $M_S(T)$:

$$\left.\frac{d}{dM}(M_0 B_J)\right|_{M=0} \overset{!}{\geq} 1 \tag{5.10}$$

From this condition, we can derive a relation for T_C. $M \to 0$ means according to (4.103)

$$M(T) \approx M_0 \frac{J+1}{3J}\, \beta J g_J \mu_B \mu_0 \lambda M = \lambda\, C\, \frac{M}{T} \tag{5.11}$$

where C is the *Curie constant* (4.130)

$$C = n\, \mu_0\, \frac{p_{eff}^2 \mu_B^2}{3k_B} \quad ; \quad p_{eff} = g_J \sqrt{J(J+1)} \tag{5.12}$$

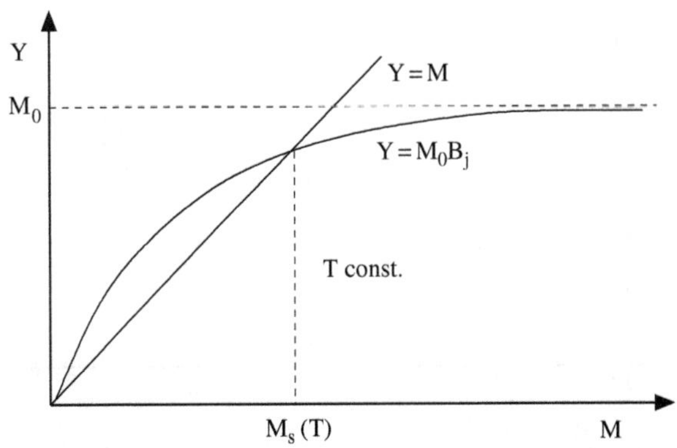

Fig. 5.4 Graphical determination of the spontaneous magnetization of a Weiss ferromagnet

The *criterion for ferromagnetism* is the following. There exists a finite temperature with

$$\frac{\lambda C}{T_C} = 1 \Leftrightarrow T_C = \lambda C \tag{5.13}$$

at which the magnetic phase transition takes place:

$$\begin{aligned} T < T_C &\rightarrow \tfrac{\lambda C}{T} > 1 \rightarrow ferromagnetism \\ T > T_C &\rightarrow \tfrac{\lambda C}{T} < 1 \rightarrow paramagnetism \end{aligned} \tag{5.14}$$

The exchange parameter λ is fixed by the experimentally measured quantity T_C.

If there exists a solution $M_S > 0$, then $-M_S$ is also a solution, so that one has a total of three solutions, namely, $M = 0, \pm M_S$. With the help of a simple thermodynamic argument, one can see that the magnetic solutions $\pm M_S$ must be stable. The free energy F takes the extremal values at the points of solution (see Problem 5.5). Further, the free energy as a function of M diverges at $\pm M_0$, since the magnetization cannot be greater than M_0 (or less than $-M_0$ as the case may be). Therefore, F has a maximum at $M = 0$ and has minima at $\pm M_S$. Therefore, the latter ones are the stable solutions (Fig. 5.5).

(c) *Low temperature behaviour*: The magnetization curve $M_S = M_S(T)$ for $T \leq T_C$ can be easily found graphically. We define

$$x = \beta J g_J \mu_B \mu_0 \lambda M_S(T)$$

Therewith we can represent the normalized spontaneous magnetization in two different ways as a function of x:

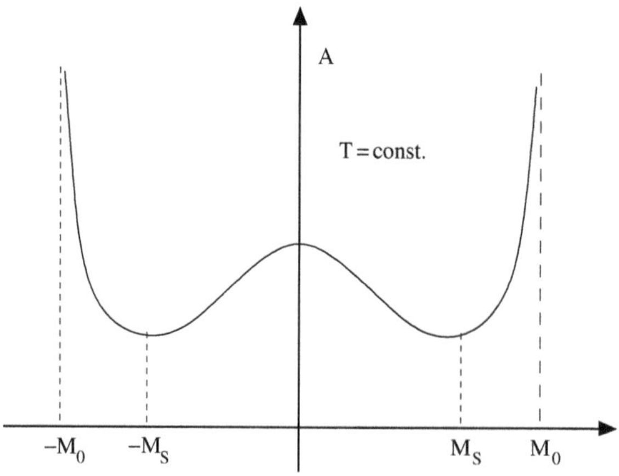

Fig. 5.5 Schematic plot of the free energy of the Weiss ferromagnet as a function of the magnetization M

$$Y_1 = \frac{M_S(T)}{M_0} = B_J(x) \quad ; \quad Y_2 = \frac{M_S(T)}{M_0} = x\frac{J+1}{3J}\frac{T}{T_C} \tag{5.15}$$

The points of intersection of the curves $Y_1(x)$ and $Y_2(x)$ give the magnetization curve (Fig. 5.6).

(d) *High temperature behaviour*: For $T > T_c$, there is no solution with $M_S \neq 0$. The system shows paramagnetic behaviour. For $T > T_C$, $M \neq 0$ can be achieved only through an external field:

$$M(T, B_0) = M_0 \ B_J \left(\beta J g_J \mu_B \mu_0 (\lambda M + \frac{1}{\mu_0} B_0) \right) \tag{5.16}$$

At high temperature ($\beta \mu_B B_0 \ll 1$) we can expand $B_J(x)$ and get

$$M(T, B_0) \approx \frac{C}{T} \left(\lambda \ M(T, B_0) + \frac{1}{\mu_0} B_0 \right)$$

which gives

$$M(T, B_0) \approx \frac{1}{\mu_0} \ B_0 \ \frac{C}{T - T_C}$$

From this follows the *Curie–Weiss* law for the susceptibility:

$$\chi(T) = \mu_0 \left(\frac{\partial M}{\partial B_0} \right)_T = \frac{C}{T - T_C} \tag{5.17}$$

The Curie–Weiss law for the high-temperature behaviour of the susceptibility is experimentally uniquely established. It can be extrapolated in order to experimentally determine the Curie temperature T_C (Fig. 5.7).

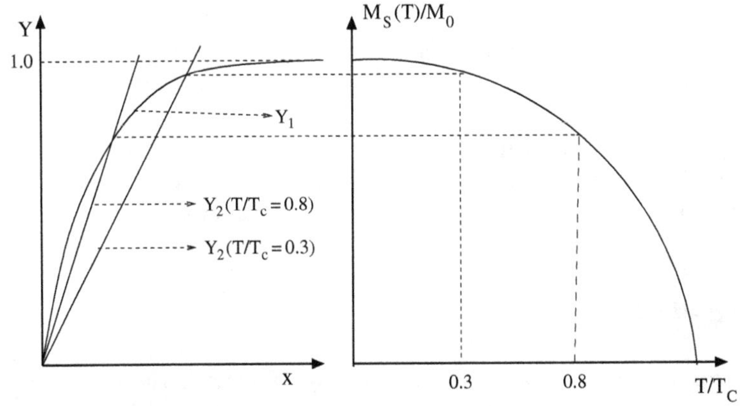

Fig. 5.6 Graphical solution of the spontaneous magnetization in the Weiss model

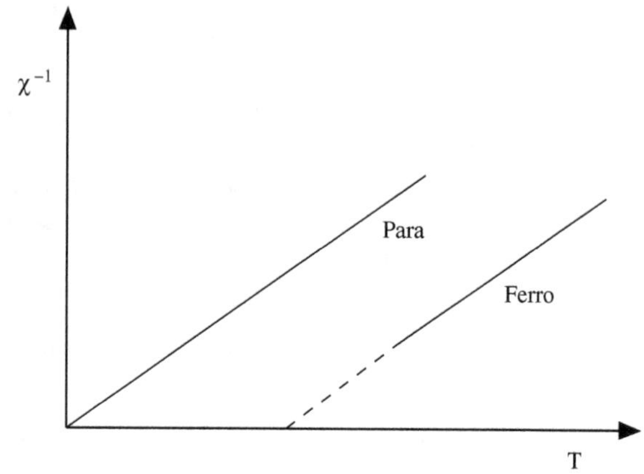

Fig. 5.7 Temperature behaviour of the inverse susceptibility of a ferromagnet and a paramagnet, respectively, at high temperatures.

5.2 Direct Exchange Interaction

The Weiss model of ferromagnetism provides, qualitatively, rather good results (magnetization curve, Curie–Weiss law). Looked into, in detail, naturally, it is too simple. Not only that, it does *not explain* how the ferromagnetic coupling comes into being, but simply *postulates* the coupling in the form of the exchange field, which is proportional to M.

The exchange interaction, which is responsible for the spontaneous magnetization, cannot be explained classically. It is of pure quantum mechanical origin, but at the same time, it is purely electrostatic in nature. It is a direct consequence of the Pauli's principle. The Pauli principle demands that the matrix elements of the electrostatic Coulomb interaction between charged particles in a solid should be constructed with fully antisymmetrized wavefunctions. Among these matrix elements, there are certain of them, which correspond in a sense, to the *exchange* of the particle indices. As we will see, these are responsible for the phenomenon of collective magnetism. That is why we can say that

origin of ferromagnets: electric fields
action of ferromagnets: magnetic fields

In explaining ferromagnetism, one can, to a good approximation, neglect the magnetic fields that are actually present in a solid (e.g. the dipole fields). We want to probe the phenomenon with a few qualitative considerations.

5.2.1 *Pauli's Principle*

We first consider a band magnet, i.e. a ferromagnetic metal. The Pauli's principle takes care that the electrons with spins parallel to each other, if they belong to the same energy band, therewith having otherwise the same set of quantum numbers, do not come too close to each other. This leads to a reduction in the Coulomb repulsion of particles of like charge, and consequently there is a gain in the potential energy ΔE_{pot}:

$$\Delta E_{pot} \approx E_{pot}(\uparrow\downarrow) - E_{pot}(\uparrow\uparrow) \tag{5.18}$$

Then, the question is, if complete spin ordering is so energetically advantageous, why is it that not all the metals are ferromagnetic? The answer lies in the kinetic energy, for which in general, ferromagnetic ordering is not advantageous because of an increase in single particle energies (Fig. 5.9 in comparison to Fig. 5.8). So there is a counteracting competition between the potential and the kinetic aspects.

For some materials of *appropriate band structure*, the reduction in the potential energy can be greater than the increase in the kinetic energy, so that an ordered state is energetically favourable compared to the disordered state. An appropriate

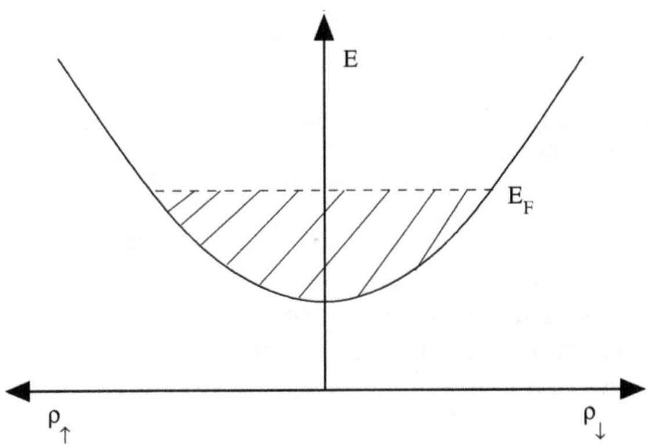

Fig. 5.8 Density of states of a paramagnetic metal, the structure of which is inconvenient to ferromagnetism

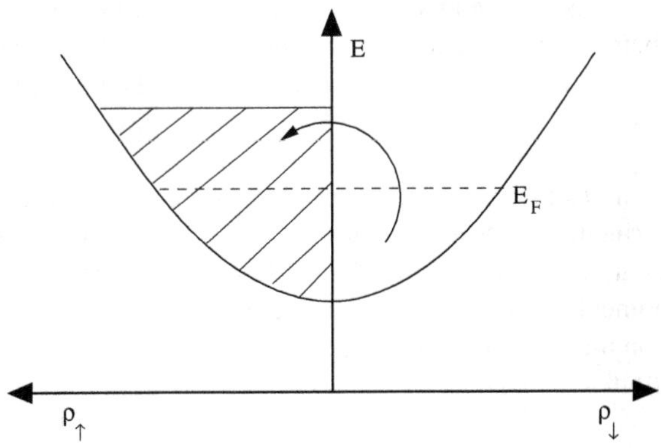

Fig. 5.9 Artificial spin polarization of metallic electrons to demonstrate the increase of single particle energy

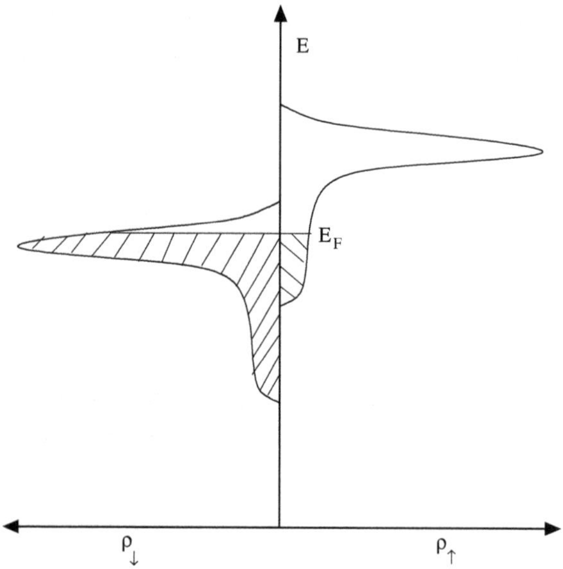

Fig. 5.10 Convenient model density of states for the appearance of band ferromagnetism in accordance with the Stoner criterion (5.20)

density of states would be such that the Fermi energy lies within a distinct peak of the density of states (Fig. 5.10). That is because, then, relatively many electrons can flip their spins without gaining too much single particle (kinetic) energy. That is the qualitative explanation for ferromagnetism. That can be summarized by the so-called *Stoner criterion*

$$U\rho(\varepsilon_F) > 1 \tag{5.19}$$

which we will later derive explicitly. This criterion says that if the intra-atomic Coulomb interaction is large and strongly screened, then, the loss in the potential energy is large being favourable for ferromagnetism, and at the same time, a sharp density of states at the Fermi level is convenient because then, relatively large number of $\downarrow$-electrons can be "converted" into $\uparrow$-electrons without increasing the kinetic energy by too large an amount.

We want to demonstrate now with a simple example of a *two-electron system*, how the Pauli's principle can lead to magnetic effects, even though, the Hamiltonian is spin independent, therefore being unable to describe a *direct* interaction between the magnetic moments. The two electrons are identical Fermions. Therefore, the total wavefunction must be antisymmetric against the interchange of particle labels. Since the Hamiltonian

$$H = \sum_{i=1}^{2} \frac{p_i^2}{2m} + V(\mathbf{r}_1, \mathbf{r}_2) \tag{5.20}$$

is spin independent, the wavefunction $|\psi\rangle$ must be a product of the space and the spin parts:

$$|\psi\rangle = |q\rangle^{(\pm)}\, |S; m_S\rangle^{(\mp)} \tag{5.21}$$

Here the space part is $|q\rangle$ and the spin part is $|S; m_S\rangle$. The spin part can be explicitly given. In a two-electron system, we have two possibilities $S = 0$ and $S = 1$. Then we can construct an *antisymmetric singlet state*

$$|0;0\rangle = \frac{1}{\sqrt{2}}(|\uparrow\downarrow\rangle - |\downarrow\uparrow\rangle) \tag{5.22}$$

and a *symmetric triplet state*

$$\begin{aligned} |1;1\rangle &= |\uparrow;\uparrow\rangle \\ |1;0\rangle &= \tfrac{1}{\sqrt{2}}(|\uparrow\downarrow\rangle + |\downarrow\uparrow\rangle) \\ |1;-1\rangle &= |\downarrow\downarrow\rangle \end{aligned} \tag{5.23}$$

The symmetry of the spin part fixes the symmetry of the corresponding space part $|q\rangle$, since the total wavefunction $|\psi\rangle$ must be antisymmetric. There are altogether four eigensolutions:

$$\begin{aligned} |\psi_1\rangle &= |q\rangle^{(+)}|0;0\rangle \\ |\psi_2(m_S)\rangle &= |q\rangle^{(-)}|1;m_S\rangle \quad ; \quad m_S = 0,\ \pm 1 \end{aligned} \tag{5.24}$$

H operates only on the space part

$$H|q\rangle^{(\pm)} = E_{\pm}|q\rangle^{(\pm)} \tag{5.25}$$

If

$$E_+ \neq E_- \tag{5.26}$$

a case about which we will learn in the next section, then spin ordering becomes automatically preferred which means there exists a *spontaneous* magnetic order. This immediately indicates that we can replace a spin-independent Hamiltonian by an effective Hamiltonian $\tilde{H}$, which, instead of operating exclusively on $|q\rangle$, operates exclusively on both electron spins. If we choose $\tilde{H}$ such that

$$\begin{aligned} \tilde{H}|0;0\rangle &= E_+|0;0\rangle \\ \tilde{H}|1;m_S\rangle &= E_-|1;m_S\rangle \end{aligned} \tag{5.27}$$

is valid, then, obviously $\tilde{H}$ and H are equivalent in their action. How should such an operator look? Let $\mathbf{s}_i$ be the spin of the ith electron ($i = 1, 2$). Then we have

$$\mathbf{s}_i^2 = \hbar^2 s_i(s_i + 1) = \hbar^2 \frac{3}{4}$$

and for the total spin

$$\begin{aligned} \mathbf{S}^2 &= (\mathbf{s}_1 + \mathbf{s}_2)^2 = S(S+1)\hbar^2 = \mathbf{s}_1^2 + \mathbf{s}_2^2 + 2\mathbf{s}_1 \cdot \mathbf{s}_2 \\ &= \frac{3}{2}\hbar^2 + 2(\mathbf{s}_1 \cdot \mathbf{s}_2) \end{aligned} \tag{5.28}$$

so that for the scalar product we get

$$\frac{1}{\hbar^2}\mathbf{s}_1 \cdot \mathbf{s}_2 = \frac{1}{2}S(S+1) - \frac{3}{4} = \begin{cases} -\frac{3}{4}\ if\ S = 0 \\ \frac{1}{4}\ if\ S = 1 \end{cases} \tag{5.29}$$

From this we see that the operator

$$\tilde{H} = \frac{1}{4}(E_+ + 3E_-) - (E_+ - E_-)\frac{1}{\hbar^2}(\mathbf{s}_1 \cdot \mathbf{s}_2) \tag{5.30}$$

gives exactly the same eigenvalues as the actual Hamiltonian H when applied to the states $|\psi_1\rangle$ and $|\psi_2(m_S)\rangle$:

$$\begin{aligned} \tilde{H}|\psi_1\rangle &= E_+|\psi_1\rangle \\ \tilde{H}|\psi_2(m_S)\rangle &= E_-|\psi_2(m_S)\rangle \end{aligned} \tag{5.31}$$

$\tilde{H}$ defined as above describes the *molecular Heisenberg model*

$$\tilde{H} = J_0 - J_{12}\mathbf{S}_1 \cdot \mathbf{S}_2 \tag{5.32}$$

$$J_{12} = \frac{1}{\hbar^2}(E_+ - E_-) \tag{5.33}$$

If $J_{12} > 0$, then the spin coupling is obviously ferromagnetic, and if $J_{12} < 0$, then it is antiferromagnetic.

5.2.2 The Heitler–London Method

In continuation of the discussion of the last section, we want to demonstrate with the help of an example, that, in fact, it is possible to have $E_+ \neq E_-$. For this purpose, we investigate, following the *Heitler–London method*, as a simple example, the two-electron system of *H_2-molecule*, in order to get at least a qualitative understanding of the *intra-atomic exchange interaction*. We note in passing that the same procedure is also followed in order to understand the covalent bonding.

Let the two protons be fixed at $\mathbf{R}_a$ and $\mathbf{R}_b$ (which means that we assume $m_a = m_b = \infty$) and let the two nuclei be in their ground state (Fig. 5.11).

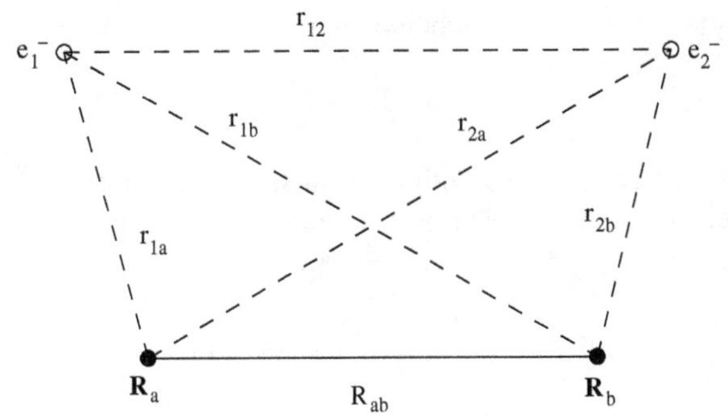

Fig. 5.11 Distances in the H_2 molecule

We split the *Hamiltonian* of the total system into an *unperturbed* part H_0 and a *perturbation* H_1:

$$H = H_0 + H_1 \tag{5.34}$$

$$H_0 = \frac{1}{2m}(p_1^2 + p_2^2) - \frac{e^2}{4\pi\varepsilon_0}\left(\frac{1}{r_{1a}} + \frac{1}{r_{2b}}\right) \tag{5.35}$$

$$H_1 = \frac{e^2}{4\pi\varepsilon_0}\left(\frac{1}{R_{ab}} + \frac{1}{r_{12}} - \frac{1}{r_{1b}} - \frac{1}{r_{2a}}\right) \tag{5.36}$$

Since H does not contain any spin parts, the eigenfunction to be determined can be factorized:

$$\begin{aligned} |\psi_s\rangle &= |q\rangle^{(+)}|0;0\rangle \\ |\psi_t\rangle &= |q\rangle^{(-)}|1;m_S\rangle \end{aligned} \tag{5.37}$$

The problem cannot be solved exactly. The unperturbed problem $H_1 = 0$ is realized by making the separation between the nuclei to be infinite. Then, we have the exactly solvable problem of two uncoupled hydrogen atoms. The solution is known from the basic quantum mechanics course:

$$\begin{aligned} \left(\frac{p_1^2}{2m} - \frac{e^2}{2\pi\varepsilon_0 r_{1a}}\right)|\phi_a^{(1)}\rangle &= E_a|\phi_a^{(1)}\rangle \\ \left(\frac{p_2^2}{2m} - \frac{e^2}{2\pi\varepsilon_0 r_{2b}}\right)|\phi_b^{(2)}\rangle &= E_b|\phi_b^{(2)}\rangle \end{aligned} \tag{5.38}$$

Then, the symmetrized space part of the "unperturbed" two-electron state must be given by

$$|q\rangle^{(\pm)} = \frac{1}{\sqrt{2}}\left(|\phi_a^{(1)}\rangle|\phi_b^{(2)}\rangle \pm |\phi_a^{(2)}\rangle|\phi_b^{(1)}\rangle\right) \equiv \frac{1}{\sqrt{2}}(|q_1\rangle \pm |q_2\rangle) \tag{5.39}$$

The last step is only a short-hand notation. We can now write

$$H_0 \, |q\rangle^{(\pm)} = (E_a + E_b) \, |q\rangle^{(\pm)} \tag{5.40}$$

When the space part is combined with the spin states, the energy $E_a + E_b$ is four-fold degenerate. The single particle wavefunctions centred at the nuclei a and b, respectively, do not overlap because $R_{ab} \to \infty$:

$$\langle \phi_a^{(1,2)} | \phi_b^{(1,2)} \rangle = \int d^3 r \phi_a^*(\mathbf{r}) \phi_b(\mathbf{r}) = 0 \tag{5.41}$$

Therefore, $|q_{1,2}\rangle$ are orthogonal to each other:

$$\langle q_1 | q_2 \rangle = 0 \tag{5.42}$$

This will not be true any more when we bring the nuclei near to each other so that the separation between them is finite, that is, when we switch on the perturbation. For $H_1 \neq 0$, the problem can be solved only approximately. We choose the variation method to solve the problem for the ground state ($E_a = E_b = E_0$). Since not only H_0, but also the total Hamiltonian is symmetric against the interchange of particles, the following *variation ansatz* appears reasonable:

$$|q\rangle = c_1 |q_1\rangle + c_2 |q_2\rangle \tag{5.43}$$

c_1, c_2 shall be real numbers. Here $|q_{1,2}\rangle$ is defined as in (5.39), but they are not orthogonal to each other any more , since, due to the finiteness of the separation between the nuclei, the so-called *overlap integral*

$$L = \langle \phi_a^{(1,2)} | \phi_b^{(1,2)} \rangle = \int d^3 r \phi_a^*(\mathbf{r}) \phi_b(\mathbf{r}) \tag{5.44}$$

is not equal to zero any more. The above ansatz neglects "polar" states of the form

$$|\phi_a^{(1)}\rangle |\phi_a^{(2)}\rangle \quad ; \quad |\phi_b^{(1)}\rangle |\phi_b^{(2)}\rangle$$

which describe situations, where both the electrons reside on the same atom. Because of the repulsive Coulomb interaction, the configurations such as these should be energetically unfavourable, and therefore may not be so important. In the case of bonding, they fix the "rest ionicity" of the covalent (homopolar) bonding.

For the variation ansatz, we further might demand the symmetry in the particle indices, so that it would be necessary to set $c_1 = \pm c_2$, and the coefficients were fixed by the normalization condition. Here, however, we will treat the c_1 and c_2 as variation parameters and through the condition

$$\frac{\partial E_V}{\partial c_{1,2}} \stackrel{!}{=} 0 \tag{5.45}$$

where

$$E_V = \frac{\langle q|H|q\rangle}{\langle q|q\rangle} = E_V(c_1, c_2) \tag{5.46}$$

We will fix the optimal c_1 and c_2. The variation energy E_V determined in this way in any case sets an upper limit to the actual ground state energy.

While calculating E_V, there appear certain characteristic integrals.

Coulomb integral:

$$V \equiv \langle q_1|H|q_1\rangle = \langle q_2|H|q_2\rangle = \int\int d^3r_1 d^3r_2 H\,|\phi_a(\mathbf{r}_1)|^2\,|\phi_b(\mathbf{r}_2)|^2 \tag{5.47}$$

Exchange integral:

$$\begin{aligned} X &\equiv \langle q_1|H|q_2\rangle = \langle q_2|H|q_1\rangle \\ &= \int\int d^3r_1 d^3r_2 \phi_a^*(\mathbf{r}_1)\phi_b^*(\mathbf{r}_2) H_1 \phi_a(\mathbf{r}_2)\phi_b(\mathbf{r}_1) \end{aligned} \tag{5.48}$$

With these definitions, we now calculate the variational energy E_V. First the denominator,

$$\begin{aligned} \langle q|q\rangle &= c_1^2\langle q_1|q_1\rangle + c_2^2\langle q_2|q_2\rangle + c_1c_2(\langle q_1|q_2\rangle + \langle q_2|q_1\rangle) \\ &= c_1^2 + c_2^2 + 2c_1c_2L^2 \end{aligned} \tag{5.49}$$

and then in similar way the numerator

$$\frac{\langle q|H_0|q\rangle}{\langle q|q\rangle} = 2E_0 \tag{5.50}$$

$$\langle q|H|q\rangle = (c_1^2 + c_2^2)V + 2c_1c_2X \tag{5.51}$$

This finally yields

$$E_V = \frac{(c_1^2 + c_2^2)V + 2c_1c_2X}{(c_1^2 + c_2^2) + 2c_1c_2L^2} \tag{5.52}$$

The variation condition (5.45) leads to

$$(c_1^2 - c_2^2)(X - VL^2) = 0 \tag{5.53}$$

The second factor is, in general, unequal to zero, so that using the normalization condition $c_1^2 + c_2^2 = 1$ we get

$$c_1 = c_2 = \frac{1}{\sqrt{2}} \tag{5.54}$$

With this, the variation ansatz (5.43) for $|q\rangle$ satisfies automatically all the symmetry requirements. With these c_1 and c_2, the ground state energy is given by

$$E_{\pm} = \frac{V \pm X}{1 \pm L^2} \tag{5.55}$$

When X and L are unequal to zero, then the condition $E_+ \neq E_-$, which is necessary for the preferred spin ordering as discussed in the last section, is realized.

In that case, we can replace the "true" Hamiltonian H by an effective *exchange Hamiltonian* $\tilde{H}$:

$$\tilde{H} = J_0 - J_{12}\mathbf{S}_1 \cdot \mathbf{S}_2 \tag{5.56}$$

$$J_{12} = \frac{1}{\hbar^2}(E_+ - E_-) = -\frac{2}{\hbar^2}\frac{VL^2 - X}{1 - L^4} \tag{5.57}$$

The sign of J_{12} depends on the relative strengths of the integrals L, V and X. In general, however, $L \ll 1$ and $X < 0$, so that J_{12} is negative, i.e. the singlet state is energetically the lowest.

$\tilde{H}$ is of the type, which we require for our magnetic problems. We thus have found a quantum mechanical mechanism, which prefers a particular spin orientation.

Postulate: *$\tilde{H}$ can be generalized to N multi-electron-atoms.*

This leads to the *Heisenberg model*

$$H = -\sum_{ij} J_{ij}\mathbf{S}_i \cdot \mathbf{S}_j \tag{5.58}$$

Discussion

1. The sign and the magnitude of the coupling constants J_{12} depend on the relative magnitudes of the integrals V, L and X. The deciding parameter is therewith the internuclear distance R_{ab} which determines the degree of overlap of the hydrogen wavefunctions $|\phi_a\rangle$ and $|\phi_b\rangle$. If we use the 1s wavefunctions of the hydrogen atom to calculate the integrals, then we get the result shown in the Fig. 5.12, which means that J_{12} is always negative or the "antiferromagnetic" singlet state is stable.
2. The Heitler–London method does not converge! This is essentially because of the non-orthogonality of the eigenstates $|\phi_{a,b}\rangle$ of hydrogen atoms centred on different nuclei. The corresponding non-orthogonality integrals appear with increasing power in the secular equation of the eigenenergies, as the number of electrons participating in the "exchange" increases. This leads to a divergence, which is called the *non-orthogonality catastrophe*.
3. Sometimes, one tries to get round the problem of (2), by using, instead of non-orthogonal, orthogonal variation states. For the states $|q_{1,2}\rangle$, we only have to require that they should be products of single particle states, which for $R_{ab} \rightarrow$

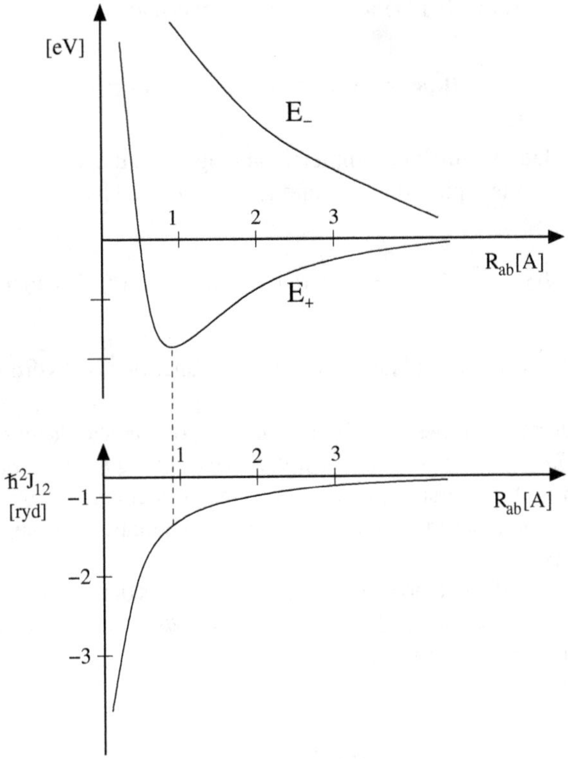

Fig. 5.12 Eigenenergies and the effective exchange integrals of the *H_2-molecule* calculated with 1s-eigenfunctions of the hydrogen atom

∞ go over to the hydrogen atom wavefunctions $|\phi_a\rangle$ and $|\phi_b\rangle$:

$$\begin{aligned} |q_{1,2}\rangle &= |u^{(1,2)}\rangle|v^{(2,1)}\rangle \\ |q\rangle^{(\pm)} &= \frac{1}{\sqrt{2}}(|q_1\rangle \pm |q_2\rangle) \end{aligned} \tag{5.59}$$

So far, we have used for $|u\rangle$ and $|v\rangle$ the hydrogen wavefunctions. Another possibility is to use the following ansatz:

$$|u\rangle = \alpha_+|\phi_a\rangle + \alpha_-|\phi_b\rangle \quad ; \quad |v\rangle = \alpha_+|\phi_b\rangle + \alpha_-|\phi_a\rangle \tag{5.60}$$

With

$$\alpha_\pm = \frac{1}{2}(1 + \langle\phi_a|\phi_b\rangle)^{-1/2} \pm \frac{1}{2}(1 - \langle\phi_a|\phi_b\rangle)^{-1/2} \tag{5.61}$$

these states indeed approach, respectively, $|\phi_a\rangle$ and $|\phi_b\rangle$ for $R_{ab} \to \infty$ because

$$\lim_{R_{ab}\to\infty} \alpha_+ \to 1 \quad ; \quad \lim_{R_{ab}\to\infty} \alpha_- \to 0 \tag{5.62}$$

One can easily show that $|u\rangle$ and $|v\rangle$ are orthonormal:

$$\langle u|u\rangle = \langle v|v\rangle = 1 \quad ; \quad \langle u|v\rangle = 0 \tag{5.63}$$

With a calculation, which is completely analogous to the one carried out above, where only $|\phi_a\rangle$ is replaced by $|u\rangle$ and $|\phi_b\rangle$ is replaced by $|v\rangle$, we get

$$L = 0\,;\; J_{12} = \frac{2}{\hbar^2}X = \frac{2}{\hbar^2}\int\int d^3r_1 d^3r_2 u^*(\mathbf{r}_1)v^*(\mathbf{r}_2)Hu(\mathbf{r}_2)v(\mathbf{r}_1) \tag{5.64}$$

This integral, when calculated with 1s-wavefunctions of hydrogen is *always positive*.
This consideration shows that the actual form of J_{12} should not be taken too seriously. *The coupling constants J_{ij} of the Heisenberg model should be treated as parameters.* It is actually pointless to want to calculate the exchange parameters J_{12} in a solid, based on a molecular model no matter how sophisticated the model may be.

4. It is important and it can be proved exactly that J_{12} is determined by the overlap integrals. *If wavefunctions of the participating electrons do not overlap, then there is no ferromagnetism.*
5. The polar states of the type

$$|\phi_a^{(1)}\rangle|\phi_a^{(2)}\rangle \; and \; |\phi_b^{(1)}\rangle|\phi_b^{(2)}\rangle$$

are not included in the variational ansatz. That means, it is implicitly assumed that, always, one electron belongs to proton a and another to proton b. As a consequence, the model is *applicable only to insulators*, and certainly not for band magnets like Fe, Co and Ni.
6. The points (4) and (5) actually exclude each other. On the one hand, the model is good, only when the concerned wavefunctions have practically no overlap, and on the other hand, if there is no overlap, there is no exchange. In reality, the coupling mechanism in magnetic insulators is therefore different, in general, of the type, called the *superexchange*, which will be described in Sect. 5.3.2. In this situation, there is, in general, a diamagnetic ion between two magnetic ions, and it mediates the coupling (prototype: MnO (Fig. 5.13)). The separation between the magnetic ions is so large that a "direct exchange" is not possible. The coupling is transmitted through the diamagnetic ion. As we will see, this again leads to a model Hamiltonian of the Heisenberg type, but of course, the interpretation of the coupling constants J_{12} will then be completely different.
The ferromagnetic metals of the class Localized Magnetism mentioned at the beginning of this chapter can also be understood only when an *indirect exchange (RKKY interaction)* mediated by the conduction electrons is invoked. This will be discussed in more detail in Sect. 5.3.1.

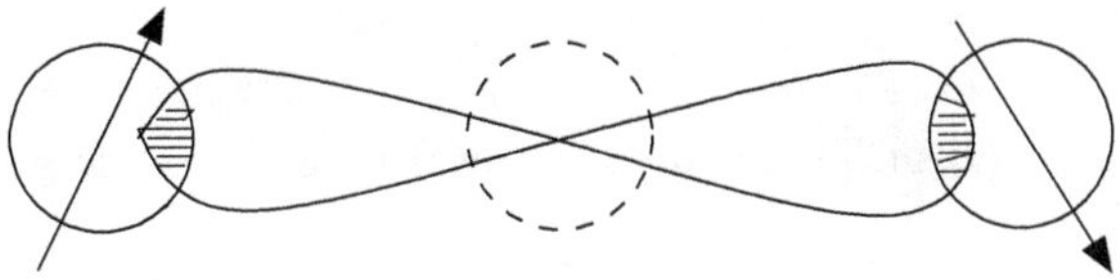

Fig. 5.13 Schematic plot which illustrates the superexchange between the Mn^{2+} ions mediated by the O^{2-} ion

5.2.3 Dirac's Vector Model

The "derivation" of the Heisenberg Hamiltonian using the Heitler–London method has the advantage that it gives a certain insight into the physical basis for the exchange interaction. The discussion in the last section, however, makes it also clear that the exchange interaction does not have a such universal character as, for example, the Coulomb's law or the Newton's axioms.

That of course does not at all mean that the whole concept of the Heisenberg Hamiltonian is questionable. There are series of other deductions, which lead to the same operator form. In this section, we want to discuss one such suggestion from Dirac, where we will show that the normal Coulomb interaction, in the first-order perturbation theory, leads to an effective Hamiltonian of the Heisenberg type. This time, however, we need not restrict ourselves to a two-electron system.

We consider an *ensemble of N indistinguishable Fermions*, which, without the presence of any interaction, occupy the single particle states

$$|\alpha_1\rangle, \ |\alpha_2, \ \rangle, \ \cdots, \ |\alpha_N\rangle$$

The α_is stand for a set of quantum numbers. The states $|\alpha_i\rangle$ can be assumed, from the beginning, to be orthogonal. Since they have to be occupied by N Fermions, they should be necessarily pairwise different from each other. A possible *unperturbed* state $|\psi\rangle$ for the whole ensemble is then,

$$|\psi\rangle = |\alpha_1^{(1)}\rangle|\alpha_2^{(2)}\rangle \cdots |\alpha_N^{(N)}\rangle \tag{5.65}$$

where the superscript enumerates the N Fermions (see Appendix A). The Hamiltonian,

$$H = H_0 + H_1 \tag{5.66}$$

and also separately H_0 and H_1 themselves, is symmetric against the particle interchanges. Therefore, the application of an arbitrary permutation $\mathcal{P}$ on the state $|\psi\rangle$ results in a new state $\mathcal{P}|\psi\rangle$ which also belongs to the same energy as $|\psi\rangle$. The new state can be written as

$$\mathcal{P}\,|\psi\rangle = |\alpha_1^{(r)}\rangle|\alpha_2^{(s)}\rangle\cdots|\alpha_N^{(z)}\rangle \tag{5.67}$$

Obviously, there are $N!$ such unperturbed states with the same energy E_0,

$$H_0\,|\psi\rangle = E_0\,|\psi\rangle$$
$$H_0\,\mathcal{P}\,|\psi\rangle = E_0\,\mathcal{P}\,|\psi\rangle,\ \cdots$$

which span the so-called "eigenspace" $\mathcal{H}_0$ corresponding to E_0. Following the usual perturbation theory for a system with degeneracy, we have to set up the *perturbation matrix* in this eigenspace. The elements of the matrix are of the type

$$\langle\psi|\mathcal{P}^{\alpha}H_1\mathcal{P}^{\beta}|\psi\rangle$$

and the eigenvalues of this matrix represent the energy corrections in the first order.

We can formally distinguish between two types of permutations, namely, the type $\mathcal{P}^{\alpha}$ which we have so far used and which acts on the superscript (particle), and the type $\mathcal{P}_{\alpha}$, which acts on the subscript, i.e. which changes the ordering of the states. Naturally, $\mathcal{P}_{\alpha}$ is meaningfully defined, only when, as was assumed in (5.65), the single particle states in the N-particle state are arranged in a particular order according to some criterion.

It is obvious that every $\mathcal{P}^{\alpha}$ commutes with every $\mathcal{P}_{\beta}$:

$$\mathcal{P}^{\alpha}\,\mathcal{P}_{\beta} = \mathcal{P}_{\beta}\,\mathcal{P}^{\alpha} \tag{5.68}$$

It is equally obvious that, for $|\psi\rangle$ given by (5.65),

$$\mathcal{P}^{\alpha}\,\mathcal{P}_{\alpha}|\psi\rangle = 1\cdot|\psi\rangle \tag{5.69}$$

This operation changes only the sequence of the factors in $|\psi\rangle$.

Since all the $|\alpha_i\rangle$ are orthogonal to each other, $|\psi\rangle$ and $\mathcal{P}|\psi\rangle$ are also orthogonal to each other, provided $\mathcal{P}$ is not the identity operator. That means

$$\langle\psi|\mathcal{P}_{\alpha}\,\mathcal{P}^{\beta}|\psi\rangle = \delta_{\alpha\beta} \tag{5.70}$$

when it is assumed that $|\psi\rangle$ is normalized.

Now, let $C(\mathcal{P})$ be some scalar quantity, which depends on the particular distribution of the N particles into the N single particle states $|\alpha_i\rangle$. Then obviously

$$C(\mathcal{P}^{\alpha}) = \sum_{\beta} C(\mathcal{P}^{\beta})\langle\psi|\mathcal{P}_{\beta}\,\mathcal{P}^{\alpha}|\psi\rangle \tag{5.71}$$

The sum runs over all the $N!$ permutations P_{β} for a given P^{α}. We now define special coefficients

$$(H_1)(\mathcal{P}) \equiv \langle\psi|H_1\mathcal{P}|\psi\rangle \tag{5.72}$$

for which, naturally, the above relation (5.71) is valid. We notice that $|\psi\rangle$ and $\mathcal{P}|\psi\rangle$ are the eigenstates of H_0 but not of H_1. Therefore, in general, $(H_1)(\mathcal{P}) \neq 0$. The perturbation H_1 is, however, symmetric against particle interchange, otherwise they would not have been "indistinguishable". As a result, for an arbitrary permutation, we have

$$H_1\mathcal{P} = \mathcal{P}H_1 \tag{5.73}$$

Then we can write

$$\begin{aligned}\langle\psi|\mathcal{P}^\alpha H_1\mathcal{P}^\beta|\psi\rangle &= \langle\psi|H_1\mathcal{P}^\alpha\mathcal{P}^\beta|\psi\rangle = (H_1)(\mathcal{P}^\alpha\mathcal{P}^\beta)\\ &= \sum_\gamma (H_1)(\mathcal{P}^\gamma)\langle\psi|\mathcal{P}_\gamma\mathcal{P}^\alpha\mathcal{P}^\beta|\psi\rangle\\ &= \sum_\gamma (H_1)(\mathcal{P}^\gamma)\langle\psi|\mathcal{P}^\alpha\mathcal{P}_\gamma\mathcal{P}^\beta|\psi\rangle\end{aligned}$$

This relation is valid for *arbitrary* states $\mathcal{P}^{\alpha,\beta}|\psi\rangle$ of the eigenspace $\mathcal{H}_0$. Therefore, we have, in this space, an *operator identity*

$$H_1 = \sum_\gamma c_\gamma\, \mathcal{P}_\gamma \tag{5.74}$$

when we interpret the *scalar* coefficient c_γ by

$$c_\gamma = (H_1)(\mathcal{P}^\gamma) \tag{5.75}$$

Therefore, the perturbation operator H_1 can be written in this space as a linear combination of permutation operators $\mathcal{P}_\gamma$.

We now want to apply this general theory to a system of electrons. Let the perturbation operator H_1 be given by the Coulomb interaction. It is important here to remember that the electron has a spin. The main consequence is not the possibility of coupling of the magnetic moment due to the spin with an external magnetic field, but the fact that according to the Pauli's principle the spin doubles the number of occupiable states. Electrons are specified by two types of variables, namely, the spin variables $\sigma_x,\ \sigma_y,\ \sigma_z$ and the position variables $x,\ y,\ z$. The two types of variables correspond to two types of permutations, $\mathcal{P}_\sigma$ for the spin variables and $\mathcal{P}_x$ for the orbit variable. The operators $\mathcal{P}_\gamma$, which were used in the general theory discussed above, encompass, naturally, the totality of all the dynamic variables:

$$\mathcal{P}_\gamma \;\rightarrow\; (\mathcal{P}_x\, \mathcal{P}_\sigma)_\gamma$$

Electrons are Fermions and so they are described by antisymmetric N-particle states $|\psi\rangle^{(-)}$:

$$(\mathcal{P}_x \cdot \mathcal{P}_\sigma)_\gamma |\psi\rangle^{(-)} = \pm |\psi\rangle^{(-)} \tag{5.76}$$

The sign + or − depends on whether the permutation is even or odd. If we restrict ourselves right from the beginning in H_0 to the appropriate antisymmetrized states $|\psi\rangle^{(-)}$, then we can treat

$$(\mathcal{P}_x \, \mathcal{P}_\sigma)_\gamma = \pm 1 \tag{5.77}$$

also as an *operator identity*. This has important consequences. Since the perturbation H_1 consists only of the usual Coulomb interaction, the Hamiltonian does not contain the spin explicitly. Therefore, we must above all investigate $(\mathcal{P}_x)_\gamma$. Because of the above operator identity, however, we can also start with the simpler problem of $(\mathcal{P}_\sigma)_\gamma$, which then uniquely fixes $(\mathcal{P}_x)_\gamma$.

We first show that $\mathcal{P}_\sigma$ can be expressed in terms of the spin operators **s**

$$\mathbf{s} = \frac{\hbar}{2}\sigma \quad , \quad \sigma = (\sigma_x, \, \sigma_y, \, \sigma_z) \tag{5.78}$$

The components of σ are the *Pauli spin matrices*:

$$\sigma_x = \begin{pmatrix} 0 & 1 \\ 1 & 0 \end{pmatrix} \; ; \; \sigma_y = \begin{pmatrix} 0 & -i \\ i & 0 \end{pmatrix} \; ; \; \sigma_z = \begin{pmatrix} 1 & 0 \\ 0 & -1 \end{pmatrix} \tag{5.79}$$

One can easily verify the following relations (Problem 2.2):

$$\sigma_i^2 = 1 \quad ; \quad i = x, \, y, \, z \tag{5.80}$$

$$\sigma_i \, \sigma_j = i \, \sigma_k \quad ; \quad (i, \, j, \, k) = (x, \, y, \, z) \text{ and cyclic} \tag{5.81}$$

$$[\sigma_i \, , \, \sigma_j]_+ = \sigma_i \, \sigma_j + \sigma_j \, \sigma_i = 2 \, \delta_{ij} \cdot 1 \tag{5.82}$$

With these relations, we can further show that (Problem 5.6)

$$\left(\sigma^{(1)} \cdot \sigma^{(2)}\right)^2 = 3 - 2 \, (\sigma^{(1)} \, \sigma^{(2)}) \tag{5.83}$$

where the superscripts are the particle indices. We now define the following operator:

$$Q_{12} = \frac{1}{2}(1 + \sigma^{(1)} \cdot \sigma^{(2)}) \tag{5.84}$$

whose square is given by

$$Q_{12}^2 = \frac{1}{4}(1 + 2\sigma^{(1)} \cdot \sigma^{(2)} + (\sigma^{(1)} \cdot \sigma^{(2)})^2) = 1 \tag{5.85}$$

In view of this relation, Q_{12} satisfies an important property of a transposition operator. From the defining equation for Q_{12}, one further sees that

$$Q_{12}\,\sigma^{(1)}_{x,y,z} = \sigma^{(2)}_{x,y,z}\,Q_{12}$$

which actually means

$$Q_{12}\,\sigma^{(1)}\,Q_{12}^{-1} = \sigma^{(2)} \;\; ; \;\; Q_{12}\,\sigma^{(2)}\,Q_{12}^{-1} = \sigma^{(1)} \;\; ; \;\; Q_{12} = Q_{12}^{-1} \tag{5.86}$$

Thus, the operator Q_{12} interchanges the spins of the particles 1 and 2 and therefore, it can be different from $(\mathcal{P}_\sigma)_{12}$, the transposition operator, at the most by a scalar factor:

$$(\mathcal{P}_\sigma)_{12} = \alpha\, Q_{12} \tag{5.87}$$

Here, because of the fact that $(\mathcal{P}_\sigma)^2_{12} = Q^2_{12} = 1$, α can be only either $+1$ or -1. The sign is decided as follows: With reference to the spin variables of the two-electron system, there are three possible symmetric states

$$\begin{aligned} &|m_S^{(1)} =\uparrow\rangle|m_S^{(2)} =\uparrow\rangle \quad ; \quad |m_S^{(1)} =\downarrow\rangle|m_S^{(2)} =\downarrow\rangle \\ &|m_S^{(1)} =\uparrow\rangle|m_S^{(2)} =\downarrow\rangle + |m_S^{(1)} =\downarrow\rangle|m_S^{(2)} =\uparrow\rangle \end{aligned} \tag{5.88}$$

and one antisymmetric state

$$|m_S^{(1)} =\uparrow\rangle|m_S^{(2)} =\downarrow\rangle - |m_S^{(1)} =\downarrow\rangle|m_S^{(2)} =\uparrow\rangle \tag{5.89}$$

Therefore, $(\mathcal{P}_\sigma)_{12}$ has the eigenvalues $1,\ 1,\ 1,\ -1$. The scalar product $\sigma^{(1)}\cdot\sigma^{(2)}$ has the eigenvalues $1,\ 1,\ 1,\ -3$, in the same sequence and so Q_{12} has the eigenvalues $1,\ 1,\ 1,\ -1$ (Problem 5.6). Therefore we must have $\alpha = +1$. Then we can write

$$(\mathcal{P}_\sigma)_{12} = \frac{1}{2}(1 + \sigma^{(1)}\cdot\sigma^{(2)}) \tag{5.90}$$

Then for the transposition operator for the position observables also we have

$$(\mathcal{P}_x)_{12} = -\frac{1}{2}(1 + \sigma^{(1)}\cdot\sigma^{(2)}) \tag{5.91}$$

The perturbation H_1, which is the Coulomb interaction of the electrons, consists of a sum of two-electron interactions. Among the matrix elements $c_\gamma = (H_1)(\mathcal{P}^\gamma)$, in the general expression for H_1 (5.74), only those will be unequal to zero, for which P^γ is either identity or it is a transposition operator for two electrons. Due to this reason, H_1 has the following form:

$$H_1 = \sum_{i<j} c_{ij}\,(\mathcal{P}_x)_{ij} + E_c \tag{5.92}$$

where E_c is a constant. In the first-order perturbation theory, the eigenvalues of the operator

$$H_1 = E_c - \frac{1}{2} \sum_{i<j} c_{ij} \left(1 + \sigma^{(i)} \cdot \sigma^{(j)}\right) \tag{5.93}$$

in the eigenspace $\mathcal{H}_0$ represent the correction to energy, which we are looking for.

In this form, H_1 provides the justification for the Heisenberg Hamiltonian. For, if we assume that, for all the pairs of electrons in the partially filled shells which are, respectively, localized at $\mathbf{R}_i$ and $\mathbf{R}_j$, the matrix elements c_{ij} are equal in the first approximation, then, we can treat all the electron spins in such a shell together as a total spin $\mathbf{S}_i$, whose index i now refers to the lattice site $\mathbf{R}_i$. By still suppressing the unimportant constants, we eventually get the *exchange operator*:

$$H = -\sum_{i,j} J_{ij}\, \mathbf{S}_i \cdot \mathbf{S}_j \tag{5.94}$$

Just as in Sect. 5.2.2, here also, H is the result of perturbation theory in the first order for the "normal" electron–electron Coulomb interaction. The coupling constants J_{ij} correspond to the classically incomprehensible exchange matrix elements $\langle \psi | H_1\ \mathcal{P}^{ij} | \psi \rangle$ of the Coulomb interaction, where $|\psi\rangle$ is a nonsymmetrized N-particle state.

5.3 Indirect Exchange Interaction

The *direct* exchange mechanism, as presented in Section 5.2 and in particular in Sect. 5.2.2, is frequently not acceptable as a coupling mechanism for magnetic materials for the reason that the separation between the magnetic ions is too large. Because of the large inter-ion separation, the overlap integrals are too small to mediate a sufficiently strong coupling. There are, however, a number of *indirect* exchange mechanisms which, within the framework of second-order perturbation theory, lead to an effective Hamiltonian of the Heisenberg type. They are different from the direct exchange because the direct exchange discussed in Sect. 5.2 is a result of perturbation theory of first order.

The concept of *indirect exchange* is not uniquely defined. In this section, we will discuss three different types of indirect coupling.

5.3.1 Rudermann–Kittel–Kasuya–Yosida (RKKY) Interaction

We begin with an indirect interaction between magnetic ions, which is mediated by quasi-free, mobile electrons of the conduction band. It is a type of coupling, which is observed, in particular in metallic 4f-systems like Gd, and therefore, it belongs to the class (Localized Magnetism) mentioned in the introduction of Chap. 5.

The idea goes back to M.A. Rudermann and C. Kittel [2], who discussed the long-range coupling between nuclear spins, on the basis of the contact hyperfine interaction between a nuclear spin and the spin of an s-conduction electron, which we have learned about in Sect. 2.8. The nuclear spin polarizes the conduction electron spin in its neighbourhood. Because of the Pauli's principle, the polarization cloud is not exclusively localized near the nuclear spin, rather, it will be a sequence of alternating "rarefactions" and "condensations". As a result, the spin polarization of the conduction electrons will have an oscillatory behaviour as a function of the distance from the polarizing nucleus. This "information" is "felt" by a neighbouring spin, from which, an effective coupling between the two nuclear spins results.

Completely analogous to this effective nuclear spin coupling, an exchange interaction between the localized electrons of partially filled electron shells of different ions in a solid and the quasi-free conduction electrons should lead to an effective coupling between the localized moments.

The idea of such a coupling mechanism in ferromagnetic metals goes to the credit of T. Kasuya [3] and K. Yosida [4]. This idea was provoked by very interesting experimental observations. When paramagnetic Mn^{2+}-ions ($(3d)^5 \Rightarrow S = 5/2,\ L = 0$) are doped into a non-magnetic Cu matrix, then, depending on the concentration of Mn, completely different phenomena are observed:

1. *Quenching* of 3d-moments.
2. *Kondo behaviour*: The conduction electrons of Cu are perturbed by the magnetic Mn-moments and this leads to an anomalous behaviour of resistivity.
3. *Spin glass behaviour*: A statistical distribution of the Mn ions in the Cu matrix leads to a coupling between the Mn ions which varies in sign and magnitude. That can mean that, the Mn spin cannot satisfy all the exchange interactions (*frustration*).
4. *Ferromagnetic ordering* of the 3d spins.
5. *Antiferromagnetic ordering*.

In order to understand such a rich variety of phenomena, the starting point is the following *model*.

Two ions located at $\mathbf{R}_i$ and $\mathbf{R}_j$, which are not directly coupled, are embedded in a "sea" of conduction electrons of a non-magnetic metal (Fig. 5.14). The conduction electrons are described by the simple Sommerfeld model (Sect. 4.1.1)

$$H_s = \sum_{\mathbf{k},\sigma} \varepsilon(\mathbf{k}) c^{\dagger}_{\mathbf{k}\sigma}\, c_{\mathbf{k}\sigma} \tag{5.95}$$

Here $c^{\dagger}_{\mathbf{k}\sigma}(c_{\mathbf{k}\sigma})$ is the creation (annihilation) operator of an electron of wavevector $\mathbf{k}$ and spin σ ($\sigma = \uparrow, \downarrow$) and energy $\varepsilon(\mathbf{k})$. The localized spins shall not directly interact with each other. Therefore, we can set the Hamiltonian that describes the interaction of the localized spins equal to zero:

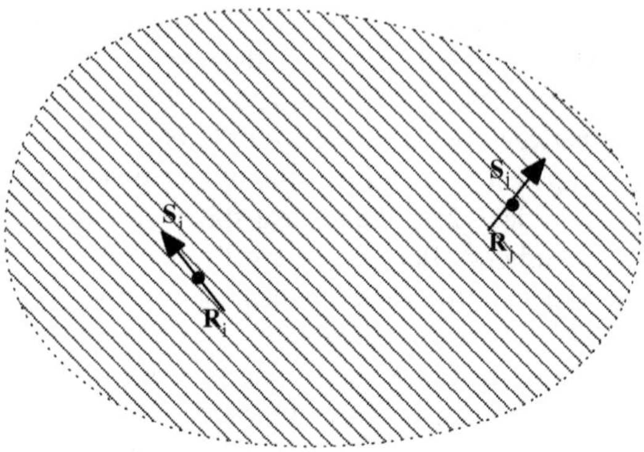

Fig. 5.14 Two localized spins $\mathbf{S}_i$ and $\mathbf{S}_j$ at the lattice sites $\mathbf{R}_i$ and $\mathbf{R}_j$ in a "sea" of conduction electrons

$$H_f \equiv 0 \tag{5.96}$$

The exchange interaction between the localized spin $\mathbf{S}_i$ and the conduction electron spin $\mathbf{s}$ is treated as the perturbation and is taken to be of Heisenberg type, as an intra-atomic exchange: J is the corresponding coupling constant.

$$H_{sf} = -J \sum_{i=1}^{2} \mathbf{s}_i \cdot \mathbf{S}_i = -J \sum_{i=1}^{2} \left\{ s_i^z S_i^z + \frac{1}{2} \left(s_i^+ S_i^- + s_i^- S_i^+ \right) \right\} \tag{5.97}$$

H_{sf} has the same structure as the hyperfine interaction H_{cont} (see Sect. 2.8) between a nuclear spin and the spin of an s-electron and therefore can be justified on the same basis (see also Problem 5.8). The electron spin operators can be expressed in terms of creation and annihilation operators:

$$s_i^z = \frac{\hbar}{2} \left(c_{i\uparrow}^\dagger c_{i\uparrow} - c_{i\downarrow}^\dagger c_{i\downarrow} \right) \tag{5.98}$$

$$s_i^+ = \hbar\, c_{i\uparrow}^\dagger c_{i\downarrow} \tag{5.99}$$

$$s_i^- = \hbar\, c_{i\downarrow}^\dagger c_{i\uparrow} \tag{5.100}$$

$c_{i\sigma}^\dagger$ ($c_{i\sigma}$) creates (annihilates) an electron of spin σ at *the position* $\mathbf{R}_i$. Using the fundamental commutation relations for Fermion operators ($[A\ ,\ B]_+ = AB + BA$ is the anticommutator, see Appendix A)

$$\left[c_{i\sigma}\ ,\ c_{j\sigma'} \right]_+ = \left[c_{i\sigma}^\dagger\ ,\ c_{j\sigma'}^\dagger \right]_+ = 0 \tag{5.101}$$

$$\left(c_{i\sigma} \right)^2 = \left(c_{i\sigma}^\dagger \right)^2 = 0 \ (\textit{Pauli's principle!}) \tag{5.102}$$

$$\left[c_{i\sigma}\ ,\ c_{j\sigma'}^\dagger \right]_+ = \delta_{ij}\delta_{\sigma\sigma'} \tag{5.103}$$

one can easily verify that the usual commutation relations for the spin operators are fulfilled (Problem 5.1):

$$\left[s_i^+ \, , \, s_j^-\right]_- = 2\,\delta_{ij}\,\hbar\,s_i^z \tag{5.104}$$

$$\left[s_i^z \, , \, s_j^\pm\right]_- = \pm\,\delta_{ij}\,\hbar\,s_i^\pm \tag{5.105}$$

One should pay attention that $[\cdots]_-$ is a commutator and $[\cdots]_+$ is an anticommutator.

It is still recommendable to transform the operators $c_{i\sigma}^\dagger$ and $c_{i\sigma}$ into the wavevector space:

$$c_{i\sigma} = \frac{1}{\sqrt{N}} \sum_{\mathbf{q}} e^{i\mathbf{q}\cdot\mathbf{R}_i} c_{\mathbf{q}\sigma} \tag{5.106}$$

$$c_{i\sigma}^\dagger = \frac{1}{\sqrt{N}} \sum_{\mathbf{q}} e^{-i\mathbf{q}\cdot\mathbf{R}_i} c_{\mathbf{q}\sigma}^\dagger \tag{5.107}$$

The perturbation operator then reads as

$$\begin{aligned} H_{sf} = -\frac{J\hbar}{2N} \sum_i \sum_{\mathbf{k},\mathbf{q}} e^{-i\mathbf{q}\cdot\mathbf{R}_i} \Big\{ S_i^z \left(c_{\mathbf{q}+\mathbf{k}\uparrow}^\dagger c_{\mathbf{k}\uparrow} - c_{\mathbf{q}+\mathbf{k}\downarrow}^\dagger c_{\mathbf{k}\downarrow} \right) \\ + S_i^+ c_{\mathbf{k}+\mathbf{q}\downarrow}^\dagger c_{\mathbf{k}\uparrow} + S_i^- c_{\mathbf{k}+\mathbf{q}\uparrow}^\dagger c_{\mathbf{k}\downarrow} \Big\} \end{aligned} \tag{5.108}$$

Without the perturbation, the conduction electrons exist in their *unpolarized* ground state. In addition, since they do not interact with each other, the unperturbed electron ground state can be written as the antisymmetrized product of single-electron states

$$|\mathbf{k}_i^{(i)}, m_{s_i}^{(i)}\rangle \equiv |\mathbf{k}_i^{(i)}\rangle |m_{s_i}^{(i)}\rangle \tag{5.109}$$

where the spin magnetic quantum number $m_{s_i}^{(i)}$ takes the values $\pm\frac{1}{2}$. $\mathbf{k}_i^{(i)}$ is a wavevector, where the superscript refers to the particle number. Further, since we want to treat the conduction electrons as s-electrons, which excludes spin–orbit interaction, we can separate the spin and the space parts. Let

$$|0; f\rangle \equiv |0\rangle |f\rangle \tag{5.110}$$

be the *"unperturbed" ground state of the total system*, where, the spin state $|f\rangle$ is specified by the relative orientation of the N spins. Since there does not exist any direct interaction between the spins, $|f\rangle$ should be a linear combination of all possible relative orientations. $|0\rangle$ symbolizes the unpolarized ground state of the conduction electrons ("filled" Fermi sphere):

$$|0\rangle = \frac{1}{N!}\sum_{\mathcal{P}}(-1)^p\,\mathcal{P}|\mathbf{k}_1^{(1)}m_{s_1}^{(1)},\ \mathbf{k}_2^{(2)}m_{s_2}^{(2)},\ \cdots\mathbf{k}_N^{(N)}m_{s_N}^{(N)}\rangle \tag{5.111}$$

$\mathcal{P}$ is the permutation operator which acts on the particle superscripts. p is the number of transpositions which build up the permutation $\mathcal{P}$.

$|0; f\rangle$ is an eigenstate of $H_0 = H_s + H_f$ and therefore it is an eigenstate of H_s:

$$H_s\ |0; f\rangle = E_0^{(0)}\ |0; f\rangle \tag{5.112}$$

We want to consider the effect of H_{sf} using the perturbation theory. First, it is easy to see that the energy correction in the first order

$$E_0^{1)} = \langle 0; f|H_{sf}|0; f\rangle \tag{5.113}$$

does not contribute. For example, we have

$$\langle 0; f|S_i^z(c_\uparrow^\dagger\ c_\uparrow - c_\downarrow^\dagger\ c_\downarrow)|0; f\rangle = 0 \tag{5.114}$$

because, without the perturbation, the electron system is not polarized. There are exactly the same number of ↑-electrons as ↓-electrons. In addition, $\langle f|S_i^z|f\rangle = 0$. Since in each of the sub-systems, in the absence of the coupling, the spin is conserved, we also have

$$\langle 0; f|S_i^+c_\downarrow\ c_\uparrow|0; f\rangle = 0 \tag{5.115}$$

$$\langle 0; f|S_i^-c_\uparrow\ c_\downarrow|0; f\rangle = 0 \tag{5.116}$$

Therefore, the s–f interaction is not noticeable in the first order:

$$E_0^{(1)} \equiv 0 \tag{5.117}$$

For the second-order correction, the following expression has to be calculated:

$$E_0^{(2)} = \sum_{(A,f)\neq(0,f)} \frac{|\langle 0; f|H_{sf}|A; f'\rangle|^2}{E_0^{(0)} - E_A^{(0)}} \tag{5.118}$$

Here

$$|A\rangle = \frac{1}{N!}\sum_{\mathcal{P}}(-1)^p\,\mathcal{P}|\mathbf{k'}_1^{(1)}m'^{(1)}_{s_1},\ \mathbf{k'}_2^{(2)}m'^{(2)}_{s_2},\ \cdots\mathbf{k'}_N^{(N)}m'^{(N)}_{s_N}\rangle \tag{5.119}$$

is an excited state of the unperturbed electron system with energy $E_A^{(0)}$.

The electron part of the perturbation operator consists exclusively of single electron operators. Due to the orthonormality of the single particle states, the matrix element splits into expressions of the form

$$\langle 0|\mathcal{O}|A\rangle \rightarrow \underbrace{\langle \mathbf{k}'m_s'|}_{from\ |0\rangle} \mathcal{O} \underbrace{|\mathbf{k}''m_s''\rangle}_{from\ |A\rangle} \tag{5.120}$$

In particular, we have

$$\begin{aligned} &\langle \mathbf{k}'m_s'|(c^\dagger_{\mathbf{k+q}\uparrow}\, c_{\mathbf{k}\uparrow} - c^\dagger_{\mathbf{k+q}\downarrow}\, c_{\mathbf{k}\downarrow})|\mathbf{k}''m_s''\rangle \\ &= \Theta(k_F - |\mathbf{k}+\mathbf{q}|)\, \Theta(|\mathbf{k}| - k_F)\delta_{\mathbf{k},\mathbf{k}''}\, \delta_{\mathbf{k+q},\mathbf{k}'}\, \frac{2}{\hbar}\, \langle m_s'|s_z|m_s''\rangle \end{aligned} \tag{5.121}$$

i.e. only those states $|\mathbf{k}''m_s''\rangle$ contribute which represent particle–hole excitations of the originally filled Fermi sphere. If we treat $|m_s\rangle$ to be a two-component spinor,

$$\begin{pmatrix}1\\0\end{pmatrix}\ or\ \begin{pmatrix}0\\1\end{pmatrix}$$

then, $\sigma_z = \frac{2}{\hbar}\, s_z$ is a 2×2 Pauli spin matrix (5.79). The step functions,

$$\Theta(x) = \begin{cases} 1\ if\ x > 0 \\ 0\ if\ x < 0 \end{cases} \tag{5.122}$$

come into play because, the state $|\mathbf{k}'m_s'\rangle$ should be part of the ground state $|0\rangle$ and the state $|\mathbf{k}''m_s''\rangle$ should be part of the excited state $|A\rangle$. In order that the matrix element needed for $E_0^{(2)}$ is unequal to zero, we must have

$$|\mathbf{k}'| = |\mathbf{k}+\mathbf{q}| \leq k_F \quad ; \quad |\mathbf{k}''| = |\mathbf{k}| > k_F$$

With the notation

$$\Theta_{\mathbf{k},\mathbf{k+q}} = \Theta(k_F - |\mathbf{k}+\mathbf{q}|)\, \Theta(|\mathbf{k}| - k_F) \tag{5.123}$$

we obtain completely analogously,

$$\begin{aligned} \langle \mathbf{k}'m_s'|c^\dagger_{\mathbf{k+q}\uparrow}\, c_{\mathbf{k}\downarrow}|\mathbf{k}''m_s''\rangle &\rightarrow \Theta_{\mathbf{k},\mathbf{k+q}}\, \delta_{\mathbf{k},\mathbf{k}''}\, \delta_{\mathbf{k}',\mathbf{k+q}}\, \frac{1}{\hbar}\, \langle m_s'|s_+|m_s''\rangle \\ \langle \mathbf{k}'m_s'|c^\dagger_{\mathbf{k+q}\downarrow}\, c_{\mathbf{k}\uparrow}|\mathbf{k}''m_s''\rangle &\rightarrow \Theta_{\mathbf{k},\mathbf{k+q}}\, \delta_{\mathbf{k},\mathbf{k}''}\, \delta_{\mathbf{k}',\mathbf{k+q}}\, \frac{1}{\hbar}\, \langle m_s'|s_-|m_s''\rangle \end{aligned}$$

The spin operators $s_\pm$ are defined as usual, in terms of the Pauli spin matrices $\sigma_x,\ \sigma_y,\ \sigma_z$ ($s_{x,\ y,\ z} = \frac{\hbar}{2}\sigma_{x,\ y,\ z}$):

$$s_+ = \frac{\hbar}{2}(\sigma_x + i\sigma_y) = \hbar\begin{pmatrix}0\ 1\\0\ 0\end{pmatrix} \quad ; \quad s_- = \frac{\hbar}{2}(\sigma_x - i\sigma_y) = \hbar\begin{pmatrix}0\ 0\\1\ 0\end{pmatrix} \tag{5.124}$$

The action of s_+ and s_- on the spinors is easy to see

$$s_+ \begin{pmatrix} 1 \\ 0 \end{pmatrix} = 0 \quad ; \quad s_+ \begin{pmatrix} 0 \\ 1 \end{pmatrix} = \hbar \begin{pmatrix} 1 \\ 0 \end{pmatrix} \tag{5.125}$$

$$s_- \begin{pmatrix} 1 \\ 0 \end{pmatrix} = \hbar \begin{pmatrix} 0 \\ 1 \end{pmatrix} \quad ; \quad s_- \begin{pmatrix} 0 \\ 1 \end{pmatrix} = 0 \tag{5.126}$$

The energy differences in the denominator of the energy correction $E_0^{(2)}$ (5.118) are obviously

$$E_0^{(0)} - E_A^{(0)} = (\varepsilon(\mathbf{k}+\mathbf{q}) - \varepsilon(\mathbf{k}))$$

We now have the perturbation correction in the second order as

$$\begin{aligned} E_0^{(2)} = &\frac{J^2}{4N^2} \sum_{\mathbf{k},\mathbf{q}} \frac{\Theta_{\mathbf{k},\mathbf{k}+\mathbf{q}}}{\varepsilon(\mathbf{k}+\mathbf{q}) - \varepsilon(\mathbf{k})} \sum_{i,j} \sum_{m'_s, m''_s} \sum_{f'} e^{-i\mathbf{q}\cdot(\mathbf{R}_i - \mathbf{R}_j)} \\ &\times \langle f | \left(2S_i^z \langle m'_s | s_z | m''_s \rangle + S_i^+ \langle m'_s | s_- | m''_s \rangle + S_i^- \langle m'_s | s_+ | m''_s \rangle \right) | f' \rangle \\ &\times \langle f' | \left(2S_j^z \langle m''_s | s_z | m'_s \rangle + S_j^+ \langle m''_s | s_- | m'_s \rangle + S_j^- \langle m''_s | s_+ | m'_s \rangle \right) | f \rangle \end{aligned} \tag{5.127}$$

This expression can be somewhat simplified by exploiting the two completeness relations

$$\sum_{f'} | f' \rangle \langle f' | = \mathbb{1} \tag{5.128}$$

$$\sum_{m''_s} | m''_s \rangle \langle m''_s | = \mathbb{1} \tag{5.129}$$

Using them, we get as an intermediate result:

$$\begin{aligned} E_0^{(2)} = &\frac{J^2}{4N^2} \sum_{\mathbf{k},\mathbf{q}} \sum_{i,j} \sum_{m'_s} \frac{\Theta_{\mathbf{k},\mathbf{k}+\mathbf{q}} e^{-i\mathbf{q}\cdot(\mathbf{R}_i - \mathbf{R}_j)}}{\varepsilon(\mathbf{k}+\mathbf{q}) - \varepsilon(\mathbf{k})} * \\ &* \Big[\langle f | \langle m'_s | \Big\{ S_i^z (4S_j^z (s_z)^2 + 2S_j^+ (s_z s_-) + 2S_j^- (s_z s_+)) \\ &+ S_i^+ (2S_j^z (s_- s_z) + S_j^+ (s_-)^2 + S_j^- (s_- s_+)) \\ &+ S_i^- (2S_j^z (s_+ s_z) + S_j^+ (s_+ s_-) + S_j^- (s_+)^2) \Big\} | m'_s \rangle | f \rangle \Big] \end{aligned} \tag{5.130}$$

One can easily see that the following relations are valid:

$$s_z^2 = \frac{\hbar^2}{4} \quad ; \quad s_+^2 = s_-^2 = 0 \tag{5.131}$$

$$s_+ \, s_- = \hbar^2 \begin{pmatrix} 1 & 0 \\ 0 & 0 \end{pmatrix} \quad ; \quad s_- \, s_+ = \hbar^2 \begin{pmatrix} 0 & 0 \\ 0 & 1 \end{pmatrix} \tag{5.132}$$

$$s_+ \, s_z = \frac{\hbar^2}{2} \begin{pmatrix} 0 & -1 \\ 0 & 0 \end{pmatrix} \quad ; \quad s_z \, s_+ = \frac{\hbar^2}{2} \begin{pmatrix} 0 & 1 \\ 0 & 0 \end{pmatrix} \tag{5.133}$$

$$s_- \, s_z = \frac{\hbar^2}{2} \begin{pmatrix} 0 & 0 \\ 1 & 0 \end{pmatrix} \quad ; \quad s_z \, s_- = \frac{\hbar^2}{2} \begin{pmatrix} 0 & 0 \\ -1 & 0 \end{pmatrix} \tag{5.134}$$

Then, while calculating the trace $\sum_{m_s'} \langle m_s' | \cdots | m_s' \rangle$, a number of terms vanish. What remains of (5.130) is

$$\begin{aligned} E_0^{(2)} &= \frac{J^2 \hbar^2}{4N^2} \sum_{\mathbf{k},\mathbf{q}} \sum_{i,j} \frac{\Theta_{\mathbf{k},\mathbf{k}+\mathbf{q}} e^{-i\mathbf{q}\cdot(\mathbf{R}_i - \mathbf{R}_j)}}{\varepsilon(\mathbf{k}+\mathbf{q}) - \varepsilon(\mathbf{k})} \\ &\times \langle f | 2 S_i^z S_j^z + S_i^+ S_j^- + S_i^- S_j^+ | f \rangle \end{aligned} \tag{5.135}$$

The sum is exactly the double of the scalar product $(\mathbf{S}_i \cdot \mathbf{S}_j)$. Therefore

$$E_0^{(2)} = \frac{J^2 \hbar^2}{2N^2} \sum_{\mathbf{k},\mathbf{q}} \sum_{i,j} \Theta_{\mathbf{k},\mathbf{k}+\mathbf{q}} \, e^{-i\mathbf{q}\cdot(\mathbf{R}_i - \mathbf{R}_j)} \frac{\langle f | \mathbf{S}_i \cdot \mathbf{S}_j | f \rangle}{\varepsilon(\mathbf{k}+\mathbf{q}) - \varepsilon(\mathbf{k})} \tag{5.136}$$

Therefore, the energy correction in the second order can obviously be considered as the eigenvalue of an *effective* Hamiltonian, which is of Heisenberg type:

$$H_f^{RKKY} = -\sum_{i,j} J_{ij}^{RKKY} \mathbf{S}_i \mathbf{S}_j \tag{5.137}$$

We have therefore shown that the electron gas mediates an indirect coupling between the localized moments.

The coupling constants J_{ij}^{RKKY} show an oscillatory behaviour as a function of the separation $|\mathbf{R}_i - \mathbf{R}_j|$:

$$J_{ij}^{RKKY} = -\frac{J^2 \hbar^2}{2N^2} \sum_{\mathbf{k},\mathbf{q}} \sum_{i,j=1}^{2} \sum_{m_s'} \Theta_{\mathbf{k},\mathbf{k}+\mathbf{q}} \frac{e^{-i\mathbf{q}\cdot(\mathbf{R}_i - \mathbf{R}_j)}}{\varepsilon(\mathbf{k}+\mathbf{q}) - \varepsilon(\mathbf{k})} \tag{5.138}$$

We want to evaluate this expression in the *effective mass approximation*:

$$\varepsilon(\mathbf{k}) = \frac{\hbar^2 k^2}{2m^*} \quad ; \quad \varepsilon_F = \frac{\hbar^2 k_F^2}{2m^*} \tag{5.139}$$

For this, we first convert the two summations into integrals:

$$\frac{1}{N^2}\sum_{\mathbf{k},\mathbf{q}} \rightarrow \frac{V^2}{N^2(2\pi)^6}\int d^3k \int d^3q$$

Taking $\mathbf{k}' = \mathbf{k} + \mathbf{q}$, we have to evaluate

$$J_{ij}^{RKKY} = \frac{m^* J^2\, V^2}{N^2\,(2\pi)^6}\int_{k'\leq k_F} d^3k' \int_{k\geq k_F} d^3k \frac{e^{-i\mathbf{q}\cdot(\mathbf{R}_i-\mathbf{R}_j)}}{k^2-k'^2} \tag{5.140}$$

In the polar coordinates, treating $\mathbf{R}_{ij} = \mathbf{R}_i - \mathbf{R}_j$ as the polar axis, the integration over the angles is straight forward:

$$\int_{-1}^{+1} dx e^{\pm ikR_{ij}x} = \frac{2\sin(kR_{ij})}{kR_{ij}}$$

Using this, we have the intermediate result

$$J_{ij}^{RKKY} = \frac{m^* J^2\, V^2}{N^2\, 4\pi^4 R_{ij}^2}\int_0^{k_F} dk'\, k' \int_{k_F}^{\infty} dk\, k \frac{\sin(k'R_{ij})\, \sin(kR_{ij})}{k^2-k'^2} \tag{5.141}$$

In the second integral, the lower limit can be set to zero because the double integral

$$\int_0^{k_F} dk'\, k' \int_0^{k_F} dk\, k \cdots = 0$$

is antisymmetric with respect to the interchange $k \leftrightarrow k'$. Further, we have (Problem 5.7)

$$\int_0^{\infty} dk\, k\, \frac{\sin(kR_{ij})}{k^2-k'^2} = \frac{\pi}{2}\cos(k'R_{ij}) \tag{5.142}$$

We still have to evaluate

$$\begin{aligned}
&\int_0^{k_F} dk'\, k' \sin(k'R_{ij})\, \cos(k'R_{ij}) = \frac{1}{2}\int_0^{k_F} dk'\, k' \sin(2k'R_{ij}) \\
&= -\frac{1}{4}\frac{d}{dR_{ij}}\int_0^{k_F} dk' \cos(2k'R_{ij}) \\
&= -\frac{1}{8}\frac{d}{dR_{ij}}\frac{1}{R_{ij}}(\sin(2k_F R_{ij}) \\
&= \frac{1}{2}k_F^4\, 4R_{ij}^2\, \frac{\sin(2k_F R_{ij}) - 2k_F R_{ij}\cos(2k_F R_{ij})}{(2R_{ij}k_F)^4}
\end{aligned}$$

We define $F(x)$ (Fig. 5.15)

$$F(x) = \frac{\sin x - x \cos x}{x^4} \tag{5.143}$$

by which we can express the *RKKY-coupling constant*:

$$J_{ij}^{RKKY} = \frac{J^2 \, k_F^6}{\varepsilon_F} \frac{\hbar^2 \, V^2}{N^2 \, (2\pi)^3} \, F(2k_F R_{ij}) \tag{5.144}$$

The exchange constant J_{ij}^{RKKY} gets an oscillatory behaviour through the function $F(x)$ as a function of the distance between the magnetic ions, i.e. depending on the separation, the interaction is either ferromagnetic or antiferromagnetic. This explains the different behaviour, listed at the beginning of this section, for the alloys CuMn, CuFe, etc., where the concentration of Mn or Fe, respectively, decides the average distance $\bar{\mathbf{R}}_{ij}$ between the magnetic ions.

Remarkable thing to notice is the relatively large range of the RKKY interaction which, for large distances goes as

$$J_{ij}^{RKKY} \sim \frac{1}{R_{ij}^3}. \tag{5.145}$$

In contrast, the direct exchange interaction decreases exponentially with the distance and therefore is of short range.

The RKKY-interaction is a second-order effect

$$J_{ij}^{RKKY} \sim J^2 \tag{5.146}$$

which very sensitively depends on the electron density $n_e = N_e/V$ of the non-magnetic matrix. Since

$$k_F = (3\pi^2 n_e)^{1/3} \quad ; \quad \varepsilon_F = \frac{\hbar^2}{2m^*}(3\pi^2 n_e)^{2/3}$$

we see that in the effective mass approximation, approximately

$$J_{ij}^{RKKY} \sim n_e^{4/3} \tag{5.147}$$

Notice, however, that the electron density also determines via k_F the period of oscillation.

5.3.2 Superexchange

The RKKY mechanism is an example for an indirect exchange in a *metallic* Heisenberg magnet, which is realized in metallic compounds of rare earths such as Gd (see class Localized Magnetism at the beginning of this chapter). Now we want to

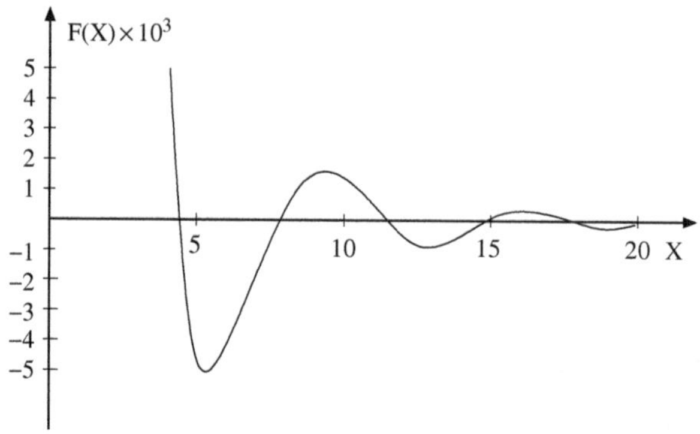

Fig. 5.15 The function $F(x)$ (5.143), which determines the oscillatory behaviour of the (indirect) RKKY exchange integrals

introduce a mechanism which is relevant to *magnetic insulators*. An interaction of the RKKY-type does not come into question, since $J_{ij}^{RKKY} \sim n_e^{4/3}$ and the density n_e of the conduction electrons in an insulator is zero. The name *superexchange* is given because of the relatively large distance over which the exchange interaction operates (Fig. 5.16). This interaction appears to be realized, in particular, in magnetic oxides or difluorides of transition metals such as,

$$MnO, \; NiO, \; MnF_2, \; FeF_2 \; CoF_2, \cdots$$

The partially filled, and therefore magnetic d-shells of Mn^{2+}, Ni^{2+}, Fe^{2+} or Co^{2+}-ions are separated from each other, in general, by more than 4 Å, so that the direct overlap of the d-wavefunctions is negligibly small. The actual exchange coupling results via non-magnetic ions like either oxygen or fluorine ions that lie in between the magnetic ions. This interaction, as a rule, is always *antiferromagnetic* for the above substances. With the help of a simple *cluster model*, we would like to understand how the diamagnetic ion can mediate a coupling between the magnetic ions.

The cluster consists of two magnetic ions such as, for example, Mn^{2+} with $S = 5/2$ located, respectively, at $\mathbf{R}_1$ and $\mathbf{R}_2$ and a diamagnetic ion such as O^{2-} located in between them at $\mathbf{R}_0$ (Fig. 5.16). The p-wavefunction of the anion overlaps ("mixes")

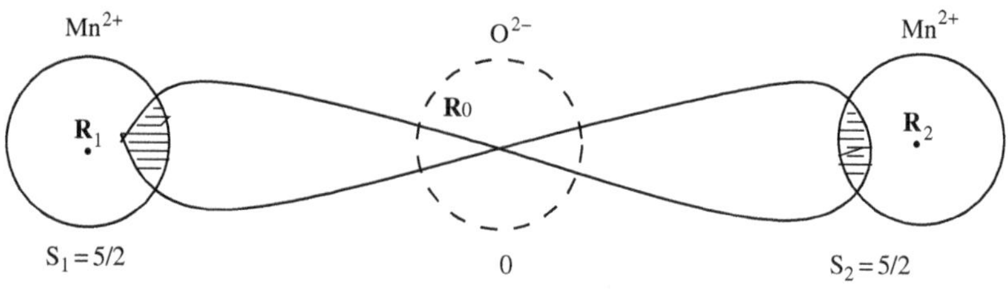

Fig. 5.16 $Mn^{2+}O^{2-}$-cluster model to explain the mechanism of superchange

very strongly with the d-wavefunction of the cations. Because of this, electronic transitions are possible, however, with the following restrictions:

1. The O^{2-} does not like to accept an extra electron, whereas it rather readily gives up an electron to the neighbouring ion.
   The Mn^{2+} ion, on the other hand, rather readily accepts an electron, but does not like to give up an electron.
2. Hund's rules (Sect. 2.1) must be satisfied. Since the 3d-shell of the Mn^{2+} is already half-filled, the electron that is "hopping" from O^{2-} to Mn^{2+} must orient its spin antiparallel to the spin of Mn^{2+}-spin.

The above mentioned facts are used to construct the following *semi-classical model*:

(a) The magnetic Mn^{2+}-ions are treated as classical spins of constant length but of variable orientation:

$$\mathbf{S}_1 \cdot \mathbf{S}_2 = S^2 \cos\vartheta \tag{5.148}$$

(b) In view of (1) and (2), only the two p-electrons of the O^{2-}-ion come into play for the hopping process. Their spins are antiparallel to each other and are oriented antiparallel to the Mn^{2+}-spins at $\mathbf{R}_1$ and $\mathbf{R}_2$, respectively.

To (b), some more remarks have to be added. $\mathbf{S}_1$ and $\mathbf{S}_2$ are oriented with respect to each other at an angle ϑ. We assume that the direction of $\mathbf{S}_1$ defines the z-axis. In view of (2), the p-electron at $\mathbf{S}_1$ is in the spin state:

$$\begin{pmatrix} 0 \\ 1 \end{pmatrix}$$

Now, how does the spin state of the electron look, when the electron "hops" to $\mathbf{R}_2$ and orients itself antiparallel to $\mathbf{S}_2$? Remember that $\mathbf{S}_2$ is at an angle ϑ with the z-axis fixed by $\mathbf{S}_1$. We find the spin state of the electron from the eigenvalue equation:

$$(\sigma \cdot \mathbf{e}) \begin{pmatrix} \chi_1 \\ \chi_2 \end{pmatrix} = \lambda \begin{pmatrix} \chi_1 \\ \chi_2 \end{pmatrix} \tag{5.149}$$

$(\sigma \cdot \mathbf{e})$ is the projection of the electron spin operator σ

$$\sigma \equiv (\sigma_x,\ \sigma_y,\ \sigma_z) \tag{5.150}$$

onto the direction $\mathbf{e}$ of $\mathbf{S}_1$:

$$\mathbf{e} = (\sin\vartheta\cos\phi,\ \sin\vartheta\sin\phi,\ \cos\vartheta) \tag{5.151}$$

Since ϕ can be chosen arbitrarily, we set $\phi = 0$. With the Pauli's spin matrices $\sigma_{x,y,z}$ given by (5.79), it immediately follows that

$$(\sigma \cdot \mathbf{e}) = \begin{pmatrix} \cos\vartheta & \sin\vartheta \\ \sin\vartheta & -\cos\vartheta \end{pmatrix} \tag{5.152}$$

The eigenvalues λ are given by the secular equation

$$det(\sigma \cdot \mathbf{e} - \lambda\,\mathbb{1}) \stackrel{!}{=} 0 = -\cos^2\vartheta + \lambda^2 - \sin^2\vartheta \tag{5.153}$$

As is to be expected, we get

$$\lambda = \pm 1 \tag{5.154}$$

The electron spin at $\mathbf{R}_2$ can also, naturally, take the values $\pm\,\hbar/2$ only. We need the eigenstate for the eigenvalue $\lambda = -1$. For this, in view of (5.149) and (5.152), we have

$$(\cos\vartheta + 1)\,\chi_1 + \sin\vartheta\,\chi_2 = 0$$

which means

$$\frac{\chi_1}{\chi_2} = -\frac{\sin\vartheta}{1+\cos\vartheta} = -\tan\frac{\vartheta}{2}$$

so that the normalized state $|\chi\rangle^{(-)}$ is given by

$$|\chi\rangle^{(-)} \equiv \begin{pmatrix} \sin\vartheta/2 \\ -\cos\vartheta/2 \end{pmatrix} \tag{5.155}$$

We now come back to our cluster model. The following are the *cluster configurations* allowed within our simple model:

1. Both the p-electrons reside at the diamagnetic anion:

$$|1\rangle = |\phi(\mathbf{r}_1 - \mathbf{R}_0)\rangle \begin{pmatrix} 0 \\ 1 \end{pmatrix} \otimes |\phi(\mathbf{r}_2 - \mathbf{R}_0)\rangle \begin{pmatrix} 1 \\ 0 \end{pmatrix} \tag{5.156}$$

   Let this state have the energy

$$E_1 = \varepsilon \tag{5.157}$$

   Note that our semiclassical model disregards the indistinguishability of the two electrons.

2. One of the p-electrons is at the cation 1:

$$|2\rangle = |\phi(\mathbf{r}_1 - \mathbf{R}_1)\rangle \begin{pmatrix} 0 \\ 1 \end{pmatrix} \otimes |\phi(\mathbf{r}_2 - \mathbf{R}_0)\rangle \begin{pmatrix} 1 \\ 0 \end{pmatrix} \tag{5.158}$$

Let the energy of this state be

$$E_2 = \varepsilon + U \tag{5.159}$$

U is something like the energy of the Coulomb interaction of the p-electron with the d-electrons of Mn^{2+}.

3. The p-electron is at the cation 2. This in comparison to (2) is not a new situation. Now, $\mathbf{S}_2$ defines the z-direction.

$$|3\rangle = |\phi(\mathbf{r}_1 - \mathbf{R}_2)\rangle \begin{pmatrix} 0 \\ 1 \end{pmatrix} \otimes |\phi(\mathbf{r}_2 - \mathbf{R}_0)\rangle \begin{pmatrix} 1 \\ 0 \end{pmatrix} \tag{5.160}$$

Let the energy of this state be

$$E_3 = \varepsilon + U \tag{5.161}$$

4. The two p-electrons reside at the two Mn^{2+} ions, one at each of them:

$$|4\rangle = |\phi(\mathbf{r}_1 - \mathbf{R}_1)\rangle \begin{pmatrix} 0 \\ 1 \end{pmatrix} \otimes |\phi(\mathbf{r}_2 - \mathbf{R}_2)\rangle \begin{pmatrix} \sin\vartheta/2 \\ -\cos\vartheta/2 \end{pmatrix} \tag{5.162}$$

Let the energy of this state be

$$E_4 = \varepsilon + U + V \tag{5.163}$$

Within the framework of the cluster model, the states $|1\rangle$ to $|4\rangle$ are considered to constitute a complete set. The eigenstate of the Hamiltonian will be a linear combination of these states.

Let us consider the matrix elements of H in this basis:

$$H_{12} = \langle 1|H|2\rangle = H_{21} = H_{13} = \langle 1|H|3\rangle = H_{31} = t \tag{5.164}$$

t is the so-called *transfer matrix element*, which is certainly small because it involves transitions, which are "virtual", i.e. energetically costly:

$$H_{24} = \langle 2|H|4\rangle = H_{42} = t\,(1\;0) \begin{pmatrix} \sin\vartheta/2 \\ -\cos\vartheta/2 \end{pmatrix} = t\sin\vartheta/2 \tag{5.165}$$

$\sin\vartheta/2$ is the probability that the $\uparrow$-p-electron orients itself at $\mathbf{R}_2$ antiparallel to $\mathbf{S}_2$. We naturally also have

$$H_{34} = \langle 3|H|4\rangle = H_{43} = t\sin\vartheta/2 \tag{5.166}$$

We consider the simultaneous hopping of both p-electrons as unlikely, i.e. $H_{14} = H_{41} \approx 0$. Furthermore, the double hopping of a p-electron from the left Mn^{2+} to the right Mn^{2+} or vice versa may be negligible, i.e. $H_{23} = H_{32} \approx 0$. In view of the above, the model Hamiltonian, in the basis $|1\rangle, \cdots |4\rangle$, has the following form:

$$H \equiv \begin{pmatrix} \varepsilon & t & t & 0 \\ t & \varepsilon + U & 0 & t \sin \vartheta/2 \\ t & 0 & \varepsilon + U & t \sin \vartheta/2 \\ 0 & t \sin \vartheta/2 & t \sin \vartheta/2 & \varepsilon + U + V \end{pmatrix} \tag{5.167}$$

We obtain the eigenvalues E from the 4×4 secular determinant

$$det(H - E) \stackrel{!}{=} 0 \tag{5.168}$$

This condition, with the following notation,

$$x = E - \varepsilon \quad ; \quad 1 - \cos \vartheta = 2 \sin^2 \frac{\vartheta}{2} \tag{5.169}$$

can be written after a lengthy but straight forward evaluation as

$$\begin{aligned} 0 \stackrel{!}{=} (x - U)[x^3 - x^2(2U + V) - x(t^2(3 - \cos \vartheta) \\ - U(U + V)) + 2t^2(U + V)]. \end{aligned} \tag{5.170}$$

One can immediately see that one of the four solutions is

$$x_4 = U \tag{5.171}$$

To solve the remaining cubic equation, we use the fact that the transfer integral t defined in (5.164) is a small quantity. So, we make a polynomial ansatz for x and arrange according to the powers of t. For $t = 0$, we have the following equation:

$$x^3 - x^2(2U + V) + xU(U + V) \stackrel{!}{=} 0 \tag{5.172}$$

whose solutions are, of course

$$x_1^{(0)} = 0 \quad ; \quad x_2^{(0)} = U \quad ; \quad x_3^{(0)} = U + V \tag{5.173}$$

Since we can take

$$U, \; V \gg t \tag{5.174}$$

and also since we are interested for $t \neq 0$, only in the ground state, we can confine ourselves to the solution which comes out of $x_1^{(0)}$. Therefore we make the ansatz

$$x_1 = \sum_{n=1}^{\infty} \alpha_n \, t^n \tag{5.175}$$

With this, we successively solve the cubic equation

$$0 = x^3 - x^2(2U + V) - x(t^2(3 - \cos\vartheta) - U(U + V)) + 2t^2(U + V) \quad (5.176)$$

From the term linear in t we get

$$0 = \alpha_1\, t\, U(U + V) \;\Leftrightarrow\; \alpha_1 = 0 \quad (5.177)$$

For the terms $\sim t^2$, it holds

$$0 = \alpha_2\, t^2\, U(U + V) + 2t^2(U + V) \;\Leftrightarrow\; \alpha_2 = -\frac{2}{U} \quad (5.178)$$

We get α_3 from the equation

$$0 = \alpha_3\, t^3\, U(U + V) \;\Leftrightarrow\; \alpha_3 = 0 \quad (5.179)$$

The terms $\sim t^4$ satisfy the equation

$$0 = -\alpha_2^2\, t^4\, (2U + V) - \alpha_2\, t^4\, (3 - \cos\vartheta) + \alpha_4\, t^4\, U(U + V)$$

Substituting for α_2 gives

$$\begin{aligned} 0 &= -\frac{4}{U^2}(2U + V) + \frac{2}{U}(3 - \cos\vartheta) + \alpha_4 U(U + V) \\ &= -\frac{2}{U^2}(U + 2V + U\cos\vartheta) + \alpha_4 U(U + V) \end{aligned}$$

From this we get

$$\alpha_4 = \frac{2}{U^2}\left(\frac{U + 2V}{U(U + V)} + \frac{\cos\vartheta}{U + V}\right) \quad (5.180)$$

The term $\sim t^5$ does not have any contribution:

$$0 = +\alpha_5\, t^5\, U(U + V) \;\Leftrightarrow\; \alpha_5 = 0 \quad (5.181)$$

Then the solution for the ground state energy is given by

$$E_0 = \left[\varepsilon - \frac{2t^2}{U} + \frac{2(U + 2V)t^4}{U^3(U + V)}\right] + 2\left(\frac{t^2}{U}\right)^2 \frac{\cos\vartheta}{U + V} + \mathcal{O}(t^6) \quad (5.182)$$

The rest $\mathcal{O}(t^6)$ contains an additive term $\sim \cos^2\vartheta$. We can express $\cos\vartheta$ in terms of the scalar product $\mathbf{S}_1 \cdot \mathbf{S}_2$ of the two spins. Then we can write E_0 as

$$E_0 = E_0^{(0)} + 2\left(\frac{t^2}{U}\right)^2 \frac{\mathbf{S}_1 \cdot \mathbf{S}_2}{S^2(U + V)} + \mathcal{O}(t^6) \quad (5.183)$$

Here $E_0^{(0)}$ is an unimportant constant

$$E_0^{(0)} = \varepsilon - \frac{2t^2}{U} + \frac{2(U+2V)t^4}{U^3(U+V)} \tag{5.184}$$

For insulators, the transfer integral t is naturally small. Therefore, the energy correction to the ground state energy up to t^5 is quite a reasonable approximation. If we accept this approximation, then the Hamiltonian of the cluster model can be replaced by the following *effective Heisenberg Hamiltonian*:

$$H_{SE} = -J_{12}^{SE}\, \mathbf{S}_1 \cdot \mathbf{S}_2 \quad ; \quad J_{12}^{SE} = -\frac{2}{S^2}\, \frac{t^4}{U^2(U+V)}. \tag{5.185}$$

We recognize that the superexchange mechanism that we have discussed here leads to an indirect *antiferromagnetic* coupling between the localized spins $\mathbf{S}_1$ and $\mathbf{S}_2$:

$$J_{12}^{SE} < 0 \tag{5.186}$$

The important thing is that even though a direct exchange between the spins is ruled out because of the large separation, the model operator obtained here is again of the Heisenberg type.

We want to close this section with a few concluding remarks:

1. The rest $\mathcal{O}(t^6)$ neglected in E_0 contains a $\cos^2\theta$ term. This leads to a correction which includes $(\mathbf{S}_1 \cdot \mathbf{S}_2)^2$ in the model Hamiltonian. Therefore, we should treat the effective exchange operator as the lowest order term in an expansion in powers of $(\mathbf{S}_1 \cdot \mathbf{S}_2)$.
2. In the literature, there exist a number of modifications and extentions to the superexchange mechanism that we have discussed here within the framework of a semiclassical cluster model. Kanamori and Goodenough [5, 6], e.g. have proposed the model Fig. 5.17:
   One of the p-electrons of the anion A^{2-} shifts to the cation C^{2+} and orients itself, according to the Hund's rule with respect to the spin there. The remaining p-electron makes the anion paramagnetic and, therefore, allows a direct exchange with the other magnetic cation. This leads to an indirect coupling of the two C-spins, which can be either ferro- or antiferromagnetic.

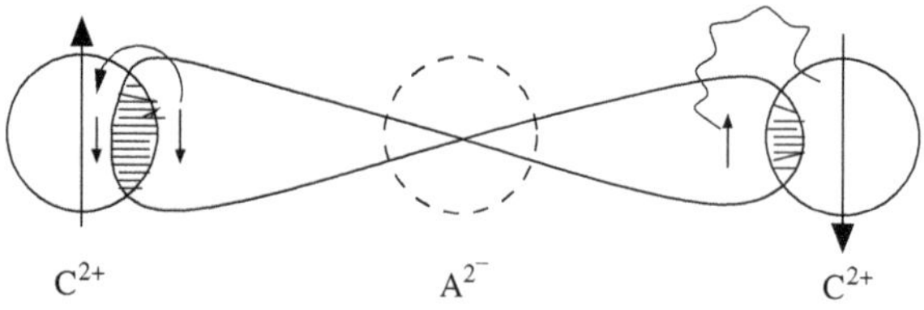

Fig. 5.17 Alternative cluster model according to J. Kanamori [5] and J.B. Goodenough [6] to explain the superexchange mechanism

5.3.3 Double Exchange

In this section, we discuss a third type of indirect exchange interaction, which leads to a model Hamiltonian which also has the characterestic scalar product $(\mathbf{S}_i \cdot \mathbf{S}_j)$. It contains, however, not only the bilinear term $(\mathbf{S}_i \cdot \mathbf{S}_j)$, but also higher powers of it. The highest power is determined by the magnitude of the localized spin:

$$H_{DE} \sim \sum_{n=0}^{2S} J_n(S)(\mathbf{S}_i \cdot \mathbf{S}_j)^n \qquad (5.187)$$

The "double exchange" is typical for systems, for which, the magnetic ion can exist in two different valence states. Jonker and van Sauten [7, 8] have discovered compounds of the form

$$(La_{1-x}M_x)MnO_3 \quad ; \quad M = Ba,\ Ca,\ Sr$$

which had unusual electric and magnetic properties. When the concentration x of the divalent nonmagnetic (!) M^{2+} becomes larger than a critical value x_c, the electrical conductivity increases and the sample, which was originally nonmagnetic, suddenly becomes ferromagnetic. Today these materials are under intensive investigation as the so-called "CMR systems" because of their "colossal magneto-resistance"-behaviour due to the simultaneous metal–insulator and ferromagnetic–paramagnetic transition [9].

When in the compound

$$La^{3+}Mn^{3+}O_3 \qquad (x = 0)$$

a trivalent La^{3+}-ion is substituted by a divalent alkaline earth M^{2+} ion, then this ion contributes only two electrons to the bonding. Therefore the required third electron must be taken from a Mn^{3+}-ion. That means, in such a compound, the manganese exists in a valence mixture of Mn^{3+} and Mn^{4+}:

$$(La^{3+}_{1-x}\ M^{2+}_x)\ Mn^{3+}_{1-x}\ Mn^{4+}_x\ O_3 \qquad (x \neq 0)$$

The sudden appearance of ferromagnetism for $x \geq x_c$ was first ascribed by Zener [10] to an electron hopping between Mn^{3+} and Mn^{4+}. This electron motion, in a way analogous to the RKKY interaction (Sect. 5.3.1), leads to an indirect coupling of the Mn spins, which, as a rule, turns out to be ferromagnetic. We will discuss this mechanism, again using a *cluster model* (Fig. 5.18)

The hopping of the "excess" electron with spin $\mathbf{s}$ does not take place directly but via the diamagnetic O^{2-} ion, which lies in between the two magnetic Mn ions. We denote the mobile electron by $1 \uparrow$ and the two O^{2-}-electrons by $2 \uparrow$ and $3 \downarrow$, respectively. Then, we can think of two equivalent transfer processes, which con-

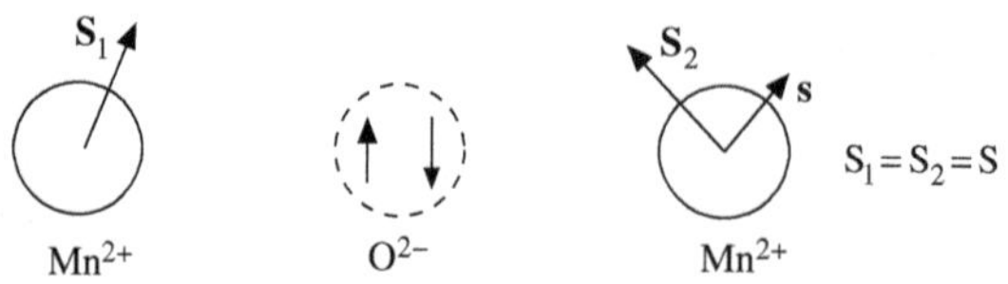

Fig. 5.18 Pictorial representation of a cluster to explain the double exchange mechanism

serve the negative charge of the oxygen atom. The two processes are symbolically represented in the following.

The first proposal was by **Zener** [10]:

$$\underbrace{Mn^{4+}}_{-}\quad\underbrace{O^{2-}}_{2\uparrow,3\downarrow}\quad\underbrace{Mn^{3+}}_{1\uparrow}\;\longrightarrow\;\underbrace{Mn^{3+}}_{2\uparrow}\quad\underbrace{O^{2-}}_{3\downarrow,1\uparrow}\quad\underbrace{Mn^{4+}}_{-}$$

Since this involves two simultaneous electron jumps, it is called the *double exchange*.

The second proposal was by **Anderson and Hasegawa** [11]:

$$\begin{aligned}\underbrace{Mn^{4+}}_{-}\quad\underbrace{O^{2-}}_{2\uparrow,3\downarrow}\quad\underbrace{Mn^{3+}}_{1\uparrow}\;&\longrightarrow\;\underbrace{Mn^{3+}}_{2\uparrow}\quad\underbrace{O^{-}}_{3\downarrow}\quad\underbrace{Mn^{3+}}_{1\uparrow}\\&\longrightarrow\;\underbrace{Mn^{3+}}_{2\uparrow}\quad\underbrace{O^{2-}}_{3\downarrow,1\uparrow}\quad\underbrace{Mn^{4+}}_{-}\end{aligned} \tag{5.188}$$

Both the processes finally achieve the same, namely, the transfer of the mobile electron from one Mn ion to the other. At the Mn site, the electron spin interacts with the localized Mn spin via the "local exchange", which was already discussed in Sect. 5.3.1 and is given by (5.97):

$$H_{sf} = -J\sum_{i=1}^{2} \mathbf{s}_i \cdot \mathbf{S}_i \tag{5.189}$$

That means, the electron should not be arranged, as in the case of superexchange (Sect. 5.3.2), according to the Hund's rule in the d-shell, but, as in the case of RKKY mechanism (Sect. 5.3.1), as a quasi-free electron interacting with a localized spin. The difference in comparison to RKKY coupling is that, now, we *cannot* treat H_{sf} as a perturbation. Since we are dealing with a *bad conductor*, the intra-atomic coupling is very much stronger than the "hopping" matrix element. Perturbation theory with respect to H_{sf} is not applicable.

The electron spin $\mathbf{s}_i$ and the localized Mn spin $\mathbf{S}_i$ can couple to give $S_{eM} = S\pm\frac{1}{2}$. The corresponding energies can be easily calculated. From

$$\mathbf{S}_{eM}^2 = \hbar^2 S_{eM}(S_{eM}+1) = (\mathbf{s}+\mathbf{S})^2 = \frac{3}{4}\hbar^2 + S(S+1)\hbar^2 + 2(\mathbf{s}\cdot\mathbf{S})$$

it follows that

$$\mathbf{s}\cdot\mathbf{S} = \frac{\hbar^2}{2}\begin{cases} -(S+1) & if\ S_{eM} = S - 1/2 \\ +S & if\ S_{eM} = S + 1/2 \end{cases} \tag{5.190}$$

The exchange interaction between the mobile electron and the magnetic ion therefore has the energies

$$\varepsilon_\alpha = -\frac{1}{2}\,\hbar^2\,J\,S \quad ; \quad \varepsilon_\beta = +\frac{1}{2}\,\hbar^2\,J\,(S+1) \tag{5.191}$$

Let the spin $\mathbf{S}_1$ define the z-axis. Then, the energies ε_α and ε_β belong to the following spin states of the mobile electron at $\mathbf{S}_1$:

$$|\alpha\rangle = \begin{pmatrix}1\\0\end{pmatrix} \quad ; \quad |\beta\rangle = \begin{pmatrix}0\\1\end{pmatrix} \tag{5.192}$$

One can easily see that $|\alpha\rangle$ and $|\beta\rangle$ are not the eigenstates of the Hamiltonian. If the angle between the spins $\mathbf{S}_1$ and $\mathbf{S}_2$ is ϑ, then the corresponding states at the second Mn ion are given by

$$\begin{aligned} |\alpha'\rangle &= \begin{pmatrix}\cos\vartheta/2\\ \sin\vartheta/2\end{pmatrix} = \cos(\vartheta/2)\,|\alpha\rangle + \sin(\vartheta/2)\,|\beta\rangle \\ |\beta'\rangle &= \begin{pmatrix}\sin\vartheta/2\\ -\cos\vartheta/2\end{pmatrix} = \sin(\vartheta/2)\,|\alpha\rangle - \cos(\vartheta/2)\,|\beta\rangle \end{aligned} \tag{5.193}$$

Within the framework of our simple cluster model, we treat the following four states as a *complete basis*:

$\vert 1\rangle = \vert 1\alpha\rangle$	Electron at atom 1 in state $\vert\alpha\rangle$
$\vert 2\rangle = \vert 1\beta\rangle$	Electron at atom 1 in state $\vert\beta\rangle$
$\vert 3\rangle = \vert 2\alpha'\rangle$	Electron at atom 2 in state $\vert\alpha'\rangle$
$\vert 4\rangle = \vert 2\beta'\rangle$	Electron at atom 2 in state $\vert\beta'\rangle$

Therewith it is implicitly assumed that the jump process of the excess electron takes place without a spin flip. The existence of the O^{2-}-ion is, within this model, no more relevant. It only catalyses the jump process. The matrix elements of the model Hamiltonian

$$H = H_0 + H_{sf}, \tag{5.194}$$

where H_0 describes the kinetic energy of the mobile electron, can be easily given, if as was done in Eq. (5.164), we introduce the transfer integral:

$$\langle 1\alpha, \beta|H|2\alpha', \beta'\rangle = \langle 1\alpha, \beta|H_0|2\alpha', \beta'\rangle = t\langle \alpha, \beta|\alpha', \beta'\rangle \tag{5.195}$$

Here we have used the fact that H_{sf} acts only locally. On the other hand, the diagonal elements of the full Hamiltonian are relatively simple ($\langle i, \alpha\beta|H_0|i, \alpha\beta\rangle = 0, \ i = 1, 2$):

$$H_{11} = \varepsilon_\alpha = -\frac{1}{2}\hbar^2\, J\, S = H_{33}\ ;\ H_{22} = \varepsilon_\beta = -\frac{1}{2}\hbar^2\, J\,(S+1) = H_{44} \tag{5.196}$$

For the non-diagonal elements, we get

$$\begin{aligned} H_{12} &= \langle\alpha|\beta\rangle = 0 = H_{21} = H_{34} = H_{43} \\ H_{13} &= t\langle\alpha|\alpha'\rangle = t\cos\vartheta/2 = H_{31} \\ H_{14} &= t\langle\alpha|\beta'\rangle = t\sin\vartheta/2 = H_{41} \\ H_{23} &= t\langle\beta|\alpha'\rangle = t\sin\vartheta/2 = H_{32} \\ H_{24} &= t\langle\beta|\beta'\rangle = -t\cos\vartheta/2 = H_{42} \end{aligned} \tag{5.197}$$

We thus have the following *Hamiltonian matrix*:

$$H \equiv \begin{pmatrix} -\frac{1}{2}\hbar^2 J\, S & 0 & t\cos\vartheta/2 & t\sin\vartheta/2 \\ 0 & \frac{1}{2}\hbar^2 J\,(S+1) & t\sin\vartheta/2 & -t\cos\vartheta/2 \\ t\cos\vartheta/2 & t\sin\vartheta/2 & -\frac{1}{2}\hbar^2 J\, S & 0 \\ t\sin\vartheta/2 & -t\cos\vartheta/2 & 0 & \frac{1}{2}\hbar^2 J\,(S+1) \end{pmatrix} \tag{5.198}$$

Since the basis $|i\rangle,\ i = 1, \cdots, 4$ does not consist of eigenstates, H is not diagonal in this representation. The eigenvalues are again obtained from the secular determinant:

$$\begin{aligned} det(H - E\ \mathbb{1}) &= \\ &= \left\{\left(E - \frac{\hbar^2 J}{4}\right)^2 - \left(t\cos\frac{\vartheta}{2} - \frac{\hbar^2 J}{2}\left(S + \frac{1}{2}\right)\right)^2 - t^2\sin^2\frac{\vartheta}{2}\right\} \\ &* \left\{\left(E - \frac{\hbar^2 J}{4}\right)^2 - \left(t\cos\frac{\vartheta}{2} + \frac{\hbar^2 J}{2}\left(S + \frac{1}{2}\right)\right)^2 - t^2\sin^2\frac{\vartheta}{2}\right\} \end{aligned} \tag{5.199}$$

Let us define

$$\gamma_\pm = \sqrt{\left(t\cos\vartheta/2 \pm \frac{1}{2}\hbar^2 J\left(S + \frac{1}{2}\right)\right)^2 + t^2\sin^2\vartheta/2} \tag{5.200}$$

Then, from the condition that the secular determinant vanishes, we get the following four eigenenergies of our model Hamiltonian:

$$\begin{aligned} E_{--} &= \tfrac{1}{4}\hbar^2 J - \gamma_- \quad ; \quad E_{-+} = \tfrac{1}{4}\hbar^2 J - \gamma_+ \\ E_{+-} &= \tfrac{1}{4}\hbar^2 J + \gamma_- \quad ; \quad E_{++} = \tfrac{1}{4}\hbar^2 J + \gamma_+ \end{aligned} \tag{5.201}$$

Just as in the case of the superexchange in Sect. 5.3.2, we will try to express the angle dependence ($\cos \vartheta/2$, $\sin^2 \vartheta/2$) in terms of the spin vectors $\mathbf{S}_1$ and $\mathbf{S}_1$. That is quite simple in a *semiclassical vector model* (Fig. 5.19):

$$|\mathbf{S}_1| = |\mathbf{S}_2| = S \quad ; \quad \cos \vartheta/2 = \frac{\frac{1}{2}|\mathbf{S}_1 + \mathbf{S}_2|}{S} = \frac{S_{tot}}{2S} \tag{5.202}$$

When the spin of the mobile electron is included, then the possible values are

$$S_0 = S_{tot} \pm \frac{1}{2} \tag{5.203}$$

Since the semiclassical model becomes correct in the limit $S \to \infty$, it is a good approximation to set

$$\cos \vartheta/2 \approx \frac{S_0}{2S} \tag{5.204}$$

A *quantum mechanically correct treatment* of the problem [11], whose details we will not go into here, leads to the same Hamiltonian as in (5.198) when we simply set

$$\cos \vartheta/2 = \frac{S_0 + \frac{1}{2}}{2S + 1} \tag{5.205}$$

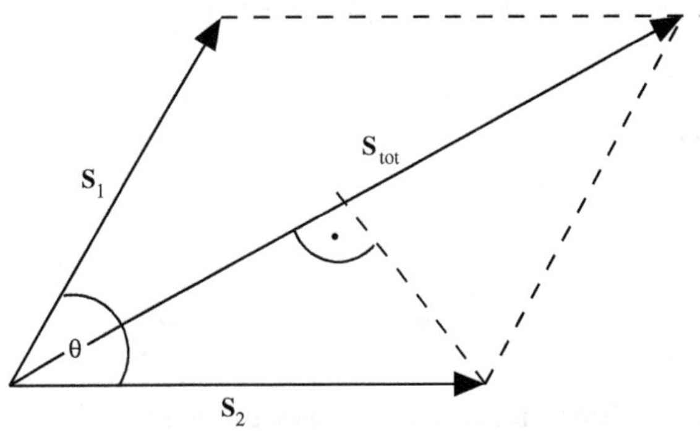

Fig. 5.19 Composition of two semiclassical spins $\mathbf{S}_1$ and $\mathbf{S}_2$ into a total spin $\mathbf{S}_{tot}$

Other than this, everything else remains unchanged. S_0 can take half-integer values between $1/2$ and $2S + 1/2$. Using this expression, we will calculate further.

The prototype materials for "double exchange" as presented at the beginning of this section are bad electric conductors, i.e. the "hopping" matrix element t is a small quantity. Therefore, we can assume $\hbar^2 J \gg t$, and then simplify the expression for $\gamma_\pm$:

$$\begin{aligned}
\gamma_\pm &= \left\{ t^2 \pm \hbar^2 J \left(S + \frac{1}{2} \right) t \cos \vartheta/2 + \frac{1}{4} (\hbar^2 J)^2 \left(S + \frac{1}{2} \right)^2 \right\}^{1/2} \\
&\approx \frac{1}{2} \hbar^2 J \left(S + \frac{1}{2} \right) \left\{ 1 \pm \frac{4t \cos \vartheta/2}{\hbar^2 J \left(S + \frac{1}{2}\right)} \right\}^{1/2} \\
&\approx \frac{1}{2} \hbar^2 J \left(S + \frac{1}{2} \right) \left\{ 1 \pm \frac{2t \cos \vartheta/2}{\hbar^2 J \left(S + \frac{1}{2}\right)} \right\} \\
&\approx \frac{1}{2} \hbar^2 J \left(S + \frac{1}{2} \right) \pm t \cos \vartheta/2
\end{aligned} \tag{5.206}$$

This gives for the energies

$$\begin{array}{ll}
E_{--} \approx -\frac{1}{2}\hbar^2 J S + t \cos \frac{\vartheta}{2} & E_{-+} \approx -\frac{1}{2}\hbar^2 J S - t \cos \frac{\vartheta}{2} \\
E_{+-} \approx \frac{1}{2}\hbar^2 J (S+1) - t \cos \frac{\vartheta}{2} & E_{++} \approx \frac{1}{2}\hbar^2 J (S+1) + t \cos \frac{\vartheta}{2}
\end{array} \tag{5.207}$$

To be concrete, we take

$$t > 0 \quad ; \quad J > 0 \tag{5.208}$$

Then, E_{-+} is the ground state energy ($\vartheta \le \pi$), on which alone, we will concentrate ourselves further on:

$$E_{-+} = E_0 \approx -\frac{1}{2}\hbar^2 J S - t\, \frac{S_0 + \frac{1}{2}}{2S + 1} \tag{5.209}$$

The rest of the effort is directed at the derivation of an effective Hamiltonian, whose ground state energy is the same as E_0:

$$\begin{aligned}
S_0 = S_{tot} + \frac{1}{2} &\Rightarrow E_0 = -\frac{1}{2}\hbar^2 J S - \frac{t}{2S+1} - t \frac{S_{tot}}{2S+1} \\
S_0 = S_{tot} - \frac{1}{2} &\Rightarrow E_0 = -\frac{1}{2}\hbar^2 J S - t \frac{S_{tot}}{2S+1}
\end{aligned}$$

The two cases are different from each other only by an unimportant constant $\frac{t}{2S+1}$. For this reason, we consider only the simpler second case. The constant $(-\frac{1}{2}\hbar^2 J S)$ is also, naturally, not important, so, we are actually looking for an effective Hamiltonian, whose eigenvalues coincide with

$$E = -t \, \frac{S_{tot}}{2S+1} \tag{5.210}$$

We first try to express S_{tot} in terms of $\mathbf{S}_1$ and $\mathbf{S}_2$:

$$\begin{aligned}
\mathbf{S}_{tot}^2 &= (\mathbf{S}_1 + \mathbf{S}_2)^2 = \hbar^2 \, S_{tot} \, (S_{tot} + 1) \\
&= \hbar^2 \left(\left(S_{tot} + \frac{1}{2} \right)^2 - \frac{1}{4} \right) \\
&= \mathbf{S}_1^2 + \mathbf{S}_2^2 + 2(\mathbf{S}_1 \cdot \mathbf{S}_2) \\
&= 2S(S+1) \, \hbar^2 + 2 \, (\mathbf{S}_1 \cdot \mathbf{S}_2)
\end{aligned}$$

With this, we get

$$S_{tot} = -\frac{1}{2} + \frac{1}{2}\sqrt{1 + 8S(S+1) + (8/\hbar^2) \, \mathbf{S}_1 \cdot \mathbf{S}_2} \tag{5.211}$$

This finally leads to the *double exchange Hamiltonian*, we are looking for, when we further ignore unimportant constants:

$$H = \frac{-t}{2(2S+1)} \sqrt{1 + 8S(S+1) + (8/\hbar^2) \, \mathbf{S}_1 \cdot \mathbf{S}_2} \tag{5.212}$$

In this form, H is *not* of Heisenberg type. If we expand the square root in a series, then, in principle, all powers of $(\mathbf{S}_1 \cdot \mathbf{S}_2)$ appear. Therefore, we can adapt the following ansatz [12]:

$$H_S = -t \sum_{n=0}^{\infty} J_n(S) \, (\mathbf{S}_1 \cdot \mathbf{S}_2)^n \tag{5.213}$$

However, not all the powers of $(\mathbf{S}_1 \cdot \mathbf{S}_2)$ are independent of each other. This we demonstrate taking two examples:

(a) $\mathbf{S} = \frac{1}{2}$

In this case, there are two possibilities:

$$S_{tot} = 0, \ 1 \tag{5.214}$$

That means

$$\frac{1}{\hbar^2}(\mathbf{S}_1 \cdot \mathbf{S}_2) = \frac{1}{2} S_{tot}(S_{tot} + 1) - S(S+1) = \begin{cases} -\frac{3}{4} \; for \; S_{tot} = 0 \\ +\frac{1}{4} \; for \; S_{tot} = 1 \end{cases} \tag{5.215}$$

With these values, we can uniquely determine the parameters α and β in the ansatz

$$\frac{1}{\hbar^4}\,(\mathbf{S}_1\cdot\mathbf{S}_2)^2=\alpha+\beta\,\frac{1}{\hbar^2}\,(\mathbf{S}_1\cdot\mathbf{S}_2) \tag{5.216}$$

$$\begin{aligned}\tfrac{9}{16}&=\alpha-\tfrac{3}{4}\,\beta\\ \tfrac{1}{16}&=\alpha+\tfrac{1}{4}\,\beta\end{aligned}$$

Then (5.216) becomes

$$\frac{1}{\hbar^4}\,(\mathbf{S}_1\cdot\mathbf{S}_2)^2=\frac{3}{16}-\frac{1}{2\hbar^2}\,(\mathbf{S}_1\cdot\mathbf{S}_2) \tag{5.217}$$

This is a relation, which we already used in Sect. 5.2.3. With the above relation, the infinite series appearing in H terminates with the linear term itself, because, all the higher powers can be expressed in terms of the linear term. This leads to a new ansatz for our model Hamiltonian:

$$H_{1/2}=-t\,(J_0(1/2)+J_1(1/2)\,(\mathbf{S}_1\cdot\mathbf{S}_2)) \tag{5.218}$$

This operator should, in view of (5.210), give the eigenvalues

$$E=-t\begin{cases}0\; for\; S_{tot}=0\\ 1/2\; for\; S_{tot}=1\end{cases} \tag{5.219}$$

This results in two equations that determine J_0 and J_1:

$$\begin{aligned}-t\,(J_0-\tfrac{3}{4}J_1\hbar^2)&=0\\ -t\,(J_0+\tfrac{1}{4}J_1\hbar^2)&=-\tfrac{1}{2}t\end{aligned}$$

from which it follows that

$$J_0(1/2)=\frac{3}{8}\quad ;\quad J_1(1/2)=\frac{1}{2\,\hbar^2} \tag{5.220}$$

Thus, for $S=1/2$, H has the usual Heisenberg structure:

$$H_{1/2}=-\frac{3}{8}\,t-\frac{1}{2}\,t\,\frac{1}{\hbar^2}\,(\mathbf{S}_1\cdot\mathbf{S}_2) \tag{5.221}$$

The procedure is completely analogous for

(b) $\mathbf{S}=\mathbf{1}$

Now the possibilities are

$$S_{tot}=0,\;1,\;2 \tag{5.222}$$

With this, we uniquely determine the coefficients α, β and γ in the ansatz

$$\frac{1}{\hbar^6}\,(\mathbf{S}_1\cdot\mathbf{S}_2)^3 = \alpha + \frac{\beta}{\hbar^2}\,(\mathbf{S}_1\cdot\mathbf{S}_2) + \frac{\gamma}{\hbar^4}\,(\mathbf{S}_1\cdot\mathbf{S}_2)^2 \tag{5.223}$$

With the analogous relation to (5.216), we get now three possible values for $(\mathbf{S}_1\cdot\mathbf{S}_2)$:

$$\frac{1}{\hbar^2}(\mathbf{S}_1\cdot\mathbf{S}_2) = \begin{cases} -2 \; for \; S_{tot} = 0 \\ -1 \; for \; S_{tot} = 1 \\ \;\;\, 1 \; for \; S_{tot} = 2 \end{cases} \tag{5.224}$$

This leads to

$$\frac{1}{\hbar^6}(\mathbf{S}_1\cdot\mathbf{S}_2)^3 = 2 + \frac{1}{\hbar^2}(\mathbf{S}_1\cdot\mathbf{S}_2) - \frac{2}{\hbar^4}(\mathbf{S}_1\cdot\mathbf{S}_2)^2$$

Thus the series for $H_{S=1}$ terminates after the quadratic term, giving the following expression:

$$H_{S=1} = -t\,\left(J_0(1) + J_1(1)\,(\mathbf{S}_1\cdot\mathbf{S}_2) + J_2(1)\,(\mathbf{S}_1\cdot\mathbf{S}_2)^2\right) \tag{5.225}$$

This operator should according to (5.210) have the eigenvalues

$$E = -\frac{t}{3}\begin{cases} 0 \; for \; S_{tot} = 0 \\ 1 \; for \; S_{tot} = 1 \\ 2 \; for \; S_{tot} = 2 \end{cases} \tag{5.226}$$

With the above relation for $\mathbf{S}_1\cdot\mathbf{S}_2$, we get the following equations to determine J_0, J_1 and J_2

$$\begin{aligned} 0 &= J_0 - 2\hbar^2 J_1 + 4\hbar^4 J_2 \\ \tfrac{1}{3} &= J_0 - \hbar^2 J_1 + \hbar^4 J_2 \\ \tfrac{2}{3} &= J_0 + \hbar^2 J_1 + \hbar^4 J_2 \end{aligned}$$

with the solutions

$$J_0(1) = \frac{5}{9} \quad ; \quad J_1(1) = \frac{1}{6\hbar^2} \quad ; \quad J_2(1) = -\frac{1}{18\hbar^4} \tag{5.227}$$

The double exchange Hamiltonian $H_{S=1}$ is now not of Heisenberg type anymore:

$$H_{S=1} = -\frac{5}{9}\,t - \frac{t}{6\hbar^2}\,(\mathbf{S}_1\cdot\mathbf{S}_2) + \frac{t}{18\hbar^4}\,(\mathbf{S}_1\cdot\mathbf{S}_2)^2 \tag{5.228}$$

For any arbitrary S, one can calculate H_S in a very similar manner. For $S > 1/2$, H_S is no more bilinear but biquadratic, bicubic and so on. The highest power of $(\mathbf{S}_1 \cdot \mathbf{S}_2)$ in H_S is $2S$. Even though the coefficients J_n decrease with n for all S, the convergence is rather slow. In case $J_n(S)$ for $n \geq 2$ are not negligible, then, H_S is not of Heisenberg type. However, the *bilinear term*, which always dominates, is *always ferromagnetic*!

In this chapter, we have learnt about a number of exchange mechanisms, which lead to model Hamiltonians, which *in principle* can describe collective magnetism. Whether a particular model shows the phase transition paramagnetism $\Leftrightarrow$ ferromagnetism or not is, for finite temperatures, not decided by the internal energy $U = \langle H \rangle$, but by the free energy $F = U - TS$. In the next chapter, special attention will be paid to the considerations, under what conditions, the models developed so far, actually give $T_{C,N} > 0$.

5.4 Problems

Problem 5.1 $c^{\dagger}_{i\sigma}$ $(c_{i\sigma})$ is the creation operator (annihilation operator) of an electron with spin σ $(\sigma = \uparrow, \downarrow)$ at the lattice site $\mathbf{R}_i$. The following Fermi commutation relations are valid:

$$\left[c_{i\sigma}, c_{j\sigma'}\right]_+ = \left[c^{\dagger}_{i\sigma}, c^{\dagger}_{j\sigma'}\right]_+ = 0$$

$$\left[c_{i\sigma}, c^{\dagger}_{j\sigma'}\right]_+ = \delta_{ij}\delta_{\sigma\sigma'}$$

$[\ldots, \ldots]_+$ is the anticommutator.

1. Show that spin operators can be defined by

$$S_i^z = \frac{\hbar}{2}\left(n_{i\uparrow} - n_{i\downarrow}\right); \qquad n_{i\sigma} = c^{\dagger}_{i\sigma}c_{i\sigma}$$

$$S_i^+ = \hbar c^{\dagger}_{i\uparrow}c_{i\downarrow}$$

$$S_i^- = \hbar c^{\dagger}_{i\downarrow}c_{i\uparrow}$$

   Verify the commutation relations for the above operators.
2. Transform the operators $c^{\dagger}_{i\sigma}$, $c_{i\sigma}$ by

$$c_{\mathbf{q}\sigma} = \frac{1}{\sqrt{N}}\sum_i \exp(-i\mathbf{q}\cdot\mathbf{R}_i)c_{i\sigma}$$

   onto the wavenumbers $\mathbf{q}$ of the first Brillouin zone. Show that $c^{\dagger}_{\mathbf{q}\sigma}$ and $c_{\mathbf{q}\sigma}$ are also Fermi operators.

Problem 5.2 Calculate the free energy and the free enthalpy for a ferromagnetic material which satisfies the Curie–Weiss law. First show that the heat capacity c_M depends only on temperature and then assume that $c_M(T)$ is known.

Problem 5.3 For the ferromagnetic material in Problem 5.2, additionally

$$c_M(T, M = 0) = \gamma T \qquad (\gamma > 0)$$

shall be valid.

1. How do $F(T, M)$, $S(T, M)$, $S(T, H)$ and $U(T, M)$ read?
2. Calculate the heat capacities c_M and c_H and also the adiabatic susceptibility χ_S.

Problem 5.4 Let the crystal lattice consist of two chemically completely equivalent sublattices 1 and 2, which by themselves order ferromagnetically below the Neél temperature T_N. Let the spontaneous magnetizations of the two sublattices be antiparallel to each other and equal in magnitude (antiferromagnet):

$$M_{1S}(T) = -M_{2S}(T) \neq 0 \ \text{ for } \ T < T_N$$

The sublattice magnetizations can be expressed in terms of the exchange fields:

$$\begin{aligned} B_A^{(1)} &= \mu_0(\lambda M_1 + \rho M_2) \\ B_A^{(2)} &= \mu_0(\rho M_1 + \lambda M_2) \end{aligned}$$

Temperature and magnetic field dependence are determined by the Brillouin function:

$$\begin{aligned} M_{1,2}(T, B_0) &= M_0 B_J \left(\beta J g_J \mu_B (B_0 + B_A^{(1,2)})\right) \\ M_0 &= \frac{N}{2V} g_J J \mu_B \\ M_{iS}(T) &= \lim_{B_0 \to 0^+} M_i(T, B_0) \quad (i = 1, 2) \end{aligned}$$

1. Express the Neél temperature in terms of the exchange constants λ and ρ.
2. Discuss the high-temperature behaviour of the susceptibility χ and determine the paramagnetic Curie temperature Θ (Curie–Weiss law).
3. Necessary condition for the antiferromagnetic structure is

$$\rho < 0$$

Discuss the behaviour of

$$-\frac{\Theta}{T_N}$$

for the cases $\lambda > 0$ and $\lambda < 0$!

Problem 5.5 Discuss the critical behaviour of the Weiss ferromagnet made up of localized moments $\mathbf{m}_i = \mathbf{m}\ \forall i$ at the lattice sites $\mathbf{R}_i$.

1. Show that with the reduced quantities

$$\hat{M} = \frac{M}{M_0}\ ; \quad b = \frac{m\mu_0 H}{k_B T}\ ; \quad \varepsilon = \frac{T - T_c}{T_c}$$
$$\left(M_0 = \frac{N}{V} m = \text{Saturation magnetization}\right)$$

the equation of state can be written as follows:

$$\hat{M} = L\left(b + \frac{3\hat{M}}{\varepsilon + 1}\right)$$

L is the Langevin function:

$$L(x) = \coth x - \frac{1}{x}$$

2. Calculate the critical exponent β of the order parameter M_S.
3. What is the value of the critical exponent δ of the critical isotherm?
4. Derive the critical exponents γ, γ' and determine the behaviour of $\frac{C}{C'}$ of the critical amplitudes.

Problem 5.6 Let σ be the Pauli spin operator with the components

$$\sigma_x = \begin{pmatrix} 0 & 1 \\ 1 & 0 \end{pmatrix} \quad ; \quad \sigma_y = \begin{pmatrix} 0 & -i \\ i & 0 \end{pmatrix} \quad ; \quad \sigma_z = \begin{pmatrix} 1 & 0 \\ 0 & -1 \end{pmatrix}$$

These have the properties proved in Problem 2.2 (5.81), (5.82) and (5.82). Let $\sigma^{(i)}$, $i = 1, 2$ be the spin operator for the particle i.

1. Show that

$$\left(\sigma^{(1)} \cdot \sigma^{(2)}\right)^2 = 3\mathbb{1} - 2\left(\sigma^{(1)} \cdot \sigma^{(2)}\right)$$

holds

2. Verify that the operator

$$Q_{12} = \frac{1}{2}\left(\mathbb{1} + \sigma^{(1)} \cdot \sigma^{(2)}\right)$$

interchanges the spins of the particles 1 and 2:

$$Q_{12}\,\sigma^{(1)}\,Q_{12}^{-1} = \sigma^{(2)}$$
$$Q_{12}\,\sigma^{(2)}\,Q_{12}^{-1} = \sigma^{(1)}$$

3. In a system of two-spin-$\frac{1}{2}$-particles, the spins can couple to give a total spin $S = 1$ and $S = 0$. The eigenstates are the symmetric triplet (5.88) and the anti-symmetric singlet (5.89). Show that these are also the eigenstates of the operator Q_{12} with the eigenvalues 1,1,1 and -1.

Problem 5.7 Calculate the integral

$$R = \int_0^\infty dx\, x \frac{\sin x}{x^2 - y^2}$$

Problem 5.8 If the Coulomb interaction of the electrons in a solid is considered to be intra-atomic, then

$$H_1 = \frac{1}{2} \sum_{i\sigma\sigma'} \sum_{\mu\mu'\nu\nu'} v(\mu\nu;\mu'\nu')\, c^\dagger_{i\mu\sigma} c^\dagger_{i\nu\sigma'} c_{i\nu'\sigma'} c_{i\mu'\sigma}$$

holds (see (8.23)). i indexes the lattice sites and μ, ν, μ', ν' are the band indices. As far as the collective magnetism is concerned all the matrix elements are not equally important. Normally one restricts oneself to the scattering processes where at the most two different bands are involved and to the so-called *direct terms* and *exchange terms* ((8.24),(8.25)):

$$\widehat{H}_1 = \frac{1}{2} \sum_{i\sigma\sigma'} \sum_{\mu\nu} \left[(1 - \delta_{\mu\nu}\delta_{\sigma\sigma'}) U_{\mu\nu} n_{i\mu\sigma} n_{i\nu\sigma'} + (1 - \delta_{\mu\nu}) J_{\mu\nu}\, c^\dagger_{i\mu\sigma} c^\dagger_{i\nu\sigma'} c_{i\mu\sigma'} c_{i\nu\sigma} \right]$$

Here $n_{i\mu\sigma} = c^\dagger_{i\mu\sigma} c_{i\mu\sigma}$ is the occupation number operator.

1. Show that $\widehat{H}_1$ can be split as follows:

$$\widehat{H}_1 = H_U + H_d + H_{ex}$$

Here H_U is an intraband Coulomb interaction corresponding to a multi-band Hubbard model (Chap. 8)

$$H_U = \frac{1}{2} \sum_{i\sigma} \sum_{\mu} U_{\mu\mu} n_{i\mu\sigma} n_{i\mu-\sigma}$$

H_d describes a direct interband Coulomb interaction

$$H_d = \frac{1}{2} \sum_{i\mu\nu}^{\mu\neq\nu} (U_{\mu\nu} - \frac{1}{2} J_{\mu\nu})\, n_{i\mu}\, n_{i\nu} \quad ; \quad n_{i\mu} = \sum_\sigma n_{i\mu\sigma}$$

H_{ex} is an interband exchange term which can also be written as spin–spin interaction:

$$H_{ex} = -\frac{4}{\hbar^2} \sum_{i\mu\nu}^{\mu \neq \nu} J_{\mu\nu}\, \mathbf{s}_{i\mu} \cdot \mathbf{s}_{i\nu}$$

The spin operator **s** is defined as in (5.78).

2. Using the result of (1), justify the Kondo-Lattice Model (sf-model) as a theoretical model to describe the local moment systems such as Gd, EuO, $Ga_{1-x}Mn_xAs$, ...

Problem 5.9 Conduction electrons, which interact with antiferromagnetically ordered localized spin system, are described by the following simplified model Hamiltonian:

$$H = H_0 + H_1$$

$$H_0 = \sum_{\mathbf{k}\sigma\alpha\beta} \varepsilon_{\alpha\beta}(\mathbf{k}) c^{\dagger}_{\mathbf{k}\sigma\alpha} c_{\mathbf{k}\sigma\beta}$$

$$H_1 = -\frac{1}{2} J \sum_{\mathbf{k}\sigma\alpha} z_\sigma \langle S^z_\alpha \rangle c^{\dagger}_{\mathbf{k}\sigma\alpha} c_{\mathbf{k}\sigma\alpha} \qquad (z_\sigma = \delta_{\sigma\uparrow} - \delta_{\sigma\downarrow})$$

$\alpha, \beta = A, B$ indicate two chemically equivalent ferromagnetic sublattices A and B. For the sublattice magnetizations

$$\langle S^z_A \rangle = -\langle S^z_B \rangle = \langle S^z \rangle$$

holds.

The Bloch energies

$$\varepsilon_{AA}(\mathbf{k}) = \varepsilon_{BB}(\mathbf{k}) \equiv \varepsilon(\mathbf{k})$$
$$\varepsilon_{AB}(\mathbf{k}) = \varepsilon^*_{BA}(\mathbf{k}) \equiv t(\mathbf{k})$$

are assumed to be known, where **k** is a wavevector in the first Brillouin zone of the *sub*lattice.

1. Calculate the eigenstates and eigenenergies of the unperturbed part H_0.
2. Using the Schrödinger perturbation theory, calculate the energy corrections in the first and second order due to H_1.
3. Repeat it with *Brillouin–Wigner* perturbation theory.
4. Compare the results of (2) and (3) with the *exact* eigenenergies.

Problem 5.10 The Weiss model of the ferromagnet starts with a single spin in a self-consistent exchange field (5.7). A reasonable extension of the model that could be investigated is the one where an exchange coupled pair of spins is in the presence of such an exchange field:

$$H = -J\,\mathbf{S}_1 \cdot \mathbf{S}_2 - \frac{g\mu_B}{\hbar}\left(S_1^z + S_2^z\right)\widehat{B}$$

Here the effective field $\widehat{B}$ consists of an external and an exchange field. The latter should again be proportional to the magnetization of the ferromagnet:

$$\widehat{B} = B_0 + B_{ex} = B_0 + \lambda\mu_0 M$$

For simplicity, all the fields are assumed to be homogeneous and oriented along the z-direction.

1. Derive the partition function of the two-spin system assuming $S_1 = S_2$.
2. For the special case of $S_1 = S_2 = \frac{1}{2}$ give the determinig equation for the magnetization. How does this result differ from the Weiss result (5.8)?

References

1. P. Weiss (1908)
2. M. A. Rudermann, C. Kittel, Phys. Rev. B **96**, 99 (1954)
3. T. Kasuya, Progr. Theoret. Phys. **16**, 45 (1956)
4. K. Yosida, Phys. Rev. **106**, 893 (1957)
5. J. Kanamori, Phys. Chem. Solids **10**, 87 (1959)
6. J. B. Goodenough, "Magnetism and the Chemical Bond", Interscience, New York, 1963, Chap. 3
7. G. H. Jonker, J. H. van Sauten, Physica **16**, 337 (1950)
8. G. H. Jonker, J. H. van Sauten, Physica **19**, 120 (1953)
9. A. Ramirez, J. Phys. Condens. Matter **9**, 8171 (1997)
10. C. Zener, Phys. Rev. **82**, 403 (1951)
11. P. W. Anderson, H. Hasegawa, Phys. Rev. **100**, 675 (1955)
12. M. Cieplak, Phys. Rev. B **18**, 3470 (1978)

Chapter 6
Ising Model

In Sect. 5.2 and 5.3, we have learnt about one of the most important models of magnetism, namely the Heisenberg model. It was conceptualized specially for magnetic insulators, but as we have argued in Sect. 5.3.1, it is also applicable to magnetic metals with localized magnetic moments. The decisive precondition for the applicability of Heisenberg model is the existence of localized permanent magnetic moments. The coupling which leads to either a direct or an indirect interaction between moments can, of course, be a variety of physical origins. From the conceptual point of view, the Heisenberg model is not applicable to the classical band magnets Fe, Co and Ni, since for these materials, the magnetism is due to the itinerant and therefore completely delocalized band electrons. Therefore for band magnets, we have to develop new concepts, which will be done in Chap. 8.

The Heisenberg model permits a few special cases if the scalar product $\mathbf{S}_i \cdot \mathbf{S}_j$ is expressed in terms of weighted components:

$$H = -\sum_{i,j} J_{ij} \left\{ \alpha \left(S_i^x \, S_j^x + S_i^y \, S_j^y \right) + \beta \, S_i^z \, S_j^z \right\}$$

$\alpha = \beta = 1$: Heisenberg model
$\alpha = 0 \, ; \; \beta = 1$: Ising model
$\alpha = 1 \, ; \; \beta = 0$: XY-model

If the localized moments are not due to pure spins $\mathbf{S}_i$ but due to the total angular momentum $\mathbf{J}_i$, then we have to, according to the Wigner–Eckart theorem (see the application example (2) in Sect. 2.5), replace $\mathbf{S}_i$ by

$$\mathbf{S}_i \rightarrow (g_J - 1) \, \mathbf{J}_i$$

where g_J is the Lande's g-factor. The structure of the Hamiltonian remains unchanged.

Before we discuss the Heisenberg model in detail in Chap. 7, we want first to investigate the Ising model in the context of the paramagnetism $\Leftrightarrow$ ferromagnetism phase transition.

W. Nolting, A. Ramakanth, *Quantum Theory of Magnetism*,
DOI 10.1007/978-3-540-85416-6_6,

6.1 The Model

Till today, the Ising model is the only somewhat realistic model of a many-body system that has shown a phase transition and has been evaluated mathematically rigorously at least in lattice dimensions of $d = 1$ and $d = 2$. Actually speaking, this model should be called "Lenz model". Around 1920, that is, before quantum mechanics was developed, Wilhelm Lenz gave his student E. Ising this problem to solve. Later the model became known as the Ising model. At that point of time the only theory of magnetism that existed was the Weiss' theory (see Sect. 5.1.2), which showed a phase transition paramagnetism $\Leftrightarrow$ ferromagnetism independent of the lattice dimensionality. Lenz expected the same result from the Ising model, which contains already rather realistic microscopic interactions. The dissertation of Ising was, therefore, disappointing on two counts. First, Ising could show that the one-dimensional model shows *no* phase transition and second, he could not solve the model in two and three dimensions.

Even today, the Ising model is treated as a *classical* model. Every one of the N lattice points on a d-dimensional ($d = 1, 2, 3$) periodic array is assigned a spin variable S_i, which is a c-number and can take only the values ± 1:

$$S_i = \pm 1 \quad ; \quad i = 1, 2, \cdots, N \tag{6.1}$$

The model is then characterized by the following Hamiltonian *function*:

$$H = -\sum_{i,j} J_{ij} S_i S_j - \mu_B B_0 \sum_i S_i \tag{6.2}$$

The possible applications of this model are extremely large. We list, in the following, the important ones:

1. *Model for magnetic insulators*: This was the original goal. The applicability is naturally restricted because of the simplification when compared to the more realistic Heisenberg model. The restriction to only the z-component of the spin operators can be justified only in magnetic systems with strongly uniaxial symmetry, where the permanent moments are confined to a particular space direction. However, there exist such systems in reality:
   $DyPO_4$, $CoCs_3Cl$, $CoRb_3Cl_5$, etc.
2. *General demonstration model of statistical mechanics*: As the simplest microscopic model which shows a phase transition, the Ising model is at the centre of the current discussion on phase transitions and critical phenomena.
3. *Model for binary alloys*: The Ising model is also useful for many non-magnetic problems. The statistical distribution of the two kinds of atoms in a lattice in the case of a binary alloy can be simulated by the two possible orientations of the Ising spins.
   $S = +1 \Leftrightarrow$ Atom type A $\quad ; \quad S_i = -1 \Leftrightarrow$ Atom type B

4. *Lattice gas model for fluids*: One divides the volume V of the fluid into small cells of volume Δv, which is roughly the volume of a molecule. One can introduce a cell variable n_i such that

$$n_i = \begin{cases} 1 & if\ cell\ i\ is\ occupied \\ 0 & if\ cell\ i\ is\ unoccupied \end{cases} \tag{6.3}$$

Then one can immediately recognize the equivalence to the Ising model:

$$S_i = 2n_i - 1 \tag{6.4}$$

The phase transition is recognized as follows. For $T > T_c$ (*gas phase*), taking any volume ΔV of macroscopic dimension (Fig. 6.1), we find a statistical distribution of the occupied cells. In contrast, for $T < T_c$, we find macroscopic regions which are either fully occupied or completely empty. This corresponds to *liquid phase*. As the model does not contain the gravitational potential, there is, of course, no interface between the "gas" and the "liquid".
5. *Model for ferroelectrics:* Ferroelectrics are characterized by a spontaneous electric polarization, which below a critical temperature T_c has two orientations. The equivalence to the Ising model is obvious.
6. *Model for biological systems:* This is a fast developing field where the Ising model is extensively applied in the area of "synergetics". We only make a mention of it. For more details one should refer to the corresponding literature.

In the following sections, we will discuss the Ising model extensively. The interest in the Ising model is because it gives a host of exact results, which is rather unusual for many-body models:

$d = 1$: Exact solution is possible

(a) with and without field B_0
(b) for nearest neighbour interactions

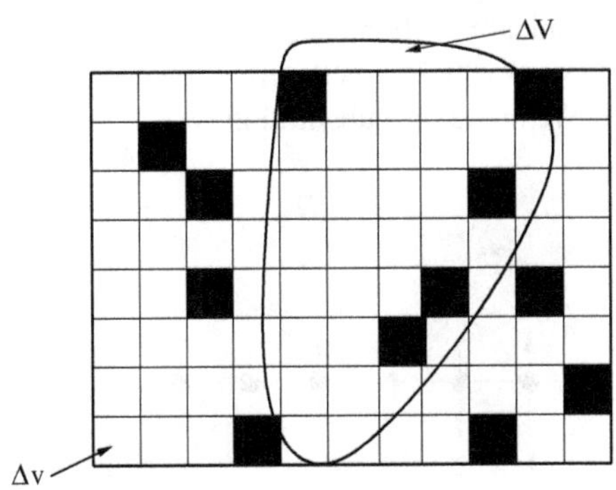

Fig. 6.1 Lattice gas model as a special Ising model

$$J_{ij} = \begin{cases} J \; if \; \mathbf{R}_i \; and \; \mathbf{R}_j \; are \; nearest \; neighbours \\ 0 \; otherwise \end{cases} \tag{6.5}$$

and for

$$J_{ij} \sim \left|\mathbf{R}_i - \mathbf{R}_j\right|^{-(d+\alpha)} \quad (\alpha > 0) \tag{6.6}$$

There is *no* phase transition ($T_c = 0$).

$d = 2$: Exact solution is possible

(a) without field
(b) for nearest neighbour interactions.

There is a phase transition ($T_c > 0$).

$d = 3$: A full exact solution is not available till now. However, there exist extrapolation methods, which can be treated for all purposes to be quasi-exact. The existence of phase transition is rigorously proved.

6.2 The One Dimensional Ising Model

6.2.1 Spontaneous Magnetization

We consider a linear chain of N spins (Fig. 6.2), first without switching on a field:

$$H = -\sum_{i=1}^{N-1} J_i \; S_i \; S_{i+1} \tag{6.7}$$

The interaction is only between the nearest neighbours: $J_{i,i+1} \rightarrow J_i$. We are interested in the question, whether a system defined by the above Hamiltonian function shows a phase transition, that is, whether there is a temperature T_c below which spontaneous magnetization sets in. To answer this question, we first calculate the canonical partition function Z_N. From this we derive the spin correlation function $\langle S_i S_{i+j} \rangle$ which directly gives the spontaneous magnetization M_s.

We make the substitution

$$j_i = \frac{J_i}{k_B T} = \beta J_i \tag{6.8}$$

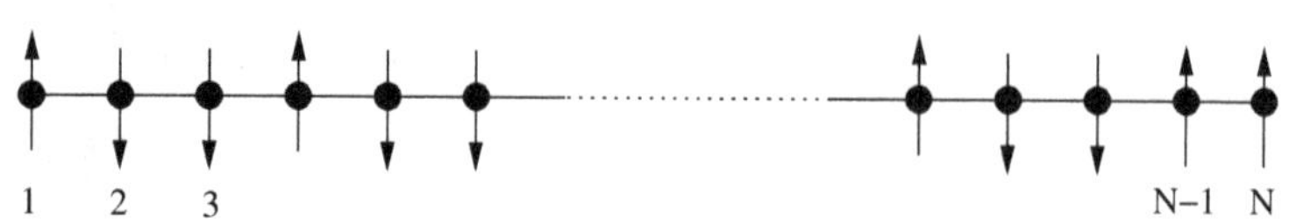

Fig. 6.2 Schematic plot of the one-dimensional Ising model consisting of localized spins

Every spin has two possible orientations. Therefore, there are 2^N possible spin orderings and so the same number of distinct states which are to be considered for the calculation of the *canonial partition function*.

$$Z_N = Z_N(j_1, \; j_2, \cdots, \; j_{N-1}) = \sum_{S_1}^{\pm 1} \sum_{S_2}^{\pm 1} \cdots \sum_{S_N}^{\pm 1} exp \left(\sum_{i=1}^{N-1} j_i S_i S_{i+1} \right) \tag{6.9}$$

To obtain a recursion relation, we extend the chain by one more spin:

$$Z_{N+1} = \sum_{S_1}^{\pm 1} \sum_{S_2}^{\pm 1} \cdots \sum_{S_N}^{\pm 1} exp \left(\sum_{i=1}^{N-1} j_i S_i S_{i+1} \right) \sum_{S_{N+1}}^{\pm 1} exp(j_N S_N S_{N+1}) \tag{6.10}$$

We see that

$$\sum_{S_{N+1}}^{\pm 1} exp\,(j_N S_N S_{N+1}) = 2\cosh(j_N S_N) = 2\cosh(j_N)$$

The last step results from the fact that S_N can take only the values ± 1 and that cosh is an even function. We now have the recursion formula that we are looking for

$$Z_{N+1} = 2\; Z_N\; cosh(j_N) \tag{6.11}$$

If Z_1 is the partition function of a single spin, then, obviously we can write,

$$Z_{N+1} = Z_1\; 2^N \prod_{i=1}^{N} \cosh(j_i) \tag{6.12}$$

A single spin has two possible orientations and no interactions. Therefore,

$$Z_1 = \sum_{S_1}^{\pm 1} e^0 = 2 \tag{6.13}$$

With this we have the *canonical partition function* of the *one-dimensional Ising model* as

$$Z_N(T) = 2^N \prod_{i=1}^{N-1} \cosh\left(\frac{J_i}{k_B T}\right) \tag{6.14}$$

Making the usual assumption

$$J_i = J \quad for\; all\; i \tag{6.15}$$

we get

$$Z_N(T) = 2^N \cosh^{N-1}(\beta J) \tag{6.16}$$

From the partition function one can derive all the physically important quantities, for example, the *spin correlation function.*

$$\langle S_i S_{i+j} \rangle = \frac{1}{Z_N} \sum_{\{S\}} S_i \; S_{i+j} exp\left(\sum_{k=1}^{N-1} j_k \; S_k \; S_{k+1} \right) \tag{6.17}$$

The summation is over all the 2^N possible spin configurations. One can now write

$$\begin{aligned}
&\frac{\partial}{\partial j_i} \frac{\partial}{\partial j_{i+1}} \cdots \frac{\partial}{\partial j_{i+j-1}} Z_N \\
&= \sum_{\{S\}} (S_i \; \underbrace{S_{i+1})(S_{i+1}}_{+1} \; S_{i+2}) \cdots \underbrace{(S_{i+j-1}}_{+1} \; S_{i+j}) * \\
&\qquad\qquad * exp\left(\sum_{k=1}^{N-1} j_k \; S_k \; S_{k+1} \right) \\
&= \sum_{\{S\}} (S_i \; S_{i+j}) exp\left(\sum_{k=1}^{N-1} j_k \; S_k \; S_{k+1} \right) \\
&= Z_N \langle S_i \; S_{i+j} \rangle
\end{aligned}$$

With the expression (6.14) for the partition function $Z_N(T)$, we can further write

$$\langle S_i \; S_{i+j} \rangle =$$

$$= \frac{2^N \cosh(j_1) \cdots \cosh(j_{i-1}) \sinh(j_i) \cdots \sinh(j_{i+j-1}) \cdots \cosh(j_{N-1})}{2^N \; \cosh(j_1) \cdots \cosh(j_{N-1})}$$

We have thus obtained the expression for the spin correlation function:

$$\langle S_i \; S_{i+j} \rangle = \prod_{r=1}^{j} \tanh\left(\beta J_{i+r-1}\right) \tag{6.18}$$

Thus, eventhough the interaction within the Ising system is of very short range (nearest neighbours), this yields a very long range correlation among the spins. If we break the Ising chain at some place i_0, that is, if we set $J_{i_0} = 0$, and if i_0 lies between i and $i + j$, then $\langle S_i \; S_{i+j} \rangle = 0$, because $\tanh(\beta J_{i_0}) = \tanh(0) = 0$. If i_0 does not lie between i and $i + j$, then the correlation function remains unaffected by the fact that the chain is broken.

In the usual case where the coupling constants $J_i = J$, the correlation function is independent of i. Only the distance between the spins plays a role:

$$\langle S_i \; S_{i+j} \rangle = \tanh^j(\beta J) \tag{6.19}$$

With the help of the correlation function, we can now answer the question, whether for $T \neq 0$, spontaneous magnetization appears in the one-dimensional Ising model. For that, we make use of the fact that two spins, infinite distance apart do not "know" of each other:

$$\langle S_i \; S_{i+j} \rangle \rightarrow \langle S_i \rangle \langle S_j \rangle \; for \; |i - j| \rightarrow \infty \tag{6.20}$$

This limiting case is, of course, meaningful only in the thermodynamic limit. As a consequence of the translational symmetry, the expectation value is same for all the spins $\langle S_i \rangle$. This provides a possible definition of the *spontaneous magnetization*:

$$M_s^2(T) = \mu_B^2 \langle S \rangle^2 = \mu_B^2 \lim_{j \to \infty} \langle S_i \; S_{i+j} \rangle \tag{6.21}$$

We know that $|\tanh x| \leq 1$, where the equality is for $x \rightarrow \pm\infty$. That leads to

$$\lim_{j \to \infty} \langle S_i \; S_{i+j} \rangle = \begin{cases} 0 \; if \; \beta \neq \infty \\ 1 \; if \; \beta = \infty \end{cases} \tag{6.22}$$

or

$$M_s(T) = \mu_B \begin{cases} 0 \; if \; T \neq 0 \\ 1 \; if \; T = 0 \end{cases} \tag{6.23}$$

The one-dimensional Ising model has a *phase transition at* $T = 0$. There is *no* spontaneous magnetization at finite temperature (Fig. 6.3).

Whether there is a phase transition or not is decided by the free energy $F = U - TS$. There are two competing tendencies, namely the orientation of the spins, which minimizes U and the statistical distribution of the spin orientations (disorder), which maximizes the entropy S. In the one-dimensional model, the tendency to

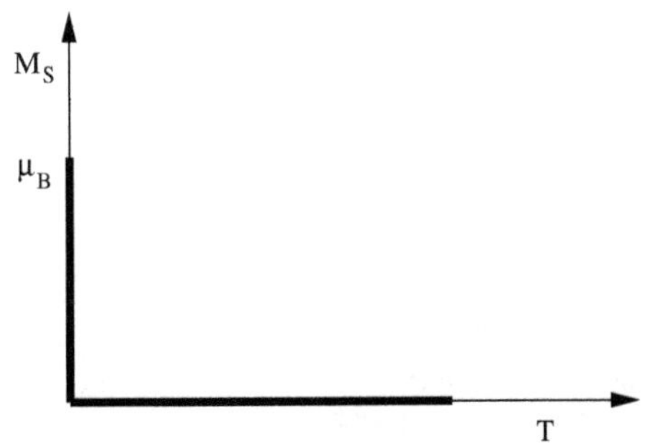

Fig. 6.3 Phase diagram of the one-dimensional Ising model

orient is always weak, as there do not exist sufficient number of nearest neighbours. Qualitatively, one can understand this as follows: Let us assume that in the one-dimensional Ising model with nearest neighbour interactions, there is a $T_c > 0$, that is, there is a phase transition. Then, in the ordered phase ($0 < T \leq T_c$), we break the chain at some place and we rotate one part of the chain into the opposite direction. For that, we require an energy $\Delta U = 2J$, because, by doing this, we are actually increasing the coupling energy only at one place, namely where we are breaking the chain. The breaking of the chain can be done at N distinct places. Therefore, there are N possibilities of realization and consequently the entropy will increase by a term proportional to $\ln N$. We see that, for $T \neq 0$, the entropy term dominates in the free energy and the free energy itself decreases. That means at $T \neq 0$, the ordered state becomes unstable against such breaking of the chain and the disordered state is always stable. Due to this reason, even for an infinitely long Ising chain with short range interaction, there cannot be a phase transition.

6.2.2 One Dimensional Ising Model in External Field

We now want to investigate the field dependence of the partition function of the one dimensional Ising model. For this, we make use of the famous *transfer matrix method*, which was introduced by Onsager in 1944 in his solution of the two-dimensional Ising model. This method will be demonstrated here for the one-dimensional case. With the substitutions

$$j = \beta J \qquad ; \qquad b = \beta \mu_B \, B_0 \tag{6.24}$$

and the assumption that the nearest neighbour interactions for all the pairs of spins are equal, we write

$$\beta H = -j \sum_{i=1}^{N} S_i \, S_{i+1} - b \sum_{i=1}^{N} S_i \tag{6.25}$$

We make use of the *periodic boundary condition* by making the chain into a ring as shown in Fig. 6.4. Then we have

Fig. 6.4 One-dimensional Ising model as a ring with periodic boundary conditions

$$S_{N+1} \equiv S_1 \tag{6.26}$$

In order to solve the problem, we now introduce the so-called *transfer function*:

$$f_{i,i+1} = exp\left(+j S_i \ S_{i+1} + \frac{1}{2} b \left(S_i + S_{i+1}\right)\right) \tag{6.27}$$

using which, we can write,

$$exp(-\beta \ H) = f_{1,2} \ f_{2,3} \cdots f_{N,1} \tag{6.28}$$

Here we have to make use of the periodic boundary condition (6.26). To the transfer function, we now define the corresponding *transfer matrix*. With the spin states

$$|S_i = +1\rangle = \begin{pmatrix} 1 \\ 0 \end{pmatrix} \quad ; \quad |S_i = -1\rangle = \begin{pmatrix} 0 \\ 1 \end{pmatrix} \tag{6.29}$$

and

$$\langle S_i | \widehat{T} | S_{i+1} \rangle = f_{i,i+1} \tag{6.30}$$

we get the transfer matrix $\widehat{T}$:

$$\widehat{T} \equiv \begin{pmatrix} e^{j+b} & e^{-j} \\ e^{-j} & e^{j-b} \end{pmatrix} \tag{6.31}$$

Exploiting the periodic boundary condition and the completeness of the spin states, the *canonical partition function* (1.86) is given by

$$\begin{aligned} Z_N(T, B_0) &= Tr\left(e^{(-\beta \ H)}\right) \\ &= \sum_{S_1}^{\pm 1} \sum_{S_2}^{\pm 1} \cdots \sum_{S_N}^{\pm 1} f_{1,2} f_{2,3} \cdots f_{N,1} \\ &= \sum_{S_1}^{\pm 1} \cdots \sum_{S_N}^{\pm 1} \langle S_1 | \widehat{T} | S_2 \rangle \langle S_2 | \widehat{T} | S_3 \rangle \cdots \langle S_N | \widehat{T} | S_1 \rangle \\ &= \sum_{S_1}^{\pm 1} \langle S_1 | \widehat{T}^N | S_1 \rangle \end{aligned} \tag{6.32}$$

If E_+ and E_- are the eigenvalues of the 2×2 matrix $\widehat{T}$, then the partition function is given by

$$Z_N(T, B_0) = Tr\left(\widehat{T}^N\right) = E_+^N + E_-^N \tag{6.33}$$

The energy eigenvalues as given by the condition

$$det(\widehat{T} - E\ 1) = 0$$

are (see Problem 6.1)

$$E_{\pm} = e^{j}\left(\cosh b \pm \sqrt{\cosh^2 b - 2\,e^{-2j}\sinh(2j)}\right) \tag{6.34}$$

Obviously, $E_+ > E_-$ so that in the thermodynamic limit, only E_+ has a role to play:

$$Z_N(T, B_0) = E_+^N\left(1 + \left(\frac{E_-}{E_+}\right)^N\right) \;\rightarrow\; E_+^N \;\; as\ N \rightarrow \infty \tag{6.35}$$

From this we can evaluate all the thermodynamically important quantities. For example,

(a) *free energy per spin*

$$\begin{aligned} f(T, B_0) &= -k_B T \lim_{N\rightarrow\infty} \frac{1}{N} \ln Z_N(T, B_0) \\ &= -k_B T \ln E_+ \\ &= -J - k_B T\ \ln\{\cosh(\beta\mu_B B_0) + \\ &+ \left.\sqrt{\cosh^2(\beta\mu_B B_0) - 2e^{2\beta J}\sinh(2\beta J)}\right\} \end{aligned} \tag{6.36}$$

From this, we get, according to (1.90), the

(b) *magnetization per spin*

$$m(T, B_0) = -\left(\frac{\partial}{\partial B_0} f(T, B_0)\right)_T \tag{6.37}$$

This yields a relatively complicated expression for $m(T, B_0)$ from which one can recognize the competition between the "ordering" tendency of the field and the "disordering" tendency of the thermal motion (see Problem 6.2):

$$m(T, B_0) = \mu_B \frac{\sinh(\beta\mu_B B_0)}{\sqrt{\cosh^2(\beta\mu_B B_0) - 2e^{2\beta J}\sinh(2\beta J)}} \tag{6.38}$$

This result makes it once more clear that in the one-dimensional Ising model spontaneous magnetization is not possible (Fig. 6.5).

$$m(T \neq 0, B_0 = 0) = 0 \tag{6.39}$$

For very large fields B_0, m goes into "saturation", $m_s = \pm\mu_B$.

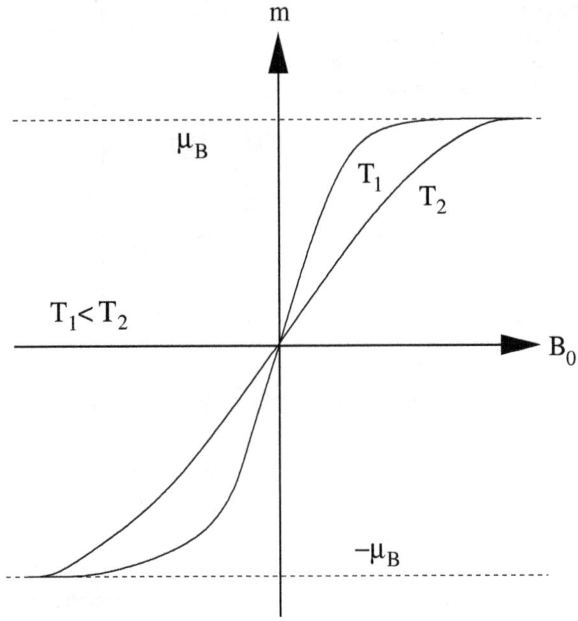

Fig. 6.5 Magnetization per spin as a function of the magnetic induction in the one-dimensional Ising model

(c) *The special case* $B_0 = 0$: In this limit the results for the Ising ring should actually coincide with those for the Ising chain (Sect. 6.2.1).
For $B_0 = 0$, the eigenenergies of the transfer matrix become

$$E_{\pm} = e^{\beta J} \pm e^{-\beta J} \tag{6.40}$$

so that the partition function takes the form

$$Z_N(T,0) \rightarrow E_+^N = 2^N \cosh^N(\beta J) \;\; for \; N \gg 1 \tag{6.41}$$

In the thermodynamic limit ($N \rightarrow \infty$), the thermodynamic potentials derived from Z_N agree with each other for the chain and the ring. For example,

$$f(T, B_0) = -k_B T \ln(2\cosh(\beta J)) \tag{6.42}$$

from which we can obtain the

(d) *entropy per spin*

$$\begin{aligned} S(T, B_0 = 0) &= -\left(\frac{\partial f}{\partial T}\right)_{B_0=0} \\ &= k_B\{\ln(2\cosh(\beta J)) - \beta J \tanh(\beta J)\} \end{aligned} \tag{6.43}$$

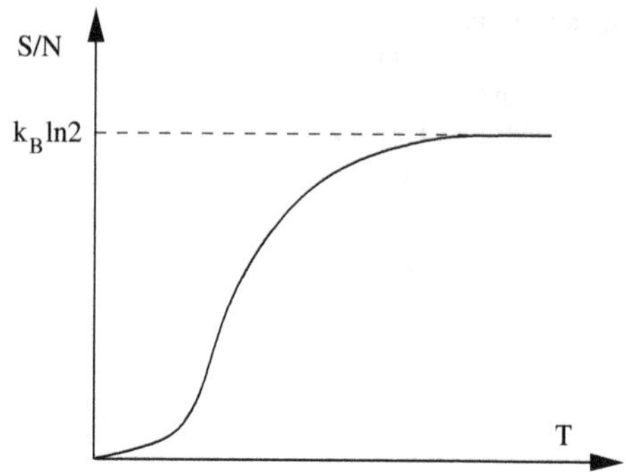

Fig. 6.6 Schematic plot of the entropy per spin as a function of temperature in the one-dimensional Ising model

One can immediately see the following limiting cases:

$$\begin{aligned} \lim_{T\to\infty}\ S &= k_B \ln 2 \\ \lim_{T\to 0}\ S &= 0 \end{aligned} \tag{6.44}$$

The second limiting case is the third law of thermodynamics. For $T \to \infty$ all the 2^N spin states are equivalent and therefore we must have $S = \frac{1}{N} k_B \ln 2^N$ (Fig. 6.6).

(e) *Specific heat*

$$C_{B_0=0} = T \left(\frac{\partial S}{\partial T}\right)_{B_0=0} = k_B \frac{\beta^2 J^2}{\cosh^2(\beta J)} \tag{6.45}$$

For $T \to {''T_c''} = 0$, the specific heat is *not* “critical”. It actually goes to zero satisfying the third law of thermodynamics (Fig. 6.7).

(f) *Susceptibility*: From (6.38) one can calculate the isothermal magnetic susceptibility (Problem 6.2):

$$\chi_T = \mu_0 \left(\frac{\partial m}{\partial B_0}\right)_{T,\ B_0\to 0} = \beta\ \mu_B^2\ \mu_0\ e^{2\beta J} \tag{6.46}$$

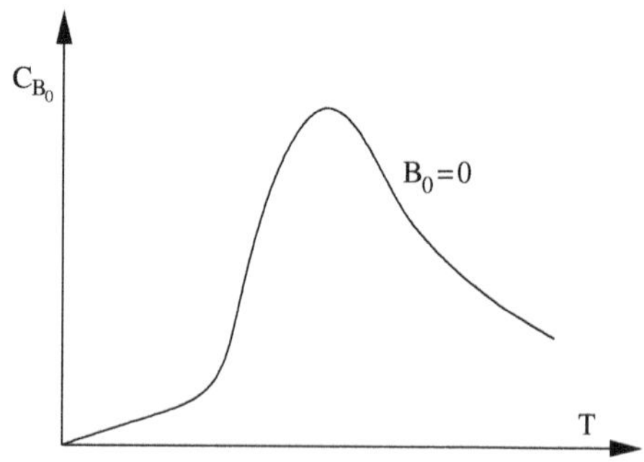

Fig. 6.7 Specific heat in zero field as a function of temperature in the one-dimensional Ising model

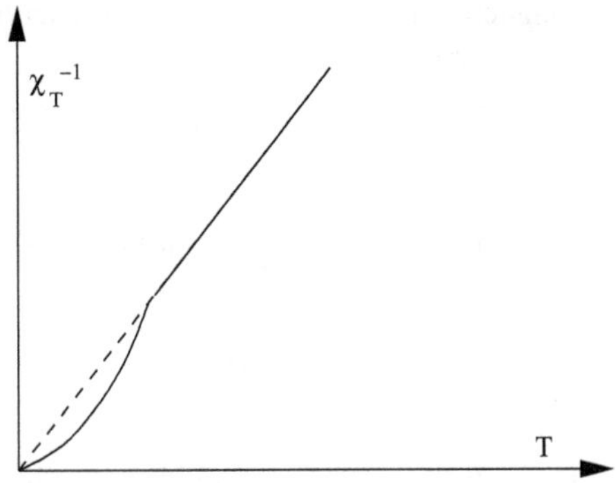

Fig. 6.8 Magnetic susceptibility of the one-dimensional Ising model as a function of temperature

The susceptibility diverges for $T \to {''T_c''} = 0$ and apparently satisfies the Curie law for high temperatures. As is to be expected for a paramagnet, χ_T is always positive (Fig. 6.8).

6.3 The Phase Transition of Two-Dimensional Ising Model

In contrast to the one-dimensional Ising model, the two-dimensional model shows a phase transition ferromagnetism $\Leftrightarrow$ paramagnetism at a finite transition temperature $T_c > 0$. In this section, we will prove the existence of the phase transition, via a chain of estimates, without actually determining T_c. We follow the idea of Peierls [1] who, long before Onsager [2] presented an exact derivation of the free energy, could demonstrate the phase transition of the two-dimensional Ising model. The original work of Peierls, however, contained a few erroneous conclusions, which were later corrected by Griffiths [3]. Now this idea of Peierls is known as the "Peierls argument".

6.3.1 The Method of Proof

Let X be a finite lattice. In connection with this we will use the following free energies:

$f(X;\ T, B_0)$: The free energy per spin for the lattice X *without* any special boundary conditions.

$\tilde{f}(X;\ T, B_0)$: The free energy per spin for the lattice X *with* special boundary conditions.

$f(T, B_0) = \lim_{X\to\infty} f(X;\ T, B_0)$: The free energy per spin in the thermodynamic limit.

The corresponding magnetizations are

$$M(X;\ T, B_0) = -\left(\frac{\partial f(X)}{\partial B_0}\right)_T \tag{6.47}$$

the magnetization of the finite lattice X *without* any boundary conditions,

$$\tilde{M}(X;\ T,\ B_0) = -\left(\frac{\partial \tilde{f}(X)}{\partial B_0}\right)_T \tag{6.48}$$

the magnetization of the finite lattice X *with* special boundary conditions,

$$M(T,\ B_0) = -\left(\frac{\partial f(T,\ B_0)}{\partial B_0}\right)_T \tag{6.49}$$

the magnetization in the thermodynamic limit.

The *spontaneous* magnetization is responsible for the phase transition.

$$M_s(X;\ T) = \lim_{B_0 \to 0} M(X;\ T,\ B_0) \tag{6.50}$$

This is the spontaneous magnetization of a *finite* system *without* boundary conditions. One shows rigorously that

$$M_s(X;\ T) \equiv 0 \tag{6.51}$$

In a finite system without any boundary conditions, there is no phase transition. However, for a finite system with appropriate boundary conditions, it is quite possible that there is a phase transition. That is,

$$\tilde{M}_S(X;\ T) = \lim_{B_0 \to 0} \tilde{M}(X;\ T,\ B_0) \tag{6.52}$$

can be unequal to zero. Actually, what decides about the phase transition is the spontaneous magnetization in the thermodynamic limit:

$$M_s(T) = \lim_{B_0 \to 0} M(T,\ B_0) \tag{6.53}$$

In the above, we have introduced all the quantities that are important for the following discussion.

The existence of the phase transition will be proved in three steps:

1. We define a finite spin lattice X with "suitable" boundary conditions and estimate the probability that a definite value ($+1$ or -1) of the spin variables S_i is realized at a temperature T.
2. We show that the spontaneous magnetization of the system X has a lower bound $m > 0$:

$$\tilde{M}_s(X;\ T) \geq m > 0, \quad if\ T < T^*(m,\ J)$$

3. We prove that m also represents the lower bound for the spontaneous magnetization in the thermodynamic limit:

$$M_s(T) \geq m > 0, \quad if\ T < T^*(m, J)$$

This will complete the proof.

6.3.2 Finite Ising Lattice with Special Boundary Conditions

The important "trick" in the derivation is that, the free energy of a finite system is calculated with so constructed boundary condition that $M_s(X;\ T) \neq 0$, which is not possible with free boundary conditions. At the same time, the boundary conditions should be chosen in such a way that they do *not* play any role in the thermodynamic limit.

Figure 6.9 represents a finite lattice X. Here we mean

$$+ \Leftrightarrow S_i = +1 \qquad\qquad - \Leftrightarrow S_i = -1$$

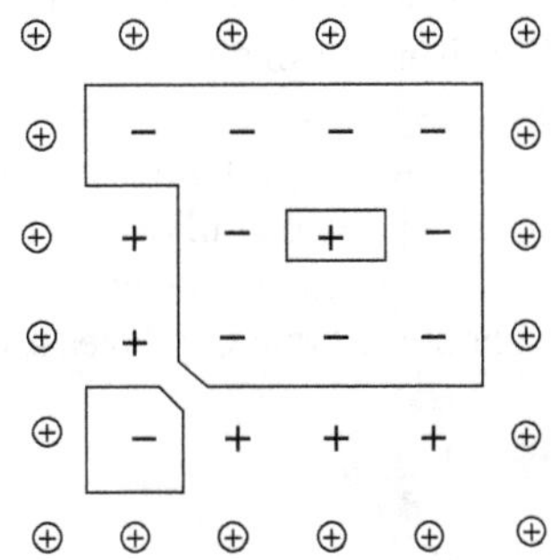

Fig. 6.9 Two-dimensional lattice of spins, oriented in positive (+) and negative (−) z-directions, with "walls" which separate (+) and (−) spins

We agree upon the *boundary condition* that the surface layer ∂X should contain only (+)-spins.

For a given spin configuration, we now introduce "walls" in the lattice. A "wall" is the line of separation between (+) and (−) sites.

When two walls intersect, at the point of intersection ("knot") there is non-uniqueness (Fig. 6.10). At these points, the walls are to be bent towards the side of the (−) places as shown in Fig. 6.10. As a consequence of the special boundary condition, the "walls" always build *closed* polygons Γ.

6.3.3 Probabilities

We define two probabilities:

$w_i(T)$: probability that at temperature T the site i is occupied by spin (−).

$W_\Gamma(T)$: probability that at temperature T, the polygon Γ exists.

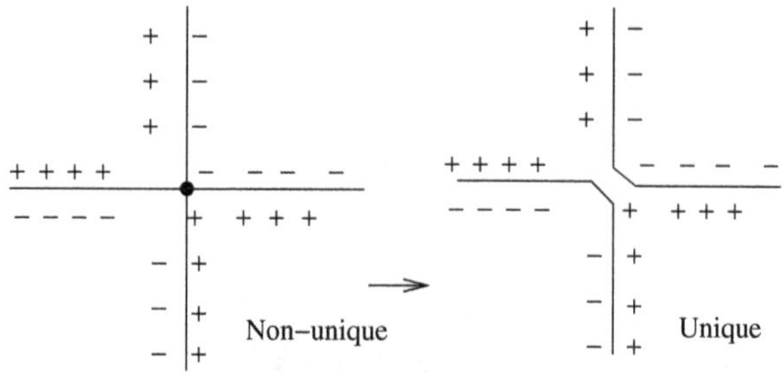

Fig. 6.10 The breakup of "knots" in the spin lattice of Fig. 6.9

Because of the special boundary condition, the $(-)$ site i lies always inside some polygon Γ. That is, i cannot be a site on the surface. The converse is of course not true, because $(+)$ place can also lie inside a polygon. Then, we can estimate

$$w_i(T) \leq \sum_{\Gamma \supset i} W_\Gamma(T) \tag{6.54}$$

The summation is over all the conceivable polygons which enclose the site i. W_Γ does not concern only the $(-)$ possibility for i. That is why the sign $\leq$.

The above estimate is surely very weak for the sites that lie deep inside the lattice X. It is increasingly sharp as the site lies more and more towards the boundary. For the sites which lie in the layer that is directly neighbouring to the surface layer, even the equality sign holds because, due to the boundary condition, these sites can lie inside a polygon only when they are themselves $(-)$-sites.

$W_\Gamma(T)$ can be calculated exactly. Let us define the lattice X' as the lattice X without the boundary layer ∂X:

$$X' = X - \partial X \tag{6.55}$$

Then there are a total of $2^{N(X')}$ possible spin configurations, since the spins in the boundary layer ∂X are fixed with $S = +1$. Let us denote by

$\sum_{\{S\}}$: sum over all the $2^{N(X')}$ spin configurations,
$\sum_{\{C\}}$: sum over all the spin configurations, in which Γ is realized.

Then, we obviously have

$$W_\Gamma(T) = \frac{\sum_{\{C\}} exp(-\beta H(C))}{\sum_{\{S\}} exp(-\beta H(S))} \tag{6.56}$$

H is the Hamiltonian function of the Ising model with nearest neighbour interactions where we set the external field $B_0 = 0$

$$H = -J \sum_{(i,j)} S_i \, S_j \tag{6.57}$$

After building the walls, the spin configuration C contains the special polygon Γ. We construct from C a new configuration C^* by reversing all the spins ($S_i \rightarrow -S_i$) within Γ. The exchange $\uparrow\uparrow \Leftrightarrow \downarrow\downarrow$ or $\uparrow\downarrow \Leftrightarrow \downarrow\uparrow$ for the nearest neighbour pairs does not change the energy of the system. In going from C to C^*, the energy of the system changes only because of the spin pairs which are separated by Γ. This is because, in C, they are antiparallel, whereas they are parallel in C^*. The change in energy per spin pair is $(-2J)$. If $|\Gamma|$ is the length of Γ measured in units of the lattice constants, then we have

$$H(C^*) = H(C) - 2J|\Gamma| \tag{6.58}$$

This gives us

$$\sum_{\{C^*\}} exp(-\beta H(C^*)) = e^{2J\beta|\Gamma|} \sum_{\{C\}} exp(-\beta H(C)) \tag{6.59}$$

From the exact relation for $W_\Gamma(T)$ (6.56), if we leave out a few of the positive definite summands in the denominator, we obtain an upper bound :

$$W_\Gamma(T) \leq \frac{\sum_{\{C\}} exp(-\beta H(C))}{\sum_{\{C^*\}} exp(-\beta H(C^*))} = e^{-2J\beta|\Gamma|} \tag{6.60}$$

With this we have obtained an intermediate result for the probability that a site is occupied by a $(-)$-spin at temperature T:

$$w_i(T) \leq \sum_{\Gamma \supset i} exp(-2\beta J|\Gamma|) \tag{6.61}$$

We will now try to perform the summation.

6.3.4 Realization Possibilities for the Polygons

We want to call two polygons to be of the *same type*, if they are related to each other by a rigid translation in the lattice (Fig. 6.11). We define:

$\rho(m, n)$: the number of realization possibilities for polygons with $2m$ vertical and $2n$ horizontal "wall units" (="rod" between nearest neighbours in the wall). In this the individual realizations shall, pairwise, be of *different type*.

It is obvious that for every polygon, the number of vertical (horizontal) individual wall units is always even. First, we want to make a rough estimate of $\rho(m, n)$. Suppose we build Γ stepwise from $2m$ vertical and $2n$ horizontal wall units. At any lattice site j, there are three independent possibilities to place the next wall unit ("backwards" is not allowed) (Fig. 6.12). Therefore, we have

$$\rho(m, n) \leq 3^{2m+2n} \tag{6.62}$$

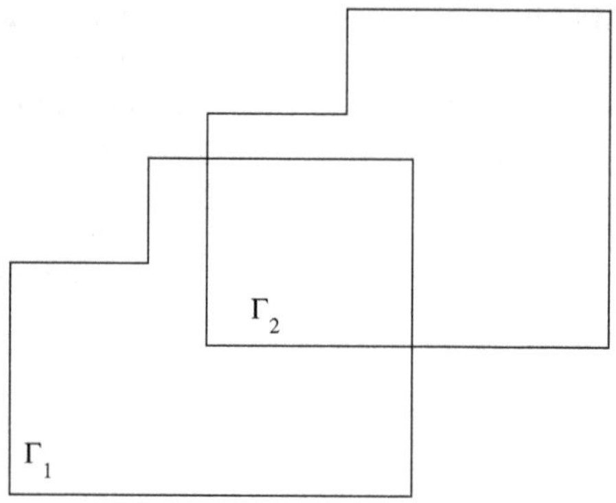

Fig. 6.11 Two polygons Γ_1 and Γ_2 of the same type

The perimeter of the polygon is $(2m + 2n)$ units. We have $\leq$ because, due to the surface, every lattice point does not have three independent possibilities. Let us denote by

$\eta_i(m, n)$: the number of possible polygons Γ built from $2m$ vertical and $2n$ horizontal wall units, and which enclose the site i.

Now let $\Gamma_i^{(0)}$ be a special $(2m, 2n)$-polygon that contains i. Let $g(\Gamma_i^{(0)})$ be the number of polygons of the same type as $\Gamma_i^{(0)}$ which also contain i. First, we want to estimate this number:

$\Gamma_i^{(0)}$ contains a maximum of $(m \cdot n)$ lattice points, that is, when it is a rectangle. That means, we can rigidly shift $\Gamma_i^{(0)}$ in a maximum of $(m \cdot n)$ ways, keeping i within the polygon.

$$g(\Gamma_i^{(0)}) \leq m \cdot n \tag{6.63}$$

We can now estimate $\eta_i(m, n)$:

$$\eta_i(m, n) \leq (m \cdot n)\, \rho(m, n) \tag{6.64}$$

This is again a very rough estimate because, in ρ on the right-hand side of the inequality, those polygons are also counted, which do not contain i at all. Using the

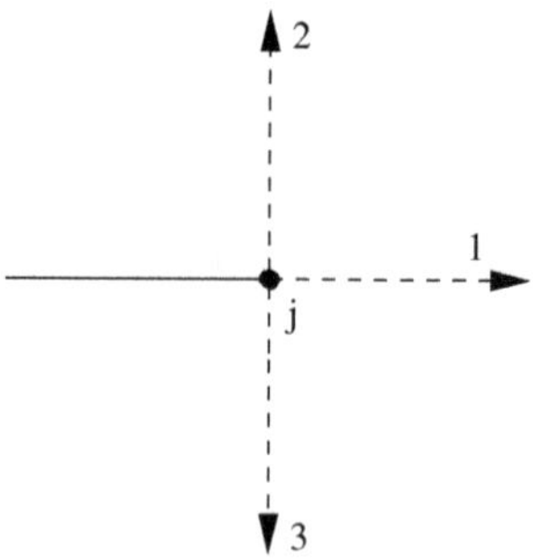

Fig. 6.12 Three allowed possibilities at site j to place the next wall unit

estimate for ρ given by (6.62) we have

$$\eta_i(m, n) \leq (m \cdot n)\, 3^{2m+2n} \tag{6.65}$$

With this, we can now further limit the probability $w_i(T)$ as given by (6.61) that the ith lattice site contains a spin $(-)$ at temperature T

$$\begin{aligned} w_i(T) &\leq \sum_{\Gamma \supset i} exp(-2\beta J|\Gamma|) \\ &\leq \sum_{m=0}^{\infty}\sum_{n=0}^{\infty} \eta_i(m, n)\, exp(-2\beta J(2m + 2n)) \end{aligned} \tag{6.66}$$

On the right-hand side there are only non-negative summands. Therefore, we can allow the sums to extend from 0 to ∞ eventhough for a finite lattice X, for every Γ, m and n must be finite and ≥ 1. In addition, we have used $|\Gamma| = 2m + 2n$. Finally, we use the estimate for $\eta_i(m, n)$ so that we get

$$w_i(T) \leq \phi(\beta, J) \tag{6.67}$$

with

$$\phi(\beta, J) = \left[\sum_{n=0}^{\infty} n \left(3e^{-2\beta J}\right)^{2n}\right]^2 \tag{6.68}$$

The series converges for $J > 0$ (ferromagnet!):

$$\begin{aligned} \phi(\beta, J) &= \left[\sum_{n=0}^{\infty} n \left(\underbrace{9e^{-4\beta J}}_{\equiv y}\right)^n\right]^2 \qquad (\beta \; such \; that \; y < 1) \\ &= \left[\left(\frac{d}{dy}\sum_{n=0}^{\infty} y^n\right) y\right]^2 = \left[y \frac{d}{dy}\left(\frac{1}{1-y}\right)\right]^2 \\ &= \frac{y^2}{(1-y)^4} \end{aligned}$$

For a sufficiently large β (i.e. small T), $\phi(\beta, J)$ can be made smaller than any arbitrary, finite value:

$$\phi(\beta, J) = \left[\frac{9e^{-4\beta J}}{\left(1 - 9e^{-4\beta J}\right)^2}\right]^2 \tag{6.69}$$

6.3.5 Magnetization of the Finite Lattice

We are now in a position to estimate the spontaneous magnetization $\tilde{M}_s(X,T)$ of the finite spin lattice:

$$\begin{aligned}\tilde{M}_s(X,T) &= \mu_B \frac{1}{N(X)} \sum_{i\in X} \langle S_i \rangle_{B_0=0} \\ &= \frac{\mu_B}{N(X)} (\langle N_+(X)\rangle - \langle N_-(X)\rangle) \\ &= \frac{\mu_B}{N(X)} (N(X)(1-w_i(T)) - N(X)w_i(T)) \\ &= \mu_B(1-2w_i(T))\end{aligned} \tag{6.70}$$

When we impose an upper bound for $w_i(T)$, then we get a lower bound for the spontaneous magnetization:

$$\tilde{M}_s(X;\,T) \geq \mu_B(1-2\phi(\beta,J)) \tag{6.71}$$

For $\beta \to \infty$, i.e. for $T \to 0$, $\phi(\beta,J)$ can be made to lie below any arbitrary finite limit. With this we complete the first part of the proof.

For every $0 < m < \mu_B$ and for every $J > 0$, there exists a critical temperature

$$T^* = T^*(m,J) \tag{6.72}$$

such that for all $T \leq T^*$ holds,

$$\tilde{M}_s(X,T) \geq m > 0$$

This is valid for a finite but in principle arbitrarily large spin lattice. Considering the fact that $M_s(X;\,T)$ for a finite system *without* boundary conditions is always zero, it must be concluded that the boundary condition for the *surface*, independent of the size of the system X does produce a *volume effect*.

6.3.6 Thermodynamic Limit

We have, so far, discussed the spontaneous magnetization $\tilde{M}_s(X,T)$ of a finite Ising lattice. For this case, we have (6.52)

$$\tilde{M}_s(X;\,T) = \lim_{B_0\to 0} \tilde{M}_s(X,T,B_0) = -\lim_{B_0\to 0} \left(\frac{\partial \tilde{f}(X)}{\partial B_0}\right)_T \tag{6.73}$$

For $T \leq T^*$, this expression is $\geq m$. For $T \neq T^*$, the thermodynamic potential $\tilde{f}(X)$ must be an analytical function, i.e. it should be continuous and differentiable any number of times. Then it follows from

$$-\left(\frac{\partial \tilde{f}(X)}{\partial B_0}\right)_{T,B_0=0} \geq m \tag{6.74}$$

for sufficiently small fields B_0:

$$\tilde{f}(X;\ T,\ B_0) - \tilde{f}(X;\ T, 0) \leq -m B_0 \qquad (T < T^*) \tag{6.75}$$

So far we have discussed only the free energy $\tilde{f}(X)$ of the finite lattice X with its special boundary conditions. We do not know the corresponding function $f(X)$ *without* the boundary conditions. Since the free energy is an extensive quantity, the two functions must coincide in the thermodynamic limit, as the boundary conditions influence only the now irrelevant surface term:

$$\begin{aligned} \tilde{f}(X;\ T,\ B_0) &\rightarrow f(T,\ B_0)\ as\ X \rightarrow \infty \\ f(X;\ T,\ B_0) &\rightarrow f(T,\ B_0)\ as\ X \rightarrow \infty \end{aligned} \tag{6.76}$$

Using this limiting case in the inequality (6.75), the right-hand side remains unaffected. Thus it holds also

$$f(T,\ B_0) - f(T,\ 0) \leq -m B_0 \qquad (T < T^*,\ B_0 \rightarrow 0) \tag{6.77}$$

Finally, it follows from this that

$$M_s(T) = \left(-\frac{\partial f}{\partial B_0}\right)_{T,B_0=0} \geq m \geq 0\ if\ T \leq T^*(m,\ J) \tag{6.78}$$

It is shown in the above that, below a critical temperature T^*, the two-dimensional Ising model has spontaneous magnetization. Therefore, the existence of a phase transition in two-dimensional Ising model is proved.

Here we have shown the phase transition only for $d = 2$. The generalization to $d = 3$ is straightforward. In place of polygons, one chooses polyhedrons. Except for that, the steps to be followed in the proof are exactly the same.

The theory presented above makes statement only about the existence of the phase transition. Because of some of the very rough estimates, naturally, we cannot give any explicit value or expression for the Curie temperature. The critical temperature

$$T_c^* = T^*(m = 0^+,\ J) \tag{6.79}$$

will be smaller than the actual T_c. T_c^* is given by the condition (6.71)

$$1 - 2\phi(\beta_c^*,\ J) \stackrel{!}{=} 0^+ \ \Leftrightarrow\ \phi(\beta_c^*,\ J) = \frac{1}{2} \tag{6.80}$$

From this it is easy to get

$$\beta_c^* J = \frac{1}{4} \ln \frac{9}{(1 + 1/\sqrt{2}) - \sqrt{1/2 + \sqrt{2}}} = 0.8314 \tag{6.81}$$

The exact solution of Onsager (19440) gives

$$\beta_c J = 0.4407 \tag{6.82}$$

That means, T_c^* is, as expected, smaller than T_c:

$$T_c^* = 0.5301\ T_c \tag{6.83}$$

The deviation, in view of the drastic estimates used, is, however, not very dramatic.

6.4 The Free Energy of the Two-Dimensional Ising Model

In the last section with the help of the *Peierls argument* we could show exactly, the existence of a phase transition in the two-dimensional Ising model. The proof was based on estimates which did not involve any complicated mathematical effort, and proved that when the temperature becomes less than a critical temperature T^*, there exists a spontaneous magnetization $M_s(T) > 0$. However, the estimates, which are partly quite rough, did not permit an exact determination of the Curie temperature T_c. In this section, we will accomplish that by a detailed discussion of the free energy of the two-dimensional Ising model [4].

The starting point is as in (6.57), the Hamiltonian function of the Ising model for the case of the nearest neighbour interaction of the spins and in the absence of an external field ($B_0 = 0$):

$$H(S) = -J \sum_{(i,j)} S_i\ S_j \tag{6.84}$$

The exact analytical derivation of the free energy for $B_0 \neq 0$ has so far not been achieved. (i, j) means the summation is over only the nearest neighbour pairs. We consider a square lattice containing N spins. Then we have to calculate the *canonical partition function*

$$Z_N(T) = \sum_{\{S\}} exp(-\beta\ H(S)) \tag{6.85}$$

The sum runs over all the 2^N spin configurations

$$S = (S_1,\ S_2,\ \cdots,\ S_N) \tag{6.86}$$

The calculation will first be done for the finite system. To find the phase transition from the free energy, we naturally need to take the thermodynamic limit. The signature of the phase transition, in case it exists, will be noticed from an "irregularity" in the free energy per spin

$$f(T) = \lim_{N\to\infty} \left[-\frac{k_B T}{N} \ln Z_N(T) \right] \tag{6.87}$$

For a better overview, we would like to divide the derivation of the free energy into a number of steps.

6.4.1 High-Temperature Expansion

Since the spin variables S_i can take only the values $+1$ or -1, it holds for an arbitrary $n \in N$:

$$(S_i\ S_j)^{2n} = 1 \qquad ; \qquad (S_i\ S_j)^{2n+1} = S_i\ S_j \tag{6.88}$$

In the series expansion of the exponential function, we can immediately see the following relation:

$$exp(\beta\ J\ S_i\ S_j) = \cosh(\beta\ J) + (S_i\ S_j)\sinh(\beta\ J)$$

We define

$$v = \tanh(\beta\ J) \tag{6.89}$$

This is a useful variable for the high-temperature expansions, since for high temperatures β and therefore v are small:

$$exp(\beta\ J\ S_i\ S_j) = \cosh(\beta\ J)(1 + v(S_i\ S_j)) \tag{6.90}$$

In a square lattice every lattice point has four nearest neighbours. Neglecting the surface effects, which anyway is allowed when we go over to the thermodynamic limit, there are $4N$ pairs of nearest neighbours out of which, however, there are two pairs of them which are equal. Thus there are $2N$ distinct pairs of nearest spin neighbours.

Using (6.90), we get a high-temperature expansion of the canonical partition function (6.85)

$$Z_N(T) = \sum_{\{S\}} exp\left(\beta\, J \sum_{(i,j)} S_i\, S_j\right)$$
$$= \cosh^{2N}(\beta\, J) \sum_{\{S\}} (1 + vS_{i_1}\, S_{j_1}) \cdots (1 + vS_{i_{2N}}\, S_{j_{2N}}) \tag{6.91}$$

After rearranging according to the powers of v, we can write

$$Z_N(T) = \cosh^{2N}(\beta\, J) \sum_{\{S\}} \left[1 + v \sum_{\nu=1}^{2N} S_{i_\nu}\, S_{j_\nu} + \right.$$
$$\left. + \; v^2 \sum_{\substack{\nu,\mu=1 \\ \nu \neq \mu}}^{2N} (S_{i_\nu}\, S_{j_\nu})(S_{i_\mu}\, S_{j_\mu}) + \cdots \right] \tag{6.92}$$

6.4.2 Spin Products as Graphs

We will symbolically indicate the *Ising interactions* between pairs of neighbouring spins by lines connecting the corresponding lattice points.

Every line carries a factor v. The lines connect only the nearest neighbours in the lattice. The points in these graphs are called the *vertices*. The order of the vertex is defined as the number of coupled interactions. Therefore, there are orders from 1 to 4 (Fig. 6.13).

On the spin products that appear in (6.92), we can make the following statement:

$$\sum_{\{S\}} (S_{i_\alpha}\, S_{j_\alpha}) \cdots (S_{i_\rho}\, S_{j_\rho}) = \begin{cases} 2^N \; if\ all\ vertices\ of\ even\ order \\ 0 \; otherwise \end{cases} \tag{6.93}$$

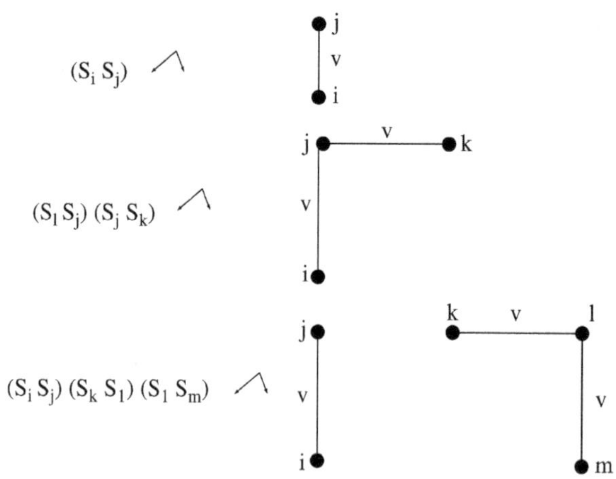

Fig. 6.13 Graphical representation of spin products

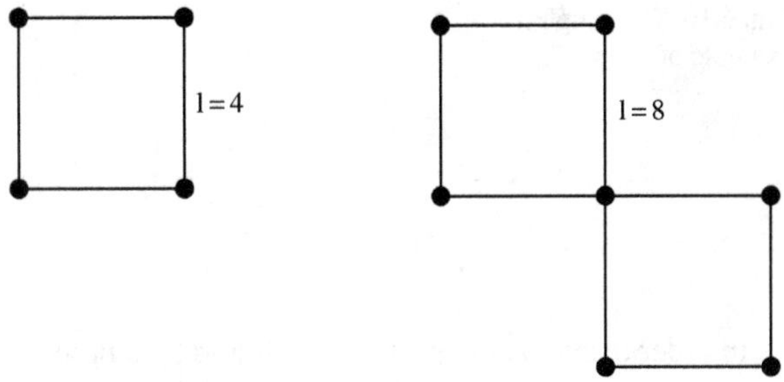

Fig. 6.14 Spin graphs with only even vertices

Since the spin variables can take values of only ± 1, an even power of a spin variable is always equal to $+1$. The sum runs over the 2^N different spin configurations. That explains the first part of (6.93). In contrast, if only one of the vertices is odd, then, there exist among the $\{S\}$ always two terms, which differ from each other only such that, at the corresponding vertex, once $S = 1$ and another time $S = -1$. Therefore, the summands compensate each other.

Using (6.93) in (6.92), we get the following intermediate result for the canonical partition function:

$$Z_N(T) = \cosh^{2N}(\beta J)\, 2^N \sum_{l=0}^{\infty} g_l v^l \tag{6.94}$$

g_l is the number of graphs made of l lines with exclusively even vertices

$$g_0 \equiv 1 \tag{6.95}$$

Now what remains is to determine g_l. The requirement that the graphs of interest must all contain even vertices says that these graphs must be made up of closed set of lines, for example, as shown in Fig. 6.14.

6.4.3 Loops

We introduce two further concepts: We denote as a *knot* a vertex of fourth order (Fig. 6.15), while a *loop* is a closed set of lines *without* knots. The simplest example is shown in Fig. 6.16.

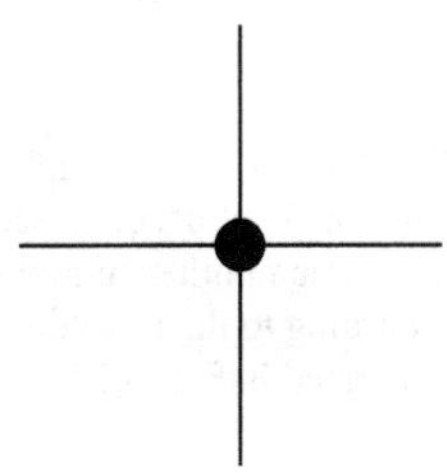

Fig. 6.15 Knot as a vertex of fourth order

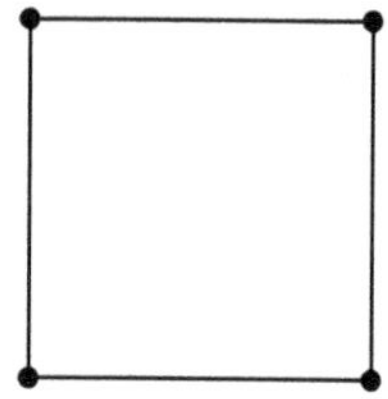

Fig. 6.16 The simplest example of a "loop"

In order to avoid ambiguities, we should fix a procedure for *dissolving knots*. The possibilities are shown in Fig. 6.17. We dissolve the knots in three different ways. We call the last possibility in Fig. 6.17 *self-intersection* and in short, denote by SI.

The described procedure of dissolving the knots, splits a graph with k knots into 3^k *families of loops* (see Fig. 6.18 for $k = 1$).

The dissolving of the knots multiplies the number of diagrams. This will be corrected by introducing a weight factor η for the loops while counting g_l:

$$\eta(loop) = (-1)^{number\ of SI} \tag{6.96}$$

$$\begin{aligned}\eta(family) &= \prod [\ \eta(loop)\ ] \\ &= (-1)^{number\ of SI\ in\ the\ family}\end{aligned} \tag{6.97}$$

Proposition: *g_l is the sum of weights of all families of loops built from a total of l lines.*

Proof:

1. Every graph without knots gives exactly *one* family of loops without any SI and therefore has a weight of $(-1)^0 = +1$ in the counting.
2. For a graph with k knots, there are $\binom{k}{j}$ possibilities to choose j knots from the k knots, which, after dissolution, have one SI. Then, there remain $(k-j)$ knots, which, after dissolution do not lead to a SI. For the latter, there are always two possibilities according to Fig. 6.17.

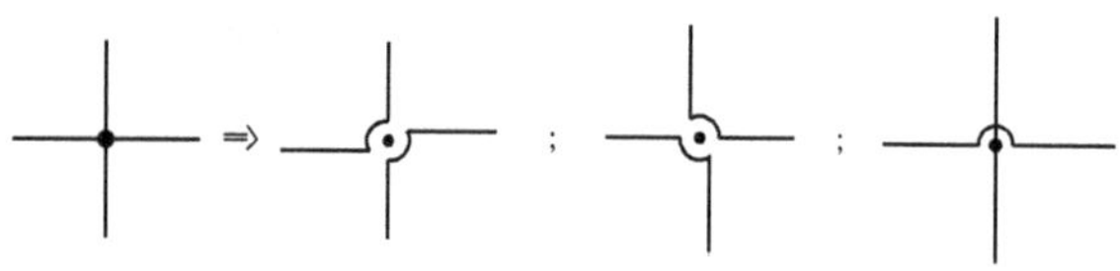

Fig. 6.17 The method of "dissolving knots"

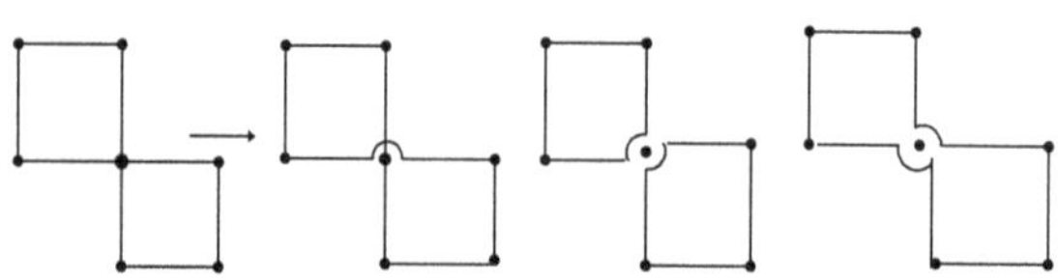

Fig. 6.18 A closed graph with one knot decomposed into three families of loops according to the procedure illustrated in Fig. 6.17

Therefore, there are $2^{k-j}\begin{pmatrix}k\\j\end{pmatrix}$ possibilities to construct a family with j SI out of a graph with k knots. Each of these families carries a weight of $(-1)^j$.

The total weight of all the families of loops constructed from a given graph with k knots is given by

$$\sum_{j=0}^{k}\begin{pmatrix}k\\j\end{pmatrix} 2^{k-j}(-1)^j = (2-1)^k = 1$$

With this, we have proved the proposition. The number of diagrams, after dissolving into families of loops, has of course increased many times. The weights, however, take care that all the families arising out of the dissolution of a given graph add up to a total weight of 1.

We now introduce a new quantity

$$D_l \equiv \textit{Sum of the weights of } \text{all } \textit{the loops with } l \textit{ lines} \tag{6.98}$$

We can connect D_l with g_l. Every family is made up of one or more loops. Therefore, we have

$$\begin{aligned} g_0 &= 1 \\ g_l &= \sum_{n=1}^{\infty}\frac{1}{n!}\sum_{\substack{l_1,\cdots,l_n \\ \sum l_i = l}} D_{l_1}\, D_{l_2}\cdots D_{l_n} \qquad (l \neq 0) \end{aligned} \tag{6.99}$$

The product of D_{l_i} captures all the possible splittings of a family with l lines into loops with the added condition $\sum l_i = 1$. Every weight, which appears only because of an interchange of D_{l_i}, corresponds to the same set of families and so, should be counted only once. This is taken care of by the factor $1/n!$. The summation over n can run up to ∞ because for $l_i < 4$, $D_{l_i} = 0$ in any case.

(6.99) as it appears is not yet quite correct. Since the summations l_i run independently, the *double occupations* of individual lines are allowed as illustrated in Fig. 6.19. Such diagrams correspond to forbidden loops, i.e. they do not represent any summand in either (6.92) or (6.94) and therefore should not contribute in our new counting. We take care of this by adapting the procedure shown in Fig. 6.20 to dissolve the double occupations. In (6.99), we allow two types of double occupations, type (a) carries a weight of (+1) since there are no SI. Type (b) has exactly one self-intersection and so carries a weight of (−1). Therefore, the total weight is zero, i.e. even though the double occupations are formally counted, they do not contribute to (6.99). From the forbidden diagram (Fig. 6.19) described above, the contributions shown in Fig. 6.21 appear.

With this procedure, now (6.99) has become unique and we get the following intermediate result:

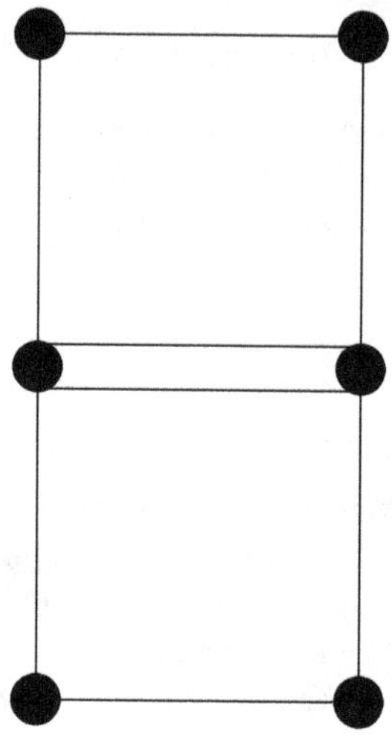

Fig. 6.19 Double occupancy of a line in a certain spin graph evaluated according to (6.99)

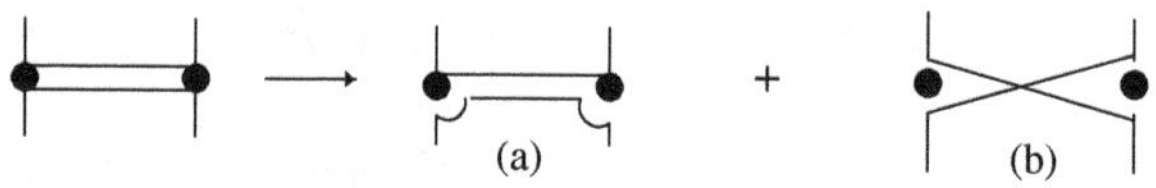

Fig. 6.20 Method of handling double occupancies of lines

$$g_l\, v^l = \sum_{n=1}^{\infty} \frac{1}{n!} \sum_{\substack{l_1,\cdots,l_n \\ \sum l_i = l}} \left(D_{l_1} v^{l_1}\right) \cdots \left(D_{l_n} v^{l_n}\right) \quad (l \neq 0) \tag{6.100}$$

For (6.99), we have to sum this expression over all l from 0 to ∞. When we do this, the l_i summations become independent of each other and their condition $\sum l_i = l$ becomes superfluous.

$$\sum_{l=0}^{\infty} g_l\, v^l = 1 + \sum_{n=1}^{\infty} \frac{1}{n!} \left\{ \sum_{m=1}^{\infty} D_m\, v^m \right\}^n = exp \left\{ \sum_{m=1}^{\infty} D_m\, v^m \right\} \tag{6.101}$$

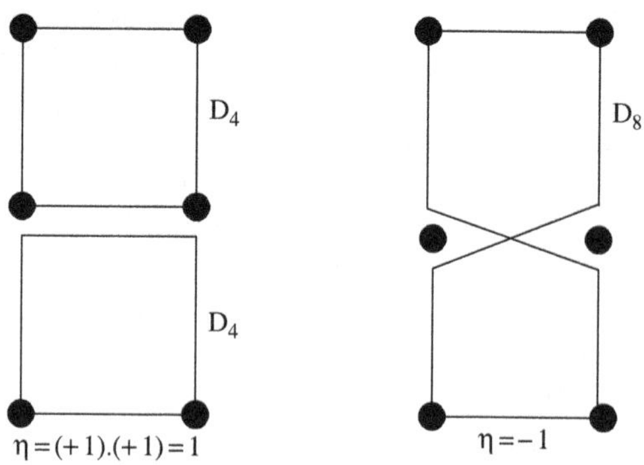

Fig. 6.21 The two possibilities of evaluating the double occupancy in the diagram of Fig. 6.19

Substituting (6.101) in (6.94) we get a further intermediate result for the canonical partition function:

$$Z_N(T) = 2^N \cosh^{2N}(\beta J) exp \left\{ \sum_{m=1}^{\infty} D_m \, v^m \right\} \tag{6.102}$$

The remaining job is to determine D_m, i.e. the weight of *all* loops which one can construct with m lines.

6.4.4 Directed Paths

The real problem is to count the self-intersections of a loop. We get round the problem by introducing directed paths: For this purpose, we define

Single step $P = (z, \alpha)$

$z = x_1 + ix_2$: starting point of the *single step*

$\alpha = 1,\ i,\ -1,\ -i$: direction of the *single step*

$z + \alpha$: end point of the *single step*

Imagine a two-dimensional Ising lattice in the complex plane. Every lattice point is represented by a complex number with real and imaginary parts x_1 and x_2 being integer numbers (Fig. 6.22).

We denote as *path from z to z′ in m steps*, a series of m single steps:

$$(z, \alpha) = P_0,\ P_1,\ P_2, \cdots,\ P_m = (z', \alpha') \tag{6.103}$$

with $z_i + \alpha_i = z_{i+1}$ and $\alpha_{i+1} \neq -\alpha_i$

With the condition $\alpha_{i+1} \neq -\alpha_i$ retracing points are forbidden. We define the *weight of the path*

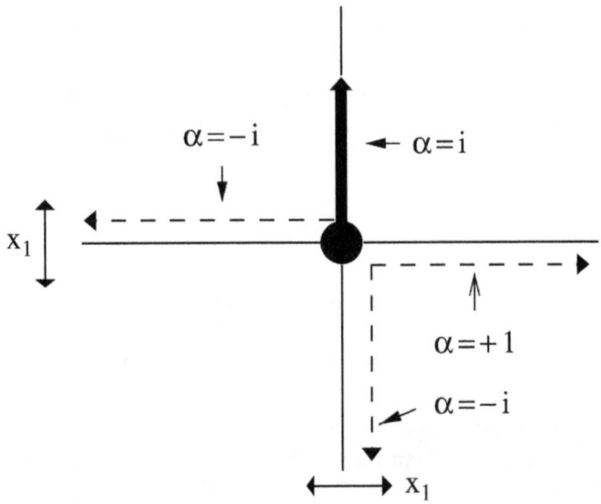

Fig. 6.22 Introduction of directed paths

$$\eta(path) = exp\left[\frac{i}{2}\left(arg\frac{\alpha_1}{\alpha_0} + \cdots + arg\frac{\alpha_m}{\alpha_{m-1}}\right)\right] \tag{6.104}$$

where

$$\frac{\alpha_{i+1}}{\alpha_i} = 1, \ \pm i \ \Leftrightarrow arg\frac{\alpha_{i+1}}{\alpha_i} = 0, \ \pm\frac{\pi}{2} \tag{6.105}$$

$arg\frac{\alpha_{i+1}}{\alpha_i}$ is the angle between the $(i+1)$st and the ith single step.

We introduce a matrix M_m whose meaning becomes clear from the definition of the matrix elements:

$$\langle P|M_m|P'\rangle = \textit{sum of the weights of all paths from P to P'} \\ \textit{in m steps} \tag{6.106}$$

The matrix element should be zero if P' is not reachable from P in m steps. If $m = m_1 + m_2$, then we can also write

$$\langle P|M_m|P'\rangle = \sum_{P''}\langle P|M_{m_1}|P''\rangle\langle P''|M_{m_2}|P'\rangle \tag{6.107}$$

This corresponds to the matrix product

$$M_m = M_{m_1}\, M_{m_2} \tag{6.108}$$

Obviously, the decomposition can be continued,

$$M_m = M_1^m \tag{6.109}$$

For N lattice points and four possibilities for α, M_1 is a $(4N \times 4N)$ matrix which contains a lot of zeros, since only those P's contribute, which are connected by a single step.

We are looking for the relation between the matrix M_m and the D_m which is required in (6.102). We expect that relation to be of the following form:

$$D_m = -\frac{1}{2m}\sum_P \langle P|M_m|P\rangle = -\frac{1}{2m}\, Tr\left(M_1^m\right) \tag{6.110}$$

This expectation will be justified in three steps:

1. D_m captures only closed paths (loops). That means, only the diagonal elements $P = P'$ can play a role.
2. In the sum over P, every one of the m loop points can be a starting point once. Further, a given loop can be run in two directions. But, such a loop should be counted only once. That is why, there is the factor $\frac{1}{2m}$.

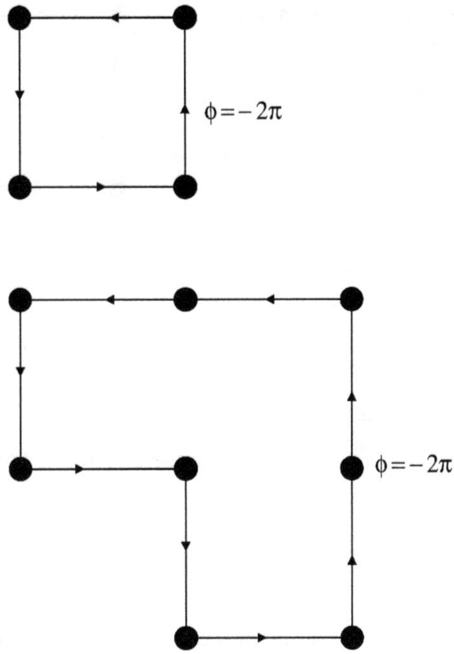

Fig. 6.23 Two diagrams with no self-intersection

3. For a closed path, the total angle of rotation is an integral multiple of 2π:

$$\phi = \arg\frac{\alpha_1}{\alpha_0} + \arg\frac{\alpha_2}{\alpha_1} + \cdots + \arg\frac{\alpha_m}{\alpha_{m-1}} = n\,2\pi \qquad n \in Z$$

That means $\eta(path) = \pm 1$.

For an even number of self-intersections, $\phi = \pm 2\pi$, and for an odd number, $\phi = 0$. Examples are given in Figs. 6.23, 6.24 and 6.25.

Comparing with (6.96), we get finally

$$\eta(path) = -\eta(loop) \tag{6.111}$$

That explains the − sign in (6.110).

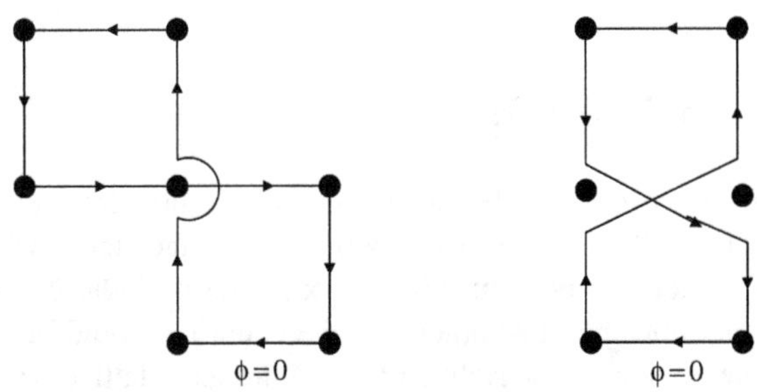

Fig. 6.24 Two diagrams with one self-intersection

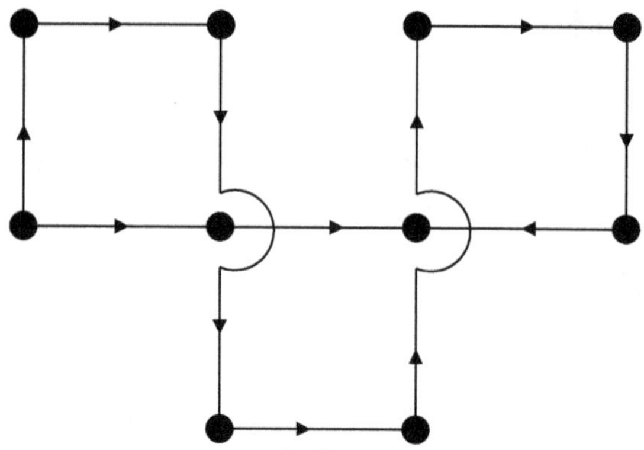

Fig. 6.25 A diagram with two self-intersections

Let $\lambda_1,\ \lambda_2, \cdots,\ \lambda_n$ be the eigenvalues of the matrix M_1. Then we have

$$Tr\left(M_1^m\right) = \sum_{j=1}^{n} \left(\lambda_j\right)^m \tag{6.112}$$

For the calculation of the canonical partition function (6.102), we require

$$\begin{aligned}\sum_{m=1}^{\infty} D_m\, v^m &= -\frac{1}{2}\sum_{j=1}^{n}\sum_{m=1}^{\infty}\frac{(\lambda_j\, v)^m}{m}\\ &= \frac{1}{2}\sum_{j=1}^{n} \ln(1 - v\lambda_j)\\ &= \ln\left\{\prod_{j=1}^{n}(1 - v\lambda_j)^{1/2}\right\}\\ &= \ln\{det(1 - vM_1)\}^{1/2}\end{aligned}$$

This leads to the following intermediate result for the canonical partition function:

$$Z_N(T) = 2^N \cosh^{2N}(\beta\, J)\{det(1 - vM_1)\}^{1/2} \tag{6.113}$$

6.4.5 *Matrix M_1*

In order to avoid the surface points, we now introduce the periodic boundary conditions. The error caused by that is a surface term, which becomes irrelevant in the thermodynamic limit ($N \to \infty$). The periodic boundary conditions are allowed only now, because otherwise the counting would be wrong. A path that runs from the left end to the right end of a finite Ising lattice would become a loop for a torus.

The elements of the matrix M_1 are easily given by

$$\langle P|M_1|P'\rangle = exp\left(\frac{i}{2}\, arg\frac{\alpha}{\alpha'}\right)\left(1-\delta_{\alpha,-\alpha'}\right)\delta_{z+\alpha,z'} \tag{6.114}$$

The first factor according to (6.104) is the weight of a single step. The second factor avoids the retrace points and the third one guarantees that only a single step from z to z' is being considered.

Due to the periodic boundary conditions, there is translational symmetry, i.e. the matrix element (6.114) for given α and α' depends only on the separation $z-z'$. Then, we can perform a Fourier transformation:

$$z = x_1 + ix_2; \quad x_i = 1, \cdots, N_i\,; \quad i = 1,\, 2,\,; \quad N = N_1 \cdot N_2 \tag{6.115}$$

$$q = q_1 - iq_2 \quad ; \quad q_i = \frac{2\pi}{N_i}(1,\, 2, \cdots, N_i) \tag{6.116}$$

The elements of $\tilde{M}_1$, the Fourier-transformed matrix of M_1, are then given by

$$\begin{aligned}\langle q\alpha|\tilde{M}_1|q'\alpha'\rangle &= \frac{1}{N}\sum_{z,z'} e^{-i\,Re(qz)}\langle z\alpha|M_1|z'\alpha'\rangle e^{i\,Re(q'z')} \\ &= \frac{1}{N}\sum_{z} e^{-i\,Re((q-q')z)} e^{i\,Re(q'\alpha)} e^{\frac{i}{2}arg\frac{\alpha}{\alpha'}}(1-\delta_{\alpha,-\alpha'}) \\ &= \delta_{q,q'} e^{i\,Re(q'\alpha)}\, e^{\frac{i}{2}arg\frac{\alpha}{\alpha'}}(1-\delta_{\alpha,-\alpha'})\end{aligned} \tag{6.117}$$

$\tilde{M}_1$ thus clearly breaks into (4 × 4) blocks along the diagonals

$$\langle q\alpha|\tilde{M}_1|q'\alpha'\rangle = \delta_{qq'}\langle\alpha|m(q)|\alpha'\rangle \tag{6.118}$$

$$\langle\alpha|m(q)|\alpha'\rangle = e^{i\ Re\ (q\alpha)} e^{\frac{i}{2}arg\frac{\alpha}{\alpha'}}\ (1-\delta_{\alpha,-\alpha'}) \tag{6.119}$$

There is no change in the determinant that is required in (6.112) after Fourier transformation. Therefore, we have

$$det(1 - vM_1) \equiv det(1 - v\tilde{M}_1) = \prod_q det(1 - vm(q)) \tag{6.120}$$

Substituting in (6.113) leads to a noticeably simple intermediate result for the canonical partition function:

$$Z_N(T) = 2^N \cosh^{2N}(\beta\, J)\{\prod_q det(1 - vm(q))\}^{1/2} \tag{6.121}$$

Now we only have to evaluate the determinant of a 4×4 matrix. Using α as the row- and α' as the column index and with the following notation:

$$\lambda = e^{i\frac{\pi}{4}} \quad ; \quad Q_1 = e^{iq_1} \quad ; \quad Q_2 = e^{iq_2} \tag{6.122}$$

the matrix in question is the following:

$$(1 - vm(q)) \equiv \begin{pmatrix} 1 - vQ_1 & -v\lambda Q_1 & 0 & -v\lambda^* Q_1 \\ -v\lambda^* Q_2 & 1 - vQ_2 & -v\lambda Q_2 & 0 \\ 0 & -v\lambda^* Q_1^* & 1 - vQ_1^* & -v\lambda Q_1^* \\ -v\lambda Q_2^* & 0 & -v\lambda^* Q_2^* & 1 - vQ_2^* \end{pmatrix} \tag{6.123}$$

It is simple to calculate the determinant of this matrix. Doing that, we get the following form for the canonical partition function:

$$Z_N(T) = 2^N \cosh^{2N}(\beta J)\{\prod_q [\,(1 + v^2)^2 - \\ -2v(1 - v^2)(\cos q_1 + \cos q_2)\,]\,\}^{1/2} \tag{6.124}$$

6.4.6 Free Energy per Spin

In the thermodynamic limit, the free energy per spin of the two-dimensional Ising model is given according to (6.87) and (6.124) by

$$-\beta f(T) = \ln 2 + 2\ln(\cosh(\beta J)) + \\ + \lim_{N\to\infty} \frac{1}{2N} \sum_{q_1,q_2} \ln\left[(1 + v^2)^2 - 2v(1 - v^2)(\cos q_1 + \cos q_2)\right] \tag{6.125}$$

We are allowed to write the double sum as a double integral:

$$\lim_{N\to\infty} \frac{1}{N} \sum_{q_1,q_2} F(q_1, q_2) \to \frac{1}{4\pi^2} \int_0^{2\pi} \int_0^{2\pi} dq_1 dq_2 \, F(q_1, q_2)$$

Further, using, according to (6.89)

$$\cosh(\beta J) = \frac{1}{\sqrt{1 - v^2}}$$

we can write

$$2\ln(\cosh(\beta J)) = \frac{1}{2}\ln(1-v^2)^{-2}$$
$$= \frac{1}{8\pi^2}\int_0^{2\pi}\int_0^{2\pi} dq_1 dq_2 \ln(1-v^2)^{-2}$$

Using this the free energy can be written as

$$-\beta f(T) = \ln 2 + \frac{1}{8\pi^2}\int_0^{2\pi}\int_0^{2\pi} dq_1 dq_2 \ln\left[\left(\frac{1+v^2}{1-v^2}\right)^2 - 2v\frac{1}{1-v^2}(\cos q_1 + \cos q_2)\right] \tag{6.126}$$

With

$$\left(\frac{1+v^2}{1-v^2}\right) = \cosh^2(2\beta J) = (1-\sinh(2\beta J))^2 + 2\sinh(2\beta J)$$

and

$$2\frac{v}{1-v^2} = \sinh(2\beta J)$$

we finally have the expression for the free energy per spin:

$$-\beta f(T) = \ln 2 + \frac{1}{8\pi^2}\int_0^{2\pi}\int_0^{2\pi} dq_1 dq_2 \ln[(1-\sinh(2\beta J))^2 + \sinh(2\beta J)(2-\cos q_1 - \cos q_2)] \tag{6.127}$$

This integral cannot be solved in a closed form.

6.4.7 Curie Temperature T_c

A phase transition must be noticeable from a non-analyticity of the free energy. Looking at (6.127), we see that such a thing is possible for $T \neq 0$ only if the argument of the logarithm becomes zero. Since both the summands in the logarithm are non-negative, both must vanish. That is, a phase transition is to be expected only if

$$q_1 = q_2 = 2\pi \tag{6.128}$$

and

$$\sinh(2\beta J) = 1 \tag{6.129}$$

simultaneously. Equation (6.129) determines the critical point T_c. This gives the already cited result (6.82):

$$\frac{J}{k_B T_c} = \frac{1}{2} \ln(1 + \sqrt{2}) = 0.44069 \tag{6.130}$$

Equation (6.130) is only an indication that a phase transition is possible. What type of phase transition this is becomes clear only by investigating the specific heat $C_{B_0=0}$ and the spontaneous magnetization M_s.

6.4.8 Specific Heat

The specific heat is mainly given by the second derivative of the free energy with respect to temperature.

$$C_{B_0=0}(T) = -T \frac{d^2 f(T)}{dT^2} \tag{6.131}$$

We want to investigate the behaviour of the specific heat around the critical point T_c. By a Taylor expansion around T_c, we get

$$\begin{aligned} \sinh(2\beta J) &\approx \sinh(2\beta_c J) + (T - T_c)\cosh(2\beta_c J)\left(-\frac{2J}{k_B T_c^2}\right) + \cdots \\ &= 1 - \frac{T - T_c}{T_c}\frac{2J}{k_B T_c}\cosh(2\beta_c J) + \cdots \\ &= 1 - c\,\varepsilon \end{aligned}$$

Here c is a number of order of magnitude 1 and

$$\varepsilon = \frac{T - T_c}{T_c} \tag{6.132}$$

Free energy can become critical if at all, only, when $q_1 = q_2 = 2\pi$. Therefore, in the neighbourhood of the critical point, we can write

$$2 - \cos q_1 - \cos q_2 \approx +\frac{1}{2}(\bar{q}_1^2 + \bar{q}_2^2) \quad (\bar{q}_i = q_i - 2\pi)$$

From (6.127), we see that the critical behaviour of the free energy is determined by the following integral:

$$I = \int_0^{\cdots} \int_0^{\cdots} d\bar{q}_1 d\bar{q}_2 \ln\left[c^2\varepsilon^2 + (1 - c\varepsilon)\frac{1}{2}(\bar{q}_1^2 + \bar{q}_2^2)\right] \tag{6.133}$$

Introducing polar coordinates

$$\bar{q}_1 = q\cos\phi \quad ; \quad \bar{q}_2 = q\sin\phi \quad ; \quad d\bar{q}_1\, d\bar{q}_2 = q\, dq\, d\phi \tag{6.134}$$

the integral I becomes

$$I = \int_0^{\pi/2} d\phi \int_0^{a(\phi)} dq\, q\, \ln\left[c^2\varepsilon^2 + \frac{1}{2}(c\varepsilon q^2)\right]$$

It is further simplified by the following substitutions

$$x = \frac{1}{2}q^2 \quad ; \quad \eta = c\,\varepsilon$$

$$I = \int_0^{\pi/2} d\phi \int_0^{b(\phi)} dx\, \ln[\eta^2 + (1-\eta)x] \tag{6.135}$$

I does not show any special behaviour as $\eta \to 0$. The free energy is therefore continuous as $T \to T_c$. This is also valid for the first derivative of f (entropy!):

$$\frac{dI}{d\eta} = \int_0^{\pi/2} d\phi \int_0^{b(\phi)} dx \frac{2\eta - x}{\eta^2 + (1-\eta)x} \overset{\eta\to 0}{\to} -\int_0^{\pi/2} d\phi\, b(\phi)$$

One does not see any critical behaviour in the first derivative. Therefore, the phase transition cannot be of first order!

For the second derivative, we have

$$\begin{aligned}\frac{d^2I}{d\eta^2} &= \int_0^{\pi/2} d\phi \int_0^{b(\phi)} dx \frac{2(\eta^2 + (1-\eta)x) - (2\eta - x)^2}{(\eta^2 + (1-\eta)x)^2} \\ &\overset{\eta\to 0}{\to} \int_0^{\pi/2} d\phi \int_0^{b(\phi)} dx \left(\frac{2}{x} - 1\right)^{\eta} \\ &= \int_0^{\pi/2} d\phi \{2\ln b(\phi) - 2\text{ “}\ln 0\text{”} - b(\phi)\}\end{aligned} \tag{6.136}$$

The second derivative thus shows a *logarithmic divergence* for $T \to T_c$. That means, it is also true for the specific heat $C_{B_0=0}$. We conclude that the two-dimensional Ising model displays a phase transition of second order at $T = T_c$. The critical exponent for the specific heat is

$$\alpha = 0 \tag{6.137}$$

corresponding to a logarithmic divergence.

6.4.9 Spontaneous Magnetization

We give here only the exact result

$$M_s(T) = \begin{cases} 0 & for\ T \geq T_c \\ (1 - x^{-4})^{1/8} & for\ T < T_c \end{cases} \tag{6.138}$$

Here is

$$x = \sinh\left(\frac{2J}{k_B T}\right) \tag{6.139}$$

Since the free energy with field $B_0 \neq 0$ could not be calculated, the magnetization was determined using (6.21). The critical exponent of the order parameter M_s is therefore given by

$$\beta = 1/8 \tag{6.140}$$

6.5 Problems

Problem 6.1 For the one-dimensional Ising model, determine the eigenvalues of the transfer matrix $\widehat{T}$ (6.31).

Problem 6.2 For the one-dimensional Ising model, calculate the magnetization (magnetic moment per particle (spin)) $m(T, B_0)$ and the isothermal susceptibility. Study the latter in particular for $B_0 \to 0$.

Problem 6.3 For the Ising model, show that the free energy $F(T,\ m)$ is an even function of the magnetic moment $m = \mu \sum_i S_i$.

Problem 6.4 A magnetic system is described by the Ising model (N localized spins).

1. Express the canonical partition function $Z_N(T, B_0)$ in terms of the moments m_l of the Hamilton function H.

$$m_l = \frac{Tr(H^l)}{Tr(\mathbb{1})} \quad ; \quad l = 1, 2, 3, \cdots$$

   What is the meaning of $Tr(\mathbb{1})$ for the Ising system?
2. Show that for the heat capacity c_{B_0} the high-temperature expansion

$$c_{B_0} = \frac{1}{k_B T^2}\left(m_2 - m_1^2\right) + \mathcal{O}(1/T^3)$$

   is valid.

Problem 6.5 Consider a spin system described by the Ising model with the magnetic moment

$$\widehat{m} = g\mu_B \sum_i S_i$$

1. Express the isothermal susceptibility χ_T by the spin correlation function $\langle S_i S_j \rangle$. For that use the *fluctuation–dissipation theorem* which was proved in Problem 1.2.
2. Use the fluctuation–dissipation theorem of part 1. to calculate the field-free susceptibility of an "open" chain of N Ising spins. Obtain χ_T as a function of $v = \tanh \beta J$.
3. Discuss the result in the thermodynamic limit $N \to \infty$ and compare it with the results of Sect. 6.2.

Problem 6.6 1. For the one-dimensional Ising model (linear open chain) without the external magnetic field, calculate the four-spin correlation function.

$$\langle S_i S_{i+1} S_j S_{j+1} \rangle$$

2. Using the result of 1. calculate the heat capacity $c_{B_0=0}$.

References

1. R. Peierls, Proc. Camb. Phil. Soc. **32**, 477 (1936)
2. L. Onsager, Phys. Rev. **65**, 117 (1944)
3. R. B. Griffiths, Phys. Rev. **136A**, 437 (1964)
4. M. L. Glasser, Am. J. Phys. **38**, 1033 (1970)

Chapter 7
Heisenberg Model

This model, which was justified in great detail in Sects. 5.2 and 5.3 can be considered to be, till today, the most intensely worked and also the best understood model of magnetism. Its region of applicability is fixed by the ideas used in the derivation given in Chap. 5. The model presumes the existence of permanent, localized moments, which interact with each other via either a direct or an indirect exchange mechanism. Therefore this model is best realized in *magnetic insulators*

$$EuO,\ EuS,\ EuTe,\ RbMnF_3,\ MnO,\ CrBr_3, \cdots$$

In addition, the *magnetic metals* in Chap. 5, where the magnetism is also due to localized moments, are successfully described by this model, so long as one is interested only in their magnetic properties.

In this chapter, we want to investigate the statements one can make, using the Heisenberg model, on the magnetic phenomena. The amount of exact solutions available till the present time is, however, much smaller than in the case of the Ising model. When comparing experiment and theory, one has to always take into account two sources of errors. The model itself is not appropriate for describing the given situation or the model is appropriate but has been solved too approximately.

7.1 Model Hamiltonian

7.1.1 Spin Operators

Many problems in handling the Heisenberg model

$$H = -\sum_{i,j} J_{ij}\, \mathbf{S}_i \cdot \mathbf{S}_j \tag{7.1}$$

W. Nolting, A. Ramakanth, *Quantum Theory of Magnetism*,
DOI 10.1007/978-3-540-85416-6_7, © Springer-Verlag Berlin Heidelberg 2009

arise due to the "inconvenient" commutation relations of the spin operators:

$$\begin{aligned}\left[S_i^x\,,\,S_j^y\right]_- &= i\,\hbar\,S_i^z\,\delta_{ij}\\ \left[S_i^y\,,\,S_j^z\right]_- &= i\,\hbar\,S_i^x\,\delta_{ij}\\ \left[S_i^z\,,\,S_j^x\right]_- &= i\,\hbar\,S_i^y\,\delta_{ij}\end{aligned} \tag{7.2}$$

The commutation relations can be formally summarized as

$$\mathbf{S}\times\mathbf{S} = i\,\hbar\,\mathbf{S} \tag{7.3}$$

We presume that the angular momentum algebra is known from the basic course on quantum mechanics and simply lists out the relations which are important for the following discussion.

We will be applying the operators

$$S^{\pm} = S^x \pm i\,S^y \tag{7.4}$$

which are known as "step up", "step down" operators and sometimes called "spin-flip" operators. For these operators, one can easily verify the following *operator identities*:

$$S_i^x = \frac{1}{2}(S_i^+ + S_i^-); \qquad S_i^y = \frac{1}{2i}(S_i^+ - S_i^-) \tag{7.5}$$

$$\left[S_i^z\,,\,S_j^{\pm}\right]_- = \pm\hbar\,\delta_{ij}\,S_i^{\pm} \tag{7.6}$$

$$\left[S_i^+\,,\,S_j^-\right]_- = 2\,\hbar\,\delta_{ij}\,S_i^z \tag{7.7}$$

$$S_i^{\pm}\,S_i^{\mp} = \hbar^2\,S(S+1) \pm \hbar\,S_i^z - \left(S_i^z\right)^2 \tag{7.8}$$

$$\prod_{m_S=-S}^{+S}(S_i^z - \hbar\,m_S) \equiv 0 \tag{7.9}$$

$$\left(S_i^+\right)^{2S+1} = \left(S_i^-\right)^{2S+1} \equiv 0 \tag{7.10}$$

The spin operators for $S = \frac{1}{2}$ form a special case. For these operators, the relations (5.79), (5.81), (5.82), (5.82), (5.83) and (5.84) are valid.

The scalar product $\mathbf{S}_i \cdot \mathbf{S}_j$ of the Heisenberg Hamiltonian (7.1) can be expressed in terms of the components as

$$\mathbf{S}_i \cdot \mathbf{S}_j = \frac{1}{2}\left(S_i^+\,S_j^- + S_i^-\,S_j^+\right) + S_i^z\,S_j^z \tag{7.11}$$

With the usual practice

$$J_{ij} = J_{ji}; \quad J_{ii} = 0 \tag{7.12}$$

the Hamiltonian of the Heisenberg model can be formulated as

$$H = -\sum_{i,j} J_{ij} \left(S_i^+ \, S_j^- + S_i^z \, S_j^z \right) \tag{7.13}$$

This form turns out to be specially useful. Using (7.6) and (7.7), in addition one can easily show that

$$\left[\mathbf{S}_i \cdot \mathbf{S}_j \, , \, \mathbf{S}_i \right]_- = i \, \hbar \, \mathbf{S}_i \times \mathbf{S}_j \quad (i \neq j) \tag{7.14}$$

and by interchanging the indices also show that

$$\left[\mathbf{S}_i \cdot \mathbf{S}_j \, , \, \mathbf{S}_j \right]_- = - \, i \, \hbar \, \mathbf{S}_i \times \mathbf{S}_j \quad (i \neq j) \tag{7.15}$$

From this it follows that the total spin $\sum_i \mathbf{S}_i$ commutes with the Heisenberg Hamiltonian and therefore is a conserved quantity.

Often, it is convenient to transform the rather cumbersome spin operators into the creation and annihilation operators of the second quantization (Appendix A). The most important possibilities of doing this will now be presented.

7.1.1.1 Pauli Operators ($S = 1/2$)

The Pauli operators are defined by the following relations:

$$\frac{1}{\hbar} \, S_i^+ = b_i \; , \quad \frac{1}{\hbar} \, S_i^- = b_i^\dagger \; , \quad \frac{1}{\hbar} \, S_i^z = \frac{1}{2} - n_i \tag{7.16}$$

b_i ($b_i^\dagger$) annihilates (creates) locally, a spin deviation. n_i is the number operator for such spin deviations.

$$n_i = b_i^\dagger \, b_i \tag{7.17}$$

The commutation relations (7.6) and (7.7) for the spin operators imply the corresponding commutation relations for the Pauli operators:

$$\left[b_i \, , \, b_j^\dagger \right]_- = (1 - 2n_i) \, \delta_{ij} \tag{7.18}$$

$$\left[b_i \, , \, b_j \right]_- = \left[b_i^\dagger \, , \, b_j^\dagger \right]_- = 0 \tag{7.19}$$

$$b_i^2 = (b_i^\dagger)^2 = 0 \tag{7.20}$$

Thus they represent neither pure Fermi nor pure Bose operators. They have the properties of Bose operators for different $(i \neq j)$ lattice sites and those of Fermi operators for the same $(i = j)$ lattice site ($[b_i \, , \, b_i^\dagger]_+ = 1$).

In terms of these new operators, the Heisenberg Hamiltonian has the following form:

$$H = -\frac{\hbar^2}{4} J_0 N - \sum_{i,j} \hbar^2 (J_{ij} - J_0 \delta_{ij}) b_j^\dagger b_i - \sum_{i,j} \hbar^2 J_{ij} n_i n_j \tag{7.21}$$

We have introduced a new quantity J_0 which is defined by

$$J_0 = \sum_i J_{ij} = \sum_j J_{ij} \tag{7.22}$$

The last term in (7.21) represents an interaction in a quasiparticle system which is defined by the other terms. When transformed into wavenumber representation, the first three terms of the Hamiltonian describe harmonic oscillators which we will interpret in Sect. 7.2.2 as *free spin waves*.

The Pauli operators are rarely used because of the restriction to $S = 1/2$ and also because, even after this restriction, the commutation relations are not very much simplified.

In Eqs. (5.99), (5.100) and (5.101) the spin operators for $S = \frac{1}{2}$ were expressed in terms of the Fermi operators $c_{i\sigma}$ and $c_{i\sigma}^\dagger$. This representation is in general much more convenient than the representation by Pauli operators.

7.1.1.2 Holstein–Primakoff Transformation (S ≥ 1/2)

Holstein and Primakoff [1] have introduced the following representation for the spin operators:

$$\frac{1}{\hbar} S_i^z = S - \widehat{n}_i \tag{7.23}$$

$$\frac{1}{\hbar} S_i^+ = \sqrt{2S}\, \phi(\widehat{n}_i)\, a_i \tag{7.24}$$

$$\frac{1}{\hbar} S_i^- = \sqrt{2S}\, a_i^\dagger\, \phi(\widehat{n}_i) \tag{7.25}$$

Here as in (7.17) n_i is defined by

$$\widehat{n}_i = a_i^\dagger \, a_i \tag{7.26}$$

and

$$\phi(\widehat{n}_i) = \sqrt{1 - \frac{\widehat{n}_i}{2S}} \tag{7.27}$$

When the eigenvalue n_i of $\widehat{n}_i$, the number of spin deviations, exceeds $2S$, the formalism becomes obviously unphysical. In actual calculations, therefore, only the physical states ($n_i = 0, 1, 2, \ldots, 2S$) have to be taken into account. This is of course a condition which cannot always be easily enforced. The definitions (7.23), (7.24) and (7.25) satisfy the commutation relations (7.6) and (7.7) if the operators $a_i^\dagger$ and a_i are treated as *Bose operators* which obey the fundamental commutation relations (Problem 7.8):

$$\left[a_i \, , \, a_j^\dagger\right]_- = \delta_{ij} \tag{7.28}$$

$$\left[a_i \, , \, a_j\right]_- = \left[a_i^\dagger \, , \, a_j^\dagger\right]_- = 0 \tag{7.29}$$

The corresponding Fourier-transformed operators

$$a_{\mathbf{q}} = \frac{1}{\sqrt{N}} \sum_i e^{-i\mathbf{q}\cdot\mathbf{R}_i} \, a_i \tag{7.30}$$

$$a_{\mathbf{q}}^\dagger = \frac{1}{\sqrt{N}} \sum_i e^{i\mathbf{q}\cdot\mathbf{R}_i} \, a_i^\dagger \tag{7.31}$$

will be interpreted later (Sect. 7.2.2) as the annihilation and creation operators for quasiparticles, the so-called *magnons*, which are the characteristic elementary excitations in magnetism. In terms of these new operators, the Heisenberg Hamiltonian has the following form:

$$\begin{aligned} H = &-N \, \hbar^2 \, S^2 \, J_0 + 2S \, \hbar^2 \, J_0 \sum_i \widehat{n}_i \\ &- 2S \, \hbar^2 \sum_{i,j} J_{ij} \, \phi(\widehat{n}_i) \, a_i \, a_j^\dagger \, \phi(\widehat{n}_j) \\ &- \hbar^2 \sum_{i,j} J_{ij} \, \widehat{n}_i \, \widehat{n}_j \end{aligned} \tag{7.32}$$

Compared with (7.21), through $\phi(n_i)$, a second interaction term comes into play. Explicit working with this Hamiltonian requires an expansion of the square root in $\phi(\widehat{n}_i)$:

$$\phi(\widehat{n}_i) = 1 - \frac{\widehat{n}_i}{4S} - \frac{\widehat{n}_i^2}{32S^2} - \frac{\widehat{n}_i^3}{128S^3} - \cdots \tag{7.33}$$

Substituting (7.33) in (7.32) leads to a series expansion of the Hamiltonian:

$$H = -N\,\hbar^2\,S^2\,J_0 + \sum_{n=1}^{\infty} H_{2n} \tag{7.34}$$

H_{2n} contains terms which are composed of $2n$ construction operators, i.e. n creation and n annihilation operators. We demand that all terms H_{2n} have the operators in the normal ordering, i.e. all the creation operators are to the left and all the annihilation operators are to the right. The possibilities for approximate treatment open up by terminating the infinite series (7.34) after a finite number of terms.

Normally, one is forced to restrict to the first two terms. Using

$$\widehat{n}_i^2 = \widehat{n}_i + (a_i^\dagger)^2\,a_i^2$$

and (7.33), we can write

$$\phi(\widehat{n}_i) = 1 - (1 - \sqrt{1 - 1/2S})\,\widehat{n}_i + \text{“}higher\ terms\text{”} \tag{7.35}$$

Then, the bilinear part is given by

$$H_2 = 2S\,\hbar^2\,J_0 \sum_i \widehat{n}_i - 2S\,\hbar^2 \sum_{i,j} J_{ij}\,a_i^\dagger\,a_j \tag{7.36}$$

and the biquadratic part by

$$\begin{aligned} H_4 = &- \hbar^2 \sum_{i,j} J_{ij}\,\widehat{n}_i\,\widehat{n}_j + 2S\,\hbar^2\,(1 - \sqrt{1 - 1/2S}) * \\ &* \sum_{i,j} J_{ij}(a_j^\dagger\,\widehat{n}_i\,a_i + a_j^\dagger\,\widehat{n}_j\,a_i) \end{aligned} \tag{7.37}$$

The disadvantage of the Holstein–Primakoff transformation is obvious. One needs a physical justification for terminating the infinite series. The simplest approximation in this sense is the so-called *spin wave approximation*

$$H_{SW} = -N\,\hbar^2\,S^2\,J_0 + H_2 \tag{7.38}$$

which will be discussed and justified later. After transformation to wavevector representation, it becomes diagonal

$$H_{SW} = -N\,\hbar^2\,S^2\,J_0 + \sum_{\mathbf{q}} \hbar\,\omega_{\mathbf{q}}\,a_{\mathbf{q}}^\dagger\,a_{\mathbf{q}} \tag{7.39}$$

and describes a system of uncoupled harmonic oscillators ("magnons") where

$$\hbar\,\omega_{\mathbf{q}} = 2S\,\hbar^2(J_0 - J(\mathbf{q})) \tag{7.40}$$

$$J(\mathbf{q}) = \frac{1}{N}\sum_{i,j} J_{ij}\, e^{i\mathbf{q}\cdot(\mathbf{R}_i - \mathbf{R}_j)} \tag{7.41}$$

The difficulties mentioned about the Holstein–Primakoff transformation do not appear in the Dyson–Maleév transformation.

7.1.1.3 Dyson–Maleév Transformation (S ≥ 1/2)

$$\frac{1}{\hbar}\,S_i^+ = \sqrt{2S}\alpha_i \tag{7.42}$$

$$\frac{1}{\hbar}\,S_i^- = \sqrt{2S}\,\alpha_i^\dagger\left(1 - \frac{\widehat{n}_i}{2S}\right) \tag{7.43}$$

$$\frac{1}{\hbar}\,S_i^z = S - \widehat{n}_i \quad (\widehat{n}_i = \alpha_i^\dagger\,\alpha_i) \tag{7.44}$$

$\alpha_i^\dagger$ and α_i are Bose operators and therefore obey (7.28) and (7.29). This transformation has the advantage that the Heisenberg Hamiltonian now has only a finite number of terms (Problem 7.11):

$$H = -N\,\hbar^2\,S^2\,J_0 + H_2 + H_4 \tag{7.45}$$

$$H_2 = 2S\,\hbar^2\,J_0\sum_i \widehat{n}_i - 2S\,\hbar^2\sum_{i,j} J_{ij}\,\alpha_i\,\alpha_j^\dagger \tag{7.46}$$

$$H_4 = -\hbar^2\sum_{i,j} J_{ij}\,\widehat{n}_i\,\widehat{n}_j + \hbar^2\sum_{i,j} J_{ij}\,\alpha_i\,\alpha_j^\dagger\,\widehat{n}_j \tag{7.47}$$

A decisive disadvantage, however, is that S_i^+ and S_i^- are no more adjoint to each other. As a result, H is not Hermitean any more!

7.1.2 Model Extensions

We want to discuss a few simple generalizations of the Heisenberg model which become necessary when, in addition to the isotropic exchange interaction, other effects play a role.

7.1.2.1 Zeeman Term H_z

When an external homogeneous magnetic field along z-direction is applied, the rotational symmetry is destroyed. The Hamiltonian has an extra term, namely, the Zeeman term:

$$H_z = -g_J \frac{\mu_B}{\hbar} B_0 \sum_{i=1}^{N} S_i^z \tag{7.48}$$

If the localized moments are pure spin moments, then g_J is naturally equal to 2. Otherwise, in place of S_i^z one has to take J_i^z and g_J is given by the Lande factor

$$g_J = 1 + \frac{J(J+1) - L(L+1) + S(S+1)}{2J(J+1)} \tag{7.49}$$

The justification for writing (7.48) in this somewhat imprecise form stems from the Wigner–Eckart theorem (2.180). Further, we assume that we are in the region of "anomalous" Zeeman effect, i.e. the field energy $\mu_B B_0$ is small compared to the spin–orbit interaction energy $\hbar^2 \Lambda(L, S)$ (2.304) so that J is still a good quantum number.

7.1.2.2 Dipole Interaction

The interaction of the localized moments with the dipole fields of the other moments, even though is not the origin of the collective magnetism as was estimated in Sect. 5.1.1, still is an important correction which leads to anisotropy effects. According to (5.4), the dipole interaction is taken into account by introducing an extra term in the Hamiltonian which is given by

$$H_D = \sum_{i,j}^{i \neq j} D_{ij} \{\mathbf{S}_i \cdot \mathbf{S}_j - 3(\mathbf{S}_i \cdot \mathbf{e}_{ij})(\mathbf{S}_j \cdot \mathbf{e}_{ij})\} \tag{7.50}$$

The coupling constants

$$D_{ij} = \frac{\mu_0}{8\pi} \frac{g_J^2 \mu_B^2}{\hbar^2 |\mathbf{R}_i - \mathbf{R}_j|^3} \tag{7.51}$$

are, as was estimated in Sect. 5.1.1, very much smaller than the exchange integrals J_{ij}. However, it should be kept in mind that the dipole interaction is of very much longer range than the exchange interaction, so that the inequality $D_{ij} \ll J_{ij}$ is valid only for not too large separations $|\mathbf{R}_i - \mathbf{R}_j|$. The vectors $\mathbf{e}_{ij}$ in (6.50) are the unit vectors defined as follows:

$$\mathbf{e}_{ij} = \frac{\mathbf{R}_i - \mathbf{R}_j}{|\mathbf{R}_i - \mathbf{R}_j|} \tag{7.52}$$

We will discuss a few consequences of the dipole interaction in Sect. 7.4.4.

7.1.2.3 Spin–Orbit Interaction

The assumption that was made in 7.1.2.1 that the replacements, respectively, in the *exchange term*

$$\mathbf{S} \rightarrow (g_J - 1)\mathbf{J} \tag{7.53}$$

and in the *Zeeman term*

$$2\, S^z \rightarrow S^z + J^z \rightarrow g_J\, J^z \tag{7.54}$$

should be made is strictly speaking not quite correct. The Hamiltonian must be extended by an explicit spin–orbit coupling term. This term, according to Sect. 2.4 or 2.10, should have the following form:

$$H_{SO} = \Lambda \sum_{i=1}^{N} \mathbf{S}_i \cdot \mathbf{L}_i \tag{7.55}$$

The meaning of Λ is same as in Sect. 2.4 (see 2.286). The assumption that Λ is an intra-atomic parameter that remains constant throughout the Wigner–Seitz cell is to be considered as a serious simplification.

In the following sections, we will not consider H_{SO} any more. We, as was done in the case of Langevin paramagnets (Sect. 4.2.2), will assume that the fine structure splitting is so strong that we are allowed to restrict our treatment only to the lowest level. In that case (7.55) becomes irrelevant.

7.1.2.4 Magnetic Anisotropy

Collective magnetism is a characteristic property of crystalline solids. It is known that in crystals, physical properties are in general anisotropic, i.e. they are different in different crystallographic directions. This is also true about a part of the magnetic properties. The important quantities such as the Curie temperature T_C and the saturation magnetization are, however, always isotropic. In contrast, the $M - B_0$ isotherms are most of the time anisotropic. For example, one speaks of *easy* and *hard* directions of magnetization.

Possible reasons for the magnetic anisotropy are

1. *the dipole interaction (7.1.2.2)* and
2. *the coupling of the electron orbits to the crystalline electric field.*

The concrete operator form of the magnetic crystal anisotropy is determined by the symmetry group of the magnetic lattice. However, some general statements can be made, because the exchange operator must remain invariant under a canonical transformation of the spin operators of the type

$$S_j^\alpha \to -S_j^\alpha \quad ; \quad i \to -i \quad (\alpha = x,\ y,\ z) \tag{7.56}$$

which retains the commutation relations unchanged. In the lowest order, the anisotropy term in the Hamiltonian should have the following form:

(a) *Uniaxial ferromagnet*

$$H_A = -\sum_{i,j} K_{ij}\ S_i^z\ S_j^z \tag{7.57}$$

(b) *Ferromagnet with cubic symmetry*

$$H_A = -\sum_{i,j,k,l} \sum_{\alpha \neq \beta}^{(x,y,z)} K_{ijkl}\ (S_i^\alpha\ S_j^\alpha)(S_k^\beta\ S_l^\beta) \tag{7.58}$$

The additional condition $\alpha \neq \beta$ is actually responsible for the anisotropy.

The diagonal term ($i = j$) in (7.57) and the term $i = j = k = l$ in (7.58) are called *single-ion anisotropy.*

(a) Uniaxial ferromagnet

$$H_A = -\sum_i K_i (S_i^z)^2 \tag{7.59}$$

(b) Ferromagnet with cubic symmetry

$$H_A = -\sum_i K_i \sum_\alpha (S_i^\alpha)^4 \tag{7.60}$$

In the last expression, the unimportant constants have been left out. From (7.58), for $i = j = k = l$ we actually have

$$\begin{aligned}
&\sum_i K_{iiii} \sum_{\alpha \neq \beta} (S_i^\alpha)^2 (S_i^\beta)^2 \\
&= \sum_i K_{iiii} \sum_{\alpha,\beta} (S_i^\alpha)^2 (S_i^\beta)^2 - \sum_i K_{iiii} \sum_\alpha (S_i^\alpha)^4 \\
&= \sum_i K_{iiii}\ \hbar^4\ (S(S+1))^2 - \sum_i K_{iiii} \sum_\alpha (S_i^\alpha)^4
\end{aligned}$$

For a few phenomena (for example, the resonance experiments), taking into account the single-ion anisotropy is already completely sufficient. The anisotropy constants are, then, in general, temperature dependent.

In a further approximation, for uniaxial anisotropy, one may frequently consider only an *effective anisotropy field* which is described in a rough approximation by

$$H_A \approx -g_J \frac{\mu_B}{\hbar} B_A \sum_i S_i^z \tag{7.61}$$

This simple ansatz is applied in particular for uniaxial antiferromagnets.

7.2 Exact Statements

In spite of the simple structure of the Hamiltonian for the Heisenberg model, there are very few exact statements possible. Without attempting to be complete, in this section, we will discuss a few of the most important exact solutions. We start with the following

7.2.1 Mermin–Wagner Theorem

This theorem states [2]

In one and two dimensions, the isotropic Heisenberg model does not have any spontaneous magnetization.

Statements and proofs of this nature have a certain tradition. For example,

Van Hove [3]: "There is no phase transition in a one dimensional gas with *hard-core* interactions of finite range".

Griffiths [4]: "The two- and three-dimensional Ising model shows a phase transition" (see Sect. 6.3).

Hohenberg [5]: "Normal superconductivity and superfluidity are impossible in one- or two-dimensional systems".

For the proof, Hohenberg used an inequality of Bogoliubov [6] which is also the starting point of Mermin and Wagner. Therefore, we want to first prove this inequality. The inequality is

$$\frac{1}{2} \beta \langle [A\,,\, A^\dagger\,]_+ \rangle \langle [[C\,,\, H]_-\,,\, C^\dagger]_- \rangle \geq |\langle [C\,,\, A]_- \rangle|^2 \tag{7.62}$$

H: Hamiltonian, $A,\ C$: arbitrary operators, $\beta = \frac{1}{k_B T}$. One should note that $[\cdots]_+$ is an anticommutator and $[\cdots]_-$ is a commutator.

The *idea* of the proof is to define an appropriate scalar product and then exploit the *Schwarz inequality*.

Let $|n\rangle$ be the eigenstates of the Hamiltonian with the eigenenergies E_n. Then, using

$$E_n = \langle n|H|n\rangle; \quad W_n = \frac{e^{-\beta\, E_n}}{Tr(e^{-\beta\, H})} \tag{7.63}$$

we can show that

$$(A, B) = \sum_{n,m}^{E_n \neq E_m} \langle n|A^\dagger|m\rangle\langle m|B|n\rangle \frac{W_m - W_n}{E_n - E_m} \tag{7.64}$$

is a *positive semidefinite scalar product.* Here A and B can be any operators. We verify the axioms:

(α) (A, B) is a complex number with

$$(A, B) = (B, A)^* \tag{7.65}$$

This is valid since $(W_m - W_n)/(E_n - E_m)$ is a real number and

$$(\langle n|B^\dagger|m\rangle\langle m|A|n\rangle)^* = \langle n|A^\dagger|m\rangle\langle m|B|n\rangle$$

(β) The linearity properties of the scalar product,

$$(A, \alpha_1 B_1 + \alpha_2 B_2) = \alpha_1(A, B_1) + \alpha_2(A, B_2) \tag{7.66}$$

with $\alpha_1,\ \alpha_2\ \in\ \mathbb{C}$ follow directly from the linearity properties of the matrix element $\langle m|B|n\rangle$.
(γ) Since $(W_m - W_n)/(E_n - E_m) \geq 0$, and $\langle n|A^\dagger|m\rangle\langle m|A|n\rangle = |\langle m|A|n\rangle|^2$ it is always true that

$$(A, A) \geq 0 \tag{7.67}$$

(δ) From $A = 0$ it naturally follows that $(A, A) = 0$. The converse is, however, not true. For example, $(H, H) = 0$ even though $H \neq 0$. Therefore, we are dealing with a "semidefinite" scalar product.

We have thus, by verifying the axioms, shown that (7.64) is a scalar product.
We now specially choose

$$B = \left[C^\dagger\,,\ H\right]_- \tag{7.68}$$

and calculate the scalar product (A, B).

$$\begin{aligned}(A, B) &= \sum_{n,m}^{E_n \neq E_m} \langle n|A^\dagger|m\rangle\langle m|[C^\dagger\,,\ H]_-|n\rangle \frac{W_m - W_n}{E_n - E_m} \\ &= \sum_{n,m} \langle n|A^\dagger|m\rangle\langle m|C^\dagger|n\rangle(W_m - W_n)\end{aligned}$$

Because of the bracket on the right, we can now include the diagonal term $E_n = E_m$ in the summation. Using the completeness relation and the definition (7.63) for W_n, we can further write

$$(A, B) = \sum_m W_m \langle m|C^\dagger A^\dagger|m\rangle - \sum_n W_n \langle n|A^\dagger C^\dagger|n\rangle$$
$$= \langle C^\dagger A^\dagger - A^\dagger C^\dagger\rangle$$

With (7.68) for B we then have

$$(A, B) = \langle [C^\dagger , A^\dagger]_-\rangle \tag{7.69}$$

If we substitute $A = B = [C^\dagger , H]_-$ in the above expression, we get

$$(B, B) = \langle [C^\dagger , [H , C]_-]_-\rangle \geq 0 \tag{7.70}$$

The inequality on the right is the general property (7.67) of the scalar product.

For the Schwarz inequality, we still require (A, A). That we get in the following way:

$$0 < \frac{W_m - W_n}{E_n - E_m}$$
$$= \left(Tr\left(e^{-\beta H}\right)\right)^{-1} \frac{e^{-\beta E_m} + e^{-\beta E_n}}{E_n - E_m} \frac{e^{-\beta E_m} - e^{-\beta E_n}}{e^{-\beta E_m} + e^{-\beta E_n}}$$
$$= \frac{W_m + W_n}{E_n - E_m} \tanh\left(\frac{\beta}{2}(E_n - E_m)\right)$$

Since $\tanh x < x$ for $x > 0$ and therefore $(E_n - E_m)^{-1} \tanh(\frac{\beta}{2}(E_n - E_m)) < \frac{\beta}{2}$ for $E_n \neq E_m$, we can see that

$$0 < \frac{W_m - W_n}{E_n - E_m} < \frac{\beta}{2}(W_n + W_m) \tag{7.71}$$

With this we can now estimate the scalar product (A, A):

$$(A, A) < \frac{1}{2}\beta \sum_{n,m}^{E_n \neq E_m} \langle n|A^\dagger|m\rangle\langle m|A|n\rangle(W_n + W_m)$$
$$\leq \frac{1}{2}\beta \sum_{n,m} \langle n|A^\dagger|m\rangle\langle m|A|n\rangle(W_n + W_m)$$
$$= \frac{1}{2}\beta \sum_n W_n(\langle n|A^\dagger A|n\rangle + \langle n|A A^\dagger|n\rangle)$$

This finally leads to the inequality

$$(A, A) \leq \frac{1}{2} \beta \langle [A , A^\dagger]_+ \rangle \tag{7.72}$$

We now substitute the expressions (7.72), (7.70) and (7.69) in the *Schwarz inequality*

$$|(A, B)|^2 \leq (A, A)(B, B) \tag{7.73}$$

and recognize that it is exactly the Bogoliubov inequality (7.62). Thus the Bogoliubov inequality is proved.

We now, using (7.62), want to find out whether the isotropic Heisenberg model gives a *spontaneous* magnetization. For that, we first calculate the magnetization of the spin system in the presence of an external homogeneous magnetic field $\mathbf{B}_0 = B_0\mathbf{e}_z$ and then set $B_0 \to 0$. The starting point for this is, according to (7.1) and (7.48), the following Hamiltonian:

$$H = -\sum_{i,j} J_{ij} \, \mathbf{S}_i \cdot \mathbf{S}_j - b \sum_i S_i^z \, e^{-i\mathbf{K}\cdot\mathbf{R}_i} \tag{7.74}$$

b is given by

$$b = g_J \, \frac{\mu_B}{\hbar} \, B_0 \tag{7.75}$$

With the help of the factor $e^{-i\mathbf{K}\cdot\mathbf{R}_i}$, we can distinguish the different spin structures. If $\mathbf{K} \equiv 0$ then it is the case of a normal ferromagnet. On the other hand, if we want to discuss, for example, the simplest antiferromagnet, namely the AB-type (sometimes called G-type), which can be considered to be made up of two ferromagnetic sub-lattices where each atom of one sub-lattice has as its nearest neighbours only atoms of the other sub-lattice, then we have to choose $\mathbf{K} = \frac{1}{2}\mathbf{Q}$. Here $\mathbf{Q}$ is the smallest reciprocal vector so that $e^{-i\mathbf{k}\cdot\mathbf{R}_i}$ is equal to $+1$ if $\mathbf{R}_i$ refers to one sub-lattice and equal to -1 if it refers to another sub-lattice.

We first calculate the *magnetization*

$$M(T, B_0) = g_J \, \frac{\mu_B}{\hbar} \sum_i e^{-i\mathbf{K}\cdot\mathbf{R}_i} \, \langle S_i^z \rangle_{T, B_0} \tag{7.76}$$

The factor $e^{-i\mathbf{K}\cdot\mathbf{R}_i}$ takes care of the fact that, in the case of an antiferromagnet, $M(T, B_0)$ represents the *sub-lattice* magnetization. The total magnetization of an antiferromagnet is naturally always equal to zero.

The question whether the Heisenberg model shows a phase transition or not is finally answered by the *spontaneous magnetization*

$$M_s(T) = \lim_{B_0 \to 0} M(T, B_0) \tag{7.77}$$

For the following analysis, we assume that the exchange integrals J_{ij} decrease sufficiently fast with the increasing distance $|\mathbf{R}_i - \mathbf{R}_j|$ so that Q defined by

$$Q = \frac{1}{N} \sum_{i,j} |\mathbf{R}_i - \mathbf{R}_j|^2 \, |J_{ij}| \tag{7.78}$$

remains finite. This is actually a weak condition which does not require that one should restrict oneself to the nearest neighbour interactions.

We are now ready to prove the *Mermin–Wagner theorem*:

For the isotropic one- and two-dimensional Heisenberg model holds

$$M_s(T) \equiv 0 \quad if \quad Q < \infty \; and \; T > 0 \tag{7.79}$$

For the proof we make use of the Bogoliubov inequality (7.62) for the following operators:

$$A = S^-(-\mathbf{k} + \mathbf{K}) \;\Leftrightarrow\; A^\dagger = S^+(\mathbf{k} - \mathbf{K}) \tag{7.80}$$

$$C = S^+(\mathbf{k}) \;\Leftrightarrow\; C^\dagger = S^-(-\mathbf{k}) \tag{7.81}$$

The spin operators in **k**-space which were introduced in the above equations are given by

$$S^\alpha(\mathbf{k}) = \sum_i S_i^\alpha e^{-i\mathbf{k}\cdot\mathbf{R}_i} \quad (\alpha = x,\ y,\ z,\ +,\ -) \tag{7.82}$$

Using (7.6) and (7.7) one can easily verify the following commutation relations:

$$\left[S^+(\mathbf{k}_1)\,,\ S^-(\mathbf{k}_2)\right]_- = 2\,\hbar\, S^z(\mathbf{k}_1 + \mathbf{k}_2) \tag{7.83}$$

$$\left[S^z(\mathbf{k}_1)\,,\ S^\pm(\mathbf{k}_2)\right]_- = \pm\,\hbar\, S^\pm(\mathbf{k}_1 + \mathbf{k}_2) \tag{7.84}$$

Using (7.80) and (7.81) we now evaluate the individual terms of the inequality (7.62).

(a) $\langle [\mathbf{C}\,,\ \mathbf{A}]_- \rangle$

$$\begin{aligned}\langle [C\,,\ A]_- \rangle &= \langle \left[S^+(\mathbf{k})\,,\ S^-(-\mathbf{k} + \mathbf{K})\right]_- \rangle \\ &= 2\,\hbar\, \langle S^z(\mathbf{K}) \rangle \\ &= 2\,\hbar \sum_i e^{-i\mathbf{K}\cdot\mathbf{R}_i} \, \langle S_i^z \rangle \end{aligned} \tag{7.85}$$

This means, from (7.76)

$$\langle [C\,,\ A]_- \rangle = \frac{2\,\hbar^2\, N}{g_J\, \mu_B}\, M(T,\, B_0) \tag{7.86}$$

(b) $\langle [\mathbf{A}\ ,\ \mathbf{A}^\dagger]_+ \rangle$

This term is easy to evaluate.

$$\begin{aligned}\sum_{\mathbf{k}} \langle [A\ ,\ A^\dagger]_+ \rangle &= \sum_{\mathbf{k}} \langle [S^-(-\mathbf{k}+\mathbf{K})\ ,\ S^+(\mathbf{k}-\mathbf{K})]_+ \rangle \\ &= \sum_{\mathbf{k}} \sum_{i,j} e^{i(\mathbf{k}-\mathbf{K})\cdot(\mathbf{R}_i-\mathbf{R}_j)} \langle S_i^- S_j^+ + S_j^+ S_i^- \rangle \\ &= 2N \sum_i \langle (S_i^x)^2 + (S_i^y)^2 \rangle \\ &\leq 2N \sum_i \langle \mathbf{S}_i^2 \rangle \end{aligned} \tag{7.87}$$

so that in (7.62) we can substitute

$$\sum_{\mathbf{k}} \langle [A\ ,\ A^\dagger]_+ \rangle \leq 2\,\hbar^2\, N^2\, S(S+1) \tag{7.88}$$

The third term requires a little more effort to evaluate.

(c) $\langle [[\mathbf{C}\ ,\ \mathbf{H}]_-\ ,\ \mathbf{C}^\dagger]_- \rangle$

First we will evaluate the commutator of S_m^+ with H using (7.74).

$$\begin{aligned}\left[S_m^+\ ,\ H\right]_- &= -\hbar \sum_i J_{im} \left\{2S_i^+ S_m^z - S_i^z S_m^+ - S_m^+ S_i^z\right\} \\ &\quad + \hbar\, b\, S_m^+\, e^{-i\mathbf{K}\cdot\mathbf{R}_m} \end{aligned} \tag{7.89}$$

Using this, we evaluate the following double commutator:

$$\begin{aligned}\left[\left[S_m^+\ ,\ H\right]_-\ ,\ S_p^-\right]_- &= 2\hbar^2 \sum_i J_{ip}\, \delta_{mp}\ \left\{S_i^+ S_p^- + 2S_i^z S_p^z\right\} \\ &\quad - 2\hbar^2\, J_{mp}\ \left\{S_m^+ S_p^- + 2S_m^z S_p^z\right\} \\ &\quad + 2\hbar^2\, b\, \delta_{mp}\, S_p^z\, e^{-i\mathbf{K}\cdot\mathbf{R}_p} \end{aligned} \tag{7.90}$$

This leads to the following intermediate result for the expectation value we are looking for:

$$\begin{aligned}&\langle \left[[C\ ,\ H]_-\ ,\ C^\dagger\right]_- \rangle \\ &= \sum_{m,p} e^{-i\mathbf{k}\cdot(\mathbf{R}_m-\mathbf{R}_p)} \langle \left[\left[S_m^+\ ,\ H\right]_-\ ,\ S_p^-\right]_- \rangle \\ &= 2\,\hbar^2\, b \sum_p \langle S_p^z \rangle\, e^{-i\mathbf{K}\cdot\mathbf{R}_p} \\ &+ 2\,\hbar^2 \sum_{m,p} J_{mp} \left(1 - e^{-i\mathbf{k}\cdot(\mathbf{R}_m-\mathbf{R}_p)}\right) \langle S_m^+ S_p^- + 2\, S_m^z\, S_p^z \rangle \end{aligned} \tag{7.91}$$

According to (7.70), this expression cannot be negative. That is also valid for the corresponding expression evaluated with $\bar{C} = S^{+}(-\mathbf{k})$ in place of $C = S^{+}(\mathbf{k})$. In this case, in (7.91) one has to only replace $\mathbf{k}$ by $(-\mathbf{k})$ on the right-hand side. Since this is also non-negative, we obtain an upper bound for the expectation value if we add it to the right-hand side of (7.91):

$$\begin{aligned}&\langle[[C\,,\,H]_-\,,\,C^\dagger]_-\rangle\\&\leq 4\,\hbar^2\,b\sum_p\langle S_p^z\rangle e^{-i\mathbf{K}\cdot\mathbf{R}_p}\\&+4\,\hbar^2\sum_{m,p}J_{mp}\,(1-\cos(\mathbf{k}\cdot(\mathbf{R}_m-\mathbf{R}_p)))\langle\mathbf{S}_m\cdot\mathbf{S}_p+S_m^zS_p^Z\rangle\end{aligned}\tag{7.92}$$

The right-hand side is positive, so that we get by use of the triangle inequality

$$\begin{aligned}&\langle[[C\,,\,H]_-\,,\,C^\dagger]_-\rangle\\&\leq 4\,\hbar^2\,b\,N\,\left|\langle S_p^z\rangle\right|+4\,\hbar^2\sum_{m,p}|J_{mp}|\,|1-\cos(\mathbf{k}\cdot(\mathbf{R}_m-\mathbf{R}_p))|\,*\\&\qquad\qquad *\left(\left|\langle\mathbf{S}_m\cdot\mathbf{S}_p\rangle\right|+\left|\langle S_m^zS_p^Z\rangle\right|\right)\\&\leq 4\,\hbar^2\,b\,N\,\left|\langle S_p^z\rangle\right|\\&+4\,\hbar^2\sum_{m,p}|J_{mp}|\,|1-\cos(\mathbf{k}\cdot(\mathbf{R}_m-\mathbf{R}_p))|\left(\hbar^2S(S+1)+\hbar^2S^2\right)\\&\leq 4\,\hbar^2\,b\,N\,\left|\langle S_p^z\rangle\right|+8\hbar^2S(S+1)\,*\\&\qquad *\sum_{m,p}|J_{mp}|\,|1-\cos(\mathbf{k}\cdot(\mathbf{R}_m-\mathbf{R}_p))|\end{aligned}\tag{7.93}$$

Therewith we have found

$$\begin{aligned}&\langle[[C\,,\,H]_-\,,\,C^\dagger]_-\rangle\\&\leq 4\,\hbar^2\,|B_0\,M(T,B_0)|\\&+8\,\hbar^2\,S(S+1)\sum_{m,p}|J_{mp}|\frac{1}{2}\,k^2|\mathbf{R}_m-\mathbf{R}_p|^2\end{aligned}\tag{7.94}$$

Using (7.78) we finally get the following inequality:

$$\langle[[C\,,\,H]_-\,,\,C^\dagger]_-\rangle\leq 4\,\hbar^2\,|B_0\,M|+4\,N\,k^2\,\hbar^4\,Q\,S(S+1)\tag{7.95}$$

We now substitute the expressions (7.86), (7.88) and (7.95) in the Bogoliubov inequality (7.62) and sum over all the wavevectors $\mathbf{k}$ of the first Brillouin zone:

$$\beta\, S(S+1) \geq \frac{M^2}{(g_J\, \mu_B)^2}\, \frac{1}{N^2} \sum_{\mathbf{k}} \frac{1}{|B_0\, M| + k^2\, \hbar^2\, N\, Q\, S(S+1)} \tag{7.96}$$

Using this inequality, we will now prove the Mermin–Wagner theorem (7.79). For that we will again convert the sum into an integral. Let the system be in a d-dimensional "volume" $V_d = L^d$ and contain $N_d = n^d$ spins. In the thermodynamic limit, let the specific volume V_d/N_d approach the constant value v_d. The integrand on the right-hand side of the inequality is positive. Therefore, we can replace the integration over the first Brillouin zone by an integration over a sphere of radius k_0 which lies completely inside the Brillouin zone:

$$S(S+1) \geq \frac{m^2\, v_d\, \Omega_d}{\beta\, (g_J\, \mu_B)^2\, (2\pi)^d} \int_0^{k_0} \frac{k^{d-1}\, dk}{|B_0\, m| + k^2\, \hbar^2\, Q\, S(S+1)} \tag{7.97}$$

Here d is the dimension of the lattice and Ω_d is the surface area of the d-dimensional "unit sphere" (see Problem 3.4)

$$\Omega_1 = 2 \quad \Omega_2 = 2\pi \quad \Omega_3 = 4\pi \tag{7.98}$$

Furthermore, we have introduced the average magnetic moment per site $m = M/N$. The integral on the right-hand side of (7.97) can be easily evaluated:

$d = 1$

$$S(S+1) \geq \frac{m^2\, v_1}{2\pi\, \beta\, (g_J\, \mu_B)^2}\, \frac{\arctan\left(k_0 \sqrt{\frac{Q\, \hbar^2\, S(S+1)}{|B_0\, m|}}\right)}{\sqrt{Q\, \hbar^2\, S(S+1)\, |B_0\, m|}} \tag{7.99}$$

We are specially interested in the behaviour of the magnetization for small fields B_0:

$$|m(T, B_0)| \leq const.\, \frac{B_0^{1/3}}{T^{2/3}} \quad \text{as } B_0 \to 0 \tag{7.100}$$

$d = 2$

For the two-dimensional lattice from (7.97) we get

$$S(S+1) \geq \frac{m^2\, v_2}{2\pi\, \beta\, (g_J\, \mu_B)^2}\, \frac{\ln\left(\frac{Q\, \hbar^2\, S(S+1) k_0^2 + |B_0\, m|}{|B_0\, m|}\right)}{2\, Q\, \hbar^2\, S(S+1)} \tag{7.101}$$

from which for small fields B_0 we get as $B_0 \to 0$

$$|m(T, B_0)| \leq const_1 \left(T\, \ln\left(\frac{const_2 + |B_0\, m|}{|B_0\, m|}\right)\right)^{-1/2} \tag{7.102}$$

From the two inequalities (7.100) and (7.102) it is clear that the spontaneous magnetization

$$m_s(T) = \lim_{B_0 \to 0} m(T, B_0) = 0 \quad \text{for } T \neq 0 \text{ and } d = 1,\ 2 \tag{7.103}$$

Thus, the *Mermin–Wagner theorem* is proved.

In the following, we add a short interpretation of the theorem:

1. The proof is valid only for $T > 0$. For $T = 0$ the inequalities (7.100) and (7.102) do not make any predictions. Just as in the case of one-dimensional Ising model (Sect. 6.2.1), for $T = 0$ spontaneous magnetization is not excluded.
2. The factor $exp(-i\mathbf{K} \cdot \mathbf{R}_i)$ which was introduced in the definition of the magnetization does not appear in the inequalities (7.100) and (7.102). Therefore, the theorem forbids even the long-range antiferromagnetic order in one- or two-dimensional spin lattices.
3. No prediction is possible for $d = 3$.
4. The theorem is valid for arbitrary spin S.
5. The theorem is valid only for the *isotropic* Heisenberg model. The proof is not valid even for a weak anisotropy. This explains the existence of a number of two-dimensional Heisenberg ferromagnets and antiferromagnets like

   $$K_2CuF_4,\ CrBr_3,\ Rb_2MnF_4, \ldots$$

   as well as high-T_c superconductors (cuprates), which show a phase transition. Distinct anisotropy in these materials makes the Mermin–Wagner theorem irrelevant to them.
6. The theorem is restricted only to the non-existence of spontaneous magnetization. It does not necessarily exclude other types of phase transitions. High-temperature expansions occasionally indicate a divergence in the isothermal magnetic susceptibility χ_T (Fig. 7.1). A divergence in χ_T at a critical temperature T^* is normally an indicator for a transition into a collectively ordered phase. In the case of two-dimensional-Heisenberg model, however, it appears that the following conditions may be simultaneously valid [7]:

   $$\begin{aligned} &M_s(T) \equiv 0 \\ &\chi_T = \lim_{B_0 \to 0} \left(\mu_0 \left(\frac{\partial M(T,B_0)}{\partial B_0} \right)_T \right) \overset{T \to T^*}{\to} \infty \end{aligned} \tag{7.104}$$

7. The Mermin–Wagner theorem can be proved also for other models like the Hubbard model, s-f (Kondo-lattice) model, X–Y model.

7.2.2 One-Magnon States of a Ferromagnet

There exist interesting analogies between the elementary excitations in a ferromagnet and an elastic solid. The oscillations of the ions in the lattice about their mean positions can be analysed in terms of the normal modes. Their amplitudes are quan-

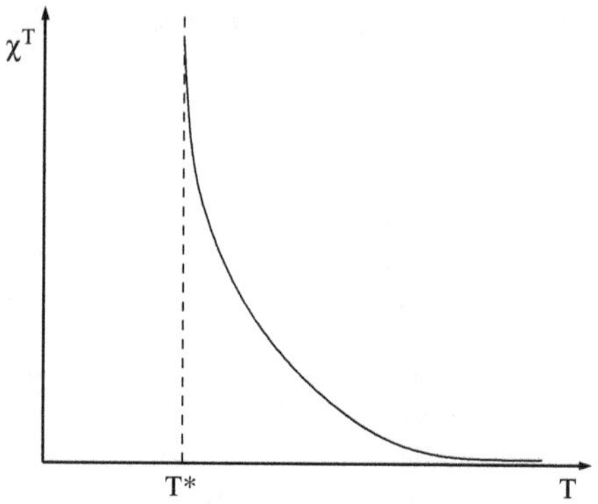

Fig. 7.1 Critical temperature behaviour of the isothermal magnetic susceptibility χ_T

tized. The quantum unit in this case is called the *phonon*. In a complete analogy, the oscillations of the ferromagnet are, following Bloch, called *spin waves* and their unit of quantization is the *magnon*. The ground state of a ferromagnet corresponds to the total alignment of all the spins. We shall see that magnons represent the spin deviations, i.e. they cause a destruction of the collective spin ordering. One-magnon states are exact eigenstates of the Heisenberg Hamiltonian. We will prove and discuss this in the following.

We use the model Hamiltonian in the form (7.13) including the Zeeman term (7.48), but we will write it in terms of wave vectors using (7.82) and (7.41):

$$\begin{aligned} H = &-\frac{1}{N}\sum_{\mathbf{k}} J(\mathbf{k})\{S^+(\mathbf{k})\, S^-(-\mathbf{k}) + S^z(\mathbf{k})\, S^z(-\mathbf{k})\} \\ &- g_J\,\frac{\mu_B}{\hbar}\, B_0\, S^z(0) \end{aligned} \tag{7.105}$$

The sum runs over all the wavevectors in the first Brillouin zone. Note that $S^+(\mathbf{k})$ and $S^-(\mathbf{k})$ are no longer adjoints and $S^z(\mathbf{k})$ is not Hermitian!

$$\left(S^+(\mathbf{k})\right)^\dagger = S^-(-\mathbf{k}); \quad \left(S^z(\mathbf{k})\right)^\dagger = S^z(-\mathbf{k})$$

Let the state $|0\rangle$ be the state where all the spins are oriented parallel to the external field. This is the ground state of the operator H (*ferromagnetic saturation*). We will first calculate its eigenvalue. The working of the spin operators on the state $|0\rangle$ is of course obvious:

$$\begin{aligned} S_i^z|0\rangle &= \hbar\, S|0\rangle \\ S^z(\mathbf{k})|0\rangle &= \hbar\, S|0\rangle \sum_i e^{-i\mathbf{k}\cdot\mathbf{R}_i} = \hbar\, N\, S|0\rangle\, \delta_{\mathbf{k},0} \end{aligned} \tag{7.106}$$

$$S_i^+|0\rangle = 0 \Rightarrow S^+(\mathbf{k})|0\rangle = 0 \tag{7.107}$$

With these relations, the working of H on $|0\rangle$ is also clear (Problem 7.6):

$$H|0\rangle = E_0|0\rangle; \quad E_0 = -N\,\hbar^2\,J_0\,S^2 - N\,g_J\,\mu_B\,B_0\,S \tag{7.108}$$

In E_0, the first term stems from the exchange interaction and the second is due to the external field. We will show that the state

$$S^-(\mathbf{k})|0\rangle$$

is also an eigenstate of H. For that, using the commutation relations (7.83) and (7.84), we will first calculate

$$\begin{aligned}
&\left[H\,,\,S^-(\mathbf{k})\right]_- \\
&= -\frac{1}{N}\sum_{\mathbf{q}} J(\mathbf{q})\left\{\left[S^+(\mathbf{q})\,,\,S^-(\mathbf{k})\right]_- S^-(-\mathbf{q})\right. \\
&+ \left.\left[S^z(\mathbf{q})\,,\,S^-(\mathbf{k})\right]_- S^z(-\mathbf{q}) + S^z(\mathbf{q})\left[S^z(-\mathbf{q})\,,\,S^-(\mathbf{k})\right]_-\right\} \\
&- \frac{1}{\hbar} g_J \mu_B B_0 \left[S^z(0)\,,\,S^-(\mathbf{k})\right]_- \\
&= -\frac{1}{N}\sum_{\mathbf{q}} J(\mathbf{q})\left\{2\hbar S^z(\mathbf{k}+\mathbf{q})S^-(-\mathbf{q}) - \hbar S^-(\mathbf{k}+\mathbf{q})S^z(-\mathbf{q})\right. \\
&- \left.\hbar S^z(\mathbf{q})S^-(\mathbf{k}-\mathbf{q})\right\} + g_J \mu_B B_0 S^-(\mathbf{k}) \\
&= -\frac{1}{N}\sum_{\mathbf{q}} J(\mathbf{q})\left\{-2\hbar^2 S^-(\mathbf{k}) + 2\hbar S^-(-\mathbf{q})S^z(\mathbf{k}+\mathbf{q})\right. \\
&- \left.\hbar S^-(\mathbf{k}+\mathbf{q})S^z(-\mathbf{q}) + \hbar^2 S^-(\mathbf{k}) - \hbar S^-(\mathbf{k}-\mathbf{q})S^z(\mathbf{q})\right\} \\
&+ g_J \mu_B B_0 S^-(\mathbf{k})
\end{aligned}$$

Therewith we can write

$$\begin{aligned}
\left[H\,,\,S^-(\mathbf{k})\right]_- &= g_J\,\mu_B\,B_0\,S^-(\mathbf{k}) \\
&- \frac{2\hbar}{N}\sum_{\mathbf{q}} J(\mathbf{q})\{S^-(-\mathbf{q})\,S^z(\mathbf{k}+\mathbf{q}) \\
&- S^-(\mathbf{k}+\mathbf{q})\,S^z(-\mathbf{q})\}
\end{aligned} \tag{7.109}$$

In view of (7.12), we have used several times

$$\frac{1}{N}\sum_{\mathbf{q}} J(\mathbf{q}) = J_{ii} = 0; \qquad J(\mathbf{q}) = J(-\mathbf{q}) \tag{7.110}$$

The commutator (7.109) is itself an operator. Using (7.106), the application of this operator on the ground state $|0\rangle$ gives

$$\left[H\,,\,S^-(\mathbf{k})\right]_-|0\rangle = \{g_J\,\mu_B\,B_0 - 2\hbar^2\,S\,(J(\mathbf{k}) - J_0)\}S^-(\mathbf{k})|0\rangle \tag{7.111}$$

Using (7.108) and (7.111) we can now write

$$\begin{aligned} & H(S^-(\mathbf{k})|0\rangle) \\ &= S^-(\mathbf{k})H|0\rangle + \left[H\,,\ S^-(\mathbf{k})\right]_-|0\rangle \\ &= \{E_0(B_0) + g_J\mu_B B_0 + 2S\hbar^2(J_0 - J(\mathbf{k}))\}S^-(\mathbf{k})|0\rangle \end{aligned} \tag{7.112}$$

Thus we see that $S^-(\mathbf{k})|0\rangle$ is indeed an eigenstate of H. This state can be normalized by

$$\begin{aligned} \langle 0|S^+(-\mathbf{k})\,S^-(\mathbf{k})|0\rangle &= \langle 0|(S^-(\mathbf{k})\,S^+(-\mathbf{k}) + 2\hbar S^z(0))|0\rangle \\ &= 2\hbar^2\,N\,S \end{aligned} \tag{7.113}$$

provided the ground state $|0\rangle$ is itself normalized. With this, we have found the following important result. The *normalized one-magnon* state

$$|\mathbf{k}\rangle = \frac{1}{\hbar\sqrt{2SN}}\,S^-(\mathbf{k})|0\rangle \tag{7.114}$$

is an eigenstate with the energy

$$E(\mathbf{k}) = E_0(B_0) + \hbar\omega(\mathbf{k}) \tag{7.115}$$

That corresponds to an *excitation energy* of

$$\hbar\omega(\mathbf{k}) = g_J\,\mu_B\,B_0 + 2S\hbar^2(J_0 - J(\mathbf{k})) \tag{7.116}$$

The energy quantum $\hbar\omega(\mathbf{k})$ is ascribed to the quasiparticle *magnon*. From the field term $g_J\mu_B B_0$, one recognizes that the total magnetic moment of the system in the state $|\mathbf{k}\rangle$ is, when compared to that in the ground state $|0\rangle$, reduced by $g_J\mu_B$. Therefore, one can ascribe spin $1\hbar$ to the quasiparticle magnon. Thus *magnons are bosons!*

One can obtain further information about this new quasiparticle by looking at the expectation value of the local operator S_i^z in the one-magnon state $|\mathbf{k}\rangle$. Using (7.114) and (7.107) one has (Problem 7.7)

$$\langle\mathbf{k}|S_i^z|\mathbf{k}\rangle = \hbar(S - \frac{1}{N}) \tag{7.117}$$

This is a remarkable result. The right-hand side is independent of i and also of $\mathbf{k}$, i.e. the spin deviation of $1\hbar$ in the state $|\mathbf{k}\rangle$ is distributed uniformly on all the lattice sites. At each lattice site, the deviation of the local spin from the totally ordered state ($\langle 0|S_i^z|0\rangle = \hbar S$ for all i) is $\hbar/N$. This fact leads inescapably to the concept of *spin waves*. This is a *collective* excitation $|\mathbf{k}\rangle$, in which all the localized spins

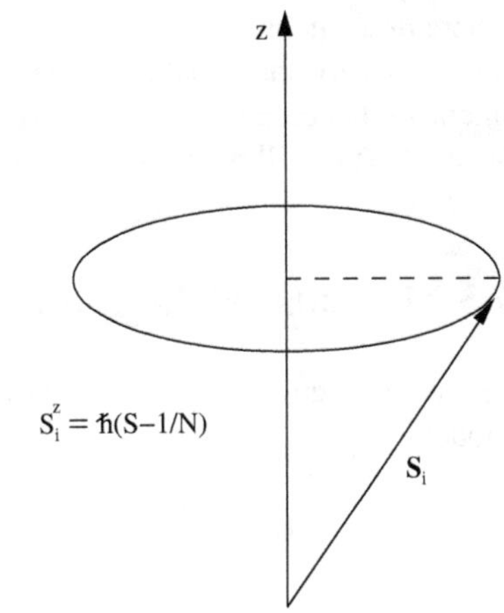

Fig. 7.2 Semiclassical picture of a local spin after one magnon has been excited

participate equally. The excitation of a spin wave means a spin deviation of one unit of angular momentum $1\hbar$ for the whole lattice. It is characterized by a propagation vector **k**. Its quantization unit is the magnon. In a picturized, quasiclassical vector model, a spin wave is represented by the precession of every localized spin about the z-axis. The angle the spin makes with the z-axis is such that the projection of the spin vector onto the z-axis is $\hbar(S - \frac{1}{N})$ (Fig. 7.2). The adjacent spin vectors have a constant phase difference determined by the propagation vector $k = 2\pi/\lambda$. With this phase difference, they build the spin *wave* which is indexed by the wavevector **k**. $\mathbf{k} = 0$ means all the spins precess about the z-axis with a phase difference of zero. From (7.116) one sees that for $B_0 \to 0$ one can excite spin waves with almost no cost in energy. That means the small thermal energy is sufficient to produce deviations from the saturation magnetization in the Heisenberg spin system by exciting spin waves. In contrast, the reversing of an individual spin costs considerably more energy (Fig. 7.3).

An obvious ansatz for the *many-magnon states* would be

$$|\psi\rangle \sim \prod_{\mathbf{k}} \left(S^-(\mathbf{k})\right)^{n_\mathbf{k}} |0\rangle \tag{7.118}$$

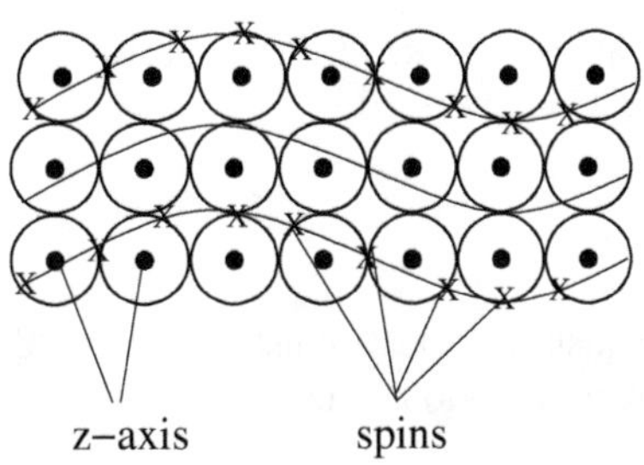

Fig. 7.3 Semiclassical model of a spin wave

where $n_{\mathbf{k}}$ is the number of the spin waves with the propagation vector **k**. However, one can easily show that these states are not eigenstates. This fact indicates that the interactions among the spin waves, i.e. among the magnons, has to be taken into account. This will be discussed in later sections.

7.3 Molecular Field Approximations

In the last section, we could treat exactly some special aspects of the Heisenberg model

$$H = -\sum_{i,j} J_{ij} \left(S_i^+ S_j^- + S_i^z S_j^z \right) - g_J \frac{\mu_B}{\hbar} B_0 \sum_i S_i^z \tag{7.119}$$

The *general* eigenvalue problem could not be solved rigorously up till now. As a result, one has to depend on approximations. The simplest approximation is constructed by linearizing the Hamiltonian. For that, the starting point is the following identity for the product of two operators A and B:

$$A \cdot B = A\langle B\rangle + \langle A\rangle B - \langle A\rangle\langle B\rangle + (A - \langle A\rangle)(B - \langle B\rangle) \tag{7.120}$$

The angular brackets symbolize again the thermodynamic average. The *molecular field approximation (MFA)* consists of the neglect of the fluctuations of the observables from their mean values:

$$A \cdot B \overset{MFA}{\longrightarrow} A\langle B\rangle + \langle A\rangle B - \langle A\rangle\langle B\rangle \tag{7.121}$$

As an operator equation this approach looks rather strange. However, since we need these operators normally in averaging processes, e.g. when calculating the magnetization, such a procedure may be roughly acceptable. Applying this approximation to the operator products in (7.119), one sees that the "spin-flip terms" $S_i^+ S_j^-$ are completely suppressed.

$$S_i^+ S_j^- \overset{MFA}{\longrightarrow} 0 \tag{7.122}$$

This is because of the conservation of the z-component of the total spin (7.14) and (7.15).

$$\langle S_i^+\rangle = \langle S_i^-\rangle = 0 \tag{7.123}$$

Equation (7.122) must be seen as the distinct disadvantage of the molecular field approximation. The approximation (7.121) leads to the following model Hamiltonian:

$$H_{MFA} = -\sum_{i,j} J_{ij} \left\{ S_i^z \langle S_j^z \rangle + \langle S_i^z \rangle S_j^z \right\} - g_J \frac{\mu_B}{\hbar} B_0 \sum_i S_i^z + D(T, B_0) \quad (7.124)$$

$D(T, B_0))$ is a temperature-dependent c-number

$$D(T, B_0) = \sum_{i,j} J_{ij} \langle S_i^z \rangle \langle S_j^z \rangle \quad (7.125)$$

which can, most of the time, be neglected. The decisive simplification achieved in (7.124) is that the original many-spin problem (7.119) is reduced to a one-spin problem. As a result, the ferro-, the ferri- or the antiferromagnet will be described as a paramagnet in an effective temperature-dependent field. Therefore, we can take over many results directly from Sect. 4.2, if we replace in the respective formulae the external field by the sum of the external and the effective fields. The exact form of the effective field is of course determined in the last analysis by the spin configuration of the Heisenberg magnet.

7.3.1 Ferromagnet

First we will consider a homogeneous ferromagnet for which case the translational symmetry is present.

$$\langle S_i^z \rangle = \langle S^z \rangle \; \forall \, i \quad (7.126)$$

This reduces the Hamiltonian to be

$$H_{MFA} = \widehat{D}(T, B_0) - \left(2 J_0 \langle S^z \rangle + g_J \frac{\mu_B}{\hbar} B_0 \right) \sum_i S_i^z \quad (7.127)$$

The c-number

$$\widehat{D}(T, B_0) = N J_0 \langle S^z \rangle^2_{T, B_0} \quad (7.128)$$

does not appear in the calculation of most of the interesting quantities such as magnetization. However, for example, if one wants to calculate the specific heat, one should not neglect this term.

In the molecular field approximation, the Heisenberg Hamiltonian has the same form as the Hamiltonian of an ideal paramagnet:

$$H_{MFA} = \widehat{D}(T, B_0) - g_J \frac{\mu_B}{\hbar} (B_A + B_0) \sum_i S_i^z \quad (7.129)$$

when one interprets B_A as an effective field, which is called the *exchange field*.

$$B_A = \frac{2J_0\,\hbar}{g_J\,\mu_B}\,\langle S^z\rangle_{T,B_0} = B_A(T,\,B_0) \tag{7.130}$$

In view of the expression for the magnetization

$$M(T,\,B_0) = n\,g_J\,\frac{\mu_B}{\hbar}\langle S^z\rangle_{T,B_0} \quad (n = N/V) \tag{7.131}$$

we notice that B_A is proportional to the magnetization

$$B_A = \mu_0\,\lambda\,M \tag{7.132}$$

where

$$\lambda = \frac{2J_0\,\hbar^2}{n\,\mu_0\,(g_J\,\mu_B)^2} \tag{7.133}$$

Thus the molecular field approximation of the Heisenberg model is fully equivalent to the phenomenological Weiss model of a ferromagnet (Sect. 5.1). Equation (7.132) is identical to (5.7). However, in addition, we now have the possibility to provide a microscopic meaning, through (7.133), to the *exchange parameter* λ.

In the following we will recover most of the results of Sect. 5.1 by deriving the physical quantities using H_{MFA}. The latter are essentially the same as those of the Weiss ferromagnet. This holds especially for the magnetization

7.3.1.1 Magnetization

The magnetization is given by (5.8) or (4.127)

$$M(T,\,B_0) = M_0\,B_s(\beta\,S\,g_J\,\mu_B(B_0 + B_A)) \tag{7.134}$$

B_s is the Brillouin function defined in (4.99) and M_0 is the *saturation magnetization*

$$M_0 = n\,S\,g_J\,\mu_B \tag{7.135}$$

In the molecular field approximation, the Heisenberg spin system shows a *phase transition* independent of the lattice dimension, i.e. there exists a critical temperature T_C below which the spontaneous magnetization $M_s(T) = M(T,\,B_0 = 0^+)$ is non-zero. The MFA therefore violates the Mermin–Wagner theorem.

$$\begin{array}{l} T > T_C : M_s(T) \equiv 0 \text{ “}paramagnetism\text{''} \\ T < T_C : M_s(T) \neq 0 \text{ “}ferromagnetism\text{''} \end{array} \tag{7.136}$$

According to the discussion in Sect. 5.1, the spontaneous magnetization satisfies the universal equation of determination (5.15)

$$\frac{M_s(T)}{M_0} = B_s\left(\frac{3S}{S+1}\,\frac{T_C}{T}\,\frac{M_s(T)}{M_0}\right) \tag{7.137}$$

The graphical solution of this equation is illustrated in Fig. 5.6. The Curie temperature, using (5.12), (5.13) and (7.133), is given by

$$k_B T_C = \frac{2}{3}\,\hbar^2\,J_0\,S(S+1) \tag{7.138}$$

T_C is therefore a direct measure of the strength of the ferromagnetic coupling between the localized moments.

The "rough" temperature dependence of the spontaneous magnetization of a ferromagnet is reproduced quite nicely by the molecular field approximation (7.137). However, a more careful look shows interesting deviations. In the critical region ($T \overset{<}{\to} T_C$, $B_0 = 0^+$), the argument of the Brillouin function (7.137) becomes very small. Therefore, we can expand as was done in (4.103).

$$B_s(x) \approx \frac{S+1}{3S}x - \frac{S+1}{3S}\,\frac{2S^2+2S+1}{30S^2}x^2$$

This gives for the spontaneous magnetization, which is also known as the *order parameter* of the ferromagnet,

$$\left(\frac{M_s}{M_0}\right)^2 \approx \frac{10}{3}\,\frac{(S+1)^2}{2S^2+2S+1}\left(\frac{T}{T_C}\right)^2\frac{T_C - T}{T_C} \tag{7.139}$$

for $T \overset{<}{\to} T_C$.

The critical exponent β of the order parameter which is defined by

$$M_s \sim (T_C - T)^\beta \;\text{ for } T \overset{<}{\to} T_C \tag{7.140}$$

in the molecular field approximation is equal to $1/2$ whereas the experimental value is $\approx 1/3$ and therefore there is a disagreement.

The low-temperature behaviour of the spontaneous magnetization is also not given quite correctly. In Sect. 7.4.1, we will derive *Bloch's $T^{3/2}$ law*,

$$\frac{M_s(T)}{M_0} \sim \left(1 - C_{3/2}\,T^{3/2}\right) \text{ for } T \to 0 \tag{7.141}$$

according to which, it can be seen that already at relatively low temperatures, the spontaneous magnetization deviates from the saturation value due to the excitation of spin waves. In the molecular field approximation, the spin waves are completely suppressed. Therefore, the deviations from saturation can appear only by the "reversal" of the individual spins. This naturally costs a lot more energy than the excitation of spin waves, which will happen with arbitrarily small energy (see Sect. 7.2.2). As a result, in the molecular field approximation, M_s, as $T \to 0$, goes into saturation exponentially.

7.3.1.2 Susceptibility

To calculate susceptibility, we will use (7.132) and (7.134). The susceptibility is given by

$$\chi(T, B_0) = \mu_0 \left(\frac{\partial M}{\partial B_0}\right)_T \tag{7.142}$$

According to (4.99), the derivative of the Brillouin function is given by

$$B_s'(x) = \frac{d}{dx} B_s(x) = \frac{1/(2S)^2}{\sinh^2\left(\frac{x}{2S}\right)} - \frac{\left(\frac{2S+1}{2S}\right)^2}{\sinh^2\left(\frac{2S+1}{2S}x\right)} \tag{7.143}$$

Using the notation

$$x = \beta\, S\, g_J\, \mu_B (B_0 + \mu_0\, \lambda\, M(T, B_0)) \tag{7.144}$$

the susceptibility χ is given by

$$\chi(T, B_0) = \mu_0\, n\, \frac{(S\, g_J\, \mu_B)^2\, B_s'(x)}{k_B T - 2J_0\, \hbar^2\, S^2\, B_s'(x)} \tag{7.145}$$

This relationship is valid in fact for all fields B_0 and for all temperatures and therefore also for $T < T_C$.

In the paramagnetic region, the *Curie–Weiss law* which was derived in Sect. 5.1 is valid:

$$M = \frac{1}{\mu_0} \chi\, B_0 \tag{7.146}$$

$$\chi(T) = \frac{C}{T - T_C} \tag{7.147}$$

C is the Curie constant which is defined in (4.130). One should remember that the linear relationship (7.146) between M and B_0 is valid only for high temperatures and small fields. Within the molecular field approximation, the expression (7.147) is exact for zero field and for all $T > T_C$. Experimentally, the linear relationship between the inverse susceptibility and temperature is valid only for very high temperatures and is unambiguously confirmed. One uses this high-temperature behaviour of the susceptibility to define a characteristic material constant: *paramagnetic Curie temperature* θ

$$\chi^{-1} \sim (T - \theta) \quad (T \gg T_C,\ B_0 = 0) \tag{7.148}$$

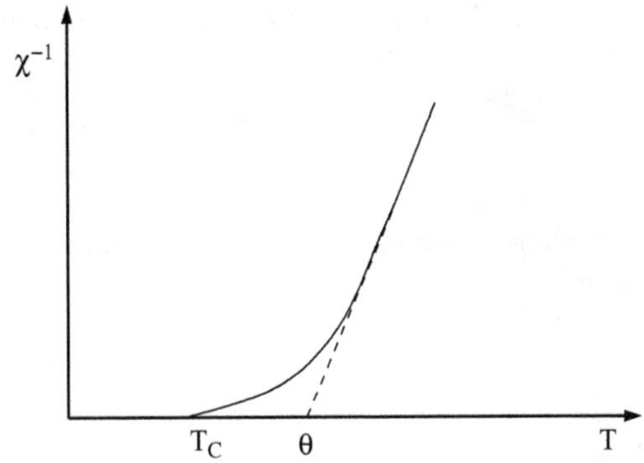

Fig. 7.4 Temperature behaviour of the inverse paramagnetic susceptibility with a zero at the Curie temperature T_C (θ, paramagnetic Curie temperature)

One obtains θ by extrapolating the high-temperature behaviour of the inverse susceptibility onto the temperature axis and finding the intercept (Fig. 7.4). In the molecular field approximation, typically,

$$\theta_{MFA} = T_C \tag{7.149}$$

Experimentally, θ is always greater than T_C.

7.3.1.3 Specific Heat

The molecular field approximation (7.129) gives the average energy per spin of a ferromagnet as

$$U = \frac{1}{N}\langle H_{MFA}\rangle = \frac{1}{N}\,\widehat{D}(T, B_0) - g_J\,\frac{\mu_B}{\hbar}\langle S^z\rangle(B_A + B_0) \tag{7.150}$$

The c-number $\widehat{D}(T, B_0)$, while calculating the specific heat, can not be neglected any more. Substituting (7.128) and (7.131) to (7.133) in (7.150), one gets for the internal energy of the ferromagnet

$$U = -\frac{1}{n}\left(\frac{1}{2}\,\mu_0\,\lambda\,M^2 + B_0\,M\right) \tag{7.151}$$

We first notice from this result that for $T > T_C$ and in zero field the magnetic contribution to the internal energy vanishes. In reality, in a ferromagnet, even above T_C, there exists a short-range ordering of the spins, which are described by the correlation functions of the type $\langle S_i^+ S_j^-\rangle$, $\langle S_i^z S_j^z\rangle$, etc., which are nonzero, even though $\langle S^z\rangle$ is already equal to zero (see Problem 6.4). According to (7.122), right from the beginning, the correlation between different lattice sites is neglected, so that the magnetic energy resulting from such short-range order cannot appear in the expression for the internal energy.

Using (7.151), we can now calculate the *specific heat*

$$C_{B_0} = T\left(\frac{\partial S}{\partial T}\right)_{B_0} \tag{7.152}$$

According to the first law of thermodynamics, the internal energy per particle is given by

$$dU = T\,dS + \frac{V}{N}\,B_0\,dM \tag{7.153}$$

The volume is to be treated as constant. S and M are functions of T and B_0. Therefore

$$\begin{aligned} dU =& \left[T\left(\frac{\partial S}{\partial T}\right)_{B_0} + \frac{V}{N}\,B_0\left(\frac{\partial M}{\partial T}\right)_{B_0}\right]dT \\ &+ \left[T\left(\frac{\partial S}{\partial B_0}\right)_T + \frac{V}{N}\,B_0\left(\frac{\partial M}{\partial B_0}\right)_T\right]dB_0 \end{aligned} \tag{7.154}$$

From this follows the generally valid relation:

$$C_{B_0} = \left(\frac{\partial U}{\partial T}\right)_{B_0} - \frac{1}{n}\,B_0\left(\frac{\partial M}{\partial T}\right)_{B_0} \tag{7.155}$$

Using (7.151), for the Heisenberg ferromagnet, we then have

$$C_{B_0} = -\frac{1}{n}\left\{2B_0\left(\frac{\partial M}{\partial T}\right)_{B_0} + \frac{\mu_0\,\lambda}{2}\left(\frac{\partial M^2}{\partial T}\right)_{B_0}\right\} \tag{7.156}$$

Since in the paramagnetic phase we have $U = 0$ and $M = 0$, we get

$$C_{B_0} \equiv 0 \quad (T > T_C,\; B_0 = 0) \tag{7.157}$$

It can be shown quite generally that, for every magnetic system at high temperatures $C_{B_0} \sim 1/T^2$ should be valid (proved as Problem 6.4). In this sense, the molecular field approximation is not correct.

For $T \stackrel{<}{\rightarrow} T_C$, the behaviour of the specific heat, particularly in the neighbourhood of the phase transition, is interesting. For $B_0 = 0$ and $T < T_C$, we can substitute (7.139) in (7.156):

$$C_{B_0=0} = 5k_B\,\frac{S(S+1)}{2S^2+2S+1}\left(\frac{T}{T_C}\right)^2\left(1+2\left(1-\frac{T_C}{T}\right)\right) \tag{7.158}$$

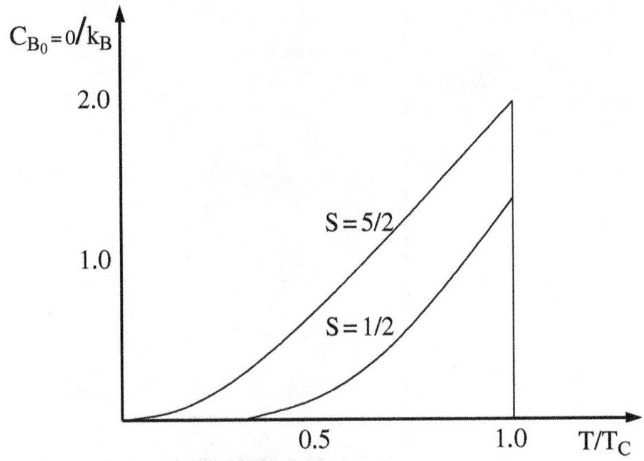

Fig. 7.5 Specific heat of the Heisenberg ferromagnet in the molecular field approximation as function of temperature

From this we see that the specific heat has a finite discontinuity at T_C which is equal to

$$\Delta C_{B_0=0}(T = T_C) = 5k_B \frac{S(S+1)}{2S^2 + 2S + 1} \tag{7.159}$$

That corresponds to a phase transition of second order in the classical Ehrenfest sense.

To calculate the general temperature dependence below T_C, we have to substitute (7.137) in (7.156). The result is displayed in Fig. 7.5, which has a certain similarity with experimental $C_{B_0} - T$ curves. In view of the greatly simplifying assumptions made in the molecular field approximation, the agreement is not expected to be quantitative. Figure 7.6 shows the temperature dependence of the specific heat of the ferromagnet $GdCl_3$ whose Curie temperature is $T_C = 2.20\ K$ [8]. The dashed line is the lattice contribution ($\sim T^3$). The difference between the continuous and dashed lines is the contribution of the ferromagnetic spin system, which is nonzero even beyond T_C due to the short-range order.

7.3.2 Antiferromagnet

We now want to apply the molecular field ansatz to the antiferromagnet. The antiferromagnet is made up of two chemically equivalent ferromagnetic sub-lattices, whose magnetizations in zero field are equal and oppositely oriented to each other at all temperatures.

$$\begin{aligned} \mathbf{M}_s(T) = \mathbf{M}_{1s}(T) + \mathbf{M}_{2s}(T) &\equiv 0 \;\; (T\ arbitrary,\ B_0 = 0) \\ \mathbf{M}_{1s}(T) \equiv -\mathbf{M}_{2s}(T) &\neq 0 \;\; \text{for } T < T_N \end{aligned} \tag{7.160}$$

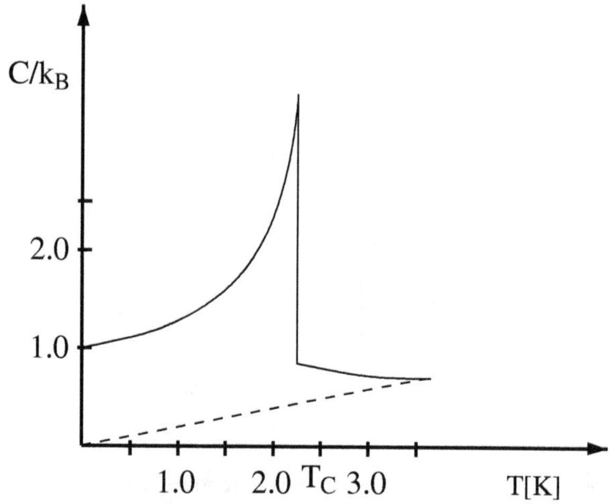

Fig. 7.6 Specific heat as function of temperature for the ferromagnet $GdCl_3$. Dashed line represents the lattice contribution

Obviously, the exchange integrals between atoms belonging to the different sub-lattices must then be negative. The antiferromagnet, in the molecular field approximation, is described by the following Hamiltonian:

$$H_{MFA} = \bar{D}(T, B_0) - g_J \frac{\mu_B}{\hbar} \sum_i (B_0 + B_A^{(i)}) S_i^z \tag{7.161}$$

The c-number $\bar{D}(T, B_0)$ is given according to (7.125) by

$$\bar{D}(T, B_0) = \sum_{i,j} J_{ij} \langle S_i^z \rangle \langle S_j^z \rangle \tag{7.162}$$

Compared to a ferromagnet, now the exchange fields are different for the two sub-lattices:

$$\frac{1}{\hbar} g_J \, \mu_B \, B_A^{(i)} = 2 \sum_j J_{ij} \langle S_j^z \rangle \tag{7.163}$$

This can be written more explicitly as

$$\begin{aligned} \tfrac{1}{\hbar} g_J \, \mu_B \, B_A^{(1)} &= 2 \, \langle S_1^z \rangle \textstyle\sum_j^{\in 1} J_{1j} + 2 \, \langle S_2^z \rangle \sum_j^{\in 2} J_{1j} \\ \tfrac{1}{\hbar} g_J \, \mu_B \, B_A^{(2)} &= 2 \, \langle S_1^z \rangle \textstyle\sum_j^{\in 1} J_{2j} + 2 \, \langle S_2^z \rangle \sum_j^{\in 2} J_{2j} \end{aligned} \tag{7.164}$$

The sub-lattice magnetization

$$M_i(T, B_0) = n_i \, g_J \, \frac{\mu_B}{\hbar} \langle S_i^z \rangle \quad n_i = \frac{N_i}{V}, \; i = 1, \, 2 \tag{7.165}$$

has to be calculated, where N_i is the number of particles in the corresponding sub-lattice. Analogously to the Weiss ferromagnet, we can introduce for the antiferromagnet also the *exchange parameters*:

$$\begin{aligned} \lambda_{11} = \lambda_{22} &= \frac{4\hbar^2}{n\,\mu_0\,(g_J\,\mu_B)^2} \sum_j^{\in 1} J_{1j} \\ \lambda_{12} = \lambda_{21} &= \frac{4\hbar^2}{n\,\mu_0\,(g_J\,\mu_B)^2} \sum_j^{\in 2} J_{1j} \end{aligned} \tag{7.166}$$

We will assume that the two sub-lattices are completely equivalent, so that, in particular, $n_i = n/2$ is valid. Then, the exchange fields have the following relationships with the sub-lattice magnetizations:

$$\begin{aligned} B_A^{(1)} &= \mu_0(\lambda_{12} M_2 + \lambda_{11} M_1) \\ B_A^{(2)} &= \mu_0(\lambda_{21} M_1 + \lambda_{22} M_2) \end{aligned} \tag{7.167}$$

We now discuss the physical properties of the Heisenberg antiferromagnet beginning with sub-lattice magnetization.

7.3.2.1 Sub-lattice Magnetization

Because of the molecular field approximation, the problem of the antiferromagnet is also converted into a single-spin problem. Therefore, in complete analogy with the ferromagnet (7.134), one gets

$$M_i(T, B_0) = M_{0i}\; B_s(\beta\; S\; g_J\; \mu_B(B_0 + B_A^{(i)})) \tag{7.168}$$

$$M_{01} = M_{02} = \frac{n}{2}\; S\; g_J\; \mu_B \tag{7.169}$$

If the field is switched off ($B_0 = 0$), we can use (7.160), as a result of which the spontaneous magnetization of the two sub-lattices must be equal and opposite:

$$M_{1s}(T) = -M_{2s}(T) \tag{7.170}$$

With (7.167), we now have to solve the following implicit equations for the sub-lattice magnetizations:

$$M_{is}(T) = M_{0i}\; B_s(\beta\; S\; g_J\; \mu_0\; \mu_B(\lambda_{11} - \lambda_{12}) M_{is}(T)) \tag{7.171}$$

Mathematically, the problem is exactly the same as the one encountered in the case of a ferromagnet. Again, one can have a graphical solution as in Sect. 5.1.2.

7.3.2.2 Neel Temperature

This is the temperature T_N at which the sub-lattice magnetization collapses. This is also derived exactly in a similar manner as in the case of a ferromagnet. According to Fig. 5.4, the necessary condition for the existence of spontaneous sub-lattice magnetization is

$$\frac{d}{d\,M_{is}} M_{0i}\; B_s \bigg|_{M_{is}=0} \geq 1$$

Using (5.11) with $M_{0i} = \frac{1}{2}M_0$, $J = S$ and $\lambda \to (\lambda_{11} - \lambda_{12})$, one gets with (5.12) and (5.13)

$$T_N = \frac{1}{2}(\lambda_{11} - \lambda_{12})C \tag{7.172}$$

C is the Curie constant (5.12).

$$k_B T_N = \frac{n}{6}\,\mu_0\,(g_J\,\mu_B)^2\; S(S+1)(\lambda_{11} - \lambda_{12}) \tag{7.173}$$

Substituting further (7.166), we finally get

$$k_B T_N = \frac{2}{3} S(S+1)\,\hbar^2 \left\{ \sum_j^{\in 1} J_{1j} - \sum_j^{\in 2} J_{1j} \right\} \tag{7.174}$$

In addition, as a natural condition for the antiferromagnetic order, we have to require

$$\sum_j^{\in 2} J_{1j} < 0 \tag{7.175}$$

We want to evaluate this expression for three typical antiferromagnetic structures.

(a) *ABAB-structure (G-type):* In this structure, there are two interpenetrating sub-lattices such that an atom in one sub-lattice has only the atoms of the other sub-lattice as its nearest neighbours. If we restrict ourselves only to the nearest and next nearest neighbour interactions J_1 and J_2, respectively, then from (7.175) it follows that

$$k_B\; T_N = \frac{2}{3}\; S(S+1)\,\hbar^2 (z_2\; J_2 - z_1\; J_1) \tag{7.176}$$

The z_1 nearest neighbours of the atom under consideration necessarily belong to the other sub-lattice and the z_2 next nearest neighbours to the same sub-lattice. In order that $T_N > 0$, in addition to $J_1 < 0$, either $J_2 > 0$ or

$$J_2 < 0 \text{ and } |z_1 J_1| > |z_2 J_2| \tag{7.177}$$

That means, J_2 can be either positive or negative. That is, the interaction within the sub-lattice, in spite of the ferromagnetic ordering, need not be ferromagnetic, but can also be antiferromagnetic.

(b) *NSNS structure:* This structure is also called the MnO structure. The (111)-planes are ferromagnetically ordered, such that the successive (111)-planes show an (NSNS) sequence of magnetization. In addition to MnO, a good example showing this structure is the antiferromagnet EuTe. In this material, the localized moments are due to the Eu^{2+} ions. These ions build an fcc lattice. One knows that in europium chalcogenides, the nearest neighbour interaction is ferromagnetic ($J_1 > 0$) and the one between the next nearest neighbours is antiferromagnetic ($J_2 < 0$). In an fcc structure, every atom has 12 nearest neighbours, out of which 6 are on the same (111)-plane, that is, in the same sub-lattice and the other 6 in the neighbouring (111)-planes. In addition there are six next nearest neighbours, all of which lie in the other sub-lattice. With this information, according to (7.174) we can write

$$\begin{aligned} k_B\, T_N &= \frac{2}{3}\, S(S+1)\, \hbar^2 (6J_1 - 6J_1 - 6J_2) \\ &= \frac{2}{3}\, S(S+1)\, \hbar^2(-6J_2) \end{aligned} \tag{7.178}$$

Because of $T_N > 0$ the next nearest neighbour interaction J_2 must be negative, i.e. antiferromagnetic. From (7.175), in addition, $6J_1 + 6J_2 < 0$ holds. This is always fulfilled for $J_1 < 0$. That means that *all* interactions are antiferromagnetic. However, $J_1 > 0$ with $J_1 < |J_2|$ is also possible. In such a case the nearest neighbour exchange would be ferromagnetic, although the system as a whole is an antiferromagnet.

(c) *NNSS structure:* An example for an antiferromagnet which shows an NNSS sequence in magnetization of the ferromagnetically ordered (111)-planes is EuSe in the temperature range between 2.8 and 4.6 K [9]. The lattice structure is same as for EuTe. The magnetic moments build an fcc lattice. Out of the 12 nearest neighbours of an arbitrarily chosen atom, 9 are in the same sub-lattice and 3 in the other sub-lattice. Out of the six next nearest neighbours, three each are in the two sub-lattices. This according to (7.174) leads to

$$\begin{aligned} k_B\, T_N &= \frac{2}{3}\, S(S+1)\, \hbar^2 (9J_1 - 3J_1 + 3J_2 - 3J_2) \\ &= \frac{2}{3}\, S(S+1)\, \hbar^2(6J_1) \end{aligned} \tag{7.179}$$

Again (7.175) also has to be fulfilled. That means $(3J_1 + 3J_2) < 0$ or $J_2 < -J_1$. Because of (7.179) the nearest neighbour exchange is ferromagnetic ($J_1 > 0$) and the next nearest neighbour exchange antiferromagnetic ($J_2 < 0$).

For any arbitrary antiferromagnet, provided it is built up of two interpenetrating sub-lattices, one can evaluate (7.174) in a similar fashion. We will now discuss the paramagnetic Curie temperature θ.

7.3.2.3 Paramagnetic Curie Temperature θ

For this, we require the high-temperature behaviour of the paramagnetic susceptibility. At high temperatures, the argument of the Brillouin function in (7.168) is small, so that we can again exploit the relation

$$B_s(x) \approx \frac{S+1}{3S} x$$

Then we get

$$M_i(T, B_0) \approx \frac{n}{6} (g_J\, \mu_B)^2\, S(S+1)\, \beta(B_0 + B_A^{(i)}) \tag{7.180}$$

The total magnetization

$$M(T, B_0) = M_1(T, B_0) + M_2(T, B_0) \tag{7.181}$$

is for $B_0 \neq 0$ of course not zero. Using (5.12) and (7.167) one sees that for antiferromagnets also the *Curie–Weiss law* is valid:

$$M(T, B_0) = \chi(T)\, \frac{1}{\mu_0}\, B_0 \tag{7.182}$$

$$\chi(T) = \frac{C}{T - \theta} \tag{7.183}$$

The paramagnetic Curie temperature θ is given by

$$k_B\, \theta = \frac{1}{2} k_B\, C(\lambda_{11} + \lambda_{12}) = \frac{2}{3} S(S+1)\, \hbar^2 \left(\sum_j^{\in 1} J_{1j} + \sum_j^{\in 2} J_{1j} \right) \tag{7.184}$$

Thus, the paramagnetic Curie temperature of an antiferromagnet satisfies formally exactly the same relation as that of a ferromagnet (7.138) which is equal to T_C in the molecular field approximation:

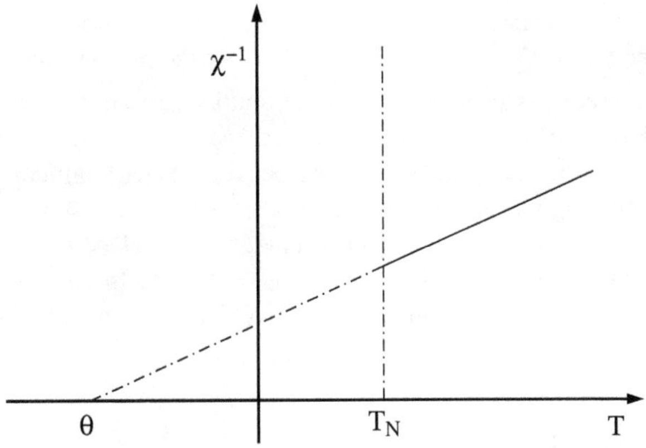

Fig. 7.7 Temperature behaviour of the inverse susceptibility of an antiferromagnet with negative paramagnetic Curie temperature

$$k_B\,\theta = \frac{2}{3}\,S(S+1)\,\hbar^2\,J_0 \tag{7.185}$$

For antiferromagnets, typically $J_0 < 0$, and therefore, the paramagnetic Curie temperature is negative (Fig. 7.7).

$$\theta < 0 \tag{7.186}$$

For the antiferromagnets EuTe and EuSe, which were already discussed, (7.185) becomes

$$k_B\theta = \frac{2}{3}\,S(S+1)\,\hbar^2(12J_1 + 6J_2) \tag{7.187}$$

With this equation and either (7.178) or (7.179), from the measurement of T_N and θ, the exchange integrals J_1 and J_2 can be determined. One finds that

$$\begin{aligned} EuTe: \hbar^2\,J_1/k_B = 0.043\text{ K};\ \hbar^2\,J_2/k_B = -0.150\text{ K} \\ EuSe: \hbar^2\,J_1/k_B = 0.073\text{ K};\ \hbar^2\,J_2/k_B = -0.011\text{ K} \end{aligned} \tag{7.188}$$

According to (7.175) and (7.184), the relationship between the two characteristic temperatures of the antiferromagnet is

$$-\frac{\theta}{T_N} = \frac{J_0}{\sum_j^{\in 2} J_{1j} - \sum_j^{\in 1} J_{1j}} \tag{7.189}$$

In most of the cases, $-\frac{\theta}{T_N} > 1$:

Material	MnO	NiO	MnF_2
$-\theta/T_N$	5.3	5.7	1.7

Since the interactions between the two sub-lattices are necessarily antiferromagnetic, i.e. $\sum_j^{\in 2} J_{1j} < 0$ (7.175), $-\theta/T_N$ can only be greater than one, when the interactions even within the sub-lattice are antiferromagnetic, i.e. $\sum_j^{\in 1} J_{1j} < 0$. By itself, each of the sub-lattices would like to order antiferromagnetically. As the antiferromagnetic coupling between the sub-lattices is stronger, a ferromagnetic coupling within the sub-lattice is forced.

There are of course antiferromagnets with $0 < -\theta/T_N < 1$. An example is EuTe which has $\theta = -4.0$ K and $T_N = 9.6$ K. In such a case, according to (7.189), $\sum_j^{\in 1} J_{1j}$ must certainly be positive. That means, in this case, the coupling within the sub-lattice is ferromagnetic.

7.3.2.4 Susceptibility for $T < T_N$

In order to calculate the susceptibility, we have to solve (7.168) for the sub-lattice magnetization for $B_0 \neq 0$ and then differentiate with respect to B_0.

Let us assume that in the absence of an external field, both the sub-lattice magnetizations $\mathbf{M}_1$ and $\mathbf{M}_2$ are, respectively, parallel and antiparallel to the z-axis. The z-axis is chosen such that it is the "easy" axis of the magnetic crystal. On the application of an external field $\mathbf{B}_0$, the magnetizations $\mathbf{M}_1$ and $\mathbf{M}_2$ will change in both direction and magnitude. One must differentiate two special cases. If the field is parallel to the easy direction, then the directions of $\mathbf{M}_1$ and $\mathbf{M}_2$ will not change but the magnitudes will. We have to expect a clear temperature dependence. On the other hand, if $\mathbf{B}_0$ lies perpendicular to the easy direction, then, to a first approximation, the magnitudes do not change but the directions will (Fig. 7.8). In this case, there will be practically no temperature dependence of the susceptibility. We want to examine these arguments a little more closely.

Let the field lie along the z-axis. In order to calculate the susceptibility, we expand the right-hand side of (7.168) in terms of the external field:

$$M_i(T, B_0) = M_{0i} \left[B_s + B_0\, \beta\, S\, g_J\, \mu_B \left(1 + \frac{\partial B_A^{(i)}}{\partial B_0} \right) B_s' + \mathcal{O}(B_0^2) \right] \tag{7.190}$$

$B_s(x)$ and $B_s'(x)$ are, respectively, odd and even functions of x (see (7.143)). Therefore, it is immaterial which of the spontaneous sub-lattice magnetizations we substitute in the argument of $B_s'(x)$:

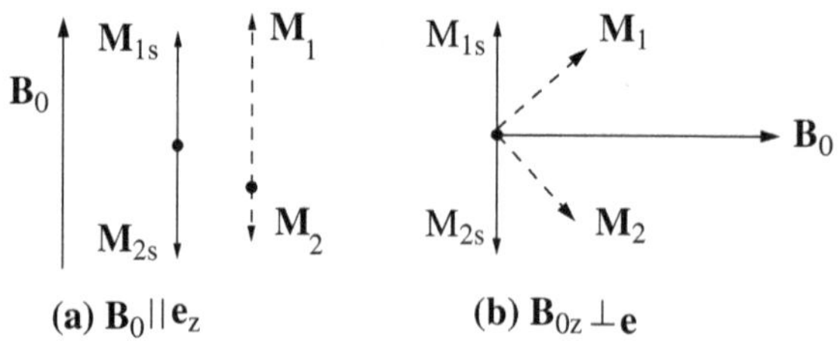

Fig. 7.8 Reaction of the sub-lattice magnetizations of an antiferromagnet to an external magnetic field $\mathbf{B}_0$. Field (**a**) parallel, (**b**) perpendicular to the easy axis

$$
\begin{aligned}
M(T, B_0) &= M_1(T, B_0) + M_2(T, B_0) \\
&= B_0\,\beta\,M_{01}\,S\,g_J\,\mu_B \left[2 + \left(\frac{\partial B_A^{(1)}}{\partial B_0}\right)_{B_0=0}\right. \\
&\left. + \left(\frac{\partial B_A^{(2)}}{\partial B_0}\right)_{B_0=0}\right] B_s'(\beta S g_J \mu_0 \mu_B(\lambda_{11} - \lambda_{12}) M_{1s}(T)) \\
&+ \mathcal{O}(B_0^2)
\end{aligned}
$$

Using this we can easily calculate the susceptibility:

$$
\chi_{\|}(T) = \mu_0 \left(\frac{\partial M(T, B_0)}{\partial B_0}\right)_{B_0=0} \tag{7.191}
$$

From (7.167), we see that

$$
\left(\frac{\partial B_A^{(1)}}{\partial B_0}\right)_{B_0=0} + \left(\frac{\partial B_A^{(2)}}{\partial B_0}\right)_{B_0=0} = \chi_{\|}(T)(\lambda_{11} + \lambda_{12}) \tag{7.192}
$$

From (7.184), we get $(\lambda_{11} + \lambda_{12}) = 2\theta/C$. Using the definitions for the saturation magnetization M_{0i} (7.169), the Curie constant C (5.12) and for the Neel temperature T_N (7.172), we get the following expression for the susceptibility:

$$
\chi_{\|}(T) = \mu_0\, n\, \frac{S^2\,(g_J \mu_B)^2\, B_s'\left(\frac{T_N}{T}\frac{3S}{S+1}\frac{M_{1s}}{M_{01}}\right)}{k_B\left(T - \frac{3S}{S+1}\theta\, B_s'\left(\frac{T_N}{T}\frac{3S}{S+1}\frac{M_{1s}}{M_{01}}\right)\right)} \tag{7.193}
$$

The spontaneous sub-lattice magnetization is determined by

$$
\frac{M_{1s}}{M_{01}} = B_s\left(\frac{T_N}{T}\frac{3S}{S+1}\frac{M_{1s}}{M_{01}}\right) \tag{7.194}
$$

With (7.143) one recognizes that

$$
\chi_{\|}(T) \overset{T\to 0}{\longrightarrow} 0 \quad \text{(exponentially)} \tag{7.195}
$$

While calculating the susceptibility $\chi_\perp$ we can use the fact that when the field is applied perpendicular to the "easy" axis, the magnitudes of the sub-lattices at first do not change but only their directions are changed. This results in a practically temperature-independent susceptibility.

$$
\chi_\perp(T) = \chi_\perp(T_N) = \chi(T_N) \tag{7.196}
$$

For $T \geq T_N$, the localized spins are "free moving", i.e. they do not have a preferential direction of orientation any more. Then the differentiation of $\chi_{\parallel}$ and $\chi_{\perp}$ is unnecessary. Both the susceptibilities are identical to the paramagnetic χ given by (7.183). With (7.173) and (7.183) substituted in (7.196), we finally have

$$\chi_{\perp} = \frac{C}{T_N - \theta} = -\frac{1}{\lambda_{12}} = n\,\mu_0\,\frac{(g_J\,\mu_B)^2}{4\,\hbar^2 \left|\sum_j^{\epsilon 2} J_{1j}\right|} \tag{7.197}$$

The typical feature of an antiferromagnet is the peak in the susceptibility at $T = T_N$ from which one can read off the exchange parameter λ_{12}. The susceptibility for an arbitrary direction can always be split up into the $\chi_{\parallel}$ and $\chi_{\perp}$ parts (Fig. 7.9). It is of course clear that the susceptibility of an antiferromagnetic powder should in any case be isotropic:

$$\chi_{powder}(T) = \frac{2}{3}\chi_{\perp}(T) + \frac{1}{3}\chi_{\parallel}(T) \tag{7.198}$$

Because of (7.195) and (7.196), we can then write

$$\frac{\chi_{powder}(0)}{\chi_{powder}(T_N)} = \frac{2}{3} \tag{7.199}$$

This relation is, in general, experimentally rather well fulfilled.

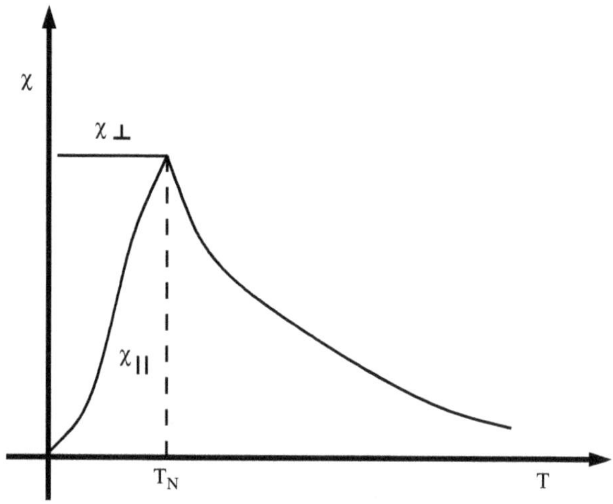

Fig. 7.9 General temperature behaviour of the susceptibility of an antiferromagnet, in case that the external field $\mathbf{B}_0 \rightarrow 0$ is parallel ($\chi_{\parallel}$) or perpendicular ($\chi_{\perp}$) to the easy axis

7.3.2.5 Specific Heat

The magnetic energy of an antiferromagnet follows almost directly from (7.161):

$$\begin{aligned}
U = \frac{1}{N}\,\langle H_{MFA}\rangle &= \frac{1}{N}\,\bar{D}(T,B_0) - \frac{1}{2}\,g_J\,\frac{\mu_B}{\hbar}\left[B_0(\langle S_1^z\rangle\right.\\
&+ \langle S_2^z\rangle) + \mu_0(\lambda_{12}M_2 + \lambda_{11}M_1)\langle S_1^z\rangle\\
&\left.+ \mu_0(\lambda_{12}M_1 + \lambda_{11}M_2)\langle S_2^z\rangle\right]\\
&= \frac{1}{N}\,D(T,B_0) - \frac{1}{n}\,[B_0(M_1+M_2)\\
&+ 2\mu_0\,\lambda_{12}M_1M_2 + \mu_0\lambda_{11}(M_1^2+M_2^2)] \qquad (7.200)
\end{aligned}$$

From (7.162), one finds for the c-number $D(T, B_0)$,

$$\frac{1}{N}\,\bar{D}(T,B_0) = \frac{1}{n}\left(\frac{1}{2}\,\mu_0\,\lambda_{11}(M_1^2+M_2^2) + \mu_0\,\lambda_{12}M_1\,M_2\right) \qquad (7.201)$$

Then, the energy per spin of an antiferromagnet is given by

$$\begin{aligned}
U(T,B_0) = -\frac{1}{n}\,[B_0(M_1+M_2) &+ \frac{1}{2}\,\mu_0\lambda_{11}(M_1^2+M_2^2)\\
&+\mu_0\,\lambda_{12}M_1\,M_2] \qquad (7.202)
\end{aligned}$$

In writing (7.200), (7.201) and (7.202) we have assumed that B_0 is parallel to the "easy" axis (z-axis).

For $B_0 = 0$ and $T < T_N$, we can exploit $M_1 = -M_2$:

$$U(T, B_0 = 0) = -\frac{\mu_0}{n}\,(\lambda_{11} - \lambda_{12})M_1^2 \qquad (7.203)$$

This has the same structure as the corresponding expression (7.151) for the ferromagnet. Because of neglect of the short-range order in the molecular field approximation, just as in the case of a ferromagnet, here also, the magnetic energy and the specific heat $C_{B_0=0}$ vanish for $T > T_N$. Analogously to (7.156), the specific heat is given by

$$C_{B_0=0} = -\frac{\mu_0}{n}(\lambda_{11} - \lambda_{12})\left(\frac{\partial M_{1s}^2}{\partial T}\right)_{B_0=0} \qquad (7.204)$$

Just as was done for (7.139), for $T \overset{<}{\rightarrow} T_N$, one can derive

$$\frac{M_{1s}^2}{M_{01}^2} \approx \frac{10}{3}\,\frac{(S+1)^2}{2S^2+2S+1}\left(\frac{T}{T_N}\right)^2\frac{T_N - T}{T_N} \qquad (7.205)$$

With (7.173) and (7.169), it then follows that

$$C_{B_0=0} = 5k_B \frac{S(S+1)}{2S^2+2S+1} \left(\frac{T}{T_N}\right)^2 \left(1 + 2\left(1 - \frac{T_N}{T}\right)\right) \tag{7.206}$$

for $T \overset{<}{\to} T_N$.

Thus the specific heat of the antiferromagnet shows exactly the same anomaly as the ferromagnet (7.159) at the phase transition point.

$$\Delta C_{B_0=0}(T = T_N) = 5k_B \frac{S(S+1)}{2S^2+2S+1} \tag{7.207}$$

The result (7.202) provides an interesting possibility to derive once again $\chi_\perp$ classically. For arbitrary direction of $\mathbf{B}_0$, (7.202) can be written as

$$\begin{aligned} U(T, B_0) = -\frac{1}{n} [\, &\mathbf{B}_0 \cdot (\mathbf{M}_1 + \mathbf{M}_2) + \frac{1}{2}\mu_0 \lambda_{11}(\mathbf{M}_1^2 + \mathbf{M}_2^2) \\ &+ \mu_0 \lambda_{12} \mathbf{M}_1 \cdot \mathbf{M}_2] \end{aligned} \tag{7.208}$$

Classically, the above equation means

$$\begin{aligned} U(T, B_0) = -\frac{1}{n} [\, &2B_0|M_1| \sin\phi + \mu_0 \lambda_{11}|M_1|^2 \\ &- \mu_0 \lambda_{12}|M_1|^2 \cos(2\phi)] \end{aligned} \tag{7.209}$$

For small fields, the angle ϕ (Fig. 7.10) will be small so that we can expand

$$\begin{aligned} U(T, B_0) \approx -\frac{1}{n} [&2B_0|M_1|\phi + \mu_0|M_1|^2 \lambda_{11} \\ &- \mu_0 |M_1|^2 \lambda_{12}(1 - 2\phi^2)] \end{aligned} \tag{7.210}$$

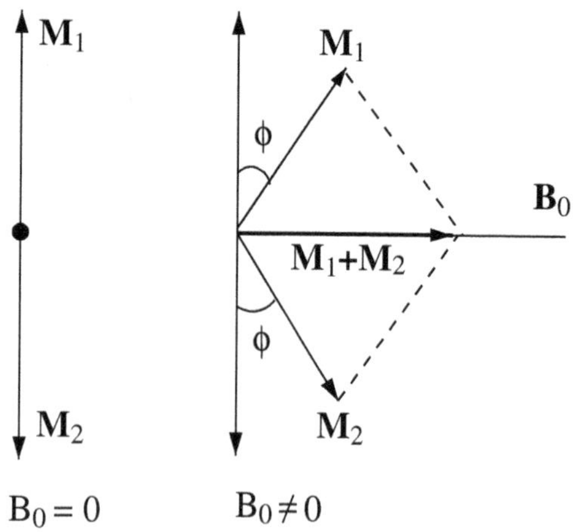

Fig. 7.10 Sub-lattice magnetizations of an antiferromagnet in an external field $\mathbf{B}_0$

The equilibrium value ϕ_0 of the angle is obtained from the condition

$$\frac{\partial U}{\partial \phi} \stackrel{!}{=} 0 \Rightarrow 0 = 2B_0|M_1| + 4\,\mu_0|M_1|^2\,\lambda_{12}\phi_0$$

With ϕ_0 given by

$$\phi_0 = -\frac{B_0}{2\mu_0\,\lambda_{12}|M_1|} \tag{7.211}$$

we can calculate the component of the magnetization parallel to the field:

$$M_{B_0} = (\mathbf{M}_1 + \mathbf{M}_2)\cdot\frac{\mathbf{B}_0}{B_0} = (|M_1| + |M_2|)\sin\phi_0 \approx 2|M_1|\phi_0 \tag{7.212}$$

With this we finally get the susceptibility $\chi_\perp$:

$$\chi_\perp = \mu_0\left(\frac{\partial M_{B_0}}{\partial B_0}\right)_T = -\frac{1}{\lambda_{12}} \tag{7.213}$$

This is exactly the result (7.197).

7.3.2.6 Spin-Flop Field

As already mentioned, the anisotropy of the antiferromagnet defines the *easy* and *hard directions* for the magnetizations. This leads to an interesting magnetic transition called the "spin-flop" in antiferromagnets (Fig. 7.11).

If the field is applied parallel to the easy direction, then first $|\mathbf{M}_1|$ increases and $|\mathbf{M}_2|$ decreases. This happens, however, only so long as the field increases up to a critical strength B_F. At this point, the magnetizations $\mathbf{M}_1$ and $\mathbf{M}_2$ jump into the direction perpendicular to the field and from then on gradually change the direction towards the direction of the field as the field is further increased.

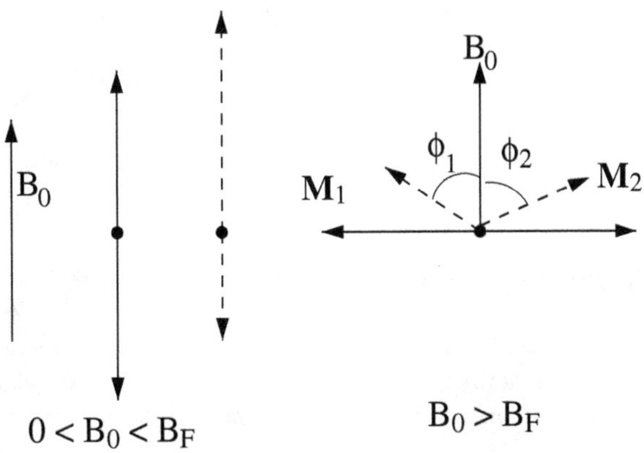

Fig. 7.11 Illustration of the spin-flop field

How does one understand this effect? We have discussed the magnetic anisotropy in Sect. 7.1.2. The "easy" axis, for which the anisotropy is responsible, can be formulated in the simplest way from (7.59), through the magnetic energy per spin U_A as follows:

$$U_{A\parallel} = K_0; \quad U_{A\perp} = K_0 + K_1 \quad (K_1 > 0) \tag{7.214}$$

In the molecular field approximation (7.208) for U, the anisotropy terms are not yet taken into account. If the magnetizations change from parallel to perpendicular direction, the last two terms in (7.208) do not change. Therefore, we can write

$$\begin{aligned} U_\parallel &= K_0 - \tfrac{1}{n} \tfrac{\chi_\parallel}{\mu_0} B_0^2 + U_0 \\ U_\perp &= K_0 + K_1 - \tfrac{1}{n} \tfrac{\chi_\perp}{\mu_0} B_0^2 + U_0 \end{aligned} \tag{7.215}$$

For $T < T_N$, $\chi_\perp > \chi_\parallel$ (Fig. 7.9). Since $K_1 > 0$, there is a critical field strength B_F defined by

$$U_\parallel(B_0 = B_F) \stackrel{!}{=} U_\perp(B_0 = B_F) \tag{7.216}$$

above which $U_\perp < U_\parallel$ so that it becomes more advantageous for the magnetization originally parallel to the field to change to the direction perpendicular to the field. From (7.215) and (7.216), we get for the *spin-flop field* B_F

$$B_F = \sqrt{\frac{n\, \mu_0\, K_1}{\chi_\perp - \chi_\parallel}} \tag{7.217}$$

In the experiments, the spin-flop becomes noticeable at low temperatures by the fact that the magnetization jumps from

$$M_\parallel = \frac{1}{\mu_0}\, \chi_\parallel\, B_0 \approx 0$$

to

$$M_\perp = \frac{1}{\mu_0}\, \chi_\perp\, B_0 = \frac{B_0}{-\mu_0\, \lambda_{12}}$$

By measuring the spin-flop field, the anisotropy constant K_1 can be experimentally determined.

It is also clear from (7.217) that *without* the anisotropy ($K_1 \to 0$), the sublattice magnetization always orients itself perpendicular to the external field. *With* anisotropy, only for $B \geq B_F$, the magnetization is decoupled from the easy direction.

7.3.3 Ferrimagnet

As the third example of the application of the molecular field approximation (7.124) of the Heisenberg model, in this section, we want to discuss the ferrimagnets which in a certain sense stay in between the ferro and antiferromagnets. Ferrimagnet, below a critical temperature, has resultant spontaneous magnetization and therefore is like a ferromagnet. On the other hand, for high temperatures, the susceptibility shows a Curie–Weiss behaviour with a negative paramagnetic Curie temperature, which is typical for an antiferromagnet. For this reason, there is no consensus in the literature regarding the naming of the transition temperature. In view of the ferromagnetic magnetization, we will call the critical temperature as the "Curie temperature" T_C, while other authors call it "Neel temperature" T_N because of the antiferromagnetic coupling mechanism and the sub-lattice structure.

By a ferrimagnet, we mean a magnet which is made up of chemically non-equivalent sub-lattices with unequal magnetic moments, so that the resulting total moment is unequal to zero. The sub-lattice magnetizations are pair-wise antiferromagnetically oriented. However, exotic spin structures such as spiral structures which are also sometimes being called ferrimagnetic will not be considered here.

Thus, ferrimagnets must be made up of at least two non-equivalent lattice sites. That is why, in general, the crystal structure of these materials is complicated. The most prominent representatives belong to the class of spinels or the perovskites with unit cells containing up to 160 atoms:

(a) *Ferrites*: $MO \cdot Fe_2O_3$
  $M =$ double-valued metal ion
(b) *Garnets*: $3M_2O_3 \cdot 5Fe_2O_3$
  $M =$ triple-valued metal ion
  (α) $M = Y$: yttrium–iron–garnet: "YIG"
  (β) $M =$ Rare earth (e.g. Gd): Rare earth–iron–garnet: "RIG"

A prominent representative of ferrites is the magnetite $Fe_3O_4 = FeO \cdot Fe_2O_3$ which has a valence mixture of Fe^{2+} and Fe^{3+}.

We want to describe here the simplest version of the ferrimagnets, namely, the *two sub-lattice model*. The evaluation within the framework of the molecular field approximation is formally identical to that of an antiferromagnet. We only have to take care that the spin and the Lande g-factors of the two sub-lattices can be different. This can, under certain conditions, lead to a nonmonotonic behaviour, even a change of sign, of the net magnetization as a function of temperature. The temperature $T_K < T_C$ at which the magnetization passes through zero is called the *compensation temperature* (Fig. 7.12). However, such compensation can, but need not necessarily, occur.

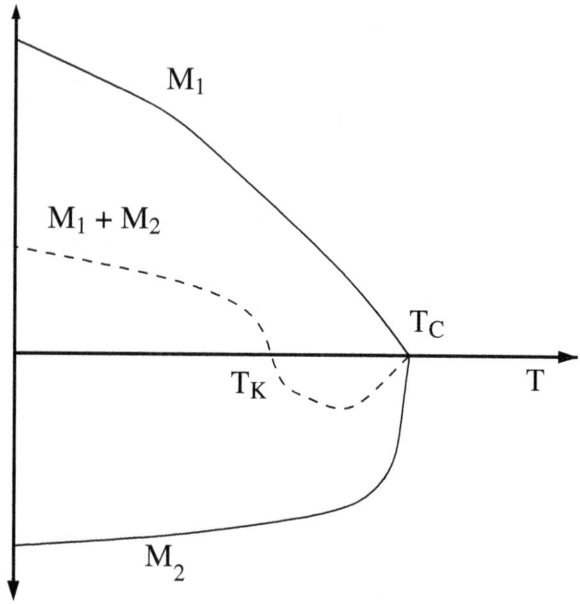

Fig. 7.12 Typical temperature behaviour of the sub-lattice and total lattice magnetizations of a ferrimagnet. T_K is the compensation temperature

7.3.3.1 Sub-lattice Magnetizations

We can practically take over all the relations that we have derived in the last section, only taking care to index certain quantities with the corresponding sub-lattice, if they are different for different sub-lattices. Then according to (7.168), we have for the sub-lattice magnetization of a ferrimagnet

$$M_i(T, B_0) = M_{0i}\; B_{S_i}\left(\beta\; S_i\; g_{J_i}\; \mu_B\left(B_0 + B_A^{(i)}\right)\right) \tag{7.218}$$

$$M_{oi} = n_i\; S_i\; g_{J_i}\; \mu_B \tag{7.219}$$

The index $i = 1,\ 2$ numbers the two sub-lattices. For the exchange fields $B_A^{(i)}$, as in (7.167), we have

$$B_A^{(1)} = \mu_0(\lambda_{12}\; M_2 + \lambda_{11}\; M_1) \tag{7.220}$$

$$B_A^{(2)} = \mu_0(\lambda_{21}\; M_1 + \lambda_{22}\; M_2) \tag{7.221}$$

Here one has to notice that even though

$$\lambda_{12} = \lambda_{21} \tag{7.222}$$

unlike in the case of antiferromagnets, we now have the possibility that

$$\lambda_{11} \neq \lambda_{22} \tag{7.223}$$

In more detail, the exchange parameters λ_{ij} are given by

$$\lambda_{12} = \lambda_{21} = \frac{2\hbar^2}{n_1\,\mu_0\,g_{J_1} g_{J_2} \mu_B^2} \sum_j^{\in 2} J_{1j} \tag{7.224}$$

$$\lambda_{11} = \frac{2\hbar^2}{n_1\,\mu_0\,(g_{J_1}\mu_B)^2} \sum_j^{\in 1} J_{1j} \tag{7.225}$$

$$\lambda_{22} = \frac{2\hbar^2}{n_2\,\mu_0\,(g_{J_2}\mu_B)^2} \sum_j^{\in 2} J_{2j} \tag{7.226}$$

Due to the antiferromagnetic coupling between the two sub-lattices we expect

$$\lambda_{12} < 0 \quad ; \quad |\lambda_{12}| \gg |\lambda_{11}|,\ |\lambda_{22}| \tag{7.227}$$

7.3.3.2 Curie Temperature

Just as in the case of a ferromagnet, T_C can be obtained by demanding that the slope of the right-hand side of (7.218) as a function of the spontaneous magnetization $M_{is} = M_i(T,0)$ be greater than or equal to one. For $B_0 = 0$ and $T \to T_C$, the argument of the Brillouin function is small and so we can apply the usual approximation

$$B_s(x) \approx \frac{S+1}{3S}\,x$$

With the two Curie constants

$$C_i = n_i\,\mu_0\,g_{J_i}\,\frac{S_i(S_i+1)}{3k_B}\,\mu_B^2 \quad i = 1,\ 2 \tag{7.228}$$

and applying Eqs. (7.218), (7.219), (7.220) and (7.221), we get

$$M_1 \approx \frac{C_1}{T}\,(\lambda_{11}\,M_1 + \lambda_{12}\,M_2) \tag{7.229}$$

$$M_2 \approx \frac{C_2}{T}\,(\lambda_{22}\,M_2 + \lambda_{12}\,M_1) \tag{7.230}$$

T_C is obtained from the condition

$$1 \stackrel{!}{=} \frac{C_1}{T_C}\left(\lambda_{11} + \lambda_{12}\,\frac{\partial M_2}{\partial M_1}\bigg|_{T=T_C}\right) \tag{7.231}$$

From (7.230) we have

$$\frac{\partial M_2}{\partial M_1}\bigg|_{T=T_C} = \frac{C_2\,\lambda_{12}}{T_C - \lambda_{22} C_2} \tag{7.232}$$

The combination of the last two equations gives a relatively complicated expression for the transition temperature:

$$T_C = \frac{1}{2}(C_1\,\lambda_{11} + C_2\,\lambda_{22}) \pm \frac{1}{2}\sqrt{(C_1\,\lambda_{11} - C_2\,\lambda_{22})^2 + 4\,C_1\,C_2\,\lambda_{12}^2} \tag{7.233}$$

The fact that T_C has to be positive allows, in view of (7.227), only the solution with the positive sign before the square root.

Before closing, we want to calculate, for the ferrimagnet, the paramagnetic susceptibility.

7.3.3.3 Paramagnetic Susceptibility

For $T > T_C$ and $B_0 \stackrel{>}{\to} 0$, we have

$$\begin{aligned} M_1(T, B_0) &\approx \tfrac{C_1}{T}(\lambda_{11}\,M_1 + \lambda_{12}\,M_2 + B_0/\mu_0) \\ M_2(T, B_0) &\approx \tfrac{C_2}{T}(\lambda_{22}\,M_2 + \lambda_{12}\,M_1 + B_0/\mu_0) \end{aligned} \tag{7.234}$$

They can be solved for M_1 and M_2 and then the total magnetization can be obtained. With the abbreviations

$$N(T) = (T - C_2\,\lambda_{22})(T - C_1\,\lambda_{11}) - C_1\,C_2\,\lambda_{12}^2 \tag{7.235}$$

$$Z(T) = (C_1 + C_2)T + C_1\,C_2(2\lambda_{12} - \lambda_{11} - \lambda_{22}) \tag{7.236}$$

for small fields B_0, we get

$$M(T, B_0) = M_1(T, B_0) + M_2(T, B_0) \approx \frac{B_0}{\mu_0}\,\frac{Z(T)}{N(T)} \tag{7.237}$$

The susceptibility

$$\chi_{T,B_0=0} = \mu_0\left(\frac{\partial M}{\partial B_0}\right)_{T,B_0=0} = \frac{Z(T)}{N(T)} \tag{7.238}$$

shows a relatively complicated temperature dependence and follows the Curie–Weiss law only for very high temperatures:

$$\chi^{-1}(T) \approx \frac{T-\theta}{C_1+C_2} \quad (T \gg T_C) \tag{7.239}$$

The paramagnetic Curie temperature θ is clearly different from T_C:

$$\theta = (C_1\,\lambda_{11} + C_2\,\lambda_{22}) + \frac{C_1\,C_2(2\lambda_{12} - \lambda_{11} - \lambda_{22})}{C_1+C_2} \tag{7.240}$$

In view of (7.227), generally, as in the case of antiferromagnets, it is to be expected that

$$\theta < 0 \tag{7.241}$$

From (7.233) and (7.235), one sees that

$$N(T_C) = 0 \tag{7.242}$$

which means

$$\chi^{-1}(T_C) = 0 \tag{7.243}$$

From the above results and those from the earlier sections, it is possible to represent schematically the inverse susceptibility of ferro-, antiferro- and ferrimagnets in the molecular field approximation by Fig. 7.13.

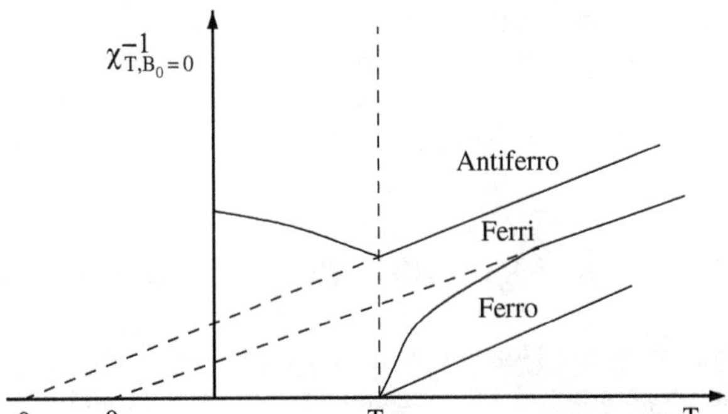

Fig. 7.13 Typical temperature behaviour of the inverse magnetic susceptibility for, respectively, antiferro-, ferri-, ferromagnets

7.4 Spin Waves

We have derived in Sect. 7.2.2 exact eigenstates of the Heisenberg Hamiltonian, which are the so-called "one-magnon states"

$$|\mathbf{k}\rangle = \frac{1}{\hbar\sqrt{2S\,N}}\,S^{-}(\mathbf{k})|0\rangle \tag{7.244}$$

with the eigenenergies

$$E(\mathbf{k}) = E_0(B_0) + g_J\,\mu_B\,B_0 + 2S\,\hbar^2\,(J_0 - J(\mathbf{k})) \tag{7.245}$$

Here $E_0(B_0)$ is the ground state energy of the Heisenberg ferromagnet

$$E_0(B_0) = -N\,\hbar^2\,J_0\,S^2 - N\,g_J\,\mu_B\,B_0\,S \tag{7.246}$$

The excitation energy

$$\hbar\,\omega(\mathbf{k}) = E(\mathbf{k}) - E_0(B_0) \tag{7.247}$$

is ascribed to the quasiparticle which is called the *magnon*. It corresponds to a spin deviation of one unit of angular momentum ($1\hbar$) which is distributed collectively among all the localized spins in the form of a spin wave.

We have introduced in Sect. 7.1.1 the Holstein–Primakoff transformation, using which the Heisenberg Hamiltonian can be expressed as an infinite series of creation and annihilation operators (7.34). If we terminate the series at the bilinear term, then we get a simplified model Hamiltonian (7.39), whose complete set of eigenstates just consists of exactly the one-magnon states (7.244). The model Hamiltonian thus obtained is called the *harmonic approximation* or the *linear spin wave approximation* of the Heisenberg model, whose predictions and the region of validity will be discussed in this section. This is naturally a *low-temperature approximation*, which is exact if no more than one single magnon is excited and therefore would certainly represent an acceptable approximation, if the number of magnons is small, so that the interaction among the magnons can be neglected. Small magnon number, on the other hand, means that the system is still very close to saturation, which in turn requires low temperatures.

7.4.1 Linear Spin Wave Theory for the Isotropic Ferromagnet

We apply the Holstein–Primakoff transformation (7.23), (7.24) and (7.25) to the spin operators of the Heisenberg Hamiltonian and obtain an infinite series of the form (7.34), which we will break after the first non-trivial term:

$$H = E_0(B_0) + \sum_{\mathbf{q}} \hbar\, \omega(\mathbf{q})\, a_{\mathbf{q}}^{\dagger}\, a_{\mathbf{q}} \tag{7.248}$$

$$\hbar\, \omega(\mathbf{q}) = g_J\, \mu_B\, B_0 + 2S\, \hbar^2\, (J_0 - J(\mathbf{q})) \tag{7.249}$$

$a_{\mathbf{q}}$ and $a_{\mathbf{q}}^{\dagger}$ were introduced in (7.30) and (7.31) as the annihilation and creation operators, respectively, for the magnons of the wavevector $\mathbf{q}$. In (7.248), the summation is over all the wavevectors $\mathbf{q}$ of the first Brillouin zone. In this approximation, H describes a system of uncoupled harmonic oscillators. The general eigenstate of this H is a product of one-magnon states (7.244):

$$|\psi\rangle = \prod_{\mathbf{q}} \left(a_{\mathbf{q}}^{\dagger}\right)^{n_{\mathbf{q}}} |0\rangle \tag{7.250}$$

Here $|0\rangle$ is the *magnon vacuum state*. $n_{\mathbf{q}}$ is the number of magnons with the wavevector $\mathbf{q}$ and therefore is not an operator but the eigenvalue of the number operator $\widehat{n}_{\mathbf{q}} = a_{\mathbf{q}}^{\dagger}\, a_{\mathbf{q}}$. The magnon vacuum state means obviously the ferromagnetic saturation. Therefore, the state $a_{\mathbf{k}}^{\dagger}|0\rangle$ is equivalent to the one-magnon state $|\mathbf{k}\rangle$ given in (7.244).

As $a_{\mathbf{q}}$ and $a_{\mathbf{q}}^{\dagger}$ are Bose operators, we have (Problem 7.13)

$$\left[a_{\mathbf{q}}\,,\, \left(a_{\mathbf{q}}^{\dagger}\right)^{n_{\mathbf{q}}}\right]_{-} = n_{\mathbf{q}} \left(a_{\mathbf{q}}^{\dagger}\right)^{n_{\mathbf{q}}-1} \tag{7.251}$$

Using this one can easily show that

$$H|\psi\rangle = \left(E_0(B_0) + \sum_{\mathbf{q}} \hbar\, \omega(\mathbf{q})\, n_{\mathbf{q}}\right) |\psi\rangle \tag{7.252}$$

where the property

$$a_{\mathbf{q}}|0\rangle = 0 \tag{7.253}$$

has been utilized (Problem 7.13). The energy states are therefore uniquely characterized by the magnon occupation numbers $n_{\mathbf{q}}$. We now know the energy eigenvalues and the eigenstates of the Hamiltonian and hence we can calculate the grand canonical partition function Ξ_0 of the magnon gas. Since the magnon number is not constant, in equilibrium at temperature T, the magnon number will be N_T, for which the free energy F is extremal:

$$\left(\frac{\partial F}{\partial T}\right)_T \Bigg|_{N=N_T} = 0 \tag{7.254}$$

On the other hand, the left-hand side is the equation for determining the chemical potential μ; that means, μ for magnons is zero. One can calculate Ξ_0 as in (3.56):

$$\begin{aligned}\Xi_0 &= e^{-\beta E_0} \prod_{\mathbf{q}} \sum_{n_{\mathbf{q}}=0}^{\infty} e^{-\beta \hbar\omega(\mathbf{q})\, n_{\mathbf{q}}} \\ &= e^{-\beta E_0} \prod_{\mathbf{q}} \frac{1}{1 - e^{-\beta\hbar\omega(\mathbf{q})}}\end{aligned} \tag{7.255}$$

From this result, we can derive all the interesting quantities, e.g. *average occupation number*

$$\langle \widehat{n}_{\mathbf{q}} \rangle = \langle a_{\mathbf{q}}^{\dagger}\, a_{\mathbf{q}} \rangle = -\frac{1}{\beta} \frac{\partial}{\partial(\hbar\omega(\mathbf{q}))} \ln \Xi_0$$

for which (3.60) is true. From (7.255) follows the expected result that $\langle \widehat{n}_{\mathbf{q}} \rangle$ is nothing but the Bose–Einstein distribution function:

$$\langle \widehat{n}_{\mathbf{q}} \rangle = \frac{1}{exp(\beta\hbar\omega(\mathbf{q})) - 1} \equiv f_{+}(\hbar\omega(\mathbf{q})) \tag{7.256}$$

The *grand canonical potential*

$$\begin{aligned}\Omega(T, B_0) &= -k_B T \ln \Xi_0 \\ &= E_0(B_0) + k_B T \sum_{\mathbf{q}} \ln\left(1 - e^{-\beta\hbar\omega(\mathbf{q})}\right)\end{aligned} \tag{7.257}$$

gives the magnetization of the spin system:

$$\begin{aligned}M(T, B_0) &= -\frac{1}{V}\left(\frac{\partial \Omega}{\partial B_0}\right)_T \\ &= g_J\, \mu_B\, S\, \frac{N}{V} - \frac{k_B T}{V} \sum_{\mathbf{q}} \frac{\beta\, g_J\, \mu_B e^{-\beta\hbar\omega(\mathbf{q})}}{1 - e^{-\beta\hbar\omega(\mathbf{q})}} \\ &= g_J\, \mu_B\, S\, \frac{N}{V}\left(1 - \frac{1}{NS} \sum_{\mathbf{q}} \langle \widehat{n}_{\mathbf{q}} \rangle\right)\end{aligned} \tag{7.258}$$

One has saturation

$$M_0 = g_J\, \mu_B\, S\, \frac{N}{V} \tag{7.259}$$

only when no magnons are excited, i.e. when $\langle\widehat{n}_{\mathbf{q}}\rangle = 0$ for all $\mathbf{q}$. Therefore

$$\frac{M_0 - M(T, B_0)}{M_0} = \frac{1}{NS}\sum_{\mathbf{q}} \frac{1}{exp(\beta\hbar\omega(\mathbf{q})) - 1} \tag{7.260}$$

We know that this expression will be more and more valid as one goes to lower and lower temperatures. Therefore, we want to investigate the behaviour for $T \to 0$ in more detail. First we will convert, as usual, the sum into an integral

$$\begin{aligned}
&\sum_{\mathbf{q}} \left(e^{\beta\hbar\omega(\mathbf{q})} - 1\right)^{-1} \\
&= \frac{V}{(2\pi)^3}\int d^3q \sum_{n=0}^{\infty} e^{-n\beta\hbar\omega(\mathbf{q})}\, e^{-\beta\hbar\omega(\mathbf{q})} \\
&= \frac{V}{(2\pi)^3}\sum_{n=1}^{\infty} e^{-n\beta g_J\mu_B B_0}\int d^3q\, e^{-2nS\beta\hbar^2(J_0 - J(\mathbf{q}))}
\end{aligned} \tag{7.261}$$

Since for low temperatures β is very large, in (7.260), specially the small magnon energies $\hbar\omega(\mathbf{q})$ play a role, i.e. those with small wavevector $|\mathbf{q}|$. In such case, we have

$$\begin{aligned}
J_0 - J(\mathbf{q}) &= \frac{1}{N}\sum_{i,j} J_{ij}\left(1 - e^{i\mathbf{q}\cdot\mathbf{R}_{ij}}\right) \\
&= \frac{1}{2N}\sum_{i,j} J_{ij}\left(\mathbf{q}\cdot\mathbf{R}_{ij}\right)^2 \equiv \frac{D}{2S\hbar^2}\, q^2
\end{aligned} \tag{7.262}$$

The linear term disappears because $\mathbf{R}_{ij} = -\mathbf{R}_{ji}$. For small $|\mathbf{q}|$, the magnon energies depend quadratically on the wavevector (see also (7.279))

$$\hbar\,\omega(\mathbf{q}) \approx g_J\,\mu_B\,B_0 + D\,q^2 \tag{7.263}$$

We can, without a great error, apply this approximation in (7.261), since in the region where (7.263) is questionable, the integrand in (7.261) is practically zero. For the same reason, one can extend the integration over the first Brillouin zone to the entire $\mathbf{q}$-space:

$$\begin{aligned}
\sum_{\mathbf{q}}\langle n_{\mathbf{q}}\rangle &= \frac{V}{2\pi^2}\sum_{n=1}^{\infty} e^{-n\beta g_J\mu_B B_0}\int_0^{\infty} dq\, q^2\, e^{-n\beta D q^2} \\
&= V\left(\frac{k_B T}{4\pi D}\right)^{3/2}\sum_{n=1}^{\infty}\frac{e^{-n\beta g_J\mu_B B_0}}{n^{3/2}}
\end{aligned} \tag{7.264}$$

We abbreviate

$$Z_m(x) = \sum_{n=1}^{\infty} \frac{e^{-nx}}{n^m} \tag{7.265}$$

For $x = 0$, this reduces to the Riemann ζ-function (3.76)

$$Z_m(0) = \zeta(m) \tag{7.266}$$

which is available in the tables ($\zeta(3/2) \approx 2.612$). With this, we have the relative magnetization of the spin system at low temperatures as

$$\frac{M_0 - M(T, B_0)}{M_0} = \frac{V}{NS}\left(\frac{k_B T}{4\pi D}\right)^{3/2} Z_{3/2}\left(\beta\, g_J\, \mu_B\, B_0\right) \tag{7.267}$$

Here we are specially interested in the spontaneous magnetization ($B_0 = 0$) for which we have derived the famous Bloch's $T^{3/2}$ law:

$$\frac{M_0 - M(T, 0)}{M_0} = C_{3/2}\, T^{3/2} \tag{7.268}$$

$$C_{3/2} = \frac{V}{NS}\left(\frac{k_B}{4\pi D}\right)^{3/2} \zeta(3/2) \tag{7.269}$$

This result of the linear spin wave theory is uniquely confirmed by experiment. We recall that the molecular field approximation predicts an exponential approach of the magnetization to its maximum value. For low temperatures, the spin wave theory provides distinctly better results for isotropic ferromagnets.

In passing, we mention that the approximation (7.262) can be easily improved by not restricting oneself to small $|\mathbf{q}|$ but only taking into account nearest neighbour interactions:

$$J(\mathbf{q}) = \frac{1}{N}\sum_{i,j} J_{ij}\, e^{i\mathbf{q}\cdot\mathbf{R}_{ij}} = \sum_{\Delta}^{n.n} J_{0\Delta}\, e^{i\mathbf{q}\cdot\mathbf{R}_{\Delta}} = z\, J_1\, \gamma_{\mathbf{q}} \tag{7.270}$$

z is the number of nearest neighbours, J_1 is the exchange integral between the nearest neighbours and $\gamma_{\mathbf{q}}$ is a structure factor

$$\gamma_{\mathbf{q}} = \frac{1}{z}\sum_{\Delta}^{n.n} e^{i\mathbf{q}\cdot\mathbf{R}_{\Delta}} \tag{7.271}$$

$\gamma_{\mathbf{q}}$ can be easily evaluated, for example, for the three cubic lattices.

(a) *Simple cubic*

$$z = 6; \quad \mathbf{R}_\Delta = \begin{cases} a(\pm 1,\ 0,\ 0) \\ a(0,\ \pm 1,\ 0) \\ a(0,\ 0,\ \pm 1) \end{cases} \tag{7.272}$$

a is the lattice constant.

$$\gamma_{\mathbf{q}}^{sc} = \frac{1}{3}(\cos(q_x a) + \cos(q_y a) + \cos(q_z a)) \tag{7.273}$$

(b) *Body centred cubic*

$$z = 8; \quad \mathbf{R}_\Delta = \frac{a}{2}\,(\pm 1,\ \pm 1,\ \pm 1) \tag{7.274}$$

$$\gamma_{\mathbf{q}}^{bcc} = \cos(\frac{1}{2}q_x a)\cos(\frac{1}{2}q_y a)\cos(\frac{1}{2}q_z a) \tag{7.275}$$

(c) *Face centred cubic*

$$z = 12; \quad \mathbf{R}_\Delta = \begin{cases} \frac{a}{2}(\pm 1,\ \pm 1,\ 0) \\ \frac{a}{2}(\pm 1,\ 0,\ \pm 1) \\ \frac{a}{2}(0,\ \pm 1,\ \pm 1) \end{cases} \tag{7.276}$$

$$\begin{aligned} \gamma_{\mathbf{q}}^{fcc} = \frac{1}{3}\ &[\cos(\frac{1}{2}q_x\ a)\ \cos(\frac{1}{2}q_y\ a) + \cos(\frac{1}{2}q_x\ a)\ \cos(\frac{1}{2}q_z\ a) \\ &+ \cos(\frac{1}{2}q_y\ a)\ \cos(\frac{1}{2}q_z\ a)] \end{aligned} \tag{7.277}$$

For small wavevectors, all the three cubic lattices hold:

$$\gamma_{\mathbf{q}} \approx 1 - \frac{1}{z}\ a^2\ q^2 \tag{7.278}$$

so that

$$\hbar\,\omega(\mathbf{q}) = g_J\ \mu_B\ B_0 + (2S\ J_1\ \hbar^2\ a^2)q^2 \tag{7.279}$$

As a consequence, the constant D introduced in (7.262) for a cubic lattice becomes

$$D = 2S\ J_1\ \hbar^2\ a^2 \tag{7.280}$$

If we calculate the relative magnetization with either (7.273) or (7.275) or (7.277) instead of with (7.262) then we get corrections to the $T^{3/2}$ term which are proportional to $T^{5/2}$, $T^{7/2}$, $\cdots$ [10]:

$$\frac{M_0 - M_s(T)}{M_0} = \frac{1}{S}\Big\{\zeta(3/2)\,t^{3/2} + \frac{3\pi}{4}\delta\,\zeta(5/2)\,t^{5/2} + \pi^2\,\delta^2\,\alpha\,\zeta(7/2)\,t^{7/2} + \cdots\Big\} \tag{7.281}$$

where t is a renormalized temperature

$$t = \frac{3k_B T}{4\pi\ S\ J_1\ \hbar^2\ z\ \delta} \tag{7.282}$$

and α and δ are structure factors.

$$\begin{aligned} \alpha_{sc} &= 33/22\ ;\ \alpha_{bcc} = 281/288\ ;\ \alpha_{fcc} = 15/16 \\ \delta_{sc} &= 1 \qquad\ ;\ \delta_{bcc} = 3\cdot 2^{-4/3}\ \ ;\ \delta_{fcc} = 2^{1/3} \end{aligned} \tag{7.283}$$

The magnetic (internal) energy of the spin wave system is calculated in a completely analogous fashion as the magnetization. With the "isotropic approximation" (7.263) and $B_0 = 0$, we first get

$$\begin{aligned} \sum_{\mathbf{q}} \hbar\omega(\mathbf{q})\ \langle n_{\mathbf{q}}\rangle &= D\sum_{\mathbf{q}} q^2\ \langle n_{\mathbf{q}}\rangle \\ &= \frac{V}{2\pi^2}\ D\sum_{n=1}^{\infty}\int_0^{\infty} dq\ q^4\ e^{-n\beta D q^2} \\ &= \frac{3V\ D}{16\pi^{3/2}}\left(\frac{k_B T}{D}\right)^{5/2}\sum_{n=1}^{\infty}\frac{1}{n^{5/2}} \end{aligned} \tag{7.284}$$

where we have exploited (7.264). This leads to the following energy expression in the region of validity of the spin wave theory:

$$U_{SW} = E_0(B_0 = 0) + \frac{3V\ \zeta(5/2)}{2(4\pi\,D)^{3/2}}\ (k_B T)^{5/2} \tag{7.285}$$

Thus the magnetic part of the specific heat also obeys a $T^{3/2}$ law,

$$C_{B_0=0}(T) = \left(\frac{\partial U_{SW}}{\partial T}\right)_{B_0=0} = \frac{15}{4}k_B\ V\ \zeta(5/2)\left(\frac{k_B T}{4\pi\,D}\right)^{3/2} \tag{7.286}$$

which is also confirmed experimentally. The improved version of the specific heat analogously to (7.281) reads as

$$C_{B_0=0}(T) = k_B\, N \left\{ \frac{15}{4} \zeta(5/2)\, t^{3/2} + \frac{105\pi}{16}\ \delta\ \zeta(7/2)\, t^{5/2} + \right.$$
$$\left. + \frac{63}{4} \pi^2\, \delta^2\, \alpha\, \zeta(9/2)\, t^{7/2} + \cdots \right\} \tag{7.287}$$

The experimental proof of the quasiparticle magnon is available from neutron scattering. Neutrons exchange energy and momentum with the phonons and magnons in the solid. The corresponding measurements provide, via the energy–momentum laws, the dispersion curves of the spin waves, i.e. the **q**-dependence of the spin wave energies. Dyson [10], by a far-reaching mathematical investigation of the Heisenberg model, confirmed the spin wave theory. He finds that the interaction among the spin waves produces correction terms, the largest one being $\sim T^4$. Therefore, the spin wave result (7.281) which includes the $T^{7/2}$ term is exact even when the interactions are taken into account. The basis for this somewhat surprising fact can be traced to the interaction term (7.47) in the Dyson–Maléev transformation, which is made up of two summands whose effects compensate each other to a large extent.

7.4.2 "Renormalized" Spin Waves

In this section, we want to discuss in a simple form the effect of interaction among spin waves (*non-linear spin wave theory*). The starting point is the Dyson–Maleev transformation (7.42), (7.43) and (7.44) which was introduced in Sect. 7.1.1. Through this transformation, the Heisenberg Hamiltonian takes the form (7.45)

$$H = E_0(B_0) + H_2 + H_4 \tag{7.288}$$

$E_0(B_0)$ is the ground state energy (7.246):

$$H_2 = \sum_{i,j} (g_J\, \mu_B\, B_0\, \delta_{ij} + 2S\, \hbar^2\, (J_0\, \delta_{ij} - J_{ij}))\, \alpha_i\, \alpha_j^\dagger \tag{7.289}$$

$$H_4 = \hbar^2 \sum_{i,j} J_{ij} (\alpha_i \alpha_j^\dagger\, \widehat{n}_j - \widehat{n}_i\, \widehat{n}_j) \tag{7.290}$$

The advantage of the Dyson–Maleev over the Holstein–Primakoff transformation lies in the fact that, in this case, the Hamiltonian consists of a finite number of terms whereas in the other case there are an infinite number of terms. The disadvantage is of course obvious from (7.42) and (7.43): H is not Hermitean any more. One can, however, show [10] that the non-Hermitean operator (7.44) contains all the eigenvalues of the original Heisenberg Hamiltonian plus an infinite number of "unphysical" eigenvalues. In the Dyson's method the unphysical eigenstates are "pushed up" so that at not too high a temperature, they do not play any role in the partition function (the contribution of a state is $\sim exp(-\beta E_n)$). The details of this method cannot be presented here but the reader is referred to the original work of Dyson [10]. We restrict ourselves here to a very much simplified treatment of the spin wave interaction. With

$$\begin{aligned} \alpha_{\mathbf{q}} &= \tfrac{1}{\sqrt{N}} \sum_i e^{-i\mathbf{q}\cdot\mathbf{R}_i}\, \alpha_i \\ \alpha_{\mathbf{q}}^{\dagger} &= \tfrac{1}{\sqrt{N}} \sum_i e^{i\mathbf{q}\cdot\mathbf{R}_i}\, \alpha_i^{\dagger} \end{aligned} \tag{7.291}$$

we will first transform the Hamiltonian into the wavevector representation. H_2 is simple:

$$H_2 = \sum_{\mathbf{q}} \hbar\, \omega(\mathbf{q})\, \alpha_{\mathbf{q}}^{\dagger}\, \alpha_{\mathbf{q}} \tag{7.292}$$

This represents the free magnon gas which was discussed in the last section. In addition we now have the interaction H_4 (Problem 7.11),

$$H_4 = \frac{\hbar^2}{N} \sum_{\mathbf{q}_1,\cdots\mathbf{q}_4} (J(\mathbf{q}_4) - J(\mathbf{q}_1 - \mathbf{q}_3))\, \delta_{\mathbf{q}_1+\mathbf{q}_2,\mathbf{q}_3+\mathbf{q}_4}\, \alpha_{\mathbf{q}_1}^{\dagger}\, \alpha_{\mathbf{q}_2}^{\dagger}\, \alpha_{\mathbf{q}_3}\, \alpha_{\mathbf{q}_4} \tag{7.293}$$

which naturally cannot be handled exactly. We will try to "diagonalize", in view of the kronecker delta, by taking only such terms for which

$$\mathbf{q}_1 = \mathbf{q}_3; \quad \mathbf{q}_2 = \mathbf{q}_4$$

or

$$\mathbf{q}_1 = \mathbf{q}_4; \quad \mathbf{q}_2 = \mathbf{q}_3$$

holds (*random phase approximation (RPA)*):

$$\begin{aligned} \tilde{H} &= \\ &= E_0(B_0) + \sum_{\mathbf{q}} \left[2S\hbar^2 (J_0 - J(\mathbf{q})) \left(1 + \frac{1}{SN}\right) + g_J \mu_B B_0 \right] \widehat{n}_{\mathbf{q}} \\ &+ \frac{\hbar^2}{N} \sum_{\mathbf{q}_1,\mathbf{q}_2} (J(\mathbf{q}_1) + J(\mathbf{q}_2) - J(\mathbf{q}_1 - \mathbf{q}_2) - J(0)) \widehat{n}_{\mathbf{q}_1} \widehat{n}_{\mathbf{q}_2} \end{aligned} \tag{7.294}$$

The approximation which leads to $\tilde{H}$ appears completely arbitrary. One can, however, show that in the space of one- and two-magnon states, $\tilde{H}$ is exact. We have treated the one-magnon states in Sect. 7.2.2. The problem of two-magnon states can also be exactly solved [11, 12].

From (7.294) one can see that the magnon occupation operator $\widehat{n}_{\mathbf{q}} = \alpha_{\mathbf{q}}^{\dagger}\, \alpha_{\mathbf{q}}$ commutes with $\tilde{H}$,

$$\left[\tilde{H}\,,\,\widehat{n}_{\mathbf{q}}\right]_- = 0 \tag{7.295}$$

so that the eigenstates of $\tilde{H}$ can be classified according to the occupation numbers $n_{\mathbf{q}}$:

$$|n_{\mathbf{q}_1}\,,\; n_{\mathbf{q}_2}\,,\; \ldots\,,\; n_{\mathbf{q}_N}\rangle$$

Let a state be given with a particular fixed magnon distribution. Then we ask ourselves what is the energy required to add an additional magnon to the system. That is relatively easy to find out. Let

$$\begin{aligned}\tilde{H}|\cdots n_{\mathbf{q}}\cdots\rangle &= E(n_{\mathbf{q}})|\cdots n_{\mathbf{q}}\cdots\rangle\\ \tilde{H}|\cdots n_{\mathbf{q}}+1\cdots\rangle &= E(n_{\mathbf{q}}+1)|\cdots n_{\mathbf{q}}+1\cdots\rangle\end{aligned} \tag{7.296}$$

Then, we can define the *renormalized spin wave energy*, renormalized by the "presence" of other magnons:

$$\hbar\tilde{\omega}(\mathbf{q}) = E(n_{\mathbf{q}}+1) - E(n_{\mathbf{q}}) \tag{7.297}$$

With

$$\left[\widehat{n}_{\mathbf{q}_1}\,,\,\alpha_{\mathbf{q}}^{\dagger}\right]_- = \delta_{\mathbf{q},\mathbf{q}_1}\,\alpha_{\mathbf{q}}^{\dagger} \tag{7.298}$$

and

$$\begin{aligned}\left[\widehat{n}_{\mathbf{q}_1}\widehat{n}_{\mathbf{q}_2}\,,\,\alpha_{\mathbf{q}}^{\dagger}\right]_- &= \delta_{\mathbf{q}\mathbf{q}_2}\widehat{n}_{\mathbf{q}_1}\alpha_{\mathbf{q}}^{\dagger} + \delta_{\mathbf{q}\mathbf{q}_1}\alpha_{\mathbf{q}}^{\dagger}\,\widehat{n}_{\mathbf{q}_2}\\ &= \delta_{\mathbf{q}\mathbf{q}_2}\alpha_{\mathbf{q}}^{\dagger}\,\widehat{n}_{\mathbf{q}_1} + \delta_{\mathbf{q}\mathbf{q}_1}\alpha_{\mathbf{q}}^{\dagger}\,\widehat{n}_{\mathbf{q}_2} + \delta_{\mathbf{q}\mathbf{q}_2}\delta_{\mathbf{q}\mathbf{q}_1}\,\alpha_{\mathbf{q}}^{\dagger}\end{aligned} \tag{7.299}$$

from (7.294) it follows that

$$\begin{aligned}\left[\tilde{H}\,,\,\alpha_{\mathbf{q}}^{\dagger}\right]_- &= \hbar\,\omega(\mathbf{q})\,\alpha_{\mathbf{q}}^{\dagger}\\ &\quad+\frac{2\hbar^2}{N}\sum_{\mathbf{q}_1}(J(\mathbf{q})+J(\mathbf{q}_1)-J(\mathbf{q}-\mathbf{q}_1)-J(0))\,\alpha_{\mathbf{q}}^{\dagger}\,\widehat{n}_{\mathbf{q}_1}\end{aligned} \tag{7.300}$$

We can now directly show that $\alpha_{\mathbf{q}}^{\dagger}|\cdots n_{\mathbf{q}}\cdots\rangle$ is an eigenstate for the magnon number $n_{\mathbf{q}}+1$. With (7.298) it holds that

$$\begin{aligned}\widehat{n}_{\mathbf{q}}\left(\alpha_{\mathbf{q}}^{\dagger}|\cdots n_{\mathbf{q}}\cdots\rangle\right) &= \left(\alpha_{\mathbf{q}}^{\dagger}\,\widehat{n}_{\mathbf{q}}+\alpha_{\mathbf{q}}^{\dagger}\right)|\cdots n_{\mathbf{q}}\cdots\rangle\\ &= \left(n_{\mathbf{q}}+1\right)\left(\alpha_{\mathbf{q}}^{\dagger}|\cdots n_{\mathbf{q}}\cdots\rangle\right)\end{aligned}$$

That means

$$\alpha_{\mathbf{q}}^{\dagger} | \cdots n_{\mathbf{q}} \cdots \rangle \sim | \cdots n_{\mathbf{q}} + 1 \cdots \rangle \tag{7.301}$$

The operation of $\tilde{H}$ on this state gives the eigenvalue $E(n_{\mathbf{q}} + 1)$:

$$\tilde{H} \left(\alpha_{\mathbf{q}}^{\dagger} | \cdots n_{\mathbf{q}} \cdots \rangle \right) = E(n_{\mathbf{q}} + 1) \left(\alpha_{\mathbf{q}}^{\dagger} | \cdots n_{\mathbf{q}} \cdots \rangle \right) \tag{7.302}$$

On the other hand, we have from (7.300)

$$\begin{aligned} \tilde{H} \left(\alpha_{\mathbf{q}}^{\dagger} | \cdots n_{\mathbf{q}} \cdots \rangle \right) &= \alpha_{\mathbf{q}}^{\dagger} \, \tilde{H} | \cdots n_{\mathbf{q}} \cdots \rangle + \left[\tilde{H} \, , \, \alpha_{\mathbf{q}}^{\dagger} \right]_{-} | \cdots n_{\mathbf{q}} \cdots \rangle \\ &= \left\{ E(n_{\mathbf{q}}) + \hbar \, \omega(\mathbf{q}) + \right. \\ &+ \frac{2\hbar^2}{N} \sum_{\mathbf{q}_1} (J(\mathbf{q}) + J(\mathbf{q}_1) - J(\mathbf{q} - \mathbf{q}_1) \\ &- \left. J(0)) n_{\mathbf{q}_1} \right\} \left(\alpha_{\mathbf{q}}^{\dagger} | \cdots n_{\mathbf{q}} \cdots \rangle \right) \end{aligned} \tag{7.303}$$

Equations (7.302) and (7.303) give via (7.297) the *renormalized spin wave energies*:

$$\hbar \, \tilde{\omega}(\mathbf{q}) = \hbar \, \omega(\mathbf{q}) + \frac{2\hbar^2}{N} \sum_{\mathbf{q}_1} (J(\mathbf{q}) + J(\mathbf{q}_1) - J(\mathbf{q} - \mathbf{q}_1) - J(0)) \, n_{\mathbf{q}_1} \tag{7.304}$$

The sum therefore represents the correction to the linear spin wave theory due to the magnon interactions. The sum contains the occupation numbers $n_{\mathbf{q}_1}$ of the magnons which can in principle be an arbitrarily large integer number which can apparently lead to unphysical states. The $\{n_{\mathbf{q}}\}$ are naturally uniquely fixed by the state $| \cdots n_{\mathbf{q}} \cdots \rangle$. We are interested in the spin wave energies which appear in the thermodynamic equilibrium. Therefore, it is reasonable to replace $n_{\mathbf{q}_1}$ by its thermodynamic expectation value which we assume is given by Bose-distribution function.

$$n_{\mathbf{q}_1} \to \langle \widehat{n}_{\mathbf{q}_1} \rangle = \{ exp(\beta \, \hbar \, \tilde{\omega}(\mathbf{q}_1)) - 1 \}^{-1} \tag{7.305}$$

This leads to an implicit equation for the "renormalized" spin wave energies:

$$\hbar \, \tilde{\omega}(\mathbf{q}) = \hbar \, \omega(\mathbf{q}) + \frac{2\hbar^2}{N} \sum_{\mathbf{q}_1} \frac{J(\mathbf{q}) + J(\mathbf{q}_1) - J(\mathbf{q} - \mathbf{q}_1) - J(0)}{exp(\beta \, \hbar \, \tilde{\omega}(\mathbf{q}_1)) - 1} \tag{7.306}$$

We want to evaluate this expression wherein, as in (7.270), we will restrict ourselves to the nearest neighbour interactions:

$$J(\mathbf{q}) = z \, J_1 \, \gamma_{\mathbf{q}}; \quad J_0 = z \, J_1 \tag{7.307}$$

By exploiting the translational symmetry of the lattice one can demonstrate that (Problem 7.17)

$$\sum_{\mathbf{q}_1} \gamma_{\mathbf{q}-\mathbf{q}_1} \langle \widehat{n}_{\mathbf{q}_1} \rangle \approx \gamma_{\mathbf{q}} \sum_{\mathbf{q}_1} \gamma_{\mathbf{q}_1} \langle \widehat{n}_{\mathbf{q}_1} \rangle \tag{7.308}$$

With this (7.306) simplifies to ($B_0 = 0$)

$$\begin{aligned} \hbar\, \tilde{\omega}(\mathbf{q}) &= 2S\, \hbar^2\, z\, J_1(1 - \gamma_{\mathbf{q}}) - \\ &- \frac{2\hbar^2}{N}\, z\, J_1(1 - \gamma_{\mathbf{q}}) \sum_{\mathbf{q}_1} (1 - \gamma_{\mathbf{q}_1}) \langle \widehat{n}_{\mathbf{q}_1} \rangle \end{aligned}$$

We define

$$A_s(T) = \frac{1}{NS} \sum_{\mathbf{q}_1} (1 - \gamma_{\mathbf{q}_1}) \langle \widehat{n}_{\mathbf{q}_1} \rangle \tag{7.309}$$

and obtain an expression for the "renormalized" spin waves which is formally identical to the one for free spin waves:

$$\hbar\, \tilde{\omega}(\mathbf{q}) = 2S\, \hbar^2\, J_0^*(T)(1 - \gamma_{\mathbf{q}}) \tag{7.310}$$

We only have to replace the exchange constant J_0 by a temperature-dependent quantity

$$J_0^*(T) = J_0(1 - A_s(T)) \tag{7.311}$$

The summation over the wavevectors of the first Brillouin zone needed in calculating $A_s(T)$ in (7.309) can be evaluated exactly for cubic lattices. We will, however, give only an estimate of $A_s(T)$ which is correct in the limit of small $|\mathbf{q}|$:

$$\hbar\, \tilde{\omega}(\mathbf{q}) \approx D(T)\, q^2 \tag{7.312}$$

$$D(T) = D(1 - A_s(T)) \tag{7.313}$$

$$D = 2S\, J_1\, \hbar^2\, a^2 \tag{7.314}$$

Then correspondingly according to (7.278)

$$1 - \gamma_{\mathbf{q}} \approx \frac{1}{z}\, a^2\, q^2 \tag{7.315}$$

holds. We substitute this in (7.309):

$$A_s(T) \approx \frac{a^2}{zNS} \sum_{\mathbf{q}} \frac{q^2}{exp(\beta D(T) q^2) - 1}$$

$$= \frac{a^2}{zNS} \frac{V}{(2\pi)^3} \int_{1st\ BZ} d^3q\, q^2 \sum_{n=1}^{\infty} e^{-n\beta D(T) q^2} \tag{7.316}$$

Because of the exponential function in the integrand, we can extend the $\mathbf{q}$-integration over the entire $\mathbf{q}$-space. This assumption is especially reasonable, because of β in the exponent, in the low-temperature region.

$$A_s(T) \approx \frac{a^2\, V}{2\pi^2\, zNS} \sum_{n=1}^{\infty} \int_0^{\infty} dq\, q^4\, e^{-n\beta D(T) q^2}$$

$$= \frac{a^2\, V}{2\pi^2\, zNS} \sum_{n=1}^{\infty} \frac{3}{8} \sqrt{\pi} \left(\frac{k_B T}{n D(T)} \right)^{5/2}$$

$$= 3 \frac{a^2\, V}{16\pi^{3/2}\, zNS} \left(\frac{k_B T}{D(T)} \right)^{5/2} \zeta(5/2) \tag{7.317}$$

This equation along with (7.313) fixes $A_s(T)$. Let us write

$$\eta = 3 \frac{a^2 V}{16\pi^{3/2} zNS} \left(\frac{k_B}{D} \right)^{5/2} \zeta(5/2)$$

to get as an intermediate result

$$A_s(T) \approx \eta \left(\frac{T}{1 - A_s(T)} \right)^{5/2}$$

For low temperatures $A_s(T)$ will be only a small correction term:

$$A_s(T) \approx \eta\, T^{5/2} \left(1 + \frac{5}{2} A_s(T) + \cdots \right)$$

$$\approx \eta\, T^{5/2} \left(1 + \frac{5}{2} \eta\, T^{5/2} + \cdots \right) = \eta\, T^{5/2} + \mathcal{O}(T^5)$$

This means for the renormalized exchange constant

$$J_0^*(T) = J_0 \left(1 - \eta\, T^{5/2} \right) \tag{7.318}$$

If this expression is used in the Bloch's $T^{3/2}$ law of the linear spin wave theory (7.268), then we obtain a T^4 term as a first correction in complete agreement with the exact theory. This can be seen as follows: The constant $C_{3/2}$ (7.269) as the pre-factor of the $T^{3/2}$ term has to be renormalized:

$$
\begin{aligned}
C_{3/2} \rightarrow C_{3/2}^{*} &= \frac{V}{NS}\left(\frac{k_B}{4\pi\, D(T)}\right)^{3/2} \zeta(3/2) \\
&\approx C_{3/2}\left(1-\eta\, T^{5/2}\right)^{-3/2} \\
&\approx C_{3/2}\left(1+\frac{3}{2}\eta\, T^{5/2}+\cdots\right) \\
&\approx C_{3/2}+C_4\, T^{5/2}+\cdots
\end{aligned}
$$

This yields for the magnetization, in complete agreement with Dyson's result,

$$
\begin{aligned}
\frac{M_0 - M_s(T)}{M_0} &= C_{3/2}T^{3/2} + C_{5/2}T^{5/2} + C_{7/2}T^{7/2} \\
&\quad + C_4 T^4 + C_{9/2}T^{9/2} + \cdots
\end{aligned}
$$

It was shown that using the renormalized spin wave theory presented here, the magnetization curves of classical ferromagnets like EuO and EuS can be almost quantitatively reproduced up to in the neighbourhood of T_C [13, 14]. In the case of europium chalcogenides, one should not restrict oneself to the nearest neighbour exchange integrals but should also consider the next nearest neighbours. Instead of (7.310) we have to use

$$
\begin{aligned}
\hbar\, \tilde{\omega}(\mathbf{q}) &= 2S\, \hbar^2\, z_1\, J_1(1 - A_1(T))(1 - \gamma_{\mathbf{q}}^{(1)}) \\
&\quad + 2S\, \hbar^2\, z_2\, J_2(1 - A_2(T))(1 - \gamma_{\mathbf{q}}^{(2)})
\end{aligned} \tag{7.319}
$$

z_1 and z_2 are the number of nearest and next nearest neighbours, respectively, and J_1 and J_2 the corresponding exchange integrals.

$$
\gamma_{\mathbf{q}}^{(i)} = \frac{1}{z_i} \sum_{\Delta_i} e^{i\mathbf{q}\cdot\mathbf{R}_{\Delta_i}} \quad (i = 1,\ 2) \tag{7.320}
$$

The sum runs over the nearest (Δ_1) or the next nearest (Δ_2) neighbours, respectively, of the lattice site under consideration.

$$
A_i(T) = \frac{1}{NS} \sum_{\mathbf{q}} \frac{1 - \gamma_{\mathbf{q}}^{(i)}}{exp(\beta\hbar\tilde{\omega}(\mathbf{q})) - 1} \quad (i = 1,\ 2) \tag{7.321}
$$

The integration over the first Brillouin zone (fcc in the case of EuO and EuS) can be exactly performed. In doing this, for EuO and EuS one has to use the following parameters (neutron scattering experiments):

$S = 7/2$

EuO: $\hbar^2 J_1/k_B = 0.625$ K EuS: $\hbar^2 J_1/k_B = 0.221$ K

$\hbar^2 J_2/k_B = 0.125$ K $\hbar^2 J_2/k_B = -0.100$ K

The agreement between the experimentally determined magnetization curves and the calculated ones using the renormalized spin wave theory and the parameters given above is practically exact in the temperature range of 0 to $0.7T_C$.

One can use either the spin wave result (7.260) or the corresponding result of the renormalized spin wave theory

$$\frac{M_0 - M_s(T)}{M_0} = \frac{1}{NS} \sum_{\mathbf{q}} \frac{1}{e^{\beta \hbar \tilde{\omega}(\mathbf{q})} - 1}$$

in order to make a rough estimate of the Curie temperature. For $T = T_C$, by definition, the magnetization is zero. If one expands the exponential function in the denominator up to the linear term, then one is left with

$$k_B\, T_C = \left\{ \frac{1}{NS} \sum_{\mathbf{q}} \frac{1}{\hbar\, \tilde{\omega}(\mathbf{q})} \right\}^{-1} \tag{7.322}$$

The sum can be explicitly evaluated for simple lattices. If one restricts the exchange interaction among the nearest neighbours, one can with (7.273), (7.275) and (7.277) obtain the following values:

$$k_B\, T_C = 2S^2\, \hbar^2\, J_0^*(T_C)\, Q \tag{7.323}$$

where the structure factor Q is given by

$$Q \equiv \left(\frac{1}{N} \sum_{\mathbf{q}} \frac{1}{1 - \gamma_{\mathbf{q}}} \right)^{-1} = \begin{cases} 0.660 : \text{sc} \\ 0.718 : \text{bcc} \\ 0.744 : \text{fcc} \end{cases}$$

For estimations this is quite a useful formula. Nevertheless, one has to be cautious about using this result since it follows from the renormalized spin wave theory, which conceptually is naturally a low-temperature approximation, and therefore the T_C obtained is questionable.

7.4.3 Harmonic Approximation for Antiferromagnets

The spin wave approximation has proved itself to be extraordinarily successful in the case of ferromagnets. The same thing cannot be expected for the cases of antiferromagnets and ferrimagnets without certain limitations. In fact, it is not even clear from the beginning whether the idea of spin waves can be taken over to the case of antiferromagnets. In contrast to the ferromagnets, the exact ground state of an antiferromagnet is known only for the one-dimensional lattice (with $S = 1/2$ or $S \to \infty$). If we accept the sub-lattice model (see Sect. 7.3.2), which has been supported by neutron scattering experiments, then the expected ground state (Neel state) is the state in which all the sub-lattices are ferromagnetically saturated. We will, however, be able to show in this section that a completely ordered state cannot be the correct ground state. With an in principle unknown ground state, one cannot talk of elementary excitations without putting some question marks on them. In spite of this, the spin wave theory, in the case of antiferromagnets also, has proved to be a useful approximate analysis. Numerical estimates, concerning the accuracy of the approximation, for a series of reference systems have established the theory to be quite reliable.

Here we want to discuss the simplest form of antiferromagnets, i.e. the ABAB structure (see Sect. 7.3.2) wherein every atom of a sub-lattice has only the atoms of the other sub-lattice as its nearest neighbours. Further, we will restrict ourselves to exchange interaction only between the nearest neighbours. Then we naturally have

$$J_1 < 0 \tag{7.324}$$

Let the external field $\mathbf{B}_0$ be along the "easy" direction, which we also take to be the z-axis. The anisotropy, which defines the "easy" direction is taken into account through an anisotropy field $\mathbf{B}_A$ (7.61):

$$\begin{aligned} H = &-\sum_{i,j}^{n.n} J_{ij}\mathbf{S}_i \cdot \mathbf{S}_j - g_J \frac{\mu_B}{\hbar}(B_0 + B_A)\sum_i^A S_i^z \\ &- g_J \frac{\mu_B}{\hbar}(B_0 - B_A)\sum_i^B S_i^z \end{aligned} \tag{7.325}$$

We will use the Holstein–Primakoff transformation (7.23), (7.24) and (7.25) of the spin operators in the *harmonic approximation*:

Sub-lattice A:

$$\frac{1}{\hbar}\,S_i^+ = \sqrt{2S}\,a_i \;\; ; \;\; \frac{1}{\hbar}\,S_i^- = \sqrt{2S}\,a_i^\dagger \;\; ; \;\; \frac{1}{\hbar}\,S_i^z = S - a_i^\dagger\,a_i \tag{7.326}$$

Sub-lattice B:

$$\frac{1}{\hbar}\,S_i^+ = \sqrt{2S}\,b_j^\dagger \;\; ; \;\; \frac{1}{\hbar}\,S_j^- = \sqrt{2S}\,b_j \;\; ; \;\; \frac{1}{\hbar}\,S_j^z = -S + b_j^\dagger\,b_j \tag{7.327}$$

All a-operators commute with all b-operators. For the sake of convenience, b_j and $b_j^\dagger$ are used for the sub-lattice B in place of the usual definition (7.24) and (7.25).

We will introduce certain abbreviations:

$$b_A = -2z\, J_1\, \hbar^2\, S + g_J\, \mu_B (B_0 + B_A) \tag{7.328}$$

$$b_B = -2z\, J_1\, \hbar^2\, S - g_J\, \mu_B (B_0 - B_A) \tag{7.329}$$

$$E_a(B_0) = -N\, g_J\, \mu_B\, S\, B_A + N\, z\, J_1\, \hbar^2\, S^2 = E_a \tag{7.330}$$

E_a is the energy of the totally ordered Neel state. The *harmonic approximation* for the operator (7.325) consists of neglecting the terms which are not bilinear in the magnon operators. By substituting (7.326) and (7.327) in (7.325) one obtains, retaining bilinear terms only

$$\begin{aligned} H = {} & E_a + b_A \sum_i^A a_i^\dagger\, a_i + b_B \sum_j^B b_j^\dagger\, b_j \\ & - 2S\hbar^2 \sum_i^A \sum_j^B J_{ij} (a_i\, b_j + a_i^\dagger\, b_j^\dagger) \end{aligned} \tag{7.331}$$

We now transform again into the wavevectors $\mathbf{q}$ which are from the first Brillouin zone of each sub-lattice. Since the two sub-lattices are identical, so are the Brillouin zones:

$$a_i = \sqrt{\frac{2}{N}} \sum_{\mathbf{q}} e^{i\mathbf{q}\cdot\mathbf{R}_i}\, a_{\mathbf{q}} \tag{7.332}$$

$$b_j = \sqrt{\frac{2}{N}} \sum_{\mathbf{q}} e^{-i\mathbf{q}\cdot\mathbf{R}_j}\, b_{\mathbf{q}} \tag{7.333}$$

$$J(\mathbf{q}) = \frac{2}{N} \sum_i^A \sum_j^B J_{ij} e^{-i\mathbf{q}\cdot(\mathbf{R}_i - \mathbf{R}_j)}$$

$$\delta_{\mathbf{q},\mathbf{q}'} = \frac{2}{N} \sum_i^A e^{i(\mathbf{q}-\mathbf{q}')\cdot\mathbf{R}_i} = \frac{2}{N} \sum_j^B e^{i(\mathbf{q}-\mathbf{q}')\cdot\mathbf{R}_j} \tag{7.334}$$

In transforming (7.325), however, we will have to remove the restriction of limiting to nearest neighbours. Then one obtains

$$H = E_a + b_A \sum_{\mathbf{q}} a_{\mathbf{q}}^\dagger\, a_{\mathbf{q}} + b_B \sum_{\mathbf{q}} b_{\mathbf{q}}^\dagger\, b_{\mathbf{q}} + \sum_{\mathbf{q}} c(\mathbf{q}) \{a_{\mathbf{q}}\, b_{\mathbf{q}} + a_{\mathbf{q}}^\dagger\, b_{\mathbf{q}}^\dagger\} \tag{7.335}$$

where

$$c(\mathbf{q}) = -2J(\mathbf{q})\,\hbar^2\,S \;\rightarrow\; -2z\,J_1\,\hbar^2\,S\,\gamma_{\mathbf{q}} \tag{7.336}$$

with $\gamma_{\mathbf{q}}$ given by (7.271). The situation is not quite as simple as in the case of ferromagnets since H is not diagonal in a's and b's. Therefore, as a next step, we look for a unitary transformation that diagonalizes H:

$$H = \tilde{E}_a(B_0) + \sum_{\mathbf{q}} \hbar\,\omega_\alpha(\mathbf{q})\,\alpha^\dagger_{\mathbf{q}}\,\alpha_{\mathbf{q}} + \sum_{\mathbf{q}} \hbar\,\omega_\beta(\mathbf{q})\,\beta^\dagger_{\mathbf{q}}\,\beta_{\mathbf{q}} \tag{7.337}$$

$\tilde{E}_a$ is the "true" ground state energy if $\omega_{\alpha,\beta}(\mathbf{q}) \geq 0$. For the "new" operators, we make the ansatz

$$\begin{aligned}\alpha_{\mathbf{q}} &= c_1\,a_{\mathbf{q}} + c_2\,b^\dagger_{\mathbf{q}}\\ \beta_{\mathbf{q}} &= d_1\,a^\dagger_{\mathbf{q}} + d_2\,b_{\mathbf{q}}\end{aligned} \tag{7.338}$$

and demand that the α's and β's fulfil the usual Bose commutation relations:

$$\left[\alpha_{\mathbf{q}}\,,\,\alpha^\dagger_{\mathbf{q}'}\right]_- = \delta_{\mathbf{q}\mathbf{q}'} \;\Leftrightarrow\; |c_1|^2 - |c_2|^2 = 1 \tag{7.339}$$

$$\left[\beta_{\mathbf{q}}\,,\,\beta^\dagger_{\mathbf{q}'}\right]_- = \delta_{\mathbf{q}\mathbf{q}'} \;\Leftrightarrow\; -|d_1|^2 + |d_2|^2 = 1 \tag{7.340}$$

$$\left[\alpha_{\mathbf{q}}\,,\,\beta_{\mathbf{q}'}\right]_- = 0 \;\Leftrightarrow\; c_1\,d_1 = c_2\,d_2 \tag{7.341}$$

With these commutation relations, one can easily see that

$$\left[\alpha_{\mathbf{q}}\,,\,H\right]_- = \hbar\,\omega_\alpha(\mathbf{q})\,\alpha_{\mathbf{q}} = \hbar\,\omega_\alpha(\mathbf{q})(c_1\,a_{\mathbf{q}} + c_2\,b^\dagger_{\mathbf{q}}) \tag{7.342}$$

According to (7.335) it also holds that

$$\begin{aligned}\left[\alpha_{\mathbf{q}}\,,\,H\right]_- &= c_1\,\left[a_{\mathbf{q}}\,,\,H\right]_- + c_2\,\left[b^\dagger_{\mathbf{q}}\,,\,H\right]_-\\ &= c_1\{b_A\,a_{\mathbf{q}} + c(\mathbf{q})\,b^\dagger_{\mathbf{q}}\}\\ &\quad + c_2\{-b_B\,b^\dagger_{\mathbf{q}} - c(\mathbf{q})\,a_{\mathbf{q}}\}\end{aligned} \tag{7.343}$$

By equating the last two equations, we get

$$\begin{aligned}&a_{\mathbf{q}}\,[c_1\,\hbar\,\omega_\alpha(\mathbf{q}) - c_1\,b_A + c_2\,c(\mathbf{q})]\\ &+ b^\dagger_{\mathbf{q}}\,[c_2\,\hbar\,\omega_\alpha(\mathbf{q}) + c_2\,b_B - c_1\,c(\mathbf{q})] = 0\end{aligned} \tag{7.344}$$

The operators $a_{\mathbf{q}}$ and $b^\dagger_{\mathbf{q}}$ act in different spaces. Therefore, each of the coefficients by itself should be equal to zero. We then obtain the following homogeneous system of equations:

$$\begin{aligned}c_1\,(\hbar\,\omega_\alpha(\mathbf{q}) - b_A) + c_2\,c(\mathbf{q}) &= 0\\ c_1\,(-c(\mathbf{q})) + c_2(\hbar\,\omega_\alpha(\mathbf{q}) + b_B) &= 0\end{aligned} \tag{7.345}$$

whose coefficients-determinant must vanish:

$$(\hbar\omega_\alpha(\mathbf{q}) - b_A)(\hbar\omega_\alpha(\mathbf{q}) - b_B) = -c^2(\mathbf{q})$$

This condition is satisfied through

$$\hbar\,\omega_\alpha(\mathbf{q}) = \frac{1}{2}\left[b_A - b_B + \sqrt{(b_A + b_B)^2 - 4c^2(\mathbf{q})}\,\right] \tag{7.346}$$

The solution with negative square root, which in principle is present, is excluded on physical grounds. This is because the excitations must always be positive even for $B_0 = 0$. However, according to (7.328) and (7.329), for $B_0 = 0$, the first summand $(b_A - b_B)$ is zero.

Completely analogously, if we start with $[\beta_\mathbf{q}\ ,\ H]_-$ instead of $[\alpha_\mathbf{q}\ ,\ H]_-$ in (7.343) we get

$$\hbar\,\omega_\beta(\mathbf{q}) = \frac{1}{2}\left[b_B - b_A + \sqrt{(b_A + b_B)^2 - 4c^2(\mathbf{q})}\,\right] \tag{7.347}$$

By substituting for the abbreviations b_A and b_B the full expressions according to (7.328) and (7.329) we get the following for the excitation energies for spin waves in an antiferromagnet:

$$\begin{aligned}\hbar\,\omega_\pm(\mathbf{q}) = &\sqrt{4S^2\hbar^4(J_0^2 - J^2(\mathbf{q})) + g_J\mu_B B_A(g_J\mu_B B_A - 4SJ_0\hbar^2)}\\ &\pm g_J\mu_B B_0\end{aligned} \tag{7.348}$$

This spectrum differs from that of a ferromagnet in a characteristic manner:

1. There are *two* spin wave branches which in the presence of a field show a constant, i.e. $\mathbf{q}$-independent splitting, and are degenerate for $B_0 = 0$.
2. The minimum spin wave energy is for $\mathbf{q} = \mathbf{0}$. This leads for $B_0 = 0$ to an *energy gap* E_g in the spin wave spectrum, which is typical for antiferromagnets.

$$E_g = \sqrt{g_J\ \mu_B\ B_A(g_J\ \mu_B\ B_A - 4S\ J_0\ \hbar^2)} \tag{7.349}$$

   E_g is determined by the anisotropy field B_A and also by the exchange J_0 ($J_0 = z\ J_1\ < 0$). As a result, E_g can be of considerable magnitude. It is obvious that this energy gap, when $k_B T$ is smaller or comparable to E_g, can influence decisively thermodynamic quantities such as the susceptibility, specific heat and sub-lattice magnetization. For $T\ < E_g/k_B$, the spin waves are *frozen* and the quantities mentioned depend exponentially on temperature.
3. From (7.348) one sees that the spin wave branch $\hbar\ \omega_-(\mathbf{q})$ can have negative values for fields

$$B_0 \geq B_0^* = E_g/g_J\ \mu_B . \tag{7.350}$$

This is a signature of the instability of the system since it is possible to have excitations of arbitrarily high order with a gain in energy. This is the collapse of the spin wave approximation. The system itself avoids this collapse by going over to a different spin configuration. B_0^* corresponds to the spin-flop field B_F (7.217) of the molecular field approximation. The sub-lattice magnetizations, which were originally either parallel or antiparallel to the easy axis, now orient themselves perpendicular to the field (see Fig. 7.11). From this one recognizes how important the anisotropy field is for the stabilization of antiferromagnets.

4. In the absence of field according to (7.348) the spin wave energy of an antiferromagnet is doubly degenerate:

$$\hbar\,\omega_0(\mathbf{q}) = \sqrt{E_g^2 + 4S^2\,\hbar^4(J_0^2 - J^2(\mathbf{q}))} \tag{7.351}$$

If we assume in addition only a small anisotropy then

$$\hbar\,\omega_0(\mathbf{q}) \approx 2S\,\hbar^2\sqrt{(J_0 + J(\mathbf{q}))(J_0 - J(\mathbf{q}))} \tag{7.352}$$

If we further restrict ourselves to small $|\mathbf{q}|$, then using (7.270) and (7.278) we can write

$$(J_0 + J(\mathbf{q})) \approx 2J_0; \quad (J_0 - J(\mathbf{q})) \approx \frac{1}{z}\,a^2\,q^2\,J_0$$

This leads to a linear $\mathbf{q}$-dependence of the spin wave energy of an antiferromagnet:

$$\hbar\,\omega(\mathbf{q}) \approx \left(2S\,\hbar^2\,|J_0|\,a\sqrt{\frac{2}{z}}\right) q \tag{7.353}$$

This is different from that of a ferromagnet, which has for small $|\mathbf{q}|$ a quadratic q-dependence. A direct consequence of the linear q-dependence is a T^3 law at low temperatures of the specific heat (Problem 7.14) which is in contrast to the $T^{3/2}$ dependence (7.286) of the specific heat of a ferromagnet. In case the anisotropy is dominant so that E_g is very large, then we again get a quadratic q-dependence:

$$\begin{aligned}\hbar\,\omega_0(\mathbf{q}) &\approx E_g + \frac{2S^2\,\hbar^4}{E_g}(J_0^2 - J^2(\mathbf{q})) \\ &\approx E_g + \left(\frac{4S^2\,J_0^2\,\hbar^4\,a^2}{z\,E_g}\right) q^2\end{aligned} \tag{7.354}$$

Then there are modifications of the T^3 behaviour.

For a further discussion of antiferromagnets, it appears meaningful to specify the transformation (7.338) more precisely. This is connected with the *Bogoliubov transformation* which is known from the theory of superconductivity. The conditions

(7.339), (7.340) and (7.341) on the coefficients $c_{1,2}$ and $d_{1,2}$ of the transformation can obviously be fulfilled through the following ansatz:

$$\begin{aligned} c_1 &= \cosh \eta_{\mathbf{q}}; \quad & c_2 &= -\sinh \eta_{\mathbf{q}} \\ d_1 &= -\sinh \eta_{\mathbf{q}}; \quad & d_2 &= \cosh \eta_{\mathbf{q}} \end{aligned} \tag{7.355}$$

$\eta_{\mathbf{q}}$ is fixed by the system of equations (7.345)

$$\tanh \eta_{\mathbf{q}} = -\frac{c_2}{c_1} = \frac{\hbar\, \omega_\alpha(\mathbf{q}) - b_A}{c(\mathbf{q})}$$

Substituting (7.346) we get

$$\begin{aligned} \tanh 2\eta_{\mathbf{q}} &= \frac{2 \tanh \eta_{\mathbf{q}}}{1 + \tanh \eta_{\mathbf{q}}} \\ &= 2c(\mathbf{q}) \frac{\hbar\omega(\mathbf{q}) - b_A}{c^2(\mathbf{q}) + (\hbar\omega(\mathbf{q}) - b_A)^2} \\ &= -\frac{2c(\mathbf{q})}{b_A + b_B} \\ &= \frac{2\hbar^2\, S\, J(\mathbf{q})}{g_J\, \mu_B\, B_A - 2\hbar^2\, S\, J_0} \end{aligned} \tag{7.356}$$

An extremely interesting quantity is the energy constant $\tilde{E}_a(B_0)$ appearing in the transformed Hamiltonian (7.337) which represents the true ground state energy of the spin system in the spin wave approximation. So far we have not calculated it explicitly. For doing this, we first use the inverse transformation to (7.338):

$$\begin{aligned} a_{\mathbf{q}} &= \bar{c}_1\, \alpha_{\mathbf{q}} + \bar{c}_2\, \beta_{\mathbf{q}}^\dagger \\ b_{\mathbf{q}} &= \bar{d}_1\, \alpha_{\mathbf{q}}^\dagger + \bar{d}_2\, \beta_{\mathbf{q}} \end{aligned} \tag{7.357}$$

$$\bar{c}_1 = \bar{d}_2 = \cosh\ \eta_{\mathbf{q}}; \quad \bar{c}_2 = \bar{d}_1 = \sinh\ \eta_{\mathbf{q}} \tag{7.358}$$

which we insert in the "original" Hamiltonian (7.335). This naturally results again in (7.337) since the transformation was conceptualized precisely for this purpose. In addition, we obtain the "new" energy constant

$$\tilde{E}_a(B_0) = E_a - \frac{N}{4}(b_A + b_B) + \frac{1}{2} \sum_{\mathbf{q}} \sqrt{(b_A + b_B)^2 - 4c^2(\mathbf{q})} = \tilde{E}_a \tag{7.359}$$

We perform the explicit transformation as Problem 7.16.

This is an extremely interesting result which shows that the ground state energy $\tilde{E}_a$ is smaller than the energy E_a of the fully ordered Neel state, in which the sub-lattice magnetizations are oriented exactly antiparallel to each other. One should recognize that the difference

$$\tilde{E}_a - E_a = -\frac{N}{2}(-2J_0\,\hbar^2\,S + g_J\,\mu_B\,B_A) + \sum_{\mathbf{q}} \sqrt{E_g^2 + 4S^2\,\hbar^4(J_0^2 - J^2(\mathbf{q}))} \leq 0 \tag{7.360}$$

is independent of field. The fully ordered Neel state is certainly not the ground state of the antiferromagnet, which has to possess some spin disorder which should be noticeable in the sub-lattice magnetizations. Therefore, as a next step, they should be calculated.

According to (7.326), for the magnetization of the sub-lattice A

$$M_A = \frac{1}{V}\frac{g_J\,\mu_B}{\hbar}\sum_{i=1}^{N/2}\langle S_i^z\rangle_A = \frac{N}{2V}\,g_J\,\mu_B\,S - \frac{1}{V}\,g_J\,\mu_B\sum_i\langle a_i^\dagger\,a_i\rangle$$

holds. When we transform into the wavenumbers, we have to evaluate

$$M_A = \frac{1}{V}\,g_J\,\mu_B\left(\frac{N}{2}\,S - \sum_{\mathbf{q}}\langle a_{\mathbf{q}}^\dagger\,a_{\mathbf{q}}\rangle\right) \tag{7.361}$$

We exploit (7.357):

$$\langle a_{\mathbf{q}}^\dagger\,a_{\mathbf{q}}\rangle = \bar{c}_1^2\langle\alpha_{\mathbf{q}}^\dagger\,\alpha_{\mathbf{q}}\rangle + \bar{c}_2^2\langle\beta_{\mathbf{q}}\;\beta_{\mathbf{q}}^\dagger\rangle + \bar{c}_1\,\bar{c}_2\left(\langle\alpha_{\mathbf{q}}^\dagger\,\beta_{\mathbf{q}}^\dagger\rangle + \langle\beta_{\mathbf{q}}\,\alpha_{\mathbf{q}}\rangle\right)$$

According to (7.337), in the linear spin wave approximation, the quasiparticles created by $\alpha_{\mathbf{q}}$ and $\beta_{\mathbf{q}}$ represent a fully decoupled system in which the particle number is conserved. That means, e.g.

$$\langle\alpha_{\mathbf{q}}^\dagger\,\beta_{\mathbf{q}}^\dagger\rangle = \langle\beta_{\mathbf{q}}\alpha_{\mathbf{q}}\rangle = 0 \tag{7.362}$$

so that

$$\langle a_{\mathbf{q}}^\dagger\,a_{\mathbf{q}}\rangle = \cosh^2\eta_{\mathbf{q}}\,\langle\alpha_{\mathbf{q}}^\dagger\,\alpha_{\mathbf{q}}\rangle + \sinh^2\eta_{\mathbf{q}}(1 + \langle\beta_{\mathbf{q}}^\dagger\,\beta_{\mathbf{q}}\rangle) \tag{7.363}$$

The α- and β-quasiparticles are non-interacting Bosons ($\mu = 0$). Therefore, the expectation values of their occupation number operators on the right-hand side of (7.363) are the Bose–Einstein distribution functions:

$$M_A = \frac{1}{V} g_J \mu_B \left[\left(\frac{N}{2} S - \sum_{\mathbf{q}} \sinh^2 \eta_{\mathbf{q}} \right) - \right.$$
$$\left. - \sum_{\mathbf{q}} \left(\frac{\cosh^2 \eta_{\mathbf{q}}}{exp(\beta \hbar \omega_\alpha(\mathbf{q})) - 1} + \frac{\sinh^2 \eta_{\mathbf{q}}}{exp(\beta \hbar \omega_\beta(\mathbf{q})) - 1} \right) \right] \tag{7.364}$$

With

$$\cosh^2 \eta_{\mathbf{q}} = \frac{1}{2} \left[1 + \left(\sqrt{1 - \tanh^2 2\eta_{\mathbf{q}}} \right)^{-1} \right]$$
$$\sinh^2 \eta_{\mathbf{q}} = \frac{1}{2} \left[-1 + \left(\sqrt{1 - \tanh^2 2\eta_{\mathbf{q}}} \right)^{-1} \right] \tag{7.365}$$

(7.356) for $\tanh 2\eta_{\mathbf{q}}$ and (7.346) and (7.347) for $\hbar\omega_\alpha$ and $\hbar\omega_\beta$, the temperature dependence of the sub-lattice magnetization in the spin wave region is fully determined. We are particularly interested in the $T = 0$ *sub-lattice magnetization*

$$M_A(T = 0) = \frac{1}{V} g_J \mu_B \left(\frac{N}{2} S - \sum_{\mathbf{q}} \sinh^2 \eta_{\mathbf{q}} \right) \tag{7.366}$$

The first summand corresponds to the complete orientation of the spins (Neel state) whereas the second corresponds to the fluctuations mentioned earlier which are not obviously visualizable. Let us set

$$M_A(T = 0) = \frac{N}{2V} g_J \mu_B (S - \sigma) \tag{7.367}$$

Then the correction σ can be easily evaluated for simple lattices if we assume the anisotropies $B_A = 0$ in (7.356):

$$\sigma = \frac{1}{N} \sum_{\mathbf{q}} \left(\frac{J_0}{\sqrt{J_0^2 - J^2(\mathbf{q})}} - 1 \right) \tag{7.368}$$

For the NaCl structure for which the whole lattice is simple cubic and the interpenetrating sub-lattices are fcc, one finds [15]

$$\sigma = 0.078 \tag{7.369}$$

That is a typical order of magnitude. The correction term is not too large but not negligible, either, especially for $S = 1/2$.

For the magnetization of the sub-lattice B (7.327) analogous to (7.361) holds:

$$M_B = \frac{1}{V}\, g_J\, \mu_B \left(-\frac{N}{2}\, S + \sum_{\mathbf{q}} \langle b^\dagger_{\mathbf{q}}\, b_{\mathbf{q}} \rangle \right) \tag{7.370}$$

where now we have to set

$$\langle b^\dagger_{\mathbf{q}}\, b_{\mathbf{q}} \rangle = \sinh^2 \eta_{\mathbf{q}} (1 + \langle \alpha^\dagger_{\mathbf{q}}\, \alpha_{\mathbf{q}} \rangle) + \cosh^2 \eta_{\mathbf{q}} \langle \beta^\dagger_{\mathbf{q}}\, \beta_{\mathbf{q}} \rangle$$

With this we get for the total magnetization

$$\begin{aligned} M(T, B_0) &= M_A(T, B_0) + M_B(T, B_0) \\ &= -\frac{1}{V}\, g_J\, \mu_B \sum_{\mathbf{q}} \left(\langle \alpha^\dagger_{\mathbf{q}}\, \alpha_{\mathbf{q}} \rangle - \langle \beta^\dagger_{\mathbf{q}}\, \beta_{\mathbf{q}} \rangle \right) \end{aligned} \tag{7.371}$$

The fluctuation terms of the two sub-lattices thus cancel out. For $B_0 = 0$, naturally $\hbar\, \omega_\alpha = \hbar\, \omega_\beta$ and therefore also $\langle \alpha^\dagger_{\mathbf{q}}\, \alpha_{\mathbf{q}} \rangle = \langle \beta^\dagger_{\mathbf{q}}\, \beta_{\mathbf{q}} \rangle$ so that the total magnetization vanishes.

At the end we now calculate the *susceptibility* of the antiferromagnet. For that we expand the magnetization up to the linear term in B_0. For $B_0 = 0$ the spin waves $\hbar\omega_\alpha$ and $\hbar\omega_\beta$ are according to (7.351) identical to $\hbar\omega_0$.

$$\begin{aligned} \langle \alpha^\dagger_{\mathbf{q}}\, \alpha_{\mathbf{q}} \rangle &= [exp(\beta \hbar \omega_\alpha(\mathbf{q})) - 1]^{-1} \\ &= \langle \alpha^\dagger_{\mathbf{q}}\, \alpha_{\mathbf{q}} \rangle^{(0)} - \beta g_J \mu_B B_0 \frac{exp(\beta\hbar\omega_0)}{(exp(\beta\hbar\omega_0) - 1)^2} + \mathcal{O}(B^2) \end{aligned}$$

A completely analogous expression is obtained for $\langle \beta^\dagger_{\mathbf{q}}\, \beta_{\mathbf{q}} \rangle$ which along with $\langle \alpha^\dagger_{\mathbf{q}}\, \alpha_{\mathbf{q}} \rangle$ we substitute in (7.371). Then the field dependence of the total magnetization reads as

$$M(T, B_0) = \frac{2}{V}\, \beta (g_J \mu_B)^2\, B_0 \sum_{\mathbf{q}} \frac{exp(\beta\hbar\omega_0(\mathbf{q}))}{(exp(\beta\hbar\omega_0(\mathbf{q})) - 1)^2} + \mathcal{O}(B_0^2) \tag{7.372}$$

With this we directly get for the susceptibility

$$\chi_\parallel(T) = \frac{2\mu_0}{V}\, \beta (g_J \mu_B)^2 \sum_{\mathbf{q}} \frac{exp(\beta\hbar\omega_0(\mathbf{q}))}{(exp(\beta\hbar\omega_0(\mathbf{q})) - 1)^2} \tag{7.373}$$

7.4.4 Harmonic Approximation for a Ferromagnet with Dipolar Interaction

In this section we want to discuss the effect of dipole interaction on the spin wave spectrum of a ferromagnet. In the last section we have seen that the fully ordered spin state is not the ground state of an antiferromagnet. In contrast, in the case of *isotropic* ferromagnet, the ground state is indicated by a complete parallel alignment of the localized spins (7.108). However, this is no more correct even for a ferromagnet as soon as we introduce anisotropy which leads to additional terms in the Hamiltonian which do not commute with the z-component of the total spin $S^z = \sum_i S_i^z$. One example for this is the dipole interaction between the localized magnetic moments, which, in particular for ferromagnets with low transition temperatures (EuS!), should not be neglected. According to (7.50), we have to take the dipole interaction into account by including the following additional term in the Hamiltonian:

$$H_D = \sum_{i,j} D_{ij} \left\{ \mathbf{S}_i \cdot \mathbf{S}_j - 3(\mathbf{S}_i \cdot \mathbf{e}_{ij})(\mathbf{S}_j \cdot \mathbf{e}_{ij}) \right\} \tag{7.374}$$

$$D_{ii} = 0; \quad D_{ij} = \frac{\mu_0}{8\pi} \frac{g_J^2 \, \mu_B^2}{\hbar^2 |\mathbf{R}_i - \mathbf{R}_j|^3} \quad (i \neq j) \tag{7.375}$$

$$\mathbf{e}_{ij} = \frac{\mathbf{R}_i - \mathbf{R}_j}{|\mathbf{R}_i - \mathbf{R}_j|} \equiv (x_{ij}, \; y_{ij}, \; z_{ij}) \tag{7.376}$$

The dipole interaction is surely of much smaller significance compared to the exchange interaction but at the same time it has much longer range. Through the second summand in (7.374) it leads to an anisotropy which can be important in many practical applications.

We split the model Hamiltonian into an isotropic (H_i) and an anisotropic part (H_a):

$$H = H_i + H_a \tag{7.377}$$

The isotropic part is different from the model discussed in Sect. 7.4.1 only by the renormalization of the coupling constants:

$$H_i = -\sum_{i,j} (J_{ij} - D_{ij})(S_i^+ \, S_j^- + S_i^z \, S_j^z) - g_J \, \frac{\mu_B}{\hbar} \, B_0 \sum_i S_i^z \tag{7.378}$$

What is new is the anisotropy part:

$$H_a = -3 \sum_{i,j} D_{ij} (\mathbf{S}_i \cdot \mathbf{e}_{ij})(\mathbf{S}_j \cdot \mathbf{e}_{ij}) \tag{7.379}$$

For the spin operators we again use the Holstein–Primakoff transformation (7.23), (7.24) and (7.25) in the "harmonic approximation":

$$\frac{1}{\hbar}\, S_i^+ = \sqrt{2S}\, a_i; \quad \frac{1}{\hbar}\, S_i^- = \sqrt{2S}\, a_i^\dagger; \quad \frac{1}{\hbar}\, S_i^z = S - a_i^\dagger\, a_i \tag{7.380}$$

Defining

$$D_0 = \sum_i D_{ij} = \sum_j D_{ij} \tag{7.381}$$

$$b_0 = g_J\, \mu_B\, B_0 + 2S\, \hbar^2\, (J_0 - D_0) \tag{7.382}$$

$$E_0(B_0) = -g_J\, \mu_B\, B_0\, N\, S - \hbar^2 (J_0 - D_0) N\, S^2 \tag{7.383}$$

we get almost directly for the isotropic part of the Hamiltonian

$$H_i = E_0(B_0) + b_0 \sum_i n_i - 2S\, \hbar^2 \sum_{i,j} (J_{ij} - D_{ij}) a_i\, a_j^\dagger \tag{7.384}$$

In obtaining this we have, in the spirit of the harmonic approximation, left out all the terms which are not bilinear in the magnon construction operators a_i, $a_i^\dagger$. The anisotropic part of the dipole interaction H_a possesses somewhat more difficulties. One finds (Problem 7.19)

$$\begin{aligned} H_a &= -3S^2\, \hbar^2 \sum_{i,j} D_{ij}\, z_{ij}^2 \\ &\quad - 3S\, \hbar^2 \sum_{i,j} D_{ij} \left[(x_{ij}^2 + y_{ij}^2) a_i^\dagger\, a_j + \frac{1}{2} (x_{ij} - i y_{ij})^2 a_i\, a_j \right. \\ &\quad \left. + \frac{1}{2} \left(x_{ij} + i y_{ij}\right)^2 a_i^\dagger\, a_j^\dagger - 2 z_{ij}^2\, n_i \right] \end{aligned} \tag{7.385}$$

The first term is a c-number which we can absorb in E_0

$$\widehat{E}_0(B_0) = E_0(B_0) - 3S^2\, \hbar^2 \sum_{i,j} D_{ij}\, z_{ij}^2 \tag{7.386}$$

A first diagonalization of the Hamiltonian is obtained by going to the wavenumbers. We define

$$B(\mathbf{q}) = -\frac{3}{2}\, S\, \hbar^2\, \frac{1}{N} \sum_{i,j} D_{ij} (x_{ij} - i y_{ij})^2 e^{i \mathbf{q} \cdot (\mathbf{R}_i - \mathbf{R}_j)} \tag{7.387}$$

$$A(\mathbf{q}) = b_0 + 6S\,\hbar^2 \sum_{i,j} D_{ij}\, z_{ij}^2 - \\ - 2S\,\hbar^2 \frac{1}{N} \sum_{i,j} \left(J_{ij} + \frac{1}{2}\, D_{ij}(1 - 3z_{ij}^2) \right) e^{i\mathbf{q}\cdot(\mathbf{R}_i - \mathbf{R}_j)} \tag{7.388}$$

and then have

$$H = \widehat{E}_0(B_0) + \sum_{\mathbf{q}} A(\mathbf{q})\, a_{\mathbf{q}}^{\dagger}\, a_{\mathbf{q}} + \\ + \sum_{\mathbf{q}} \left(B(\mathbf{q})\, a_{\mathbf{q}}\, a_{-\mathbf{q}} + B^*(\mathbf{q})\, a_{\mathbf{q}}^{\dagger}\, a_{-\mathbf{q}}^{\dagger} \right) \tag{7.389}$$

Because of the anisotropy $B(\mathbf{q})$, H is not yet diagonal. Just as in the case of antiferromagnets we now take H through the Bogoliubov transformation to the form

$$H = E_1(B_0) + \sum_{\mathbf{q}} \hbar\, \omega_D(\mathbf{q})\, \alpha_{\mathbf{q}}^{\dagger}\, \alpha_{\mathbf{q}} \tag{7.390}$$

For that, for the new operators we make the ansatz:

$$\alpha_{\mathbf{q}} = c_{\mathbf{q}}\, a_{\mathbf{q}} + d_{\mathbf{q}}\, a_{-\mathbf{q}}^{\dagger} \tag{7.391}$$

The fulfilment of the fundamental Bose commutator rules is ensured by the following conditions on the coefficients $c_{\mathbf{q}}$ and $d_{\mathbf{q}}$:

$$\begin{aligned} |c_{\mathbf{q}}|^2 - |d_{\mathbf{q}}|^2 &= 1 \\ c_{\mathbf{q}}\, d_{-\mathbf{q}} - d_{\mathbf{q}}\, c_{-\mathbf{q}} &= 0 \end{aligned} \tag{7.392}$$

Further calculational procedure is similar to that followed for determining the spin wave energies of antiferromagnets.

$$\begin{aligned} \left[\alpha_{\mathbf{q}}\,,\, H\right]_- &= \hbar\, \omega_D(\mathbf{q})\, \alpha_{\mathbf{q}} = \hbar\, \omega_D(\mathbf{q})(c_{\mathbf{q}}\, a_{\mathbf{q}} + d_{\mathbf{q}}\, a_{-\mathbf{q}}^{\dagger}) \\ &= c_{\mathbf{q}} \left[a_{\mathbf{q}}\,,\, H\right]_- + d_{\mathbf{q}} \left[a_{-\mathbf{q}}^{\dagger}\,,\, H\right]_- \\ &= c_{\mathbf{q}} \left(A(\mathbf{q})\, a_{\mathbf{q}} + B^*(\mathbf{q})\, a_{-\mathbf{q}}^{\dagger} + B^*(-\mathbf{q})\, a_{-\mathbf{q}}^{\dagger} \right) + \\ &+ d_{\mathbf{q}} \left(-A(-\mathbf{q})\, a_{-\mathbf{q}}^{\dagger} - B(-\mathbf{q})\, a_{\mathbf{q}} - B(\mathbf{q}) a_{\mathbf{q}} \right) \end{aligned}$$

With $B(\mathbf{q}) = B(-\mathbf{q})$ and $A(\mathbf{q}) = A(-\mathbf{q})$ it further follows that

$$a_{\mathbf{q}} \left\{ c_{\mathbf{q}}\, \hbar\, \omega_D(\mathbf{q}) - c_{\mathbf{q}}\, A(\mathbf{q}) + 2d_{\mathbf{q}}\, B(\mathbf{q}) \right\} + \\ + a_{-\mathbf{q}}^{\dagger} \left\{ d_{\mathbf{q}}\, \hbar\, \omega_D(\mathbf{q}) - 2c_{\mathbf{q}}\, B^*(\mathbf{q}) + d_{\mathbf{q}}\, A(\mathbf{q}) \right\} = 0$$

This equation can only be satisfied when each coefficient by itself is equal to zero. This gives a homogeneous system of equations:

$$\begin{aligned} c_{\mathbf{q}}\,(\hbar\,\omega_D(\mathbf{q}) - A(\mathbf{q})) + d_{\mathbf{q}}\,2B(\mathbf{q}) &= 0 \\ c_{\mathbf{q}}\,2B^*(\mathbf{q}) + d_{\mathbf{q}}(-\hbar\,\omega_D(\mathbf{q}) - A(\mathbf{q})) &= 0 \end{aligned} \tag{7.393}$$

whose solvability condition

$$-(\hbar\,\omega_D(\mathbf{q}))^2 + A^2(\mathbf{q}) - 4|B(\mathbf{q})|^2 = 0$$

determines the spin wave energies:

$$\hbar\,\omega_D(\mathbf{q}) = \sqrt{A^2(\mathbf{q}) - 4|B(\mathbf{q})|^2} \tag{7.394}$$

Due to the long range of the dipole interaction, the lattice sums needed to calculate $A(\mathbf{q})$ and $B(\mathbf{q})$ are not easy to perform. However, it is in principle quite possible to do it for not too complicated lattices. We will not attempt to do it here in detail [16].

The energy constant $E_1(B_0)$, which obviously according to (7.390) represents the ground state energy of a ferromagnet with dipole interaction, provides many interesting conclusions. First we obtain using (7.393) with (7.392) for the coefficients of the transformation (7.393)

$$|c_{\mathbf{q}}|^2 = \frac{1}{2}\left(\frac{A(\mathbf{q})}{\hbar\,\omega_D(\mathbf{q})} + 1\right); \quad |d_{\mathbf{q}}|^2 = \frac{1}{2}\left(\frac{A(\mathbf{q})}{\hbar\,\omega_D(\mathbf{q})} - 1\right) \tag{7.395}$$

where the phase remains free for the moment. This will be uniquely determined from the condition that the Hamiltonian (7.389) takes the form (7.390) after the transformation. For that we invert the transformation (7.391)

$$a_{\mathbf{q}} = c_{\mathbf{q}}^*\,\alpha_{\mathbf{q}} - d_{\mathbf{q}}\,\alpha_{-\mathbf{q}}^\dagger \tag{7.396}$$

and use this in (7.389). H will be diagonal only when we fix the phase as follows:

$$B(\mathbf{q}) = |B(\mathbf{q})|\,e^{i\phi}; \quad c_{\mathbf{q}} = |c_{\mathbf{q}}|\,e^{-i\phi/2}; \quad d_{\mathbf{q}} = |d_{\mathbf{q}}|\,e^{i\phi/2} \tag{7.397}$$

With this we finally get the energy constant $E_1(B_0)$ in (7.390):

$$E_1(B_0) = \widehat{E}_0(B_0) - \frac{1}{2}\sum_{\mathbf{q}}\{A(\mathbf{q}) - \hbar\,\omega_D(\mathbf{q})\} \tag{7.398}$$

$\widehat{E}_0(B_0)$ is the energy of the fully ordered ferromagnet. From

$$E_1(B_0) < \widehat{E}_0(B_0) \tag{7.399}$$

we must conclude as in the case of an antiferromagnet that due to the anisotropy part in the dipole interaction (7.379), the fully ordered spin state is not the ground state of a ferromagnet any more.

Finally we have also described a ferromagnet with dipole interaction by a system of non-interacting bosons. For the quantities such as the partition function, the average occupation density, the grand canonical potential and the internal energy, formally the same relations are valid as for an isotropic ferromagnet. We only have to replace $\hbar\omega(\mathbf{q})$ by $\hbar\omega_D(\mathbf{q})$ everywhere:

$$\Xi(T, B_0) = exp(-\beta\, E_1(B_0)) \prod_{\mathbf{q}} \frac{1}{1 - exp(\beta\hbar\omega_D(\mathbf{q}))} \tag{7.400}$$

$$\langle n(\mathbf{q})\rangle = [exp(\beta\hbar\omega_D(\mathbf{q})) - 1]^{-1} \tag{7.401}$$

$$\begin{aligned} \Omega(T, B_0) = {} & E_1(B_0) + \\ & + k_B T \sum_{\mathbf{q}} \ln(1 - exp(-\beta\hbar\omega_D(\mathbf{q}))) \end{aligned} \tag{7.402}$$

$$U = E_1(B_0) + \sum_{\mathbf{q}} \hbar\omega_D(\mathbf{q})\langle n(\mathbf{q})\rangle \tag{7.403}$$

While calculating the *magnetization*

$$M(T, B_0) = -\frac{1}{V}\left(\frac{\partial\Omega}{\partial B_0}\right)_T \tag{7.404}$$

we have to pay attention to the fact that $A(\mathbf{q})$ depends on B_0 whereas $B(\mathbf{q})$ does not. With (7.383), (7.386) and (7.398), the following holds for the energy constant:

$$\frac{\partial E_1(B_0)}{\partial B_0} = -g_J\,\mu_B\,N\,S - \frac{1}{2}\,g_J\,\mu_B \sum_{\mathbf{q}} \left\{1 - \frac{A(\mathbf{q})}{\hbar\,\omega_D(\mathbf{q})}\right\} \tag{7.405}$$

We can split the magnetization into an isotropic and an anisotropic part

$$M(T, B_0) = M_i(T, B_0) + M_a(T, B_0) \tag{7.406}$$

M_i is the magnetization of the isotropic ferromagnet:

$$M_i(T, B_0) = M_0\left(1 - \frac{1}{NS}\sum_{\mathbf{q}}\langle n_{\mathbf{q}}\rangle\right) \tag{7.407}$$

M_0 denotes the magnetization of the fully ordered spin system

$$M_0 = g_J\,\mu_B\,\frac{N}{V}\,S \tag{7.408}$$

$M_a(T, B_0)$ expresses the influence of the anisotropy:

$$M_a = g_J\,\mu_B\,\frac{1}{V}\sum_{\mathbf{q}}\left(1 - \frac{A(\mathbf{q})}{\hbar\,\omega_D(\mathbf{q})}\right)\left(\frac{1}{2} + \langle n_{\mathbf{q}}\rangle\right) \tag{7.409}$$

M_a is obviously a negative quantity. The dipole interaction thus produces a partial demagnetization.

For the special case $D_{ij} = 0$, naturally, we recover the *old* result of the isotropic ferromagnet, because then $B(\mathbf{q}) \equiv 0$ and $A(\mathbf{q}) = \hbar\,\omega(\mathbf{q})$. That means according to (7.394) $\hbar\,\omega_D(\mathbf{q}) = \hbar\,\omega(\mathbf{q})$ and with that $M_a \equiv 0$.

7.5 Thermodynamics of $S = 1/2$ Ferromagnet

We have so far investigated in general the not exactly solvable Heisenberg model by using two approximate methods, namely the molecular field approximation in Sect. 7.3 and the spin wave approximation in Sect. 7.4. While the spin wave picture is restricted to low temperatures, the molecular field approximation though valid at all temperatures is too coarse for many purposes. In this section, we want to extract detailed information about the thermodynamics of the Heisenberg model with the help of the method of double-time Green's functions (see Appendix B). We will restrict our treatment to ferromagnets.

7.5.1 Tyablikov Decoupling

We consider a system of localized magnetic moments described by the spin operators $\mathbf{S}_i, \mathbf{S}_j$ in a homogeneous, time-independent magnetic field B_0:

$$H = -\sum_{i,j} J_{ij}(S_i^+\,S_j^- + S_i^z\,S_j^z) - g_J\,\frac{\mu_B}{\hbar}\,B_0\sum_i S_i^z \tag{7.410}$$

We are primarily interested in the magnetization of the system,

$$M(T, B_0) = \frac{1}{V}\,g_J\,\frac{\mu_B}{\hbar}\sum_i \langle S_i^z\rangle_{T,B_0} \tag{7.411}$$

for which we need the thermodynamic expectation value of the spin operator S_i^z. Here we first assume that the localized spins are of magnitude

$$S = \frac{1}{2} \tag{7.412}$$

This is, as we will justify more precisely, a special case which is decisively simpler to handle than the general case of $S \geq 1/2$. The relation (7.8), which is valid in general

$$S_i^{\pm} S_i^{\mp} = \hbar^2 S(S+1) \pm \hbar S_i^z - \left(S_i^z\right)^2 \tag{7.413}$$

simplifies for $S = 1/2$ to

$$S_i^{\pm} S_i^{\mp} = \hbar (\hbar S \pm S_i^z) \tag{7.414}$$

since in this case due to (7.9)

$$\left(S_i^z\right)^2 = \frac{\hbar^2}{4} \mathbb{1} \tag{7.415}$$

represents an operator identity.

The relation (7.414) suggests that the following retarded Green's function should be studied:

$$\begin{aligned} G_{ij}^{ret}(t,t') &= \langle\langle S_i^+(t)\,;\, S_j^-(t')\rangle\rangle^{ret} \\ &= -i\,\theta(t-t')\langle\left[S_i^+(t)\,,\, S_j^-(t')\right]_-\rangle \end{aligned} \tag{7.416}$$

For the energy-dependent Fourier transform (B.18)

$$G_{ij}^{ret}(E) = \int_{-\infty}^{+\infty} d(t-t')\, e^{\frac{i}{\hbar}E(t-t')} G_{ij}^{ret}(t-t') \tag{7.417}$$

the simple algebraic equation of motion holds (see B.83):

$$E\, G_{ij}^{ret}(E) = \hbar \,\langle\left[S_i^+\,,\, S_j^-\right]_-\rangle + \langle\langle\left[S_i^+\,,\, H\right]_-\,;\, S_j^-\rangle\rangle_E^{ret} \tag{7.418}$$

For further evaluation we need the commutators on the right-hand side. For that we use the relations (7.6) and (7.7):

$$\left[S_i^+\,,\, S_j^-\right]_- = 2\hbar\, \delta_{ij}\, S_i^z \tag{7.419}$$

$$\left[S_i^+\,,\, H\right]_- = -2\hbar \sum_m J_{im}(S_m^+ S_i^z - S_i^+ S_m^z) + g_J\, \mu_B\, B_0\, S_i^+ \tag{7.420}$$

With this the equation of motion (7.418) becomes

$$\begin{aligned} (E - g_J\, \mu_B\, B_0)\, G_{ij}^{ret}(E) &= 2\hbar^2\, \delta_{ij}\, \langle S_i^z\rangle - \\ - 2\hbar \sum_m J_{im} &\left(\langle\langle S_i^z S_m^+\,;\, S_j^-\rangle\rangle_E^{ret} - \langle\langle S_m^z S_i^+\,;\, S_j^-\rangle\rangle_E^{ret}\right) \end{aligned} \tag{7.421}$$

This equation is still exact. It cannot, however, be solved exactly since there appear "higher" Green's functions on the right-hand side. It appears natural to set up again the equations of motion for these Green's functions. This leads to *still higher* Green's functions so that eventually one has an infinite chain ("hierarchy") of equations of motions, which, with the help of approximate methods, have to be somehow decoupled.

We want to apply the decoupling already on (7.421) by using an approximation similar to the molecular field approximation. We replace the operator S_i^z by its thermodynamic expectation value $\langle S_i^z \rangle$. As a c-number, this can be brought out of the Green's function. In the literature, this is known as the *Tyablikov approximation* or the *RPA decoupling* (random phase approximation) [17]:

$$\begin{aligned} \langle\langle S_i^z \, S_m^+ \, ; \, S_j^- \rangle\rangle &\rightarrow \langle S_i^z \rangle \langle\langle S_m^+ \, ; \, S_j^- \rangle\rangle_E^{ret} \\ \langle\langle S_m^z \, S_i^+ \, ; \, S_j^- \rangle\rangle &\rightarrow \langle S_m^z \rangle \langle\langle S_i^+ \, ; \, S_j^- \rangle\rangle_E^{ret} \end{aligned} \tag{7.422}$$

This decoupling procedure is the weak point of the theory since there is no direct justification for doing this. The only justification is the results obtained with this method.

We can further assume translational symmetry:

$$\langle S_i^z \rangle = \langle S_m^z \rangle = \langle S^z \rangle \tag{7.423}$$

and with (7.421) and (7.422) have the following simplified equation of motion:

$$\begin{aligned} &(E - g_J \, \mu_B \, B_0 - 2\hbar \, \langle S^z \rangle \, J_0) \, G_{ij}^{ret}(E) \\ &= 2\hbar^2 \, \delta_{ij} \, \langle S^z \rangle - 2\hbar \, \langle S^z \rangle \sum_m J_{im} \, G_{mj}^{ret}(E) \end{aligned} \tag{7.424}$$

which can be solved by Fourier transforming to the wavenumber representation.

$$G_{\mathbf{q}}^{ret}(E) = \frac{1}{N} \sum_{i,j} G_{ij}^{ret}(E) \, e^{i\mathbf{q}\cdot(\mathbf{R}_i - \mathbf{R}_j)} \tag{7.425}$$

$$\delta_{ij} = \frac{1}{N} \sum_{\mathbf{q}} e^{-i\mathbf{q}\cdot(\mathbf{R}_i - \mathbf{R}_j)} \tag{7.426}$$

With this we get

$$G_{\mathbf{q}}^{ret}(E) = \frac{2\hbar^2 \, \langle S^z \rangle}{E - E(\mathbf{q}) + i0^+} \tag{7.427}$$

The poles of the Green's function correspond to the elementary excitations of the spin system (see Appendix B):

$$E(\mathbf{q}) = 2\hbar \langle S^z \rangle (J_0 - J(\mathbf{q})) + g_J \mu_B B_0 \tag{7.428}$$

Typical for the quasiparticle concept of the many-body theory is the temperature dependence of the elementary excitations which in the present case is manifested in the expectation value $\langle S^z \rangle$. For $T = 0$

$$\langle S^z \rangle_{T=0} = \hbar S \tag{7.429}$$

holds so that (7.428) reproduces exactly the spin wave result (7.249). One expects the temperature dependence intuitively on the basis of interactions since at higher temperatures, it should be easier to excite spin waves, i.e. spin deviations.

The solution as it stands in (7.427) is not yet complete since we still do not know the expectation value $\langle S^z \rangle$.

7.5.2 Spontaneous Magnetization

The spectral density $S_{\mathbf{q}}(E)$ (B.65) according to (7.427) has the simple form

$$\begin{aligned} S_{\mathbf{q}}(E) &= -\frac{1}{\pi} \, Im \, G^{ret}_{\mathbf{q}}(E + i0^+) \\ &= 2\hbar \langle S^z \rangle \, \delta(E - E(\mathbf{q})) \end{aligned} \tag{7.430}$$

where the Dirac identity (B.92) has been used. With the help of the fundamental spectral theorem (B.95) one easily gets the expectation value $\langle S_j^- \, S_i^+ \rangle$ ($D \equiv 0 \; for \; B_0 > 0$):

$$\begin{aligned} \langle S_j^- \, S_i^+ \rangle &= \frac{1}{N} \sum_{\mathbf{q}} e^{-i\mathbf{q}\cdot(\mathbf{R}_i - \mathbf{R}_j)} \frac{1}{\hbar} \int_{-\infty}^{+\infty} \frac{S_{\mathbf{q}}(E)}{e^{\beta E} - 1} \\ &= \frac{1}{N} \sum_{\mathbf{q}} e^{-i\mathbf{q}\cdot(\mathbf{R}_i - \mathbf{R}_j)} \frac{2\hbar \langle S^z \rangle}{e^{\beta E(\mathbf{q})} - 1} \end{aligned} \tag{7.431}$$

For $i = j$ in particular we have

$$\langle S_i^- \, S_i^+ \rangle = 2\hbar \langle S^z \rangle \frac{1}{N} \sum_{\mathbf{q}} \frac{1}{exp(\beta E(\mathbf{q})) - 1} \tag{7.432}$$

At this stage we make use of the simplification declared above, namely we limit ourselves to $S = 1/2$. Then we can use (7.414):

$$\langle S_i^- \, S_i^+ \rangle = \hbar^2 \, S - \hbar \langle S_i^z \rangle \tag{7.433}$$

so that we get with (7.432)

$$\langle S^z \rangle = \hbar\, S \left(1 + \frac{2}{N} \sum_{\mathbf{q}} \frac{1}{exp(\beta E(\mathbf{q})) - 1} \right)^{-1} \tag{7.434}$$

Along with (7.428) we have found two equations which permit a self-consistent determination of $\langle S^z \rangle$. Numerical evaluation is done most conveniently by using the following relation which is equivalent to (7.434):

$$\frac{\hbar\, S}{\langle S^z \rangle} = \frac{V}{N(2\pi)^3} \int d^3q \coth \left[\frac{2\hbar \langle S^z \rangle (J_0 - J(\mathbf{q})) + g_J \mu_B B_0}{2k_B T} \right] \tag{7.435}$$

The integration is performed over the first Brillouin zone of the lattice. There is no difficulty in numerically evaluating this equation for all temperatures; however, one can analytically discuss this in certain temperature regions.

First, (7.435) permits to derive an explicit equation for the Curie temperature. For

$$B_0 = 0^+; \quad T \overset{<}{\to} T_C \leftrightarrow \langle S^z \rangle \overset{>}{\to} 0$$

the argument of coth is small. Therefore, we can terminate the expansion

$$\coth x = \frac{1}{x} + \sum_{n=1}^{\infty} (-1)^{n-1} \frac{2^{2n}\, B_n}{(2n)!}\, x^{2n-1} \tag{7.436}$$

(B_n = Bernoulli numbers) already after the first term and obtain

$$k_B\, T_C = \left\{ \frac{1}{NS} \sum_{\mathbf{q}} \frac{1}{\hbar^2 (J_0 - J(\mathbf{q}))} \right\}^{-1} \tag{7.437}$$

a relation which formally agrees with the result obtained from the spin wave theory (7.322) for $S = 1/2$.

We now want to evaluate (7.435) for *low temperatures*. For small T, the argument of coth is very large. Therefore, it is convenient to expand

$$\coth x = 1 + 2 \sum_{m=1}^{\infty} exp(-2mx) \tag{7.438}$$

In addition, $\langle S^z \rangle \approx \hbar\, S$, so that the following ansatz is meaningful:

$$\frac{\langle S^z \rangle}{\hbar\, S} = \frac{1}{1 + 2\phi} \tag{7.439}$$

where according to (7.434) and (7.438)

$$\phi = \frac{1}{N}\sum_{\mathbf{q}}\langle n_{\mathbf{q}}\rangle = \sum_{m=1}^{\infty} exp(-\beta\; m\; g_J\; \mu_B\; B_0) *$$

$$* \frac{V}{N(2\pi)^3}\int d^3q\; exp(-m\;\beta(2\hbar\;\langle S^z\rangle(J_0 - J(\mathbf{q})))) \tag{7.440}$$

Comparing this expression with (7.261) which is the analogous term of the linear spin wave theory, we see that the only difference is that in the integrand S is replaced by $\langle S\rangle/\hbar$. This does not influence the q-integration in any way. Therefore, we can take over directly either (7.264) or (7.267)

$$\phi = \frac{V}{N}\left(\frac{k_B T}{4\pi\; D^*}\right)^{3/2} Z_{3/2}(\beta\; g_J\; \mu_B\; B_0) \tag{7.441}$$

$$D^* = 2\;\langle S^z\rangle\; J_1\;\hbar\; a^2 \tag{7.442}$$

The function $Z_m(x)$ is defined in (7.265). We have restricted ourselves here again to the nearest neighbour interactions ($J_{ij} \rightarrow J_1$).

For low temperatures ϕ must be a small quantity so that an expansion of (7.439) in powers of ϕ appears meaningful:

$$\frac{\langle S^z\rangle}{\hbar\; S} = 1 - 2\phi + (2\phi)^2 - \cdots \tag{7.443}$$

which we in a first approximation can terminate after the first term. This gives for the magnetization (7.411)

$$M(T, B_0) = \frac{N}{V}\; g_J\; \frac{\mu_B}{\hbar}\;\langle S^z\rangle \approx \frac{N}{V}\; g_J\;\mu_B\; S(1 - 2\phi)$$

The factor in front of the bracket is exactly the saturation magnetization M_0 (7.259) so that one finds the following expression for the deviation of magnetization from its $T = 0$ value:

$$\frac{M_0 - M(T, B_0)}{M_0} \approx 2\;\frac{V}{N}\left(\frac{k_B T}{4\pi\, D^*}\right)^{3/2} Z_{3/2}(\beta g_J \mu_B B_0) \tag{7.444}$$

One gets the spontaneous magnetization $M_s(T)$ from this by setting $B_0 = 0$. Then on the right-hand side we have instead of the function $Z_{3/2}$ the Riemann ζ-function (7.266) $\zeta(3/2)$. D^* still contains the magnetization itself (7.442). If one restricts oneself to terms of the order of $T^{3/2}$, then we can replace D^* by D (7.280), i.e. in D^* $\langle S^z\rangle$ by $\hbar\; S$. Then we exactly have the Bloch's $T^{3/2}$ law (7.268) for $S = 1/2$ which is reproduced correctly by the Tyablikov approximation discussed here.

In order to find the higher powers in temperature, we can practically take over (7.281); we only have to replace the reduced temperature t by

$$\tau = \frac{\hbar\, S}{\langle S^z \rangle}\, t = \frac{3k_B T}{4\pi\, \hbar\, \langle S^z \rangle\, z\, J_1\, \delta} \tag{7.445}$$

If we make the same assumptions as in Sect. 7.4.1 for the free spin waves, i.e. restrict ourselves to cubic lattices with exchange interactions only between nearest neighbours, then the following holds:

$$\phi = \zeta(3/2)\, \tau^{3/2} + \frac{3\pi\,\delta}{4} \zeta(5/2)\, \tau^{5/2} + \pi^2\, \delta^2\, \alpha^2\, \zeta(7/2)\, \tau^{7/2} + \cdots \tag{7.446}$$

δ and α are the structure factors given in (7.283). Inserting (7.446) in (7.443) we get an equation for $\langle S^z \rangle$ which has to be solved self-consistently. The leading temperature terms can be easily given:

$$\begin{aligned}\frac{\langle S^z \rangle}{\hbar\, S} &= 1 - 2\zeta(3/2)\, t^{3/2} - \frac{3\pi\,\delta}{2}\, \zeta(5/2)\, t^{5/2} \\ &\quad - 2\zeta(3/2)\, t^3 - 2\pi^2\, \delta^2\, \alpha\, \zeta(7/2)\, t^{7/2} - \cdots\end{aligned} \tag{7.447}$$

Except for the t^3 term it agrees with the exact spin wave result of Dyson [10].

We want to finally investigate the *high-temperature behaviour of the magnetization in external field* ($\mathbf{B_0} \neq \mathbf{0}$). For that we expand coth in (7.435) whose argument is small for high temperatures. We first abbreviate

$$b(T) = \frac{g_J\, \mu_B\, B_0}{2k_B T}; \quad a_{\mathbf{q}}(T) = \frac{2\hbar\, \langle S^z \rangle\, (J_0 - J(\mathbf{q}))}{2k_B T} \tag{7.448}$$

$$t_b = \tanh b(T); \quad t_a = \tanh a_{\mathbf{q}}(T) \tag{7.449}$$

With this the integrand in (7.435) can be written as

$$\begin{aligned}\coth(b(T) + a_{\mathbf{q}}(T)) &= \frac{1 + t_a\, t_b}{t_a + t_b} \\ &= \frac{1}{t_b}\left(1 - (1 - t_b^2)\frac{t_a}{t_b + t_a}\right) \\ &= \frac{1}{t_b}\left(1 + (1 - t_b^2)\sum_{n=1}^{\infty}(-1)^n \left(\frac{t_a}{t_b}\right)^n\right)\end{aligned} \tag{7.450}$$

We insert this in (7.435)

$$\frac{\hbar\, S}{\langle S^z \rangle} = \frac{1}{t_b}\left(1 + (1 - t_b^2)\sum_{n=1}^{\infty} \frac{(-1)^n}{t_b^n} \frac{V}{N(2\pi)^3} \int d^3q\, t_a^n\right) \tag{7.451}$$

For t_a we apply the usual expansion of the hyperbolic tangent:

$$t_a = \sum_{n=1}^{\infty} (-1)^{n-1} \frac{2^{2n}(2^{2n}-1)B_n}{(2n)!} (a_{\mathbf{q}}(T))^{2n-1} \tag{7.452}$$

Before we insert (7.452) in (7.451) we introduce a few more abbreviations:

$$\sigma = \frac{\langle S^z \rangle}{\hbar S} = 2\frac{\langle S^z \rangle}{\hbar} \tag{7.453}$$

$$\tau = \frac{2k_B T}{z\,\hbar^2\, J_1} \tag{7.454}$$

$$a_{\mathbf{q}}(T) = \frac{\sigma}{\tau}(1-\gamma_{\mathbf{q}}) \tag{7.455}$$

$$c_m = \frac{V}{N(2\pi)^3} \int d^3q\,(1-\gamma_{\mathbf{q}})^m \tag{7.456}$$

The values of c_m can easily be calculated (Problem 7.18):

$$c_1 = 1\,,\ c_2 = 1 + \frac{1}{z}\,,\ c_3 = 1 + \frac{3}{z}\,,\ \ldots \tag{7.457}$$

For the presumed high temperatures, we consider $a_{\mathbf{q}}$ to be a small quantity,

$$t_a \approx \frac{\sigma}{\tau}(1-\gamma_{\mathbf{q}}) - \frac{1}{3}\left(\frac{\sigma}{\tau}\right)^3 (1-\gamma_{\mathbf{q}})^3 + \cdots \tag{7.458}$$

and expand (7.451) up to the quadratic term in $1/\tau$

$$\sigma^{-1} = \frac{1}{t_b}\left(1 - c_1\frac{1-t_b^2}{t_b}\left(\frac{\sigma}{\tau}\right) + c_2\frac{1-t_b^2}{t_b^2}\left(\frac{\sigma}{\tau}\right)^2 + \cdots\right) \tag{7.459}$$

This equation can be solved by successive iteration:

$$\sigma_{(0)} = t_b = \tanh(g_J\,\mu_B\,B_0/2k_B T) \tag{7.460}$$

In the zeroth approximation we obtain the result (4.127) of the ideal paramagnet (non-interacting local moments in external magnetic field) for $S = 1/2$. The further corrections are produced by the exchange interaction of the spins:

$$\sigma_{(1)} = t_b\left(1 + (1-t_b^2)\frac{1}{\tau}\right) \tag{7.461}$$

In the next step, by inserting $\sigma_{(1)}$ on the right-hand side of (7.459) we obtain the solution which is correct up to the order of $1/\tau^2$:

$$\frac{\langle S^z \rangle}{\hbar\, S} = t_b + t_b(1 - t_b^2)\left(\frac{z\, J_1\, \hbar^2}{2k_B T}\right)$$
$$+\, t_b \left(\frac{z-1}{z} - 2t_b^2\right)\left(\frac{z\, J_1\, \hbar^2}{2k_B T}\right)^2 + \cdots \qquad (7.462)$$

For very high temperatures one can assume the exchange interaction energy $z\, J_1\, \hbar^2$ to be very small compared to the thermal energy $k_B T$. Then, one can handle the Heisenberg model perturbatively [18]. A very good agreement with (7.462) is found. This shows that the RPA approximation (7.422), which appears to be completely arbitrary at first sight, leads to a useful approximate solution of the Heisenberg model for $S = 1/2$ both for low temperatures (7.447) and for high temperatures (7.462). Therefore, even for intermediate temperatures one can expect it to have certain reliability.

7.5.3 Thermodynamic Potentials

In this section, we want to derive the internal and free energies of the Heisenberg ferromagnet. The internal energy U is nothing but the thermodynamic expectation value of the Hamiltonian H:

$$U = \langle H \rangle = -\sum_{i,j} J_{ij}\left(\langle S_i^+\, S_j^- \rangle + \langle S_i^z\, S_j^z \rangle\right) - g_J\, \frac{\mu_B}{\hbar}\, B_0\, N\, \langle S^z \rangle \qquad (7.463)$$

The expectation values $\langle S_i^+\, S_j^- \rangle$ and $\langle S^z \rangle$ were derived and discussed in the last section. What is new now is the term $\langle S_i^z\, S_j^z \rangle$ which can also be expressed in terms of the spectral density $S_{\mathbf{q}}(E)$ (7.430).

With the operator identities valid for $S = 1/2$

$$S_i^-\, S_i^z = \frac{\hbar}{2} S_i^-; \quad S_i^-\, S_i^+ = \hbar^2\, S - \hbar\, S_i^z \qquad (7.464)$$

the following follows from (7.420):

$$S_i^-\left[S_i^+\, ,\, H\right]_- = -\hbar^2 \sum_j J_{ij}\, S_i^-\, S_j^+ - 2\hbar^2 \sum_j J_{ij}\, S_i^z\, S_j^z$$
$$+\, 2\hbar^3\, S \sum_j J_{ij}\, S_j^z + g_J\, \mu_B\, B_0(\hbar^2\, S - \hbar\, S_i^z)$$

Averaging and summing over all the lattice sites lead to

$$
\begin{aligned}
-\sum_{i,j} J_{ij} \langle S_i^z \; S_j^z \rangle = & \frac{1}{2\hbar^2} \sum_i \langle S_i^- \left[S_i^+ \, , \, H \right]_- \rangle \\
& + \frac{1}{2} \sum_{i,j} J_{ij} \langle S_i^- \; S_j^+ \rangle \\
& - J_0 \, \hbar \, S \, N \, \langle S^z \rangle - \frac{N}{2\hbar^2} \, g_J \, \mu_B \, B_0 \, \langle S_i^- \; S_i^+ \rangle
\end{aligned}
$$

Substituting this in (7.463) we get the following intermediate result:

$$
\begin{aligned}
U = E_0(B_0) + & \frac{1}{2} \sum_{i,j} \left((J_0 \, \delta_{ij} - J_{ij}) + \frac{\delta_{ij}}{\hbar^2} \, g_J \, \mu_B \, B_0 \right) \langle S_i^- \; S_j^+ \rangle \\
& + \frac{1}{2\hbar^2} \sum_i \langle S_i^- \left[S_i^+ \, , \, H \right]_- \rangle
\end{aligned} \tag{7.465}
$$

$E_0(B_0)$ is the ground state energy (7.246) of the ferromagnet. With (7.431), we write this expression in terms of wavenumbers: as

$$
\begin{aligned}
U = E_0(B_0) + & \frac{1}{2\hbar^3} \sum_{\mathbf{q}} \int_{-\infty}^{+\infty} dE \, \frac{\hbar \, \omega(\mathbf{q}) \; S_{\mathbf{q}}(E)}{exp(\beta E) - 1} \\
& + \frac{1}{2\hbar^2} \sum_i \langle S_i^- \left[S_i^+ \, , \, H \right]_- \rangle
\end{aligned} \tag{7.466}
$$

$\hbar\omega(\mathbf{q})$ are the free spin waves for $S = 1/2$:

$$
\hbar\omega(\mathbf{q}) = \hbar^2 \left(J_0 - J(\mathbf{q}) \right) + g_J \, \mu_B \, B_0 \tag{7.467}
$$

The remaining expectation value in (7.466) can also be expressed in terms of $S_{\mathbf{q}}(E)$:

$$
\begin{aligned}
& \frac{1}{2\hbar^2} \sum_i \langle S_i^- \left[S_i^+ \, , \, H \right]_- \rangle \\
& = \frac{i\hbar}{2\hbar^2} \sum_i \frac{d}{dt} \langle S_i^-(t') \; S_i^+(t) \rangle \Big|_{t=t'} \\
& = \frac{i\hbar}{2\hbar^2} \sum_{\mathbf{q}} \frac{1}{\hbar} \frac{d}{dt} \int_{-\infty}^{+\infty} dE \, \frac{S_{\mathbf{q}}(E)}{exp(\beta E) - 1} e^{-\frac{i}{\hbar} \, E(t-t')} \Big|_{t=t'} \\
& = \frac{1}{2\hbar^3} \sum_{\mathbf{q}} \int_{-\infty}^{+\infty} dE \, \frac{E \; S_{\mathbf{q}}(E)}{exp(\beta E) - 1}
\end{aligned} \tag{7.468}
$$

In writing this equation, in the first step the equation of motion for time-dependent Heisenberg operators and in the second step the spectral theorem (B.95) have been used. Equations (7.466) and (7.468) finally give the following *exact* relation for the internal energy U:

$$U = E_0(B_0) + \frac{1}{2\hbar^3} \sum_{\mathbf{q}} \int_{-\infty}^{+\infty} dE \, \frac{(E + \hbar\,\omega(\mathbf{q}))\, S_{\mathbf{q}}(E)}{exp(\beta E) - 1} \tag{7.469}$$

This equation can easily be evaluated using (7.430) for $S = 1/2$ ferromagnets in random phase approximation:

$$U = E_0(B_0) + \frac{\langle S^z \rangle}{\hbar} \sum_{\mathbf{q}} \frac{\hbar^2 (J_0 - J(\mathbf{q}))\left(1 + \frac{2\langle S^z \rangle}{\hbar}\right) + 2g_J\,\mu_B\,B_0}{e^{\beta E(\mathbf{q})} - 1} \tag{7.470}$$

We have defined the quasiparticle energies $E(\mathbf{q})$ in (7.428).

In Appendix B, the following relationship between the internal energy and the free energy is derived (B.198):

$$F(T) = U(0) - T \int_0^T dT' \, \frac{U(T') - U(0)}{T'^2} \tag{7.471}$$

Substituting (7.470) gives a formally simple expression ($U(0) = E_0(B_0)$):

$$F(T) = E_0(B_0) + T\,\frac{1}{\hbar} \sum_{\mathbf{q}} \int_0^T dT' \, \frac{\langle S^z \rangle_{T'}}{T'^2} \, \frac{E(\mathbf{q}) + \hbar\,\omega(\mathbf{q})}{exp(\beta E(\mathbf{q})) - 1} \tag{7.472}$$

in which, however, $E(\mathbf{q})$ and $\langle S^z \rangle$ are temperature dependent, so that it is not possible to carry out the T-integration analytically.

7.6 Thermodynamics of $S \geq 1/2$ Ferromagnets

7.6.1 Green's Functions

The central quantity in our discussions is the magnetization $M(T, B_0)$ (7.411), which, up to some scalar factors, is determined by the expectation value $\langle S^z \rangle$ of the z-component of the spin operator. We therefore need a Green's function, with whose help it is possible to evaluate this expectation value in the simplest possible manner. For the special case of $S = 1/2$ which was discussed in the last section (7.433)

$$\langle S_i^- \, S_i^+ \rangle = \hbar^2 \, S - \hbar \, \langle S^z \rangle \tag{7.473}$$

holds, so that the appropriate Green's function is

$$G_{ij}(E) = \langle\langle S_i^+ \, ; \, S_j^- \rangle\rangle_E^{ret} \tag{7.474}$$

which, via the spectral theorem (B.95), fixes the expectation value $\langle S_i^- \, S_i^+ \rangle$. Even though the latter is valid also for $S > 1/2$, its relationship with $\langle S^z \rangle$ is not so simple as in (7.473). For $S > 1/2$, according to (7.8) the following holds:

$$\langle S_i^- \, S_i^+ \rangle = \hbar^2 \, S(S+1) - \hbar \, \langle S_i^z \rangle - \langle (S_i^z)^2 \rangle \tag{7.475}$$

The term $\langle (S_i^z)^2 \rangle$ creates difficulties. We must look for equations which connect the powers of the spin operators $(S_i^z)^n$ with each other. For this purpose, the Green's function (7.474) is not suitable. Tyablikov suggested the following:

$$G_{ij}^{(n)}(E) = \langle\langle S_i^+ \, ; \, (S_j^z)^n \, S_j^- \rangle\rangle_E^{ret} \tag{7.476}$$

We will see later that we can limit ourselves to $n = 0, \, 1, \, \ldots, \, 2S - 1$. Since the operator before the semicolon is the same as in $G_{ij}(E)$ in (7.474), which was the Green's function used for the special case $S = 1/2$, in the equation of motion (7.421) only the inhomogeneity on the right-hand side will change.

$$\begin{aligned}(E - g_J \, \mu_B \, B_0) G_{ij}^{(n)}(E) &= \hbar \langle \left[S_i^+ \, , \, (S_j^z)^n \, S_j^- \right]_- \rangle \\ - 2\hbar \sum_m J_{im} &\left\{ \langle\langle S_i^z \, S_m^+ \, ; \, (S_j^z)^n S_j^- \rangle\rangle - \langle\langle S_m^z \, S_i^+ \, ; \, (S_j^z)^n S_j^- \rangle\rangle \right\}\end{aligned} \tag{7.477}$$

We will apply the same RPA decoupling as for the case of $S = 1/2$ (7.422). Since the operator to the right of the semicolon in the Green's function is inactive so far as the equation of motion and its decoupling are concerned, we can directly write down formally the result analogous to (7.427):

$$G_{\mathbf{q}}^{(n)}(E) = \frac{\hbar \, \langle \left[S_i^+ \, , \, (S_i^z)^n \, S_i^- \right]_- \rangle}{E - E(\mathbf{q})} \tag{7.478}$$

The quasiparticle energies are exactly the same as in (7.428). For the associated spectral density the following now holds:

$$S_{\mathbf{q}}^{(n)}(E) = \hbar \langle \left[S_i^+ \, , \, (S_i^z)^n \, S_i^- \right]_- \rangle \, \delta(E - E(\mathbf{q})) \tag{7.479}$$

With the help of the spectral theorem (B.95) it follows analogously to (7.434)

$$\langle (S_i^z)^n \, S_i^- \, S_i^+ \rangle = \langle \left[S_i^+ \, , \, (S_i^z)^n \, S_i^- \right]_- \rangle \, \frac{1}{N} \sum_{\mathbf{q}} \left(e^{\beta E(\mathbf{q})} - 1 \right)^{-1} \tag{7.480}$$

In the next section we will see how one can indeed derive predictions on $\langle S^z \rangle$ from this expression for an arbitrary spin S.

The selection of (7.476) as a suitable choice of the Green's function is not the only possibility. Here we mention another suggestion from Tahir-Kheli and ter

Haar [19]:

$$G_{ij}^{(n)}(E) = \langle\langle S_i^+ \, ; \, (S_j^-)^n \, (S_j^+)^{n-1} \rangle\rangle_E^{ret} \tag{7.481}$$

The operator to the left of the semicolon is again the same as in (7.474) and (7.476). Applying once again the RPA decoupling (7.422), we can immediately give the result that follows from (7.481):

$$\langle (S_i^-)^n \, (S_i^+)^n \rangle = \langle \left[S_i^+ \, , \, (S_i^-)^n \, (S_i^+)^{n-1} \right]_- \rangle \, \phi(S) \tag{7.482}$$

where, as in (7.439), we define

$$\phi(S) = \frac{1}{N} \sum_{\mathbf{q}} \left(e^{\beta E(\mathbf{q})} - 1 \right)^{-1} \tag{7.483}$$

7.6.2 Spontaneous Magnetization

We will try using (7.480) and (7.482) to determine $\langle S^z \rangle$ for an arbitrary spin. We start with the Tyablikov result (7.480). With (7.475) for the left-hand side the following holds:

$$\langle (S_i^z)^n \, S_i^- \, S_i^+ \rangle = \hbar^2 \, S(S+1) \langle (S_i^z)^n \rangle - \hbar \, \langle (S_i^z)^{n+1} \rangle - \langle (S_i^z)^{n+2} \rangle \tag{7.484}$$

We need this equation only for $n = 0, \; 1, \; \ldots, \; 2S - 1$, since because of the operator identity (7.9),

$$\prod_{m_s = -S}^{+S} (S_i^z - \hbar \, m_s) = 0 \tag{7.485}$$

the chain of equations (7.484) automatically terminates. For $n = 2S - 1$ the highest power of S_i^z on the right-hand side turns out to be $(S_i^z)^{2S+1}$. This, however, can be due to (7.485) expressed by $(S_i^z)^n$ with $n = 0, \; 1, \; \ldots, \; 2S$:

$$\langle (S_i^z)^{2S+1} \rangle = \sum_{n=0}^{2S} \alpha_n(S) \, \langle (S_i^z)^n \rangle \tag{7.486}$$

The $\alpha_n(S)$ are numbers which can easily be determined from (7.485) (see Problem 7.3).

We now need the commutator on the right-hand side of Eq. (7.480). One can easily prove by induction that

$$S_i^+ \, (S_i^z)^n = (S_i^z - \hbar)^n \, S_i^+ \tag{7.487}$$

With this one calculates without any difficulty the required commutator:

$$\begin{aligned}&\left[S_i^+ \,,\, (S_i^z)^n\, S_i^-\right]_- = 2\hbar\, S_i^z(S_i^z-\hbar)^n\\&+((S_i^z-\hbar)^n-(S_i^z)^n)(\hbar^2\, S(S+1)-\hbar\, S_i^z-(S_i^z)^2)\end{aligned} \tag{7.488}$$

Finally substituting (7.484) and (7.488) in (7.480) one obtains the following system of equations:

$$\begin{aligned}&\hbar^2\, S(S+1)\,\langle (S_i^z)^n\rangle - \hbar\,\langle (S_i^z)^{n+1}\rangle - \langle (S_i^z)^{n+2}\rangle\\&=[2\hbar\,\langle S_i^z(S_i^z-\hbar)^n\rangle + \langle((S_i^z-\hbar)^n-(S_i^z)^n)(\hbar^2\, S(S+1)\\&-\hbar\, S_i^z-(S_i^z)^2)\rangle]\,\phi(S)\\&(n=0,\;1\,2,\;\ldots,\;2S-1)\end{aligned} \tag{7.489}$$

This along with (7.485) is a system of $(2S+1)$ equations for the $(2S+1)$ expectation values $\langle (S_i^z)^m\rangle$ $(m=1,\;\ldots,\;2S+1)$, which naturally can in principle be solved. One should, however, pay attention that in ϕ, according to (7.483), there is $E(\mathbf{q})$ due to which $\phi(S)$ again depends on $\langle S^z\rangle$. We want to investigate two examples $S=1/2$ and $S=1$ a little further:

(a) $S=1/2$
In this case it follows from (7.485) that

$$(S_i^z)^2=\frac{\hbar^2}{4} \tag{7.490}$$

We need (7.489) only for $n=0$:

$$\frac{3}{4}\,\hbar^2-\hbar\,\langle S^z\rangle-\langle (S^z)^2\rangle = 2\hbar\,\langle S^z\rangle\,\phi(1/2). \tag{7.491}$$

From this follows the result (7.434) or (7.439) which was already discussed in Sect. 7.5 for the $S=1/2$ ferromagnets:

$$\langle S^z\rangle_{S=1/2}=\frac{\hbar}{2}\,(1+2\phi(1/2))^{-1} \tag{7.492}$$

Thus the special case $S=1/2$ is contained in the theory presented here.

(b) S=1
In this case it follows from (7.485) that

$$(S_i^z)^3=\hbar^2\, S_i^z \tag{7.493}$$

The system of equations now has to be evaluated for $n=0$ and $n=1$:

$$2\hbar^2-\hbar\,\langle S_i^z\rangle-\langle (S_i^z)^2\rangle = 2\hbar\,\langle S_i^z\rangle\,\phi(1) \tag{7.494}$$

$$
\begin{aligned}
&2\hbar^2 \langle S_i^z \rangle - \hbar \langle (S_i^z)^2 \rangle - \langle (S_i^z)^3 \rangle \\
&= \phi(1)[2\hbar \langle (S_i^z)^2 \rangle - 2\hbar^2 \langle S_i^z \rangle - 2\hbar^3 + \hbar^2 \langle S_i^z \rangle + \hbar \langle (S_i^z)^2 \rangle]
\end{aligned}
\tag{7.495}
$$

This can be solved for $\langle S_i^z \rangle$ without any difficulty:

$$
\langle S_i^z \rangle_{S=1} = \hbar \frac{1 + 2\phi(1)}{1 + 3\phi(1) + 3\phi^2(1)} \tag{7.496}
$$

We also get the expectation value of $(S_i^z)^2$:

$$
\langle (S_i^z)^2 \rangle_{S=1} = 2\hbar^2 \frac{\phi(1)}{1 + 3\phi(1)} + \hbar \langle S_i^z \rangle \frac{1 + \phi(1)}{1 + 3\phi(1)} \tag{7.497}
$$

In this manner we can evaluate stepwise $\langle S_i^z \rangle$ for any arbitrary spin S. For $S > 1$, it is possible to give the solution in a general form [20]:

$$
\langle S_i^z \rangle_S = \hbar \frac{[S - \phi(S)][1 + \phi(S)]^{2S+1} + [1 + S + \phi(S)][\phi(S)]^{2S+1}}{[1 + \phi(S)]^{2S+1} - [\phi(S)]^{2S+1}} \tag{7.498}
$$

This result actually follows from the Tyablikov result (7.480). Before we continue our discussion for different temperature regions, we want to show that the result of Tahir-Kheli and ter Haar (7.482) also leads exactly to the same predictions. For the expectation values on the left-hand side of (7.482) we use the relations (Problem 7.2)

$$
\left[(S_i^-)^n \, , \, S_i^z\right]_- = n \, \hbar \, (S_i^-)^n \tag{7.499}
$$

$$
\left[(S_i^-)^n \, , \, (S_i^z)^2\right]_- = n^2 \, \hbar^2 \, (S_i^-)^n + 2n \, \hbar \, S_i^z \, (S_i^-)^n \tag{7.500}
$$

With this we can show that

$$
\begin{aligned}
(S_i^-)^n \, (S_i^+)^n &= (S_i^-)^{n-1} (\hbar^2 \, S(S+1) - \hbar \, S_i^z - (S_i^z)^2) \, (S_i^+)^{n-1} \\
&= [\hbar^2 \, S(S+1) - n(n-1) \, \hbar^2 - (2n-1) \, \hbar \, S_i^z \\
&\quad - (S_i^z)^2] (S_i^-)^{n-1} \, (S_i^+)^{n-1}
\end{aligned}
\tag{7.501}
$$

holds, from which eventually the following relation results:

$$
\begin{aligned}
(S_i^-)^n \, (S_i^+)^n = \prod_{p=1}^{n} &[\hbar^2 \, S(S+1) - (n-p)(n-p+1)\hbar^2 \\
&- (2n - 2p + 1)\hbar \, S_i^z - (S_i^z)^2]
\end{aligned}
\tag{7.502}
$$

$$(n \geq 1)$$

The commutator on the right-hand side of (7.482) now remains to be evaluated:

$$\left[S_i^+ \,,\, (S_i^-)^n \,(S_i^+)^{n-1}\right]_- = \left[S_i^+ \,,\, (S_i^-)^n\right]_- \,(S_i^+)^{n-1}$$

One can prove by induction (Problem 7.2)

$$\left[S_i^+ \,,\, (S_i^-)^n\right]_- = (2n\,\hbar\, S_i^z + \hbar^2\, n(n-1))\,(S_i^-)^{n-1} \tag{7.503}$$

With this we obtain the following recursion formula:

$$\begin{aligned}\left[S_i^+ \,,\, (S_i^-)^n \,(S_i^+)^{n-1}\right]_- & \\ &= n\,\hbar(2S_i^z + \hbar\,(n-1))(S_i^-)^{n-1}\,(S_i^+)^{n-1} \\ &\qquad (n \geq 1)\end{aligned} \tag{7.504}$$

For **n** $>$ **1**, we can express the remaining operator product on the right-hand side using (7.502):

$$\begin{aligned}\left[S_i^+ \,,\, (S_i^-)^n \,(S_i^+)^{n-1}\right]_- &= 2\hbar\, S_i^z\,, \quad \text{for } \mathbf{n} = \mathbf{1} \\ \left[S_i^+ \,,\, (S_i^-)^n \,(S_i^+)^{n-1}\right]_- &= (2n\,\hbar\, S_i^z + n(n-1)\hbar^2) * \\ * \prod_{p=1}^{n-1} [\hbar^2\, S(S+1) - (n-1-p)(n-p)\hbar^2 & \\ - (2n-2p-1)\hbar\, S_i^z - (S_i^z)^2]\,, \quad \text{for } \mathbf{n} \geq \mathbf{2} &\end{aligned} \tag{7.505}$$

Substituting (7.502) and (7.505) in (7.482), we obtain with (7.485) again a complete system of equations for determining $\langle S^z \rangle$. This will be demonstrated with a couple of examples:

(a) S=1/2
In this case it is $n = 1$ so that from (7.485) and (7.502) and (7.505) it directly follows that

$$\frac{3}{4}\hbar^2 - \hbar\,\langle S^z \rangle - \langle (S^z)^2 \rangle = 2\hbar\,\langle S^z \rangle\,\phi(1/2) \tag{7.506}$$

This equation is identical with (7.491) and therefore reproduces the result (7.492) for $\langle S^z \rangle_{S=1/2}$. Thus both the Green's functions (7.476) and (7.481) lead to the same physical results for $S = 1/2$.

(b) S=1
The equations that follow for $n = 1$ and $n = 2$ from (7.482) and (7.502) are

$$2\hbar^2 - \hbar\,\langle S^z \rangle - \langle (S^z)^2 \rangle = 2\hbar\,\langle S^z \rangle\,\phi(1) \tag{7.507}$$

$$\begin{aligned}&\langle -6\hbar^3\, S^z + \hbar^2\, (S^z)^2 + 4\hbar (S^z)^3 + (S^z)^4 \rangle \\ &= \langle 4\hbar^4 + 6\hbar^3\, S^z - 6\hbar^2\, (S^z)^2 - 4\hbar\, (S^z)^3 \rangle\, \phi(1)\end{aligned} \tag{7.508}$$

In addition, from (7.485) we get the following operator identities;

$$(S^z)^3 = \hbar^2\, S^z \tag{7.509}$$

$$(S^z)^4 = \hbar^2\, (S^z)^2 \tag{7.510}$$

so that (7.508) becomes

$$-\,\hbar^2\, \langle S^z \rangle + \hbar\, \langle (S^z)^2 \rangle = (2\hbar^3 + \hbar^2\, \langle S^z \rangle - 3\hbar\, \langle (S^z)^2 \rangle)\, \phi(1) \tag{7.511}$$

(7.507) and (7.511) agree exactly with (7.494) and (7.495) so that for $\langle S^z \rangle_{S=1}$ also the earlier result (7.496) is reproduced. One can also show this from the general result (7.498) for $S > 1$ so that the two Green's functions (7.476) and (7.481) are completely equivalent. Therefore, we can continue further discussions from (7.492), (7.496) and (7.498).

7.6.2.1 Low-Temperature Region

At low temperatures, the expansion (7.446) for $\phi(S)$ is valid. Further, $\phi(S)$ should itself be a small quantity so that we can expand (7.492), (7.496) and (7.498)

$$\langle S^z \rangle_{1/2} = \frac{\hbar}{2} - \hbar\, \phi(1/2) + 2\hbar\, \phi^2(1/2) + \mathcal{O}(\phi^3) \tag{7.512}$$

$$\langle S^z \rangle_{S \geq 1} = \hbar\, S - \hbar\, \phi(S) + \mathcal{O}(\phi^3) \tag{7.513}$$

(7.512) agrees with (7.443); therefore, we obtain exactly the same result (7.447) again for $S = 1/2$. Therefore, here only the case $S \geq 1$ is interesting. With (7.446) in (7.513) we first get

$$\begin{aligned}\langle S^z \rangle_{S \geq 1} = \hbar\, S - \hbar \Bigg(& \zeta(3/2)\, \tau^{3/2} + \frac{3\pi\, \delta}{4}\, \zeta(5/2)\, \tau^{5/2} \\ & + \pi^2\, \delta^2\, \alpha^2\, \zeta(7/2)\, \tau^{7/2} + \cdots \Bigg)\end{aligned} \tag{7.514}$$

α and δ are the structure factors given in (7.283). The reduced temperature τ is defined in (7.445):

$$\tau = \frac{\hbar\, S}{\langle S^z \rangle}\, t = \frac{3k_B T}{4\pi\, \hbar\, \langle S^z \rangle\, z\, J_1\, \delta}$$

That means (7.447)

$$\tau^{n/2} = t^{n/2} \left(\frac{\hbar\, S}{\langle S^z \rangle} \right)^{n/2} = \left(1 - \frac{\zeta(3/2)}{S}\, t^{3/2} + \cdots \right)^{-n/2} t^{n/2}$$
$$\approx \left(1 + \frac{n}{2S}\, \zeta(3/2)\, t^{3/2} \right) t^{n/2} \tag{7.515}$$

Therefore, the correct low-temperature expansion of $\langle S^z \rangle$ reads as

$$\langle S^z \rangle_{S \geq 1} = \hbar \left[S - \zeta(3/2)\, t^{3/2} - \frac{3\pi\, \delta}{4}\, \zeta(5/2)\, t^{5/2} - \frac{3}{2S}\, \zeta^2(3/2)\, t^3 - \pi^2\, \delta^2\, \alpha\, \zeta(7/2)\, t^{7/2} + \cdots \right] \tag{7.516}$$

A comparison with the Dyson solution [10] leads to an exact agreement in the t^0, $t^{3/2}$, $t^{5/2}$ and $t^{7/2}$ terms. As in the $S = 1/2$ solution (7.447), we obtain in addition a t^3 term. In the case $S = 1/2$ also this correction term originates from the term $(2\hbar\, \phi^2(1/2))$ in the expansion (7.512). Since in the expansion (7.513) for $S \geq 1$ a ϕ^2 term does not appear, here $\phi(S)$ itself is responsible for the t^3 correction. This error arises mainly due to the imprecise RPA decoupling (7.422), which according to (7.428) leads to spin wave energies, whose first temperature correction behaves as $T^{3/2}$. One can see this directly when one inserts (7.516) in (7.428).

$$E(\mathbf{q}) = A(S) + B(S)\, T^{3/2} + \cdots \tag{7.517}$$

But we know that according to (7.318)

$$E(\mathbf{q}) = A(S) + B(S)\, T^{5/2} + \cdots \tag{7.518}$$

must hold.

7.6.2.2 Critical Region ($\mathbf{T \stackrel{<}{\to} T_C}$)

In the neighbourhood of the transition temperature $\phi(S)$ is no more a small quantity but a rather large quantity. Therefore, we expand (7.498) in powers of $1/\phi$:

$$\langle S^z \rangle_S = \hbar\, \frac{[S - \phi(S)][1/\phi + 1]^{2S+1} + [1 + S + \phi]}{[1/\phi + 1]^{2S+1} - 1}$$
$$= \frac{\hbar}{(2S+1)/\phi} \left\{ (S - \phi) \left(1 + \frac{2S+1}{\phi} + \frac{S(2S+1)}{\phi^2} + \frac{(2S+1)2S(2S-1)}{6\phi^3} + \cdots \right) + 1 + S + \phi \right\}$$

$$\langle S^z \rangle_S = \hbar \, \frac{S(S+1)}{3\phi(S)} + \mathcal{O}(1/\phi^3) \tag{7.519}$$

$\phi(S)$ still contains $\langle S^z \rangle$ as one recognizes from (7.483):

$$1 + 2\phi(S) = \frac{1}{N} \sum_{\mathbf{q}} \coth \frac{\sigma(1-\gamma_{\mathbf{q}})}{\tau} \tag{7.520}$$

Here we have used the abbreviations (7.453) and (7.454):

$$\sigma = \frac{\langle S^z \rangle}{\hbar \, S}; \quad \tau = \frac{k_B T}{z \, J_1 \, \hbar^2 \, S} \tag{7.521}$$

Because of σ the argument of coth in (7.520) is small. Therefore, as in (7.436), we expand and use the relation (7.456):

$$1 + 2\phi(S) = c_{-1} \, \frac{\tau}{\sigma} + c_1 \, \frac{\sigma}{3\tau} - c_3 \, \frac{\sigma^3}{45\tau^3} + \cdots \tag{7.522}$$

c_1, c_3 are simple numbers from (7.457). The constant c_{-1} can be calculated for simple cubic lattices [19]:

$$\begin{aligned} c_{-1}^{sc} &= 1.51638 \\ c_{-1}^{bcc} &= 1.39320 \\ c_{-1}^{fcc} &= 1.34466 \end{aligned} \tag{7.523}$$

From (7.522) one can further estimate

$$\phi^{-1}(S) = \frac{2\sigma}{\tau \, c_{-1}} \left(1 + \frac{\sigma}{\tau \, c_{-1}} - \frac{\sigma^3}{3\tau^3 \, c_{-1}} + \cdots \right) \tag{7.524}$$

We substitute this in (7.519) and let $T \to T_C$, i.e. drive $\langle S^z \rangle \to 0$. This provides a simple equation for the Curie temperature T_C:

$$\begin{aligned} k_B T_C &= k_B \, \theta_C / c_{-1} \\ k_B \, \theta_C &= \tfrac{2}{3} \, S(S+1) \, z \, J_1 \, \hbar^2 \end{aligned} \tag{7.525}$$

Except for the factor $1/c_{-1}$ this is exactly the molecular field result (7.138) for the Curie temperature θ_C. Since according to (7.523) $1/c_{-1}$ is always less than one, the Curie temperature calculated here is less than that in the molecular field approximation. This is in agreement with experimental observations.

7.6.2.3 High-Temperature Region

For temperatures well above T_C, we are interested primarily in the *magnetic susceptibility* χ. For very high temperatures and not too strong fields, we can assume

$$\phi(S) \gg 1 \tag{7.526}$$

and use the expression (7.519) for the expectation value $\langle S^z \rangle$. For $B_0 = 0$ naturally $\phi(S) = \infty$ and consequently $\langle S^z \rangle = 0$. From the magnetization

$$M(T, B_0) = \frac{N}{V} g_J \mu_B \frac{1}{\hbar} \langle S^z \rangle \approx \frac{N}{V} g_J \mu_B \frac{2}{3} S(S+1) \frac{1}{2\phi(S)} \tag{7.527}$$

we get the susceptibility

$$\begin{aligned} \chi_T &= \mu_0 \left(\frac{\partial M}{\partial B_0} \right)_{T,(B_0=0)} \\ &= \frac{N}{V} g_J \mu_B \frac{2}{3} S(S+1) \left(\frac{\partial}{\partial B_0} \frac{1}{2\phi(S)} \right)_T (B_0 = 0) \end{aligned} \tag{7.528}$$

According to (7.459), for high-temperature region the following holds:

$$\begin{aligned} 2\phi(S) &\approx 1 + 2\phi(S) \\ &= \frac{1}{t_b} \left(1 - \frac{1 - t_b^2}{t_b} \left(\frac{\sigma}{\tau} \right) + \frac{z+1}{z} \frac{1 - t_b^2}{t_b^2} \left(\frac{\sigma}{\tau} \right)^2 + \cdots \right) \end{aligned} \tag{7.529}$$

Here

$$t_b = \tanh \left(\frac{g_J \mu_B B_0}{2k_B T} \right) \tag{7.530}$$

$$\frac{\sigma}{\tau} = \frac{z \hbar J_1 \langle S^z \rangle}{k_B T} = \frac{\theta_C}{T} \frac{1}{2\phi(S)} \tag{7.531}$$

Substituting these in (7.529), we get an equation for $\phi(S)$,

$$\begin{aligned} 2\phi(S) \approx \frac{1}{t_b} \bigg(1 &- \frac{1 - t_b^2}{t_b} \left(\frac{\theta_C}{T} \right) \frac{1}{2\phi(S)} \\ &+ \frac{z+1}{z} \frac{1 - t_b^2}{t_b^2} \left(\frac{\theta_C}{T} \right)^2 \frac{1}{4\phi^2(S)} + \cdots \bigg) \end{aligned} \tag{7.532}$$

which can be solved successively up to an arbitrary degree of accuracy. In the zeroth and first order the following holds:

$$\left(\frac{1}{2\phi(S)}\right)^{(0)} = t_b \Rightarrow \left(\frac{1}{2\phi(S)}\right)^{(1)} = t_b \left(1 + (1 - t_b^2)\left(\frac{\theta_C}{T}\right)\right) \tag{7.533}$$

Substituting this in (7.529) we obtain an expression which is exact up to the quadratic term in $1/T$:

$$\frac{1}{2\phi(S)} = t_b \left(1 + (1 - t_b^2)\left(\frac{\theta_C}{T}\right) + \right.$$
$$\left. + (1 - t_b^2)\left(\frac{z-1}{z} - 2t_b^2\right)\left(\frac{\theta_C}{T}\right)^2 + \cdots\right) \tag{7.534}$$

The derivative at $B_0 = 0$

$$\left(\frac{\partial}{\partial B_0}\frac{1}{2\phi(S)}\right)_{B_0=0} = \frac{g_J\,\mu_B}{2k_B T}\left(1 + \frac{\theta_C}{T} + \frac{z-1}{z}\left(\frac{\theta_C}{T}\right)^2 + \cdots\right)$$

fixes in (7.528) the susceptibility

$$\chi_T = \mu_0 \frac{N}{V}(g_J\,\mu_B)^2 \frac{S(S+1)}{3k_B T} *$$
$$* \left(1 + \frac{\theta_C}{T} + \frac{z-1}{z}\left(\frac{\theta_C}{T}\right)^2 + \mathcal{O}(1/T^3)\right) \tag{7.535}$$

If we terminate the series just after the second term and use

$$1 + \frac{\theta_C}{T} \approx \left(1 - \frac{\theta_C}{T}\right)^{-1}$$

we get the Curie–Weiss law (7.147) of the molecular field theory, which in any case should be valid at sufficiently high temperatures.

In summary one can say that the RPA decoupling (7.422) leads to reliable results both for $S = 1/2$ and also for $S > 1/2$. Therefore, it possesses a kind of belated justification. It, however, remains unsatisfactory that one cannot offer direct physical justification for the decoupling (7.422). From this point of view, the Callen decoupling, which will be discussed in the next section, represents a distinct improvement.

7.6.3 The Callen Method

In order to calculate the expectation value $\langle S^z \rangle$, which fixes the temperature and field dependence of the magnetization, the choice of a suitable Green's function is decisive. For the special case of $S = 1/2$ the choice (7.416) is unique, but not for $S \geq 1$. The difficulties that arise for higher spins were discussed in detail in connection with the Green's functions (7.476) and (7.481). From the Green's functions defined there, with the help of the spectral theorem, it was possible to derive a system of $2S$ equations, from which $\langle S^z \rangle$ could be determined. As the Green's functions cannot be calculated exactly, naturally, the solution depends on the approximations used. In all the methods discussed up to now, the RPA decoupling (7.422) was used. The results obtained in this way are quite satisfying. In high-temperature region, we get the Curie–Weiss law, for the Curie temperature T_C a clear improvement compared to the molecular field approximation and in spin wave region a correct reproduction of the $T^{3/2}$, $T^{5/2}$ and $T^{7/2}$ terms of the Dyson theory. What is disturbing is the T^3 term which is clearly arising due to the approximation used.

The method that will be discussed now, which was suggested by Callen [21], has advantages from two points of view as compared to the procedure of the last section. First, the decoupling of the chain of equations of motion can be physically better justified than the RPA decoupling (7.422), and second, the disturbing T^3 term for $S > 1/2$ in the low-temperature expansion disappears. In particular for large spin, the Callen method represents an improvement over the theories of the last section.

Starting point is the following parametrized Green's function:

$$G_{ij}^{(a)}(E) = \langle\langle S_i^+ \, ; \, exp(a\, S_j^z)\, S_j^- \rangle\rangle_E^{ret} \tag{7.536}$$

The operator before the semicolon is the same as the one in the Green's functions (7.416), (7.476) and (7.481) used earlier. As a result, the equation of motion changes in comparison to those of the other functions only through the inhomogeneity on the right-hand side,

$$\eta(a) \equiv \langle \left[S_i^+ \, , \, exp(a\, S_i^z)\, S_i^- \right]_- \rangle \tag{7.537}$$

The operator to the right of the semicolon $B_j = exp(a\, S_j^z)\, S_j^-$ remains, for the moment, inactive. With (7.420), it directly follows that

$$\begin{aligned}(E - g_J\, \mu_B\, B_0) G_{ij}^{(a)}(E) &= \hbar\, \eta(a)\, \delta_{ij} \\ - 2\hbar \sum_m J_{im} &\left\{ \langle\langle S_i^z\, S_m^+ \, ; \, B_j \rangle\rangle_E^{ret} - \langle\langle S_m^z\, S_i^+ \, ; \, B_j \rangle\rangle_E^{ret} \right\}\end{aligned} \tag{7.538}$$

The RPA decoupling (7.422) consist in replacing in the higher Green's functions on the right-hand side of this equation of motion the operators S_i^z and S_m^z by their thermodynamic expectation values $\langle S^z \rangle$. Here we will use a different procedure, which will first be explained for the special case of $S = 1/2$. For this case, there are three possible representations for the operator S_i^z:

$$S_i^z = \frac{1}{2\hbar}\left(S_i^+ S_i^- - S_i^- S_i^+\right) \tag{7.539}$$

$$S_i^z = \hbar\, S - \frac{1}{\hbar}\, S_i^- S_i^+ \tag{7.540}$$

$$S_i^z = -\hbar\, S + \frac{1}{\hbar}\, S_i^+ S_i^- \tag{7.541}$$

We multiply the first equation with $(1-\alpha)$ and the second with α and add the two equations:

$$S_i^z = \alpha\, \hbar\, S + \frac{1}{2\hbar}(1-\alpha)\, S_i^+ S_i^- - \frac{1}{2\hbar}(1+\alpha)\, S_i^- S_i^+ \tag{7.542}$$

α is a parameter which is still to be fixed. Substituting (7.542) in (7.538) for S_i^z and S_m^z, one obtains Green's functions of the following form:

$$\begin{array}{ll} \langle\langle S_i^+ S_i^- S_m^+ \,;\, B_j\rangle\rangle \,; & \langle\langle S_i^- S_i^+ S_m^+ \,;\, B_j\rangle\rangle \\ \langle\langle S_m^+ S_m^- S_i^+ \,;\, B_j\rangle\rangle \,; & \langle\langle S_m^- S_m^+ S_i^+ \,;\, B_j\rangle\rangle \end{array} \tag{7.543}$$

We will simplify them through a *symmetric decoupling*, which at first is arbitrary, taking into account only spin conservation.

$$\langle\langle S_i^+ S_i^- S_m^+ \,;\, B_j\rangle\rangle \Rightarrow \langle S_i^+ S_i^-\rangle\langle\langle S_m^+ \,;\, B_j\rangle\rangle + \langle S_i^- S_m^+\rangle\langle\langle S_i^+ \,;\, B_j\rangle\rangle \tag{7.544}$$

$$\langle\langle S_i^- S_i^+ S_m^+ \,;\, B_j\rangle\rangle \Rightarrow \langle S_i^- S_i^+\rangle\langle\langle S_m^+ \,;\, B_j\rangle\rangle + \langle S_i^- S_m^+\rangle\langle\langle S_i^+ \,;\, B_j\rangle\rangle \tag{7.545}$$

$$\langle\langle S_m^+ S_m^- S_i^+ \,;\, B_j\rangle\rangle \Rightarrow \langle S_m^+ S_m^-\rangle\langle\langle S_i^+ \,;\, B_j\rangle\rangle + \langle S_m^- S_i^+\rangle\langle\langle S_m^+ \,;\, B_j\rangle\rangle \tag{7.546}$$

$$\langle\langle S_m^- S_m^+ S_i^+ \,;\, B_j\rangle\rangle \Rightarrow \langle S_m^- S_m^+\rangle\langle\langle S_i^+ \,;\, B_j\rangle\rangle + \langle S_m^- S_i^+\rangle\langle\langle S_m^+ \,;\, B_j\rangle\rangle \tag{7.547}$$

If we substitute the ansatz (7.542) in the "higher" Green's functions of the equation of motion (7.538) and decouple according to (7.544), (7.545), (7.546) and (7.547) we get

$$\langle\langle S_i^z S_m^+ \,;\, B_j\rangle\rangle \Rightarrow \langle S^z\rangle\langle\langle S_m^+ \,;\, B_j\rangle\rangle - \frac{\alpha}{\hbar}\langle S_i^- S_m^+\rangle\langle\langle S_i^+ \,;\, B_j\rangle\rangle \tag{7.548}$$

$$\langle\langle S_m^z S_i^+ \,;\, B_j\rangle\rangle \Rightarrow \langle S^z\rangle\langle\langle S_i^+ \,;\, B_j\rangle\rangle - \frac{\alpha}{\hbar}\langle S_m^- S_i^+\rangle\langle\langle S_m^+ \,;\, B_j\rangle\rangle \tag{7.549}$$

The next step is to physically fix the still undetermined parameter α. The idea is to choose α such that the error arising out of the decoupling (7.548) and (7.549) remains as small as possible. One recognizes from (7.542) that for $\alpha = 0$, S_i^z is represented by (7.539). The operator combination $\frac{1}{2\hbar}(S_i^+ \, S_i^- - S_i^- \, S_i^+)$ gives the fluctuation of S_i^z around zero. On the other hand our decoupling captures these combinations $S_i^+ \, S_i^-$ and $S_i^- \, S_i^+$. $\alpha = 0$ would have been a good choice, had these fluctuations been small, so that the consequent "decoupling error" is not so grave. Therefore

$$\alpha = 0 \quad \text{for} \quad \langle S^z \rangle \approx 0$$

For $\alpha = 1$ (7.542) becomes the representation (7.540) for S^z. In this case, the operator product $S_i^- \, S_i^+$ gives the deviation of S_i^z from the value $\hbar \, S$. Thus, $\alpha = 1$ would be the choice if these deviations are small:

$$\alpha = 1 \quad \text{for} \quad \langle S^z \rangle \approx \hbar \, S$$

With $\alpha = -1$ we get (7.541) from (7.542). With a corresponding justification, we can choose

$$\alpha = -1 \quad \text{for} \quad \langle S^z \rangle \approx -\hbar \, S$$

Obviously, these three limiting cases will be simultaneously satisfied by the following ansatz:

$$\alpha_{S=1/2} = \frac{\langle S^z \rangle}{\hbar \, S} \tag{7.550}$$

The above considerations are valid exclusively for the case $S = 1/2$. We will now generalize to an arbitrary value of S. The relations analogous to (7.539), (7.540) and (7.541) read as

$$S_i^z = \frac{1}{2\hbar} \, (S_i^+ \, S_i^- - S_i^- \, S_i^+) \tag{7.551}$$

$$S_i^z = \hbar \, S(S+1) - \frac{1}{\hbar} \, (S_i^z)^2 - \frac{1}{\hbar}(S_i^- \, S_i^+) \tag{7.552}$$

$$S_i^z = -\hbar \, S(S+1) + \frac{1}{\hbar} \, (S_i^z)^2 + \frac{1}{\hbar}(S_i^+ \, S_i^-) \tag{7.553}$$

We again multiply the first equation with $(1 - \alpha)$, the second with α and add both:

$$\begin{aligned} S_i^z &= \alpha \, \hbar \, S(S+1) + \frac{1}{2\hbar}(1-\alpha) \, S_i^+ \, S_i^- \\ &- \frac{1}{2\hbar} \, (1+\alpha) \, S_i^- \, S_i^+ - \frac{\alpha}{\hbar} \, (S_i^z)^2 \end{aligned} \tag{7.554}$$

Substituting this in the "higher" Green's functions of the equation of motion (7.538) and applying the symmetric decouplings (7.544), (7.545), (7.546) and (7.547) we again obtain exactly the same expressions (7.548) and (7.549) if we, as an additional approximation, neglect the fluctuations of the operator $(S_i^z)^2$.

The physical fixing of the parameter α is unfortunately not quite as plausible as for the case of $S = 1/2$. For example, we cannot simply assume that the operator product $S_i^- S_i^+$ determines the fluctuation of S_i^z around the value $+\hbar\, S$. We choose

$$\alpha = \frac{1}{2S} \frac{\langle S^z \rangle}{\hbar\, S} \tag{7.555}$$

and justify it in the following way:

1. α should reproduce the ansatz (7.550) for $S = 1/2$.
2. (7.551) for arbitrary S is identical to (7.539) for $S = 1/2$, so that with the same justification as for (7.550), it must be assumed that $\alpha \approx 0$ for $\langle S^z \rangle \approx 0$.
3. For low temperatures, $\langle S^z \rangle \approx \hbar\, S$, i.e. the spin wave approximation is valid:

$$S_i^+ \approx \hbar\sqrt{2S}\, a_i; \quad S_i^- \approx \hbar\sqrt{2S}\, a_i^\dagger$$

Then according to (7.23) the following holds:

$$\begin{aligned} \langle S^z \rangle &= \hbar\, S - \hbar\, \langle a_i^\dagger\, a_i \rangle \\ &\approx \hbar\, S - \frac{1}{2S\,\hbar} \langle S_i^-\, S_i^+ \rangle \end{aligned} \tag{7.556}$$

When we write (7.554) as

$$\langle S^z \rangle = \alpha\hbar S(S+1) - \frac{\alpha}{\hbar}\langle S_i^-\, S_i^+ \rangle + (1-\alpha)\langle S^z \rangle - \frac{\alpha}{\hbar}\langle (S^z)^2 \rangle$$

and substitute on the right-hand side $\langle (S_i^z)^2 \rangle \approx \hbar^2\, S^2$ and $\langle S_i^z \rangle \approx \hbar\, S$, because of the low temperatures we get

$$\langle S^z \rangle \approx \hbar S - \frac{\alpha}{\hbar}\langle S_i^- S_i^+ \rangle$$

With (7.556) we are therefore forced to assume $\alpha \approx \frac{1}{2S}$.

Obviously (7.555) satisfies all the three criteria. With (7.555) in (7.548) and (7.549), the decoupling is completely determined, so that we can now solve the approximated equation of motion (7.538).

$$
\begin{aligned}
(E - g_J\,\mu_B\,B_0)G_{ij}^{(a)}(E) = \hbar\,\eta(a) - \\
&- 2\hbar \sum_m J_{im} \left\{ G_{mj}^{(a)}(E) \Big(\langle S^z \rangle \right. \\
&+ \frac{\langle S^z \rangle}{2S^2\,\hbar^2} \langle S_m^-\,S_i^+ \rangle \Big) \\
&\left. - G_{ij}^{(a)}(E) \left(\langle S^z \rangle + \frac{\langle S^z \rangle}{2S^2\,\hbar^2} \langle S_i^-\,S_m^+ \rangle \right) \right\}
\end{aligned}
\tag{7.557}
$$

We solve this equation by Fourier transforming, wherein we use the following abbreviation:

$$
p(\mathbf{k}, a) = \frac{1}{N} \sum_{i,j} e^{i\mathbf{k}\cdot(\mathbf{R}_i - \mathbf{R}_j)} \langle e^{aS_j^z}\,S_j^-\,S_i^+ \rangle \tag{7.558}
$$

With this one finds the following solution:

$$
G_{\mathbf{q}}^{(a)}(E) = \frac{1}{N} \sum_{i,j} G_{ij}^{(a)}(E)\, e^{i\mathbf{q}\cdot(\mathbf{R}_i - \mathbf{R}_j)} = \frac{\hbar\,\eta(a)}{E - E(\mathbf{q})} \tag{7.559}
$$

$$
\begin{aligned}
E(\mathbf{q}) = 2\hbar\,\langle S^z \rangle\,(J_0 - J(\mathbf{q})) - \\
- \frac{\langle S^z \rangle}{\hbar^2\,S^2} \frac{1}{N} \sum_{\mathbf{k}} (J(\mathbf{k} + \mathbf{q}) - J(\mathbf{k})) p(\mathbf{k}, 0) + g_J\,\mu_B\,B_0
\end{aligned}
\tag{7.560}
$$

As a consequence of the changed decoupling method, the quasiparticle energies $E(\mathbf{q})$ are different from the result (7.428) that was obtained with the help of the RPA decoupling (7.422) by the second term in (7.560). This correction contains the correlation $p(\mathbf{k}, a)$, which we must now try to fix.

For the spectral density $S_{\mathbf{q}}^{(a)}$, we find using (B.94) along with (7.559)

$$
S_{\mathbf{q}}^{(a)}(E) = \hbar\,\eta(a)\,\delta(E - E(\mathbf{q})) \tag{7.561}
$$

Then from the spectral theorem (B.95), it follows immediately for $q \neq 0$

$$
p(\mathbf{q}, a) = \frac{\eta(a)}{exp(\beta E(\mathbf{q})) - 1} \tag{7.562}
$$

We now look for connection with the expectation value $\langle S^z \rangle$ which is what we are actually interested in. From (7.537) it follows that

$$
\eta(0) = 2\hbar\,\langle S^z \rangle \tag{7.563}
$$

Further, according to (7.558) the following holds:

$$\bar{p}(0) = \frac{1}{N}\sum_{\mathbf{q}} p(\mathbf{q}, 0) = \langle S_i^- \; S_i^+ \rangle = \hbar^2 \; S(S+1) - \hbar \; \langle S^z \rangle - \langle (S_i^z)^2 \rangle \tag{7.564}$$

Here again the term $\langle (S_i^z)^2 \rangle$ is disturbing. We eliminate this by looking for and exploiting a functional relationship among $\bar{p}(a)$, $\eta(a)$ and the parameter a. For this purpose we derive a differential equation for

$$\Omega(a) = \langle exp(aS^z) \rangle \tag{7.565}$$

solve this and build

$$\langle S^z \rangle = \frac{d}{da}\Omega(a)\bigg|_{a=0} \tag{7.566}$$

First let us determine $\eta(a)$. With (7.487),

$$\left[S_i^+ \, , \, (S_i^z)^n\right]_- = \left\{(S_i^z - \hbar)^n - (S_i^z)^n\right\} \; S_i^+ \tag{7.567}$$

follows

$$\begin{aligned}\left[S_i^+ \, , \, e^{aS_i^z}\right]_- &= \sum_n \frac{1}{n!} \, a^n \left[S_i^+ \, , \, (S_i^z)^n\right]_- \\ &= (e^{-a\hbar} - 1) e^{aS_i^z} \; S_i^+ \end{aligned}\tag{7.568}$$

With this we can now give the expectation value $\eta(a)$ defined in (7.537):

$$\begin{aligned}\eta(a) &= 2\hbar \, \langle e^{aS^z} S^z \rangle + (e^{-a\hbar} - 1)\langle e^{aS^z} S^+ \; S^- \rangle \\ &= \hbar^2 \; S(S+1)(e^{-a\hbar} - 1)\Omega(a) + (e^{-a\hbar} + 1)\hbar \frac{d}{da}\Omega(a) - \\ &\quad -(e^{-a\hbar} - 1) \, \frac{d^2}{da^2}\Omega(a)\end{aligned}\tag{7.569}$$

In addition, from (7.562) and (7.564) we find

$$\begin{aligned}\bar{p}(a) &= \frac{1}{N}\sum_{\mathbf{q}} p(\mathbf{q}, a) = \langle e^{aS^z} \; S^- \; S^+ \rangle \\ &= \hbar^2 \; S(S+1)\Omega(a) - \hbar \frac{d}{da}\Omega(a) - \frac{d^2}{da^2}\Omega(a)\end{aligned}\tag{7.570}$$

Due to (7.562) we can combine the two equations:

$$\bar{p}(a) = \eta(a) \, \phi(S) \tag{7.571}$$

$\phi(S)$ is defined as in (7.483) but with the "new" quasiparticle energies $E(\mathbf{q})$ given by (7.560). We eventually get the following differential equation for $\Omega(a)$:

$$\frac{d^2\Omega}{da^2} + \frac{(1+\phi) + \phi\, e^{-a\hbar}}{(1+\phi) - \phi\, e^{-a\hbar}}\, \hbar\, \frac{d\Omega}{da} - \hbar^2\, S(S+1)\,\Omega = 0 \tag{7.572}$$

This is analogous to the $2S$ coupled equations (7.489) or (7.505) obtained by the procedures of Tyablikov and Tahir-Kheli and ter Haar discussed in the last section. In order to solve this differential equation, we require two boundary conditions. One follows from the definition (7.565) for $\Omega(a)$:

$$\Omega(0) = 1 \tag{7.573}$$

The second boundary condition we derive from (7.485):

$$\prod_{m_s=-S}^{+S} \left(\frac{d}{da} - \hbar\, m_s \right) \Omega(a) \bigg|_{a=0} \equiv D_s\, \Omega(0) = 0. \tag{7.574}$$

One defines as an ansatz

$$\omega(x, a) = \frac{e^{\hbar\, a\, x}}{(1+\phi)\, e^{\hbar\, a} - \phi} \tag{7.575}$$

and then one gets by substituting in (7.572) that

$$x^2 - x - S(S+1) = 0$$

must hold and therefore

$$\omega(-S, a) \ \text{and} \ \omega(S+1, a)$$

are particular solutions of the differential equation (Problem 7.20). The general solution then reads as

$$\Omega(a) = \alpha\, \omega(-S, a) + \beta\, \omega(S+1, a) \tag{7.576}$$

The two boundary conditions (7.573) and (7.574) demand that

$$\begin{aligned} &\alpha + \beta = 1 \\ &\alpha\, D_s\, \omega(-S, 0) + \beta\, D_s\, \omega(S+1, 0) = 0 \end{aligned}$$

and with that fix α and β:

$$\Omega(a) = \frac{\omega(-S, a)\, D_s\, \omega(S+1, 0) - \omega(S+1, a)\, D_s\, \omega(-S, 0)}{D_s\, \omega(S+1, 0) - D_s\, \omega(-S, 0)} \tag{7.577}$$

This expression can further be somewhat reformulated. With the substitution

$$y = e^{\hbar\, a}$$

from (7.574) and (7.575) it follows that

$$D_s\, \omega(x, a) = -\frac{\hbar^{2S+1}}{\phi} \prod_{m_s=-S}^{+S} \left(y\frac{d}{dy} - m_s \right) \frac{y^x}{1 - \left(\frac{1+\phi}{\phi}\right) y}$$

For the fraction on the right-hand side we insert the corresponding series expansion, perform the differentiation and then set $a = 0$, i.e. $y = 1$:

$$\begin{aligned} D_s\, \omega(x, 0) &= -\frac{\hbar^{2S+1}}{\phi} \sum_{p=0}^{\infty} \left(\frac{1+\phi}{\phi} \right)^p \prod_{m_s=-S}^{+S} (p + x - m_s) \\ &= -\frac{\hbar^{2S+1}}{\phi} \left(\frac{1+\phi}{\phi} \right)^{-x} \sum_{p=x}^{\infty} \left(\frac{1+\phi}{\phi} \right)^p \prod_{m_S=-S}^{+S} (p - m_S) \end{aligned} \tag{7.578}$$

If we choose $x = -S$ then the first $(2S + 1)$ summands vanish because for $-S \leq p \leq +S$ always one of the factors in the product is zero:

$$D_s\omega(-S, 0) = -\frac{\hbar^{2S+1}}{\phi} \left(\frac{1+\phi}{\phi} \right)^S \sum_{p=S+1}^{\infty} \left(\frac{1+\phi}{\phi} \right)^p \prod_{m_s} (p - m_s)$$

From this we recognize (put $x = S + 1$ in (7.578))

$$D_s\, \omega(-S, 0) = \left(\frac{1+\phi}{\phi} \right)^{2S+1} D_s\, \omega(S + 1, 0) \tag{7.579}$$

With this the final solution of the differential equation (7.572) reads as

$$\Omega(a) = \frac{e^{-\hbar a S}\, \phi^{2S+1} - e^{\hbar a(S+1)}(1 + \phi)^{2S+1}}{[\phi^{2S+1} - (1 + \phi)^{2S+1}][(1 + \phi)\, e^{\hbar a} - \phi]} \tag{7.580}$$

The derivative of $\Omega(a)$ with respect to a at $a = 0$ according to (7.566) provides the expectation value $\langle S^z \rangle$ that we are looking for:

$$\langle S^z \rangle = \hbar\, \frac{(1 + \phi)^{2S+1}\, (S - \phi) + \phi^{2S+1}\, (S + 1 + \phi)}{(1 + \phi)^{2S+1} - \phi^{2S+1}}. \tag{7.581}$$

This result agrees formally exactly with the result (7.498) of the earlier discussed methods of Tyablikov and Tahir-Kheli and ter Haar. The differences lie in the $\phi(S)$ due to the different decoupling methods. We will summarize once again the formal solution of the Callen theory presented in this section:

(a) $\langle S^z \rangle$ from (7.581) as a function of ϕ

(b) $\phi(S) = \frac{1}{N}\sum_{\mathbf{q}}(exp(\beta E(\mathbf{q})) - 1)^{-1}$

(c) $E(\mathbf{q}) = g_J\ \mu_B\ B_0 + 2\hbar\langle S^z\rangle(J_0 - J(\mathbf{q})) - \frac{\langle S^z\rangle}{N\ \hbar^2\ S^2}\sum_{\mathbf{k}}(J(\mathbf{k}+\mathbf{q}) - J(\mathbf{k}))\ p(\mathbf{k},0)$

(d) $p(\mathbf{k},0) = \frac{2\hbar\ \langle S^z\rangle}{exp(\beta E(\mathbf{k}))-1}$

Equations $(a) - (d)$ build a complete system of equations which can be solved self-consistently for $\langle S^z\rangle$ and $E(\mathbf{q})$.

With the usual restriction to nearest neighbour interactions and a reformulation analogous to (7.308) we can write the quasiparticle energies $E(\mathbf{q})$ as follows:

$$E(\mathbf{q}) = g_J\ \mu_B\ B_0 + 2\hbar\ \langle S^z\rangle(J_0 - J(\mathbf{q}))\left(1 + \frac{\langle S^z\rangle}{S^2}\ Q\right) \tag{7.582}$$

They differ from the result (7.428) of the simpler RPA procedure of the earlier section by a temperature-dependent correction factor:

$$Q = \frac{1}{N\hbar\ J_0}\sum_{\mathbf{k}}\frac{J(\mathbf{k})}{exp(\beta E(\mathbf{k})) - 1} \tag{7.583}$$

The evaluation of the formal solution proceeds exactly in the same manner as in Sect. 7.6.2 and therefore will not be presented here in detail. One can use exactly the same series expansion as was done there. Modification of the results are only due to the factor $1 + (\langle S^z\rangle/S^2)Q)$ in the "new" quasiparticle energies (7.582). The results can be summarized as in the following [21]:

(a) *Low-temperature region*: The expansion of $\langle S^z\rangle$ in powers of the reduced temperature t (7.282) gives terms in t^0, $t^{3/2}$, $t^{5/2}$ and $t^{7/2}$ which agree with the exact Dyson solution. That also happens to be the case with the solution (7.516) of the methods of Tahir-Kheli and ter Haar and Tyablikov. The disturbing fact there is the presence of an additional t^3 term. This does not appear in the Callen solution for $S > 1/2$. Here the next higher terms are proportional to t^4, $t^{3S+3/2}$, $t^{3S+5/2}$, ... The $t^{3S+3/2}$ and $t^{3S+5/2}$ terms are apparently the consequences of the decoupling method. They reproduce the disturbing t^3 term for $S = 1/2$. The t^4 term differs slightly only in the coefficient from the exact Dyson result.
On the whole one can say that the solution of the Callen procedure, in the low-temperature region and for large S, represents a clear improvement over the

RPA methods of the preceding section. For $S = 1/2$, no improvements are recognizable.

(b) *Critical region*: Here also the expansions and estimates are completely analogous to those in Sect. 7.6.2. There we had in Eq. (7.525)

$$(k_B T_C)^{RPA} = k_B\, \theta_C / c_{-1}$$

where c_{-1} is a structure-dependent number of the order of magnitude of 1 (7.523). θ_C is the Curie temperature of the molecular field approximation. The method discussed in this section leads to

$$k_B T_C = (k_B T_C)^{RPA}\, \frac{(4S+1)c_{-1} - (S+1)}{3c_{-1}\, S} \tag{7.584}$$

The correction factor is somewhat larger than 1 so that the Curie temperature T_C in the Callen method lies higher than that in RPA.

(c) *High-temperature region*: The high-temperature expansion of the susceptibility is obtained with the same series representation as in the RPA theory (7.535) in Sect. 7.6.2. The results differ from each other in the coefficient of the $1/T^2$ term:

$$\chi_T = \mu_0\, \frac{N}{V}\, (g_J\, \mu_B)^2\, \frac{S(S+1)}{3k_B T} \left\{ 1 + \frac{\theta_C}{T} \right.$$
$$\left. + \left(1 - \frac{2S-1}{3z\, S}\right)\left(\frac{\theta_C}{T}\right)^2 + \mathcal{O}(1/T^2) \right\} \tag{7.585}$$

Precisely as in (7.535), the first two terms reproduce the exact $1/t$ expansion [22]. For very high temperatures, they lead to the Curie–Weiss law (7.147) of the molecular field theory.

7.7 Problems

Problem 7.1 Let $\mathbf{S}_i$, $\mathbf{S}_j$ be the spin operators corresponding to magnetic moments at the sites $\mathbf{R}_i$, $\mathbf{R}_j$. As is well known

$$S_j^{\pm} = S_j^x \pm i S_j^y$$

Prove that

1. $\left[S_i^z,\, S_j^{\pm}\right]_- = \pm\hbar\delta_{ij} S_i^{\pm}$
2. $\left[S_i^+,\, S_j^-\right]_- = 2\hbar\delta_{ij} S_i^z$
3. $S_i^{\pm} S_i^{\mp} = \hbar^2 S(S+1) \pm \hbar S_i^z - \left(S_i^z\right)^2$
4. $\mathbf{S}_i \cdot \mathbf{S}_j = \frac{1}{2}\left(S_i^+ S_j^- + S_i^- S_j^+\right) + S_i^z S_j^z$

5. Due to the fact that $J_{ij} = J_{ji}$; $J_{ii} = 0$, the Heisenberg Hamiltonian can also be written as

$$H = -\sum_{i,j} J_{ij} \left(S_i^+ S_j^- + S_i^z S_j^z \right)$$

Problem 7.2 Verify the following commutators for the spin operators:

1. $\left[(S_i^-)^n, S_i^z\right]_- = n\hbar(S_i^-)^n$; $n = 1, 2, \ldots$
2. $\left[(S_i^-)^n, (S_i^z)^2\right]_- = n^2\hbar^2(S_i^-)^n + 2n\hbar S_i^z(S_i^-)^n$; $n = 1, 2, \ldots$
3. $\left[S_i^+, (S_i^-)^n\right]_- = (2n\hbar S_i^z + \hbar^2 n(n-1))(S_i^-)^{n-1}$; $n = 1, 2, \ldots$

Problem 7.3 For the z-component of the local spin operator S_i^z, with real coefficients $\alpha_n(S)$ the following holds:

$$\langle (S_i^z)^{2S+1} \rangle = \sum_{n=0}^{2S} \alpha_n(S) \langle (S_i^z)^n \rangle$$

Calculate these coefficients for

$$S = \frac{1}{2}, \ 1, \ \frac{3}{2}, \ 2, \ \frac{7}{2}.$$

Problem 7.4 EuSe is an antiferromagnet between 2.8 and 4.6 K with a NNSS sequence of ferromagnetically ordering (111)-planes. Within the framework of molecular field approximation, calculate the Neél temperature T_N and the paramagnetic Curie temperature Θ as functions of the exchange integrals between the nearest (J_1) and next nearest neighbours (J_2).

Problem 7.5 1. Calculate the eigenvalues of a tridiagonal matrix:

$$A = \begin{pmatrix} a\, b\, 0\, 0\, 0 \cdots 0\, 0 \\ b\, a\, b\, 0\, 0 \cdots 0\, 0 \\ 0\, b\, a\, b\, 0 \cdots 0\, 0 \\ \cdots\cdots \\ \cdots\cdots \\ 0\, 0\, 0\, 0\, 0 \cdots b\, a \end{pmatrix}$$

2. Using the result of 1. determine the Curie temperature of a *Heisenberg film* of d monolayers in the framework of molecular field approximation. How does T_C vary with the thickness d of the film?

Problem 7.6 Let $|0\rangle$ be the state in which all the spins are oriented parallel to the external field $\mathbf{B}_0 = B_0\mathbf{e}_z$ (*ferromagnetic saturation*). Show that this state is an eigenstate of the Heisenberg Hamiltonian and calculate the corresponding energy eigenvalue $E_0(B_0)$.

Problem 7.7 Let

$$|\mathbf{k}\rangle = \frac{1}{\hbar\sqrt{2SN}} S^{-}(\mathbf{k})\,|0\rangle$$

be the normalized one-magnon state (7.114). Here $|0\rangle$ is the (normalized) magnon vacuum (*ferromagnetic saturation*). Show that for the expectation value of the local spin operator in the state $|\mathbf{k}\rangle$ the following holds:

$$\langle \mathbf{k}|\, S_i^z\, |\mathbf{k}\rangle = \hbar \left(S - \frac{1}{N} \right)$$

Problem 7.8 1. Show that the Holstein–Primakoff transformation preserves the commutation relations of the spin operators:

$$\left[S_i^{+}, S_j^{-}\right]_{-} = 2\hbar\delta_{ij} S_i^z$$
$$\left[S_i^{z}, S_j^{\pm}\right]_{-} = \pm\hbar\delta_{ij} S_i^{\pm}$$

2. Verify $\mathbf{S}_i^2 = \hbar^2 S(S+1)\mathbb{1}$

Problem 7.9 Repeat the consideration of Problem 7.8 for the Dyson–Maléev transformation.

Problem 7.10 Show that the spin wave approximation

$$S^{+}(\mathbf{k}) \approx \hbar\sqrt{2SN} a_{\mathbf{k}}$$
$$S^{-}(\mathbf{k}) \approx \hbar\sqrt{2SN} a^{+}_{-\mathbf{k}}$$
$$S^{z}(\mathbf{k}) = \hbar S N \delta_{\mathbf{k},0} - \hbar \sum_{\mathbf{q}} a^{+}_{\mathbf{q}} a_{\mathbf{k}+\mathbf{q}}$$

preserves the fundamental commutation relations of the spin operators.

Problem 7.11 Using Dyson–Maléev transformation, transform the Heisenberg Hamiltonian into wavenumber representation.

Problem 7.12 In the spin wave approximation, calculate the internal energy and heat capacity of a ferromagnet at low temperatures.

Problem 7.13 In the linear spin wave approximation the Heisenberg Hamiltonian reads as

$$H = E_0(B_0) + \sum_{\mathbf{q}} \hbar\omega(\mathbf{q}) a^{\dagger}_{\mathbf{q}} a_{\mathbf{q}}$$

$a_{\mathbf{q}}^{\dagger}$ and $a_{\mathbf{q}}$ are Bose operators for creation and annihilation of magnons with wavenumber $\mathbf{q}$. $\hbar\omega(\mathbf{q})$ are the magnon energies and $E_0(B_0)$ is the field-dependent constant of Eq. (7.246).

1. Show that

$$\left[\hat{n}_{\mathbf{q}}, (a_{\mathbf{q}}^{\dagger})^p\right]_- = p(a_{\mathbf{q}}^{\dagger})^p$$

   holds where $\hat{n}_{\mathbf{q}} = a_{\mathbf{q}}^{\dagger} a_{\mathbf{q}}$ and $p = 0, 1, 2, \ldots$
2. Show that the product state

$$|\psi\rangle = \prod_{\mathbf{q}} (a_{\mathbf{q}}^{\dagger})^{n_{\mathbf{q}}} |0\rangle$$

   is an eigenstate of H and find the eigenvalue. $|0\rangle$ is the magnon vacuum and $n_{\mathbf{q}}$ is the eigenvalue of $\hat{n}_{\mathbf{q}}$ which is the number of magnons with wavenumber $\mathbf{q}$.

Problem 7.14 Calculate the internal energy and the heat capacity of an antiferromagnet (ABAB structure) at low temperature in the spin wave approximation.

Problem 7.15 What is the temperature dependence shown by the sub-lattice magnetization of an antiferromagnet at low temperature? Assume the anisotropic contributions to be negligible. Apply the spin wave approximation for an ABAB antiferromagnet.

Problem 7.16 Calculate the ground state energy $\hat{E}_a$ of an antiferromagnet (ABAB structure). For that apply in the spin wave approximation the Bogoliubov transformation discussed in Sect. 7.4.3:

$$H = \hat{E}_a + \sum_{\mathbf{q}} \left\{ E_\alpha(\mathbf{q}) \alpha_{\mathbf{q}}^{+} \alpha_{\mathbf{q}} + E_\beta(\mathbf{q}) \beta_{\mathbf{q}}^{+} \beta_{\mathbf{q}} \right\}$$

(Notation as in Sect. 7.4.3!)

Problem 7.17 Consider Heisenberg spins on a lattice with translational symmetry. Let the exchange integrals J_{ij} be restricted to nearest neighbours $\mathbf{R}_i$ and $\mathbf{R}_j$. Then the wavenumber dependence of the magnon properties is determined by the structure factor

$$\gamma_{\mathbf{q}} = \frac{1}{z_1} \sum_{\Delta_1} e^{i\mathbf{q}\cdot\mathbf{R}_{\Delta_1}}$$

Here z_1 is the number of the nearest neighbours, $\mathbf{R}_{\Delta_1}$ is a vector from the point of consideration to one of the nearest neighbouring lattice sites and $\mathbf{q}$ is a vector in the first Brillouin zone. Show that from the translational symmetry it follows that

$$\sum_{\mathbf{q}_1} \gamma_{\mathbf{q}-\mathbf{q}_1} \langle \hat{n}_{\mathbf{q}_1} \rangle = \gamma_{\mathbf{q}} \sum_{\mathbf{q}_1} \gamma_{\mathbf{q}_1} \langle \hat{n}_{\mathbf{q}_1} \rangle$$

Here, one has

$$\begin{aligned} \hat{n}_{\mathbf{q}} &= a_{\mathbf{q}}^{\dagger} a_{\mathbf{q}} \\ a_{\mathbf{q}} &= \frac{1}{\sqrt{N}} \sum_i a_i e^{-i\mathbf{q}\cdot\mathbf{R}_i} \\ a_i &= \frac{1}{\sqrt{N}} \sum_{\mathbf{q}} a_{\mathbf{q}} e^{i\mathbf{q}\cdot\mathbf{R}_i} \end{aligned}$$

Problem 7.18 Let the structure factors $\gamma_{\mathbf{q}}$ be defined as in Problem 7.17. For high-temperature expansions of magnetization, expressions like the following are important:

$$c_m = \frac{1}{N} \sum_{\mathbf{q}} \left(1 - \gamma_{\mathbf{q}}\right)^m$$

Calculate c_1, c_2 and c_3 for a primitive cubic lattice.

Problem 7.19 For a system of localized spins, the dipole interaction

$$\begin{aligned} H_D &= \sum_{i,j} D_{ij} \mathbf{S}_i \cdot \mathbf{S}_j + H_a \\ H_a &= -3 \sum_{i,j} D_{ij} \left(\mathbf{S}_i \cdot \mathbf{e}_{ij}\right) \left(\mathbf{S}_j \cdot \mathbf{e}_{ij}\right) \end{aligned}$$

represents an anisotropic addition to the Heisenberg exchange. The symbols D_{ij} and $\mathbf{e}_{ij}$ are defined in (7.375) and (7.376).

Express H_a for low temperatures in the spin wave approximation. Check this with Eq. (7.385).

Problem 7.20 The Callen method for an approximate solution of the spontaneous magnetization in Heisenberg model uses the definition (7.565)

$$\Omega(a) = \langle exp(aS^z) \rangle$$

Für diese ergibt sich die Bestimmungsgleichung (7.572):

$$\frac{\mathrm{d}^2\Omega}{\mathrm{d}a^2} + \frac{(1+\varphi) + \varphi e^{-a\hbar}}{(1+\varphi) - \varphi e^{-a\hbar}} \hbar \frac{\mathrm{d}\Omega}{\mathrm{d}a} - \hbar^2 S(S+1)\Omega = 0$$

The solution uses the boundary conditions

$$\Omega(0) = 1$$

$$\prod_{m_S=-S}^{+S} \left(\frac{\mathrm{d}}{\mathrm{d}a} - \hbar m_s \right) \Omega(a)|_{a=0} \equiv \mathfrak{D}_S \Omega(0) = 0$$

Show that

$$\omega(x, a) = \frac{e^{\hbar a x}}{(1+\varphi)e^{\hbar a} - \varphi}$$

is a solution of the differential equation and find the possible values of x. What is the general solution?

References

1. T. Holstein, H. Primakoff, Phys. Rev. **58**, 1048 (1940)
2. N. D. Mermin, H. Wagner, Phys. Rev. Lett. **17**, 1133 (1966)
3. L. Van Hove, Physica **16**, 137 (1950)
4. R. B. Griffiths, Phys. Rev. A **136**, 437 (1964)
5. C. Hohenberg, Phys. Rev. **158**, 383 (1967)
6. N. N. Bogoliubov, Phys. Abhandl. Sowjetunion **6**, 1, 113, 229 (1962)
7. H. E. Stanley, Phys. Rev. **164**, 709 (1967)
8. G. Garton et al., J. Appl. Phys. **34**, 1083 (1963)
9. R. Griessen et al., Solid State Commun. **9**, 2219 (1971)
10. F. J. Dyson, Phys. Rev. **102**, 1217–1230 (1956)
11. M. Wortis, Phys. Rev. **132**, 85 (1963)
12. M. Wortis, Phys. Rev. **138A**, 1126 (1965)
13. L. Passell et al., Phys. Rev. **B14**, 4897, 4908, 4923 (1976)
14. H. G. Bohn et al., Phys. Rev. B **22**, 5447 (1980)
15. P. W. Anderson, Phys. Rev. **86**, 694 (1952)
16. G. T. Rado, H. Suhl, "Magnetism I", Academic Press, New York, 1963, p. 235
17. N. N. Bogoliubov, S. V. Tyablikov, Dokl. Akad. Nauk USSR **126**, 53 (1959)
18. R. B. Stinchcombe et al., Phys. Rev. **130**, 155 (1962)
19. R. A. Tahir-Kheli, D. ter Haar, Phys. Rev. **127**, 88 (1962)
20. E. Praveczki, Phys. Lett. **6**, 147 (1963)
21. H. B. Callen, Phys. Rev. **130**, 890 (1963)
22. H. A. Brown, J. M. Luttinger, Phys. Rev. **100**, 685 (1962)

Chapter 8
Hubbard Model

8.1 Introduction

The Heisenberg model which was discussed in detail in Chap. 7, cannot be applied to all classes of ferro-, ferri- or antiferromagnets. Its region of validity is immediately clear from the derivation in Chap. 5. Heisenberg model presumes the existence of permanent *localized* magnetic moments, which, as a consequence of either a direct or an indirect exchange interaction, order themselves collectively below a critical temperature T^*. Therefore, Heisenberg model is a good model for most of the magnetic insulators but, conceptually, it is totally inappropriate for magnetic metals such as Fe, Co and Ni (*band magnets*). In order to describe such substances, one has to use a completely different ansatz. The model which is most often used in this connection is due to J. Hubbard [1]. It was simultaneously proposed by J. Kanamori [2] and by M.C. Gutzwiller [3]. This model is the central point of this chapter.

Band magnets are primarily characterized by the fact that one and the same group of electrons are responsible for both magnetism and electrical conduction. Simple and very successful (with limitations of course) models for metallic solids are the Sommerfeld model (Sect. 3.3) and the Jellium model (Sect. 4.1.3). Describing collective magnetism is however beyond the scope of both the models. An important simplication made in these models is not taking into account the atomic structure of the solid. For this reason, these models are suitable to describe electrons in broad energy bands, e.g. electrons in the conduction bands of alkali metals. The qualitative discussion in Sect. 5.2.1 makes it clear that the structure of the density of states of the concerned energy bands should play not an unimportant role in deciding the possibility of collective magnetism. Added to this is the observation that band magnetism appears primarily in the narrow bands of the transition metals. The assumption made in the Jellium model, that the ions in the solid are represented only by a homogeneously smeared out positive charge, is not a good starting point for the theory of band magnetism. A consequence of this assumption is that the conduction electrons have a constant probability to exist anywhere in the entire crystal. *Narrow energy band*, on the other hand, means a relatively small mobility of the electrons and therefore sharp peaks around the lattice points for the probability of finding the electrons. Using plane waves to describe band electrons as is done in the Jellium model is obviously not appropriate.

W. Nolting, A. Ramakanth, *Quantum Theory of Magnetism*,
DOI 10.1007/978-3-540-85416-6_8,

8.2 Model for Band Magnets

8.2.1 Solid as a Many-Body System

A solid is made up of a large number of particles such as atoms, molecules, clusters, which are interacting with each other. Ultimately, they are positively charged nuclei and negatively charged electron shells. One denotes as *core electrons*, the electrons which are close to the nucleus and are tightly bound to it, and in general, form closed shells. The core electrons have little influence on the typical solid state properties. *Valence*- and *conduction electrons* which stem as a rule, from partially filled shells, are relatively free to move about (itinerant) and are responsible for the bonding of the solid. They therefore influence the solid state properties in a decisive way. One defines

$$Lattice\ ion = nucleus + core\ electrons$$

That leads to a first modelling of a solid:

$$Solid:$$
$$Interacting\ lattice\ ions\ and\ valence\ electrons$$

The Hamiltonian, therefore, has the structure

$$H = H_e + H_i + H_{ei} \tag{8.1}$$

H_e is the electronic part:

$$H_e = T_e + V_{ee} = \sum_{i=1}^{N_e} \frac{\mathbf{p}_i^2}{2m} + \frac{1}{2}\sum_{i,j}^{i \neq j} \frac{e^2}{4\pi\varepsilon_0 \mid \mathbf{r}_i - \mathbf{r}_j \mid} \tag{8.2}$$

$\mathbf{r}_i$ and $\mathbf{p}_i$ denote the position and momentum of the ith electron.
H_i describes the ionic subsystem:

$$H_i = T_i + V_{ii} = \sum_{\alpha=1}^{N_i} \frac{\mathbf{P}_\alpha^2}{2M_\alpha} + \frac{1}{2}\sum_{\alpha,\beta}^{\alpha \neq \beta} \frac{Z_\alpha Z_\beta e^2}{4\pi\varepsilon_0 \mid \mathbf{R}_\alpha - \mathbf{R}_\beta \mid} \tag{8.3}$$

$\mathbf{R}_\alpha$ and $\mathbf{P}_\alpha$ are the position and momentum of the ion α. The inter-ion interaction term can be separated into a part $V_{ii}^{(0)}$, which represents the rigid lattice $\mathbf{R}_\alpha^{(0)}$, e.g. is responsible for the binding energy, and a part V_p, which describes the lattice dynamics ("phonons"):

$$V_{ii} = V_{ii}^{(0)} + V_p \tag{8.4}$$

The interaction between the electronic and ionic subsystems is described by

$$H_{ei} = -\sum_{i=1}^{N_e}\sum_{\alpha=1}^{N_i} \frac{Z_\alpha e^2}{4\pi\varepsilon_0 \mid \mathbf{r}_i - \mathbf{R}_\alpha \mid} = V_{ei}^{(0)} + V_{ep} \tag{8.5}$$

Here $V_{ei}^{(0)}$ denotes the periodic lattice potential, in which the electrons in the solid move, and V_{ep} is the electron–phonon interaction term. Up till now our model is still very general. However, which parts are really important for describing band magnetism?

8.2.2 Electrons in Narrow Energy Bands

We want to first list a few experimental facts in order to find the essential features around which an appropriate theoretical model can be built.

1. Band ferromagnetism is produced apparently by electrons in *relatively narrow* energy bands, e.g. in $3d$-bands of the transition metals.
2. The states in the *relatively broad* $(s, p, \cdots)$ bands can hybridize with the states in the narrow d-bands, so that, e.g. the number of electrons per lattice site may become a non-integer. Further, the electrostatic potential of the ions "seen" by the d-electrons and also the inter- and intra-atomic interaction potentials among the d-electrons are screened by the substantially "faster" $(s, p, \cdots)$-electrons so that there is a certain renormalization of the corresponding matrix elements.
3. It appears that the lattice *dynamics* (V_{ep}) is relatively unimportant for band magnetism whereas the lattice *structure* (V_{ei}) plays an important role.

Therefore, in the first modelling of a *band ferromagnetic* solid, we will restrict ourselves to interacting d-electrons in a *rigid* ion lattice with matrix elements renormalized by $(s, p, \cdots)$-electrons:

$$H = H_0 + H_1 \tag{8.6}$$

H_1 is the Coulomb interaction of the d-electrons which is to be considered later, while H_0 is the kinetic energy T_e^d of the d-electrons and their interaction with the periodic lattice potential $V_{ei}^{0,d}$.

$$H_0 = T_e^d + V_{ei}^{(0,d)} \tag{8.7}$$

We want to represent H in second quantized form and for that purpose ask ourselves which is the appropriate single-particle basis. Figure 8.1 shows the schematic behaviour of the potential and the average probability of finding an atomic electron in an isolated single ion. If N such ions are brought near each other so that the interionic distance is finite, then the potential hills will overlap and there is a possibility of hopping of the electron with a finite tunnelling probability. For a *narrow* energy band, this tunnelling probability is quite small so that the probability $\mid \psi \mid^2$ of the

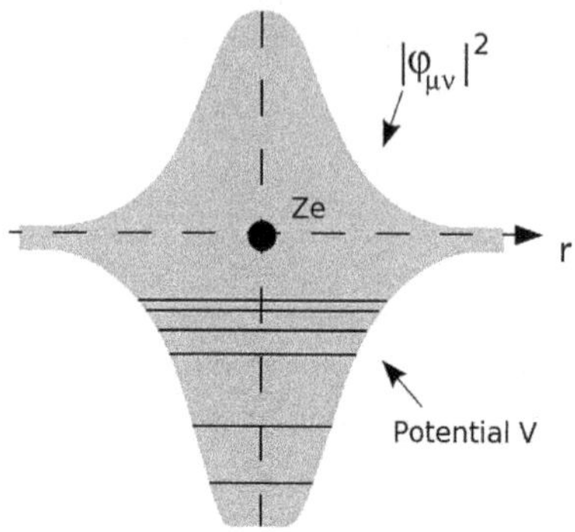

Fig. 8.1 Schematic plot of the potential V and the probability density $|\varphi_{\mu\sigma}|^2$ for an electron in the Z-fold positively charged single ion

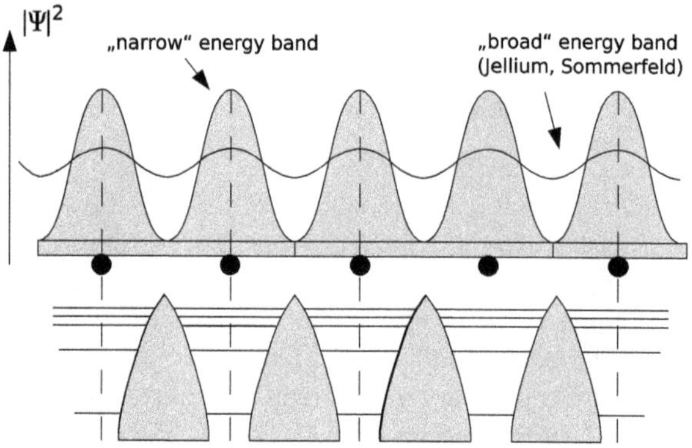

Fig. 8.2 Schematic plot of the lattice potential V and the probability density $|\varphi_{\mu\sigma}|^2$ for an electron in the narrow band of a solid

d-electrons has distinct maxima about the nuclei (see Fig. 8.2). Thus the electron still has a strong reference to the atomic structure. Therefore, the natural choice of the basis needed for second quantization could be the atomic wavefunctions

$$\varphi_{\mu\sigma}(\mathbf{r}-\mathbf{R}_i) = \langle \mathbf{r} \mid \varphi_{\mu\sigma}^{(i)} \rangle \tag{8.8}$$

$$\mid \varphi_{\mu\sigma}(\mathbf{k})\rangle = \frac{1}{\sqrt{N_i}} \sum_{j=1}^{N_i} e^{i\mathbf{k}\cdot\mathbf{R}_j} \mid \varphi_{\mu\sigma}^{(j)} \rangle \tag{8.9}$$

μ denotes the orbital and σ is as usual the spin projection. $\mathbf{R}_i$ is the position of the nucleus. $\{\varphi_{\mu\sigma}\}$ represents in the subspace of interest a complete one-particle basis. That means for the one-particle part of the Hamiltonian:

$$H_0 = \sum_{ij\sigma} \sum_{\mu\nu} T_{ij}^{\mu\nu} c_{i\mu\sigma}^{\dagger} c_{j\nu\sigma} \tag{8.10}$$

The *hopping* integrals $T_{ij}^{\mu\nu}$ are expressed in terms of the atomic wavefunctions by

$$T_{ij}^{\mu\nu} = \int d^3r\, \varphi_{\mu\sigma}^{*}(\mathbf{r}-\mathbf{R}_i) \left(-\frac{\hbar^2}{2m}\Delta + V_{ei}^{(0,d)} \right) \varphi_{\nu\sigma}(\mathbf{r}-\mathbf{R}_j) \tag{8.11}$$

After Fourier transforming we obtain the (μ, ν)-element of the $(2l + 1) \times (2l + 1)$-matrix $\widehat{T}_{\mathbf{k}}$:

$$T_{\mathbf{k}}^{\mu\nu} = \frac{1}{N_i} \sum_{i,j} T_{ij}^{\mu\nu} \, e^{-i\mathbf{k}\cdot(\mathbf{R}_i - \mathbf{R}_j)} \tag{8.12}$$

The eigenstates of $\widehat{T}_{\mathbf{k}}$,

$$\widehat{T}_{\mathbf{k}} \mid u_{m\mathbf{k}\sigma}\rangle = \varepsilon_m(\mathbf{k}) \mid u_{m\mathbf{k}\sigma}\rangle \tag{8.13}$$

are used to build the unitary matrix

$$U_{\mathbf{k}\sigma}^{m\mu} = \langle u_{m\mathbf{k}\sigma} \mid \varphi_{\mu\sigma}(\mathbf{k})\rangle \tag{8.14}$$

which diagonalizes $\widehat{T}_{\mathbf{k}}$:

$$\widehat{U}_{\mathbf{k}\sigma}\, \widehat{T}_{\mathbf{k}}\, \widehat{U}_{\mathbf{k}\sigma}^{\dagger} = \begin{pmatrix} \varepsilon_1(\mathbf{k}) & 0 & \cdots 0 & 0 \\ 0 & \varepsilon_2(\mathbf{k}) & \cdots 0 & 0 \\ & \cdots\cdots & \cdots\cdots & \\ 0 & 0 & \cdots 0 & \varepsilon_{2l+1}(\mathbf{k}) \end{pmatrix} \tag{8.15}$$

The diagonal elements $\varepsilon_m(\mathbf{k})$ are the *Bloch energies*, which provide us the *band structure* of the non-interacting electron system. m indexes the sub-bands whose number is the same as the number of atomic orbitals.

How do the construction operators transform themselves?

$$\begin{aligned} c_{\mathbf{k}m\sigma}^{\dagger} \mid 0\rangle = \mid u_{m\mathbf{k}\sigma}\rangle &= \sum_{\mu} \mid \varphi_{\mu\sigma}(\mathbf{k})\rangle\langle\varphi_{\mu\sigma}(\mathbf{k}) \mid u_{m\mathbf{k}\sigma}\rangle \\ &= \sum_{\mu} \left(U_{\mathbf{k}\sigma}^{m\mu}\right)^* \mid \varphi_{\mu\sigma}(\mathbf{k})\rangle \\ &= \frac{1}{\sqrt{N_i}} \sum_{j\mu} e^{i\mathbf{k}\cdot\mathbf{R}_j} \left(U_{\mathbf{k}\sigma}^{m\mu}\right)^* c_{j\mu\sigma}^{\dagger} \mid 0\rangle \end{aligned}$$

so that for the creation operators holds:

$$c_{\mathbf{k}m\sigma}^{\dagger} = \frac{1}{\sqrt{N_i}} \sum_{j\mu} e^{i\mathbf{k}\cdot\mathbf{R}_j} \left(U_{\mathbf{k}\sigma}^{m\mu}\right)^* c_{j\mu\sigma}^{\dagger} \tag{8.16}$$

$$c_{j\mu\sigma}^{\dagger} = \frac{1}{\sqrt{N_i}} \sum_{\mathbf{k}m} e^{-i\mathbf{k}\cdot\mathbf{R}_j}\, U_{\mathbf{k}\sigma}^{m\mu}\, c_{\mathbf{k}m\sigma}^{\dagger} \tag{8.17}$$

By taking adjoint one obtains the corresponding annihilation operators.

Substituting these transformation formulas in (8.10) and exploiting the fact that the columns and rows of the matrix $\widehat{U}_{\mathbf{k}\sigma}$ are orthonormal, one obtains the one-particle part of the Hamiltonian (Problem 8.1):

$$H_0 = \sum_{\mathbf{k}m\sigma} \varepsilon_m(\mathbf{k})\, c^{\dagger}_{\mathbf{k}m\sigma}\, c_{\mathbf{k}m\sigma}\,. \tag{8.18}$$

The number of sub-bands corresponds to the number of given orbitals.

The *Coulomb interaction* which is still remaining must also naturally be formulated in the one-particle basis already fixed by H_0. Since there is no spin dependence, for the Coulomb matrix element at the moment holds:

$$v(i\mu\sigma_1, j\nu\sigma_2; l\mu'\sigma_4, k\nu'\sigma_3) = \delta_{\sigma_1\sigma_4}\,\delta_{\sigma_2\sigma_3}\, v(i\mu, j\nu; l\mu', k\nu') \tag{8.19}$$

where

$$\begin{aligned} v(i\mu, j\nu; l\mu', k\nu') = \int\int d^3r d^3r' \varphi^*_{\mu\sigma}(\mathbf{r}-\mathbf{R}_i)\varphi^*_{\nu\sigma'}(\mathbf{r}'-\mathbf{R}_j) * \\ * \frac{e^2}{4\pi\varepsilon_0 \mid \mathbf{r}-\mathbf{r}' \mid} \varphi_{\mu'\sigma}(\mathbf{r}-\mathbf{R}_l)\varphi_{\nu'\sigma'}(\mathbf{r}'-\mathbf{R}_k) \end{aligned} \tag{8.20}$$

The interaction term in the model Hamiltonian, therefore, has the following form:

$$H_1 = \frac{1}{2}\sum_{ijkl}\sum_{\mu\mu'\nu\nu'}\sum_{\sigma\sigma'} v(i\mu, j\nu; l\mu', k\nu') c^{\dagger}_{i\mu\sigma} c^{\dagger}_{j\nu\sigma'} c_{k\nu'\sigma'} c_{l\mu'\sigma} \tag{8.21}$$

Up to now everything is still exact, except that we are limiting ourselves to the subspace fixed by the atomic wavefunctions, namely the d-states. After all, we have chosen only one special complete one-particle basis.

As a first *simplification*, one can exploit the fact that the wavefunctions centred at different lattice sites have a very small overlap so that the *intra-atomic* matrix elements surely dominate:

$$v(\mu\nu; \mu'\nu') = v(i\mu, i\nu; i\mu', i\nu') \tag{8.22}$$

All the other matrix elements are neglected:

$$H_1 = \frac{1}{2}\sum_{i\sigma\sigma'}\sum_{\mu\mu'\nu\nu'} v(\mu\nu; \mu'\nu') c^{\dagger}_{i\mu\sigma} c^{\dagger}_{i\nu\sigma'} c_{i\nu'\sigma'} c_{i\mu'\sigma} \tag{8.23}$$

That is, the interacting partners belong to the same lattice site! One finds that not all the matrix elements are equally important. The particularly important ones are

(a) *direct terms*: These are analogous to classical Coulomb repulsion ($\mu = \mu'$; $\nu = \nu'$):

$$U_{\mu\nu} \equiv v(\mu\nu;\ \mu\nu) \tag{8.24}$$

(b) *Exchange terms*: These have no classical analogue ($\mu = \nu'$; $\nu = \mu'$):

$$J_{\mu\nu} \equiv v(\mu\nu;\ \nu\mu) \tag{8.25}$$

The other matrix elements, i.e. other index combinations, on the one hand are smaller compared to the above terms and on the other, appear to not to influence the magnetic properties in any decisive manner.

$$H_1 = \frac{1}{2}\sum_{i\sigma\sigma'}\sum_{\mu\nu}\Big[(1-\delta_{\mu\nu}\delta_{\sigma\sigma'})U_{\mu\nu}\, n_{i\mu\sigma}n_{i\nu\sigma'} + (1-\delta_{\mu\nu})J_{\mu\nu}c^{\dagger}_{i\mu\sigma}c^{\dagger}_{i\nu\sigma'}c_{i\mu\sigma'}c_{i\nu\sigma}\Big] \tag{8.26}$$

$n_{i\mu\sigma} = c^{\dagger}_{i\mu\sigma}c_{i\mu\sigma}$ is the occupation number operator.

Inspite of the various simplifications, this operator along with H_0 appears to be a realistic starting point for describing band magnets. This is particularly the case when the one-particle energies $\varepsilon_m(\mathbf{k})$, $T_{ij}^{\mu\nu}$ and the matrix elements are suitably renormalized in order to take into account the effects, that are not covered by the model, at least in an average manner.

8.2.3 Hubbard Model

If we are primarily interested in a *qualitative* understanding of magnetism, one could assume as a first step that the d-band degeneracy is not so decisive. As a further simplification we then agree to limit ourselves to "narrow" *s-bands*($\rightarrow\ \mu = \nu = 1$).

The indexing of the orbitals is then superfluous. Then we have the model Hamiltonian named after Hubbard (Fig. 8.3):

$$H = \sum_{ij\sigma} T_{ij}c^{\dagger}_{i\sigma}c_{j\sigma} + \frac{1}{2}U\sum_{i\sigma} n_{i\sigma}n_{i-\sigma} \tag{8.27}$$

Does this model still have a relation to reality? The *model components*

- kinetic energy
- Coulomb interaction
- Pauli principle
- lattice structure

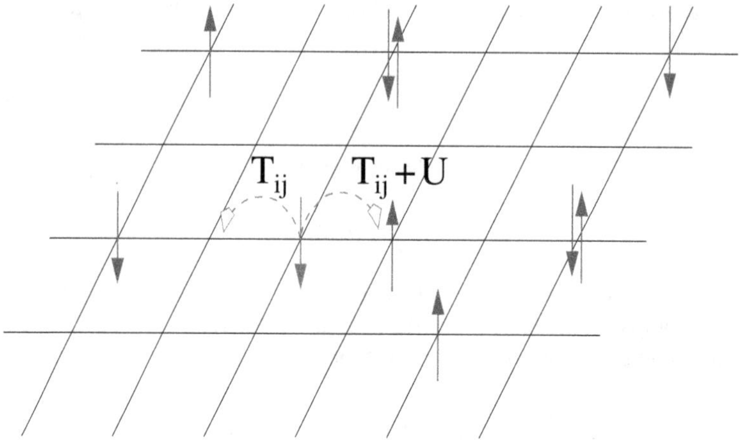

Fig. 8.3 Correlated fermions on a lattice

surely represent a minimal set in order to study the electronic correlations on a lattice.

Let us recall the salient *model simplifications*:

- *one* orbital per atom ($\rightarrow$ s-band)
- only *intra-atomic* electron–electron interaction
- *one* atom per unit cell
- one-particle basis built out of *atomic* wavefunctions

The important *model parameters*, through which the model systems differ from each other are

- the strength of the Coulomb coupling U
- bandwidth W of the Bloch band
- band occupation: $n = \sum_{i\sigma} \langle n_{i\sigma} \rangle$; $0 \le n \le 2$
- crystal structure

Is it really allowed to restrict oneself to intra-atomic Coulomb matrix element U? J. Hubbard has, in his original work [1], given the numbers for the *matrix elements* which are calculated using atomic Ni wavefunctions:

$$\begin{aligned} U = v(ii;ii) &\approx 20\,eV \\ v(ij;ij) &\approx 6\,eV \\ v(ii;ij) &\approx 0.5\,eV \\ v(ij;ik) &\approx 0.1\,eV \\ v(ij;ji) &\approx 0.025\,eV \\ v(ii;jj) &\approx 0.025\,eV \end{aligned}$$

The indices $i,\ j,\ k$ refer to nearest neighbours in the lattice. Due to screening effects, the interatomic matrix elements can further reduce by up to 20%. One should, however, note that the terms which contain the nearest neighbour indices should be multiplied by the number of nearest neighbours. A logical *model extension* which takes into account the influence of $v(ij;ij)$ would be

$$H = \sum_{ij\sigma} T_{ij} c_{i\sigma}^{\dagger} c_{j\sigma} + \frac{1}{2} U \sum_{i\sigma} n_{i\sigma} n_{i-\sigma} + \frac{1}{2} \sum_{ij\sigma\sigma'}^{n.N.} Q_{ij} n_{i\sigma} n_{j\sigma'} \tag{8.28}$$

The extra term for the nearest neighbour interactions, according to the available approximate investigations, appear to reinforce the tendency towards a ferromagnetic ground state. However, the generalization to a multi-band model as in (8.26) should be more important. We will, therefore, not discuss the model extension (8.28) any further.

What are the most important *applications of the model* today?

- electronic properties of narrow energy band solids (transition metals)
- band magnetism (Fe, Co, Ni, etc.)
- metal–insulator transitions (Mott transitions)
- high-temperature superconductivity
- general concepts of Statistical Mechanics

8.3 Stoner Model

The interaction term in the Hubbard Hamiltonian (8.27) prevents, at least in the general case, an exact solution of the corresponding many-body problem. As a result, one is constrained to resort to approximations. We begin with the simplest ansatz, namely the molecular field approximation. Just as the Weiss model of the ferromagnet (Sect. 5.1.2) represents the molecular field approximation of the Heisenberg model (7.1), one can treat the Stoner model as the molecular field approximation of the Hubbard model. Just as Weiss could not have been aware of the Heisenberg model, Stoner's original proposal of his model was before the Hubbard model was introduced. Stoner could demonstrate a band magnetic phase transition by constructing a self-consistent phenomenological *exchange field*.

8.3.1 Stoner Ansatz (Ferromagnet)

We use the molecular field approximation in the form (7.121) for the interaction term of the Hubbard Hamiltonian:

$$n_{i\sigma} n_{i-\sigma} \Longrightarrow \langle n_{i\sigma} \rangle n_{i-\sigma} + n_{i\sigma} \langle n_{i-\sigma} \rangle - \langle n_{i\sigma} \rangle \langle n_{i-\sigma} \rangle \tag{8.29}$$

With this the operator (8.27) becomes a one-particle operator:

$$H \to H_S = \sum_{ij\sigma} \left(T_{ij} + U\langle n_{i-\sigma}\rangle\delta_{ij}\right) c^{\dagger}_{i\sigma} c_{j\sigma} + D_S(T) \tag{8.30}$$

The relatively unimportant c-number

$$D_S(T) = -\frac{1}{2}U\sum_{i\sigma}\langle n_{i\sigma}\rangle\langle n_{i-\sigma}\rangle \tag{8.31}$$

will not play any role in the following considerations. Since we are concerned here exclusively with ferro- and paramagnetism, we can exploit translational symmetry. That means the expectation values of the occupation number operators are independent of the lattice site:

$$\langle n_{i\sigma}\rangle = n_\sigma \;\forall\, i \tag{8.32}$$

Then the Stoner approximation of the Hubbard model reads as

$$H_S = \sum_{ij\sigma}\left(T_{ij} + U\, n_{-\sigma}\delta_{ij}\right) c^{\dagger}_{i\sigma} c_{j\sigma} + D_S(T) \tag{8.33}$$

$$= \sum_{\mathbf{k}\sigma} \varepsilon_\sigma c^{\dagger}_{\mathbf{k}\sigma} c_{\mathbf{k}\sigma} + D_S(T) \tag{8.34}$$

$$D_S(T) = -NU\, n_\sigma\, n_{-\sigma} \tag{8.35}$$

After transforming into **k**-space, the Stoner Hamiltonian becomes diagonal and describes non-interacting Fermions with renormalized one-particle energies:

$$\varepsilon_\sigma(\mathbf{k}) = \varepsilon(\mathbf{k}) + U\, n_{-\sigma} \tag{8.36}$$

$n_{-\sigma}$ has to be determined self-consistently. We further introduce

$$'\textbf{polarization}' \;\; m(T, n) = n_\uparrow - n_\downarrow \tag{8.37}$$

and

$$'\textbf{particle density}' \;\; n = n_\uparrow + n_\downarrow \tag{8.38}$$

and write the Stoner energies as

$$\varepsilon_\sigma(\mathbf{k}) = \left(\varepsilon(\mathbf{k}) + \frac{1}{2}U\, n\right) - z_\sigma \frac{1}{2}Um\;;\; (z_\sigma = \delta_{\sigma\uparrow} - \delta_{\sigma\downarrow}) \tag{8.39}$$

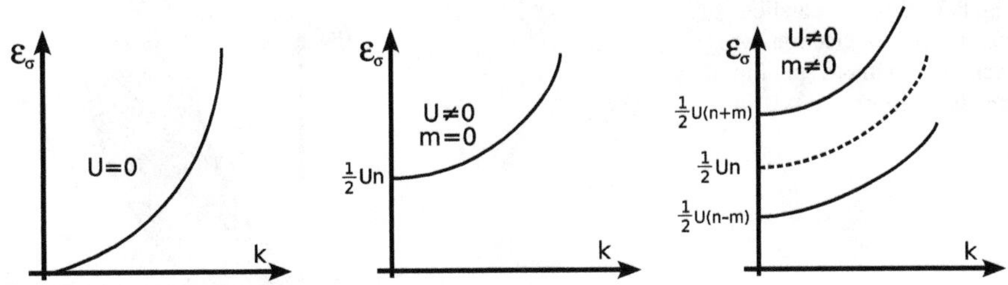

Fig. 8.4 Stoner energies as functions of the wavevector for (**a**) $U = 0$, (**b**) $U \neq 0, m = 0$, (**c**) $U \neq 0, m \neq 0$

The analogy to the Weiss ferromagnet (Sect. 5.1.2) becomes obvious when we define as *(spin-dependent) exchange field*

$$B_\sigma^{ex} = \frac{U}{\mu_B} n_{-\sigma} = \frac{U}{2\mu_B}(n - z_\sigma m) \tag{8.40}$$

$$\rightarrow \varepsilon_\sigma(\mathbf{k}) = \varepsilon(\mathbf{k}) + \mu_B B_\sigma^{ex} \tag{8.41}$$

As soon as the exchange field exhibits a real spin dependence, the Bloch dispersion $\varepsilon(\mathbf{k})$ splits into two spin dispersions which are rigidly shifted with respect to each other by the *exchange splitting* $\Delta E_{ex} = Um$. The splitting is temperature dependent and vanishes above T_C.

8.3.2 Stoner Excitations

According to (8.39) the Stoner model permits the following electronic excitations (Fig. 8.4):

$$\hbar\omega_{\mathbf{k}}^{\sigma\sigma'}(\mathbf{q}) = \varepsilon_{\sigma'}(\mathbf{k}+\mathbf{q}) - \varepsilon_\sigma(\mathbf{k}) = \varepsilon(\mathbf{k}+\mathbf{q}) - \varepsilon(\mathbf{k}) - \frac{1}{2}U\, m(z_{\sigma'} - z_\sigma) \tag{8.42}$$

If such an excitation takes place within the same spin band, i.e. without a spin flip ($\sigma = \sigma'$), then, because of the rigid band splitting, this is identical to an excitation in the free system. For a spherically symmetric Fermi body ($\varepsilon(\mathbf{k}) = \hbar^2 k^2/2m^*$), the *excitation continuum* lies then in between the following curves (Fig. 8.5):

$$\hbar\omega_{\mathbf{k}}^{max}(\mathbf{q}) = \frac{\hbar^2}{2m^*}(q^2 + 2k_F q) \tag{8.43}$$

$$\hbar\omega_{\mathbf{k}}^{min}(\mathbf{q}) = \begin{cases} \frac{\hbar^2}{2m^*}(q^2 - 2k_F q) & \text{if } q > 2k_F \\ 0 & \text{otherwise} \end{cases} \tag{8.44}$$

Excitations with spin flip mean transitions between the two spin bands:

$$\hbar\omega_{\mathbf{k}}^{\uparrow\downarrow} = \varepsilon(\mathbf{k}+\mathbf{q}) - \varepsilon(\mathbf{k}) + U\, m \tag{8.45}$$

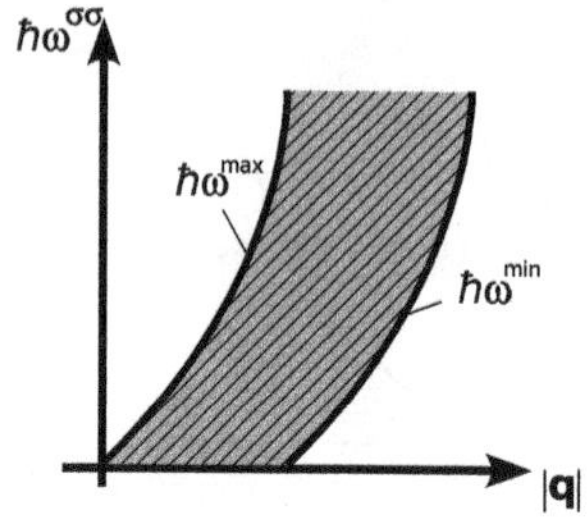

Fig. 8.5 "Stoner excitation continuum" for electron excitations without spin flip (*shaded region*)

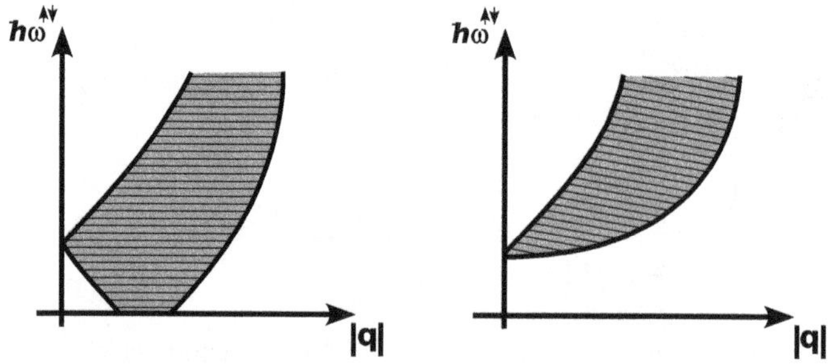

Fig. 8.6 "Stoner excitation continuum" for electron excitations with spin flip (*shaded region*): (**a**) $Um < \varepsilon_F$; (**b**) $Um > \varepsilon_F$

If $U\,m$ is greater than the Fermi energy ε_F then it is called *strong ferromagnetism* (Fig. 8.6b) and if $U\,m < \varepsilon_F$ it is called *weak ferromagnetism* (Fig. 8.6a).

8.3.3 Magnetic Phase Transition

A finite magnetization $m \neq 0$ is possible only if there is a preferential spin orientation in the relevant partially filled band. According to (8.39), this is the case only when $U \neq 0$ and $m \neq 0$. Thus, m must be determined self-consistently. This can be further understood from the *quasiparticle density of states* $\rho_\sigma(E)$ which for the *Stoner electrons* is not significantly different from the *Bloch density of states* $\rho_0(E)$ of the free electrons

$$\rho_0(E) = \frac{1}{N}\sum_{\mathbf{k}} \delta(E - \varepsilon(\mathbf{k})) \tag{8.46}$$

$$\rho_\sigma(E) = \frac{1}{N}\sum_{\mathbf{k}} \delta(E - \varepsilon_\sigma(\mathbf{k})) = \rho_0(E - Un_{-\sigma}) \tag{8.47}$$

It is, of course, dependent on temperature and particle density.

For $U = 0$, the two spin bands are naturally degenerate (Fig. 8.7a). This is also the case for $U \neq 0$, $m = 0$ (Fig. 8.7b). The two sub-bands are together shifted by

$\frac{1}{2}Un$. On the other hand, if $m \neq 0$ (Fig. 8.7c) there is a rigid spin-dependent shift. The $\uparrow$-band is shifted by $\frac{1}{2}U(n-m)$ and the $\downarrow$-band by $\frac{1}{2}U(n+m)$. When the two sub-bands are filled up to the common Fermi energy, then there will be more $\uparrow$- than $\downarrow$-electrons resulting in a non-zero polarization $m \neq 0$. One should pay attention that in order to get a spin-dependent shift (*exchange splitting*) one has to assume $m \neq 0$ in the first place. Therefore, m has to be determined self-consistently.

We therefore ask the question, under what conditions is

$$m(T, n) \neq 0?$$

In order to answer that we need the spin-dependent occupation numbers n_σ and $n_{-\sigma}$. If $f_-(E)$ is the Fermi function then holds

$$n_\sigma = \int_{-\infty}^{+\infty} dE f_-(E)\rho_\sigma(E) = \frac{1}{N}\sum_{\mathbf{k}} \left(e^{\beta(\varepsilon(\mathbf{k})+Un_{-\sigma}-\mu)} + 1\right)^{-1} \tag{8.48}$$

If we replace σ by $-\sigma$ in this equation, we obtain a second equation which along with (8.48) constitute a system of implicit equations for n_σ and $n_{-\sigma}$ or for n and m. As an abbreviation we define

$$g(\beta, n, m; \varepsilon(\mathbf{k})) = \left[\cosh\left(\beta(\varepsilon(\mathbf{k}) + \frac{1}{2}Un - \mu)\right) + \cosh\left(\frac{1}{2}\beta Um\right)\right]^{-1} \tag{8.49}$$

so that for the particle density, which fixes the chemical potential, we get

$$n = \frac{1}{N}\sum_{\mathbf{k}} \left(e^{-\beta(\varepsilon(\mathbf{k})-\mu+\frac{1}{2}Un)} + \cosh(\frac{1}{2}\beta Um)\right) \cdot g(\beta, n, m; \varepsilon(\mathbf{k})) \tag{8.50}$$

The temperature dependence on the right-hand side is only formal. It is compensated by the chemical potential. n is naturally temperature independent. This, however, is *not* the case for the polarization m:

$$m(T, n) = \sinh(\frac{1}{2}\beta Um)\frac{1}{N}\sum_{\mathbf{k}} g(\beta, n, m; \varepsilon(\mathbf{k})) \tag{8.51}$$

Since the **k**-dependence plays a role only through the Bloch energies $\varepsilon(\mathbf{k})$, the **k**-summation can be replaced by a simple energy integration over the Bloch density of states (8.46), which, any way we always assume to be known:

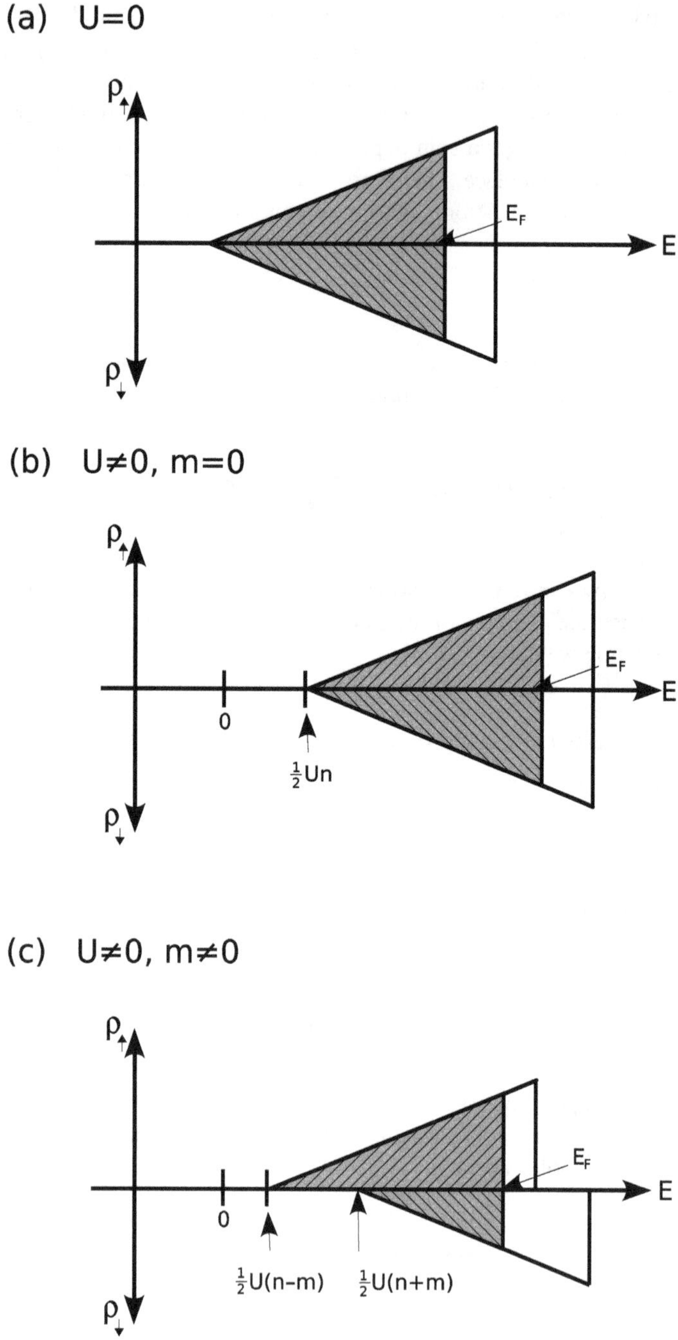

Fig. 8.7 Schematic plot of the spin-dependent quasiparticle density of states of the Stoner model: (**a**) $U = 0$, (**b**) $U \neq 0, m = 0$, (**c**) $U \neq 0, m \neq 0$

$$m(T,n) = \sinh(\frac{1}{2}\beta U m) * \\ * \int_{-\infty}^{+\infty} dx \frac{\rho_0(x)}{\cosh\left(\beta(x+\frac{1}{2}Un-\mu)\right)+\cosh\left(\frac{1}{2}\beta U m\right)} \tag{8.52}$$

We want to examine this expression in a little more detail.

1. *Paramagnetism* ($m = 0$)

   - Since $\sinh(0) = 0$, we have $m = 0$ so that paramagnetism is always a solution.
   - In case an additional ferromagnetic solution $m_s > 0$ exists, so since $\sinh(-x) = -\sinh(x)$, $-m_s < 0$ is also a solution. Then there are three mathematical solutions to (8.52) and the free energy $F(T,m)$ must decide which of them is the stable one. With the same justification as for the Weiss ferromagnet discussed in Sect. 5.1.2 (see Fig. 5.5), we conclude that the ferromagnetic solution, when it exists, is the stable one.

2. *High-temperature behaviour*: For $T \to \infty$, i.e. for $\beta \to 0$, we can expand the hyperbolic functions in (8.52):

$$\sinh(x) = x + \frac{1}{3!}x^3 + \cdots$$
$$\cosh(x) = 1 + \frac{1}{2!}x^2 + \cdots$$

   and obtain

$$m \approx \frac{1}{2}\beta U m \int_{-\infty}^{+\infty} dx \, \frac{\rho_0(x)}{2} = \frac{1}{4}\beta U m \tag{8.53}$$

   Thus at high temperatures, only the non-magnetic solution $m = 0$ exists. That means the system is paramagnetic.

3. *Low-temperature behaviour*: If at all a ferromagnetic solution appears, then because of 2., it can be only below a critical temperature T_C.

$$T \leq T_C \quad \Leftrightarrow \quad m \geq 0 \tag{8.54}$$

   In the neighbourhood of T_C (8.52) then simplifies to

$$1 \simeq \frac{1}{2}\beta_C U \int_{-\infty}^{+\infty} dx \, \frac{\rho_0(x)}{1+\cosh\left(\beta_C(x+\frac{1}{2}Un-\mu)\right)} \tag{8.55}$$

$$\left(\beta_C = \frac{1}{k_B T_C}\right)$$

In order to find a criterion for ferromagnetism, let us consider the *most unfavourable situation*: $T_C = 0^+$. This means $\beta_C \to \infty$.

$$1 \simeq U \int_{-\infty}^{+\infty} dx\, \rho_0(x) \lim_{\beta_C \to \infty} \frac{1}{2} \frac{\beta_C}{1 + \cosh\left(\beta_C(x + \frac{1}{2}Un - \mu)\right)} \tag{8.56}$$

We use the following representation of the δ-function which we prove as Problem 8.5:

$$\delta(x) = \lim_{\alpha \to \infty} \frac{1}{2} \frac{\alpha}{1 + \cosh(\alpha x)} \tag{8.57}$$

and exploit

$$\mu(T = 0^+) = \varepsilon_F + \frac{1}{2}Un \tag{8.58}$$

where ε_F is the Fermi energy of the free system. Then, we can read off from (8.56) the following *criterion for ferromagnetism*

$$1 \leq U\, \rho_0(\varepsilon_F) \tag{8.59}$$

In case the so-called *Stoner criterion* is fulfilled, a *spontaneous* polarization of the band electrons appears. We have, already in the qualitative discussion of band ferromagnetism in Sect. 5.2.1, more or less *guessed* this criterion. It is qualitatively confirmed by the observed trends in the periodic table.

- In a *row* of the transition metals in the periodic table the number of electrons in the d-shell increases from left to right and therefore also the $\rho(\varepsilon_F)$ for typical d-density of states.
- In a column of the periodic table, the total electron number increases from top to bottom and therefore also the screening of the Coulomb interaction, which means U becomes smaller.

According to Stoner criterion, ferromagnetism should be stabler as we go from left to right in a row and from bottom to top in a column. As a result good ferromagnets should be found in the top right corner of the transition metals. This is in fact roughly qualitatively confirmed.

4. *Curie Temperature*: If the Stoner criterion is fulfilled, then the numerical evaluation of the transcendental equation (8.52) for $m(T, n)$ gives the well-known and typical temperature dependence (Fig. 8.8).
Let us estimate the order of magnitude of the transition temperature to be expected from (8.55). For that we choose a particularly simple Bloch density of states (Fig. 8.9).

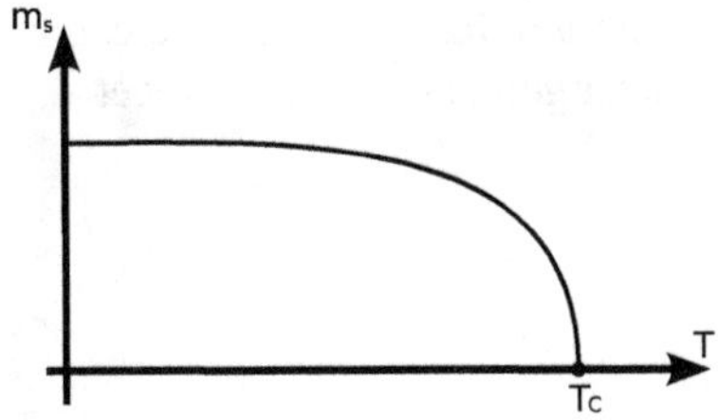

Fig. 8.8 Spontaneous magnetization in the Stoner model

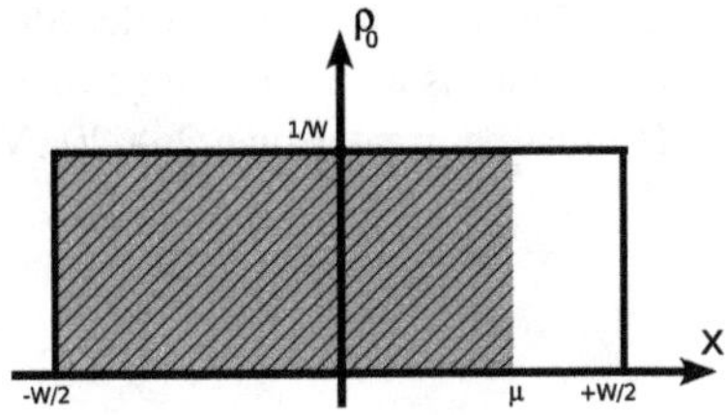

Fig. 8.9 Rectangular Bloch density of states

$$\rho_0(x) = \begin{cases} \frac{1}{W} \text{ if } -\frac{W}{2} \le x \le +\frac{W}{2} \\ 0 \quad \text{otherwise} \end{cases} \tag{8.60}$$

We can approximate $\mu(T_C) \simeq \varepsilon_F + \frac{1}{2}Un$ and then have from (8.56):

$$\begin{aligned} 1 &\simeq \frac{1}{2}\beta_C U \frac{1}{W} \int_{-\frac{W}{2}}^{+\frac{W}{2}} dx \, \frac{1}{1+\cosh(\beta_C(x-\varepsilon_F))} \\ &= \frac{1}{4}\beta_C U \frac{1}{W} \int_{-\frac{W}{2}}^{+\frac{W}{2}} dx \, \frac{1}{\cosh^2(\frac{1}{2}\beta_C(x-\varepsilon_F))} \\ &= \frac{U}{2W} \int_{-\frac{W}{2}}^{+\frac{W}{2}} dx \, \frac{d}{dx} \tanh(\frac{1}{2}\beta_C(x-\varepsilon_F)) \\ &= \frac{U}{2W} \left[\tanh\left(\frac{1}{2}\beta_C(\frac{W}{2}-\varepsilon_F)\right) \right. \\ &\qquad \left. - \tanh\left(\frac{1}{2}\beta_C(-\frac{W}{2}-\varepsilon_F)\right) \right] \end{aligned}$$

Through the Fermi energy the Curie temperature is obviously dependent on the particle number also. In order to have an estimate let us consider the special case of the half-filled band so that we can set $\varepsilon_F = 0$. Then we obtain

$$\begin{aligned} \frac{W}{U} &= \tanh\left(\frac{W}{4k_B T_C}\right) \\ \rightarrow k_B T_C &= \frac{W}{4\, arctanh(\frac{W}{U})} \end{aligned}$$

Already for $\frac{W}{U} < 0.5$, we can replace $arctanh(x)$ to a good approximation by its argument x so that we get

$$k_B T_C \simeq \frac{U}{4} \tag{8.61}$$

Is it realistic? Because of the Stoner criterion, we must anyway have $U > 2W$. Band widths are in the range of a few electron volts ($W = 1 - 10\,\text{eV}$). Therefore U must be greater than $2 - 20\,\text{eV}$. That means

$$T_C \geq 10^4 K \tag{8.62}$$

which is obviously an absurd result. The conclusion is that the molecular field approximation of the Hubbard model clearly overestimates the possibility and stability of ferromagnetism which appears to be a general characteristic of molecular field approximations.

5. *Critical behaviour*: In the neighbourhood of T_C, $m \stackrel{>}{\to} 0$ holds so that we can expand the implicit equation (8.52) as follows:

$$\begin{aligned}
m &= \left(\frac{1}{2}\beta U m + \frac{1}{6}\frac{(\beta U m)^3}{8} + \cdots\right) * \\
&* \int_{-\infty}^{+\infty} dx \frac{\rho_0(x)}{1 + \frac{1}{2}\frac{(\beta U m)^2}{4} + \cdots + \cosh\left(\beta(x + \frac{1}{2}Un - \mu)\right)} \\
&= \left(\frac{1}{2}\beta U m + \frac{1}{6}\frac{(\beta U m)^3}{8} + \cdots\right) * \\
&* \int_{-\infty}^{+\infty} dx \frac{\rho_0(x)}{1 + \cosh\left(\beta(x + \frac{1}{2}Un - \mu)\right)} * \\
&* \left(1 - \frac{1}{2}\frac{(\beta U m)^2}{4}\frac{1}{1 + \cosh\left(\beta(x + \frac{1}{2}Un - \mu)\right)} + \cdots\right)
\end{aligned}$$

For abbreviation, we define the surely non-critical quantity J:

$$J = \int_{-\infty}^{+\infty} dx \frac{\rho_0(x)}{\left(1 + \cosh\left(\beta(x + \frac{1}{2}Un - \mu)\right)\right)^2}$$

so that along with (8.56) we have

$$
\begin{aligned}
m &\simeq \left(\frac{1}{2}\beta U m + \frac{1}{48}(\beta U m)^3 + \cdots\right) * \\
&\quad * \left(\frac{2}{\beta_C U} - \frac{1}{8}(\beta U m)^2 \cdot J\right) \\
\rightarrow 1 &\simeq \frac{\beta}{\beta_C} + m^2 \left(\frac{\beta^3 U^2}{24\beta_C} - \frac{\beta^3 U^3}{16} \cdot J + \cdots\right) + \cdots
\end{aligned}
$$

Since $T \leq T_C$, the quantity in the parenthesis must necessarily be negative $(-\gamma^{-2})$. Thus the polarization m (magnetization) shows a power law behaviour:

$$m \sim \gamma \left(\frac{T_C}{T} - 1\right)^{\frac{1}{2}} \tag{8.63}$$

As in the case of the molecular field approximation of the Heisenberg ferromagnet (Sect. 7.3.1), for the critical exponent of the order parameter, we get the value which is expected from classical theories:

$$\beta = \frac{1}{2} \tag{8.64}$$

6. *Summary:* In conclusion we can state that the Stoner model describes the ground state properties qualitatively quite well but at finite temperatures exhibits a number of unrealistic features. The Curie temperature is extremely overestimated. Ferromagnetism is permitted in all lattice dimensions which is a violation of the Mermin–Wagner theorem which is valid also for the Hubbard model as we will show later. The Stoner model is furthermore, of course, overstrained to account for collective excitations (spin waves). Therefore, the molecular field approximation of the Hubbard model is surely too rough for quantitative comparison with experiment. In particular, the electronic correlations, which are responsible for many phenomena, are suppressed.

8.3.4 Static Susceptibility

An important indicator for magnetic phase transitions is the static susceptibility. If one calculates this for a paramagnetic system, then its singularities signal the instabilities of the paramagnetic state against ferromagnetism. As we are going to use the susceptibility at various places later, we shall first derive a general expression for the susceptibility χ of an interacting electron system so that we can apply it to the Stoner model. We can, from χ, derive criteria for the possibility of spontaneous magnetization.

According to (1.68), the static susceptibility is defined as follows:

$$\chi = \mu_0 \mu_B \cdot \frac{N}{V} \cdot \bar{\chi} \tag{8.65}$$

$$\bar{\chi} = \left(\frac{\partial m}{\partial B_0} \right)_{T, B_0 \to 0} \tag{8.66}$$

where

$$m = n_\uparrow - n_\downarrow = 2n_\uparrow - n = n - 2n_\downarrow \tag{8.67}$$

is the dimensionless magnetization per lattice site. We want to first express $\bar{\chi}$ through the one-electron spectral density $S_{\mathbf{k}\sigma}(E)$. According to spectral theorem (B.95) holds

$$n_\sigma = \frac{1}{\hbar} \frac{1}{N} \sum_{\mathbf{k}} \int_{-\infty}^{+\infty} dE \, f_-(E) S_{\mathbf{k}\sigma}(E - \mu) \tag{8.68}$$

$$m = \frac{1}{\hbar N} \sum_{\mathbf{k},\sigma} z_\sigma \int_{-\infty}^{+\infty} dE \, f_-(E) S_{\mathbf{k}\sigma}(E - \mu) \tag{8.69}$$

With (8.66) it follows:

$$\bar{\chi} = \frac{1}{N\hbar} \sum_{\mathbf{k}} \int_{-\infty}^{+\infty} dE \left\{ \frac{\partial f_-(E)}{\partial B_0} \sum_\sigma z_\sigma S_{\mathbf{k}\sigma}(E - \mu) + \right.$$
$$\left. + f_-(E) \sum_\sigma z_\sigma \frac{\partial}{\partial B_0} S_{\mathbf{k}\sigma}(E - \mu) \right\}_{B_0 = 0} \tag{8.70}$$

In the following, we will always calculate $\bar{\chi}$ for the *paramagnetic phase*. Therefore,

$$\left\{ \sum_\sigma z_\sigma S_{\mathbf{k}\sigma}(E - \mu) \right\}_0 = 0 \tag{8.71}$$

In (8.70) there remains only the second summand:

$$\bar{\chi} = \frac{1}{N\hbar} \sum_{\mathbf{k},\sigma} \int_{-\infty}^{+\infty} dE \, f_-(E) z_\sigma \left\{ \frac{\partial}{\partial B_0} S_{\mathbf{k}\sigma}(E - \mu) \right\}_0 \tag{8.72}$$

How does the spectral density in the presence of an external field B_0 look? First, the energies contain the extra Zeeman term $(-z_\sigma \mu_B B_0)$. In addition all the equal-time correlations still get an implicit field dependence. In the case which is frequently encountered, namely that these equal-time correlations are exclusively occupation numbers $n_{-\sigma}$ (or $n_\sigma = n - n_{-\sigma}$), we can write

$$S_{\mathbf{k}\sigma}(E) \rightarrow S_{\mathbf{k}\sigma}(E + z_\sigma \mu_B B_0;\, n_{-\sigma}(B_0)) \tag{8.73}$$

That means in (8.72)

$$\begin{aligned}(\partial_B S_{\mathbf{k}\sigma}(E-\mu))_0 &= z_\sigma \mu_B\, (\partial_E S_{\mathbf{k}\sigma}(E-\mu))_0 \\ &\quad + (\partial_B n_{-\sigma})_0 (\partial_{n_{-\sigma}} S_{\mathbf{k}\sigma}(E-\mu)_0)\end{aligned} \tag{8.74}$$

Now we have

$$\partial_B\, n = 0 \;\leftrightarrow\; \partial_B\, n_\sigma = -\partial_B\, n_{-\sigma} \tag{8.75}$$

so that

$$(\partial_B\, n_{-\sigma})_0 = -\frac{1}{2} z_\sigma \bar{\chi} = -(\partial_B n_\sigma)_0 \tag{8.76}$$

For (8.72), the first summand in (8.74) gives the contribution:

$$\begin{aligned}2\mu_B \frac{1}{N\hbar} \sum_{\mathbf{k}} \int\limits_{-\infty}^{+\infty} dE\; (f_-(E) \partial_E S_{\mathbf{k}\sigma}(E-\mu))_0 &= \\ = -2\mu_B \frac{1}{N\hbar} \sum_{\mathbf{k}} \int\limits_{-\infty}^{+\infty} dE\; \left(f_-'(E) S_{\mathbf{k}\sigma}(E-\mu)\right)_0\end{aligned}$$

In the integration by parts performed here, the term integrated out vanishes at the two limits $\pm\infty$. We define

$$\begin{aligned}Z &= -\frac{1}{N\hbar} \sum_{\mathbf{k}} \int\limits_{-\infty}^{+\infty} dE\; \left\{S_{\mathbf{k}\sigma}(E-\mu) f_-'(E)\right\}_0 \\ &= -\int\limits_{-\infty}^{+\infty} dE\; \left\{\rho_\sigma(E) f_-'(E)\right\}_0\end{aligned} \tag{8.77}$$

$$\widehat{N} = \frac{1}{2N\hbar} \sum_{\mathbf{k},\sigma} \int\limits_{-\infty}^{+\infty} dE\; \left(f_-(E)\, \partial_{n_{-\sigma}} S_{\mathbf{k}\sigma}(E-\mu)\right)_0 \tag{8.78}$$

The static susceptibility then reads as

$$\bar{\chi} = 2\mu_B \cdot \frac{Z}{1+\widehat{N}} \tag{8.79}$$

In the interaction-free case ($U = 0$),

$$S^{(0)}_{\mathbf{k}\sigma} = \hbar\delta(E + \mu - \varepsilon(\mathbf{k}))$$

$\widehat{N}$ vanishes and what remains is the Pauli susceptibility (4.2):

$$\bar{\chi}_0 = 2\mu_B Z_0 = -2\mu_B \int\limits_{-\infty}^{+\infty} dE\, \rho_0(E) f'_-(E) \tag{8.80}$$

The denominator $(1 + \widehat{N})$ in (8.79) expresses essentially the influence of the electron–electron interaction. Naturally the numerator also changes by switching on of the interaction.

Now we want to calculate the susceptibility for the Stoner model, which is the molecular field approximation of the Hubbard model.

With the quasiparticle energies (8.36)

$$\varepsilon_\sigma(\mathbf{k}) = \varepsilon(\mathbf{k}) + Un_{-\sigma} \tag{8.81}$$

the spectral density simply reads as

$$\begin{aligned} S_{\mathbf{k}\sigma}(E) &= \hbar\delta(E + \mu - \varepsilon_\sigma(\mathbf{k})) \\ \Leftrightarrow \rho_\sigma(E) &= \rho_0(E - Un_{-\sigma}) \end{aligned} \tag{8.82}$$

So that for (8.77) and (8.78) because of

$$\begin{aligned} \widehat{N}_S &= -\frac{U}{N}\sum_{\mathbf{k}} \int_{-\infty}^{+\infty} dE\, f_-(E)\frac{d}{dE}\delta\left(E - \frac{1}{2}Un - \varepsilon(\mathbf{k})\right) \\ &= +U \int_{-\infty}^{+\infty} f'_-(E)\,\rho_0\left(E - \frac{1}{2}Un\right) \end{aligned}$$

we get the following expression:

$$Z_S = -\frac{1}{U}\widehat{N}_S = -\int\limits_{-\infty}^{+\infty} d\eta\, f'_-\left(\eta + \frac{1}{2}Un\right)\rho_0(\eta) \tag{8.83}$$

Since in the Stoner model the quasiparticle bands are only rigidly shifted by $\frac{1}{2}Un$ with respect to the free Bloch bands, it holds

$$\mu = \mu^{(0)} + \frac{1}{2}Un \tag{8.84}$$

where $\mu^{(0)}$ is the chemical potential of the ideal Fermi gas. With this Z_S becomes identical to Z_0 in (8.80) and we recover the well-known RPA result for the static susceptibility:

$$\bar{\chi}_S = 2\mu_B \frac{Z_0}{1 - UZ_0} \tag{8.85}$$

Therefore, the paramagnetic state becomes unstable against ferromagnetism if

$$1 \stackrel{!}{=} UZ_0 \tag{8.86}$$

or expressed in more detail

$$\begin{aligned} 1 &\stackrel{!}{=} -U \int_{-\infty}^{+\infty} dE\, \rho_0(E) \partial_E f_-(E) \\ &= \frac{\beta_C U}{4} \int_{-\infty}^{+\infty} dE \frac{\rho_0(E)}{\cosh^2(\frac{1}{2}\beta_c(E - \mu^{(0)}))} \\ \Leftrightarrow 1 &\stackrel{!}{=} \frac{1}{2}\beta_c U \int_{-\infty}^{+\infty} dE \frac{\rho_0(E)}{1 + \cosh(\beta_c(E - \mu^{(0)}))} \end{aligned} \tag{8.87}$$

This equations determines T_c. We have already derived it in (8.55) in another way. For the *most unfavourable case* $T_c = 0^+$ ($\beta_c \to \infty$), with (8.57) and $\mu^{(0)}(T = 0) = \varepsilon_F$ we get the *Stoner criterion*:

$$1 \leq U\rho_0(\varepsilon_F) \tag{8.88}$$

Thus the static susceptibility as far as the magnetic phase transition is concerned makes the same statement as the directly calculated magnetization curve.

8.4 Exact Statements and General Properties

Without claiming any completeness, in this section, we will derive a few exact results which are possible for the Hubbard model. For a model which cannot be rigorously exactly solved in the general case, exactly calculable special cases are of extreme value in assessing the reliability of the unavoidable approximations. Apart from that, a lot of physical information about the model is contained in these exact results.

8.4.1 Mermin–Wagner Theorem

The aim of the following considerations is to derive the condition for spontaneous magnetization in the Hubbard model. Can the model describe band ferromagnetism? The Stoner model certainly predicts ferromagnetism but as a molecular field approximation, it is *too favourable* for spontaneous ferromagnetic ordering. Therefore, in this first subsection we will prove the Mermin–Wagner theorem [4] which we have already inspected in Sect. 7.2.1. The statement made in that section that the isotropic Heisenberg model in one and two dimensions does *not* show any spontaneous collective magnetism, can also be proved in the case of the Hubbard model. Since the proof follows similar lines as for the Heisenberg model, we will skip to a large extent the calculational details.

We consider the interacting electron system in the presence of a homogeneous magnetic field $\mathbf{B}_0 = B_0\mathbf{e}_z$:

$$H = \sum_{ij\sigma} T_{ij} c_{i\sigma}^{\dagger} c_{j\sigma} + \frac{1}{2} U \sum_{i\sigma} n_{i\sigma} n_{i-\sigma} - \mu_B B_0 \sum_{i\sigma} z_\sigma n_{i\sigma} \tag{8.89}$$

$z_\sigma = (\delta_{\sigma\uparrow} - \delta_{\sigma\downarrow})$ is only a sign factor. We use the spin operators introduced in (5.98), (5.99) and (5.100) (which are dimensionless here!):

$$\sigma_i^z = \frac{1}{2}(n_{i\uparrow} - n_{i\downarrow}) \; ; \; \sigma_i^+ = c_{i\uparrow}^{\dagger} c_{i\downarrow} \; ; \; \sigma_i^- = c_{i\downarrow}^{\dagger} c_{i\uparrow} \tag{8.90}$$

We also need the wavevector-dependent Fourier-transformed operators:

$$\begin{aligned} \sigma^\alpha(\mathbf{k}) = \sum_i \sigma_i^\alpha \, e^{-i\mathbf{k}\cdot\mathbf{R}_i} \quad &; \quad \sigma_i^\alpha = \frac{1}{N} \sum_{\mathbf{k}} \sigma^\alpha(\mathbf{k}) \, e^{i\mathbf{k}\cdot\mathbf{R}_i} \\ \alpha = \pm, x, y, z & \end{aligned} \tag{8.91}$$

whose commutation relations can easily be calculated from those for the position-dependent spin operators:

$$\left[\sigma^+(\mathbf{k}_1), \, \sigma^-(\mathbf{k}_2)\right]_- = 2\sigma^z(\mathbf{k}_1 + \mathbf{k}_2) \tag{8.92}$$

$$\left[\sigma^z(\mathbf{k}_1), \, \sigma^\pm(\mathbf{k}_2)\right]_- = \pm\sigma^\pm(\mathbf{k}_1 + \mathbf{k}_2) \tag{8.93}$$

We now rewrite the Hamiltonian (8.89) as far as possible in terms of these operators. For that we use

$$\begin{aligned} \sigma_i \cdot \sigma_i &= \sigma_i^+ \sigma_i^- + (\sigma_i^z)^2 - \sigma_i^z = \\ &= c_{i\uparrow}^{\dagger} c_{i\downarrow} c_{i\downarrow}^{\dagger} c_{i\uparrow} + \frac{1}{4}(n_{i\uparrow} - n_{i\downarrow})^2 - \frac{1}{2}(n_{i\uparrow} - n_{i\downarrow}) = \\ &= \frac{3}{4} n_{i\uparrow} + \frac{3}{4} n_{i\downarrow} - \frac{3}{2} n_{i\uparrow} n_{i\downarrow} \end{aligned}$$

Here we have exploited the relation $n_{i\sigma}^2 = n_{i\sigma}$ which is valid for Fermions. With the operator for the total electron number

$$\widehat{N} = \sum_{i\sigma} n_{i\sigma} \tag{8.94}$$

obviously holds

$$\sum_i \sigma_i^2 = \frac{3}{4}\widehat{N} - \frac{3}{2}\sum_i n_{i\uparrow}n_{i\downarrow} \tag{8.95}$$

so that the Hamiltonian (8.89) reads as

$$H = \sum_{ij\sigma} T_{ij} c_{i\sigma}^\dagger c_{j\sigma} - \frac{2}{3}U\sum_i \sigma_i^2 + \frac{1}{2}U\widehat{N} - 2\mu_B B_0 \sum_i \sigma_i^z \tag{8.96}$$

After Fourier transformation it becomes

$$H = \sum_{\mathbf{k}\sigma} \varepsilon(\mathbf{k}) c_{\mathbf{k}\sigma}^\dagger c_{\mathbf{k}\sigma} - \frac{2}{3N}U\sum_{\mathbf{k}} \sigma(\mathbf{k})\cdot\sigma(-\mathbf{k}) + \frac{1}{2}U\widehat{N} - 2\mu_B B_0 \sigma^z(\mathbf{0}) \tag{8.97}$$

In the following, we will need both the versions (8.96) and (8.97). In addition, we will assume that the *hopping* integrals T_{ij} fall off sufficiently fast with increasing distance R_{ij}:

$$Q = \frac{1}{N}\sum_{ij} \mid T_{ij} \mid \left(\mathbf{R}_i - \mathbf{R}_j\right)^2 < \infty \tag{8.98}$$

As T_{ij} are overlap integrals, this is only a very weak condition.

Our aim is to calculate the magnetization:

$$M(T, B_0) = 2\mu_B \frac{1}{N}\sum_i \langle \sigma_i^z \rangle = \frac{2\mu_B}{N}\langle \sigma^z(\mathbf{0})\rangle \tag{8.99}$$

From this we obtain the *spontaneous* magnetization $M_S(T)$ by taking the limit $B_0 \to 0$. We can, without any loss of generality, assume that $M(T, B_0) \geq 0$.

As in Sect. 7.2.1, the starting point for the proof is the *Bogoliubov inequality* (7.62):

$$\left|\langle [C, A]_-\rangle\right|^2 \leq \frac{1}{2}\beta\langle\left[A, A^\dagger\right]_+\rangle\langle\left[[C, H]_-, C^\dagger\right]_-\rangle \tag{8.100}$$

Here A and B are arbitrary operators; H is the Hamiltonian of the system and as usual $\beta = \frac{1}{k_B T}$. We recall that from the derivation of this formula via the Schwarz inequality, we see that the expectation value of the double commutator on the right-hand side must be positive definite.

In order to estimate the magnetization with the help of the Bogoliubov inequality, it is found that the following operators are convenient:

$$C = \sigma^{+}(\mathbf{k}) \leftrightarrow C^{\dagger} = \sigma^{-}(-\mathbf{k}) \tag{8.101}$$

$$A = \sigma^{-}(-\mathbf{k}) \tag{8.102}$$

We will now evaluate stepwise the individual terms in the inequality (8.100). The left-hand side is simple:

$$\langle [C,\ A]_{-} \rangle = 2\,\langle \sigma^{z}(\mathbf{0}) \rangle = \frac{N}{\mu_B} M(T,\ B_0) \tag{8.103}$$

Further holds

$$\begin{aligned}
\sum_{\mathbf{k}} \langle [A,\ A^{\dagger}]_{+} \rangle &= \sum_{\mathbf{k}} \langle [\sigma^{-}(-\mathbf{k}),\ \sigma^{+}(\mathbf{k})]_{+} \rangle \\
&= N \sum_{i} \langle \sigma_i^{-} \sigma_i^{+} + \sigma_i^{+} \sigma_i^{-} \rangle \\
&= N \sum_{i} \langle c_{i\downarrow}^{\dagger} c_{i\uparrow} c_{i\uparrow}^{\dagger} c_{i\downarrow} + c_{i\uparrow}^{\dagger} c_{i\downarrow} c_{i\downarrow}^{\dagger} c_{i\uparrow} \rangle \\
&= N \sum_{i} \langle n_{i\downarrow}(1 - n_{i\uparrow}) + n_{i\uparrow}(1 - n_{i\downarrow}) \rangle \\
&= N \sum_{i} \langle (n_{i\uparrow} - n_{i\downarrow})^2 \rangle
\end{aligned}$$

In the last step we have again used $n_{i\sigma}^2 = n_{i\sigma}$. Since the electrons move in an s-band, the thermodynamic expectation value on the right-hand side is certainly not greater than one. So we have

$$\sum_{\mathbf{k}} \langle [A,\ A^{\dagger}]_{+} \rangle \le N^2 \tag{8.104}$$

In the inequality (8.100), we divide by the positive double commutator and sum over all the wavevectors. Then with (8.103) and (8.104) we get the intermediate result

$$\frac{M^2(T,\ B_0)}{\mu_B^2} \sum_{\mathbf{k}} \frac{1}{\langle [[\sigma^{+}(\mathbf{k}),\ H]_{-},\ \sigma^{-}(-\mathbf{k})]_{-} \rangle} \le \frac{1}{2}\beta \tag{8.105}$$

What remains is the evaluation of the double commutator, which needs somewhat more effort. We leave it as an excercise to the reader to verify the correctness of the following commutators (Problem 8.6):

$$\left[\sigma^+(\mathbf{k}),\ \sum_{mn\sigma} T_{mn} c^\dagger_{m\sigma} c_{n\sigma}\right]_- = \sum_{mn} T_{mn}\left(e^{-i\mathbf{k}\cdot\mathbf{R}_m} - e^{-i\mathbf{k}\cdot\mathbf{R}_n}\right) c^\dagger_{m\uparrow} c_{n\downarrow}$$

$$\left[\sigma^+(\mathbf{k}),\ \sum_{\mathbf{p}} \sigma(\mathbf{p})\cdot\sigma(-\mathbf{p})\right]_- = 0$$

$$\left[\sigma^+(\mathbf{k}),\ \widehat{N}\right]_- = 0$$

$$\left[\sigma^+(\mathbf{k}),\ \sigma^z(\mathbf{0})\right]_- = -\sigma^+(\mathbf{k})$$

These can be summarized into (Problem 8.6):

$$\left[\sigma^+(\mathbf{k}),\ H\right]_- = 2\mu_B B_0\, \sigma^+(\mathbf{k}) + \sum_{mn} T_{mn}\left(e^{-i\mathbf{k}\cdot\mathbf{R}_m} - e^{-i\mathbf{k}\cdot\mathbf{R}_n}\right) c^\dagger_{m\uparrow} c_{n\downarrow} \tag{8.106}$$

Finally, the double commutator needed in (8.105) is found to be

$$\begin{aligned}&\left[\left[\sigma^+(\mathbf{k})\,,\ H\right]_-,\ \sigma^-(-\mathbf{k})\right]_- \\ &= 4\mu_B B_0\, \sigma^z(\mathbf{0}) + \sum_{mn,\sigma} T_{mn}\left(e^{iz_\sigma \mathbf{k}\cdot(\mathbf{R}_n-\mathbf{R}_m)} - 1\right) c^\dagger_{m\sigma} c_{n\sigma}\end{aligned} \tag{8.107}$$

From this we have to build the thermodynamic expectation value. In doing this we can exploit the fact that due to translational symmetry and the triangular inequality it must hold

$$|\ \langle c^\dagger_{m\sigma} c_{n\sigma}\rangle\ | \le \frac{1}{N}\sum_{\mathbf{k}} |\ e^{i\mathbf{k}\cdot(\mathbf{R}_m-\mathbf{R}_n)}\ ||\ \langle n_{\mathbf{k}\sigma}\rangle\ |$$

Since surely $|\ \langle n_{\mathbf{k}\sigma}\rangle\ | \le 1$, we can assume

$$|\ \langle c^\dagger_{m\sigma} c_{n\sigma}\rangle\ | \le 1$$

This we use for the estimate in (8.107)

$$\begin{aligned}&\langle\left[\left[\sigma^+(\mathbf{k}),\ H\right]_-,\ \sigma^-(-\mathbf{k})\right]_-\rangle \\ &\le 4\mu_B B_0\ |\ \langle\sigma^z(\mathbf{0})\rangle\ | + \sum_{mn,\sigma} |\ T_{mn}\ ||\ e^{iz_\sigma \mathbf{k}\cdot(\mathbf{R}_n-\mathbf{R}_m)} - 1\ | \\ &\le 4\mu_B B_0\ |\ \langle\sigma^z(\mathbf{0})\rangle\ | + 2\sum_{mn} |\ T_{mn}\ ||\ \cos(\mathbf{k}\cdot(\mathbf{R}_m - \mathbf{R}_n) - 1\ | \\ &\le 4\mu_B B_0\ |\ \langle\sigma^z(\mathbf{0})\rangle\ | + k^2 N Q\end{aligned}$$

With this we have found

$$\langle\Big[\big[\sigma^+(\mathbf{k}),\ H\big]_-,\ \sigma^-(-\mathbf{k})\Big]_-\rangle \le k^2 N Q + 2N B_0 M(T, B_0) \tag{8.108}$$

This we substitute in the inequality (8.105):

$$\beta \ge \frac{M^2}{\mu_B^2}\frac{1}{N}\sum_{\mathbf{k}}\frac{1}{B_0 M + \frac{1}{2}Qk^2} \tag{8.109}$$

The sum we evaluate in the thermodynamic limit

$$\frac{1}{N_d}\sum_{\mathbf{k}} \rightarrow \frac{v_d}{(2\pi)^d}\int d^d k \tag{8.110}$$

Here d is the dimension of the system and $v_d = V_d/N_d < \infty$ is the finite volume per particle. The summand on the right-hand side of the inequality (8.109) is positive definite. We can therefore instead of integrating over the full Brillouin zone, limit ourselves to integrate over a sphere of radius k_0 which lies completely inside the Brillouin zone. By doing this the inequality is only further strengthened.

$$\beta \ge \frac{M^2}{\mu_B^2}\frac{v_d\Omega_d}{(2\pi)^d}\int_0^{k_0}\frac{dk\,k^{d-1}}{B_0 M + \frac{1}{2}Qk^2} \tag{8.111}$$

Because of the above assumption, the integration over the angle can be directly performed. Ω_d is the surface of the d-dimensional unit sphere (Problem 3.4):

$$\Omega_1 = 2\ ;\ \Omega_2 = 2\pi\ ;\ \Omega_3 = 4\pi\ ;\ \Omega_d = \frac{2\pi^{\frac{d}{2}}}{\Gamma(\frac{d}{2})} \tag{8.112}$$

We will evaluate the inequality (8.111) explicitly:

- $d = 1$: With the standard integral

$$\int \frac{dx}{a^2x^2 + b^2} = \frac{1}{ab}\arctan\frac{ax}{b} + c$$

and $\arctan 0 = 0$ follows

$$\beta \ge \frac{M^2}{\mu_B^2}\frac{v_1}{2\pi}\frac{1}{\sqrt{\frac{1}{2}Q B_0 M}}\arctan\left(k_0\sqrt{\frac{Q}{2B_0 M}}\right)$$

For $B_0 \to 0$, arctan (...) approaches $\pi/2$. So that the temperature and field dependence of the magnetization for small fields ($B_0 \to 0$) is given by

$$M(T; B_0) \overset{<}{\to} const. \cdot \frac{B_0^{1/3}}{T^{2/3}} \tag{8.113}$$

This, however, means that the spontaneous magnetization vanishes for finite temperatures:

$$M_S(T) \equiv 0 \quad \text{for } T \neq 0 \text{ and } d = 1 \tag{8.114}$$

In one-dimensional Hubbard lattice there is no ferromagnetism!

- $d = 2$: We now use the integral

$$\int \frac{dx\, x}{a^2x^2 + b^2} = \frac{1}{2a^2} \ln c(a^2x^2 + b^2) \tag{8.115}$$

So that we get the following inequality

$$\beta \geq \frac{M^2}{\mu_B^2} \frac{v_2}{2\pi} \frac{1}{Q} \ln \left(\frac{B_0 M + \frac{1}{2} Q k^2}{B_0 M} \right) \tag{8.116}$$

That means for $B_0 \to 0$:

$$M(T; B_0) \overset{<}{\to} \frac{const.}{\sqrt{T \cdot (-\ln B_0 M)}} \tag{8.117}$$

Thus for finite temperatures, in two-dimensional lattice also there is no spontaneous magnetization:

$$M_S(T) \equiv 0 \text{ for } T \neq 0 \text{ and } d = 2 \tag{8.118}$$

In two-dimensional Hubbard lattice there is no spontaneous magnetization!

8.4.2 The Infinitely Narrow Band

Ferromagnetism is certainly a phenomenon of *strong coupling* U/W. This is already indicated by the Stoner criterion (8.59). Perhaps something can be learnt from the extreme case of infinitely narrow band ($W \to 0$), which can be handled mathematically rigorously.

$$W \to 0 : \varepsilon(\mathbf{k}) \equiv T_0 \; \forall\, \mathbf{k} \tag{8.119}$$

$$T_{ij} = T_0 \, \delta_{ij} \tag{8.120}$$

We understand this to be a limiting process, i.e. an energy *band* of the solid with vanishing dispersion. This, however, does not mean an isolated atom! In order to solve this problem we use the equation of motion method of the Green's functions (see Appendix B). The starting point is the retarded one-electron Green's function in the position- (Wannier-) representation:

$$G_{ij\sigma}(E) = \langle\langle c_{i\sigma}; c^\dagger_{j\sigma}\rangle\rangle_E \tag{8.121}$$

For the equation of motion (B.83) we need the commutator of $c_{i\sigma}$ with the (grand canonical) Hubbard Hamiltonian (8.27):

$$\mathcal{H} = H - \mu\widehat{N} = \sum_{ij\sigma}(T_{ij} - \mu\delta_{ij})c^\dagger_{i\sigma}c_{j\sigma} + \frac{1}{2}U\sum_{i\sigma} n_{i\sigma}n_{i-\sigma} \tag{8.122}$$

$\widehat{N}$ is the particle number operator. One finds

$$[c_{i\sigma}, \mathcal{H}]_- = \sum_m (T_{im} - \mu\delta_{im})c_{m\sigma} + Un_{i-\sigma}c_{i\sigma} \tag{8.123}$$

The second summand on the right-hand side leads to a *higher* Green's function:

$$\Gamma_{ilm;j\sigma}(E) = \langle\langle c^\dagger_{i-\sigma}c_{l-\sigma}c_{m\sigma}; c^\dagger_{j\sigma}\rangle\rangle_E \tag{8.124}$$

So that the equation of motion reads as

$$(E + \mu)G_{ij\sigma}(E) = \hbar\delta_{ij} + \sum_m T_{im}G_{mj\sigma}(E) + U\,\Gamma_{iii;j\sigma}(E) \tag{8.125}$$

A direct solution of the problem is not possible because of the higher Green's function on the right-hand side. Therefore, we write down the equation of motion for the higher Green's function also. For that we need

$$\begin{aligned}
&\left[n_{i-\sigma}c_{i\sigma}, \mathcal{H}_0\right]_- \\
&= \sum_{lm\sigma'}(T_{lm} - \mu\delta_{lm})\left[n_{i-\sigma}c_{i\sigma}, c^\dagger_{l\sigma'}c_{m\sigma'}\right]_- \\
&= \sum_{lm\sigma'}(T_{lm} - \mu\delta_{lm})\Big(\delta_{il}\delta_{\sigma\sigma'}n_{i-\sigma}c_{m\sigma'} - \delta_{il}\delta_{\sigma-\sigma'}c^\dagger_{i-\sigma}c_{i\sigma}c_{m\sigma'} \\
&\quad -\delta_{im}\delta_{\sigma-\sigma'}c^\dagger_{l\sigma'}c_{i-\sigma}c_{i\sigma}\Big) \\
&= \sum_m (T_{im} - \mu\delta_{im})\left(n_{i-\sigma}c_{m\sigma} + c^\dagger_{i-\sigma}c_{m-\sigma}c_{i\sigma} - c^\dagger_{m-\sigma}c_{i-\sigma}c_{i\sigma}\right)
\end{aligned}$$

$$
\begin{aligned}
&\left[n_{i-\sigma} c_{i\sigma}, H_1\right]_- \\
&= \frac{1}{2} U \sum_{m\sigma'} \left[n_{i-\sigma} c_{i\sigma}, n_{m\sigma'} n_{m-\sigma'}\right]_- \\
&= \frac{1}{2} U \sum_{m\sigma'} n_{i-\sigma} \left[c_{i\sigma}, n_{m\sigma'} n_{m-\sigma'}\right]_- \\
&= \frac{1}{2} U \sum_{m\sigma'} n_{i-\sigma} \left(\delta_{im} \delta_{\sigma\sigma'} c_{m\sigma'} n_{m-\sigma'} + \delta_{im} \delta_{\sigma-\sigma'} n_{m\sigma'} c_{m-\sigma'}\right) \\
&= U\, c_{i\sigma} n_{i-\sigma}
\end{aligned}
$$

In the last step, we have used the relation $n_{i\sigma}^2 = n_{i\sigma}$ valid for Fermions. Then, we have the equation of motion

$$
(E + \mu - U)\, \Gamma_{iii;j\sigma} = \hbar \delta_{ij} n_{-\sigma} + \sum_m T_{im} \left\{\Gamma_{iim;j\sigma} + \Gamma_{imi;j\sigma} - \Gamma_{mii;j\sigma}\right\} \tag{8.126}
$$

Up to this point everything is still exact where, however, we have assumed again translational symmetry ($\langle n_{i-\sigma} \rangle = n_{-\sigma}\ \forall i$).

We will now concentrate on the limiting case of infinitely narrow band for which we can exploit (8.119) and (8.120) so that (8.126) simplifies to

$$
(E + \mu - T_0 - U)\, \Gamma_{iii;j\sigma}(E) = \hbar \delta_{ij}\, n_{-\sigma}
$$

and can be solved easily:

$$
\Gamma_{iii;j\sigma}(E) = \delta_{ij} \frac{\hbar n_{-\sigma}}{E + \mu - T_0 - U + i0^+} \tag{8.127}
$$

We substitute the solution (8.127) in (8.125) and obtain for the one-particle Green's function of the Hubbard model being necessarily "local" in the zero-bandwidth limit:

$$
G_{ii\sigma}(E) = \hbar \left(\frac{1 - n_{-\sigma}}{E + \mu - T_0 + i0^+} + \frac{n_{-\sigma}}{E + \mu - T_0 - U + i0^+} \right) \tag{8.128}
$$

$G_{ii\sigma}(E)$ thus has two poles representing the possible excitation energies measured with respect to the chemical potential μ:

$$
E_{1\sigma} = T_0 = E_{1-\sigma} \tag{8.129}
$$

$$
E_{2\sigma} = T_0 + U = E_{2-\sigma} \tag{8.130}
$$

The quasiparticle energies are spin independent but the respective spectral weights are spin dependent:

$$
\alpha_{1\sigma} = 1 - n_{-\sigma}\,,\quad \alpha_{2\sigma} = n_{-\sigma} \tag{8.131}
$$

They are a measure of the probability that a test electron of spin σ encounters a $(-\sigma)$-electron at a lattice site ($\alpha_{2\sigma}$) or it does not ($\alpha_{1\sigma}$). If the electron finds such an interaction partner then it costs an extra Coulomb energy U. The Bloch *band*, which becomes a contracted N-fold degenerate level T_0 in the limit (8.119), is split into two infinitely narrow sub-bands at T_0 and $T_0 + U$ with *degrees of degeneracies*,

$$g_{1\sigma} = \alpha_{1\sigma} N = (1 - n_{-\sigma})N \ , \ \ g_{2\sigma} = \alpha_{2\sigma} N = n_{-\sigma} N \tag{8.132}$$

Due to lack of dispersion, in this special case, the quasiparticle density of states is identical to the local one-electron spectral density

$$\begin{aligned}\rho_\sigma(E) &= \frac{1}{N\hbar}\sum_i S_{ii\sigma}(E-\mu) = \frac{1}{\hbar}S_{ii\sigma}(E-\mu)\\ &= -\frac{1}{\pi}ImG_{ii\sigma}(E-\mu) = \hbar\sum_{j=1}^{2}\alpha_{j\sigma}\delta(E-E_{j\sigma})\end{aligned}$$

$$\rho_\sigma(E) = (1-n_{-\sigma})\delta(E-T_0) + n_{-\sigma}\delta(E-T_0-U) \tag{8.133}$$

The two δ-functions represent the two infinitely narrow subbands (Fig. 8.10).

The Green's function (8.128) for the infinitely narrow band ($W = 0$) is sometimes also called, somewhat less appropriately, *atomic limit*. It is interesting to compare it with the general structure of Green's functions (B.149):

$$G_{ii\sigma}(E) = \hbar\left(E + \mu - T_0 - \Sigma_\sigma^{W=0}(E) + i0^+\right)^{-1} \tag{8.134}$$

That implies the *atomic-limit self-energy*

$$\Sigma_\sigma^{W=0}(E) = Un_{-\sigma}\,\frac{E+\mu-T_0}{E+\mu-T_0-U(1-n_{-\sigma})} \tag{8.135}$$

which we will come across again in connection with the so-called *Hubbard-I* solution (Sect. 8.5.1) and as a result of the *interpolation method* (Sect. 8.5.2)

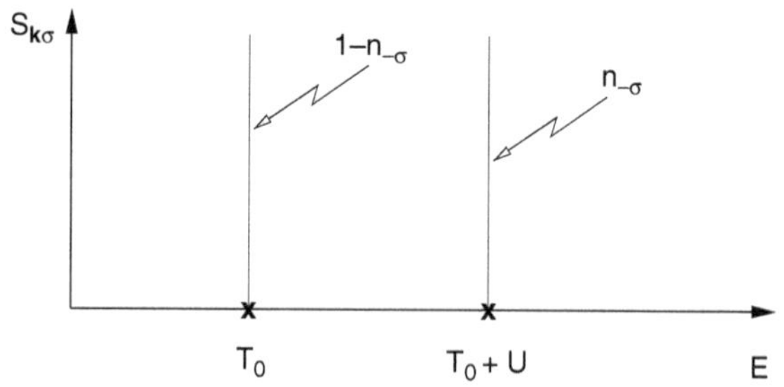

Fig. 8.10 Spectral density (=quasiparticle density of states) of the Hubbard model for the zero-bandwidth case

In order to completely determine the quasiparticle density of states or the Green's function, we still need the expectation value $n_{-\sigma} \equiv \langle n_{i-\sigma} \rangle$, which has to be self-consistently determined with the help of the spectral theorem (B.95). We can couple this job with the question:

Does a finite spontaneous magnetization exist?

From the spectral theorem we have

$$n_{-\sigma} = \frac{1}{\hbar} \int_{-\infty}^{+\infty} dE \, \frac{S_{ii-\sigma}(E)}{\exp(\beta E) + 1}$$

$$n_{-\sigma} = (1 - n_\sigma) f_-(T_0) + n_\sigma f_-(T_0 + U) \tag{8.136}$$

Here again $f_-(E)$ is the Fermi function. By reversing the spin in (8.136) we get the corresponding equation for n_σ which can in turn be substituted in (8.136). Then we finally get

$$n_{-\sigma} = \frac{f_-(T_0)}{1 - (f_-(T_0 + U) - f_-(T_0))} = n_\sigma \tag{8.137}$$

In the limiting case of infinitely narrow band it therefore always holds

$$n_\sigma = n_{-\sigma} = \frac{n}{2} \tag{8.138}$$

Ferromagnetism is excluded in this limiting case.

We will finally calculate the static susceptibility for the limiting case of infinitely narrow band. With (8.133), the spectral density is available exactly:

$$S_\sigma(E) = \hbar(1 - n_{-\sigma})\delta(E + \mu - T_0) + \hbar n_{-\sigma}\delta(E + \mu - T_0 - U) \tag{8.139}$$

Now we need that for the paramagnetic electron system:

$$n_\sigma = n_{-\sigma} = \frac{1}{2} n \tag{8.140}$$

Further, we will assume a half-filled *band* ($n = 1$). From (8.137) one reads off (Problem 8.7):

$$\mu(n = 1) = T_0 + \frac{1}{2} U \tag{8.141}$$

Equation (8.139) in (8.77) and (8.78) gives

$$N_{W=0} = -f_{-}(T_0) + f_{-}(T_0 + U)$$
$$Z_{W=0} = -\frac{1}{2}(f'_{-}(T_0) + f'_{-}(T_0 + U))$$

These two expressions can easily be evaluated (Problem 8.8):

$$N_{W=0} = -\tanh\left(\frac{1}{4}\beta U\right) \tag{8.142}$$

$$Z_{W=0} = \frac{\beta}{4\cosh^2\left(\frac{1}{4}\beta U\right)} \tag{8.143}$$

So that the susceptibility $\bar{\chi}_{W=0}$ is given by

$$\bar{\chi}_{W=0} = \frac{1}{2}\beta\mu_B\left(1 + \tanh\left(\frac{1}{4}\beta U\right)\right) \tag{8.144}$$

$\bar{\chi}_{W=0}$ can never be singular for $T > 0$, i.e. the paramagnetic state is always stable. As we already know, spontaneous magnetization is not possible in the atomic Hubbard model. On the other hand, the important Curie law (1.72) of the paramagnet is obviously fulfilled.

8.4.3 The Two-Site Model

In the last section, we have seen that the infinitely narrow *band* T_0, because of the Coulomb interaction U splits into two sub*bands*. We want now, with the help of a very simple model, gain certain insight about how the kinetic energy modifies this picture. What happens when the electrons are no more *frozen* on their lattice sites but can hop from site to site? To answer this we consider a simple system made up of two lattice sites so that it can hold a maximum of four s-electrons. This system is described by the following simplified Hamiltonian:

$$H = \sum_{\sigma}\left[t(c^{\dagger}_{1\sigma}c_{2\sigma} + c^{\dagger}_{2\sigma}c_{1\sigma}) + \frac{1}{2}U(n_{1\sigma}n_{1-\sigma} + n_{2\sigma}n_{2-\sigma})\right] \tag{8.145}$$

We will again first assume $t = 0$. The possible energy levels with their degeneracies are depicted in Fig. 8.11 ($T_0 = 0$). The degeneracies occur for the total of five possible configurations mainly because of the equivalence of the two lattice sites.

We are interested in the quasiparticle density of states which is very sensitively dependent on the electron concentration. We want to determine the quasiparticle density of states for the special case that the system contains exactly one electron with spin $(-\sigma)$. According to (B.188) it holds

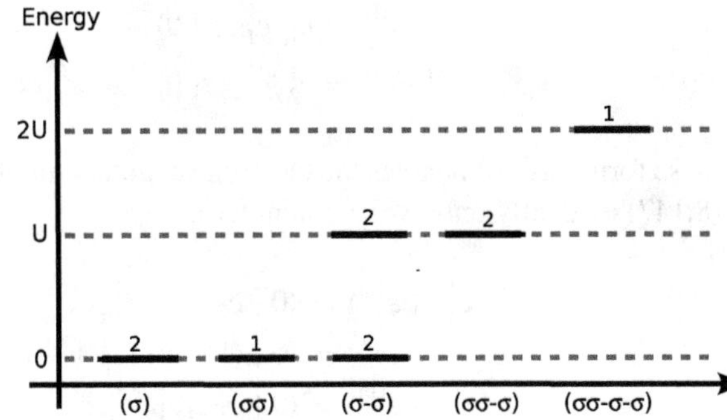

Fig. 8.11 Energy levels and corresponding degrees of degeneracy for the two-site Hubbard model in the case of zero hopping

$$\rho_{\sigma}^{(-\sigma)}(E) = \frac{1}{\hbar} S_{11\sigma}^{(-\sigma)} = S_{22\sigma}^{(-\sigma)} =$$
$$= \frac{1}{2\pi\hbar} \int_{-\infty}^{+\infty} d(t-t') \langle \left[c_{1\sigma}(t), c_{1\sigma}^{\dagger}(t') \right]_{+} \rangle \exp(\frac{i}{\hbar} E(t-t')) \tag{8.146}$$

The averaging is over the energy eigenstates of the system, which contains exactly one $(-\sigma)$-electron. That is, there is no σ-electron. Therefore clearly a σ-electron cannot be annihilated. As a result, only one term in the anticommutator in (8.146) is non-zero. With simple reformulations which correspond to those used for the spectral representations in Appendix B (e.g. (B.86)), one gets the following spectral representation of the quasiparticle density of states (8.146):

$$\rho_{\sigma}^{(-\sigma)}(E) = \frac{1}{Z} \sum_{nm} | \langle E_m^{(2)} | c_{1\sigma}^{\dagger} | E_n^{(1)} \rangle |^2 e^{-\beta E_n^{(1)}} \delta\left(E - (E_m^{(2)} - E_n^{(1)})\right) \tag{8.147}$$

$| E_n^{(i)} \rangle$ denotes an i-particle energy state. Z is the canonical partition function. Because of (8.146) we could have also chosen $c_{2\sigma}^{\dagger}$ instead of $c_{1\sigma}^{\dagger}$. For $t = 0$, (8.147) can easily be evaluated. The following one-particle states and one-particle energies are possible:

$$| \varepsilon_1^{(1)} \rangle = c_{1-\sigma}^{\dagger} | 0 \rangle \leftrightarrow \varepsilon_1^{(1)} = 0 \tag{8.148}$$
$$| \varepsilon_2^{(1)} \rangle = c_{2-\sigma}^{\dagger} | 0 \rangle \leftrightarrow \varepsilon_2^{(1)} = 0 \tag{8.149}$$

For the two-particle system consisting of a σ- and a $(-\sigma)$-electron, there are four possible eigenstates:

$$| \varepsilon_1^{(2)} \rangle = c_{1\sigma}^{\dagger} c_{2-\sigma}^{\dagger} | 0 \rangle \leftrightarrow \varepsilon_1^{(2)} = 0 \tag{8.150}$$
$$| \varepsilon_2^{(2)} \rangle = c_{2\sigma}^{\dagger} c_{1-\sigma}^{\dagger} | 0 \rangle \leftrightarrow \varepsilon_2^{(2)} = 0 \tag{8.151}$$

$$| \varepsilon_3^{(2)} \rangle = c_{1\sigma}^{\dagger} c_{1-\sigma}^{\dagger} \mid 0 \rangle \; \leftrightarrow \; \varepsilon_3^{(2)} = U \tag{8.152}$$

$$| \varepsilon_4^{(2)} \rangle = c_{2\sigma}^{\dagger} c_{2-\sigma}^{\dagger} \mid 0 \rangle \; \leftrightarrow \; \varepsilon_4^{(2)} = U \tag{8.153}$$

In these formulas $| \, 0 \rangle$ denotes the electron vacuum. Out of the eight matrix elements in (8.147) evidently only two are non-zero:

$$\begin{aligned} \langle \varepsilon_1^{(2)} \mid c_{1\sigma}^{\dagger} \mid \varepsilon_2^{(1)} \rangle &= \langle 0 \mid c_{2-\sigma} c_{1\sigma} c_{1\sigma}^{\dagger} c_{2-\sigma}^{\dagger} \mid 0 \rangle \\ &= \langle 0 \mid (1 - n_{2-\sigma})(1 - n_{1\sigma}) \mid 0 \rangle = 1 \end{aligned} \tag{8.154}$$

$$\begin{aligned} \langle \varepsilon_3^{(2)} \mid c_{1\sigma}^{\dagger} \mid \varepsilon_1^{(1)} \rangle &= \langle 0 \mid c_{1-\sigma} c_{1\sigma} c_{1\sigma}^{\dagger} c_{1-\sigma}^{\dagger} \mid 0 \rangle \\ &= \langle 0 \mid (1 - n_{1-\sigma})(1 - n_{1\sigma}) \mid 0 \rangle = 1 \end{aligned} \tag{8.155}$$

Application of the occupation number operator on the electron vacuum $|0\rangle$ results naturally in zero. We require in addition the partition function Z:

$$Z\{-\sigma\} = \langle \varepsilon_1^{(1)} \mid e^{\beta H} \mid \varepsilon_1^{(1)} \rangle + \langle \varepsilon_2^{(1)} \mid e^{\beta H} \mid \varepsilon_2^{(1)} \rangle = 2 \tag{8.156}$$

Substituting (8.148) to (8.156) in (8.147) we get

$$\rho_\sigma^{-\sigma}(E; t = 0) = \frac{1}{2}\delta(E) + \frac{1}{2}\delta(E - U) \tag{8.157}$$

This according to (8.139) is of course the expected result.

How does the density of states change when electron *hopping* is allowed?

$$t \neq 0 \, , \; t \ll U \;\; ('narrow\ band') \tag{8.158}$$

One should expect the degeneracy of the levels in Fig. 8.11 to be lifted. Because of $t \neq 0$, the states in (8.148) to (8.153) are no more the eigenstates. But we can use them as complete basis to represent the energy matrix which is to be diagonalized. One can expect that the degeneracy of the levels will be lifted.

For the one-particle system ($-\sigma$-electron) H is a (2×2)-matrix,

$$H^{(1)} = \begin{pmatrix} 0 & t \\ t & 0 \end{pmatrix} \tag{8.159}$$

which can easily be diagonalized. One finds the following eigenstates and eigenvalues (see Problem 8.9):

$$| E_1^{(1)} \rangle = \frac{1}{\sqrt{2}} \left(c_{1-\sigma}^{\dagger} - c_{2-\sigma}^{\dagger} \right) | \, 0 \rangle \; ; \; E_1^{(1)} = -t \tag{8.160}$$

$$| E_2^{(1)} \rangle = \frac{1}{\sqrt{2}} \left(c_{1-\sigma}^{\dagger} + c_{2-\sigma}^{\dagger} \right) | \, 0 \rangle \; ; \; E_2^{(1)} = +t \tag{8.161}$$

For the two-particle system $(\sigma, -\sigma)$ H is a (4×4)-matrix. With the basis states (8.150), (8.151), (8.152) and (8.153) one gets

$$H^{(2)} = \begin{pmatrix} 0 & 0 & t & t \\ 0 & 0 & t & t \\ t & t & U & 0 \\ t & t & 0 & U \end{pmatrix} \tag{8.162}$$

This matrix too can be exactly diagonalized (Problem 8.10). From

$$\det(H^{(2)} - E \cdot \mathbb{1}) = E(E-U)(E(E-U) - 4t^2) = 0$$

we get the eigenenergies

$$E_1^{(2)} = E_- \,,\; E_2^{(2)} = 0 \,,\; E_3^{(2)} = U,$$
$$E_4^{(2)} = E_+ \; with \; E_\pm = \frac{1}{2}U \pm \sqrt{\frac{1}{4}U^2 + 4t^2} \tag{8.163}$$

The corresponding eigenstates are (Problem 8.10)

$$\mid E_1^{(2)}\rangle = \frac{1}{\sqrt{2(1+\gamma_-^2)}}\Big(c_{1\sigma}^\dagger c_{2-\sigma}^\dagger + c_{2\sigma}^\dagger c_{1-\sigma}^\dagger + \gamma_- \sum_{i=1}^{2} c_{i\sigma}^\dagger c_{i-\sigma}^\dagger\Big) \mid 0\rangle \tag{8.164}$$

$$\mid E_2^{(2)}\rangle = \frac{1}{\sqrt{2}}\left(c_{1\sigma}^\dagger c_{2-\sigma}^\dagger - c_{2\sigma}^\dagger c_{1-\sigma}^\dagger\right) \mid 0\rangle \tag{8.165}$$

$$\mid E_3^{(2)}\rangle = \frac{1}{\sqrt{2}}\left(c_{1\sigma}^\dagger c_{1-\sigma}^\dagger - c_{2\sigma}^\dagger c_{2-\sigma}^\dagger\right) \mid 0\rangle \tag{8.166}$$

$$\mid E_4^{(2)}\rangle = \frac{1}{\sqrt{2(1+\gamma_+^2)}}\Big(c_{1\sigma}^\dagger c_{2-\sigma}^\dagger + c_{2\sigma}^\dagger c_{1-\sigma}^\dagger + \gamma_+ \sum_{i=1}^{2} c_{i\sigma}^\dagger c_{i-\sigma}^\dagger\Big) \mid 0\rangle \tag{8.167}$$

$$\gamma_\pm = \frac{1}{2t} E_\pm \tag{8.168}$$

When the electron *hopping* is switched on, indeed, the degeneracy which was present at $t = 0$ is completely lifted (Fig. 8.12).

With the states (8.160) and (8.161) for the one-electron system and (8.164), (8.165), (8.166) and (8.167) for the two-electron system, we can calculate the matrix elements required for the quasiparticle density of states (8.147). The partition func-

Fig. 8.12 Energy levels for the two-site Hubbard model for the case that there is a single $-\sigma$ electron and for the case that there are a σ and a $-\sigma$ electron

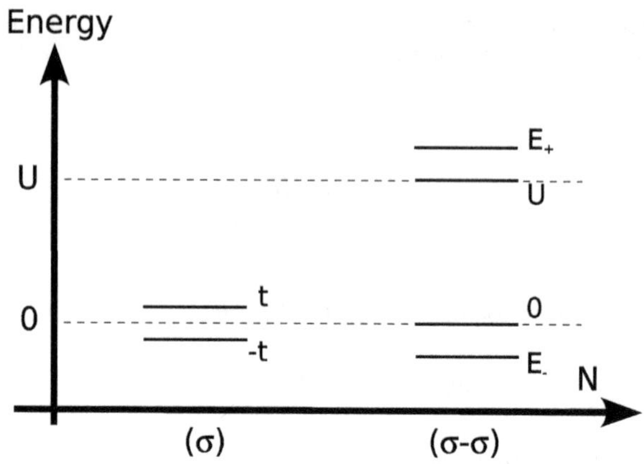

tion Z which is also to be substituted is given with (8.160) and (8.161) by

$$Z = e^{\beta t} + e^{-\beta t} \tag{8.169}$$

The quasiparticle density of states consists of a total of eight terms (Problem 8.10):

$$\rho_{\sigma}^{(-\sigma)}(E) = \rho_{t\sigma}^{(-\sigma)}(E) + \rho_{U\sigma}^{(-\sigma)}(E) \tag{8.170}$$

$$\begin{aligned}\rho_{t\sigma}^{(-\sigma)}(E) = \frac{1}{4(1+e^{-2\beta t})} & \left\{ \frac{(1-\gamma_-)^2}{1+\gamma_-^2}\delta(E-(E_- + t)) \right. \\ & + \delta(E-t) \\ & + \frac{(1+\gamma_-)^2}{1+\gamma_-^2} e^{-2\beta t}\delta(E-(E_- - t)) \\ & \left. + e^{-2\beta t}\delta(E+t) \right\}\end{aligned} \tag{8.171}$$

$$\begin{aligned}\rho_{U\sigma}^{(-\sigma)}(E) = \frac{1}{4(1+e^{-2\beta t})} & \left\{ \frac{(1-\gamma_+)^2}{1+\gamma_+^2}\delta(E-(E_+ + t)) \right. \\ & + \delta(E-(U+t)) \\ & + \frac{(1+\gamma_+)^2}{1+\gamma_+^2} e^{-2\beta t}\delta(E-(E_+ - t)) \\ & \left. + e^{-2\beta t}\delta(E-(U-t)) \right\}\end{aligned} \tag{8.172}$$

With the assumption $t \ll U$, the two parts of the quasiparticle density of states $\rho_{t\sigma}^{(-\sigma)}(E)$ and $\rho_{U\sigma}^{(-\sigma)}(E)$ are energetically distinctly separated. The four poles of $\rho_{t\sigma}^{(-\sigma)}(E)$ lie near the energy 0, constituting a *lower sub-band* and the four poles of $\rho_{U\sigma}^{(-\sigma)}(E)$ constituting a *upper sub-band* lie near the energy U. Thus the spectrum clearly consists of two quasiparticle sub-bands which are separated from each other by about U.

We obtain the statistical weights of the individual sub-bands when we integrate $\rho_{\sigma}^{(-\sigma)}(E)$ over the energy range which includes, respectively, either only the lower or the upper sub-band:

$$A_{t(U)} \equiv \int dE\, \rho_{t(U)\sigma}^{(-\sigma)}(E) \tag{8.173}$$

One finds with (8.171) and (8.172)

$$A_t = \frac{1}{2} + \frac{tE_+}{4t^2 + E_+^2} \frac{1 - e^{-2\beta t}}{1 + e^{-2\beta t}} \tag{8.174}$$

$$A_U = \frac{1}{2} - \frac{tE_+}{4t^2 + E_+^2} \frac{1 - e^{-2\beta t}}{1 + e^{-2\beta t}} \tag{8.175}$$

With this the sum rule

$$A_t + A_U = \int_{-\infty}^{+\infty} dE\, \rho_{\sigma}^{(-\sigma)}(E) = 1 \tag{8.176}$$

is obviously fulfilled. Equations (8.174) and (8.175) contain for $t = 0$ the special case of the infinitely narrow band: $A_t^{t=0} = A_U^{t=0} = \frac{1}{2}$ (8.157). Otherwise, for finite *hopping* this value is attained only for $T \to \infty$. For finite temperatures a weight transfer between the sub-bands occurs which is decisively induced by the *hopping integral* t.

Finally, we want to consider for our simple two-site model one more important special case, namely the situation where there is one electron per atom (*half-filled s-band*). For the sake of simplicity, we set $T = 0$. In the case of the infinitely narrow band ($t = 0$), there are two infinitely narrow sub-*bands* out of which the lower one is occupied and the upper one is empty. What happens when the *hopping* is switched on? We have seen above that the weights of the sub-bands which correspond to the area under the partial densities of states change with t. That would mean that quasiparticle states shift from one sub-band to the other. For the special case, that we are now considering, consequently the sub-bands would no longer be full and empty, respectively. The system would change from an *insulator* to a *metal*. We will, however, see that at $T = 0$, and for one electron per lattice site, the weights $A_{t,U}$ are *independent* of t so that in this special case the Hubbard system remains an insulator.

Because of $T = 0$, we can perform the average required for the quasiparticle density of states with the ground state of the two-electron system. This state according to Fig. 8.12 is $| E_1^{(2)}\rangle$ with the energy E_-:

$$\rho_\sigma^{(\sigma,-\sigma)}(E) = \sum_n | \langle E_n^{(3)} | c_{1\sigma}^\dagger | E_1^{(2)}\rangle |^2 \, \delta(E - (E_n^{(3)} - E_-)) + \sum_m | \langle E_m^{(1)} | c_{1\sigma} | E_1^{(2)}\rangle |^2 \, \delta(E - (E_- - E_m^{(1)})) \tag{8.177}$$

We have calculated the one-electron states $| E_m^{(1)}\rangle$ in (8.160) and (8.161) and the two-electron states $| E_1^{(2)}\rangle$ in (8.164), (8.165), (8.166), (8.167) and (8.168). The three-electron states $| E_n^{(3)}\rangle$ which have not yet been determined can almost be *guessed*. In Problem 8.11 we show that

$$| E_1^{(3)}\rangle = \frac{1}{\sqrt{2}}\left(c_{1-\sigma}^\dagger - c_{2-\sigma}^\dagger\right) c_{1\sigma}^\dagger c_{2\sigma}^\dagger \, | 0\rangle \tag{8.178}$$

$$| E_2^{(3)}\rangle = \frac{1}{\sqrt{2}}\left(c_{1-\sigma}^\dagger + c_{2-\sigma}^\dagger\right) c_{1\sigma}^\dagger c_{2\sigma}^\dagger \, | 0\rangle \tag{8.179}$$

are the eigenstates with energies:

$$E_1^{(3)} = U - t \; ; \; E_2^{(3)} = U + t \tag{8.180}$$

Now we have everything to evaluate $\rho_\sigma^{(\sigma,-\sigma)}(E)$:

$$\begin{aligned}\rho_\sigma^{(\sigma,-\sigma)}(E) &= \frac{(1-\gamma_-)^2}{4(1+\gamma_-^2)}\{\delta\,(E - (U - t - E_-)) + \delta\,(E - (E_- + t))\} \\ &+ \frac{(1+\gamma_-)^2}{4(1+\gamma_-^2)}\{\delta\,(E - (U + t - E_-)) + \delta\,(E - (E_- - t))\}\end{aligned} \tag{8.181}$$

Because of $t \ll U$, we have $E_- \approx t$; i.e. now also we have the total spectrum split into two sub-*bands* which are separated by about U. In this special case the weights of the two sub-*bands* are equal and importantly are independent of t:

$$A_t = A_U = \frac{1}{4(1+\gamma_-^2)}\left((1-\gamma_-)^2 + (1+\gamma_-)^2\right) = \frac{1}{2} \tag{8.182}$$

They have the same value as in the atomic limit (8.157). The independence from t is, of course, valid only for this special case in which the system contains exactly one electron per lattice site. The lower *sub-band* is fully occupied and the upper one is empty. This corresponds to an insulator since the electrons in the lower *sub-band* must overcome an energy gap in order to reach unoccupied states. This feature transfers, as we shall see, to the full three-dimensional situation and represents a characteristic correlation effect. The strong Coulomb interaction sees to it that the Hubbard crystal is an insulator even though the s-band is only half-filled. According to the elementary band theory such a system should show metallic behaviour.

In the next section also we will consider the special properties of the half-filled band.

8.4.4 The Exactly Half-Filled Band

Sometimes, one can gain valuable physical information when one transforms the actual Hamiltonian into an equivalent effective operator. An interesting possibility in this context is the Hubbard model for the special case of an exactly half-filled energy band. In the Hubbard model the system is considered to be a lattice of atoms with a single orbital which can be occupied by a maximum of two electrons with opposite spins. That means

$$half - filled\ band \ \leftrightarrow\ n = 1$$

In the special case of infinitely narrow band ($W \to 0$), in the ground state, every lattice point is occupied by exactly one electron. Then the only variable is the electron spin ($\sigma = \uparrow, \downarrow$). Therefore, the ground state energy is 2^N-fold degenerate. The ground state does not have any double occupation. If now the *hopping is switched on*, the band electrons in the limit of strong coupling

$$W \ll U\ \ (strong - coupling\ limit)$$

remain localized to a large extent. However, *virtual* changes of sites as in the case of superexchange of the Heisenberg model (Sect. 5.3.2) cause an *indirect* coupling between the electron spins on different lattice sites. These changes of sites are called *virtual* because they are associated with energy cost.

With the help of an elementary perturbation theory, we want to show, for the half-filled band, in the strong-coupling regime, an equivalence of the Hubbard and Heisenberg models. For that we treat the electron hopping as perturbation.

$$H = H_0 + H_1 \tag{8.183}$$

$$H_0 = T_0 \sum_{i\sigma} n_{i\sigma} + \frac{1}{2} U \sum_{i\sigma} n_{i\sigma} n_{i-\sigma} \tag{8.184}$$

$$H_1 = \sum_{ij\sigma}^{i \neq j} T_{ij} c_{i\sigma}^{\dagger} c_{j\sigma} \tag{8.185}$$

We consider only the ground state. All the eigenvalues and eigenstates of H_0 are characterized by d = *number of doubly occupied lattice sites* The states with the same d are highly degenerate since there are many ways in which N_σ σ-electrons can be arranged on N lattice sites. Let the corresponding indexing be through Greek letters $\alpha, \beta, \gamma \cdots$.

$$H_0 \mid d\alpha\rangle^{(0)} = E_d^{(0)} \mid d\alpha\rangle^{(0)} = (NT_0 + dU) \mid d\alpha\rangle^{(0)} \tag{8.186}$$

$\mid 0\alpha\rangle^{(0)}$ is the ground state which is 2^N-fold degenerate due to $n = 1$. First order perturbation theory requires the solution of the secular equation

$$\det\left({}^{(0)}\langle 0\alpha' \mid H_1 \mid 0\alpha\rangle^{(0)} - E_0^{(1)}\delta_{\alpha\alpha'}\right) \stackrel{!}{=} 0 \tag{8.187}$$

with 2^N solutions $E_{0\alpha}^{(1)}$. One can easily see that since every summand of H_1 produces an empty and a doubly occupied site, it holds

$${}^{(0)}\langle d\alpha' \mid H_1 \mid 0\alpha\rangle^{(0)} \neq 0 \text{ at the most for } d = 1 \tag{8.188}$$

Because of this the perturbation matrix in (8.187) consists of elements which are all zeros. Therefore, all the energy corrections in the first order $E_{0\alpha}^{(1)}$ vanish; the degeneracy remains unaffected.

Second order perturbation theory requires the solution of the following system of equations:

$$\sum_\alpha c_\alpha \left\{ \sum_{d\gamma}^{d\neq 0} \frac{{}^{(0)}\langle 0\alpha' \mid H_1 \mid d\gamma\rangle^{(0)} \,.^{(0)}\langle d\gamma \mid H_1 \mid 0\alpha\rangle^{(0)}}{E_0^{(0)} - E_d^{(0)}} - E_0^{(2)}\delta_{\alpha\alpha'} \right\} \stackrel{!}{=} 0 \tag{8.189}$$

We interpret this as eigenvalue equation of an "effective" *Hamiltonian*, whose eigenvalue corresponds to the second order correction and whose matrix elements are given by

$$\begin{aligned}
&{}^{(0)}\langle 0\alpha' \mid H_1 \sum_{d\gamma}^{d\neq 0} \frac{\mid d\gamma\rangle^{(0)}\, {}^{(0)}\langle d\gamma \mid}{E_0^{(0)} - E_d^{(0)}} H_1 \mid 0\alpha\rangle^{(0)} = \\
&= -\frac{1}{U}\, {}^{(0)}\langle 0\alpha' \mid H_1 \left(\sum_{d\gamma} \mid d\gamma\rangle^{(0)}\, {}^{(0)}\langle d\gamma \mid \right) H_1 \mid 0\alpha\rangle^{(0)} = \\
&= -\frac{1}{U}\, {}^{(0)}\langle 0\alpha' \mid H_1^2 \mid 0\alpha\rangle^{(0)}
\end{aligned} \tag{8.190}$$

In the first step we have used (8.188) so that

$$\left(E_d^{(0)} - E_0^{(0)}\right) \longrightarrow \left(E_1^{(0)} - E_0^{(0)}\right) = U$$

and the limitation $d \neq 0$ can be done away with. The second step follows from the completeness relation of the *unperturbed* states $| \, d\gamma\rangle^{(0)}$. The effective Hamiltonian then reads as

$$H_{eff} = P_0 \left(-\frac{H_1^2}{U}\right) P_0 \tag{8.191}$$

where P_0 is the projection operator onto the $d = 0$ subspace. What have we gained from this? In order to see that, we express this effective operator in terms of spin operators. We first substitute (8.185):

$$H_{eff} = -\frac{1}{U} P_0 \left(\sum_{ij\sigma}^{i\neq j} \sum_{mn\sigma'}^{m\neq n} T_{ij} T_{mn} c_{i\sigma}^{\dagger} c_{j\sigma} c_{m\sigma'}^{\dagger} c_{n\sigma'} \right) P_0 \tag{8.192}$$

Because of P_0 in the multiple sum only the terms $i = n$ and $j = m$ contribute. Therefore, what remains is

$$\begin{aligned} H_{eff} &= -\frac{1}{U} P_0 \left(\sum_{ij\sigma\sigma'}^{i\neq j} T_{ij} T_{ji} c_{i\sigma}^{\dagger} c_{j\sigma} c_{j\sigma'}^{\dagger} c_{i\sigma'} \right) P_0 \\ &= -\frac{1}{U} P_0 \left(\sum_{ij\sigma\sigma'}^{i\neq j} | \, T_{ij} \, |^2 \, c_{i\sigma}^{\dagger} c_{i\sigma'} (\delta_{\sigma\sigma'} - c_{j\sigma'}^{\dagger} c_{j\sigma}) \right) P_0 \\ &= -\frac{1}{U} P_0 \left(\sum_{ij\sigma}^{i\neq j} | \, T_{ij} \, |^2 \, (n_{i\sigma} - n_{i\sigma} n_{j\sigma} - c_{i\sigma}^{\dagger} c_{i-\sigma} c_{j-\sigma}^{\dagger} c_{j\sigma}) \right) P_0 \end{aligned} \tag{8.193}$$

We now use the spin operators from (5.98) and (5.101), which, as was shown in Problem 5.1, fulfil the conventional commutation relations.

$$S_i^z = \frac{\hbar}{2} \sum_{\sigma} z_{\sigma} n_{i\sigma} \quad (z_{\sigma} = \delta_{\sigma\uparrow} - \delta_{\sigma\downarrow})$$

$$S_i^{\sigma} = \hbar c_{i\sigma}^{\dagger} c_{i-\sigma} \quad (S_i^{\uparrow} \equiv S_i^{+} \, ; \; S_i^{\downarrow} \equiv S_i^{-})$$

It then holds:

$$
\begin{aligned}
P_0\left\{S_i^z S_j^z\right\} &= \frac{\hbar^2}{4}\sum_{\sigma\sigma'} z_\sigma z_{\sigma'} P_0\{n_{i\sigma}n_{j\sigma'}\}P_0 \\
&= \frac{\hbar^2}{4}\sum_{\sigma}\left\{P_0\{n_{i\sigma}n_{j\sigma}\}P_0 - P_0\{n_{i\sigma}n_{j-\sigma}\}P_0\right\} \\
&= \frac{\hbar^2}{4}\sum_{\sigma}\left\{P_0\{n_{i\sigma}n_{j\sigma}\}P_0 - P_0\{n_{i\sigma}(1-n_{j\sigma}\}P_0\right\} \\
&= \frac{\hbar^2}{2}P_0\left\{\sum_\sigma n_{i\sigma}n_{j\sigma}\right\}P_0 - \frac{\hbar^2}{4}P_0\left\{\sum_\sigma n_{i\sigma}\right\}P_0 \\
&= \frac{\hbar^2}{2}P_0\left\{\sum_\sigma n_{i\sigma}n_{j\sigma}\right\}P_0 - \frac{\hbar^2}{4}P_0^2
\end{aligned}
$$

As an intermediate result we have:

$$
P_0\left\{\sum_\sigma n_{i\sigma}n_{j\sigma}\right\}P_0 = P_0\left\{\frac{2}{\hbar^2}S_i^z S_j^z + \frac{1}{2}\mathbb{1}\right\}P_0 \tag{8.194}
$$

where in particular

$$
P_0\left\{\sum_\sigma n_{i\sigma}\right\}P_0 = P_0\mathbb{1}P_0
$$

has been used, which is correct of course only for our special case $n = 1$. Directly from the above definition of the spin operators it further follows:

$$
\begin{aligned}
P_0\left\{\sum_\sigma c_{i\sigma}^\dagger c_{i-\sigma}c_{j-\sigma}^\dagger c_{j\sigma}\right\}P_0 &= P_0\left\{\frac{1}{\hbar^2}\sum_\sigma S_i^\sigma S_j^{-\sigma}\right\}P_0 = \\
&\overset{i\neq j}{=} P_0\left\{\frac{2}{\hbar^2}\left(S_i^x S_j^x + S_i^y S_j^y\right)\right\}P_0
\end{aligned}
\tag{8.195}
$$

Substituting (8.194) and (8.195) in the effective operator (8.193) and using the (actually unimportant) constant,

$$
\eta = -\frac{1}{2U}\sum_{ij}^{i\neq j} \mid T_{ij}\mid^2 \tag{8.196}
$$

one recognizes that H_{eff} has the same structure as the *Heisenberg Hamiltonian*:

$$H_{eff} = P_0 \left\{ \eta - \sum_{ij}^{i \neq j} J_{ij} \mathbf{S}_i \cdot \mathbf{S}_j \right\} P_0 \tag{8.197}$$

The "exchange integrals"

$$J_{ij} = -\frac{2}{\hbar^2} \frac{|\, T_{ij} \,|^2}{U} \tag{8.198}$$

are always negative, and therefore mediate an *antiferromagnetic coupling* between the electron spins. Therefore, for half-filled band ($n = 1$), within the framework of Hubbard model, antiferromagnetism is to be expected, at least in the strong-coupling regime U/W. This will be confirmed also by later analysis. The Hubbard model in this special case is equivalent to Heisenberg model, where in this special case, the exchange integrals acquire a well-defined microscopic meaning.

According to (8.193) the second order perturbation theory contains the jump processes of the electron from lattice site $\mathbf{R}_i$ to lattice site $\mathbf{R}_j$ and back (see Fig. 8.13). According to (8.193), these jump processes lead to an energy *gain*. The jump probability is proportional to T_{ij} and is maximum between the nearest neighbours since T_{ij} is as a rule of *short range*. The more are the jumps, the larger is the decrease in the energy. In a "saturated" ferromagnet the *virtual* hopping is forbidden due to the Pauli's principle. In contrast, in an antiferromagnet, not all the neighbouring spins are parallel so that *virtual hopping* is allowed. This is a qualitative explanation for the negative exchange integrals (8.198). Ultimately it is of course the free energy that decides the stability of a magnetic phase.

8.4.5 Strong-Coupling Regime

We have seen in Sect. 8.4.2 that for infinitely narrow band case ($W \to 0$) the density of states and the spectral density of the Hubbard model is made up of two δ-functions at T_0 and $T_0 + U$. In this case the number of doubly occupied sites d is a conserved quantity because

$$\left[n_{i\sigma} n_{i-\sigma}, H_0 \right]_- = 0$$

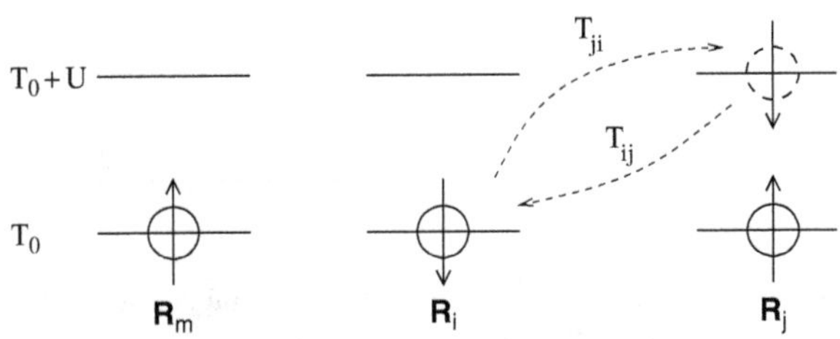

Fig. 8.13 Virtual electron hopping in the strongly correlated Hubbard model at half-filling ($n = 1$)

It is equal to zero in the ground state of a less than half-filled band ($n < 1$). Ferromagnetism turned out to be impossible. What happens now when the *hopping is switched on*? For this we will assume

$$W \neq 0 \qquad W \ll U \qquad (strong - coupling\ limit) \tag{8.199}$$

We can get the first indications from the discussion on the two-site model in Sect. 8.4.3. The kinetic energy lifts the degeneracies of both the quasiparticle levels. The new energies, however, group around the energies T_0 and $T_0 + U$, indicating the formation of two quasiparticle sub-bands. Therefore, for the full lattice one should expect:

- Smearing out of the δ-peaks in the spectral densities.
- Shifting of the centres of gravity of the peaks.
- Shifting of the spectral weight($\cong$ area under the respective peak).
- Appearance of *satellite peaks* at energies

$$T_0 + pU\ ;\ p = -1, -2, \cdots ; p = 2, 3, \cdots$$

as a consequence of *multiple processes* (see Fig. 8.14).

By *multiple processes* we mean the situations in which, when an additional electron is introduced, because of the finite hopping probability in the Hubbard system, the number of double occupations d, which is no more a conserved quantity, changes. Large energy shifts are involved in this so that such processes in the *strong-coupling limit* (8.199) should be relatively improbable. Therefore, the corresponding satellite peaks will carry relatively small spectral weights.

Even in the limit (8.199) the difficult many-body problem of the Hubbard model cannot be exactly solved. But nevertheless, certain rigorously valid statements can

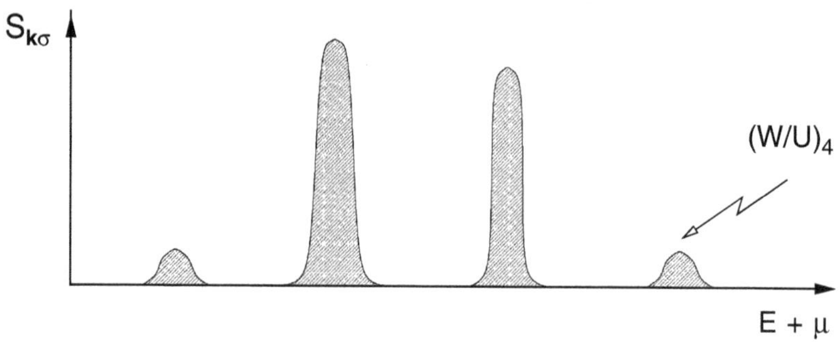

Fig. 8.14 Schematic representation of the single-electron spectral density of the Hubbard model in the strong-coupling regime as function of energy ($W \ll U$)

be made. A few of these are due to the pioneering work of Harris and Lange [5], which we will briefly mention, without going into the detailed presentation of partly very involved derivations. For more details, the reader is referred to the original work [5].

The main idea is based on the splitting of the construction operators,

$$c_{i\sigma} = \sum_p c_{i\sigma,p} \;\; ; \;\; c^{\dagger}_{i\sigma} = \sum_p c^{\dagger}_{i\sigma,p} \tag{8.200}$$

in such a way that the *new* operators couple the states with each other, whose $W \to 0$-energies are different by $p\,U$. That means

$$\begin{aligned} \langle E_b \mid c_{i\sigma,p} \mid E_a \rangle &= 0 \\ \langle E_a \mid c^{\dagger}_{i\sigma,p} \mid E_b \rangle &= 0 \\ if\ not \;\; : \;\; E_a - E_b &= p\,U + \mathcal{O}(W) \end{aligned} \tag{8.201}$$

This ansatz is meaningful only in the strong-coupling regime where the spectral density has the structure shown in Fig. 8.14. One should note that $c_{i\sigma,p}$ and $c^{\dagger}_{i\sigma,p}$ are not adjoint of each other. Rather what holds is $c^{\dagger}_{i\sigma,p} = (c_{i\sigma,-p})^{\dagger}$. As an example let us consider the simple special case $W = 0$. In the decomposition

$$c_{i\sigma} = n_{i-\sigma} c_{i\sigma} + (1 - n_{i-\sigma}) c_{i\sigma} \tag{8.202}$$

the first summand annihilates a doubly occupied site and the second annihilates a singly occupied one. That means

$$\begin{aligned} c_{i\sigma,0} &= (1 - n_{i-\sigma}) c_{i\sigma} \\ c_{i\sigma,-1} &= n_{i-\sigma} c_{i\sigma} \\ c_{i\sigma,p} &= 0 \;\; for\ p \neq 0, -1 \end{aligned} \tag{8.203}$$

Analogously it follows from

$$c^{\dagger}_{i\sigma} = n_{i-\sigma} c^{\dagger}_{i\sigma} + (1 - n_{i-\sigma}) c^{\dagger}_{i\sigma} \tag{8.204}$$

the separation:

$$\begin{aligned} c^{\dagger}_{i\sigma,0} &= (1 - n_{i-\sigma}) c^{\dagger}_{i\sigma} \\ c^{\dagger}_{i\sigma,+1} &= n_{i-\sigma} c^{\dagger}_{i\sigma} \\ c^{\dagger}_{i\sigma,p} &= 0 \;\; for\ p \neq 0, 1 \end{aligned}$$

For finite bandwidth but with $W \ll U$, the one-electron spectral density also can be correspondingly decomposed:

$$S_{ij\sigma}(E) = \sum_p S_{ij\sigma,p}(E) \tag{8.205}$$

where $S_{ij\sigma,p}(E)$ is non-zero only in a narrow energy region around $T_0 + pU$. Because of the assumption (8.199) no *mixed* p, p'-terms appear. The spectral density peak at $T_0 + pU$ is completely described by the *partial* spectral density $S_{ij\sigma,p}(E)$, which is formally defined exactly in the same way as the *normal* spectral density $S_{ij\sigma}(E)$, except that the *new* construction operators of (8.200) appear

$$S_{ij\sigma,p}(E) = \frac{1}{2\pi}\int_{-\infty}^{+\infty} d(t-t')\, e^{\frac{i}{\hbar}E(t-t')}\left\langle\left[c_{i\sigma,-p}(t),\, c_{j\sigma,p}^{\dagger}(t')\right]\right\rangle \tag{8.206}$$

Then we can define the *quite normal* spectral moments corresponding to these *partial* spectral densities, which provide information about the individual spectral density peaks:

$$\begin{aligned} m_{ij\sigma,p}^{(n)} &= \frac{1}{\hbar}\int_{-\infty}^{+\infty} dE\,(E - T_0 - pU)^n S_{ij\sigma,p}(E) = \\ &= \left(i\hbar\frac{\partial}{\partial t} - T_0 - pU\right)^n \left\langle\left[c_{i\sigma,-p}(t),\, c_{j\sigma,p}^{\dagger}(t')\right]\right\rangle\Big|_{t=t'} \\ n &= 0, 1, 2, \cdots \end{aligned} \tag{8.207}$$

The $n = 0$ moment gives the spectral weight (area under the curve) of the corresponding peak while the $n = 1$ moment fixes the centre of gravity.

The main question now is, how does one find the p-separation (8.200) of the construction operators? Harris and Lange have developed a unitary transformation on the new construction operators, which has exactly the required properties, and with these they could, via the spectral moments (8.207), analyse the individual peaks of the spectral density. The procedure can in principle be iterated to any required order in W/U.

This confirms the expectation of Fig. 8.14 that in *strong-coupling regime* (8.199), the (wavenumber-dependent) spectral density of the Hubbard model is made up of two main peaks at T_0 and $T_0 + U$ and additional satellite peaks near the energies $T_0 + pU$. The spectral weights of the satellite peaks, however, decrease rapidly with the increase in distance from the main peaks. Already the immediately neighbouring peaks ($p = -1, +2$) have spectral weights of the order of $(W/U)^4$, so that in the strong-coupling regime they are relatively unimportant. Therefore, for $U \gg W$, the spectral density has mainly a two-peak structure. The exact shape of the peaks is not known, but using the first two spectral moments one can make statements regarding the positions of the centres of gravity of the peaks $T_{1,2\sigma}(\mathbf{k})$ and the corresponding spectral weights $\alpha_{1,2\sigma}(\mathbf{k})$:

$$T_{1\sigma}(\mathbf{k}) = T_0 + (1 - n_{-\sigma})(\varepsilon(\mathbf{k}) - T_0) + n_{-\sigma} B_{\mathbf{k}-\sigma} + \mathcal{O}\left(\frac{W}{U}\right)^4 \tag{8.208}$$

$$T_{2\sigma}(\mathbf{k}) = T_0 + U + n_{-\sigma}(\varepsilon(\mathbf{k}) - T_0) + (1 - n_{-\sigma}) B_{\mathbf{k}-\sigma} + \mathcal{O}\left(\frac{W}{U}\right)^4 \tag{8.209}$$

$$\alpha_{1\sigma}(\mathbf{k}) = (1 - n_{-\sigma}) + \mathcal{O}\left(\frac{W}{U}\right) = 1 - \alpha_{2\sigma}(\mathbf{k}) \tag{8.210}$$

We have again assumed translational symmetry. Therefore, the thermodynamic expectation values of the occupation numbers are as in (8.32) independent of the lattice site index. Basically, the centres of gravity of the peaks can also possess a spin dependence which results in an additional exchange splitting of the main peaks (Fig. 8.15), which is a fundamental basic requirement for ferromagnetism. The question, whether such a correlation-induced spin asymmetry appears or not, as the following investigation will show, is to a large extent determined by the *band correction* $B_{\mathbf{k}-\sigma}$:

$$B_{\mathbf{k}-\sigma} = B_{-\sigma} + F_{\mathbf{k}-\sigma} \tag{8.211}$$

The wavevector independent, local term $B_{-\sigma}$ shall be called in the following as *band shift*. It describes a *correlated electron hopping*:

$$n_{-\sigma}(1 - n_{-\sigma}) B_{-\sigma} = \frac{1}{N} \sum_{ij}^{i \neq j} T_{ij} \langle c_{i-\sigma}^{\dagger} c_{j-\sigma} (2n_{i\sigma} - 1) \rangle \tag{8.212}$$

For $U \gg W$ and for less than half-filled bands, double occupations are very improbable, so that the second term in the parenthesis dominates over the first term. Then we see that the possible shift of the σ-spectrum correlates with the negative kinetic energy of the $(-\sigma)$-electron.

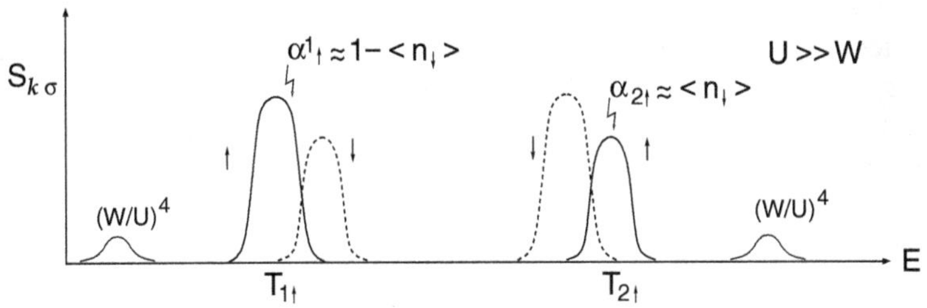

Fig. 8.15 Exchange-split single-electron spectral density of the Hubbard model in the strong-coupling regime ($U \gg W$) as function of energy

The **k**-dependent part $F_{\mathbf{k}-\sigma}$ of the band correction (8.211), which we will in the following call *bandwidth correction* consists of terms which can be easily interpreted, namely a density correlation, a double-hopping term and a spin-flip correlation:

$$\begin{aligned} n_{-\sigma}(1-n_{-\sigma})F_{\mathbf{k}-\sigma} = & \sum_{ij}^{i\neq j} T_{ij} e^{-i\mathbf{k}(\mathbf{R}_i-\mathbf{R}_j)}\Big((\langle n_{i-\sigma} n_{j-\sigma}\rangle - n_{-\sigma}^2) + \\ & + \langle c_{j-\sigma}^{\dagger} c_{j\sigma}^{\dagger} c_{i-\sigma} c_{i\sigma}\rangle + \langle c_{j\sigma}^{\dagger} c_{j-\sigma} c_{i-\sigma}^{\dagger} c_{i\sigma}\rangle \Big) \end{aligned} \quad (8.213)$$

Just as the band shift $B_{-\sigma}$, $F_{\mathbf{k}-\sigma}$ also vanishes in the limit of infinitely narrow band ($W \to 0$) and has in addition no direct influence on the centres of gravity $T_{1,2\sigma}$ of the two quasiparticle sub-bands (*Hubbard bands*), which are obtained from the two main peaks of the spectral density (Fig. 8.15) after summing over the wavevectors of the first Brillouin zone:

$$T_{1\sigma} = \frac{1}{N}\sum_{\mathbf{k}} T_{1\sigma}(\mathbf{k}) = T_0 + n_{-\sigma} B_{-\sigma} \quad (8.214)$$

$$T_{2\sigma} = \frac{1}{N}\sum_{\mathbf{k}} T_{2\sigma}(\mathbf{k}) = T_0 + U + (1-n_{-\sigma}) B_{-\sigma} \quad (8.215)$$

Here we have used

$$\frac{1}{N}\sum_{\mathbf{k}} F_{\mathbf{k}-\sigma} = 0 \quad (8.216)$$

This is because the **k**-summation in the bandwidth correction results in a Kronecker delta δ_{ij}, so that the sum over i and j vanishes. The bandwidth correction should be of less importance than the band shift $B_{-\sigma}$ as far as the possibility of ferromagnetism in the Hubbard model is concerned. However, $F_{\mathbf{k}-\sigma}$ can lead to a spin-dependent bandwidth correction and in this sense can compete with the pre-factors of $(\varepsilon(\mathbf{k}) - T_0)$ in (8.208) and (8.209). This has to be still investigated.

For later concrete evaluations it will turn out important and also advantageous that the spin-dependent band shift can be exactly expressed in terms of the one-electron spectral density or the corresponding Green's function, even though it is made up of *higher* correlation functions. We prove this as Problem 8.12:

$$\begin{aligned} n_{-\sigma}(1-n_{-\sigma})B_{-\sigma} = & \frac{1}{N\hbar}\sum_{\mathbf{k}} (\varepsilon(\mathbf{k}) - T_0) * \\ & * \int_{-\infty}^{+\infty} dE\, f_{-}(E)\left(\frac{2}{U}(E-\varepsilon(\mathbf{k})) - 1\right) * \\ & * S_{\mathbf{k}-\sigma}(E-\mu) \end{aligned} \quad (8.217)$$

$f_-(E)$ is the Fermi function. Unfortunately, the bandwidth correction $F_{\mathbf{k}-\sigma}$ cannot be expressed in a similar manner in terms of one-electron spectral density.

8.4.6 Spectral Moments

The spectral moments (Sect. B.3.4) are of great value in the context of assessing unavoidable approximations and also for fixing free parameters in respective trial functions.

$$M_{ij\sigma}^{(n)} = \frac{1}{\hbar}\int_{-\infty}^{+\infty} dE\, E^n\, S_{ij\sigma}(E)\,;\; n = 0, 1, 2, \cdots \tag{8.218}$$

Using the equations of motion of the Heisenberg operators in terms of which the spectral density is constructed, one can derive an equivalent expression for the moments (B.100):

$$M_{ij\sigma}^{n} = \langle \Big[\underbrace{[\cdots[c_{i\sigma}, \mathcal{H}]_- \cdots, \mathcal{H}]_-}_{(n-p)-fold},\, \underbrace{[\mathcal{H}, \cdots [\mathcal{H}, c_{j\sigma}^\dagger]_- \cdots]_-}_{p-fold} \Big]_+ \rangle \tag{8.219}$$

p is an integer between 0 and n. With $\mathcal{H}$ from (8.122), one can calculate, in principle, all the moments exactly, that, too, independently (!) of the corresponding spectral density. In practice, however, only a few moments can be determined without prohibitively large effort. In addition, with increasing order n, there appear more and more *higher* expectation values in the spectral moments which are unknown and cannot be self-consistently determined using the spectral density.

However, with the help of the moments, certain general statements can still be made which will be listed out here without proof:

- *Centre of gravity of the energy spectrum:* The spin-dependent centre of gravity of the total energy spectrum,

$$T_\sigma = \frac{1}{\hbar N}\sum_{\mathbf{k}}\int_{-\infty}^{+\infty} dE\, E S_{\mathbf{k}\sigma}(E-\mu) = \int_{-\infty}^{+\infty} dE\, E\rho_\sigma(E) \tag{8.220}$$

  are correctly recovered, if the first two wavevector-dependent moments

$$M_{\mathbf{k}\sigma}^{(n)}\,;\; n = 0, 1$$

  are fulfilled.
- *Hubbard bands:* Typical for the strong-coupling regime of the Hubbard model, as we will later see and also as it was already indicated in Sects. 8.4.2 and 8.4.3, is the splitting of the original Bloch density of states into two quasiparticle subbands (Fig. 8.16) with an energy separation of U.
  A necessary condition for that is the first three moments should be fulfilled.

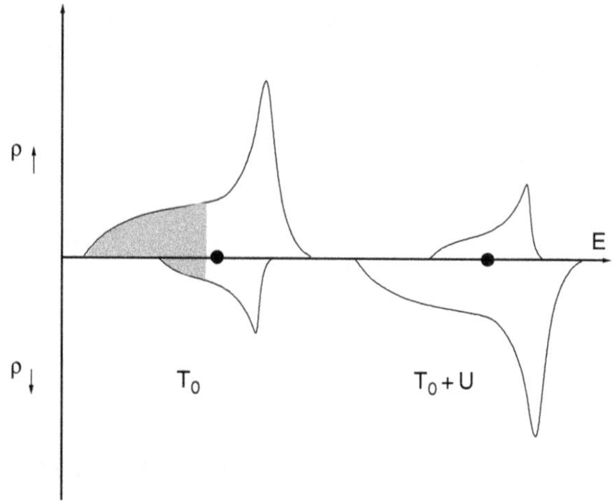

Fig. 8.16 Schematic representation of the strong-coupling quasiparticle sub-bands of the Hubbard model (Hubbard bands)

$$M_{\mathbf{k}\sigma}^{(n)} \; ; \; n = 0, 1, 2$$

The condition is however not sufficient. The self-consistent second order perturbation theory in U for infinite lattice dimensions ($d = \infty$) satisfies the first three moments but does not give the above-mentioned *Hubbard splitting*.

- *"Strong-coupling" behaviour:* The correct behaviour in the strong-coupling regime according to the theory of Sect. 8.4.5 is based on two conditions. First, for $W \ll U$, the *Hubbard bands* (Fig. 8.16) must exist. One can show that this is guaranteed by every theory which correctly reproduces the limit of infinitely narrow band (Sect. 8.4.2). The second condition is that the first four spectral moments

$$M_{\mathbf{k}\sigma}^{(n)} \; ; \; n = 0, 1, 2, 3$$

must be satisfied. These two conditions guarantee the correct centres of gravity (8.208) and (8.209) of the spin-dependent spectral density and the correct spectral weights (8.210) of both the main peaks.

The calculation of the first four spectral moments requires some effort, but does not encounter any principal difficulties (Problem 8.13):

$$M_{\mathbf{k}\sigma}^{(0)} = 1 \tag{8.221}$$

$$M_{\mathbf{k}\sigma}^{(1)} = \varepsilon(\mathbf{k}) - \mu + Un_{-\sigma} \tag{8.222}$$

$$M_{\mathbf{k}\sigma}^{(2)} = (\varepsilon(\mathbf{k}) - \mu)^2 + 2(\varepsilon(\mathbf{k}) - \mu)Un_{-\sigma} + U^2 n_{-\sigma} \tag{8.223}$$

$$\begin{aligned} M_{\mathbf{k}\sigma}^{(3)} &= (\varepsilon(\mathbf{k}) - \mu)^3 + 3(\varepsilon(\mathbf{k}) - \mu)^2 Un_{-\sigma} + (\varepsilon(\mathbf{k}) - \mu)U^2 n_{-\sigma}(2 + n_{-\sigma}) \\ &\quad + U^3 n_{-\sigma} + U^2 n_{-\sigma}(1 - n_{-\sigma})(T_0 + B_{\mathbf{k}-\sigma} - \mu) \end{aligned} \tag{8.224}$$

The fourth moment ($n = 3$) contains the important *band correction* $B_{\mathbf{k}-\sigma}$, which is made up of the *higher* correlation functions (8.211) and proves to be one of the decisive factors for the possibility of spontaneous ferromagnetism.

8.4.7 High-Energy Expansions

There exists a very close connection between the spectral moments and the high-energy expansion of the one-electron Green's function whose spectral representation (B.91)

$$G_{\mathbf{k}\sigma}(E) = \int_{-\infty}^{+\infty} dE' \, \frac{S_{\mathbf{k}\sigma}(E')}{E - E'} \tag{8.225}$$

can also be written as follows:

$$\begin{aligned} G_{\mathbf{k}\sigma}(E) &= \frac{1}{E} \int_{-\infty}^{+\infty} dE' \, \frac{S_{\mathbf{k}\sigma}(E')}{1 - \frac{E'}{E}} = \\ &= \frac{1}{E} \sum_{n=0}^{\infty} \int_{-\infty}^{+\infty} dE' \left(\frac{E'}{E} \right)^n S_{\mathbf{k}\sigma}(E') \end{aligned}$$

In this expression one recognizes the spectral moments (8.218), so that the following important relationship holds:

$$G_{\mathbf{k}\sigma}(E) = \hbar \sum_{n=0}^{\infty} \frac{M_{\mathbf{k}\sigma}^{(n)}}{E^{n+1}} \tag{8.226}$$

A completely analogous high-energy expansion can be performed also for the self-energy:

$$\Sigma_{\mathbf{k}\sigma}(E) = \sum_{m=0}^{\infty} \frac{C_{\mathbf{k}\sigma}^{(m)}}{E^m} \tag{8.227}$$

The coefficients $C_{\mathbf{k}\sigma}^{(m)}$ are determined using the generally valid Dyson equation (B.152):

$$E \, G_{\mathbf{k}\sigma}(E) = \hbar + (\varepsilon(\mathbf{k}) - \mu + \Sigma_{\mathbf{k}\sigma}(E)) \, G_{\mathbf{k}\sigma}(E)$$

Here we insert the high-energy expansions (8.226) and (8.227):

$$\sum_{n=0}^{\infty} \frac{\hbar}{E^n} M_{\mathbf{k}\sigma} = \hbar + (\varepsilon(\mathbf{k}) - \mu) \sum_{n=0}^{\infty} \frac{\hbar}{E^{n+1}} M_{\mathbf{k}\sigma}^{(n)} + \sum_{m,n=0}^{\infty} \frac{\hbar}{E^{m+n+1}} C_{\mathbf{k}\sigma}^{(m)} M_{\mathbf{k}\sigma}^{(n)}$$

Comparing the coefficients of the same powers of $\frac{1}{E}$, one reads off:

$$C_{\mathbf{k}\sigma}^{(0)} = M_{\mathbf{k}\sigma}^{(1)} - \varepsilon(\mathbf{k}) + \mu \tag{8.228}$$

$$C_{\mathbf{k}\sigma}^{(1)} = M_{\mathbf{k}\sigma}^{(2)} - \left(M_{\mathbf{k}\sigma}^{(1)}\right)^2 \tag{8.229}$$

$$C_{\mathbf{k}\sigma}^{(2)} = M_{\mathbf{k}\sigma}^{(3)} - 2M_{\mathbf{k}\sigma}^{(2)} M_{\mathbf{k}\sigma}^{(1)} + \left(M_{\mathbf{k}\sigma}^{(1)}\right)^3 \tag{8.230}$$

$$C_{\mathbf{k}\sigma}^{(3)} = M_{\mathbf{k}\sigma}^{(4)} - 2M_{\mathbf{k}\sigma}^{(3)} M_{\mathbf{k}\sigma}^{(1)} + 3M_{\mathbf{k}\sigma}^{(2)} \left(M_{\mathbf{k}\sigma}^{(1)}\right)^2 - \left(M_{\mathbf{k}\sigma}^{(2)}\right)^2 - \left(M_{\mathbf{k}\sigma}^{(1)}\right)^4 \tag{8.231}$$

We substitute the moments (8.221), (8.222), (8.223) and (8.224) here and obtain specially for the Hubbard model:

$$C_{\mathbf{k}\sigma}^{(0)} = U\, n_{-\sigma} \tag{8.232}$$

$$C_{\mathbf{k}\sigma}^{(1)} = U^2\, n_{-\sigma}(1 - n_{-\sigma}) \tag{8.233}$$

$$C_{\mathbf{k}\sigma}^{(2)} = U^2 n_{-\sigma}(1 - n_{-\sigma})(T_0 + B_{\mathbf{k}-\sigma} - \mu) + U^3 n_{-\sigma}(1 - n_{-\sigma})^2 \tag{8.234}$$

Every analytical, normally of course approximate expression for the self-energy can be tested with the help of these exact relations by an expansion with respect to $\frac{1}{E}$. If the moments up to $m = 3$ or the coefficients $C_{\mathbf{k}\sigma}^{(m)}$ up to $m = 2$ are satisfied, then one can be sure that the *strong-coupling* behaviour of Sect. 8.4.5 is correctly reproduced.

It is quite instructive to use these general results for a first estimate of the electronic self-energy. In the high-energy region one can write

$$\Sigma_{\mathbf{k}\sigma}(E) = C_{\mathbf{k}\sigma}^{(0)} + \frac{C_{\mathbf{k}\sigma}^{(1)}}{E} + \frac{C_{\mathbf{k}\sigma}^{(2)}}{E^2} + \cdots = C_{\mathbf{k}\sigma}^{(0)} + \frac{C_{\mathbf{k}\sigma}^{(1)}}{E}\left(1 + \frac{C_{\mathbf{k}\sigma}^{(2)}}{C_{\mathbf{k}\sigma}^{(1)} E} + \cdots\right)$$

$$\approx C^{(0)}_{\mathbf{k}\sigma} + \frac{C^{(1)}_{\mathbf{k}\sigma}}{E - \frac{C^{(2)}_{\mathbf{k}\sigma}}{C^{(1)}_{\mathbf{k}\sigma}}}$$

$$= Un_{-\sigma} + \frac{U^2 n_{-\sigma}(1 - n_{-\sigma})}{E + \mu - T_0 - B_{\mathbf{k}-\sigma} - U(1 - n_{-\sigma})}$$

So that we have found an approximate expression for the self-energy:

$$\Sigma_{\mathbf{k}\sigma}(E) \approx U\, n_{-\sigma} \frac{E + \mu - T_0 - B_{\mathbf{k}-\sigma}}{E + \mu - T_0 - B_{\mathbf{k}-\sigma} - U(1 - n_{-\sigma})} \tag{8.235}$$

This expression has formally the same structure as the $W \to 0$-self-energy (8.135) when we set the band correction $B_{\mathbf{k}-\sigma}$ equal to zero. We will encounter this result later once more and show it to be identical to the approximate self-energy which results from the physically justifiable *Spectral density ansatz (SDA)* (Sect. 8.5.6). The approximation is particularly attractive if one has the grounds to make a local ansatz ($\Sigma_{\mathbf{k}\sigma} \to \Sigma_\sigma$; $B_{\mathbf{k}-\sigma} \to B_{-\sigma}$; $F_{\mathbf{k}-\sigma} \approx 0$) acceptable, for example, in the case of infinite lattice dimensions ($d \to \infty$). In that case, the correlation functions $n_{-\sigma}$ and $B_{-\sigma}$ appearing in (8.235) can be exactly expressed in terms of the one-electron spectral density (8.217) and therefore in terms of the self-energy. We then have a closed system of equations which can be solved self-consistently. The result will certainly have the correct behaviour in the regime of strong coupling .

8.4.8 Weak-Coupling Regime

So far we discussed only the strong-coupling regime because ferromagnetism is expected above all in systems with strong coupling ($U \gg W$). Therefore, it is interesting to question whether the correct *weak-coupling* behaviour and the right *low-energy* properties can influence the *strong-coupling* phenomenon of ferromagnetism in a decisive way. Perturbation theories should be successful in the limit

$$U/W \ll 1$$

In order to investigate this limit we want to use a diagrammatic representation.

In Fig. 8.17 the *full*-Matsubara propagator, the *free*-Matsubara propagator and the electron self-energy are diagrammatically represented. Here E_n is the Fermionic Matsubara energy:

$$E_n = (2n + 1)\frac{\pi}{\beta} \ ; \ n = \cdots, -1, 0, 1. \cdots \tag{8.236}$$

The Dyson equation for the one-electron Green's function has then the diagrammatic structure shown in Fig. 8.18:

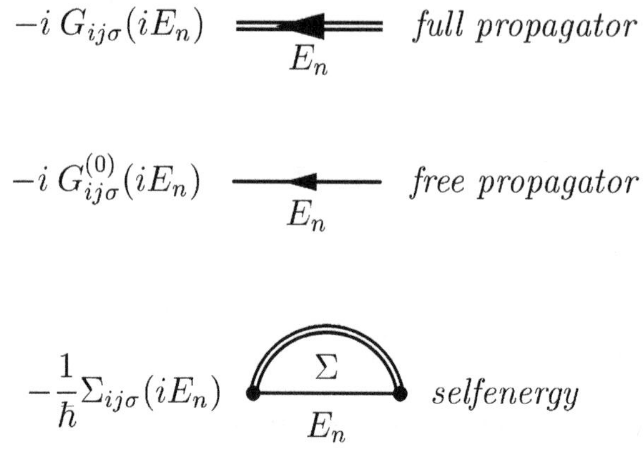

Fig. 8.17 Diagrammatic parts of the Dyson equation

Fig. 8.18 Diagrammatic Dyson equation

$$G_{ij\sigma}(iE_n) = G^{(0)}_{ij\sigma}(iE_n) + \sum_{lm} G^{(0)}_{il\sigma}(iE_n)\frac{1}{\hbar}\Sigma_{lm\sigma}(iE_n)G_{mj\sigma}(iE_n) \tag{8.237}$$

Actually to make contact with experiment it is necessary to consider the retarded Green's function, which, however, cannot be handled diagrammatically. But it can be obtained from the Matsubara function in a simple way by taking the limit

$$i\,E_n \;\rightarrow\; E + i0^+$$

One can show that the self-energy is the sum of all the *dressed skeleton* diagrams. By *skeleton* diagram we mean a self-energy diagram which is built up of only (free) propagators which contain no self-energy parts. A self-energy part is a diagram part of a one-electron Matsubara function which is connected by two propagators with the rest of the diagram. If in a *skeleton* diagram the free propagators are replaced by full ones then we get a *dressed skeleton* diagram.

- *Hartree-Fock approximation:* In the lowest order for the self-energy, one has to evaluate the diagrams shown in Fig. 8.19.
  The Fock-part vanishes in the case of Hubbard model because there have to be oppositely oriented spins at the two ends of the vertex. The Hartree-part gives

$$\Sigma^{HF}_{ij\sigma} \equiv \delta_{ij}\,U\langle n_{i-\sigma}\rangle \tag{8.238}$$

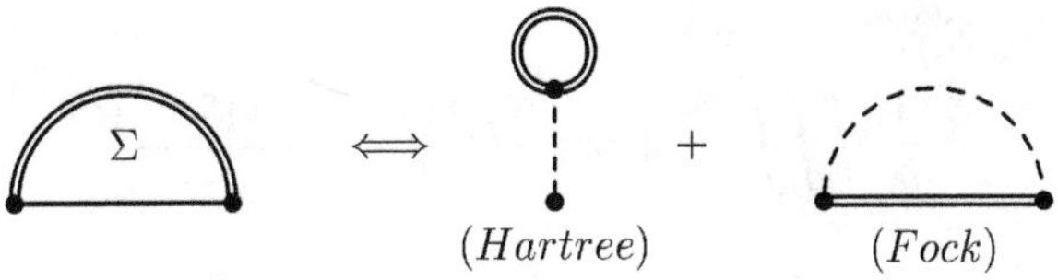

Fig. 8.19 Hartree–Fock self-energy diagrams

In this approximation, the Hubbard model is identical to the Stoner model of Sect. 8.3 which, as we have seen, overestimates the possibility of ferromagnetism. The Hartree–Fock approximation satisfies the first two spectral moments and therefore according to (8.220) gives the correct spin-dependent centres of gravity T_σ of the total energy spectrum. Actually (8.238) is not yet a real solution since $\langle n_{i-\sigma}\rangle \equiv n_{-\sigma}$ has to be self-consistently determined with the help of the spectral theorem either from the *full* one-particle Matsubara function or from the corresponding spectral density. For $U \to 0$, (8.238) is exact.

- *Second order self-consistent perturbation theory:* According to the diagrammatic perturbation theory, which we cannot develop in detail here, the self-energy diagrams shown in Fig. 8.20 have to be evaluated.
  One can easily see that the second and the fourth diagrams do not contribute in the special case of the Hubbard model. Inclusion of the third diagram gives the correct *weak-coupling* expansion of the self-energy:

$$\Sigma_{ij\sigma}^{wc} = \delta_{ij}\, U\, n_{-\sigma} + U^2 \Sigma_{ij\sigma}^{(SOC)}(E) + \mathcal{O}(U^3) \tag{8.239}$$

With this, the second order contribution can be calculated with the help of the conventional rules of the diagram technique, with some effort, but without any principal problems to give

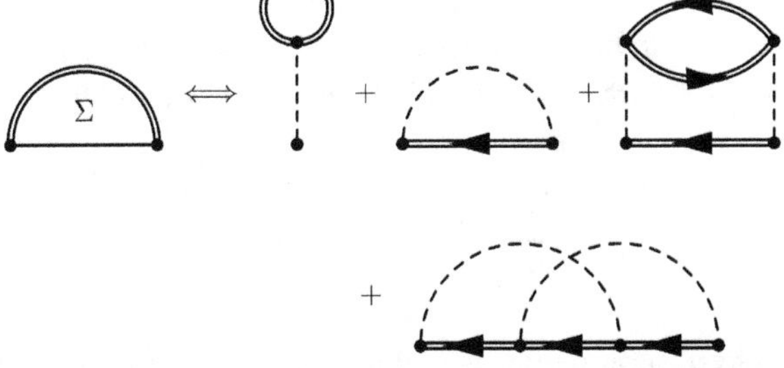

Fig. 8.20 Skeleton diagrams of the electronic self-energy up to second order in the interaction

$$\Sigma_{ij\sigma}^{(SOC)}(iE) =$$
$$= \frac{1}{\hbar^3} \iiint dx\, dy\, dz\, \frac{S_{ij\sigma}(x)\, S_{ji-\sigma}(y) S_{ij-\sigma}(z)}{iE - x + y - z} *$$
$$* (f(x)f(-y)f(z) + f(-x)f(y)f(-z)) \tag{8.240}$$

Here $f(x)$ is the Fermi function for $\mu = 0$ $(f(-x) = 1 - f(x))$. S is the full spectral density, which has to be calculated self-consistently. The self-energy contribution in second order comes out to be non-local, energy dependent and in general complex. In case that it is claimed to be exact up to terms in U^2, the spectral density on the right-hand side can be interpreted as that of the interaction-free or of the Stoner model ($S_{\mathbf{k}\sigma}^{(1)}(E)$) (Hartree–Fock approximation (8.238)):

$$S_{ij\sigma}^{(0)}(E) = \frac{\hbar}{N} \sum_{\mathbf{k}} \delta(E + \mu - \varepsilon(\mathbf{k}))\, e^{i\mathbf{k}\cdot(\mathbf{R}_i - \mathbf{R}_j)} \tag{8.241}$$

$$S_{ij\sigma}^{(1)}(E) = S_{ij\sigma}^{(0)}(E - Un_{-\sigma}^{(1)}) \tag{8.242}$$

The upper index (1) indicates that the particle density $n_{-\sigma}^{(1)}$ has to be determined self-consistently within the framework of the Hartree–Fock approximation. If in (8.240) the *full (free)* spectral density is used then it is called *self-consistent (non-self-consistent)* second order perturbation theory. On the other hand, if $S_{ij\sigma}^{(1)}(E)$ is used then it is called *perturbation theory around Hartree–Fock.*

8.4.9 Infinite Dimensions

It is possible to make interesting and wide ranging exact statements about the many-body problem of the Hubbard model in infinite lattice dimensions $d \to \infty$ [6]. Even though at first it appears exotic, it is of great direct relevance. Naturally the limit $d \to \infty$ has to be taken in a *physically meaningful* sense. For example, the kinetic energy must be *appropriately* rescaled. If the hopping integral between the nearest neighbours were taken to be $t = const.$, then the kinetic energy per particle would grow without limit for $d \to \infty$ whereas the potential energy remains constant.

$$H = -t \sum_{<ij>\sigma} c_{i\sigma}^{\dagger} c_{j\sigma} + \frac{1}{2} U \sum_{i\sigma} n_{i\sigma}\, n_{i-\sigma} \tag{8.243}$$

As a result, the physical properties would be trivially identical to those of an ideal Fermi gas. Therefore, one demands that

$$\sqrt{\frac{1}{N}\sum_{\mathbf{k}}\varepsilon^2(\mathbf{k})} = finite \ \ for \ d \to \infty \tag{8.244}$$

That means for a hypercubic lattice, e.g. for that each lattice site has $2d$ nearest neighbours:

$$\frac{1}{N}\sum_{\mathbf{k}}\varepsilon^2(\mathbf{k}) = \sum_j T_{ij}T_{ji} = 2dt^2 \stackrel{!}{\to} finite$$

This is possible only for

$$t \propto \frac{1}{\sqrt{d}} \tag{8.245}$$

We now want to exploit this fact in order to see how the corresponding Bloch density of states looks like.

$$\rho_0(E) = \frac{1}{N}\sum_{\mathbf{k}}\delta(E - \varepsilon(\mathbf{k})) = \frac{a^d}{(2\pi)^d}\int d^dk\ \delta(E - \varepsilon(\mathbf{k})) \tag{8.246}$$

a is the lattice constant and $\mathbf{k}$ is the d-dimensional wavevector.

$$\varepsilon(\mathbf{k}) = -2t\sum_{i=1}^{d}\cos(k_i a) \tag{8.247}$$
$$\mathbf{k} = (k_1, k_2, \cdots, k_d)\ ; \ -\pi \le k_i a \le +\pi$$

Substituting (8.247) in (8.246), one obtains after Fourier transformation:

$$\rho_0(E) = \left(\prod_{i=1}^{d}\int_{-\pi}^{+\pi}\frac{dk_i\ a}{2\pi}\right)\delta\left(E + 2t\sum_{j=1}^{d}\cos(k_j a)\right) =$$
$$= \frac{1}{2\pi\hbar}\int_{-\infty}^{+\infty}d\tau\ e^{\frac{i}{\hbar}E\tau}\prod_{i=1}^{d}\left(\int_{-\pi}^{+\pi}\frac{dx_i}{2\pi}\exp\left(2\frac{i}{\hbar}t\cos x_i\cdot\tau\right)\right) \tag{8.248}$$

The x_i-integral can be estimated ($t \propto d^{-\frac{1}{2}},\ d \to \infty$) as follows:

$$\int_{-\pi}^{+\pi}\frac{dx_i}{2\pi}\exp\left(2\frac{i}{\hbar}t\cos x_i\cdot\tau\right) =$$
$$= \int_{-\pi}^{+\pi}\frac{dx_i}{2\pi}\left(1 + 2\frac{i}{\hbar}t\,\tau\cos x_i - \frac{2}{\hbar^2}t^2\tau^2\cos^2 x_i + \cdots\right)$$

$$
\begin{aligned}
&= 1 - \frac{t^2}{\hbar^2}\tau^2 + O(t^4) \\
&\approx \exp\left(-\frac{t^2}{\hbar^2}\tau^2\right)
\end{aligned}
\tag{8.249}
$$

Here we have used the definite integral

$$
\frac{1}{2\pi}\int_{-\pi}^{+\pi} dx_i \, \cos^2 x_i = \frac{1}{2}
$$

Then we arrive at the following expression for the Bloch density of states:

$$
\rho_0(E) \approx \frac{1}{2\pi\hbar}\int_{-\infty}^{+\infty} d\tau \exp\left(\frac{i}{\hbar}E\tau - \frac{t^2}{\hbar^2}\tau^2 d\right)
$$

The integral can be performed elementarily.

$$
\begin{aligned}
\rho_0(E) &\approx \frac{1}{2\pi\hbar}\exp(-\frac{E^2}{4t^2 d})\int_{-\infty}^{+\infty} d\tau \exp\left(-\left(\frac{t}{\hbar}\tau\sqrt{d} - \frac{i}{2}\frac{E}{t\sqrt{d}}\right)^2\right) \\
&= \frac{1}{2\pi\hbar}\exp(-\frac{E^2}{4t^2 d})\frac{\hbar}{t\sqrt{d}}\int_{-\infty}^{+\infty} dy \exp(-y^2) \\
&= \frac{1}{2t\sqrt{d\pi}}\exp(-\frac{E^2}{4t^2 d})
\end{aligned}
$$

Thus for a hypercubic lattice in the case of infinite dimensions for the Bloch density of states holds

$$
\rho_0^{(\infty)}(E) = \frac{1}{t^*\sqrt{2\pi}}\exp(-\frac{E^2}{2t^{*2}})
\tag{8.250}
$$

For a non-trivial density of states, one must therefore require

$$
t^* = t\sqrt{2d} \stackrel{!}{=} finite
\tag{8.251}
$$

This agrees with (8.245). With this scaling it is guaranteed that the competition between kinetic and potential energies and also the physically relevant correlation effects are properly taken into account.

Now we want to discuss an important consequence of this scaling. By construction, the kinetic energy per particle is a finite quantity:

$$
\frac{1}{N}\langle H_0\rangle = -t\sum_{<ij>\sigma}\langle c_{i\sigma}^{\dagger}c_{j\sigma}\rangle^{(0)} \stackrel{!}{=} finite
\tag{8.252}
$$

The sum runs over the $2d$ nearest neighbours of the ith lattice site. Because of (8.251) it must therefore hold

$$\langle c_{i\sigma}^{\dagger} c_{j\sigma} \rangle^{(0)} = \mathcal{O}\left(\frac{1}{\sqrt{d}}\right) \quad i \neq j \tag{8.253}$$

Because of the spectral theorem, this order of magnitude is reflected in the spectral density and the Green's function:

$$S_{ij\sigma}^{(0)} = \mathcal{O}\left(\frac{1}{\sqrt{d}}\right) \; ; \; G_{ij\sigma}^{(0)} = \mathcal{O}\left(\frac{1}{\sqrt{d}}\right) \quad i \neq j \tag{8.254}$$

The effect on the self-energy is obvious [6] as one can see from the diagram presented in Fig. 8.21.

The three propagators that appear in the diagram contribute in the form

$$\sum_j G_{ij\sigma}^{(0)}(E) G_{ij-\sigma}^{(0)}(E) G_{ji-\sigma}^{(0)}(E) = \mathcal{O}\left(d \cdot \left(\frac{1}{\sqrt{d}}\right)^3\right) = \mathcal{O}\left(\frac{1}{\sqrt{d}}\right)$$

which obviously vanish for $d \to \infty$. That is valid for all the non-diagonal terms in the self-energy [7]. With this we obtain the important result that the self-energy of the Hubbard model in the case of infinite dimensions is local:

$$\Sigma_{ij\sigma}(E) = \Sigma_{\sigma}(E)\,\delta_{ij} \; ; \; d \to \infty \tag{8.255}$$

This exact result can be used in many ways as we will see later. The next section discusses a special application.

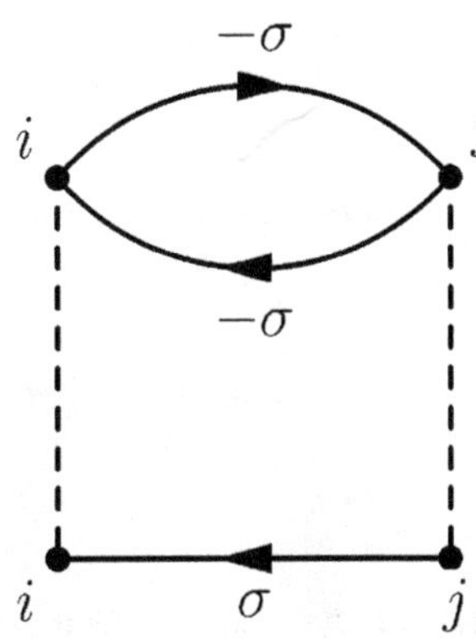

Fig. 8.21 Self-energy diagram of second order

8.4.10 Effective "impurity"-Problem

In this section, we want to assume that the self-energy of the Hubbard model is local, which is shown in the last section to be exact in the case of infinite dimensions $d \to \infty$. There are indications that even for finite dimension ($d = 3$), a local self-energy can be an acceptable approximation. Then the Dyson equation for the local propagator simplifies to:

$$G_{ii\sigma}(iE_n) = G^{(0)}_{ii\sigma}(iE_n) + \sum_j G^{(0)}_{ij\sigma}(iE_n)\frac{1}{\hbar}\Sigma_\sigma(iE_n)G_{ji\sigma}(iE_n) \tag{8.256}$$

The diagrammatic representation of the Dyson equation is shown in Fig. 8.22. We now define a new Matsubara function which is different from the old one (8.256) by removing the diagonal term $j = i$ from the sum:

$$F_{li\sigma}(iE_n) = G^{(0)}_{li\sigma}(iE_n) + \sum_j^{j\neq i} G^{(0)}_{lj\sigma}(iE_n)\frac{1}{\hbar}\Sigma_\sigma(iE_n)F_{ji\sigma}(iE_n) \tag{8.257}$$

We want to connect this function with $G_{ii\sigma}(iE_n)$. For that we consider

$$\begin{aligned} F_{li\sigma}(iE_n) = G^{(0)}_{li\sigma}(iE_n) &+ \sum_j G^{(0)}_{lj\sigma}(iE_n)\frac{1}{\hbar}\Sigma_\sigma(iE_n)F_{ji\sigma}(iE_n) \\ &- G^{(0)}_{li\sigma}(iE_n)\frac{1}{\hbar}\Sigma_\sigma(iE_n)F_{ii\sigma}(iE_n) \end{aligned} \tag{8.258}$$

We perform Fourier transformation of $G^{(0)}_{li\sigma}(iE_n)$ and $F_{li\sigma}(iE_n)$, but not $F_{ii\sigma}(iE_n)$, so that we get the following relation:

$$G^{(0)}_{\mathbf{k}\sigma}(iE_n) = \frac{F_{\mathbf{k}\sigma}(iE_n)}{1 + \frac{1}{\hbar}\Sigma_\sigma(iE_n)F_{\mathbf{k}\sigma}(iE_n) - \frac{1}{\hbar}\Sigma_\sigma(iE_n)F_{ii\sigma}(iE_n)} \tag{8.259}$$

We substitute this in the *normal* Dyson equation (8.237)

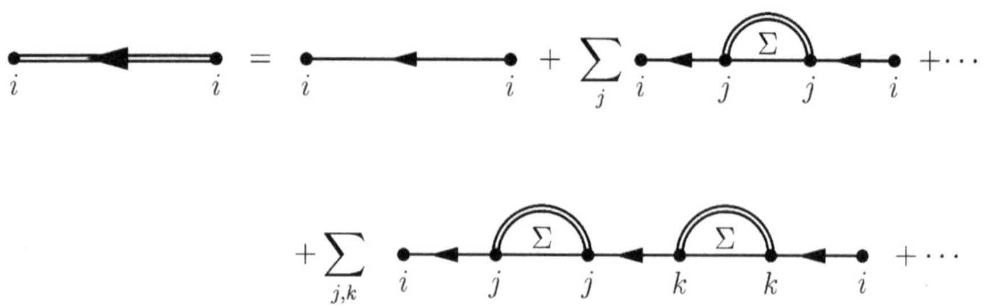

Fig. 8.22 Dyson equation for the local propagator

$$G_{\mathbf{k}\sigma}(iE_n) = G^{(0)}_{\mathbf{k}\sigma}(iE_n) + G^{(0)}_{\mathbf{k}\sigma}(iE_n)\frac{1}{\hbar}\Sigma_\sigma(iE_n)G_{\mathbf{k}\sigma}(iE_n) \tag{8.260}$$

so that what remains is

$$G_{\mathbf{k}\sigma}(iE_n) = F_{\mathbf{k}\sigma}(iE_n) + F_{ii\sigma}(iE_n)\frac{1}{\hbar}\Sigma_\sigma(iE_n)G_{\mathbf{k}\sigma}(iE_n) \tag{8.261}$$

After summing over all the wavenumbers, it follows ($G_{ii\sigma} = 1/N\sum_{\mathbf{k}} G_{\mathbf{k}\sigma}$):

$$G_{ii\sigma}(iE_n) = F_{ii\sigma}(iE_n) + F_{ii\sigma}(iE_n)\frac{1}{\hbar}\Sigma_\sigma(iE_n)G_{ii\sigma}(iE_n) \tag{8.262}$$

We represent the propagator $F_{ii\sigma}(iE_n)$ by a wavy line as in Fig. 8.23, so that (8.262) can be represented as shown in Fig. 8.24.

We will consider now the self-energy diagrams for the special case where only the scattering centre at $\mathbf{R}_i$ is *switched on*. Then we obtain *impurity self-energy diagrams* (ISD) shown in Fig. 8.25. These define a special diagram class C_σ:

Fig. 8.23 Graphical representation of the local propagator from (8.262)

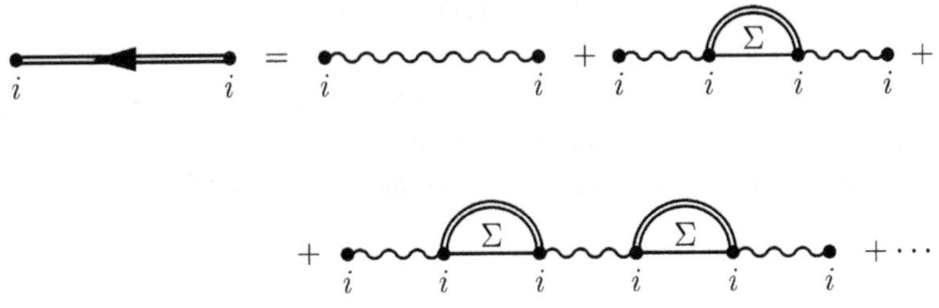

Fig. 8.24 Modified Dyson equation for the local propagator

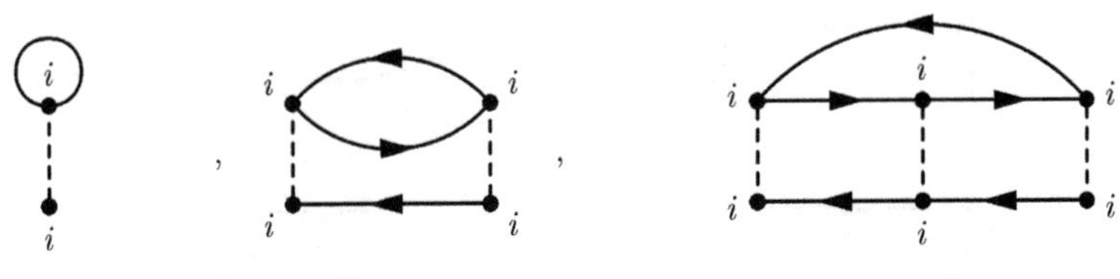

Fig. 8.25 Impurity self-energy diagrams

$$\Sigma_\sigma^{ISD}(iE_n) \equiv C_\sigma\left[\left\{G_{ii\sigma}^{(0)}(iE_n)\right\}, U\right] \tag{8.263}$$

C_σ is independent of the special way of realizing the diagrams through the *free* propagator $G_{ii\sigma}(iE_n)$.

We now investigate, which of the diagrams contribute to the $d \to \infty$-self-energy of the Hubbard model. Naturally, these are, first, all the *local diagrams* for which all the lattice indices belong to the same site i. These are actually the diagrams shown in Fig. 8.25. We have, however, also to count certain non-local diagrams which acquire the non-locality from a self-energy insertion as, for example shown in Fig. 8.26.

Their contribution does not vanish for $d \to \infty$, since the two Matsubara functions,

$$G_{ij\sigma}^{(0)}(iE_n) \cdot G_{ji\sigma}^{(0)}(iE_n) \sim \mathcal{O}\left(\frac{1}{d}\right)$$

are outweighed by the extra summation,

$$\sum_j \to \mathcal{O}(d)$$

Obviously we obtain all the self-energy diagrams when we replace in the ISD diagrams in Fig. 8.25 the *free* propagator by the *modified* propagator of Fig. 8.23.

According to (8.263) this means

$$\Sigma_\sigma(iE_n) = C_\sigma\left[\{F_{ii\sigma}(iE_n)\}, U\right] \tag{8.264}$$

With this we have expressed the $d \to \infty$-self-energy of the Hubbard model in terms of the diagrammatic functional C_σ of the *impurity* scattering (Fig. 8.27). This result forms the basis of the *Dynamical mean-field theory* (DMFT), which will be discussed later.

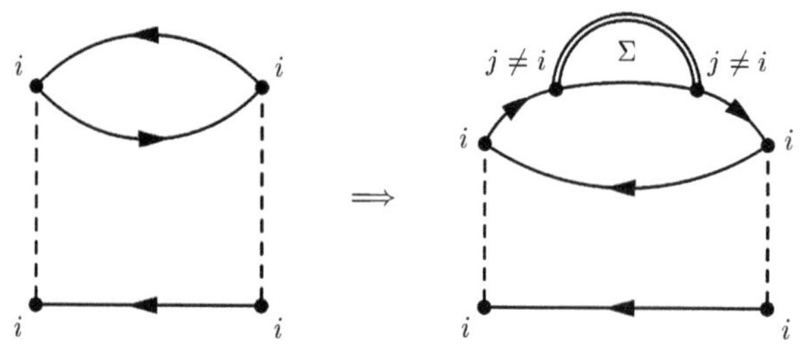

Fig. 8.26 Local self-energy diagram with non-locality contribution by a self-energy insertion

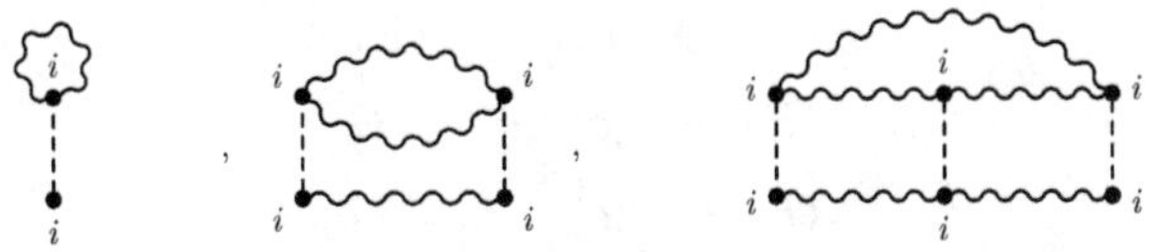

Fig. 8.27 Local self-energy diagrams of the $d \to \infty$ Hubbard model

8.5 Magnetism and Electronic Correlations

In this section we want to investigate, within the framework of the Hubbard model, the possibility of band ferromagnetism, as is observed in the classical ferromagnets Fe, Co and Ni. The decisive question in this context will be whether the Hubbard model is at all capable of describing a spontaneous magnetization of itinerant band electrons? One has to be circumspect about the prediction made by the Hartree–Fock approximation of the Hubbard (Stoner model) with respect to spontaneous ferromagnetism in view of the fact that molecular field approximations always grossly overestimate the possibility of ferromagnetism.

We have discussed in Sect. 8.4.2 the limiting case of the infinitely narrow band which one can imagine to be *realized* by lattice sites which are infinitely apart from each other. The main result of this case is that the original atomic level T_0 is split due to the Coulomb interaction U into two quasiparticle levels T_0 and $T_0 + U$ with formally spin-dependent spectral weights $1 - n_{-\sigma}$ and $n_{-\sigma}$. Now it has to be inspected what happens when the ions of the solid are brought nearer to each other so that the separation between them is finite so that the atomic wavefunctions begin to overlap and consequently the *hopping* probability of the band electrons is non-zero. From the two-site model discussed in Sect. 8.4.3 we can already get certain trends. From this we can expect that at least in the case of strong coupling ($U \gg W$, Sect. 8.4.5) the two atomic quasiparticle levels spread out into two quasiparticle sub-bands which are clearly separated from each other by about U and which have densities of states strongly dependent on temperature and band occupation. This feature is to be investigated in more detail which will be done from now on, based certainly necessarily on approximate theories.

8.5.1 Hubbard-I Approximation

The first proposal for a theory that takes into account electronic correlations was developed by Hubbard himself in his pioneering work [1]. He used the equation of motion method for the (retarded) Green's function. The starting point is the exact equation of motion (8.125) of the one-electron Green's function $G_{ij\sigma}(E)$ (8.121). This is not exactly solvable due to the presence of the *higher* Green's function $\Gamma_{iii;j\sigma}(E)$ (8.124). The same is true for the equation of motion (8.126) of $\Gamma_{iii;j\sigma}(E)$. Hubbard proposed an approximation consisting of a *mean field* decoupling (7.120) of the Green's function on the right-hand side of (8.126) for $i \neq m$:

$$\Gamma_{iim;j\sigma}(E) \rightarrow n_{-\sigma}\, G_{mj\sigma}(E) \tag{8.265}$$

$$\Gamma_{imi;j\sigma}(E) \rightarrow \langle c^{\dagger}_{i-\sigma} c_{m-\sigma}\rangle\, G_{ij\sigma}(E) \tag{8.266}$$

$$\Gamma_{mii;j\sigma}(E) \rightarrow \langle c^{\dagger}_{m-\sigma} c_{i-\sigma}\rangle\, G_{ij\sigma}(E) \tag{8.267}$$

In this decoupling, the particle number and spin are conserved. On substituting in (8.126), the terms (8.266) and (8.267)

$$\sum_{m}^{m\neq i} T_{im}\left(\Gamma_{imi;j\sigma}(E) - \Gamma_{mii;j\sigma}(E)\right) \rightarrow$$
$$\rightarrow G_{ij\sigma}(E) \sum_{m} T_{im}\left(\langle c^{\dagger}_{i-\sigma} c_{m-\sigma}\rangle - \langle c^{\dagger}_{m-\sigma} c_{i-\sigma}\rangle\right)$$

do not contribute due to the assumed translational symmetry:

$$\begin{aligned}
&\sum_{m} T_{im}\left(\langle c^{\dagger}_{i-\sigma} c_{m-\sigma}\rangle - \langle c^{\dagger}_{m-\sigma} c_{i-\sigma}\rangle\right) = \\
&= \frac{1}{N}\sum_{im} T_{im}\left(\langle c^{\dagger}_{i-\sigma} c_{m-\sigma}\rangle - \langle c^{\dagger}_{m-\sigma} c_{i-\sigma}\rangle\right) = \\
&= \frac{1}{N}\sum_{im} (T_{im} - T_{mi})\,\langle c^{\dagger}_{i-\sigma} c_{m-\sigma}\rangle = \\
&= 0
\end{aligned} \tag{8.268}$$

Then what remains as the equation of motion for $\Gamma_{iii;j\sigma}(E)$ is

$$(E + \mu - T_0 - U)\, \Gamma_{iii;j\sigma}(E) = \hbar\delta_{ij}\, n_{-\sigma} + n_{-\sigma} \sum_{m}^{m\neq i} T_{im} G_{mj\sigma}(E)$$

This can easily be solved for $\Gamma_{iii;j\sigma}(E)$ and is then substituted in the equation of motion (8.125) for $G_{ij\sigma}(E)$:

$$(E + \mu - T_0)\, G_{ij\sigma}(E) = \left(\hbar\delta_{ij} + \sum_{m}^{m\neq i} T_{im} G_{mj\sigma}(E)\right) *$$
$$* \left(1 + \frac{U\, n_{-\sigma}}{E + \mu - T_0 - U}\right)$$

The complete solution is obtained by Fourier transformation to wavenumbers.

$$\left[(E+\mu-T_0)-(\varepsilon(\mathbf{k})-T_0)\frac{E+\mu-T_0-U(1-n_{-\sigma})}{E+\mu-T_0-U}\right]G_{\mathbf{k}\sigma}(E)$$
$$=\hbar\frac{E+\mu-T_0-U(1-n_{-\sigma})}{E+\mu-T_0-U}$$

$$\Rightarrow\left[\frac{(E+\mu-T_0)(E+\mu-T_0-U)}{E+\mu-T_0-U(1-n_{-\sigma})}-(\varepsilon(\mathbf{k})-T_0)\right]G_{\mathbf{k}\sigma}(E)=\hbar$$

Let us recall the general structure of one-electron Green's function,

$$G_{\mathbf{k}\sigma}(E)=\frac{\hbar}{E+\mu-\varepsilon(\mathbf{k})-\Sigma_{\mathbf{k}\sigma}^{HI}(E)+i\,0^+} \tag{8.269}$$

then we have a self-energy which agrees exactly with the limiting case of the infinitely narrow band (8.135):

$$\Sigma_{\mathbf{k}\sigma}^{HI}(E)\equiv\Sigma_{\sigma}^{HI}(E)=U\,n_{-\sigma}\frac{E+\mu-T_0}{E+\mu-T_0-U(1-n_{-\sigma})} \tag{8.270}$$

A certain *physical support* of the *Hubbard decouplings* (8.265), (8.266), and (8.267), which at first glance appear completely arbitrary, is derived from the fact that it reproduces both the limiting cases

- $U\to 0$ *band limit*
- $W\to 0$ *atomic limit*
- satisfies the spectral moments $M_{\mathbf{k}\sigma}^{(0,1,2)}$

One should note that the Hartree–Fock solution (8.47) (*Stoner model*),

$$\Sigma_{\mathbf{k}\sigma}^{(HF)}(E)\equiv\Sigma_{\sigma}^{(HF)}\equiv U\,n_{-\sigma} \tag{8.271}$$

satisfies the band limit and the first two moments $M_{\mathbf{k}\sigma}^{(0,1)}$ only.

In contrast, the fourth moment $M_{\mathbf{k}\sigma}^{(3)}$ is violated by the Hubbard-I approximation which means that the correct *strong-coupling* behaviour cannot be expected from it. From the point of view of the possibility of ferromagnetism, this proves to be a serious limitation. Before we analyse this in more detail, we want to reproduce the Hubbard-I solution with the help of a simple interpolation method, which does not need the problematic decouplings and therefore has a little more to say about the real worth of the Hubbard approximation.

8.5.2 Interpolation Method

The *Hubbard problem* can be exactly solved in two limits, namely the *band limit* ($U = 0$) and the limit of infinitely narrow band ($W \to 0$, Sect. 8.4.2). We want to construct a solution which interpolates between these limits.

- *Band limit:* The solution for the one-electron Green's function is known:

$$G_{\mathbf{k}}^{(0)}(E) = \frac{\hbar}{E + \mu - \varepsilon(\mathbf{k})} \tag{8.272}$$

This can be formally written as a Dyson equation,

$$G_{\mathbf{k}}^{(0)}(E) = G_{00}(E) + G_{00}(E)\frac{1}{\hbar}\Sigma_{\mathbf{k}}^{(0)}\, G_{\mathbf{k}}^{(0)}(E) \tag{8.273}$$

where the corresponding *free* Green's function $G_{00}(E)$ should be interpreted as the one for the infinitely narrow band:

$$G_{00}(E) = \frac{\hbar}{E + \mu - T_0} \tag{8.274}$$

Then the *self-energy of the band limit* is given by

$$\Sigma_{\mathbf{k}}^{(0)} = \varepsilon(\mathbf{k}) - T_0 \tag{8.275}$$

- *Atomic limit:* The solution again is known from (8.134) and can be written as Dyson equation as follows:

$$G_{\sigma}^{W=0}(E) = G_{00}(E) + G_{00}(E)\frac{1}{\hbar}\Sigma_{\sigma}^{W=0}(E)\, G_{\sigma}^{W=0}(E) \tag{8.276}$$

- *"Full" problem:* This is not exactly solvable. Formal Dyson equations can be written with the two limiting cases as the *free parts*, then of course with different self-energies:

$$G_{\mathbf{k}\sigma}(E) = G_{\mathbf{k}}^{(0)}(E) + G_{\mathbf{k}}^{(0)}(E)\frac{1}{\hbar}\Sigma_{\mathbf{k}\sigma}^{U=0}(E)G_{\mathbf{k}\sigma}(E) \tag{8.277}$$

$$G_{\mathbf{k}\sigma}(E) = G_{\sigma}^{W=0}(E) + G_{\sigma}^{W=0}(E)\frac{1}{\hbar}\Sigma_{\mathbf{k}\sigma}^{W=0}(E)G_{\mathbf{k}\sigma}(E) \tag{8.278}$$

The self-energies $\Sigma_{\mathbf{k}\sigma}^{W=0}(E)$ and $\Sigma_{\mathbf{k}\sigma}^{U=0}(E)$ are, however, not exactly known. It appears reasonable to use in (8.277) a self-energy approximation, which is convincing in the $W = 0$ limit, since in the other limit ($U = 0$) the *free* function is already correct. In contrast, in the ansatz (8.278) the self-energy is to be so chosen that it is useful in the $U = 0$ limit, since in this case the *free* function already takes care for the correct $W = 0$ behaviour. Therefore, we replace in (8.277) the self-energy by the exact expression (8.135) of the infinitely narrow band:

$$\Sigma_{\mathbf{k}\sigma}^{U=0}(E) \to \Sigma_{\sigma}^{W=0}(E) = U\, n_{-\sigma} \frac{E + \mu - T_0}{E + \mu - T_0 - U(1 - n_{-\sigma})} \tag{8.279}$$

and in (8.278) the self-energy by the one of the band limit (8.275):

$$\Sigma_{\mathbf{k}\sigma}^{W=0}(E) \to \Sigma_{\mathbf{k}}^{(0)} = \varepsilon(\mathbf{k}) - T_0 \tag{8.280}$$

These replacements give in both the cases exactly the same result:

$$G_{\mathbf{k}\sigma}(E) \approx \left[\left(G_{\mathbf{k}}^{(0)}(E)\right)^{-1} - \frac{1}{\hbar}\Sigma_{\sigma}^{W=0}(E)\right]^{-1}$$

$$G_{\mathbf{k}\sigma}(E) \approx \left[\left(G_{\sigma}^{W=0}(E)\right)^{-1} - \frac{1}{\hbar}\Sigma_{\mathbf{k}}^{(0)}\right]^{-1}$$

This results in both cases in the Hubbard-I solution (8.269). The decoupling method proposed by Hubbard (8.265), (8.266), and (8.267), which at first sight appears to be somewhat arbitrary, can therefore be considered to be an *interpolation method* which *interpolates* between the two limiting cases (8.275) and (8.279). The result obtains further weight from the fact that both the *interpolation paths* lead to the same approximate result.

8.5.3 Correlation Effects and Ferromagnetism

The self-energy (8.270) of the Hubbard-I solution is real and wavenumber independent and therefore describes quasiparticles of infinite lifetime. From (8.269) and with the help of the Dirac identity (B.92), we get for the spectral density

$$S_{\mathbf{k}\sigma}(E) = \hbar\,\delta\left(E + \mu - \varepsilon(\mathbf{k}) - \Sigma_{\sigma}^{HI}(E)\right) \tag{8.281}$$

Then the quasiparticle density of states $\rho_\sigma(E)$ can be expressed in terms of Bloch density of states of the non-interacting system $\rho_0(E)$:

$$\rho_\sigma(E) = \frac{1}{N}\sum_{\mathbf{k}} S_{\mathbf{k}\sigma}(E - \mu) = \\ = \rho_0\left(E - \Sigma_{\sigma}^{HI}(E - \mu)\right) \tag{8.282}$$

An important speciality of the Hubbard-I solution and also of the $W = 0$ self-energy is that the function $\left(E - \Sigma_\sigma^{HI}(E - \mu)\right)$ has a singularity at

$$E = E_{0\sigma} = T_0 + U(1 - n_{-\sigma}) \tag{8.283}$$

As discussed in detail in Appendix B, such a singularity always leads to a splitting of the energy band for each spin direction into two quasiparticle sub-bands. (Fig. 8.28). Thus the solution (8.282) reproduces the main results of the two-site model (Sect. 8.4.3). The Coulomb interaction U causes the splitting, whose origin can be understood as follows. If the electron is moving in the upper sub-band, then it hops mainly onto such lattice points where, there already exists another electron with oppositely oriented spin. On the other hand, the electron in the lower sub-band prefers the lattice sites which are empty. This leads to an energy separation of U between the two sub-bands. For comparison, if the Stoner model (Hartree–Fock approximation (8.271)) is considered, one recognizes that the energy- and wavenumber-independent self-energy $Un_{-\sigma}$ neither deforms nor splits the spin bands. For $n_\uparrow \neq n_\downarrow$, however, a shift of the spin-bands relative to each other takes place (Fig. 8.29).

The one-electron Green's function (8.269), as can be easily seen, represents a two-pole function:

$$G_{\mathbf{k}\sigma}(E) = \hbar \sum_{j=1}^{2} \frac{\alpha_{j\sigma}(\mathbf{k})}{E + \mu - E_{j\sigma}(\mathbf{k})} \tag{8.284}$$

Here the (real) quasiparticle energies are given by ($j = 1, 2$):

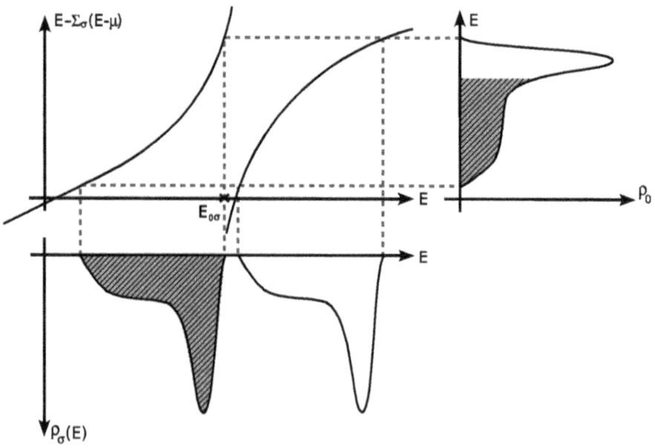

Fig. 8.28 Schematic plot of the energy band splitting into two quasiparticle sub-bands according to the Hubbard-I solution (8.282)

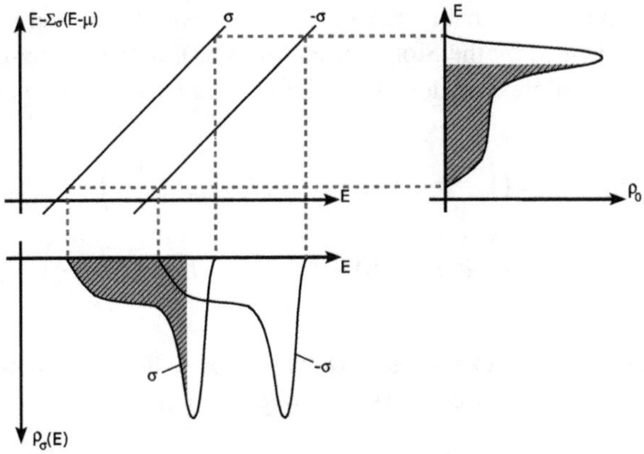

Fig. 8.29 As comparison to Fig. 8.28, a schematic plot of the origin of the energy shift of the spin sub-bands of the Hubbard model in the Hartree–Fock approximation (Stoner model)

$$E_{j\sigma}(\mathbf{k}) = \frac{1}{2}\Big(T_0 + \varepsilon(\mathbf{k}) + \\ + U + (-1)^j \sqrt{(T_0 - \varepsilon(\mathbf{k}) + U)^2 + 4U\, n_{-\sigma}(\varepsilon(\mathbf{k}) - T_0)}\Big) \tag{8.285}$$

and the corresponding spectral weights by

$$\alpha_{1\sigma}(\mathbf{k}) = \frac{E_{1\sigma}(\mathbf{k}) - T_0 - U\,(1 - n_{-\sigma})}{E_{1\sigma}(\mathbf{k}) - E_{2\sigma}(\mathbf{k})} = 1 - \alpha_{2\sigma}(\mathbf{k}) \tag{8.286}$$

Thus the spectral density is a sum of two weighted δ-functions:

$$S_{\mathbf{k}\sigma}(E) = \hbar \sum_{j=1}^{2} \alpha_{j\sigma}(\mathbf{k})\delta\left(E + \mu - E_{j\sigma}(\mathbf{k})\right) \tag{8.287}$$

The band splitting inducing singularity $E_{0\sigma}$ (8.283), however, is also responsible for a serious shortcoming of the Hubbard-I solution. From physical grounds one should expect that with decreasing U/W, the originally separate sub-bands should gradually overlap. But, from (8.283) one sees that for arbitrarily small couplings there is always a singularity and therewith a band gap.

8.5.4 Criterion for Ferromagnetism

The actually interesting question is about the possibility of ferromagnetism in a system of band electrons. It is reasonable to assume that for a given band occupation $n = n_\sigma + n_{-\sigma}$, for weak coupling U, the paramagnetic state is preferred and ferromagnetism, if at all, probably can appear only beyond a critical value of $U_c(n)$. This is already indicated by the Stoner criterion (8.59). In the *strong-coupling limit* $U \gg W$, the quasiparticle energies (8.285) of the *Hubbard-I solution* simplify to

$$E_{1\sigma}(\mathbf{k}) = (1 - n_{-\sigma})\varepsilon(\mathbf{k}) + n_{-\sigma}T_0 + \mathcal{O}\left(\frac{W}{U}\right) \tag{8.288}$$

$$E_{2\sigma}(\mathbf{k}) = U + n_{-\sigma}\varepsilon(\mathbf{k}) + (1 - n_{-\sigma})T_0 + \mathcal{O}\left(\frac{W}{U}\right) \tag{8.289}$$

As a comparison with the exact result (8.210) shows, the corresponding spectral weights turn out to be correct in the *strong-coupling limit*:

$$\alpha_{1\sigma}(\mathbf{k}) \approx 1 - n_{-\sigma}\ ;\ \alpha_{2\sigma}(\mathbf{k}) \approx n_{-\sigma} \tag{8.290}$$

In this limit it is meaningful to define, for the two energetically separated *Hubbard bands*, partial density of states which can be connected to the *free* Bloch density of states in a simple way:

$$\rho_\sigma^{lower}(E) \approx \rho_0\left(\frac{E - n_{-\sigma}T_0}{1 - n_{-\sigma}}\right) \tag{8.291}$$

$$\rho_\sigma^{upper}(E) \approx \rho_0\left(\frac{E - U - (1 - n_{-\sigma})T_0}{n_{-\sigma}}\right) \tag{8.292}$$

The two bands have spin-independent centres of gravity:

$$T_{1\sigma} \equiv \int dE\, E\, \rho_\sigma^{lower}(E) \equiv \frac{1}{N}\sum_{\mathbf{k}} E_{1\sigma}(\mathbf{k}) = T_0 = T_{1-\sigma} \tag{8.293}$$

$$T_{2\sigma} \equiv \int dE\, E\, \rho_\sigma^{upper}(E) \equiv \frac{1}{N}\sum_{\mathbf{k}} E_{2\sigma}(\mathbf{k}) = T_0 + U = T_{2-\sigma} \tag{8.294}$$

The *strong-coupling* result of the *Hubbard-I* theory is schematically shown in Fig. 8.16. The spin independence of the centres of gravity turns out to be the decisive disadvantage for the possibility of ferromagnetism. A comparison with (8.214) and (8.215) also shows that the *Hubbard-I* solution does *not* have the correct *strong-coupling* behaviour.

Ferromagnetism presumes a spontaneous preferential spin orientation, i.e. $n_\uparrow \neq n_\downarrow$, which in turn requires a spin asymmetry of the sub-bands in Fig. 8.28. Formally, of course the quasiparticle energies (8.288) and (8.289) and also the spectral weights (8.290) are spin dependent so that such an asymmetry is in principle conceivable. But it must be the result of a self-consistent calculation. For the average occupation number per lattice site and spin, according to the spectral theorem (B.95) holds for $T = 0$:

$$n_\sigma(T=0) = \int_{-\infty}^{E_F} dE\rho_\sigma(E) = \int_{-\infty}^{E_F} dE\rho_0(E - \Sigma_\sigma(E-\mu)) \tag{8.295}$$

E_F is the Fermi energy. If there is a solution with $n_\sigma^* \neq n_{-\sigma}^*$, then on symmetry grounds, in addition to $n_\sigma^* < n/2$, $n - n_\sigma^*$ should also be a solution. That means there exist either none or two magnetic solutions. Strictly speaking, four, six, ..., solutions are conceivable which, however, we are excluding here. From the fact that two magnetic solutions must be available, we want to derive a criterion for ferromagnetism. In Fig. 8.30 the two positive definite sides of (8.295) are plotted formally as functions of n_σ.

The left-hand side is simply the bisectrix. Since paramagnetism $n_\sigma = n_{-\sigma} = n/2$ is always a mathematical solution, the right-hand side of (8.295) must intersect the bisectrix exactly at three places as is shown by the continuous lines in Fig. 8.30. The dashed line does not come into question since it leads to further solutions at the most at $n_\sigma^* = 0$. As the coupling U decreases, the two magnetic solutions collapse to the point $n_\sigma = \frac{n}{2}$. For $U \leq U_c(n)$, the curve representing the right-hand side of (8.295) touches the straight line at $n_\sigma = \frac{n}{2}$, which means at that point it has a slope equal to 1. We then establish a *criterion for ferromagnetism* to be that the right-hand side of (8.295) as a function of n_σ should have a slope greater than 1 at $n_\sigma = \frac{n}{2}$:

$$1 < \left[\frac{d}{dn_\sigma}\int_{-\infty}^{E_F} dE\rho_\sigma(E)\right]_{n_\sigma=\frac{n}{2}} \tag{8.296}$$

This is still rather generally valid. We evaluate the criterion for the Hubbard solution (8.284) under the condition $n < 1$ and $U \gg W$ and therefore we can use the

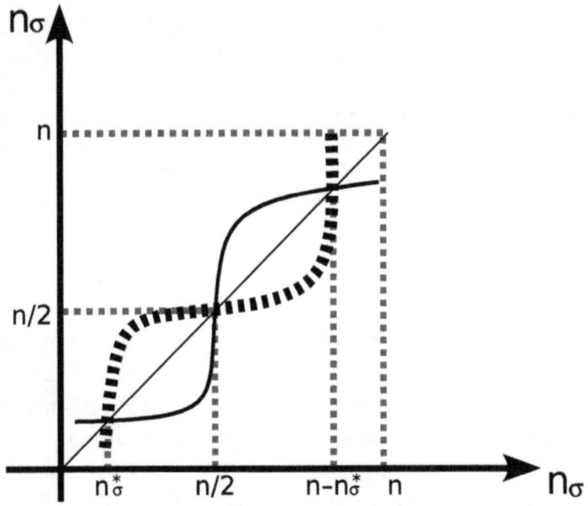

Fig. 8.30 Schematic representation of the two sides of (8.295) as functions of n_σ to derive a criterion for ferromagnetism in the correlated electron system

expression (8.291) for the quasiparticle density of states:

$$1 < \left[\frac{d}{dn_\sigma}\int_{-\infty}^{E_F} dE\rho_0\left(\frac{E-(n-n_\sigma)T_0}{1-n+n_\sigma}\right)\right]_{n_\sigma=\frac{n}{2}} \tag{8.297}$$

With the substitution

$$\eta_\sigma \equiv \frac{E-(n-n_\sigma)T_0}{1-n+n_\sigma}\;;\;\eta_{F\sigma}=\eta_\sigma(E=E_F)$$

which results in

$$dE=(1-n+n_\sigma)d\eta_\sigma$$

the criterion (8.297) reads as

$$1 < \left[\frac{d}{dn_\sigma}(1-n+n_\sigma)\int_{-\infty}^{\eta_{F\sigma}} d\eta_\sigma\rho_0(\eta_\sigma)\right]_{n_\sigma=\frac{n}{2}} =$$

$$= \left[\int_{-\infty}^{\eta_{F\sigma}} d\eta_\sigma\rho_0(\eta_\sigma)+(1-n+n_\sigma)\rho_0(\eta_{F\sigma})\frac{d\eta_{F\sigma}}{dn_\sigma}\right]_{n_\sigma=\frac{n}{2}} =$$

$$= \frac{n}{2-n}+(1-\frac{1}{2}n)\rho_0\left(\frac{E_F-\frac{n}{2}T_0}{1-\frac{n}{2}}\right)\left(\frac{d\eta_{F\sigma}}{dn_\sigma}\right)_{n_\sigma=\frac{n}{2}}$$

Finally with

$$\left(\frac{d\eta_{F\sigma}}{dn_\sigma}\right)_{n_\sigma=\frac{n}{2}} = \frac{T_0-E_F}{(1-\frac{n}{2})^2}$$

the *Hubbard-I criterion for ferromagnetism* is given by

$$1-n+(E_F-T_0)\rho_0\left(\frac{2E_F-nT_0}{2-n}\right) \overset{!}{<} 0 \tag{8.298}$$

In comparison to the Stoner criterion (8.59), this is considerably sharper. In particular, it puts a condition also on the band occupation n. Since $(1-n)$ and the density of states ρ_0 are non-negative quantities, for the appearance of ferromagnetism, it is necessary that $(E_F-T_0) < 0$. Since according to (8.293) T_0 coincides with the centre of gravity of the lower band, the criterion can be fulfilled only for small band occupations. The origin for this apparently implausible condition should be searched in the missing of a (formal) spin dependence of the centres of gravity of the bands (8.293) and (8.294). On the other hand, this is what is to be expected from the exact *strong-coupling* result (8.214) and (8.215). We will consider the significance of spin-dependent band shift in detail in the next section.

Before that let us investigate the criterion (8.298), as an example, for the simple rectangular density of states shown in Fig. 8.9.

$$\rho_0(x) = \begin{cases} \frac{1}{W} \; if \; T_0 - \frac{W}{2} \leq x \leq T_0 + \frac{W}{2} \\ 0 \quad otherwise \end{cases}, \tag{8.299}$$

The Stoner criterion (8.59) demands only $U > W$ and gives an extremely high Curie temperature (8.61). In contrast, according to (8.298), we have to require

$$1 - n + \frac{E_F - T_0}{W} \overset{!}{\leq} 0 \tag{8.300}$$

Here one has to note that the Fermi energy E_F is, for an interacting electron system, a function of the band occupation, which, however, can be easily determined here:

$$n_\sigma \;\to\; \frac{n}{2} = \int_{E_u}^{E_F} dE \, \frac{1}{W} = \frac{1}{W}(E_F - E_u) \;\; (T = 0)$$

E_u is the lower band edge for which holds:

$$T_0 - \frac{W}{2} \overset{!}{=} \frac{E_u - \frac{n}{2} T_0}{1 - \frac{n}{2}} \;\to\; E_u = T_0 - \frac{W}{2}(1 - \frac{n}{2})$$

That means

$$E_F = \frac{n}{2} W + E_u = T_0 + \frac{W}{2}\left(\frac{3}{2}n - 1\right)$$

If this is substituted in (8.300), we get the unsatisfiable condition

$$2 < n \tag{8.301}$$

Thus in contrast to the Stoner model, in the *Hubbard-I* theory, there is no ferromagnetism for the rectangular density of states.

8.5.5 Static Susceptibility and Ferromagnetism

Before we elaborate the obviously very important spin-dependent band shift, we want to make further checks on the possibility of ferromagnetism in the *Hubbard-I* theory by calculating the static paramagnetic susceptibility χ. For this we assume a slightly less than half-filled band and a strong Coulomb interaction.

$$n \le 1 \qquad ; \qquad U >> W \tag{8.302}$$

Then the Fermi energy is definitely in the lower quasiparticle band (8.291). Hence we can limit ourselves to the quasiparticle density of states of the lower sub-band (8.291):

$$\rho_\sigma^{lower}(E) = \rho_0 \left(\frac{E - n_{-\sigma} T_0}{1 - n_{-\sigma}} \right) \tag{8.303}$$

ρ_0 is as usual the Bloch density of states. Therewith the numerator (8.77) of the static susceptibility reads as

$$\begin{aligned} Z_H &= - \int_{-\infty}^{+\infty} dE \, (\rho_\sigma(E) f'_-(E))_0 \\ &= - \int_{-\infty}^{+\infty} dE \, \rho_0 \left(\frac{2E - nT_0}{2 - n} \right) f'_-(E) \end{aligned} \tag{8.304}$$

For $(T = 0)$, $f'_-(E)$ is a δ-function:

$$f'_-(E; T = 0) = -\delta(E - E_F) \tag{8.305}$$

That means

$$Z_H(T = 0) = \rho_0 \left(\frac{2E_F - nT_0}{2 - n} \right) \tag{8.306}$$

For the denominator of the susceptibility we need

$$\begin{aligned} \frac{\partial}{\partial n_{-\sigma}} S_{\mathbf{k}\sigma}^{lower}(E - \mu) = &-\hbar\delta(E - (1 - n_{-\sigma})\varepsilon(\mathbf{k}) - n_{-\sigma} T_0) + \\ &+ \hbar(1 - n_{-\sigma})(\varepsilon(\mathbf{k}) - T_0) * \\ &* \delta'(E - (1 - n_{-\sigma})\, \varepsilon(\mathbf{k}) - n_{-\sigma} T_0) \end{aligned}$$

that gives in (8.78):

$$N_H = -\frac{1}{N}\sum_{\mathbf{k}} \int_{-\infty}^{+\infty} dE\, f_-(E)\delta\left(E - \left(1 - \frac{n}{2}\right)\varepsilon\left(\mathbf{k}\right) - \frac{n}{2}T_0\right) +$$

$$+ \frac{1}{N}\sum_{\mathbf{k}}\left(1 - \frac{n}{2}\right) \int_{-\infty}^{+\infty} dE\, f_-(E)(\varepsilon(\mathbf{k}) - T_0) *$$

$$* \, \delta'\left(E - \left(1 - \frac{n}{2}\right)\varepsilon\left(\mathbf{k}\right) - \frac{n}{2}T_0\right)$$

$$= -\frac{2}{2-n}\int_{-\infty}^{+\infty} dE\, f_-(E)\rho_0\left(\frac{2E - nT_0}{2-n}\right) -$$

$$- \int_{-\infty}^{+\infty} dE\, f_-'(E)\left(\frac{2E - nT_0}{2-n} - T_0\right)\rho_0\left(\frac{2E - nT_0}{2-n}\right)$$

We finally get

$$N_H = -\frac{n}{2-n} - \frac{2}{2-n}\int_{-\infty}^{+\infty} dE\, f_-'(E)(E - T_0)\rho_0\left(\frac{2E - nT_0}{2-n}\right) \tag{8.307}$$

With this the static paramagnetic susceptibility is completely determined:

$$\bar{\chi}_H = -\mu_B \int_{-\infty}^{+\infty} dE\, f_-'(E)\rho_0\left(\frac{2E - nT_0}{2-n}\right)(2-n) *$$

$$* \left[1 - n - \int_{-\infty}^{+\infty} dE\, f_-'(E)(E - T_0)\rho_0\left(\frac{2E - nT_0}{2-n}\right)\right]^{-1} \tag{8.308}$$

The instability of the paramagnetism against ferromagnetic ordering is given by the zero of the denominator:

$$0 = 1 - n - \int_{-\infty}^{+\infty} f_-'(E)(E - T_0)\rho_0\left(\frac{2E - nT_0}{2-n}\right) \tag{8.309}$$

For the most *unfavourable* case of $T_c = 0^+$, we can use (8.305). That gives the following *criterion for ferromagnetism*:

$$0 = 1 - n + (E_F - T_0)\rho_0\left(\frac{2E_F - nT_0}{2-n}\right) \tag{8.310}$$

This result agrees with (8.298). Therefore, we can take over the discussion presented there and avoid repetition.

8.5.6 Spin-Dependent Band Shift

The approximate evaluation of the Hubbard model as discussed in the preceding section led to the criterion (8.298) for the existence of ferromagnetism, which is obviously only very difficult to fulfil as was demonstrated for a special case. The possible reason for that is clear from Fig. 8.16. The centres of gravity of the ↑- and ↓-spectra coincide so that a spontaneous magnetization, which presumes a preferential spin orientation, appears to be possible only for small band occupation. The *Hubbard-I* solution satisfies the first three spectral moments but not the fourth. Therefore, it is not correct in the *strong-coupling* regime, which happens to be the regime where ferromagnetism is expected in the first place. Therefore, the appearance of a spin-dependent band shift will be essential for a spontaneous ordering of electron moments.

A distinct improvement from this point of view is provided by *moment method* which is conceptually very simple and non-perturbative. It consists of two steps [8]: First, one tries to guess the general structure of the spectral density, i.e. one makes a *spectral density ansatz (SDA)*, whereby, one is guided by exactly solvable limits, exact spectral representations, sum rules and also by *plausibility* and *intuition*. A comparison with already existing reliable theories also certainly helps. The zero-bandwidth limit (Sect. 8.4.2), the two-site model (Sect. 8.4.3), the exact *strong-coupling* behaviour and also the approximate *Hubbard-I* solution (8.287) altogether make a two-pole ansatz [9] meaningful:

$$S_{\mathbf{k}\sigma}(E) = \hbar \sum_{j=1}^{2} \alpha_{j\sigma}(\mathbf{k}) \delta\left(E + \mu - E_{j\sigma}(\mathbf{k})\right) \tag{8.311}$$

The spectral weights $\alpha_{j\sigma}(\mathbf{k})$ and the quasiparticle energies $E_{j\sigma}(\mathbf{k})$ are treated in the beginning as free parameters.

In the second step of the procedure, using the relation (8.218), these free parameters are fitted with the exactly calculated spectral moments of the Hubbard model (8.221), (8.222), (8.223), and (8.224). After simple reformulations, one obtains real quasiparticle energies ($j = 1, 2$)

$$\begin{aligned} E_{j\sigma}(\mathbf{k}) = \frac{1}{2}\Big(& T_0 + B_{\mathbf{k}-\sigma} + \varepsilon(\mathbf{k}) + U + (-1)^j \, * \\ * & \sqrt{(T_0 + B_{\mathbf{k}-\sigma} - \varepsilon(\mathbf{k}) + U)^2 + 4U\, n_{-\sigma}\left(\varepsilon(\mathbf{k}) - T_0 - B_{\mathbf{k}-\sigma}\right)}\Big) \end{aligned} \tag{8.312}$$

with spectral weights of the form

$$\alpha_{1\sigma}(\mathbf{k}) = \frac{E_{1\sigma}(\mathbf{k}) - T_0 - B_{\mathbf{k}-\sigma} - U(1 - n_{-\sigma})}{E_{1\sigma}(\mathbf{k}) - E_{2\sigma}(\mathbf{k})} = 1 - \alpha_{2\sigma}(\mathbf{k}) \tag{8.313}$$

This result implies the SDA self-energy:

$$\Sigma_{\mathbf{k}\sigma}^{SDA}(E) = U n_{-\sigma} \frac{E + \mu - T_0 - B_{\mathbf{k}-\sigma}}{E + \mu - T_0 - B_{\mathbf{k}-\sigma} - U(1 - n_{-\sigma})} \tag{8.314}$$

The SDA self-energy has the same structure as the one of *Hubbard-I* solution, what is new is only the term $B_{\mathbf{k}-\sigma}$, which we already got to know in Sect. 8.4.5 as the "band correction" (8.211). This is made up of two terms, namely the *band shift* $B_{-\sigma}$ (8.212) and the *band width correction* $F_{\mathbf{k}-\sigma}$ (8.213). It is interesting to note that the SDA self-energy agrees exactly with the expression (8.235) which is obtained as the leading term of the general high-energy expansion. With this it is particularly clear that the solution (8.314) will have the correct *strong-coupling* behaviour. By construction, the SDA solution satisfies the first four spectral moments.

Because of the presence of $B_{\mathbf{k}-\sigma}$ and also because of $n_{-\sigma}$, the many-body problem with (8.314) is not yet completely solved. The important *band shift* $B_{-\sigma}$ can be expressed according to (8.217) exactly by the one-electron Green's function or spectral density and therefore does not need any further approximation. The wavenumber-dependent *band width correction* $F_{\mathbf{k}-\sigma}$, on the other hand, cannot be fixed directly using the one-electron functions. Investigations, which we do not want to go into in detail here, have however shown that this part of the band correction, as far as ferromagnetism is concerned, in contrast to the band shift, does not play such a decisive role. Therefore, one might assume that in a first approximation it can be neglected.

However, going beyond this, it would be consistent to determine the "higher" correlation functions appearing in $F_{\mathbf{k}-\sigma}$ also by a moment method. Here we will briefly discuss the method developed in [10]. First, it can be shown that in case of translational symmetry and restriction to nearest neighbour electron hopping, the wavenumber dependence of the bandwidth correction can be separated into

$$n_{-\sigma}(1 - n_{-\sigma})F_{\mathbf{k}-\sigma} = (\varepsilon(\mathbf{k}) - T_0) \sum_{i=1}^{3} F_{-\sigma}^{(i)} \tag{8.315}$$

According to (8.213) the three summands have the meaning

$$F_{-\sigma}^{(1)} = \langle n_{i-\sigma} n_{j-\sigma} \rangle - n_{-\sigma}^2 \quad density\ correlation \tag{8.316}$$

$$F_{-\sigma}^{(2)} = -\langle c_{j\sigma}^{\dagger} c_{j-\sigma}^{\dagger} c_{i-\sigma} c_{i\sigma} \rangle \quad double-hopping\ correlation \tag{8.317}$$

$$F_{-\sigma}^{(3)} = -\langle c_{j\sigma}^{\dagger} c_{i-\sigma}^{\dagger} c_{j-\sigma} c_{i\sigma} \rangle \quad spin-flip\ correlation \tag{8.318}$$

i and j number the nearest neighbour lattice sites. As a result, the expectation values are not explicitly dependent on the lattice sites. We will demonstrate the procedure on the correlation $F_{-\sigma}^{(3)}$. We can first write

$$F^{(3)}_{-\sigma} = -\sum_l \delta_{jl} \langle c^\dagger_{l\sigma} c^\dagger_{j+\Delta-\sigma} c_{j-\sigma} c_{j+\Delta\sigma} \rangle \tag{8.319}$$

Here the index Δ represents the lattice vector that connects two neighbouring lattice sites $\mathbf{R}_i$ and $\mathbf{R}_j$ with each other. Due to the assumed translational symmetry, $F^{(3)}_{-\sigma}$ is the same expression for all the nearest neighbours. We will now introduce a *suitable, higher* spectral density,

$$S^{(3)}_{jl\sigma}(t,t') = \frac{1}{2\pi} \langle \left[c^\dagger_{j+\Delta-\sigma} c_{j-\sigma} c_{j+\Delta\sigma}(t),\, c^\dagger_{l\sigma}(t') \right]_+ \rangle \tag{8.320}$$

whose Fourier transform $S^{(3)}_{\mathbf{k}\sigma}(E)$ using the spectral theorem (B.95) gives the correlation that we are looking for

$$F^{(3)}_{-\sigma} = -\frac{1}{N} \sum_{\mathbf{k}} \int_{-\infty}^{+\infty} dE f_-(E) S^{(3)}_{\mathbf{k}\sigma}(E-\mu) \tag{8.321}$$

The poles of $S^{(3)}_{\mathbf{k}\sigma}(E)$, as can be seen from the definition (8.320), belong to the one-particle excitation spectrum of the Hubbard system. By comparing the spectral representation (B.86) of $S^{(3)}_{\mathbf{k}\sigma}(E)$ with that of the starting function (8.311), one can to a good approximation assume that the pole structure of the two functions is the same. Therefore, the difference should be only in the spectral weights. That is why in analogy to (8.311) it is reasonable to make the following ansatz for the *higher* spectral density:

$$S^{(3)}_{\mathbf{k}\sigma}(E) = \hbar \sum_{j=1}^{2} \beta_{j\sigma}(\mathbf{k})\, \delta(E + \mu - E_{j\sigma}(\mathbf{k})) \tag{8.322}$$

where the quasiparticle energies $E_{j\sigma}(\mathbf{k})$ are the same as in (8.311) or (8.312). Thus, only the spectral weights $\beta_{j\sigma}(\mathbf{k})$ are the unknowns. They are fixed by the first two exactly calculated spectral moments of the *higher* spectral density $S^{(3)}_{\mathbf{k}\sigma}(E)$ and the correlation $F^{(3)}_{-\sigma}$ is then determined through (8.321).

The correlation terms $F^{(1)}_{-\sigma}$ and $F^{(2)}_{-\sigma}$ are handled in an analogous way, i.e. with two-pole ansatz similar to (8.322), for suitably chosen *higher* spectral densities. It should not remain unmentioned that the same procedure can be applied also for the higher correlation $\langle n_{i\sigma} c^\dagger_{i-\sigma} c_{j-\sigma} \rangle$ that appears in the *band shift* $B_{-\sigma}$ (8.212) which leads to the exact result (8.217). This can be seen as a strong support for the method.

Figure 8.31 shows a typical result of the SDA for a strongly coupled band electron system on a bcc lattice. In the figure the quasiparticle density of states is plotted as a function of energy for different band occupations n and at temperature $T = 0$. For band occupations $n \leq 0.55$ there does not exist any ferromagnetic solution ($n_\uparrow \neq n_\downarrow$). There is no spin asymmetry of the density of states ($\rho_\sigma(E) \equiv \rho_{-\sigma}(E)$). We, however, observe a splitting into the so-called *Hubbard bands* which are shifted

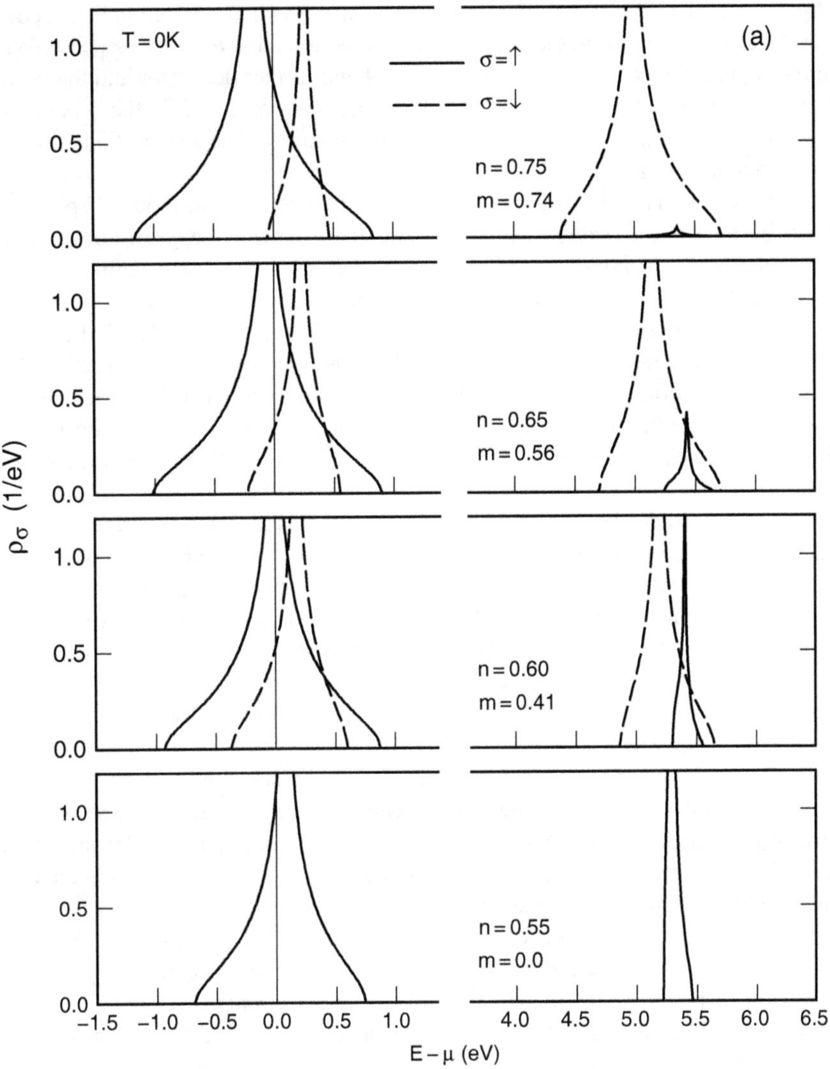

Fig. 8.31 Quasiparticle density of states in the SDA for the Hubbard model as function of the energy for various band occupations. Parameters: bcc lattice, $U = 5\,eV$, $W = 2\,eV$, $T = 0K$. *Solid lines* for $\sigma = \uparrow$, *broken lines* for $\sigma = \downarrow$

with respect to each other by about U and whose physical meaning has already been discussed. We want to call this splitting the *quasiparticle splitting*. The areas under the partial density of states scale roughly with the probability that a propagating electron finds at a lattice point an interaction partner ($n_{-\sigma}$, upper band) or does not ($1 - n_{-\sigma}$, lower band). For $n > 0.58$ an additional spin-splitting sets in for each of the sub-bands (*exchange splitting*). The lower $\downarrow$-band becomes narrower and shifts to higher energies. The states are occupied by electrons up to a common Fermi edge. Therefore, the system shows a spontaneous spin polarization $m = n_{\uparrow} - n_{\downarrow} \neq 0$. With increasing particle density, the width of the $\downarrow$-band further decreases and the centre of gravity moves to higher energies. For band occupations $n \geq 0.78$ the $\downarrow$-band lies completely above the Fermi edge and is therefore empty. The system finds itself in *ferromagnetic saturation*.

Obviously we have to differentiate between two correlation-induced splittings. This is because in the *strong-coupling* regime, there appears the *quasiparticle splitting* of the *Hubbard bands* practically for all the parameter constellations. Added to this, under certain conditions, there is an *exchange splitting*, which creates the spontaneous magnetization in the ferromagnetic phase. A more detailed analysis shows that a spontaneous polarization of the band electrons sets in exactly when the band correction $B_{\mathbf{k}-\sigma}$ (8.211), in particular the band shift $B_{-\sigma}$ (8.212), has a real spin dependence. This term is missing in the *Hubbard-I* solution (8.269) which again explains why ferromagnetism is extremely difficult to realize in that theory.

The quasiparticle density of states shows a characteristic temperature dependence, from which finally the magnetization curve, typical for ferromagnets, results. The density of states for a band occupation $n = 0.65$ is plotted in Fig. 8.32 for different temperatures. According to Fig. 8.31 (second picture from above), for this particle concentration, the system is ferromagnetic and at $T = 0$ near to saturation ($m = 0.56$). With increasing temperature, the $\downarrow$-band moves to lower energies and becomes broader such that at $T = 550K$ it coincides with the $\uparrow$-band. At this point, the spin asymmetry and exchange splitting vanish. The system is in its paramagnetic phase. By multiplying the spin-dependent density of states with the Fermi function and integrating, we obtain the spin-dependent average occupation numbers and therewith the electron polarization m. Dividing by the total particle density (m/n) results in a temperature dependence which is similar to the magnetization curve of Fig. 8.8 with an obviously realistic Curie temperature.

These results can again be verified partly by the static susceptibility (8.65). In Fig. 8.33, the inverse paramagnetic susceptibility for an sc lattice is plotted as a function of band occupation for different coupling strengths U/W. The zeroes provide us the instabilities of the paramagnetic state against ferromagnetism. The appearance of two zeroes, i.e. two instabilities, at first glance seems to be a special feature of the SDA. How far, possibly, this is a property of the Hubbard model itself is not clear. It turns out that always the solution starting at the lower band occupation is stable. This approaches, in contrast to the second solution, saturation $m = n$ for higher band occupations.

An interesting point is the influence of the *non-locality* of the electronic self-energy (8.314) which is given by the wavenumber-dependent *bandwidth correction* $F_{\mathbf{k}-\sigma}$ in (8.315). Part (b) of Fig. 8.33 gives us an idea. The critical coupling U/W increases from about 4 to about 14. For a primitive cubic lattice, apparently the *bandwidth correction* cannot be neglected. Detailed investigation, however, has shown that the influence of $F_{\mathbf{k}-\sigma}$ on the magnetic stability considerably decreases with increasing coordination number (sc$\rightarrow$bcc$\rightarrow$fcc). What in principle is decisive

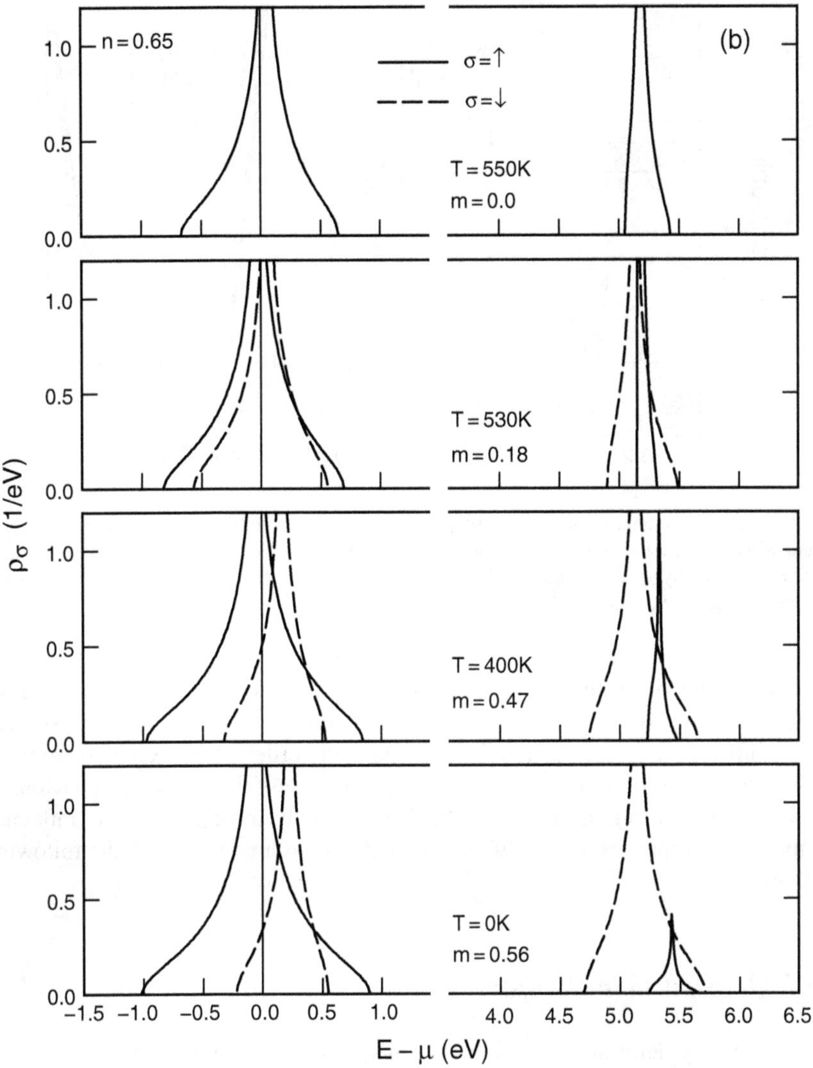

Fig. 8.32 Quasiparticle density of states in the SDA for the Hubbard model as function of the energy for various temperatures. Parameters: bcc lattice, $U = 5\,eV$, $W = 2\,eV$, $n = 0.65$. *Solid lines* for $\sigma = \uparrow$, *broken lines* for $\sigma = \downarrow$

for the possibility of ferromagnetism in the Hubbard model is the spin-dependent *band shift* $B_{-\sigma}$.

The SDA (*spectral density approach*), presented in this section, gives a qualitatively convincing picture of the band ferromagnetism of correlated electrons. It shows that the spin-dependent *band shift* and the lattice structure are the most important components for a ferromagnetic ordering. At the same time it is also very

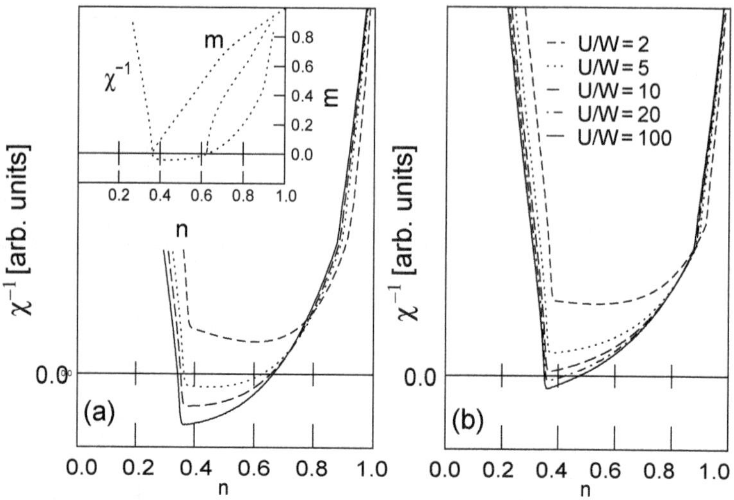

Fig. 8.33 Inverse paramagnetic static susceptibility χ^{-1} for a sc lattice as a function of the bandoccupation for various values of the coupling U/W at $T = 0K$. The *curves* are found within the SDA [11].(**a**) System with the full **k**-dependent self-energy (8.275). (**b**) System with a local self-energy, i.e. for $F_{\mathbf{k}-\sigma} \equiv 0$ (8.276)

clear, where the disadvantages of the method lie. First, by ansatz (8.311), the quasiparticle energies (8.312) are real. Quasiparticle damping, which can be expected to destabilize to a certain degree the ferromagnetic ordering, is excluded from the beginning . In addition, the SDA result does not possess the expected behaviour in the weak-coupling regime (Sect. 8.4.8). How strongly these two factors influence the magnetic properties of the Hubbard model will be investigated in the following sections.

8.5.7 *Quasiparticle Damping*

More or less by definition, the SDA self-energy (8.314) is a real quantity. It has a pole at

$$E = T_0 + B_{\mathbf{k}-\sigma} + U(1 - n_{-\sigma}) - \mu$$

which leads to a δ-peak in the imaginary part of the self-energy $Im\ \Sigma_{\mathbf{k}\sigma}(E + i0^+)$. This, of course, has no influence since the pole always lies in the band gap. As a result, the quasiparticles are stable from the beginning. One can see in this a serious shortcoming of the moment method of the last section. Therefore, one should look for a theory which retains the advantages of the SDA and at the same time, in addition, allows for finite lifetimes of quasiparticles. The question we are interested in this section is the influence of quasiparticle damping on the magnetic stability of the Hubbard model. As a preparation for this, we start with an *alloy analogy* of the Hubbard model suggested by Hubbard himself [12].

The starting point is the idea that the σ-electron propagates through a crystal with $(-\sigma)$-electrons frozen at certain lattice sites which are statistically distributed over the lattice. When the σ-electron lands at a certain lattice site there are two possibilities. It either finds a $(-\sigma)$-electron there or it does not. This can be understood as motion through a *fictitious* binary alloy. The two alloy components have atomic levels $E_{1\sigma}$ and $E_{2\sigma}$ and appear with *concentrations* $x_{1\sigma}$ and $x_{2\sigma}$ $(x_{1\sigma} + x_{1\sigma} = 1)$. The *coherent potential approximation* (CPA) is a standard method with which a configurational average over the statistically distributed positions of the *frozen* $(-\sigma)$-electron can be performed [13]. The resulting self-energy of the σ-electron obeys the following equation:

$$0 = \sum_{p=1}^{2} x_{p\sigma} \frac{E_{p\sigma} - \Sigma_\sigma(E) - T_0}{1 - \frac{1}{\hbar} G_\sigma(E)(E_{p\sigma} - \Sigma_\sigma(E) - T_0)}$$

$$G_\sigma(E) = \frac{1}{N} \sum_{\mathbf{k}} G_{\mathbf{k}\sigma}(E) = \int_{-\infty}^{+\infty} dx \frac{\rho_0(x)}{E - \Sigma_\sigma(E) - x} \tag{8.323}$$

$\rho_0(x)$ is again the Bloch density of states of the non-interacting electron system. The CPA is a *single-site* approximation and therefore the self-energy is wavenumber independent. Now to be concrete, we have to specify the fictitious alloy. The first choice (*conventional alloy analogy* (CAA)) would be the use of the results of the infinitely narrow band case ((8.129), (8.130), and (8.131), *atomic limit*)

$$\begin{aligned} E_{1\sigma} &= T_0 \quad &; x_{1\sigma} &= 1 - n_{-\sigma} \\ E_{2\sigma} &= T_0 + U &; x_{2\sigma} &= n_{-\sigma} \end{aligned} \tag{8.324}$$

Substituting this in (8.323) one gets a self-energy $\Sigma_\sigma^{CAA}(E)$, which is in general complex and therefore includes quasiparticle damping. On the other hand, it could be shown that this self-energy excludes in any case spontaneous ferromagnetism. This is a crude contradiction to the result of SDA of the last section. Then one has to ask oneself whether the damping (finite lifetime) of the quasiparticles which was neglected in SDA has really such a destabilizing effect on collective magnetism. This is a question which has to be taken seriously. It could be shown [14] that in infinite dimensional lattice (Sect. 8.4.9), CPA is an exact (!) theory of the alloy problem. On the other hand, the CAA self-energy violates the correct *strong-coupling* behaviour (8.227) as well as the limit of weak coupling (8.239). If the CPA theory is exact for the alloy problem at least for $d \to \infty$, then the contradiction can be resolved only by assuming that the *conventional* alloy analogy (8.324) is not the correct starting point. In particular, the picture of *frozen* $(-\sigma)$-electrons is surely not acceptable.

The form of the fictitious alloy is indeed by no means predetermined. Therefore, we will now treat the energies $E_{1,2\sigma}$ and the concentrations $x_{1,2\sigma}$ of the still to be found "optimal" fictitious alloy, as free parameters. In order to fix them, we will substitute the high-energy expansions (8.226) and (8.227) in the CPA equation and gather the powers of $\frac{1}{E}$. In view of the single-site aspect of CPA, we limit the band correction $B_{\mathbf{k}-\sigma}$ to its local part $B_{-\sigma}$. This gives the following determining equations for the energies and concentrations:

$$\begin{aligned}
&\sum_{p=1}^{2} x_{p\sigma} = 1 \\
&\sum_{p=1}^{2} x_{p\sigma}(E_{p\sigma} - T_0) = U\, n_{-\sigma} \\
&\sum_{p=1}^{2} x_{p\sigma}(E_{p\sigma} - T_0)^2 = U^2\, n_{-\sigma} \\
&\sum_{p=1}^{2} x_{p\sigma}(E_{p\sigma} - T_0)^3 = U^3\, n_{-\sigma} + U^2\, B_{-\sigma}\, n_{-\sigma}(1 - n_{-\sigma})
\end{aligned} \tag{8.325}$$

When the energies and concentrations of a *modified alloy analogy* (MAA) are determined from these relations, it is automatically guaranteed that the first four spectral moments are fulfilled from which a correct *strong-coupling* behaviour results. One finds

$$E_{1\sigma}^{MAA} = T_0 + \frac{1}{2}\left(U + B_{-\sigma} - \sqrt{(U + B_{-\sigma})^2 - 4U n_{-\sigma} B_{-\sigma}}\right) \tag{8.326}$$

$$E_{2\sigma}^{MAA} = T_0 + \frac{1}{2}\left(U + B_{-\sigma} + \sqrt{(U + B_{-\sigma})^2 - 4U n_{-\sigma} B_{-\sigma}}\right) \tag{8.327}$$

$$x_{1\sigma}^{MAA} = \frac{E_{1\sigma}^{MAA} - T_0 - B_{-\sigma} - U(1 - n_{-\sigma})}{E_{1\sigma}^{MAA} - E_{2\sigma}^{MAA}} = 1 - x_{2\sigma}^{MAA} \tag{8.328}$$

Interestingly, the weights and the energies coincide with the results (8.312) and (8.313) of SDA if the band energies $\varepsilon(\mathbf{k})$ and the band corrections $B_{\mathbf{k}-\sigma}$ are replaced by their corresponding centres of gravity T_0 and $B_{-\sigma}$, respectively. Therefore, one can understand the following choice as *optimal alloy analogy* $p = 1, 2$:

$$E_{p\sigma}^{MAA} = \left(E_{p\sigma}^{SDA}(\mathbf{k})\right)_{\varepsilon(\mathbf{k})=T_0} = f_p(T_0, U, n_{-\sigma}, B_{-\sigma}) \tag{8.329}$$

$$x_{p\sigma}^{MAA} = \left(\alpha_{p\sigma}^{SDA}(\mathbf{k})\right)_{\varepsilon(\mathbf{k})=T_0} = g_p(T_0, U, n_{-\sigma}, B_{-\sigma}) \tag{8.330}$$

Substituting this alloy analogy in the CPA equation (8.323) one obtains the MAA self-energy for all energies E with which all the interesting quantities can be derived.

- As a CPA result, MAA includes quasiparticle damping ($Im\Sigma_\sigma^{MAA}(E) \not\equiv 0$) without sacrificing the advantages of SDA.
- The thermodynamic averages $n_{-\sigma}$ and $B_{-\sigma}$ can be determined self-consistently with the help of the spectral theorem (8.136) and (8.217) and providing therewith, in principle, particle density-, temperature- and spin-dependent atomic data (8.329) and (8.330) of the two components of the alloy. The important thing here is that through $B_{-\sigma}$ in some sense the itineracy of the $(-\sigma)$-electrons is included in the calculation (*correlated electron hopping*). Contrary to the conventional alloy analogy CAA (8.324) they are not seen as *frozen* on the lattice sites.
- By definition the first four spectral moments are fulfilled so that the correct *strong-coupling* behaviour is taken care of.
- The *strong-coupling* behaviour can also be tested with the help of an exact result of the general CPA theory [15] in the so-called *split-band regime* $U \gg W$. According to this the spectral density should consist of two separated peaks with centres of gravity at

$$T_{p\sigma}^{CPA} = E_{p\sigma} + x_{p\sigma}(\varepsilon(\mathbf{k} - T_0) \quad ; \; p = 1, 2 \tag{8.331}$$

  Substituting (8.329) and (8.330) in this equation we get for $U \gg W$ the correct result (8.208) and (8.209) when we further replace $B_{\mathbf{k}-\sigma}$ by $B_{-\sigma}$. This can be seen as a strong support for the *modified alloy analogy*.
- For us the most important fact, however, is that in contrast to CAA, the MAA allows *spontaneous* ferromagnetism!

Figures 8.34 and 8.35 show the typical results of the MAA theory for two different temperatures for the spectral density $S_{\mathbf{k}\sigma}(E)$ of a strongly correlated electron system ($U/W = 5$) on an fcc lattice. As Bloch density of states a *tight-binding* version [16] is chosen. For less than half-filled bands ($n < 1$) the system is paramagnetic; a spontaneous spin asymmetry does not appear. The band occupation used in Figs. 8.34 and 8.35, $n = 1.6$ allows ferromagnetism provided the Coulomb interaction U exceeds a critical value. As already observed in the results of SDA (Figs. 8.31 and 8.32), two different splittings appear. The spectral density consists for each $\mathbf{k}$-vector of a high-energy and a low-energy peaks (*quasiparticle splitting*), which are separated from each other by about U. The finite widths of the peaks are consequences of the quasiparticle damping, which displays a clear energy-, wavenumber-, spin- and temperature-dependence. The spectral weight of the low-energy peak, given by the area under the curve, scales with the probability that the $(\mathbf{k}, \sigma)$-electron propagating in the more than half-filled band finds an empty lattice site. On the other hand, the weight of the upper peak corresponds to the probability that the electron meets an electron with opposite spin. We have already established in connection with the SDA results (Figs. 8.31 and 8.32) that this *quasiparticle splitting* is not tied to the ferromagnetic phase. It represents the correct *strong-coupling* behaviour (Fig. 8.15).

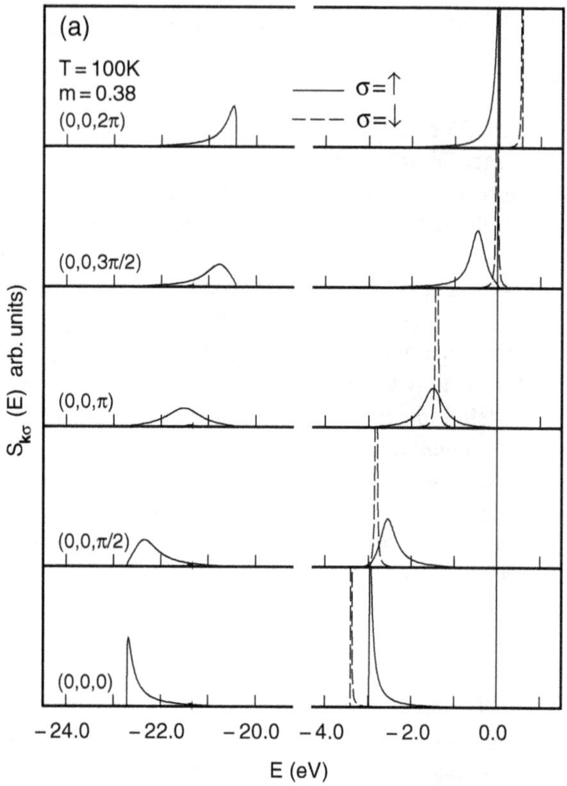

Fig. 8.34 Spectral density as a function of energy for an fcc lattice calculated within the MAA at the temperature $T = 100\text{K}$ for different **k**-vectors equidistant along the (0, 0, 1) direction of the 1. Brillouin zone. Further parameters: $n = 1.6$, $U = 20\,eV$, $W = 4\,eV$. *Vertical line* indicates the position of the chemical potential

Ferromagnetism appears exactly then, when in addition to the *quasiparticle splitting* there sets in an extra *exchange splitting* of each of the two spectral density peaks. At low temperatures as in Fig. 8.34 ($T = 100\text{K}$), the system is practically in ferromagnetic saturation ($m = 2 - n$), i.e. almost all $\uparrow$-states are occupied so that a $\downarrow$-electron finds an interacting partner at almost every lattice site. Consequently, the low-energy peak of the $\downarrow$ spectral density vanishes. The high-energy peak is very sharp indicating a long-lived quasiparticle. A $\downarrow$-hole has no chance to be scattered by a $\uparrow$-hole. At higher temperatures as in Fig. 8.35, due to partial demagnetization, i.e. due to a finite hole density in the $\uparrow$-spectrum, there appears again a low-energy $\downarrow$-peak.

The exchange splitting has an interesting dependence on wavenumber. At the upper end of the spectrum (X-point: $(0, 0, 2\pi)$) we find a *normal* splitting, i.e. the $\downarrow$-peak lies above the $\uparrow$-peak. At the lower end (Γ-point: (0,0,0)) it is exactly the opposite, namely the $\downarrow$-peak lies below the $\uparrow$-peak. The quasiparticle dispersions intersect each other as functions of **k**. This is the result of two competing correlation effects. One leads to a spin-dependent exchange shift of the centres of gravity of the quasiparticle subbands (8.214) and (8.215) with respect to each other in the *normal* sense and the other to a spin-dependent bandwidth reduction, which overcompensates the *normal* shift at the band bottom.

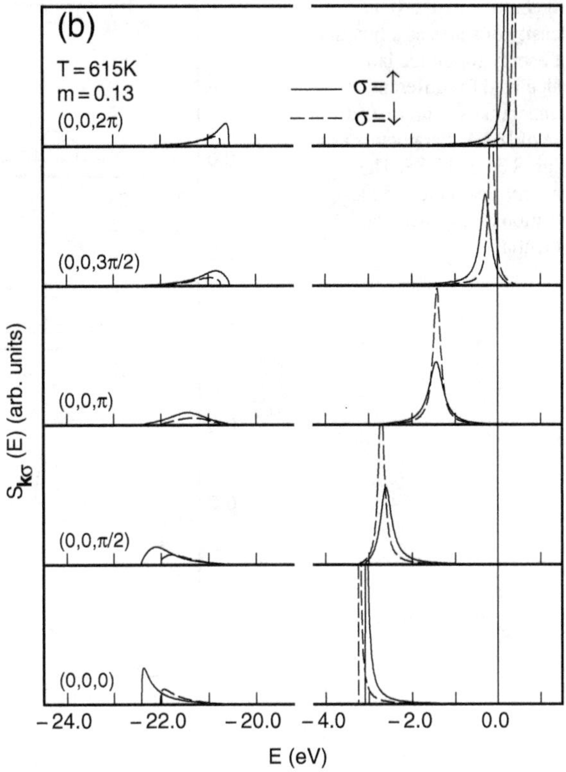

Fig. 8.35 Spectral density as a function of energy for an fcc lattice calculated within the MAA at the temperature $T = 615\text{K}$ for different **k**-vectors equidistant along the (0, 0, 1) direction of the 1. Brillouin zone. Further parameters: $n = 1.6$, $U = 20\,eV$, $W = 4\,eV$. *Vertical line* indicates the position of the chemical potential

An example of the temperature dependence of the quasiparticle density of states is demonstrated in Fig. 8.36. The qualitative interpretation is the same as that for the SDA results presented in Fig. 8.32 and therefore need not be repeated. On the whole the curves are *rounder* compared to those in SDA due to quasiparticle damping. We will see in Sect. 8.5.10 that this damping leads to an appreciable destabilization of ferromagnetism compared to the SDA. A main reason for this is a finite overlap of the spectral density peaks which clearly weakens the self-consistently calculated ferromagnetic solution compared to the paramagnetic solution. In summary one may say that because of the inclusion of the quasiparticle damping, the MAA is a systematic improvement of the SDA without loosing its advantage (spin-dependent band shift!). The second shortcoming of the SDA mentioned above, namely the incorrect *weak-coupling* behaviour, however, remains even in the MAA. This will be a point to ponder in the next section.

8.5.8 Dynamical Mean Field Theory

Neither the SDA self-energy nor the MAA self-energy satisfy the correct *weak-coupling* behaviour (8.239). Therefore, we now search for a method which in a

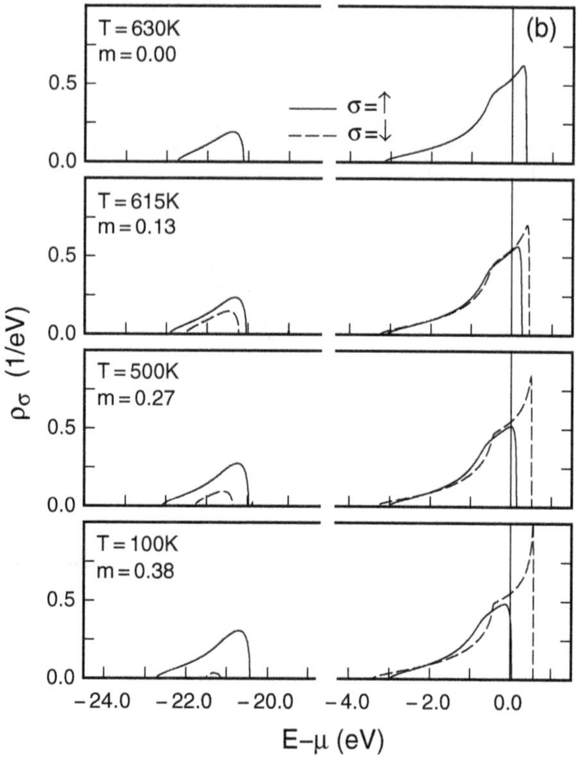

Fig. 8.36 Quasiparticle density of states as a function of energy for an fcc lattice calculated for different temperatures T up to T_C by use of MAA. Parameters as in Figs. 8.34 and 8.35. The *vertical line* indicates the position of the chemical potential

reasonable way interpolates between the rigorous results of the weak and strong coupling. This is achieved by the recently developed and extraordinarily successful *dynamical mean field theory* (DMFT) [17]. First, we start with a wavenumber-independent (*local*) self-energy. According to (8.255) this is exact in the case of a lattice of infinite dimension d. Let us assume that this is an acceptable starting point even for finite dimensions. We have shown for this case in Sect. 8.4.10 (8.264) that the self-energy in the Hubbard model can be expressed by the diagrammatic functional C_σ of an *impurity* scattering (8.263). This fact will now be exploited together with an investigation of a well-known *impurity* problem [18].

The starting point is the *"single-impurity" Anderson model* (SIAM). The model describes a partially filled energy band which hybridizes with a localized correlated impurity (d) level and therefore is described by the following Hamiltonian:

$$
\begin{aligned}
H = &\sum_{\sigma}(\varepsilon_d - \mu)n_{d\sigma} + \sum_{\mathbf{k}\sigma}(\varepsilon(\mathbf{k}) - \mu)n_{\mathbf{k}\sigma} + \\
&+ \sum_{\mathbf{k}\sigma}\left(V_{\mathbf{k}d}c^{\dagger}_{d\sigma}c_{\mathbf{k}\sigma} + V_{d\mathbf{k}}c^{\dagger}_{\mathbf{k}\sigma}c_{d\sigma}\right) + \frac{1}{2}U\sum_{\sigma}n_{d\sigma}n_{d-\sigma} \qquad (8.332)
\end{aligned}
$$

$c^{\dagger}_{d\sigma}(c_{d\sigma})$ is the creation (annihilation) operator of an electron with spin σ in a single localized d-level (*impurity level*) and correspondingly $c^{\dagger}_{\mathbf{k}\sigma}(c_{\mathbf{k}\sigma})$ for a band electron with the wavevector $\mathbf{k}$. $n_{d\sigma} = c^{\dagger}_{d\sigma}c_{d\sigma}$ is the occupation number operator for the impurity level and $n_{\mathbf{k}\sigma} = c^{\dagger}_{\mathbf{k}\sigma}c_{\mathbf{k}\sigma}$ for the band states. $V_{\mathbf{k}d}(= V^{*}_{d\mathbf{k}})$ is the hybridization matrix element and U is the Coulomb repulsion on the d-level.

The equation of motion of the *d-impurity* Green's function

$$G_{d\sigma}(E) = \langle\langle c_{d\sigma}\,;\, c^{\dagger}_{d\sigma}\rangle\rangle \tag{8.333}$$

is formally solved by introducing a corresponding d-self-energy $\Sigma_{d\sigma}(E)$ using

$$\left\langle\!\!\left\langle \left[c_{d\sigma}\,,\, \frac{1}{2}U\sum_{\sigma'} n_{d\sigma'}n_{d-\sigma'} \right]_{-} ;\, c^{\dagger}_{d\sigma} \right\rangle\!\!\right\rangle \equiv \Sigma_{d\sigma}(E)G_{d\sigma}(E) \tag{8.334}$$

and by defining a *hybridization function*

$$\Delta(E) = \sum_{\mathbf{k}} \frac{|V_{\mathbf{k}d}|^2}{E + \mu - \varepsilon(\mathbf{k})} \tag{8.335}$$

which absorbs the band energies $\varepsilon(\mathbf{k})$ and the hybridization matrix element $V_{\mathbf{k}d}$ (Problem 8.16):

$$G_{d\sigma}(E) = \frac{\hbar}{E + \mu - \varepsilon_d - \Delta(E) - \Sigma_{d\sigma}(E)} \tag{8.336}$$

With the respective *free* Green's function

$$G^{(0)}_{d\sigma}(E) = \frac{\hbar}{E + \mu - \varepsilon_d - \Delta(E)} \tag{8.337}$$

Equation (8.336) can also be written as Dyson equation:

$$G_{d\sigma}(E) = G^{(0)}_{d\sigma}(E) + G^{(0)}_{d\sigma}(E)\frac{1}{\hbar}\Sigma_{d\sigma}(E)G_{d\sigma}(E) \tag{8.338}$$

This equation has the same structure as the version (8.262) for the Hubbard model. The corresponding self-energy diagrams belong to the class C_σ (8.263) of *impurity* diagrams

$$\Sigma_{d\sigma}(E) = C_\sigma\left[\left\{G^{(0)}_{d\sigma}(E)\right\}\,,\, U\right]\,, \tag{8.339}$$

provided it is assumed that U has the same meaning as in (8.263).

We now choose in (8.337) the hybridization function $\Delta(E)$ such that it holds

$$G^{(0)}_{d\sigma}(E) \stackrel{!}{\equiv} F_{ii\sigma}(E) \tag{8.340}$$

$F_{ii\sigma}(E)$ is defined in (8.257) ($l = i$). One can show that (8.340) can be realized only when the chemical potential μ is same for both the systems and formally one sets

$$\varepsilon_d = T_0 \tag{8.341}$$

Comparing (8.264) with (8.339) and (8.340), we get

$$\Sigma_\sigma(E) \equiv \Sigma_{d\sigma}(E) \tag{8.342}$$

and then due to (8.338) and (8.262):

$$G_{ii\sigma}(E) \equiv G_{d\sigma}(E) \tag{8.343}$$

With this the Hubbard model is ascribed to the simpler SIAM.

We want to summarize the results once more. Equation (8.262) can be written as follows:

$$F^{-1}_{ii\sigma}(E) = G^{-1}_{ii\sigma}(E) + \frac{1}{\hbar}\Sigma_\sigma(E)$$

With (8.340) this means

$$\left(G^{(0)}_{d\sigma}\right)^{-1} = \frac{1}{\hbar}(E + \mu - \varepsilon_d - \Delta(E)) = G^{-1}_{ii\sigma}(E) + \frac{1}{\hbar}\Sigma_\sigma(E)$$

and leads with (8.341) to the following self-consistent equation:

$$\Delta(E) = E + \mu - T_0 - \Sigma_\sigma(E) - \hbar G^{-1}_{ii\sigma}(E) \tag{8.344}$$

Due to the **k**-independence of the self-energy the propagator $G_{ii\sigma}(E)$ can be expressed in terms of the in general known *free* Bloch density of states $\rho_0(E)$.

$$G_{ii\sigma}(E) = \hbar \int_{-\infty}^{+\infty} dx \frac{\rho_0(x)}{E + \mu - x - \Sigma_\sigma(E)} \tag{8.345}$$

The DMFT solves the many-body problem of the Hubbard model finally through the following selfconsistency cycle:

- Choose a starting value for $\Sigma_\sigma(E)$, e.g., $\Sigma_\sigma(E) \equiv 0$!
- Using (8.345) calculate $G_{ii\sigma}(E)$!
- Determine $\Delta(E)$ using (8.344)

- Solve the SIAM problem (8.336) with the obtained $\Delta(E)$, i.e. find $\Sigma_{d\sigma}(E)$!
- Use (8.342) for a new starting value of $\Sigma_\sigma(E)$!

The remaining job lies in the solution of SIAM in the fourth step. This is certainly not a trivial problem but very much better worked out than the original Hubbard problem. With the help of Quantum Monte Carlo methods [19] it is possible to find a numerically essentially exact solution of the SIAM which can be used in the DMFT cycle to solve the Hubbard problem [20]. In the next section, as an example, we want to discuss an analytical theory of the SIAM problem, then apply it to the Hubbard model via DMFT to finally obtain statements about band ferromagnetism in the Hubbard model.

8.5.9 Modified Perturbation Theory

We are looking for a theory of the *single-impurity* Anderson model, which on implanting in the DMFT procedure reproduces the correct *weak-coupling* behaviour of the Hubbard model (Sect. 8.4.8) without loosing the advantages of the methods discussed in Sects. 8.5.6 and 8.5.7. That means a self-energy has to be found which reproduces the exact results in the strong as well as weak-coupling regimes. Most interesting question here is how far does the *weak-coupling* physics influence the possibility of band ferromagnetism in the Hubbard model.

A first step in answering this consists of a second order perturbation theory for the SIAM

$$\Sigma_{d\sigma}^{SOPT}(E) = Un_{d-\sigma} + U^2\Sigma_{d\sigma}^{(SOC)} \tag{8.346}$$

The procedure is same as described in Sect. 8.4.8 for the Hubbard model. Figure 8.37 shows the corresponding second order skeleton diagram for SIAM. Evaluation of this diagram gives an expression which is completely analogous to (8.240), the only difference being one has to replace the *single-electron* spectral density of the Hubbard model by the *impurity* spectral density of the Anderson model:

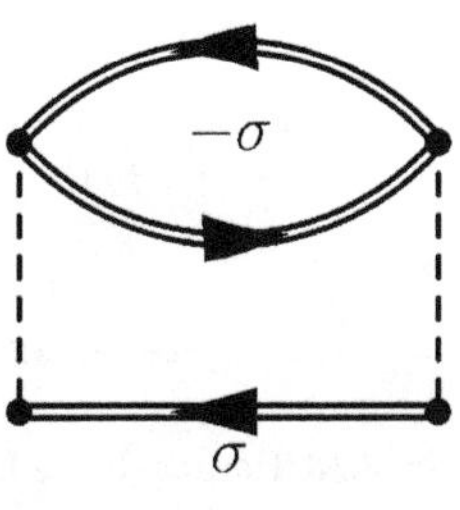

Fig. 8.37 Second order contribution to the self-energy of the single-impurity Anderson model

$$\Sigma_{d\sigma}^{(SOC)}(E) = \frac{1}{\hbar^3} \iiint dx\, dy\, dz \frac{S_{d\sigma}^{(1)}(x)\, S_{d-\sigma}^{(1)}(y) S_{d-\sigma}^{(1)}(z)}{E - x + y - z} *$$

$$* (f(x)f(-y)f(z) + f(-x)f(y)f(-z)) \tag{8.347}$$

It was found convenient to choose here the perturbation theory *around Hartree–Fock* (see (8.242)). The corresponding spectral density $S_{d\sigma}^{(1)}$ is determined from the reduced Dyson equation presented in Fig. 8.38 which takes into account only the skeleton diagrams of first order. Combining this perturbation theory result already with the DMFT, leads to *iterative perturbation theory* (ITP) [18], from which one can show that for the half-filled band ($n = 1$) it gives excellent result. For $n \neq 1$, however, it turns out to be insufficient particularly from the point of view of collective magnetic ordering. The *strong-coupling* behaviour is not correctly reproduced by this approximation. More promising is therefore the following interpolating self-energy ansatz [21]:

$$\Sigma_{d\sigma}(E) = U n_{d-\sigma} + \frac{a_\sigma U^2 \Sigma_{d\sigma}^{(SOC)}(E)}{1 - b_\sigma U^2 \Sigma_{d\sigma}^{(SOC)}(E)} \tag{8.348}$$

A high-energy expansion for the SIAM, analogous to the one for the Hubbard model in Sect. 8.4.7, is used in order to fix the unknown parameters a_σ and b_σ. The explicit calculation shows that the first two spectral moments are automatically satisfied by the ansatz so that the third and fourth moments are needed for fixing the parameters. One finds

$$a_\sigma = \frac{n_{d-\sigma}(1 - n_{d-\sigma})}{n_{d-\sigma}^{(1)}(1 - n_{d-\sigma}^{(1)})} \tag{8.349}$$

$$b_\sigma = \frac{B_{d-\sigma} - B_{d-\sigma}^{(1)} - (\mu - \tilde{\mu}) + U(1 - 2n_{d-\sigma})}{U^2 n_{d-\sigma}^{(1)}(1 - n_{d-\sigma}^{(1)})} \tag{8.350}$$

Here $B_{d\sigma}$ is a spin-dependent band shift,

$$n_{d\sigma}(1 - n_{d\sigma})(B_{d\sigma} - \varepsilon_d) = \sum_{\mathbf{k}} V_{\mathbf{k}d} \langle c_{\mathbf{k}\sigma}^\dagger c_{d\sigma}(2n_{d-\sigma} - 1)\rangle \tag{8.351}$$

Fig. 8.38 Hartree–Fock–Dyson equation for the impurity Green's function of the single-impurity Anderson model

completely analogous to the one for the Hubbard model (8.212). How this higher correlation function can be expressed by the *single-impurity* Green's function is shown in Problem 8.17:

$$n_{d\sigma}(1-n_{d\sigma})(B_{d\sigma}-\varepsilon_d) = \frac{1}{\pi\hbar} Im \int dE f_-(E)\Delta(E-\mu) * \\ * \left(1-\frac{2}{U}\Sigma_{d\sigma}(E-\mu)\right) G_{d\sigma}(E-\mu) \quad (8.352)$$

$B^{(1)}_{d\sigma}$ is the *mean-field* analogue:

$$n^{(1)}_{d\sigma}(1-n^{(1)}_{d\sigma})(B^{(1)}_{d\sigma}-\varepsilon_d) = \left(2n^{(1)}_{d-\sigma}-1\right)\sum_{\mathbf{k}} V_{\mathbf{k}d}\langle c^{\dagger}_{\mathbf{k}\sigma}c_{d\sigma}\rangle^{(1)} \quad (8.353)$$

The occupation number $n^{(1)}_{d\sigma}$ is to be correspondingly interpreted. The parameter $\tilde{\mu}$ is further introduced in order to fulfil the Luttinger sum rule [22], about which we will not discuss further here. It is eventually fixed by the condition

$$n^{(1)}_{d\sigma} = n_{d\sigma} \quad (8.354)$$

The *modified perturbation theory* presented here has a number of advantages so that the approximation, inspite of the seemingly arbitrarily proposed ansatz (8.348), appears to be quite reliable.

1. Correct *weak-coupling* behaviour up to U^2 terms.
2. Correct Fermi liquid behaviour for arbitrary band occupations n (not investigated in detail here!).
3. Luttinger theorem ($T = 0$) is satisfied for band occupations not far from half-filling ($n = 1$).
4. Appearance of a *Kondo resonance* for low temperatures.
5. $W \to 0$ limit exact for all n.
6. Correct behaviour for strong coupling.
7. Spontaneous band ferromagnetism possible!

Figure 8.39 shows a typical result of the MPT for the local spectral density $1/\hbar S_{ii\sigma}(E) = \rho_\sigma(E+\mu)$, calculated for a hypercubic lattice with the density of states (8.250) and band occupation $n = 0.94$ for different temperatures. The energy units are so chosen that in (8.251) $t^* = 1/\sqrt{2}$ is valid. One recognizes clearly the two *strong-coupling* structures (*Hubbard bands*) separated from each other by about U. Striking however is a resonant quasiparticle structure at the chemical potential μ. It is very revealing to observe that the resonance begins to appear exactly when the static susceptibility starts to deviate from the Curie-like behaviour. This indicates a suppression of the average local moment due to the antiferromagnetic correlations of the itinerant electrons at the Fermi edge. The striking structure at μ, which did not appear in the other methods discussed so far, can therefore be interpreted as

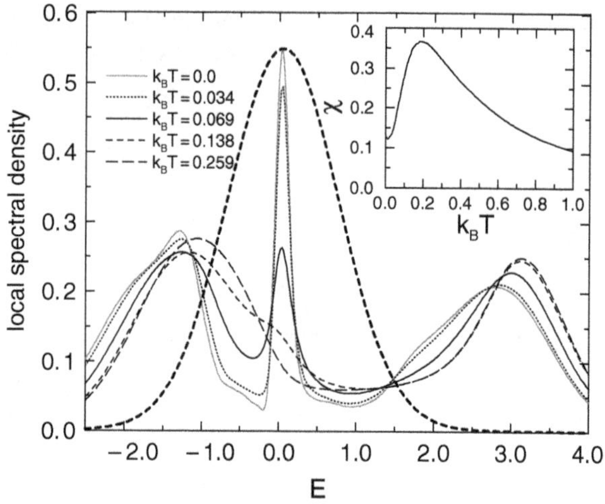

Fig. 8.39 Spectral density $1/\hbar S_{ii\sigma}(E) = \rho_\sigma(E + \mu)$ for a hypercubic lattice (8.250), $n = 0.94$ and U=4 (energy unit $t^* = 1/\sqrt{2}$ (8.251)) as function of the energy and for various temperatures T. (MPT, [21]). *Thin dashed line:* $\rho_0(E + \mu)$, $U = 0$. Inset: Static susceptibility χ as function of the temperature

Kondo-type resonance. Apparently this can be captured only by such theories of the Hubbard model which possess the correct *weak-coupling behaviour*. For temperatures at which χ has the Curie-like behaviour (e.g. $k_BT = 0.259$ in Fig. 8.39) there is no Kondo resonance at the Fermi edge. On the other hand with decreasing temperatures, the resonance grows larger approaching the value of the *free* density of states (local spectral density) at $T = 0$. Although not discussed in detail here, it means according to [22] that the theory (MPT) satisfies the important Luttinger sum rule: $S_{ii\sigma}(E = 0) \stackrel{!}{=} S^{(0)}_{ii\sigma}(E = 0)$ $(d = \infty)$. Assessing from the number of the exact results to the Hubbard model reproduced, the MPT appears to be an optimal analytical approach to the many-body problem of the Hubbard model. In the next section, we want to compare the results of MPT for band ferromagnetism with those of the theories discussed earlier.

8.5.10 Curie Temperature, Magnetization and Static Susceptibility

All the theories discussed in the earlier sections exclude spontaneous band ferromagnetism for the hypercubic lattice with the density of states (8.250) shown in Fig. 8.39. It appears that a distinctly more asymmetric density of states is necessary for collective magnetism. Therefore, we take a $d = 3$ fcc density of states generalized to $d = \infty$, which was also used in the Quantum Monte Carlo calculation in [20]:

$$\rho_0(E) = \frac{\exp\left(-\frac{1}{2}\left(1 - \frac{\sqrt{2}E}{t^*}\right)\right)}{t^*\sqrt{\pi\left(1 - \frac{\sqrt{2}E}{t^*}\right)}} \tag{8.355}$$

The energy units are so fixed that it holds

$$t^* = t\sqrt{2d(d-1)} \overset{!}{=} 1 \tag{8.356}$$

For this density of states, all the theories presented predict ferromagnetism only for more than half-filled band. That is why in Fig. 8.40 by band *filling* we mean the hole concentration. The corresponding hole density of states is given from (8.355) by $t^* \to -t^*$. The figure shows the Curie temperature as a function of hole concentration n. The electron concentration in this case is $2 - n$. The results of the analytical methods SDA, MAA and MPT are compared with the numerically essentially exact Quantum Monte Carlo results which can hold as a yardstick of the reliability of the proposed theories. The T_C curves for all the methods look qualitatively similar. The role of the spin-dependent band shift $B_{-\sigma}$ (8.212) and (8.217), which was discussed earlier, becomes strikingly clear when we compare the results displayed for the individual theories with the results of the corresponding approximations which differ exactly by the missing of such a band shift from SDA, MAA and MPT. The counterpart of SDA in this sense is the Hubbard-I solution (Sect. 8.5.1) which allows ferromagnetism only for a very asymmetric density of states. For the density of states (8.355) it indeed has ferromagnetic solutions but in a very narrow range of n values. The band shift $B_{-\sigma}$, which takes finite values in SDA apparently causes a drastic increase of the ferromagnetic stability. It becomes still more clear if we compare the *modified alloy analogy* (MAA) with its $B_{-\sigma} \equiv 0$ counterpart. This is the conventional alloy analogy CAA (8.323) and (8.324) which excludes ferromagnetism in all cases, i.e. for all coupling strengths U and band occupations n [23].

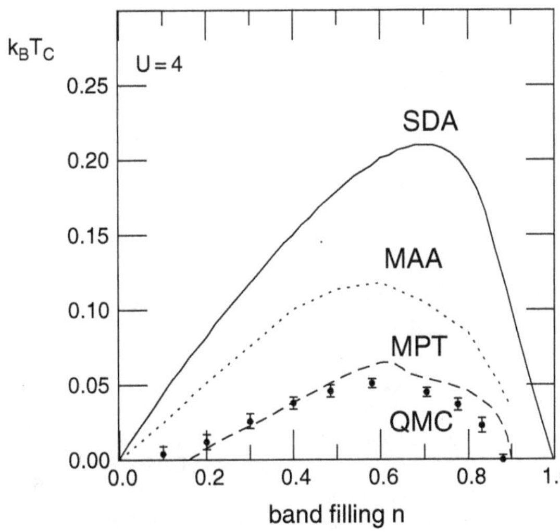

Fig. 8.40 Curie temperature as a function of band "filling" (hole concentration) for an fcc-type $d = \infty$-lattice (8.355). The points with error bars are the Quantum Monte Carlo results from [20]

The comparison of SDA and MAA gives information about the quasiparticle damping, which is present in MAA but not in SDA. The ferromagnetic coupling seems to be considerably weakened by the finite lifetimes of the quasiparticles. The Curie temperatures are distinctly lower. The correct capture of the *weak-coupling* behaviour in the MPT leads to a further reduction of T_C. The possible reason could be the tendency to screen the effective magnetic moments mediated by the Kondo resonance. The Curie temperatures of MPT are nearest to those of QMC. We conclude that the *strong-coupling* phenomenon of ferromagnetism is also influenced not inappreciably by the *weak-coupling* aspects.

Figure 8.41 demonstrates the qualitative equivalence of the theories of the Hubbard model that are discussed here, by considering the spontaneous magnetization (spin polarization of the band electrons) and the inverse static susceptibility as functions of reduced temperature T/T_C. All the magnetization curves can be well approximated by Brillouin function and reach saturation at $T = 0$ (a small deviation for MPT). There appear, at least for the parameters chosen, exclusively second order phase transitions at critical point. The static paramagnetic susceptibility follows for all theories the Curie–Weiss law: $\chi = C(T - \Theta)^{-1}$, where in all cases the paramag-

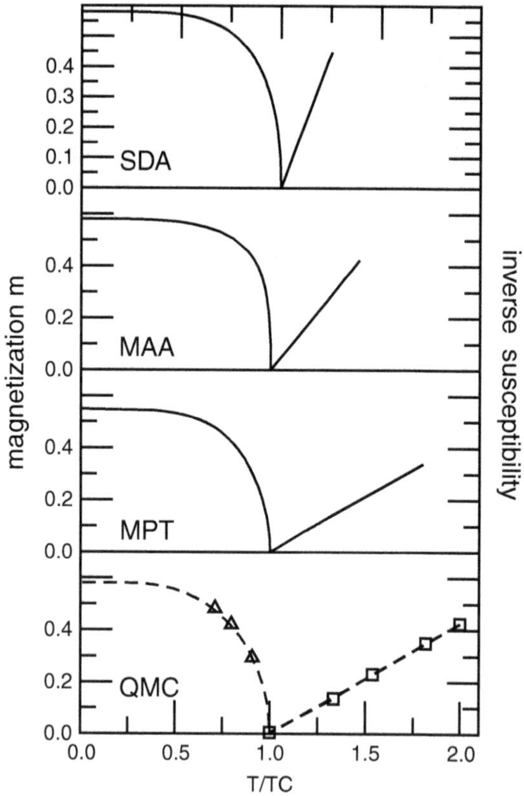

Fig. 8.41 Electron spin polarization m and inverse static susceptibility χ^{-1} as function of the reduced temperature T/T_C. *Bottom figure*: QMC results, *dashed line* for $T/T_C \leq 1$: $S = 1/2$-Brillouin function. Parameters: $U = 4, n = 0.58$

netic Curie temperature Θ coincides with the Curie temperature T_C. Even the Curie constant C itself is roughly the same (≈ 0.5) for the different theories.

The foregoing considerations have shown that band ferromagnetism is possible in the Hubbard model depending on the following central parameters:

lattice structure,
band occupation,
Coulomb interaction,
temperature.

Our comparison of many analytical procedures has shown as important factors for magnetic stability, in a positive sense, a self-consistent, spin-dependent band shift, which through an effective exchange splitting leads to a ferromagnetic phase, and in a negative sense, a finite lifetime of the quasiparticles which may drastically reduce the effective ferromagnetic coupling strength. An additional inclusion of the correct *weak-coupling* properties (*Fermi fluid*) eventually provides a convincing description of band ferromagnetism in the Hubbard model.

8.6 Problems

Problem 8.1 Consider non-interacting electrons in a periodic crystal potential:

$$H_0 = \sum_{ij\sigma} \sum_{\mu\nu} T_{ij}^{\mu\nu} c_{i\mu\sigma}^{+} c_{j\nu\sigma} \qquad \mu,\ \nu : \ band\ indices$$

After Fourier transformation for the (μ, ν)-elememt of the $(2l+1)\times(2l+1)$-matrix $\hat{T}_\mathbf{k}$ holds

$$T_\mathbf{k}^{\mu\nu} = \frac{1}{N_i} \sum_{i,j} T_{ij}^{\mu\nu} e^{-i\mathbf{k}(\mathbf{R}_i - \mathbf{R}_j)} \qquad N_i \ : \ number\ of\ lattice\ sites$$

The eigenstates of $\hat{T}_\mathbf{k}$

$$\hat{T}_\mathbf{k}\ |u_{m\mathbf{k}\sigma}\rangle = \varepsilon_m(\mathbf{k})\ |u_{m\mathbf{k}\sigma}\rangle$$

build the unitary matrix $\hat{U}_{\mathbf{k}\sigma}$, which diagonalizes $\hat{T}_\mathbf{k}$.

$$\hat{U}_{\mathbf{k}\sigma}\ \hat{T}_\mathbf{k}\ \hat{U}_{\mathbf{k}\sigma}^{\dagger} = \begin{pmatrix} \varepsilon_1(\mathbf{k}) & \dots & 0 & 0 \\ 0 & \varepsilon_2(\mathbf{k}) & \cdots & 0 \\ \vdots & & \ddots & \vdots \\ 0 & 0 & \dots & \varepsilon_{2l+1}(\mathbf{k}) \end{pmatrix}$$

Then the construction operators are transformed as follows:

$$c_{j\mu\sigma} = \frac{1}{\sqrt{N!}} \sum_{\mathbf{k}m} e^{i\mathbf{k}\cdot\mathbf{R}_j} \left(U_{\mathbf{k}\sigma}^{m\mu}\right)^* c_{\mathbf{k}m\sigma}$$

Then prove that

$$H_0 = \sum_{\mathbf{k}m\sigma} \varepsilon_m(\mathbf{k}) c_{\mathbf{k}m\sigma}^{\dagger} c_{\mathbf{k}m\sigma}$$

Problem 8.2 How does the Hubbard Hamiltonian appear in the Bloch representation?

Problem 8.3 Check whether the Stoner approximation (Sect. 8.3.1) correctly reproduces the exact limiting cases of the band limit ($U \to 0$) and the infinitely narrow band limit ($\varepsilon(\mathbf{k}) \to T_0 \ \forall \mathbf{k}$).

Problem 8.4 Consider a band antiferromagnet within the framework of the Stoner model:

$$H = \sum_{\mathbf{k}\sigma,\alpha,\beta} \varepsilon_{\sigma}^{\alpha\beta}(\mathbf{k}) c_{\mathbf{k}\alpha\sigma}^{\dagger} c_{\mathbf{k}\beta\sigma}$$

Let the antiferromagnet be made up of two ferromagnetically ordered sub-lattices A and B:

$$\begin{aligned}
\varepsilon_{\sigma}^{\alpha\alpha}(\mathbf{k}) &= \varepsilon(\mathbf{k}) + \frac{1}{2}Un - \frac{1}{2}z_{\sigma}Um_{\alpha} - \mu \\
\varepsilon_{\sigma}^{AB}(\mathbf{k}) &= t(\mathbf{k}) = \varepsilon_{\sigma}^{BA*}(\mathbf{k}) \\
m_A &= -m_B = m = n_{A\uparrow} - n_{A\downarrow} \\
n &= n_{\alpha\uparrow} + n_{\alpha\downarrow}
\end{aligned}$$

1. Calculate the quasiparticle energies. Show that they are independent of electron spin σ.
2. Determine the quasiparticle density of states!

Problem 8.5 Verify the following representation of the δ-function:

$$\delta(x) = \frac{1}{2} \lim_{\beta\to\infty} \frac{\beta}{1 + \cosh(\beta x)}, \qquad (\beta > 0)$$

Problem 8.6 Let

$$\sigma^{\alpha}(\mathbf{k}) = \sum_i \sigma_i^{\alpha} e^{-i\mathbf{k}\cdot\mathbf{R}_i}, \qquad (\alpha = \pm, z, x, y)$$

be the electron spin operators:

$$\sigma_i^z = \frac{1}{2}(n_{i\uparrow} - n_{i\downarrow}); \qquad \sigma_i^+ = c_{i\uparrow}^+ c_{i\downarrow}; \qquad \sigma_i^- = c_{i\downarrow}^+ c_{i\uparrow}$$

Calculate for the Hubbard model in a homogeneous external field $\mathbf{B}_0 = B_0 \mathbf{e}_z$ the following commutators:

1. $\left[\sigma^+(\mathbf{k})\,,\, H\right]_-$
2. $\left[\left[\sigma^+(\mathbf{k})\,,\, H\right]_-\,,\, \sigma^-(-\mathbf{k})\right]_-$

Problem 8.7 Calculate the chemical potential μ of the Hubbard model at half-filling in the zero-bandwidth limit (Sect. 8.4.2)!

Problem 8.8 Using (8.79) calculate the static susceptibility of the Hubbard model at half-filling ($n = 1$) in the zero-bandwidth limit.

Problem 8.9 Consider the two-site model (Sect. 8.4.3). The system contains exactly one σ-electron which can hop in either direction between the two sites. Calculate the energy eigenvalues and the eigenstates of the electron.

Problem 8.10 Let the two-site Hubbard model

$$H = t \sum_\sigma \left(c_{1\sigma}^\dagger c_{2\sigma} + c_{2\sigma}^\dagger c_{1\sigma} \right) + \frac{1}{2} U \sum_{i=1}^{2} \sum_\sigma n_{i\sigma} n_{i-\sigma}$$

be occupied by two electrons with opposite spins. Use the exact eigenstates for $t = 0$ as the basis states:

$$\left|\varepsilon_1^{(2)}\right\rangle = c_{1\sigma}^\dagger c_{2-\sigma}^\dagger \left|0\right\rangle$$
$$\left|\varepsilon_2^{(2)}\right\rangle = c_{2\sigma}^\dagger c_{1-\sigma}^\dagger \left|0\right\rangle$$
$$\left|\varepsilon_3^{(2)}\right\rangle = c_{1\sigma}^\dagger c_{1-\sigma}^\dagger \left|0\right\rangle$$
$$\left|\varepsilon_4^{(2)}\right\rangle = c_{2\sigma}^\dagger c_{2-\sigma}^\dagger \left|0\right\rangle \qquad \langle 0 \mid 0 \rangle = 1$$

1. Show that the states $\left|\varepsilon_i^{(2)}\right\rangle$ are orthonormal.
2. Calculate the Hamiltonian matrix

$$H_{ij}^{(2)} = \left\langle \varepsilon_i^{(2)} \right| H \left| \varepsilon_j^{(2)} \right\rangle; \qquad i, j = 1 \ldots 4$$

3. Calculate the eigenvalues and eigenstates of the two-site Hamiltonian!
4. Using the above results calculate the quasiparticle density of states (8.147):

$$\rho_{\sigma}^{(-\sigma)}(E) = \frac{1}{Z}\sum_{n,m}\left|\left\langle\varepsilon_m^{(2)}\right| c_{1\sigma}^{\dagger}\left|\varepsilon_n^{(1)}\right\rangle\right|^2 e^{-\beta E_n^{(1)}} *$$
$$*\delta\left(E - \left(E_m^{(2)} - E_n^{(1)}\right)\right)$$

Problem 8.11 1. Consider the two-site Hubbard model in the case of two σ-electrons and one $(-\sigma)$-electron. Show that

$$\left|E_1^{(3)}\right\rangle = \frac{1}{\sqrt{2}}\left(c_{1-\sigma}^{\dagger} - c_{2-\sigma}^{\dagger}\right) c_{1\sigma}^{\dagger} c_{2\sigma}^{\dagger}\left|0\right\rangle$$
$$\left|E_2^{(3)}\right\rangle = \frac{1}{\sqrt{2}}\left(c_{1-\sigma}^{\dagger} + c_{2-\sigma}^{\dagger}\right) c_{1\sigma}^{\dagger} c_{2\sigma}^{\dagger}\left|0\right\rangle$$

are the eigenstates and calculate the corresponding eigenvalues.
2. Calculate the density of states $\rho_{\sigma}^{(\sigma,-\sigma)}(E)$ for the case where the two-site model is occupied by one σ- and one $(-\sigma)$-elctron.

Problem 8.12 Particularly important for the possibility of ferromagnetism in Hubbard model is the spin-dependent band shift:

$$n_{-\sigma}(1 - n_{-\sigma})B_{-\sigma} = \frac{1}{N}\sum_{i,j}^{i\neq j} T_{ij}\left\langle c_{i-\sigma}^{\dagger} c_{j-\sigma}(2n_{i\sigma} - 1)\right\rangle$$

Show that this higher correlation $\left\langle c_{i-\sigma}^{\dagger} c_{j-\sigma}(2n_{i\sigma} - 1)\right\rangle$ is expressible in terms of the one-electron spectral density $S_{\mathbf{k}\sigma}(E)$.

Problem 8.13 Calculate the first four one-electron spectral moments of the Hubbard model. Compare your results with (8.221), (8.222), (8.223) and (8.224).

Problem 8.14 The determinant built by the one-electron spectral moments $M_{\mathbf{k}\sigma}^{(n)}$

$$\Delta_{\mathbf{k}\sigma}^{(r)} = \begin{vmatrix} M_{\mathbf{k}\sigma}^{(0)} & \cdots & M_{\mathbf{k}\sigma}^{(r)} \\ \vdots & & \vdots \\ M_{\mathbf{k}\sigma}^{(r)} & \cdots & M_{\mathbf{k}\sigma}^{(2r)} \end{vmatrix}$$

allows statements to be made about the structure of the one-electron spectral density [24]. That is exactly an n-pole function,

$$S_{\mathbf{k}\sigma}(E) = \hbar\sum_{i=1}^{n}\alpha_{\mathbf{k}\sigma}^{(i)}\delta(E + \mu - E_{i\sigma}(\mathbf{k}))$$

when

$$\Delta_{\mathbf{k}\sigma}^{(n)} = 0$$
$$\Delta_{\mathbf{k}\sigma}^{(r)} \neq 0 \quad for\ 0 \leq r \leq n-1$$

Using this theorem, prove that for the case of the Stoner model, the spectral density is a one-pole function.

Problem 8.15 Consider the Hubbard model in the limit of infinitely narrow band:

$$T_{ij} = T_0 \delta_{ij} \leftrightarrow \varepsilon(\mathbf{k}) = T_0 \ \forall \mathbf{k}$$

1. Show that for the one-electron spectral moments holds
$$M_{ii\sigma}^{(n)} = (T_0 - \mu)^n + \left[(T_0 + U - \mu)^n - (T_0 - \mu)^n \right] n_{-\sigma}$$
$$n = 0, 1, 2, \cdots$$
2. Use the theorem from Problem 8.14 to show that the one-electron spectral density is a two-pole function, i.e. a linear combination of two δ-functions.
3. Calculate the quasiparticle energies and their spectral weights.

Problem 8.16 Determine the Green's function

$$G_{d\sigma}(E) = \left\langle\!\left\langle c_{d\sigma}; c_{d\sigma}^\dagger \right\rangle\!\right\rangle$$

of the Anderson model (SIAM). Show that it holds

$$G_{d\sigma}(E) = \frac{\hbar}{E + \mu - \varepsilon_d - \Delta(E) - \Sigma_{d\sigma}(E)}$$

where $\Delta(E)$ is the hybridization function defined in (8.335) and $\Sigma_{d\sigma}(E)$ is the self-energy fixed by (8.334).

Problem 8.17 Show that the spin-dependent band shift of the Anderson impurity model

$$n_{d\sigma}(1 - n_{d\sigma})(B_{d\sigma} - \varepsilon_d) = \sum_{\mathbf{k}} V_{\mathbf{k}d} \left\langle c_{\mathbf{k}\sigma}^\dagger c_{d\sigma} (2n_{d-\sigma} - 1) \right\rangle$$

can be expressed in terms of the impurity Green's function $G_{d\sigma}$ as follows:

$$n_{d\sigma}(1 - n_{d\sigma})(B_{d\sigma} - \varepsilon_d) = -\frac{1}{\pi\hbar} Im \int_{-\infty}^{+\infty} dE\ f_-(E) \Delta(E - \mu) *$$
$$* \ G_{d\sigma}(E - \mu) \left(\frac{2}{U} \Sigma_{d\sigma}(E - \mu) - 1 \right)$$

The notations correspond to the ones in Sects. 8.5.8 and 8.5.9.

References

1. J. Hubbard, Proc. R. Soc. London **A276**, 238 (1963)
2. J. Kanamori, Progr. Theor. Phys. **A30**, 275 (1963)
3. M. C. Gutzwiller, Phys. Rev. Lett. **A10**, 59 (1963)
4. N. D. Mermin, H. Wagner, Phys. Rev. Lett. **17**, 1133 (1966)
5. A. B. Harris, R. V. Lange, Phys. Rev. **157**, 295 (1967)
6. W. Metzner, D. Vollhardt, Phys. Rev. Lett. **62**, 324 (1989)
7. E. Muller-Hartmann, Z. Phys. **74**, 507 (1987)
8. W. Nolting, W. Borgiel, Phys. Rev. B **39**, 6962 (1989)
9. L. M. Roth, Phys. Rev. **184**, 451 (1969)
10. T. Hermann, W. Nolting, J. Magn. Magn. Mat. **170**, 253 (1997)
11. T. Hermann, W. Nolting, Solid State Commun. **103**, 351 (1997)
12. J. Hubbard, Proc. R. Soc. London **A281**, 401 (1964)
13. H. Fukuyama, H. Ehernreich, Phys. Rev. B **7**, 3266 (1973)
14. R. Vlamming, D. Vollhardt, Phys. Rev. B **45**, 4637 (1992)
15. B. Velicky, S. Kirkpatrick, H. Ehrenreich, Phys. Rev. **175**, 747 (1968)
16. R. Jellitto, J. Phys. Chem. Solids **30**, 609 (1969)
17. G. Kotliar, D. Vollhardt, Phys. Today **57**, 53 (2004)
18. A. Georges, G. Kotliar, Phys. Rev. B **45**, 6479 (1992)
19. J. E. Hirsch, R. M. Fye, Phys. Rev. Lett. **56**, 2521 (1986)
20. M. Ulmke, Eur. Phys. J. B **1**, 301 (1998)
21. M. Potthoff, T. Wegner, W. Nolting, Phys. Rev. B **55**, 16132 (1997)
22. E. Müller-Hartmann, Z. Phys. **74**, 507 (1989)
23. J. Schneider, V. Drchal, Phys. Stat. Sol. (b) **68**, 207 (1975)
24. A. Lonke, J. Math. Phys. **12**, 2422 (1971)

Appendix A
Second Quantization

An exact description of an interacting many-body system requires the solving of the corresponding many-body Schrödinger equations. The formalism of second quantization leads to a substantial simplification in the description of such a many-body system, but it should be noted that it is only a reformulation of the original Schrödinger equation but not yet a solution. The essential step in the second quantization is the introduction of so-called *creation* and *annihilation* operators. By doing this, we eliminate the need for the laborious construction, respectively, of the symmetrized or the anti-symmetrized N-particle wavefunctions from the single-particle wavefunctions. The entire statistics is contained in *fundamental commutation relations* of these operators. Forces and interactions are expressed in terms of these "creation" and "annihilation" operators.

How does one handle an N-particle system? In case the particles are distinguishable, that is, if they are enumerable, then the method of description follows directly the general postulates of quantum mechanics:

$\mathcal{H}_1^{(i)}$: Hilbert space of the ith particle with the orthonormal basis $\{|\phi_\alpha^{(i)}\rangle\}$:

$$\langle\phi_\alpha^{(i)}|\phi_\beta^{(i)}\rangle = \delta_{\alpha\beta} \tag{A.1}$$

$\mathcal{H}_N$: Hilbert space of the N-particle system

$$\mathcal{H}_N = \mathcal{H}_1^{(1)} \otimes \mathcal{H}_1^{(2)} \otimes \cdots \otimes \mathcal{H}_1^{(N)} \tag{A.2}$$

with the basis $\{|\phi_N\rangle\}$:

$$|\phi_N\rangle = |\phi_{\alpha_1}^{(1)}\phi_{\alpha_2}^{(2)}\cdots\phi_{\alpha_N}^{(N)}\rangle = |\phi_{\alpha_1}^{(1)}\rangle\, |\phi_{\alpha_2}^{(2)}\rangle \cdots |\phi_{\alpha_N}^{(N)}\rangle \tag{A.3}$$

An arbitrary N-particle state $|\psi_N\rangle$,

$$|\psi_N\rangle = \int\sum_{\alpha_1\cdots\alpha_N} c(\alpha_1\cdots\alpha_N)|\phi_{\alpha_1}^{(1)}\phi_{\alpha_2}^{(2)}\cdots\phi_{\alpha_N}^{(N)}\rangle \tag{A.4}$$

underlies the same statistical interpretation as in the case of a 1-particle system. The dynamics of the N-particle system results from the formally unchanged Schrödinger equation:

W. Nolting, A. Ramakanth, *Quantum Theory of Magnetism*,
DOI 10.1007/978-3-540-85416-6_BM2, © Springer-Verlag Berlin Heidelberg 2009

$$i\hbar \frac{\partial |\psi_N\rangle}{\partial t} = \widehat{H}\ |\psi_N\rangle \tag{A.5}$$

The handling of the many-body problem in quantum mechanics, in the case of distinguishable particles, confronts exactly the same difficulties as in the classical physics, simply because of the large number of degrees of freedom. There are no extra, typically quantum mechanical complications.

A.1 Identical Particles

What are "identical particles"? To avoid misunderstandings let us strictly separate "particle properties" from "measured quantities of particle observables". "Particle properties" as e.g. mass, spin, charge, magnetic moment, etc. are in principle unchangeable intrinsic characteristics of the particle. The "measured values of particle observables" as e.g. position, momentum, angular momentum, spin projection, etc., on the other hand, can always change with time. We define "identical particles" in the quantum mechanical sense as particles which agree in all their particle properties. *Identical particles* are therefore particles, which behave exactly in the same manner under the same physical conditions, i.e. no measurement can differentiate one from the other. Identical particles also exist in classical mechanics. However, if the initial conditions are known, the state of the system for all times is determined by the Hamilton's equations of motion. Thus the particles are always identifiable! In quantum mechanics, on the contrary, there is the *principle of indistinguishability*, which says that, identical particles are intrinsically indistinguishable. In quantum mechanics, in some sense, identical particles lose their individuality. This originates from the fact that there do not exist sharp particle trajectories (only "spreading" wave packets!). The regions, where the probability of finding the particle is unequal to zero, overlap for different particles. Any question whose answer requires the observation of one single particle is physically meaningless.

We face now the problem, that, essentially for calculational purposes, we cannot avoid, to number the particles. Then, however, the numbering must be so that all physically relevant statements, that are made, are absolutely invariant under a change of this particle labelling. How to manage this is the subject of the following considerations.

We introduce the *permutation operator* $\mathcal{P}$, which, in the N-particle state, interchanges the particle indices:

$$\mathcal{P}|\phi_{\alpha_1}^{(1)}\phi_{\alpha_2}^{(2)}\cdots\phi_{\alpha_N}^{(N)}\rangle = |\phi_{\alpha_1}^{(i_1)}\phi_{\alpha_2}^{(i_2)}\cdots\phi_{\alpha_N}^{(i_N)}\rangle \tag{A.6}$$

Every permutation operator can be written as a product of *transposition operators* P_{ij}. On application of P_{ij} to a state of identical particles, results, according to the principle of indistinguishability, in a state, which, at the most, is different from the initial state by an unimportant phase factor $\lambda = exp(i\eta)$:

$$P_{ij}|\cdots\phi_{\alpha_i}^{(i)}\cdots\phi_{\alpha_j}^{(j)}\cdots\rangle = |\cdots\phi_{\alpha_i}^{(j)}\cdots\phi_{\alpha_j}^{(i)}\cdots\rangle$$

$$= \lambda\,|\cdots\phi_{\alpha_i}^{(i)}\cdots\phi_{\alpha_j}^{(j)}\cdots\rangle \tag{A.7}$$

Since $P_{ij}^2 = 1$, it is necessary that $\lambda = \pm 1$. Therefore, a system of identical particles must be either symmetric or antisymmetric against interchange of particles! This defines two different Hilbert spaces:

$\mathcal{H}_N^{(+)}$: the space of symmetric states $|\psi_N\rangle^{(+)}$

$$P_{ij}\,|\psi_N\rangle^{(+)} = |\psi_N\rangle^{(+)} \tag{A.8}$$

$\mathcal{H}_N^{(-)}$: the space of antisymmetric states $|\psi_N\rangle^{(-)}$

$$P_{ij}\,|\psi_N\rangle^{(-)} = -|\psi_N\rangle^{(-)} \tag{A.9}$$

In these spaces, P_{ij} are Hermitian and unitary!

What are the properties the observables must have for a system of identical particles? They must necessarily depend on the coordinates of *all* the particles

$$\widehat{A} = \widehat{A}(1,\ 2,\ \cdots,\ N) \tag{A.10}$$

and must commute with all the transpositions (permutations)

$$\left[P_{ij}\,,\ \widehat{A}\right]_- = 0 \tag{A.11}$$

This is valid specially for the Hamiltonian H and therefore also for the time evolution operator

$$U(t, t_0) = exp\left(-\frac{i}{\hbar}H(t-t_0)\right) \quad ; \quad (H \neq H(t)) \tag{A.12}$$

$$\left[P_{ij}\,,\ U\right]_- = 0 \tag{A.13}$$

That means the symmetry character of an N-particle state remains unchanged for all times!

Which Hilbert space out of $\mathcal{H}^{(+)}$ and $\mathcal{H}^{(-)}$ is applicable for which type of particles is established in relativistic quantum field theory. We will take over the *spin-statistics theorem* from there, without any proof.

$\mathcal{H}_{\mathbf{N}}^{(+)}$: Space of symmetric states of N identical particles with *integral spin*. These particles are called *Bosons*.

$\mathcal{H}_{\mathbf{N}}^{(-)}$: Space of antisymmetric states of N identical particles with *half-integral spin*. These particles are called *Fermions*.

A.2 Continuous Fock Representation

A.2.1 Symmetrized Many-Particle States

Let $\mathcal{H}_N^{(\varepsilon)}$ be the Hilbert space of a system of N identical particles. Here

$$\varepsilon = \begin{cases} + : \; Bosons \\ - : \; Fermions \end{cases} \tag{A.14}$$

Let $\widehat{\phi}$ be a 1-particle observable (or a set of 1-particle observables) with a *continuous* spectrum, ϕ_α being a particular eigenvalue corresponding to the 1-particle eigenstate $|\phi_\alpha\rangle$:

$$\widehat{\phi} \, |\phi_\alpha\rangle = \phi_\alpha \, |\phi_\alpha\rangle \tag{A.15}$$

$$\langle \phi_\alpha | \phi_\beta \rangle = \delta(\phi_\alpha - \phi_\beta) \;\; (\equiv \delta(\alpha - \beta)) \tag{A.16}$$

$$\int d\phi_\alpha |\phi_\alpha\rangle\langle\phi_\alpha| = \mathbb{1} \text{ in } \mathcal{H}_1 \tag{A.17}$$

A basis of $\mathcal{H}_N^{(\varepsilon)}$ is constructed from the following *(anti-) symmetrized N-particle states*:

$$|\phi_{\alpha_1} \cdots \phi_{\alpha_N}\rangle^{(\varepsilon)} = \frac{1}{N!} \sum_{\mathcal{P}} \varepsilon^p \, \mathcal{P} \left\{ |\phi_{\alpha_1}^{(1)}\rangle |\phi_{\alpha_2}^{(2)}\rangle \cdots |\phi_{\alpha_N}^{(N)}\rangle \right\} \tag{A.18}$$

p is the number of transpositions in $\mathcal{P}$. The summation runs over all the possible permutations $\mathcal{P}$. The sequence of the 1-particle states in the ket on the left-hand side of (A.18) is called the *standard ordering*. It is arbitrary, but has to be fixed right at the beginning. Interchange of two 1-particle symbols on the left means only a constant factor ε on the right.

One can easily prove the following relations for the above introduced N-particle states:

Scalar product:

$$\begin{aligned} &{}^{(\varepsilon)}\langle \phi_{\beta_1} \cdots | \phi_{\alpha_1} \cdots \rangle^{(\varepsilon)} \\ &= \frac{1}{N!} \sum_{\mathcal{P}_\alpha} \varepsilon^{p_\alpha} \mathcal{P}_\alpha \left\{ \delta(\phi_{\beta_1} - \phi_{\alpha_1}) \cdots \delta(\phi_{\beta_N} - \phi_{\alpha_N}) \right\} \end{aligned} \tag{A.19}$$

The index α shall indicate that $\mathcal{P}_\alpha$ permutes only the ϕ_α's.

Completeness relation:

$$\int \cdots \int d\phi_{\beta_1} \cdots d\phi_{\beta_N} |\phi_{\beta_1} \cdots \rangle^{(\varepsilon)(\varepsilon)} \langle \phi_{\beta_1} \cdots | = \mathbb{1} \tag{A.20}$$

This leads to a formal representation of an *observable of the N-particle system*, which will be used later a few times:

$$\widehat{A} = \int \cdots \int d\phi_{\alpha_1} \cdots d\phi_{\alpha_N}\, d\phi_{\beta_1} \cdots d\phi_{\beta_N} \, * \\ * \, |\phi_{\alpha_1} \cdots\rangle^{(\varepsilon)}\, {}^{(\varepsilon)}\langle \phi_{\alpha_1} \cdots |\, \widehat{A}\, |\phi_{\beta_1} \cdots\rangle^{(\varepsilon)}\, {}^{(\varepsilon)}\langle \phi_{\beta_1} \cdots | \tag{A.21}$$

A.2.2 Construction Operators

We want to build up the basis states of $\mathcal{H}_N^{(\varepsilon)}$, step by step, from the vacuum state $|0\rangle$ ($\langle 0|0\rangle = 1$) with the help of the operator

$$c^\dagger_{\phi_\alpha} \equiv c^\dagger_\alpha$$

These operators are defined by their action on the states:

$$c^\dagger_{\alpha_1}|0\rangle = \sqrt{1}|\phi_{\alpha_1}\rangle^{(\varepsilon)} \quad \in \mathcal{H}_1^{(\varepsilon)} \tag{A.22}$$

$$c^\dagger_{\alpha_2}|\phi_{\alpha_1}\rangle^{(\varepsilon)} = \sqrt{2}|\phi_{\alpha_2}\phi_{\alpha_1}\rangle^{(\varepsilon)} \quad \in \mathcal{H}_2^{(\varepsilon)} \tag{A.23}$$

Or, in general

$$c^\dagger_\beta \underbrace{|\phi_{\alpha_1} \cdots \phi_{\alpha_N}\rangle^{(\varepsilon)}}_{\in \mathcal{H}_N^{(\varepsilon)}} = \sqrt{N+1}\, \underbrace{|\phi_\beta \phi_{\alpha_1} \cdots \phi_{\alpha_N}\rangle^{(\varepsilon)}}_{\in \mathcal{H}_{N+1}^{(\varepsilon)}} \tag{A.24}$$

$c^\dagger_\beta$ is called the *creation operator*. It *creates* an extra particle in the N-particle state. The relation (A.24) is obviously reversible:

$$|\phi_{\alpha_1}\, \phi_{\alpha_2}\, \phi_{\alpha_3} \cdots \phi_{\alpha_N}\rangle^{(\varepsilon)} = \frac{1}{\sqrt{N!}} c^\dagger_{\alpha_1}\, c^\dagger_{\alpha_2} \cdots c^\dagger_{\alpha_N}|0\rangle \tag{A.25}$$

The N-particle state $|\phi_{\alpha_i} \cdots\rangle^{(\varepsilon)}$ can be built up from the vacuum state $|0\rangle$ by applying a sequence of N creation operators, where the order of the operators is to be strictly obeyed.

For a product of two creation operators, it follows from (A.24)

$$c^\dagger_{\alpha_1}\, c^\dagger_{\alpha_2}|\phi_{\alpha_3} \cdots \phi_{\alpha_N}\rangle^{(\varepsilon)} = \sqrt{N(N-1)}|\phi_{\alpha_1}\, \phi_{\alpha_2}\, \phi_{\alpha_3} \cdots \phi_{\alpha_N}\rangle^{(\varepsilon)} \tag{A.26}$$

If the sequence of the operators is reversed, then we have

$$\begin{aligned} & c^\dagger_{\alpha_2}\, c^\dagger_{\alpha_1}|\phi_{\alpha_3} \cdots \phi_{\alpha_N}\rangle^{(\varepsilon)} \\ &= \sqrt{N(N-1)}|\phi_{\alpha_2}\, \phi_{\alpha_1}\, \phi_{\alpha_3} \cdots \phi_{\alpha_N}\rangle^{(\varepsilon)} \\ &= \varepsilon\sqrt{N(N-1)}|\phi_{\alpha_1}\, \phi_{\alpha_2}\, \phi_{\alpha_3} \cdots \phi_{\alpha_N}\rangle^{(\varepsilon)} \end{aligned} \tag{A.27}$$

The last step follows because of (A.18). Since the states in (A.26) and (A.27) are basis states, by comparing, we can therefore read off the following operator identity:

$$\left[c^{\dagger}_{\alpha_1}\,,\,c^{\dagger}_{\alpha_2}\right]_{-\varepsilon} = c^{\dagger}_{\alpha_1}\,c^{\dagger}_{\alpha_2} - \varepsilon\,c^{\dagger}_{\alpha_2}c^{\dagger}_{\alpha_1} = 0 \tag{A.28}$$

The creation operators commute in the case of Bosons ($\varepsilon = +1$) and anticommute ($\varepsilon = -1$) in the case of Fermions.

We now consider the operator c_α which is the adjoint of $c^{\dagger}_{\alpha}$. Because of (A.24) and (A.25), we can write

$$\begin{aligned} {}^{(\varepsilon)}\langle\phi_{\alpha_1}\cdots\phi_{\alpha_N}|\,c_\gamma &= \sqrt{N+1}\,{}^{(\varepsilon)}\langle\phi_\gamma\,\phi_{\alpha_1}\cdots\phi_{\alpha_N}| \\ {}^{(\varepsilon)}\langle\phi_{\alpha_1}\cdots\phi_{\alpha_N}| &= \frac{1}{\sqrt{N!}}\,\langle 0|\,c_{\alpha_N}\cdots c_{\alpha_2}\,c_{\alpha_1} \end{aligned} \tag{A.29}$$

The meaning of c_γ is made clear by the following consideration:

$$\begin{aligned} &\underbrace{{}^{(\varepsilon)}\langle\phi_{\beta_2}\cdots\phi_{\beta_N}}_{\in\mathcal{H}^{(\varepsilon)}_{N-1}}|c_\gamma|\underbrace{\phi_{\alpha_1}\cdots\phi_{\alpha_N}\rangle^{(\varepsilon)}}_{\in\mathcal{H}^{(\varepsilon)}_{N}} \\ &= \sqrt{N}\,{}^{(\varepsilon)}\langle\phi_\gamma\phi_{\beta_2}\cdots\phi_{\beta_N}|\phi_{\alpha_1}\cdots\phi_{\alpha_N}\rangle^{(\varepsilon)} \\ &= \frac{\sqrt{N}}{N!}\sum_{\mathcal{P}_\alpha}\varepsilon^{p_\alpha}\mathcal{P}_\alpha\left\{\delta(\phi_\gamma-\phi_{\alpha_1})\delta(\phi_{\beta_2}-\phi_{\alpha_2})\cdots\right. \\ &\qquad\left.\cdots\,\delta(\phi_{\beta_N}-\phi_{\alpha_N})\right\} \end{aligned}$$

Here we have used in the first step (A.29) and in the second step (A.19). We can further rewrite the right-hand side

$$\begin{aligned} {}^{(\varepsilon)}\langle\cdots|c_\gamma|\cdots\rangle^{(\varepsilon)} &= \frac{1}{\sqrt{N}}\frac{1}{(N-1)!}\,* \\ *&\left\{\delta(\phi_\gamma-\phi_{\alpha_1})\sum_{\mathcal{P}_\alpha}\varepsilon^{p_\alpha}\mathcal{P}_\alpha(\delta(\phi_{\beta_2}-\phi_{\alpha_2})\cdots\delta(\phi_{\beta_N}-\phi_{\alpha_N}))\right. \\ &+\varepsilon\delta(\phi_\gamma-\phi_{\alpha_2})\sum_{\mathcal{P}_\alpha}\varepsilon^{p_\alpha}\mathcal{P}_\alpha(\delta(\phi_{\beta_2}-\phi_{\alpha_1})\delta(\phi_{\beta_3}-\phi_{\alpha_3})\cdots\delta(\phi_{\beta_N}-\phi_{\alpha_N})) \\ &+\cdots+ \\ &+\varepsilon^{N-1}\delta(\phi_\gamma-\phi_{\alpha_N})\sum_{\mathcal{P}_\alpha}\varepsilon^{p_\alpha}\mathcal{P}_\alpha(\delta(\phi_{\beta_2}-\phi_{\alpha_1})\delta(\phi_{\beta_3}-\phi_{\alpha_2}) \\ &\quad\left.\cdots\delta(\phi_{\beta_N}-\phi_{\alpha_{N-1}}))\right\} \\ &=\frac{1}{\sqrt{N}}\left\{\delta(\phi_\gamma-\phi_{\alpha_1})^{(\varepsilon)}\langle\phi_{\beta_2}\cdots\phi_{\beta_N}|\phi_{\alpha_2}\cdots\phi_{\alpha_N}\rangle^{(\varepsilon)}\right. \\ &+\varepsilon\delta(\phi_\gamma-\phi_{\alpha_2})^{(\varepsilon)}\langle\phi_{\beta_2}\cdots\phi_{\beta_N}|\phi_{\alpha_1}\phi_{\alpha_3}\cdots\phi_{\alpha_N}\rangle^{(\varepsilon)} \\ &+\cdots+ \\ &\left.+\varepsilon^{N-1}\delta(\phi_\gamma-\phi_{\alpha_N})^{(\varepsilon)}\langle\phi_{\beta_2}\cdots\phi_{\beta_N}|\phi_{\alpha_1}\phi_{\alpha_2}\cdots\phi_{\alpha_{N-1}}\rangle^{(\varepsilon)}\right\} \end{aligned}$$

With this, the action of c_γ is clear, since ${}^{(\varepsilon)}\langle\phi_{\beta_2}\cdots\phi_{\beta_N}|$ is an arbitrary $(N-1)$-particle bra basis-state:

$$\begin{aligned}
&c_\gamma|\phi_{\alpha_1}\cdots\phi_{\alpha_N}\rangle^{(\varepsilon)} = \\
&= \frac{1}{\sqrt{N}}\{\delta(\phi_\gamma-\phi_{\alpha_1})|\phi_{\alpha_2}\cdots\phi_{\alpha_N}\rangle^{(\varepsilon)} + \\
&+\cdots+ \\
&+\varepsilon^{N-1}\delta(\phi_\gamma-\phi_{\alpha_N})|\phi_{\alpha_1}\cdots\phi_{\alpha_{N-1}}\rangle^{(\varepsilon)}\}
\end{aligned} \tag{A.30}$$

c_γ *annihilates* a particle in the state $|\phi_\gamma\rangle$ and that is why it is called the *annihilation operator*. From (A.28), it immediately follows that

$$\left[c_{\alpha_1}\,,\,c_{\alpha_2}\right]_{-\varepsilon} = -\varepsilon\left(\left[c^\dagger_{\alpha_1}\,,\,c^\dagger_{\alpha_2}\right]_{-\varepsilon}\right)^\dagger = 0 \tag{A.31}$$

The annihilation operators commute in the case of Bosons ($\varepsilon=-1$) and anticommute ($\varepsilon=+1$) in the case of Fermions.

From (A.24) and (A.30) one can show that (Problem A.1)

$$(c_\beta\,c^\dagger_\gamma - \varepsilon c^\dagger_\gamma\,c_\beta)|\phi_{\alpha_1}\cdots\phi_{\alpha_N}\rangle^{(\varepsilon)} = \delta(\phi_\beta-\phi_\gamma)|\phi_{\alpha_1}\cdots\phi_{\alpha_N}\rangle^{(\varepsilon)}$$

From this it follows that

$$\left[c_\beta\,,\,c^\dagger_\gamma\right]_{-\varepsilon} = \delta(\phi_\beta-\phi_\gamma) \tag{A.32}$$

(A.28), (A.31) and (A.32) are the three *fundamental commutation rules* of the construction operators a_γ and $a^\dagger_\gamma$.

A.2.3 Many-Body Operators

We start with the formal ("spectral") representation of the N-particle observable $\widehat{A}$ as given in (A.21). Using (A.25) and (A.29), we can write ($d\phi_{\alpha_i}\equiv d\alpha_i$)

$$\begin{aligned}
\widehat{A} &= \frac{1}{N!}\int\cdots\int d\alpha_1\cdots d\alpha_N d\beta_1\cdots d\beta_N \,* \\
&* \,c^\dagger_{\alpha_1}\cdots\,c^\dagger_{\alpha_N}|0\rangle\,{}^{(\varepsilon)}\langle\phi_{\alpha_1}\cdots|\widehat{A}|\phi_{\beta_1}\cdots\rangle^{(\varepsilon)}\langle 0|c_{\beta_N}\cdots c_{\beta_1}
\end{aligned} \tag{A.33}$$

Normally, such an operator consists of 1-particle and 2-particle parts:

$$\widehat{A} = \sum_{i=1}^{N}\widehat{A}^{(1)}_i + \frac{1}{2}\sum_{i,j}^{i\neq j}\widehat{A}^{(2)}_{ij} \tag{A.34}$$

Using this, we will calculate the matrix elements in (A.33). We start with the 1-particle part:

$$
{}^{(\varepsilon)}\langle\phi_{\alpha_1}\cdots|\sum_{i=1}^{N}\widehat{A}_i^{(1)}|\phi_{\beta_1}\cdots\rangle^{(\varepsilon)} = \frac{1}{(N!)^2}\sum_{\mathcal{P}_\alpha}\sum_{\mathcal{P}_\beta}\varepsilon^{p_\alpha+p_\beta} *
$$

$$
* \left\{\langle\phi_{\alpha_N}^{(N)}|\cdots\langle\phi_{\alpha_1}^{(1)}|\right\}\left(\mathcal{P}_\alpha^\dagger\sum_{i=1}^{N}\widehat{A}_i^{(1)}\mathcal{P}_\beta\right)\left\{|\phi_{\beta_1}^{(N)}\rangle\cdots|\phi_{\beta_N}^{(N)}\rangle\right\} \tag{A.35}
$$

One can easily see that every summand in the double sum $\sum_{\mathcal{P}_\alpha}\sum_{\mathcal{P}_\beta}$ gives the same contribution, since every permuted ordering of $\{|\phi_\alpha\rangle\}$ or $\{|\phi_\beta\rangle\}$ can be restored to the standard ordering by renumbering the integration variables in (A.33). In order to bring back the creation and annihilation operators, which have been reindexed in the process, into the "correct" sequence, we require, according to (A.28) and (A.31), a factor $\varepsilon^{p_\alpha+p_\beta}$, which along with the corresponding factor in the above equation (A.35) gives a factor $+1$. Thus for (A.33), we need (A.35) only in the following simplified form:

$$
{}^{(\varepsilon)}\langle\phi_{\alpha_1}\cdots|\sum_{i=1}^{N}\widehat{A}_i^{(1)}|\phi_{\beta_1}\cdots\rangle^{(\varepsilon)} \Rightarrow
$$

$$
\left\{\langle\phi_{\alpha_N}^{(N)}|\cdots\langle\phi_{\alpha_1}^{(1)}|\right\}\sum_{i=1}^{N}\widehat{A}_i^{(1)}\left\{|\phi_{\beta_1}^{(1)}\rangle\cdots|\phi_{\beta_N}^{(N)}\rangle\right\} \tag{A.36}
$$

Substituting this in (A.33), we get

$$
\begin{aligned}
\sum_{i=1}^{N}\widehat{A}_i^{(1)} &= \frac{1}{N!}\int\cdots\int d\alpha_1\cdots d\beta_N c_{\alpha_1}^\dagger\cdots c_{\alpha_N}^\dagger|0\rangle * \\
&* \left\{\langle\phi_{\alpha_1}^{(1)}|\widehat{A}_1^{(1)}|\phi_{\beta_1}^{(1)}\rangle\langle\phi_{\alpha_2}^{(2)}|\phi_{\beta_2}^{(2)}\rangle\cdots\langle\phi_{\alpha_N}^{(N)}|\phi_{\beta_N}^{(N)}\rangle\right. \\
&+ \left.\langle\phi_{\alpha_1}^{(1)}|\phi_{\beta_1}^{(1)}\rangle\langle\phi_{\alpha_2}^{(2)}|\widehat{A}_2^{(1)}|\phi_{\beta_2}^{(2)}\rangle\cdots\langle\phi_{\alpha_N}^{(N)}|\phi_{\beta_N}^{(N)}\rangle+\cdots\right\} * \\
&* \langle 0|c_{\beta_N}\cdots c_{\beta_1} \\
&= \frac{1}{N!}\int\cdots\int d\alpha_1\cdots d\alpha_N c_{\alpha_1}^\dagger\cdots c_{\alpha_N}^\dagger|0\rangle * \\
&* \left\{\int d\beta_1\langle\phi_{\alpha_1}^{(1)}|\widehat{A}_1^{(1)}|\phi_{\beta_1}^{(1)}\rangle\langle 0|c_{\alpha_N}\cdots c_{\alpha_2}c_{\beta_1}\right. \\
&+ \left.\int d\beta_2\langle\phi_{\alpha_2}^{(2)}|\widehat{A}_2^{(1)}|\phi_{\beta_2}^{(2)}\rangle\langle 0|c_{\alpha_N}\cdots c_{\alpha_3}c_{\beta_2}c_{\alpha_1}+\cdots\right\}
\end{aligned}
$$

$$= \frac{1}{N} \int \int d\alpha_1 d\beta_1 \langle \phi^{(1)}_{\alpha_1} | \widehat{A}^{(1)}_1 | \phi^{(1)}_{\beta_1} \rangle a^{\dagger}_{\alpha_1} \left\{ \frac{1}{(N-1)!} \right.$$

$$\left. \int \cdots \int d\alpha_2 \cdots d\alpha_N c^{\dagger}_{\alpha_2} \cdots c^{\dagger}_{\alpha_N} |0\rangle \langle 0| c_{\alpha_N} \cdots c_{\alpha_2} \right\} c_{\beta_1}$$

$$+ \frac{1}{N} \int \int d\alpha_2 d\beta_2 \langle \phi^{(2)}_{\alpha_2} | \widehat{A}^{(1)}_2 | \phi^{(2)}_{\beta_2} \rangle c^{\dagger}_{\alpha_2} \left\{ \frac{1}{(N-1)!} \right.$$

$$\int \cdots \int d\alpha_1 d\alpha_3 \cdots d\alpha_N c^{\dagger}_{\alpha_1} c^{\dagger}_{\alpha_3} \cdots c^{\dagger}_{\alpha_N} |0\rangle$$

$$\left. \langle 0| c_{\alpha_N} \cdots c_{\alpha_3} c_{\alpha_1} \right\} c_{\beta_2} \varepsilon^2 + \cdots \tag{A.37}$$

The factor ε^2 in the last line stems from the (anti)commutation $c^{\dagger}_{\alpha_2} \leftrightarrow c^{\dagger}_{\alpha_1}$ and $c^{\dagger}_{\alpha_1} \leftrightarrow c^{\dagger}_{\beta_2}$. Due to the analogous rearrangements all the other terms get a factor ε^{2m} which in any case is equal to $+1$. The term in each of the curly brackets is the identity for the Hilbert space $\mathcal{H}^{(\varepsilon)}_{N-1}$ as given in (A.20). Therefore what remains is

$$\sum_{i=1}^{N} \widehat{A}^{(1)}_i = \frac{1}{N} \sum_{i=1}^{N} \int \int d\alpha_i d\beta_i \langle \phi^{(i)}_{\alpha_i} | \widehat{A}^{(1)}_i | \phi^{(i)}_{\beta_i} \rangle c^{\dagger}_{\alpha_i} c_{\beta_i} \tag{A.38}$$

The matrix elements are naturally the same for all the N identical particles. Therefore, the factor $1/N$ and the summation cancel out:

$$\sum_{i=1}^{N} \widehat{A}^{(1)}_i = \int \int d\alpha d\beta \langle \phi_{\alpha} | \widehat{A}^{(1)} | \phi_{\beta} \rangle c^{\dagger}_{\alpha} c_{\beta} \tag{A.39}$$

The remaining matrix element is in general easy to calculate. On the right-hand side, we do not have the particle number any more. It is of course implicitly present because, in between $c^{\dagger}_{\alpha}$ and c_{β}, there appears in principle an identity $\mathbb{1}$, corresponding to the Hilbert space $\mathcal{H}^{(\varepsilon)}_{N-1}$.

For the two-particle part of the operator $\widehat{A}$, at first, exactly the same considerations are valid which led us from (A.35) to (A.36). Therefore, we can use in (A.33)

$${}^{(\varepsilon)}\langle \phi_{\alpha_1} \cdots | \frac{1}{2} \sum_{i,j}^{i \neq j} \widehat{A}^{(2)}_{ij} | \phi_{\beta_1} \cdots \rangle^{(\varepsilon)} \Rightarrow$$

$$\left\{ \langle \phi^{(N)}_{\alpha_N} | \cdots \langle \phi^{(1)}_{\alpha_1} | \right\} \frac{1}{2} \sum_{i,j}^{i \neq j} \widehat{A}^{(2)}_{ij} \left\{ | \phi^{(1)}_{\beta_1} \rangle \cdots | \phi^{(N)}_{\beta_N} \rangle \right\} \tag{A.40}$$

Substituting this we are left with

$$\frac{1}{2}\sum_{i,j}^{i\neq j}\widehat{A}_{ij}^{(2)} = \frac{1}{2}\frac{1}{N!}\int\cdots\int d\alpha_1\cdots d\beta_N c_{\alpha_1}^{\dagger}\cdots c_{\alpha_N}^{\dagger}|0\rangle *$$

$$* \left\{ \langle\phi_{\alpha_2}^{(2)}|\langle\phi_{\alpha_1}^{(1)}|\widehat{A}_{12}^{(2)}|\phi_{\beta_1}^{(1)}\rangle|\phi_{\beta_2}^{(2)}\rangle\langle\phi_{\alpha_3}^{(3)}|\phi_{\beta_3}^{(3)}\rangle\cdots\langle\phi_{\alpha_N}^{(N)}|\phi_{\beta_N}^{(N)}\rangle\right.$$

$$\left. + \cdots \right\}\langle 0|c_{\beta_N}\cdots c_{\beta_1}$$

$$= \frac{1}{2}\frac{1}{N!}\int\cdots\int d\alpha_1\cdots d\alpha_N d\beta_1 d\beta_2 c_{\alpha_1}^{\dagger}c_{\alpha_2}^{\dagger}\cdots c_{\alpha_N}^{\dagger}|0\rangle *$$

$$* \langle\phi_{\alpha_2}^{(2)}|\langle\phi_{\alpha_1}^{(1)}|\widehat{A}_{12}^{(2)}|\phi_{\beta_1}^{(1)}\rangle|\phi_{\beta_2}^{(2)}\rangle\langle 0|c_{\alpha_N}\cdots c_{\alpha_3}c_{\beta_2}c_{\beta_1} + \cdots$$

$$= \frac{1}{2N(N-1)}\int\cdots\int d\alpha_1 d\alpha_2 d\beta_1 d\beta_2 *$$

$$* \langle\phi_{\alpha_1}^{(1)}\phi_{\alpha_2}^{(2)}|\widehat{A}_{12}^{(2)}|\phi_{\beta_1}^{(1)}\phi_{\beta_2}^{(2)}\rangle c_{\alpha_1}^{\dagger}c_{\alpha_2}^{\dagger} *$$

$$* \left\{\frac{1}{(N-2)!}\int\cdots\int d\alpha_3\cdots d\alpha_N c_{\alpha_3}^{\dagger}\cdots c_{\alpha_N}^{\dagger}|0\rangle\right.$$

$$\left.\langle 0|c_{\alpha_N}\cdots c_{\alpha_3}\right\} c_{\beta_2}c_{\beta_1} + \cdots$$

The curly bracket now gives the identity in the Hilbert space $\mathcal{H}_{N-2}^{(\varepsilon)}$, so that we have

$$\frac{1}{2}\sum_{i,j}^{i\neq j}\widehat{A}_{ij}^{(2)} = \frac{1}{2N(N-1)}\sum_{i,j}^{i\neq j}\int\cdots\int d\alpha_i\, d\alpha_j\, d\beta_i\, d\beta_j \times$$

$$\times \langle\phi_{\alpha_i}^{(i)}\phi_{\alpha_j}^{(j)}|\widehat{A}_{ij}^{(2)}|\phi_{\beta_i}^{(i)}\phi_{\beta_j}^{(j)}\rangle\, c_{\alpha_i}^{\dagger}\, c_{\alpha_j}^{\dagger}\, c_{\beta_j}\, c_{\beta_i} \tag{A.41}$$

In a system of identical particles, naturally, all the summands on the right-hand side give identical contributions. Therefore, we have

$$\frac{1}{2}\sum_{i,j}^{i\neq j}\widehat{A}_{ij}^{(2)} = \frac{1}{2}\int\cdots\int d\alpha_1\, d\alpha_2\, d\beta_1\, d\beta_2 *$$

$$* \langle\phi_{\alpha_1}^{(1)}\phi_{\alpha_2}^{(2)}|\widehat{A}_{12}^{(2)}|\phi_{\beta_1}^{(1)}\phi_{\beta_2}^{(2)}\rangle\, c_{\alpha_1}^{\dagger}\, c_{\alpha_2}^{\dagger}\, c_{\beta_2}\, c_{\beta_1} \tag{A.42}$$

The remaining matrix element on the right-hand side can be built either with unsymmetrized two-particle states

$$\langle\phi_{\alpha_1}\phi_{\alpha_2}| = \langle\phi_{\alpha_1}|\langle\phi_{\alpha_2}| \quad ; \quad |\phi_{\beta_1}\phi_{\beta_2}\rangle = |\phi_{\beta_1}\rangle|\phi_{\beta_2}\rangle \tag{A.43}$$

or also with the symmetrized states

$$|\phi_{\beta_1}\phi_{\beta_2}\rangle^{(\varepsilon)} = \frac{1}{2!}\left\{|\phi_{\beta_1}^{(1)}\rangle|\phi_{\beta_2}^{(2)}\rangle + \varepsilon|\phi_{\beta_1}^{(2)}\rangle|\phi_{\beta_2}^{(1)}\rangle\right\} \tag{A.44}$$

What have we achieved? We have, through (A.25) and (A.29), replaced the laborious building up of symmetrized products from the single-particle states, by the application of products of construction operators to the vacuum state $|0\rangle$. The operation is quite simple, for example,

$$c_\alpha|0\rangle = 0 \tag{A.45}$$

and the whole statistics is taken care by the fundamental commutation relations (A.28), (A.31) and (A.32). The N-particle observables can also be expressed by the construction operators, (A.39) and (A.42) where the remaining matrix elements can be, usually, easily calculated with single-particle states. Note that the choice of the single-particle basis $\{|\phi_\alpha\rangle\}$ is absolutely arbitrary, a great advantage with respect to practical purposes. We will demonstrate this in (A.4) with a few examples.

The theory developed so far is valid for continuous as well as discrete single-particle spectra. Only δ-functions have to be replaced by Kronecker-δs and integrals by summations in the case of discrete spectra.

A.3 Discrete Fock Representation (Occupation Number Representation)

Let $\mathcal{H}_N^{(\varepsilon)}$ be again the Hilbert space of a system of N identical particles. Now, let $\widehat{\phi}$ be a single-particle observable with *discrete* spectrum. In principle, the same considerations are valid as in (A.2).

A.3.1 Symmetrized Many-Particle States

We will use the following *(anti-) symmetrized N-particle states* as the basis of $\mathcal{H}_N^{(\varepsilon)}$:

$$|\phi_{\alpha_1}\cdots\phi_{\alpha_N}\rangle^{(\varepsilon)} = C_\varepsilon \sum_{\mathcal{P}} \varepsilon^p \mathcal{P}\left\{|\phi_{\alpha_1}^{(1)}\rangle\cdots|\phi_{\alpha_N}^{(N)}\rangle\right\} \tag{A.46}$$

Up to the still to be determined normalization constant C_ε, this definition agrees with the corresponding definition (A.18) of the continuous case. However, now for the single-particle states, we have

$$\langle\phi_\alpha|\phi_\beta\rangle = \delta_{\alpha\beta} \quad ; \quad \sum_\alpha |\phi_\alpha\rangle\langle\phi_\alpha| = \mathbb{1} \text{ in } \mathcal{H}_1 \tag{A.47}$$

One can see that (A.46) can be written for Fermions ($\varepsilon = -1$), as a determinant

$$|\phi_{\alpha_1}\cdots\phi_{\alpha_N}\rangle^{(-)} = C_- \begin{vmatrix} |\phi^{(1)}_{\alpha_1}\rangle & |\phi^{(2)}_{\alpha_1}\rangle & \cdots & |\phi^{(N)}_{\alpha_1}\rangle \\ |\phi^{(1)}_{\alpha_2}\rangle & |\phi^{(2)}_{\alpha_2}\rangle & \cdots & |\phi^{(N)}_{\alpha_2}\rangle \\ \cdot & \cdot & \cdot & \cdot \\ \cdot & \cdot & \cdot & \cdot \\ \cdot & \cdot & \cdot & \cdot \\ |\phi^{(1)}_{\alpha_N}\rangle & |\phi^{(2)}_{\alpha_N}\rangle & \cdots & |\phi^{(N)}_{\alpha_N}\rangle \end{vmatrix} \tag{A.48}$$

This is known as the *Slater determinant*.

In case two sets of quantum numbers are equal, say, $\alpha_i = \alpha_j$, then, two rows of the Slater determinant are identical, which means, the determinant is equal to zero. Consequently, the probability that such a situation exists for a system of identical Fermions is zero. This is exactly the statement of the *Pauli's principle!*

We define

n_i = *occupation number*, i.e. the frequency with which the state $|\phi_{\alpha_i}\rangle$ appears in the N-particle state $|\phi_{\alpha_1}\cdots\phi_{\alpha_N}\rangle^{(-)}$.

Naturally, we have

$$\sum_i n_i = N \tag{A.49}$$

where the values n_i can be

$$\begin{aligned} n_i &= 0,\ 1 && \mathit{for\ Fermions} \\ n_i &= 0,\ 1,\ 2,\ \cdots && \mathit{for\ Bosons} \end{aligned} \tag{A.50}$$

First, we want to fix the normalization constant, which we assume to be real:

$$\begin{aligned} 1 &\stackrel{!}{=} {}^{(\varepsilon)}\langle\phi_{\alpha_1}\cdots|\phi_{\alpha_1}\cdots\rangle^{(\varepsilon)} \\ &= C_\varepsilon^2 \sum_{\mathcal{P}}\sum_{\mathcal{P}'} \varepsilon^{p+p'} \left\{\langle\phi^{(N)}_{\alpha_N}|\cdots\right\} \mathcal{P}^\dagger\, \mathcal{P}' \left\{|\phi^{(1)}_{\alpha_1}\rangle\cdots\right\} \end{aligned} \tag{A.51}$$

In the case of Fermions ($\varepsilon = -1$), every state is occupied only once. Therefore, in the sum, only the terms with $\mathcal{P} = \mathcal{P}'$ are unequal to zero and due to (A.47), each term is exactly equal to 1. Therefore, we get

$$C_- = \frac{1}{\sqrt{N!}} \quad (\mathit{Fermions}) \tag{A.52}$$

In the case of Bosons ($\varepsilon = +1$), the summands are then unequal to zero, when $\mathcal{P}$ differs from $\mathcal{P}'$ at the most by such transpositions for which only the groups of n_i identical single-particle states $|\phi_i\rangle$ are interchanged among themselves. There are naturally $n_i!$ such possibilities. With this argument, we get for Bosons

$$C_+ = (N!\, n_1!\, n_2! \cdots n_i! \cdots)^{-1/2} \quad (Bosons) \tag{A.53}$$

Obviously, a symmetrized basis state can be completely specified by giving the occupation numbers. This permits the representation by the *Fock states*

$$\begin{aligned} &|N;\, n_1\, n_2 \cdots n_i \cdots n_j \cdots\rangle^{(\varepsilon)} \equiv |\phi_{\alpha_1} \cdots \phi_{\alpha_N}\rangle^{(\varepsilon)} \\ &= C_\varepsilon \sum_{\mathcal{P}} \varepsilon^P \mathcal{P} \left\{ \underbrace{|\phi_{\alpha_1}^{(1)}\rangle|\phi_{\alpha_1}^{(2)}\rangle \cdots}_{n_1} \underbrace{\cdots |\phi_{\alpha_i}^{(r)}\rangle|\phi_{\alpha_i}^{(r+1)}\rangle \cdots}_{n_i} \cdots \right\} \end{aligned} \tag{A.54}$$

One has to give *all* the occupation numbers, even those with $n_i = 0$. Completeness and orthonormality follow from the corresponding relations for the symmetrized states:

$$^{(\varepsilon)}\langle N; \cdots n_i \cdots | \tilde{N}; \cdots \tilde{n}_i \cdots \rangle^{(\varepsilon)} = \delta_{N\tilde{N}} \prod_i \delta_{n_i \tilde{n}_i} \tag{A.55}$$

$$\underbrace{\sum_{n_1} \sum_{n_2} \cdots \sum_{n_i} \cdots}_{(\sum_i n_i = N)} |N; \cdots n_i \cdots\rangle^{(\varepsilon)}\, {}^{(\varepsilon)}\langle N; \cdots n_i \cdots | = \mathbb{1} \tag{A.56}$$

A.3.2 Construction Operators

Up to normalization constants, we define the construction operators exactly in the same manner as we have done for the case of continuous spectrum in (A.2.2):

$$\begin{aligned} c_{\alpha_r}^\dagger |\tilde{N}; \cdots \tilde{n}_r \cdots\rangle^{(\varepsilon)} &= c_{\alpha_r}^\dagger |\phi_{\alpha_1} \cdots \phi_{\alpha_N}\rangle^{(\varepsilon)} \\ &= \sqrt{n_r + 1}\, |\phi_{\alpha_r} \phi_{\alpha_1} \cdots \phi_{\alpha_N}\rangle^{(\varepsilon)} \\ &= \varepsilon^{N_r} \sqrt{n_r + 1}\, |\phi_{\alpha_1} \cdots \phi_{\alpha_r} \cdots \phi_{\alpha_N}\rangle^{(\varepsilon)} \\ &= \varepsilon^{N_r} \sqrt{n_r + 1}\, |N + 1; \cdots n_r + 1 \cdots\rangle^{(\varepsilon)} \end{aligned} \tag{A.57}$$

Here, N_r is the number of transpositions necessary to bring the single-particle state $|\phi_{\alpha_r}\rangle$ to the "correct" place:

$$N_r = \sum_{i=1}^{r-1} n_i \tag{A.58}$$

Equation (A.57) does not, as yet, contain the Pauli's principle in a correct way. The operation of the so-called *creation operator* is precisely defined as follows.

$$\begin{aligned}
&\textit{Bosons}: \\
&c^{\dagger}_{\alpha_r}|N;\cdots n_r\cdots\rangle^{(+)} = \sqrt{n_r+1}\,|N+1;\cdots n_r+1\cdots\rangle^{(+)} \\
&\textit{Fermions}: \\
&c^{\dagger}_{\alpha_r}|N;\cdots n_r\cdots\rangle^{(-)} = (-1)^{N_r}\,\delta_{n_r,0}\,|N+1;\cdots n_r+1\cdots\rangle^{(-)}
\end{aligned} \tag{A.59}$$

Any N-particle state can be built up by repeated application of the creation operators on the vacuum state $|0\rangle$:

$$|N;n_1\,n_2\cdots\rangle^{(\varepsilon)} = \prod_p^{\sum n_p=N} \frac{1}{\sqrt{n_p!}}\left(c^{\dagger}_{\alpha_p}\right)^{n_p}\varepsilon^{N_p}|0\rangle \tag{A.60}$$

The *annihilation operator* is again defined as the adjoint of the creation operator:

$$c_{\alpha_r} = \left(c^{\dagger}_{\alpha_r}\right)^{\dagger} \tag{A.61}$$

The action of this operator becomes clear from the following:

$$\begin{aligned}
&{}^{(\varepsilon)}\langle N;\cdots n_r\cdots|c_{\alpha_r}|\bar{N};\cdots\bar{n}_r\cdots\rangle^{(\varepsilon)} \\
&= \varepsilon^{N_r}\sqrt{n_r+1}\;{}^{(\varepsilon)}\langle N+1;\cdots n_r+1\cdots|\bar{N};\cdots\bar{n}_r\cdots\rangle^{(\varepsilon)} \\
&= \varepsilon^{N_r}\sqrt{n_r+1}\,\delta_{N+1,\bar{N}}\left(\delta_{n_1,\bar{n}_1}\cdots\delta_{n_r+1,\bar{n}_r}\cdots\right) \\
&= \varepsilon^{\bar{N}_r}\sqrt{\bar{n}_r}\,\delta_{N,\bar{N}-1}\,\left(\delta_{n_1,\bar{n}_1}\cdots\delta_{n_r,\bar{n}_r-1}\cdots\right) \\
&= \varepsilon^{\bar{N}_r}\,\sqrt{\bar{n}_r}\;{}^{(\varepsilon)}\langle N;\cdots n_r\cdots|\bar{N}-1;\cdots\bar{n}_r-1\cdots\rangle^{(\varepsilon)}
\end{aligned}$$

$\bar{N}_r$ is defined as in (A.58). Since ${}^{(\varepsilon)}\langle N_1\cdots N_r\cdots|$ is an arbitrary bra-basis state, we have to conclude

$$c_{\alpha_r}|N;\cdots n_r\cdots\rangle^{(\varepsilon)} = \varepsilon^{N_r}\sqrt{n_r}|N-1;\cdots n_r-1\cdots\rangle^{(\varepsilon)} \tag{A.62}$$

or, more specifically,

$$\begin{aligned}
&\textit{Bosons}: \\
&c_{\alpha_r}|N;\cdots n_r\cdots\rangle^{(+)} = \sqrt{n_r}|N-1;\cdots n_r-1\cdots\rangle^{(+)} \\
&\textit{Fermions}: \\
&c_{\alpha_r}|N;\cdots n_r\cdots\rangle^{(-)} = \delta_{n_r,1}(-1)^{N_r}|N-1;\cdots n_r-1\cdots\rangle^{(-)}
\end{aligned} \tag{A.63}$$

With (A.59) and (A.63), one can easily prove three *fundamental commutation rules* (Problem A.2):

$$\begin{aligned}\left[c_{\alpha_r}\,,\, c_{\alpha_s}\right]_{-\varepsilon} &= \left[c^{\dagger}_{\alpha_r}\,,\, c^{\dagger}_{\alpha_s}\right]_{-\varepsilon} = 0 \\ \left[c_{\alpha_r}\,,\, c^{\dagger}_{\alpha_s}\right]_{-\varepsilon} &= \delta_{rs}\end{aligned} \tag{A.64}$$

We further introduce two special operators, namely, the *occupation number operator*:

$$\widehat{n}_r = c^{\dagger}_{\alpha_r}\, c_{\alpha_r} \tag{A.65}$$

and the *particle number operator*:

$$\widehat{N} = \sum_r \widehat{n}_r \tag{A.66}$$

The Fock states are the eigenstates of $\widehat{n}_r$ as well as of $\widehat{N}$. One can easily show with (A.59) and (A.63) that

$$\widehat{n}_r |N; \cdots n_r \cdots \rangle^{(\varepsilon)} = n_r\, |N; \cdots n_r \cdots \rangle^{(\varepsilon)} \tag{A.67}$$

Thus $\widehat{n}_r$ refers to the number of particles occupying the rth single-particle state. The eigenvalue of $\widehat{N}$ is the total number of particles N:

$$\begin{aligned}\widehat{N} |N; \cdots n_r \cdots \rangle^{(\varepsilon)} &= \left(\sum_r \widehat{n}_r\right) |N; \cdots n_r \cdots \rangle^{(\varepsilon)} \\ &= N |N; \cdots n_r \cdots \rangle^{(\varepsilon)}\end{aligned} \tag{A.68}$$

The following relations are valid for both Bosons and Fermions (Problem A.3):

$$\begin{aligned}\left[\widehat{n}_r\,,\, c^{\dagger}_s\right]_- &= \delta_{rs}\, c^{\dagger}_r \quad ; \quad \left[\widehat{n}_r\,,\, c_s\right]_- = -\delta_{rs}\, c_r \\ \left[\widehat{N}\,,\, c^{\dagger}_s\right]_- &= c^{\dagger}_s \quad ; \quad \left[\widehat{N}\,,\, c_s\right]_- = -c_s\end{aligned} \tag{A.69}$$

In order to transform a general operator $\widehat{A}$ (A.34) with a discrete spectrum into the formalism of second quantization, we have to follow the same procedure as was done in the case of the continuous spectrum. We only have to replace integrations by summations and delta functions by Kronecker deltas. Thus, we have expressions analogous to (A.39) and (A.42):

$$\begin{aligned}\widehat{A} \equiv & \sum_{r,r'} \langle \phi_{\alpha_r} | \widehat{A}^{(1)} | \phi_{\alpha_{r'}} \rangle c^{\dagger}_{\alpha_r}\, c_{\alpha_{r'}} \\ & + \frac{1}{2} \sum_{\substack{r,r' \\ s,s'}} \langle \phi^{(1)}_{\alpha_r} \phi^{(2)}_{\alpha_s} | \widehat{A}^{(2)} | \phi^{(1)}_{\alpha_{r'}} \phi^{(2)}_{\alpha_{s'}} \rangle c^{\dagger}_{\alpha_r}\, c^{\dagger}_{\alpha_s}\, c_{\alpha_{s'}}\, c_{\alpha_{r'}}\end{aligned} \tag{A.70}$$

In contrast to the case of continuous spectrum, here, the matrix elements must be calculated with non-symmetrized two-particle states. The reason for this is the different normalization used here.

A.4 Examples

In this section, we want to transform a few of the most often used operators from the first to second quantized form.

A.4.1 Bloch Electrons

We consider electrons in a rigid ion-lattice. The electrons interact with the lattice potential but do not interact with each other:

$$H_0 = H_{e,kin} + H_{ei}^{(0)} = \sum_{i=1}^{N_e} h_0^{(i)} \tag{A.71}$$

$H_{e,kin}$ is the operator of the kinetic energy

$$H_{e,kin} = \sum_{i=1}^{N_e} \frac{p_i^2}{2m} \tag{A.72}$$

N_e is the number of electrons which interact with the rigid ion-lattice via $H_{ei}^{(0)}$:

$$H_{ei}^{(0)} = \sum_{i=1}^{N_e} v(\mathbf{r}_i) \quad ; \quad v(\mathbf{r}_i) = \sum_{\alpha=1}^{N} V_{ei}(\mathbf{r}_i - \mathbf{R}_\alpha) \tag{A.73}$$

N is the number of lattice atoms, whose equilibrium positions are given by $\mathbf{R}_\alpha$. $v(\mathbf{r}_i)$ has the same periodicity as the lattice:

$$v(\mathbf{r}_i) = v(\mathbf{r}_i + \mathbf{R}_\alpha) \tag{A.74}$$

H_0 is obviously a single-particle operator. The eigenvalue equation for

$$h_0 = \frac{p^2}{2m} + v(\mathbf{r}) \tag{A.75}$$

defines the *Bloch function* $\psi_{\mathbf{k}}(\mathbf{r})$ and the *Bloch energy* $\varepsilon(\mathbf{k})$:

$$h_0\, \psi_{\mathbf{k}}(\mathbf{r}) = \varepsilon(\mathbf{k})\, \psi_{\mathbf{k}}(\mathbf{r}) \tag{A.76}$$

For $\psi_{\mathbf{k}}(\mathbf{r})$, we make the usual ansatz

$$\psi_{\mathbf{k}}(\mathbf{r}) = u_{\mathbf{k}}(\mathbf{r})\, e^{i\mathbf{k}\cdot\mathbf{r}} \tag{A.77}$$

where the amplitude function $u_{\mathbf{k}}(\mathbf{r})$ has the periodicity of the lattice. The Bloch functions build a complete orthonormal set:

$$\int d^3r\ \psi^*_{\mathbf{k}}(\mathbf{r})\psi_{\mathbf{k}'}(\mathbf{r}) = \delta_{\mathbf{k}\mathbf{k}'} \tag{A.78}$$

$$\sum_{\mathbf{k}}^{1st\ B.Z.} \psi^*_{\mathbf{k}}(\mathbf{r})\psi_{\mathbf{k}}(\mathbf{r}') = \delta(\mathbf{r}-\mathbf{r}') \tag{A.79}$$

Neither H_0 nor h_0 contains spin terms. Therefore the complete solutions are

$$|\mathbf{k}\sigma\rangle \ \Leftrightarrow\ \langle\mathbf{r}|\mathbf{k}\sigma\rangle = \psi_{\mathbf{k}\sigma}(\mathbf{r}) = \psi_{\mathbf{k}}(\mathbf{r})\chi_\sigma \tag{A.80}$$

$$\chi_\uparrow = \begin{pmatrix}1\\0\end{pmatrix}\ ;\ \chi_\downarrow = \begin{pmatrix}0\\1\end{pmatrix} \tag{A.81}$$

We define

$c^\dagger_{\mathbf{k}\sigma}$ ($c_{\mathbf{k}\sigma}$) - creation (annihilation) operator of a Bloch electron

Since H_0 is a single-particle operator, according to (A.70), we can write

$$H_0 = \sum_{\substack{\mathbf{k},\sigma\\ \mathbf{k}',\sigma'}} \langle\mathbf{k}\sigma|h_0|\mathbf{k}'\sigma'\rangle\ c^\dagger_{\mathbf{k}\sigma}\ c_{\mathbf{k}'\sigma'} \tag{A.82}$$

The matrix element is given by

$$\begin{aligned}\langle\mathbf{k}\sigma|h_0|\mathbf{k}'\sigma'\rangle &= \int d^3r \langle\mathbf{k}\sigma|\mathbf{r}\rangle\langle\mathbf{r}|h_0|\mathbf{k}'\sigma'\rangle\\ &= \varepsilon(\mathbf{k}')\int d^3r\ \psi^*_{\mathbf{k}\sigma}(\mathbf{r})\ \psi_{\mathbf{k}'\sigma'}(\mathbf{r})\\ &= \varepsilon(\mathbf{k})\ \delta_{\mathbf{k}\mathbf{k}'}\ \delta_{\sigma\sigma'}\end{aligned} \tag{A.83}$$

So that we can finally write

$$H_0 = \sum_{\mathbf{k},\sigma} \varepsilon(\mathbf{k})\ c^\dagger_{\mathbf{k}\sigma}\ c_{\mathbf{k}\sigma} \tag{A.84}$$

The Bloch operators satisfy the fundamental commutation relations of Fermion operators:

$$\begin{aligned} \left[c_{\mathbf{k}\sigma}\,,\,c_{\mathbf{k}'\sigma'}\right]_+ &= \left[c^\dagger_{\mathbf{k}\sigma}\,,\,c^\dagger_{\mathbf{k}'\sigma'}\right]_+ = 0 \\ \left[c_{\mathbf{k}\sigma}\,,\,c^\dagger_{\mathbf{k}'\sigma'}\right]_+ &= \delta_{\mathbf{k}\mathbf{k}'}\,\delta\sigma\sigma' \end{aligned} \tag{A.85}$$

In the special case, where one can neglect the crystal structure, e.g., as in the Jellium model, the Bloch functions become plane waves

$$\psi_{\mathbf{k}}(\mathbf{r}) \Rightarrow \frac{1}{\sqrt{V}}e^{i\mathbf{k}\cdot\mathbf{r}}; \qquad \varepsilon(\mathbf{k}) \Rightarrow \frac{\hbar^2k^2}{2m} \quad : (v(\mathbf{r}) \equiv const) \tag{A.86}$$

A.4.2 Wannier Electrons

The representation using *Wannier functions*

$$w_\sigma(\mathbf{r}-\mathbf{R}_i) = \frac{1}{\sqrt{N}}\sum_{\mathbf{k}}^{1st\ B.Z.} e^{-i\mathbf{k}\cdot\mathbf{R}_i}\,\psi_{\mathbf{k}\sigma}(\mathbf{r}) \tag{A.87}$$

is a special, frequently used position representation. The typical property of the Wannier function is its strong concentration around the respective lattice site $\mathbf{R}_i$:

$$\int d^3r\; w^*_{\sigma'}(\mathbf{r}-\mathbf{R}_i)\,w_\sigma(\mathbf{r}-\mathbf{R}_j) = \delta_{\sigma\sigma'}\,\delta_{ij} \tag{A.88}$$

We define

$c^\dagger_{i\sigma}$ $(c_{i\sigma})$ - creation (annihilation) operator of an electron with spin σ in a Wannier state at the lattice site $\mathbf{R}_i$

These construction operators satisfy the following commutation relations:

$$\left[c_{i\sigma}\,,\,c_{j\sigma'}\right]_+ = \left[c^\dagger_{i\sigma}\,,\,c^\dagger_{j\sigma'}\right]_+ = 0; \quad \left[c_{i\sigma}\,,\,c^\dagger_{j\sigma'}\right]_+ = \delta_{\sigma\sigma'}\,\delta_{ij} \tag{A.89}$$

In this basis, H_0 looks as follows:

$$H_0 = \sum_{i,j,\sigma} T_{ij}c^\dagger_{i\sigma}\,c_{j\sigma'} \tag{A.90}$$

$$T_{ij} = \int d^3r\; w^*_\sigma(\mathbf{r}-\mathbf{R}_i)\,h_0\,w_\sigma(\mathbf{r}-\mathbf{R}_j) \tag{A.91}$$

T_{ij} is known as the *hopping integral*. In this form, H_0 describes, more transparently, the hopping of an electron of spin σ from the lattice site $\mathbf{R}_j$ to the lattice site $\mathbf{R}_i$. The connection with the Bloch representation (A.4.1) is easy to establish by using (A.78) and (A.87):

$$T_{ij} = \frac{1}{N} \sum_{\mathbf{k}}^{1st\ B.Z.} \varepsilon(\mathbf{k}) e^{i\mathbf{k}\cdot(\mathbf{R}_i - \mathbf{R}_j)} \tag{A.92}$$

$$c_{i\sigma} = \frac{1}{\sqrt{N}} \sum_{\mathbf{k}}^{1st\ B.Z.} e^{i\mathbf{k}\cdot\mathbf{R}_i} c_{\mathbf{k}\sigma} \tag{A.93}$$

A.4.3 Density Operator

The operator for the electron density

$$\widehat{\rho}(\mathbf{r}) = \sum_{i=1}^{N_e} \delta(\mathbf{r} - \widehat{\mathbf{r}}_i) \tag{A.94}$$

is another example for a single-particle operator. It should be noticed that the electron position $\widehat{\mathbf{r}}_i$ is an operator but not the variable $\mathbf{r}$:

$$\widehat{\rho}(\mathbf{r}) = \sum_{\mathbf{k}\sigma,\ \mathbf{k}'\sigma'} \langle \mathbf{k}\sigma | \delta(\mathbf{r} - \widehat{\mathbf{r}}_i) | \mathbf{k}'\sigma' \rangle c^{\dagger}_{\mathbf{k}\sigma}\, c_{\mathbf{k}'\sigma'} \tag{A.95}$$

The matrix element is given by

$$\begin{aligned} \langle \mathbf{k}\sigma | \delta(\mathbf{r} - \widehat{\mathbf{r}}') | \mathbf{k}'\sigma' \rangle &= \int d^3r''\ \langle \mathbf{k}\sigma | \delta(\mathbf{r} - \widehat{\mathbf{r}}') | \mathbf{r}'' \rangle \langle \mathbf{r}'' | \mathbf{k}'\sigma' \rangle \\ &= \delta_{\sigma\sigma'} \int d^3r''\ \delta(\mathbf{r} - \mathbf{r}'') \langle \mathbf{k}\sigma | \mathbf{r}'' \rangle \langle \mathbf{r}'' | \mathbf{k}'\sigma \rangle \\ &= \delta_{\sigma\sigma'}\ \psi^*_{\mathbf{k}\sigma}(\mathbf{r})\ \psi_{\mathbf{k}'\sigma}(\mathbf{r}) \end{aligned} \tag{A.96}$$

If one restricts oneself to plane waves ($v(\mathbf{r}) = const.$), then

$$\langle \mathbf{k}\sigma | \delta(\mathbf{r} - \widehat{\mathbf{r}}') | \mathbf{k} + \mathbf{q}\sigma' \rangle = \delta_{\sigma\sigma'} \frac{1}{V} e^{i\mathbf{q}\cdot\mathbf{r}} \tag{A.97}$$

Using this in (A.95) we get

$$\widehat{\rho}(\mathbf{r}) = \frac{1}{V} \sum_{\mathbf{k}\mathbf{q}\sigma} c^{\dagger}_{\mathbf{k}\sigma}\, c_{\mathbf{k}+\mathbf{q}\sigma} e^{i\mathbf{q}\cdot\mathbf{r}} \tag{A.98}$$

From the above equation, we therefore find the Fourier component of the density operator:

$$\widehat{\rho}_{\mathbf{q}} = \sum_{\mathbf{k}\sigma} c^{\dagger}_{\mathbf{k}\sigma}\, c_{\mathbf{k}+\mathbf{q}\sigma} \tag{A.99}$$

A.4.4 Coulomb Interaction

Now we have to deal with a two-particle operators:

$$H_C = \frac{1}{2}\frac{e^2}{4\pi\varepsilon_0}\sum_{i,j}^{i\neq j}\frac{1}{\left|\widehat{\mathbf{r}}_i - \widehat{\mathbf{r}}_j\right|} \tag{A.100}$$

How does this operator look in the formalism of second quantization? We again choose the momentum representation:

$$\begin{aligned} H_C = {} & \frac{e^2}{8\pi\varepsilon_0} * \\ & * \sum_{\substack{\mathbf{k}_1\cdots\mathbf{k}_4 \\ \sigma_1\cdots\sigma_4}} \langle(\mathbf{k}_1\sigma_1)^{(1)}(\mathbf{k}_2\sigma_2)^{(2)}|\frac{1}{\left|\widehat{\mathbf{r}}^{(1)} - \widehat{\mathbf{r}}^{(2)}\right|}|(\mathbf{k}_3\sigma_3)^{(1)}(\mathbf{k}_4\sigma_4)^{(2)}\rangle * \\ & * c^{\dagger}_{\mathbf{k}_1\sigma_1}\, c^{\dagger}_{\mathbf{k}_2\sigma_2}\, c_{\mathbf{k}_4\sigma_4}\, c_{\mathbf{k}_3\sigma_3} \end{aligned} \tag{A.101}$$

Since the operator itself is independent of spin, the matrix element is surely nonzero only for $\sigma_1 = \sigma_3$ and $\sigma_2 = \sigma_4$. Therefore, we are left with the following matrix element:

$$\begin{aligned} v(\mathbf{k}_1\cdots\mathbf{k}_4) &= \tfrac{e^2}{4\pi\varepsilon_0}\langle\mathbf{k}_1^{(1)}\mathbf{k}_2^{(2)}|\tfrac{1}{|\widehat{\mathbf{r}}^{(1)}-\widehat{\mathbf{r}}^{(2)}|}|\mathbf{k}_3^{(1)}\mathbf{k}_4^{(2)}\rangle \\ &= \tfrac{e^2}{4\pi\varepsilon_0}\int\int d^3r_1 d^3r_2\, \langle\mathbf{k}_1^{(1)}\mathbf{k}_2^{(2)}|\tfrac{1}{|\widehat{\mathbf{r}}^{(1)}-\widehat{\mathbf{r}}^{(2)}|}|\mathbf{r}_1^{(1)}\mathbf{r}_2^{(2)}\rangle\langle\mathbf{r}_1^{(1)}\mathbf{r}_2^{(2)}|\mathbf{k}_3^{(1)}\mathbf{k}_4^{(2)}\rangle \\ &= \tfrac{e^2}{4\pi\varepsilon_0}\int\int d^3r_1 d^3r_2\, \tfrac{1}{|\mathbf{r}_1-\mathbf{r}_2|}\langle\mathbf{k}_1^{(1)}\mathbf{k}_2^{(2)}|\mathbf{r}_1^{(1)}\mathbf{r}_2^{(2)}\rangle\langle\mathbf{r}_1^{(1)}\mathbf{r}_2^{(2)}|\mathbf{k}_3^{(1)}\mathbf{k}_4^{(2)}\rangle \\ &= \tfrac{e^2}{4\pi\varepsilon_0}\int\int d^3r_1 d^3r_2\, \tfrac{1}{|\mathbf{r}_1-\mathbf{r}_2|}\psi^*_{\mathbf{k}_1}(\mathbf{r}_1)\psi^*_{\mathbf{k}_2}(\mathbf{r}_2)\psi_{\mathbf{k}_3}(\mathbf{r}_1)\psi_{\mathbf{k}_4}(\mathbf{r}_2) \end{aligned} \tag{A.102}$$

Translational symmetry demands that we must have

$$\mathbf{k}_1 + \mathbf{k}_2 = \mathbf{k}_3 + \mathbf{k}_4 \tag{A.103}$$

Thus we have obtained the following expression for the Coulomb interaction:

$$H_C = \frac{1}{2}\sum_{\substack{\mathbf{kpq} \\ \sigma\sigma'}} v(\mathbf{k},\ \mathbf{p},\ \mathbf{q})\, c^{\dagger}_{\mathbf{k}+\mathbf{q}\sigma}\, c^{\dagger}_{\mathbf{p}-\mathbf{q}\sigma'}\, c_{\mathbf{p}\sigma'} c_{\mathbf{k}\sigma} \tag{A.104}$$

with

$$v(\mathbf{k}, \mathbf{p}, \mathbf{q}) = \frac{e^2}{4\pi\varepsilon_0} \int\int d^3r_1\, d^3r_2\, \psi^*_{\mathbf{k}+\mathbf{q}}(\mathbf{r}_1)\, \psi_{\mathbf{p}-\mathbf{q}}(\mathbf{r}_2) \, * \\ * \, \frac{1}{|\mathbf{r}_1 - \mathbf{r}_2|} \psi_{\mathbf{k}}(\mathbf{r}_1)\, \psi_{\mathbf{p}}(\mathbf{r}_2) \tag{A.105}$$

A.5 Problems

Problem A.1 Let $|\varphi_{\alpha_1} \cdots \varphi_{\alpha_N}\rangle^{(\varepsilon)}$ be an (anti) symmetrized N-particle basis state for the case of a continuous one-particle spectrum. Then show that

$$\left(c_\beta c_\gamma^\dagger - \varepsilon c_\gamma^\dagger c_\beta\right) |\varphi_{\alpha_1} \cdots \varphi_{\alpha_N}\rangle^{(\varepsilon)} = \delta(\varphi_\beta - \varphi_\gamma) |\varphi_{\alpha_1} \cdots \varphi_{\alpha_N}\rangle^{(\varepsilon)}$$

Problem A.2 Prove the fundamental commutation relations

$$\left[c_{\alpha_r}, c_{\alpha_s}\right]_{\mp} = \left[c_{\alpha_s}^\dagger, c_{\alpha_r}^\dagger\right]_{\mp} = 0$$
$$\left[c_{\alpha_r}, c_{\alpha_s}^\dagger\right]_{\mp} = \delta_{r,s}$$

for the creation and annihilation operators for the Bosons and Fermions in the discrete Fock space.

Problem A.3 $c_{\varphi_\alpha} \equiv c_\alpha$ und $c_{\varphi_\alpha}^\dagger \equiv c_\alpha^\dagger$ are the annihilation and creation operators for one-particle states $|\varphi_\alpha\rangle$ of an observable $\widehat{\Phi}$ with discrete spectrum. With the help of the fundamental commutation relations for Bosons and Fermions, calculate the following commutators:

$$1.)\ \left[\widehat{n}_\alpha, c_\beta^\dagger\right]_- \qquad 2.)\ \left[\widehat{n}_\alpha, c_\beta\right]_- \qquad 3.)\ \left[\widehat{N}, c_\alpha\right]_-$$

Here $\widehat{N} = \sum_\alpha \widehat{n}_\alpha = \sum_\alpha c_\alpha^\dagger c_\alpha$ is the particle number operator.

Problem A.4 Under the same assumptions as in Problem A.3 for Fermions, prove the following relations:

1. $(c_\alpha)^2 = 0\,;\ \left(c_\alpha^\dagger\right)^2 = 0$
2. $(\widehat{n}_\alpha)^2 = \widehat{n}_\alpha$
3. $c_\alpha\, \widehat{n}_\alpha = c_\alpha\,;\ c_\alpha^\dagger\, \widehat{n}_\alpha = 0$
4. $\widehat{n}_\alpha\, c_\alpha = 0\,;\ \widehat{n}_\alpha\, c_\alpha^\dagger = c_\alpha^\dagger$

Problem A.5 Let $|0\rangle$ be the normalized vacuum state. Let $c_\alpha^\dagger$ und c_α be the creation and annihilation operators for a particle in one-particle state $|\varphi_\alpha\rangle$. Using the fundamental commutation relations derive the relation

$$\langle 0|\, c_{\beta_N} \cdots c_{\beta_1} c_{\alpha_1}^\dagger \cdots c_{\alpha_N}^\dagger \,|0\rangle = \sum_{\mathcal{P}_\alpha} (\pm)^{p_\alpha} \mathcal{P}_\alpha \left[\delta(\beta_1, \alpha_1)\delta(\beta_2, \alpha_2) \cdots \delta(\beta_N, \alpha_N)\right]$$

$\mathcal{P}_\alpha$ is the permutation operator that operates on the indices α_i.

Problem A.6 For the occupation number density operator calculate the commutators

$$1)\ \left[\widehat{n}_\alpha,\, c_\beta^\dagger\right]_- \ ;\ 2.)\ \left[\widehat{n}_\alpha,\, c_\beta\right]_-$$

Is there a difference for Bosons and Fermions?

Problem A.7 The anti-symmetrized basis states $\left|\varphi_{\alpha_1} \cdots \varphi_{\alpha_N}\right\rangle^{(\pm)}$ of $\mathcal{H}_N^{(\pm)}$ are built from *continuous* one-particle basis states. They are the eigenstates of the particle number operator $\widehat{N}$. Then show that

$$1)\ c_\beta^\dagger \left|\varphi_{\alpha_1} \cdots \varphi_{\alpha_N}\right\rangle^{(\pm)} \quad 2)\ c_\beta \left|\varphi_{\alpha_1} \cdots \varphi_{\alpha_N}\right\rangle^{(\pm)}$$

are also eigenstates of $\widehat{N}$ and calculate the corresponding eigenvalues.

Problem A.8 A system of N electrons in volume $V = L^3$ interact among themselves via the Coulomb interaction

$$V_2 = \frac{1}{2} \sum_{i,j}^{i \neq j} V_2^{(i,j)}; \quad V_2^{(i,j)} = \frac{e^2}{4\pi\varepsilon_0} \frac{1}{\left|\widehat{\mathbf{r}}_i - \widehat{\mathbf{r}}_j\right|}$$

$\widehat{\mathbf{r}}_i$ und $\widehat{\mathbf{r}}_j$ are, respectively, the position operators of the ith and jth electron. Formulate the Hamiltonian of the system in second quantization. Use as the one-particle basis plane waves which have discrete wavevectors $\mathbf{k}$ as a consequence of the periodic boundary conditions on $V = L^3$

Problem A.9 Show that the Hamiltonian calculated in Problem A.8 of the interacting N-electron system

$$H_N = \sum_{\mathbf{k}\sigma} \varepsilon_0(\mathbf{k})\, c_{\mathbf{k}\sigma}^\dagger c_{\mathbf{k}\sigma} + \frac{1}{2} \sum_{\substack{\mathbf{kpq} \\ \sigma\sigma'}} v_0(\mathbf{q})\, c_{\mathbf{k}+\mathbf{q}\sigma}^\dagger c_{\mathbf{p}-\mathbf{q}\sigma'}^\dagger c_{\mathbf{p}\sigma'} c_{\mathbf{k}\sigma}$$

commutes with the particle number operator

$$\widehat{N} = \sum_{\mathbf{k}\sigma} c_{\mathbf{k}\sigma}^\dagger c_{\mathbf{k}\sigma}$$

What is the physical meaning of this?

Problem A.10 Two identical particles move in a one-dimensional infinite potential well:

$$V(x) = \begin{cases} 0 & \text{for } 0 \le x \le a, \\ \infty & \text{for } x < 0 \text{ and } x > a \end{cases} \tag{A.106}$$

Calculate the energy eigenfunctions and energy eigenvalues of the two-particle system if they are (a) Bosons and (b) Fermions. What is the ground state energy in the case $N \gg 1$ for Bosons and for Fermions?

Problem A.11 There is a system of non-interacting identical Bosons or Fermions described by

$$H = \sum_{i=1}^{N} H_1^{(i)}$$

The one-particle operator $H_1^{(i)}$ has a discrete non-degenerate spectrum:

$$H_1^{(i)}|\varphi_r^{(i)}\rangle = \epsilon_r|\varphi_r^{(i)}\rangle; \quad \langle\varphi_r^{(i)}|\varphi_s^{(i)}\rangle = \delta_{rs}$$

$|\varphi_r^{(i)}\rangle$ are used to build the Fock states $|N; n_1, n_2, \ldots\rangle^{(\epsilon)}$. The general state of the system is described by the un-normalized density matrix ρ, for which the grand canonical ensemble (variable particle number) holds

$$\rho = \exp[-\beta(H - \mu\hat{N})]$$

1. How does the Hamiltonian read in second quantization?
2. Show that the grand canonical partition is given by

$$\Xi(T, V, \mu) = \mathrm{Sp}\rho = \begin{cases} \prod_i \{1 - \exp[-\beta(\epsilon_i - \mu)]\}^{-1} & \text{Bosons,} \\ \prod_i \{1 + \exp[-\beta(\epsilon_i - \mu)]\} & \text{Fermions.} \end{cases}$$

3. Calculate the expectation value of the particle number:

$$\langle\hat{N}\rangle = \frac{1}{\Xi}\mathrm{Tr}(\rho\hat{N})$$

4. Calculate the internal energy:

$$U = \langle H\rangle = \frac{1}{\Xi}\mathrm{Tr}(\rho H)$$

5. Calculate the average occupation number of the ith one-particle state

$$\langle \hat{n}_i \rangle = \frac{1}{\Xi} \mathrm{Tr}(\rho a_i^\dagger a_i)$$

and show that

$$U = \sum_i \epsilon_i \langle \hat{n}_i \rangle; \quad \langle \hat{N} \rangle = \sum_i \langle \hat{n}_i \rangle$$

are valid.

Appendix B
The Method of Green's Functions

B.1 Linear Response Theory

We want to introduce the Green's functions using a concrete physical context: *How does a physical system respond to an external perturbation?* The answer is provided by the so-called *response functions*. Well-known examples are the magnetic or electric susceptibility, the electrical conductivity, the thermal conductivity, etc. These are completely expressed by a special type of Green's function.

B.1.1 Kubo Formula

Let the Hamiltonian

$$H = H_0 + V_t \tag{B.1}$$

consist of two parts. V_t is the *perturbation* which is the interaction of the system with a possibly time-dependent external field. H_0 is the Hamiltonian of the *field-free* but certainly interacting particle system. Thus H_0 is already in general not exactly solvable. Let the perturbation be operating through a scalar field F_t which couples to the observable $\widehat{B}$:

$$V_t = \widehat{B} \cdot F_t \tag{B.2}$$

The restriction made now to a *scalar* field can be easily removed later. One should note that $\widehat{B}$ is an operator, whereas F_t is a c-number.

Now let $\widehat{A}$ be a not explicitly time-dependent observable. The interesting question is the following:

How does the measured quantity $\langle \widehat{A} \rangle$ react to the perturbation V_t?

Without the field holds

$$\langle \widehat{A} \rangle_0 = Tr(\rho_0 \widehat{A}) \tag{B.3}$$

$$\rho_0 = \frac{\exp(-\beta H_0)}{Tr\,(\exp(-\beta H_0))} \tag{B.4}$$

where ρ_0 is the statistical operator of the field-free system, at the moment in the canonical ensemble. $\beta = 1/k_B T$ is as usual the reciprocal temperature.

On the other hand, *with* field holds:

$$\langle\widehat{A}\rangle_t = Tr(\rho_t \widehat{A}) \tag{B.5}$$

$$\rho_t = \frac{\exp(-\beta H)}{Tr\,(\exp(-\beta H))} \tag{B.6}$$

where ρ_t is now the statistical operator of the particle system in the presence of the external field F_t. Thus a measure of the response of the system to the perturbation could be the change in the value of $\langle\widehat{A}\rangle$ under the influence of the external field:

$$\Delta A_t = \langle\widehat{A}\rangle_t - \langle\widehat{A}\rangle_0 \tag{B.7}$$

To calculate this we need ρ_t. In Schrödinger picture ρ_t satisfies the equation of motion

$$i\hbar\dot{\rho}_t = [H_0 + V_t,\ \rho_t]_- \tag{B.8}$$

with the boundary condition that the field for $t \to -\infty$ is switched off:

$$\lim_{t\to-\infty} \rho_t = \rho_0 \tag{B.9}$$

Because of H_0, the Schrödinger picture is not convenient for perturbational approaches. More advantageous would be the Dirac picture:

$$\rho_t^D(t) = e^{\frac{i}{\hbar}H_0 t}\rho_t e^{-\frac{i}{\hbar}H_0 t} \tag{B.10}$$

One should note the two different time dependences. The lower index t denotes the possible explicit time dependence caused by the field. On the contrary, the argument of the statistical operator represents the dynamic time dependence. The equation of motion now is determined solely by the perturbation:

$$i\hbar\dot{\rho}_t^D(t) = \left[V_t^D(t),\ \rho_t^D(t)\right]_- \tag{B.11}$$

The boundary condition follows from (B.9) and (B.10) where ρ_0 commutes with H_0:

$$\lim_{t\to-\infty} \rho_t^D(t) = \rho_0 \tag{B.12}$$

We can now formally integrate (B.11):

$$\rho_t^D(t) = \rho_0 - \frac{i}{\hbar}\int_{-\infty}^{t} dt' \left[V_{t'}^D(t'),\, \rho_{t'}^D(t')\right]_- \tag{B.13}$$

This integral equation can be solved iteratively up to arbitrary accuracy in the perturbation V_t. For small perturbations one can restrict oneself to the first non-trivial step (*linear response*). That means after back-transformation to Schrödinger picture

$$\rho_t \approx \rho_0 - \frac{i}{\hbar}\int_{-\infty}^{t} dt' e^{-\frac{i}{\hbar}H_0 t} \left[V_{t'}^D(t'),\, \rho_0\right]_- e^{\frac{i}{\hbar}H_0 t} \tag{B.14}$$

We now calculate $\langle\widehat{A}\rangle_t$ approximately by substituting ρ_t in (B.5):

$$\begin{aligned}
\langle\widehat{A}\rangle_t &= \langle\widehat{A}\rangle_0 - \frac{i}{\hbar}\int_{-\infty}^{t} dt' Tr\left(e^{-\frac{i}{\hbar}H_0 t}\left[V_{t'}^D(t'),\, \rho_0\right]_- e^{\frac{i}{\hbar}H_0 t}\cdot\widehat{A}\right)\\
&= \langle\widehat{A}\rangle_0 - \frac{i}{\hbar}\int_{-\infty}^{t} dt' Tr\left(V_{t'}^D(t')\,\rho_0\widehat{A}^D(t) - \rho_0 V_{t'}^D(t')\widehat{A}^D(t)\right)\\
&= \langle\widehat{A}\rangle_0 - \frac{i}{\hbar}\int_{-\infty}^{t} dt' Tr\left(\rho_0\left[\widehat{A}^D(t),\, V_{t'}^D(t')\right]_-\right)
\end{aligned}$$

In the reformulation, we many times used the cyclic invariance of trace. We thus obtain the reaction of the system to external perturbation:

$$\Delta A_t = -\frac{i}{\hbar}\int_{-\infty}^{t} dt' F_{t'}\left\langle\left[\widehat{A}^D(t),\, \widehat{B}^D(t')\right]_-\right\rangle_0 \tag{B.15}$$

As a consequence of linear response, the reaction of the system is determined by an expectation value for the field-free system. One should note that the Dirac representation of the operators $\widehat{A}$ and $\widehat{B}$ here correspond to the Heisenberg representation *without* field ($H_0 \to H$).

One defines: *"double-time, retarded Green's function"*

$$G_{AB}^{ret}(t,t') = \langle\langle\widehat{A}(t);\widehat{B}(t')\rangle\rangle^{ret} = -i\Theta(t-t')\left\langle\left[\widehat{A}(t),\,\widehat{B}(t')\right]_-\right\rangle \tag{B.16}$$

where Θ is the Heaviside step function. The operators are in the field-free Heisenberg representation. The properties of the retarded Green's function will be discussed in more detail in the following. Here, we recognize that it describes the reaction of the system in terms of the observable $\widehat{A}$ when the external perturbation couples to the observable $\widehat{B}$

$$\Delta A_t = \frac{1}{\hbar}\int_{-\infty}^{+\infty} dt' F_{t'}\, G_{AB}^{ret}(t,t') \tag{B.17}$$

Later we will assume that the (field-free) Hamiltonian H is not explicitly dependent on time. In that case one can show that the Green's function does not depend on two times but only on the time difference. Then it is called *homogeneous in time*:

$$G^{ret}_{AB}(t,t') \rightarrow G^{ret}_{AB}(t-t')$$

In such a situation one has the Fourier transformation:

$$\begin{aligned} G^{ret}_{AB}(E) &= \langle\langle \widehat{A}; \widehat{B} \rangle\rangle^{ret}_{E} \\ &= \int_{-\infty}^{+\infty} d(t-t')\, G^{ret}_{AB}(t-t')\, e^{\frac{i}{\hbar}E(t-t')} \end{aligned} \tag{B.18}$$

$$G^{ret}_{AB}(t-t') = \frac{1}{2\pi\hbar}\int_{-\infty}^{+\infty} dE\, G^{ret}_{AB}(E)\, e^{-\frac{i}{\hbar}E(t-t')} \tag{B.19}$$

In the following, all the time/energy functions will be transformed in this way. In particular, for the delta function holds:

$$\delta(E-E') = \frac{1}{2\pi\hbar}\int_{-\infty}^{+\infty} dt\, e^{-\frac{i}{\hbar}(E-E')t} \tag{B.20}$$

$$\delta(t-t') = \frac{1}{2\pi\hbar}\int_{-\infty}^{+\infty} dE\, e^{\frac{i}{\hbar}E(t-t')} \tag{B.21}$$

We write the perturbing field as a Fourier integral:

$$F_t = \frac{1}{2\pi\hbar}\int_{-\infty}^{+\infty} dE\, F(E)\, e^{-\frac{i}{\hbar}(E+i0^+)t} \tag{B.22}$$

The additional $i0^+$ takes care of fulfilling the boundary condition (B.9). Substituting this expression now in (B.17) we finally obtain the *Kubo formula*:

$$\Delta A_t = \frac{1}{2\pi\hbar^2}\int_{-\infty}^{+\infty} dE\, F(E)\, G^{ret}_{AB}(E+i0^+)\, e^{-\frac{i}{\hbar}(E+i0^+)t} \tag{B.23}$$

In the following sections we will present a few applications of this important formula.

B.1.2 Magnetic Susceptibility

We consider a magnetic induction $\mathbf{B}_t$ which is homogeneous in space and oscillating in time as *perturbation* which couples to the total magnetic moment $\mathbf{m}$ of a localized spin system (e.g. Heisenberg model) (Fig. B.1):

$$\mathbf{m} = \sum_i \mathbf{m}_i \tag{B.24}$$

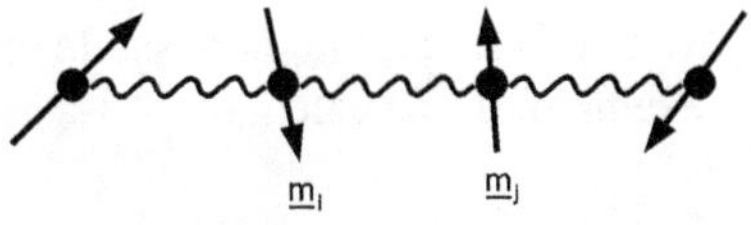

Fig. B.1 Interacting localized moments

i indexes the lattice sites at which the localized moments (spins) are present. The perturbation term then reads

$$V_t = -\mathbf{m} \cdot \mathbf{B}_t = -\frac{1}{2\pi\hbar} \sum_{\alpha}^{(x,y,z)} \int_{-\infty}^{+\infty} dE\, m^\alpha B^\alpha(E)\, e^{-\frac{i}{\hbar}(E+i0^+)t} \tag{B.25}$$

Interesting here is the reaction of the magnetization of the moment system:

$$\mathbf{M} = \frac{1}{V} \langle \mathbf{m} \rangle = \frac{1}{V} \sum_i \langle \mathbf{m}_i \rangle \tag{B.26}$$

It can be approximately calculated with the help of the Kubo formula (B.17):

$$\Delta M_t^\beta = M_t^\beta - M_0^\beta = -\frac{1}{V\hbar} \int_{-\infty}^{+\infty} dt' \sum_\alpha B_{t'}^\alpha \langle\langle m^\beta(t);\, m^\alpha(t') \rangle\rangle \tag{B.27}$$

Here M_0 is the field-free magnetization which is unequal zero only for ferromagnets. One defines "magnetic susceptibility tensor"

$$\chi_{ij}^{\alpha\beta}(t,t') = -\frac{\mu_0}{V\hbar} \langle\langle m_i^\beta(t);\, m_j^\alpha(t') \rangle\rangle \tag{B.28}$$

This function, which is so important for magnetism, is thus a retarded Green's function:

$$\Delta M_t^\beta = \frac{1}{\mu_0} \sum_{i,j} \sum_\alpha \int_{-\infty}^{+\infty} dt'\, \chi_{ij}^{\alpha\beta}(t,t') B_{t'}^\alpha \tag{B.29}$$

If the susceptibility is homogeneous in time, $\chi_{ij}^{\alpha\beta}(t,t') \equiv \chi_{ij}^{\alpha\beta}(t-t')$, which we want to assume, then energy representation is meaningful:

$$\Delta M_t^\beta = \frac{1}{2\pi\hbar\mu_0} \sum_{i,j} \sum_\alpha \int_{-\infty}^{+\infty} dE\, \chi_{ij}^{\alpha\beta}(E+i0^+) B^\alpha(E) e^{-\frac{i}{\hbar}(E+i0^+)t} \tag{B.30}$$

Of particular interest are

- "longitudinal" susceptibility:

$$\chi_{ij}^{zz}(E) = -\frac{\mu_0}{V\hbar} \langle\langle m_i^z;\, m_j^z \rangle\rangle_E \tag{B.31}$$

With this function it is possible to make statements about *magnetic stability*. For example, if one calculates

$$\chi_{\mathbf{q}}^{zz}(E) = \frac{1}{N} \sum_{i,j} \chi_{ij}^{zz}(E) e^{i\mathbf{q}\cdot(\mathbf{R}_i - \mathbf{R}_j)} \tag{B.32}$$

for the *paramagnetic* moment system, then in the limit $(\mathbf{q}, E) \to 0$ the singularities

$$\left[\lim_{(\mathbf{q},E)\to 0} \chi_{\mathbf{q}}^{zz}(E) \right]^{-1} \overset{!}{=} 0$$

give the instabilities of the paramagnetic state against ferromagnetic ordering. At these points even an infinetesimal, symmetry breaking field produces a finite magnetization. The poles of the function $\chi_{\mathbf{q}}^{zz}(E)$ therefore describe the phase transition *para-ferromagnetism*.

- "transverse" susceptibility:

$$\chi_{ij}^{+-}(E) = -\frac{\mu_0}{V\hbar} \langle\langle m_i^+; m_j^- \rangle\rangle_E \tag{B.33}$$

Here holds

$$m_j^{\pm} = m_j^x \pm i\, m_j^y$$

For this function also the poles are interesting. They represent *resonances* or *eigenoscillations*. One obtains from them the energies of *spin waves* or *magnons*:

$$\left(\chi_{\mathbf{q}}^{+-}(E)\right)^{-1} \overset{!}{=} 0 \iff E = \hbar\omega(\mathbf{q}) \tag{B.34}$$

Thus the linear response theory is not only an approximate procedure for weak perturbations but also provides valuable information about the unperturbed system.

B.1.3 Dielectric Function

We want to discuss another application of the linear response theory. Into a system of quasifree conduction electrons (metal), an external charge density is introduced. Because of Coulomb repulsion, the charge carriers react to the *perturbation*. This leads to changes in the density of metal electrons in such a way that it causes effectively a more or less strong *screening* of the perturbing charge. This is described by

$$\varepsilon(\mathbf{q}, E) : \textit{dielectric function}$$

We first want to handle the problem classically and undertake the necessary quantization later. For the external charge density we write

$$\rho_{ext}(\mathbf{r},t) = \frac{1}{2\pi\hbar V}\int_{-\infty}^{+\infty} dE \sum_{\mathbf{q}} \rho_{ext}(\mathbf{q},E)\, e^{i\mathbf{q}\cdot\mathbf{r}}\, e^{-\frac{i}{\hbar}(E+i0^+)t} \tag{B.35}$$

A corresponding expression should hold for the conduction electrons:

$$-e\rho(\mathbf{r}) = -\frac{e}{V}\sum_{\mathbf{q}} \rho_{\mathbf{q}}\, e^{i\mathbf{q}\cdot\mathbf{r}} \tag{B.36}$$

The interaction of the conduction electrons with the external charge density then reads as

$$V_t = \frac{-e}{4\pi\varepsilon_0}\int\int d^3r\, d^3r'\, \frac{\rho(\mathbf{r})\,\rho_{ext}(\mathbf{r}',t)}{|\mathbf{r}-\mathbf{r}'|} \tag{B.37}$$

One shows (Problem 4.3)

$$\int\int d^3r\, d^3r'\, \frac{e^{i(\mathbf{q}\cdot\mathbf{r}+\mathbf{q}'\cdot\mathbf{r}')}}{|\mathbf{r}-\mathbf{r}'|} = \frac{4\pi V}{q^2}\,\delta_{\mathbf{q},-\mathbf{q}'} \tag{B.38}$$

Using this it holds with the Coulomb potential

$$v_0(\mathbf{q}) = \frac{1}{V}\frac{e^2}{\varepsilon_0 q^2} \tag{B.39}$$

after simple reformulation:

$$V_t = \frac{1}{2\pi\hbar}\int_{-\infty}^{+\infty} dE\, e^{-\frac{i}{\hbar}(E+i0^+)t} \sum_{\mathbf{q}} \frac{v_0(\mathbf{q})}{-e}\,\rho_{-\mathbf{q}}\,\rho_{ext}(\mathbf{q},E) \tag{B.40}$$

We assume that without the external perturbation the charge densities of the conduction electrons and the positively charged ions of the solid exactly compensate each other. Then we have for the total charge density

$$\rho_{tot}(\mathbf{r},t) = \rho_{ext}(\mathbf{r},t) + \rho_{ind}(\mathbf{r},t) \tag{B.41}$$

where $\rho_{ind}(\mathbf{r},t)$ is the charge density induced by the displacement of charges. With the Fourier transformed Maxwell equations

$$i\mathbf{q}\cdot\mathbf{D}(\mathbf{q},E) = \rho_{ext}(\mathbf{q},E) \tag{B.42}$$

$$i\mathbf{q}\cdot\mathbf{F}(\mathbf{q},E) = \frac{1}{\varepsilon_0}\left(\rho_{ext}(\mathbf{q},E) + \rho_{ind}(\mathbf{q},E)\right) \tag{B.43}$$

where **F** is the electric field strength and **D** is the electric displacement, and the material equation

$$\mathbf{D}(\mathbf{q}, E) = \varepsilon_0 \varepsilon(\mathbf{q}, E)\, \mathbf{F}(\mathbf{q}, E) \tag{B.44}$$

holds for the induced charge density:

$$\rho_{ind}(\mathbf{q}, E) = \left[\frac{1}{\varepsilon(\mathbf{q}, E)} - 1\right] \rho_{ext}(\mathbf{q}, E) \tag{B.45}$$

We now translate the formulas obtained classically so far into quantum mechanics. When we do this, the electron density becomes the density *operator*

$$\rho_{\mathbf{q}} = \sum_{\mathbf{k},\sigma} c^{\dagger}_{\mathbf{k}\sigma} c_{\mathbf{k}+\mathbf{q}\sigma} \quad ; \quad \rho_{-\mathbf{q}} = \rho^{\dagger}_{\mathbf{q}} \tag{B.46}$$

Then the interaction energy also becomes an operator. According to (B.40) it holds

$$V_t = \sum_{\mathbf{q}} \rho^{\dagger}_{\mathbf{q}} \widetilde{F}_t(\mathbf{q}) \tag{B.47}$$

The perturbing field

$$-e\widetilde{F}_t(\mathbf{q}) = \frac{v_0(\mathbf{q})}{2\pi\hbar} \int_{-\infty}^{+\infty} dE\, \rho_{ext}(\mathbf{q}, E)\, e^{-\frac{i}{\hbar}(E+i0^+)t} \tag{B.48}$$

however, remains naturally, a *c*-number. How does the induced charge density (operator!) react to the perturbing field produced by the external charge?

$$\langle \rho_{ind}(\mathbf{q}, t)\rangle = -e\left\{\langle \rho_{\mathbf{q}}\rangle_t - \langle \rho_{\mathbf{q}}\rangle_0\right\} = -e\Delta\left(\rho_{\mathbf{q}}\right)_t \tag{B.49}$$

We now use the Kubo formula (B.17)

$$\Delta\left(\rho_{\mathbf{q}}\right)_t = \frac{1}{\hbar}\sum_{\mathbf{q}'} \int_{-\infty}^{+\infty} dt'\, \widetilde{F}_{t'}(\mathbf{q}') \langle\langle \rho_{\mathbf{q}}(t);\, \rho^{\dagger}_{\mathbf{q}'}(t')\rangle\rangle \tag{B.50}$$

We assume a system with translational symmetry so that we can use

$$\langle\langle \rho_{\mathbf{q}}(t);\, \rho^{\dagger}_{\mathbf{q}'}(t')\rangle\rangle \to \delta_{\mathbf{q},\mathbf{q}'}\, \langle\langle \rho_{\mathbf{q}}(t);\, \rho^{\dagger}_{\mathbf{q}}(t')\rangle\rangle$$

Then we have

$$\langle \rho_{ind}(\mathbf{q}, t)\rangle = \frac{-e}{\hbar} \int_{-\infty}^{+\infty} dt'\, \widetilde{F}_{t'}(\mathbf{q}) \langle\langle \rho_{\mathbf{q}}(t);\, \rho^{\dagger}_{\mathbf{q}}(t')\rangle\rangle \tag{B.51}$$

or after Fourier transformation

$$\langle \rho_{ind}(\mathbf{q}, E) \rangle = \frac{v_0(\mathbf{q})}{\hbar} \rho_{ext}(\mathbf{q}, E) \langle\langle \rho_{\mathbf{q}}; \rho_{\mathbf{q}}^{\dagger} \rangle\rangle_{E+i0^+} \tag{B.52}$$

We compare this with the classical result (B.45) and then we can represent the dielectric function by a retarded Green's function

$$\frac{1}{\varepsilon(\mathbf{q}, E)} = 1 + \frac{1}{\hbar} v_0(\mathbf{q}) \langle\langle \rho_{\mathbf{q}}; \rho_{\mathbf{q}}^{\dagger} \rangle\rangle_{E+i0^+} \tag{B.53}$$

The following two limiting cases are interesting:

- $\varepsilon(\mathbf{q}, E) \gg 1 : \Rightarrow \quad \langle \rho_{ind}(\mathbf{q}, E) \rangle \approx -\rho_{ext}(\mathbf{q}, E)$, almost complete screening of the perturbing charge.
- $\varepsilon(\mathbf{q}, E) \to 0 : \Rightarrow$ arbitrarily small perturbations produce finite density oscillations $\Rightarrow$ "resonances", i.e. collective eigenoscillations of the electron system.

The poles of the Green's function $\langle\langle \rho_{\mathbf{q}}; \rho_{\mathbf{q}}^{\dagger} \rangle\rangle_{E+i0^+}$ are just the energies of the so-called "plasmons".

B.2 Spectroscopies and Spectral Densities

An additional important motivation for the study of Green's functions is their close connection to "elementary excitations" of the system, which are directly observable by appropriate spectroscopies. Thus certain Green's functions provide a direct access to experiment. This is more directly valid for another fundamental function, namely the so-called "spectral density" which has a close relation to the Green's functions. Figure B.2 shows in a schematic form, which elementary processes are involved in four well-known spectroscopies for the determination of electronic structure. The *photoemission (PES)* and the *inverse photoemission (IPE)* are the so-called one-particle spectroscopies since the system (solid) contains after the excitation process one particle more (less) as compared to before the process. In the *photoemission* the energy $\hbar\omega$ of the photon is absorbed by an electron in a (partially) occupied energy band. This gain in energy makes the electron to leave the solid. Analysis of the kinetic energy of the photoelectron leads to conclusions about the occupied states of the concerned energy band. Then the transition operator $Z_{-1} = c_\alpha$ corresponds to the annihilation operator c_α, if the particle occupies the one-particle state $|\alpha\rangle$ before the excitation process. In the *inverse photoemission*, a somewhat reverse process takes place. An electron is shot into the solid and it lands in an unoccupied state $|\beta\rangle$ of a partially filled energy band. The energy released is emitted as a photon $\hbar\omega$ which is analysed. Now the system contains one electron more than before the process. This corresponds to the transition operator $Z_{+1} = c_\beta^{\dagger}$. PES and IPE are in some ways complementary spectroscopies. The former provides

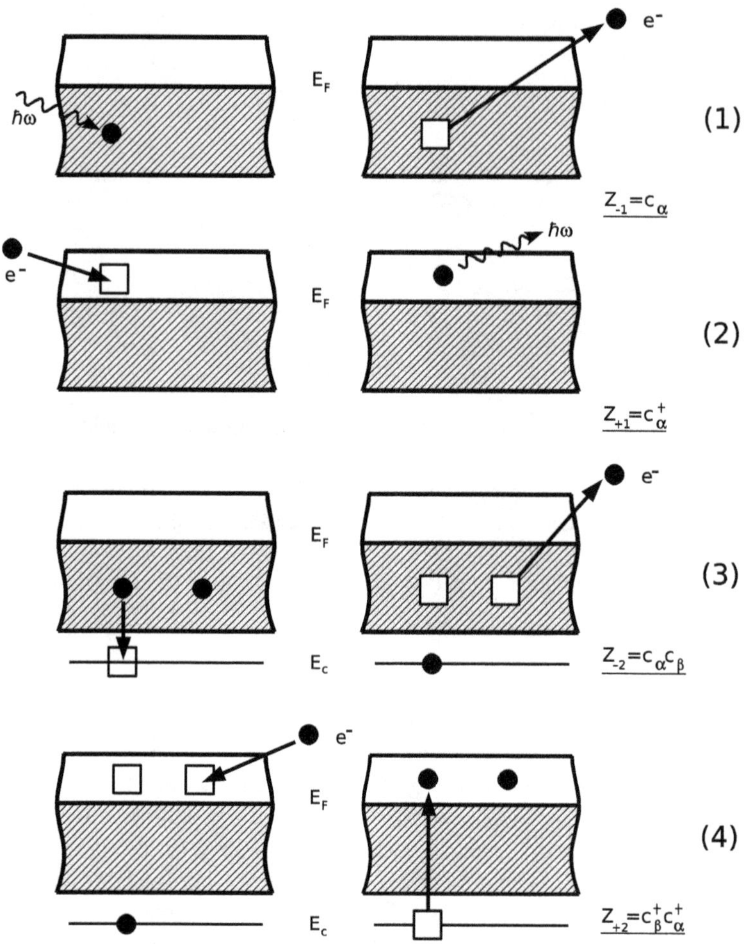

Fig. B.2 Elementary processes relevant for four different spectroscopies: 1. photoemission (PES), 2. inverse photoemission (IPE), 3. Auger electron spectroscopy (AES), and 4. appearance-potential spectroscopy (APS). Z_j is the transition operator, where j means the change in the electron numbers due to the respective excitation

information regarding the occupied and the latter about the unoccupied states of the energy band.

Auger-electron spectroscopy (AES) and *appearance potential spectroscopy (APS)* are two-particle spectroscopies. The starting situation for AES is characterized by the existence of a hole in a deep lying core state. An electron of the partially filled band drops down into this core state and transfers the energy to another electron of the same band so that the latter can leave the solid. The analysis of the kinetic energy of the emitted electron provides information about the energy structure of the occupied states (two-particle density of states). The system (energy band) contains two particles less than before the process: $Z_{-2} = c_\alpha c_\beta$. Almost the reverse process is exploited in APS. An electron lands in an unoccupied state of an energy band. The energy released is used in exciting a core electron to come up and occupy another unoccupied state of the same energy band. The subsequent *de-exciting* process can be analyzed to gain information about the unoccupied states of the energy band. In this case then the system (energy band) contains two electrons more after the process as compared to before the process. Therefore the transition operator is $Z_{+2} = c_{\beta}^{\dagger} c_{\alpha}^{\dagger}$. AES and APS are obviously complementary two-particle spectroscopies.

We now, using simple arguments, want to estimate the associated intensities for the individual processes.

- The system under investigation is described by the Hamiltonian:

$$\mathcal{H} = H - \mu \widehat{N} \tag{B.54}$$

  Here μ is the chemical potential and $\widehat{N}$ is the particle number operator. Here we use $\mathcal{H}$ instead of H since in the following we want to perform averages in grand canonical ensemble. This is necessary because the above discussed transition operators change the particle number. However, H and $\widehat{N}$ should commute which means they possess a common set of eigenstates:

$$H \,|E_n(N)\rangle = E_n(N)\,|E_n(N)\rangle \;\; ; \;\; \widehat{N}\,|E_n(N)\rangle = N\,|E_n(N)\rangle$$

  Then $\mathcal{H}$ satisfies the eigenvalue equation

$$\mathcal{H}\,|E_n(N)\rangle = (E_n(N) - \mu N)\,|E_n(N)\rangle \to E_n\,|E_n\rangle \tag{B.55}$$

  In order to save writing effort, in the following, so long as it does not create any confusion, we will write the short forms i.e. for the eigenvalue instead of $(E_n(N) - \mu N)$ write E_n and for the eigenstates instead of $|E_n(N)\rangle$ simply write $|E_n\rangle$. However, the actual dependence of the states and eigenenergies on the particle number should always be kept in mind.
- The system at temperature T finds itself in an eigenstate $|E_n\rangle$ of the Hamiltonian $\mathcal{H}$ with the probability

$$\frac{1}{\Xi} \exp(-\beta E_n)$$

  where Ξ is the grand canonical partition function

$$\Xi = Tr\,(\exp(-\beta\mathcal{H}) \tag{B.56}$$

- The transition operator Z_r induces transitions between the states $|E_n\rangle$ and $|E_m\rangle$ with the probability

$$|\langle E_m|Z_r|E_n\rangle|^2 \qquad r = \pm 1, \pm 2$$

- The *intensity* of the measured elementary process corresponds to the total number of transitions with excitation energies between E and $E + dE$:

$$I_r(E) = \frac{1}{\Xi} \sum_{m,n} e^{-\beta E_n} \, |\langle E_m | Z_r | E_n \rangle|^2 \, \delta(E - (E_m - E_n)) \tag{B.57}$$

  If the excitation energies are sufficiently dense, as is anyway the case for a solid, then $I_r(E)$ is a continuous function of the energy E.
- At this point we neglect certain secondary effects, which of course are important for a quantitative analysis of the respective experiments, but are not decisive for the actually interesting processes. This is valid, for example, in PES and AES for the fact that the photoelectron leaving the solid can still couple to the residual system (*sudden approximation*). In addition, the matrix elements for the transition from band level into vacuum are not taken into account here. However, the *bare line forms* of the mentioned spectroscopies should be correctly described by (B.57).

One should note that for the transition operator holds

$$Z_r = Z^{\dagger}_{-r} \tag{B.58}$$

That means complementary spectroscopies are related to each other in some way. This is now investigated in more detail:

$$\begin{aligned}
I_r(E) &= \frac{1}{\Xi} \sum_{m,n} e^{\beta E} e^{-\beta E_m} \, |\langle E_m | Z_r | E_n \rangle|^2 \, \delta(E - (E_m - E_n)) \\
&= \frac{1}{\Xi} \sum_{n,m} e^{\beta E} e^{-\beta E_n} \, |\langle E_n | Z_r | E_m \rangle|^2 \, \delta(E - (E_n - E_m)) \\
&= \frac{e^{\beta E}}{\Xi} \sum_{n,m} e^{-\beta E_n} \, |\langle E_m | Z_{-r} | E_n \rangle|^2 \, \delta((-E) - (E_m - E_n))
\end{aligned}$$

In the second step only the summation indices n and m are interchanged; the last transition uses (B.58). We then have derived a "symmetry relation" for the *complementary* spectroscopies:

$$I_r(E) = e^{\beta E} \, I_{-r}(-E) \tag{B.59}$$

We now define the *spectral density* which is important for the following:

$$\frac{1}{\hbar} S_r^{(\pm)}(E) = I_{-r}(E) \mp I_r(-E) = \left(e^{\beta E} \mp 1\right) I_r(-E) \tag{B.60}$$

The freedom in the sign will be explained later. From (B.59) and (B.60) one recognizes that the intensities of the *complementary* spectroscopies are determined by the same spectral density in a simple manner:

$$\hbar\, I_r(E) = \frac{1}{e^{-\beta E} \mp 1}\, S_r^{(\pm)}(-E) \tag{B.61}$$

$$\hbar\, I_{-r}(E) = \frac{e^{\beta E}}{e^{\beta E} \mp 1}\, S_r^{(\pm)}(E) \tag{B.62}$$

Thus the just introduced spectral density is closely related to the intensities of spectroscopies. Therefore we want to further investigate this function by performing a Fourier transformation into the time domain:

$$\begin{aligned}
&\frac{1}{2\pi\hbar}\int_{-\infty}^{+\infty} dE e^{-\frac{i}{\hbar}E(t-t')} I_{-r}(E) \\
&= \frac{1}{2\pi\hbar}\frac{1}{\Xi}\sum_{m,n} e^{-\beta E_n}\, e^{-\frac{i}{\hbar}(E_m-E_n)(t-t')} \langle E_m|Z_{-r}|E_n\rangle\, \langle E_n|Z^{\dagger}_{-r}|E_m\rangle \\
&= \frac{1}{2\pi\hbar}\frac{1}{\Xi}\sum_{m,n} e^{-\beta E_n} \langle E_m|e^{\frac{i}{\hbar}\mathcal{H}t'} Z_{-r} e^{-\frac{i}{\hbar}\mathcal{H}t'}|E_n\rangle \,* \\
&\qquad\qquad * \langle E_n|e^{\frac{i}{\hbar}\mathcal{H}t} Z_r e^{-\frac{i}{\hbar}\mathcal{H}t}|E_m\rangle \\
&= \frac{1}{2\pi\hbar}\frac{1}{\Xi}\sum_{m,n} e^{-\beta E_n} \langle E_n|Z_r(t)|E_m\rangle\, \langle E_m|Z_r^{\dagger}(t')|E_n\rangle \\
&= \frac{1}{2\pi\hbar}\frac{1}{\Xi}\sum_{n} e^{-\beta E_n} \langle E_n|Z_r(t)Z_r^{\dagger}(t')|E_n\rangle \\
&= \frac{1}{2\pi\hbar}\langle Z_r(t)\, Z_r^{\dagger}(t')\rangle
\end{aligned}$$

Completely analogously one finds

$$\frac{1}{2\pi\hbar}\int_{-\infty}^{+\infty} dE e^{-\frac{i}{\hbar}E(t-t')} I_r(-E) = \frac{1}{2\pi\hbar}\langle Z_r^{\dagger}(t')\, Z_r(t)\rangle$$

That means with (B.60) for the double-time spectral density

$$\begin{aligned}
S_r^{(\eta)}(t,t') &= \frac{1}{2\pi\hbar}\int_{-\infty}^{+\infty} dE e^{-\frac{i}{\hbar}E(t-t')}\, S_r^{(\eta)}(E) \\
&= \frac{1}{2\pi}\,\langle\big[Z_r(t),\; Z_r^{\dagger}(t')\big]_{-\eta}\rangle
\end{aligned} \tag{B.63}$$

Here $\eta = \pm$ is at the moment only an arbitrary sign factor. $[\cdots,\cdots]_{-\eta}$ is either a commutator or an anti-commutator:

$$\big[Z_r(t),\; Z_r^{\dagger}(t')\big]_{-\eta} = Z_r(t)\, Z_r^{\dagger}(t') - \eta\, Z_r^{\dagger}(t')\, Z_r(t) \tag{B.64}$$

We have shown that spectral density in (B.63) is of central importance for the intensities of spectroscopies. In addition it can also be shown that a generalization of spectral density to arbitrary operators $\widehat{A}$ and $\widehat{B}$ is closely related to the retarded Green's functions introduced in (B.16). This is also true for the other types of Green's functions to be defined in the next section. The "spectral density"

$$S_{AB}^{(\eta)}(t,t') = \frac{1}{2\pi}\langle\big[\widehat{A}(t),\ \widehat{B}(t')\big]_{-\eta}\rangle \tag{B.65}$$

has the same place of importance as the Green's functions in the many-body theory.

B.3 Double-Time Green's Functions

B.3.1 Definitions and Equations of Motion

In order to construct the full Green's function formalism the retarded Green's function introduced earlier is not sufficient. One needs two other types:
Retarded Green's function

$$G_{AB}^{ret}(t,t') = \langle\langle A(t);\ B(t')\rangle\rangle^{ret} = -i\Theta(t-t')\langle\big[A(t),\ B(t')\big]_{-\eta}\rangle \tag{B.66}$$

Advanced Green's function

$$G_{AB}^{ad}(t,t') = \langle\langle A(t);\ B(t')\rangle\rangle^{ad} = +i\Theta(t'-t)\langle\big[A(t),\ B(t')\big]_{-\eta}\rangle \tag{B.67}$$

Causal Green's function

$$G_{AB}^{c}(t,t') = \langle\langle A(t);\ B(t')\rangle\rangle^{c} = -i\langle T_\eta\left\{A(t)B(t')\right\}\rangle \tag{B.68}$$

The operators here are in their time-dependent Heisenberg picture:

$$X(t) = e^{\frac{i}{\hbar}\mathcal{H}t}\, X\, e^{-\frac{i}{\hbar}\mathcal{H}t} \tag{B.69}$$

where, just as in the derivation of the spectral density in the last section, the transformation should be carried out with the grand canonical Hamiltonian $\mathcal{H} = H - \mu\widehat{N}$. In doing this we assume that $\mathcal{H}$ is not explicitly dependent on time. The averages are performed in the grand canonical ensemble:

$$\langle XY\rangle = \frac{1}{\Xi}\, Tr\left(e^{-\beta\mathcal{H}}\, XY\right) \tag{B.70}$$

Ξ is the grand canonical partition function (B.56). The choice of the sign $\eta = \pm$ is arbitrary and appears based on the convenience in a given situation. If $\widehat{A}$ and $\widehat{B}$ are pure Fermi (Bose) operators, then $\eta = -$ $(\eta = +)$ is convenient but not necessarily required. The brackets in (B.66) and (B.67) are anti-commutators in the former case and commutators in the latter case as was fixed in (B.64).

The definition (B.68) of the causal Green's function contains the *Wick's time ordering operator*:

$$T_\eta \{A(t)B(t')\} = \Theta(t-t')\, A(t)B(t') + \eta\, \Theta(t'-t)\, B(t')A(t) \tag{B.71}$$

The step function Θ

$$\Theta(t-t') = \begin{cases} 1 \text{ for } t > t' \\ 0 \text{ for } t < t' \end{cases} \tag{B.72}$$

is not defined for equal times. This is reflected in the Green's functions and so has to be considered later. Because of the averaging in (B.70), the Green's functions are also temperature dependent. As a result, later, an unusual but convenient relationship between the temperature and time variables will be established.

In addition to the Green's functions (B.66), (B.67) and (B.68), the spectral density (B.65) is of equal significance.

We now prove "retrospectively" that if $\mathcal{H}$ is not explicitly dependent on time, then the Green's functions and the spectral density are homogeneous in time:

$$\frac{\partial \mathcal{H}}{\partial t} = 0 \quad \curvearrowright \quad G_{AB}^{ret,ad,c}(t,t') = G_{AB}^{ret,ad,c}(t-t') \tag{B.73}$$

$$S_{AB}(t,t') = S_{AB}(t-t') \tag{B.74}$$

The proof is complete provided the homogeneity of the "correlation functions"

$$\langle A(t)\, B(t') \rangle \;\; ; \;\; \langle B(t')\, A(t) \rangle$$

is proved. This is achieved by using the cyclic invariance of trace:

$$\begin{aligned} Tr\left(e^{-\beta\mathcal{H}} A(t)\, B(t')\right) &= Tr\left(e^{-\beta\mathcal{H}} e^{\frac{i}{\hbar}\mathcal{H}t}\, A\, e^{-\frac{i}{\hbar}\mathcal{H}t}\, e^{\frac{i}{\hbar}\mathcal{H}t'}\, B\, e^{-\frac{i}{\hbar}\mathcal{H}t'}\right) \\ &= Tr\left(e^{-\frac{i}{\hbar}\mathcal{H}t'} e^{-\beta\mathcal{H}} e^{\frac{i}{\hbar}\mathcal{H}t}\, A\, e^{-\frac{i}{\hbar}\mathcal{H}t}\, e^{\frac{i}{\hbar}\mathcal{H}t'}\, B\right) \\ &= Tr\left(e^{-\beta\mathcal{H}} e^{-\frac{i}{\hbar}\mathcal{H}t'}\, e^{\frac{i}{\hbar}\mathcal{H}t}\, A\, e^{-\frac{i}{\hbar}\mathcal{H}t}\, e^{\frac{i}{\hbar}\mathcal{H}t'}\, B\right) \\ &= Tr\left(e^{-\beta\mathcal{H}}\, A(t-t')\, B\right) \end{aligned}$$

Thus the homogeneity is shown as

$$\langle A(t)\, B(t') \rangle = \langle A(t-t')\, B(0) \rangle \tag{B.75}$$

Analogously one finds

$$\langle B(t')\, A(t) \rangle = \langle B(0)\, A(t-t') \rangle \tag{B.76}$$

For the approximate determination of a Green's function, one can use the respective *equation of motion* which is directly derived from the equation of motion of the corresponding time-dependent Heisenberg operators. With

$$\frac{\partial}{\partial t}\Theta(t-t') = \delta(t-t') = -\frac{\partial}{\partial t'}\Theta(t-t') \tag{B.77}$$

one gets formally the same equation of motion for all the three Green's functions:

$$\begin{aligned} i\hbar\frac{\partial}{\partial t}G_{AB}^{ret,ad,c}(t,t') &= \hbar\delta(t-t')\langle[A,\ B]_{-\eta}\rangle \\ &\quad + \langle\langle[A,\ \mathcal{H}]_{-}(t);\ B(t')\rangle\rangle^{ret,ad,c} \end{aligned} \tag{B.78}$$

The boundary conditions are, however, different:

$$G_{AB}^{ret}(t,t') = 0 \quad \text{for} \quad t < t' \tag{B.79}$$

$$G_{AB}^{av}(t,t') = 0 \quad \text{for} \quad t > t' \tag{B.80}$$

$$G_{AB}^{c}(t,t') = \begin{cases} -i\langle A(t-t')\,B(0)\rangle & \text{for } t > t' \\ -i\eta\langle B(0)\,A(t-t')\rangle & \text{for } t < t' \end{cases} \tag{B.81}$$

The boundary conditions for the causal function are quite unmanageable. For that reason this function does not play any role in the equation of motion method. That is why it will not be considered here any more.

On the right-hand side of the equation of motion (B.78) appears a *higher* Green's function as $[A,\ \mathcal{H}]_{-}(t)$ is itself a time-dependent operator. In some special cases $[A,\ \mathcal{H}]_{-}(t) \propto A(t)$ holds. Then an exact solution of the equation of motion is directly possible. However, in general, such a proportionality does not hold. Then the *higher* Green's function satisfies its own equation of motion of the form

$$\begin{aligned} i\hbar\frac{\partial}{\partial t}\langle\langle[A,\ \mathcal{H}]_{-}(t);\ B(t')\rangle\rangle^{ret,ad} &= \\ = \hbar\delta(t-t')\langle\big[[A,\ \mathcal{H}]_{-},\ B\big]_{-\eta}\rangle &+ \\ +\langle\langle\big[[A,\ \mathcal{H}]_{-},\ \mathcal{H}\big]_{-}(t);\ B(t')\rangle\rangle^{ret,ad} \end{aligned} \tag{B.82}$$

On the right-hand side appears a *still higher* Green's function for which again another equation of motion can be written. This leads to an infinite *chain of equations of motion* which, at some stage, has to be decoupled physically meaningfully.

Going from the time domain into energy domain also does not change anything for this chain of equations. However, with (B.18) and (B.21), one gets a pure algebraic equation which can possibly be of advantage:

$$E\langle\langle A;\ B\rangle\rangle_E^{ret,ad} = \hbar\langle[A,\ B]_{-\eta}\rangle + \langle\langle[A,\ \mathcal{H}]_{-};\ B\rangle\rangle_E^{ret,ad} \tag{B.83}$$

The boundary conditions (B.79) and (B.80) manifest themselves as different analytical behaviours of the Green's functions in the complex E-plane. This shall be investigated in detail in the next section.

B.3.2 Spectral Representations

In order to learn more about the analytical properties of these functions, with the help of the eigenvalues and eigenstates (B.55) of the Hamiltonian $\mathcal{H}$, we derive the spectral representations of the retarded and advanced Green's functions. The eigenstates constitute a complete orthonormal system:

$$\langle E_n | E_m \rangle = \delta_{nm} \;\; ; \;\; \sum_n |E_n\rangle\langle E_n| = \mathbb{1} \tag{B.84}$$

Using this, we first rewrite the correlation functions $\langle A(t)\, B(t')\rangle$, $\langle B(t')\, A(t)\rangle$:

$$\begin{aligned}
\Xi \cdot \langle A(t)\, B(t')\rangle &= Tr\left\{e^{-\beta\mathcal{H}} A(t)\, B(t')\right\} \\
&= \sum_n \langle E_n | e^{-\beta\mathcal{H}} A(t)\, B(t') | E_n\rangle \\
&= \sum_{n,m} \langle E_n | e^{-\beta\mathcal{H}} A(t) | E_m\rangle\langle E_m | B(t') | E_n\rangle \\
&= \sum_{n,m} e^{-\beta E_n} \langle E_n | A | E_m\rangle\langle E_m | B | E_n\rangle e^{\frac{i}{\hbar}(E_n - E_m)(t-t')} \\
&= \sum_{n,m} e^{-\beta E_n} e^{\beta(E_n - E_m)} \langle E_n | B | E_m\rangle\langle E_m | A | E_n\rangle\, * \\
&\qquad\qquad * e^{-\frac{i}{\hbar}(E_n - E_m)(t-t')}
\end{aligned}$$

In the last step we simply interchanged the summation indices n and m. Completely analogously one finds the other correlation function

$$\Xi \cdot \langle B(t')\, A(t)\rangle = \sum_{n,m} e^{-\beta E_n} \langle E_n | B | E_m\rangle\langle E_m | A | E_n\rangle e^{-\frac{i}{\hbar}(E_n - E_m)(t-t')}$$

Comparing these expressions with the Fourier representation of the spectral density

$$\begin{aligned}
S_{AB}(t, t') &= \frac{1}{2\pi}\langle\left[A(t),\; B(t')\right]_{-\eta}\rangle \\
&= \frac{1}{2\pi\hbar}\int_{-\infty}^{+\infty} dE\, e^{-\frac{i}{\hbar}E(t-t')} S_{AB}(E)
\end{aligned} \tag{B.85}$$

we get the important "spectral representation of the spectral density"

$$S_{AB}(E) = \frac{\hbar}{\Xi} \sum_{n,m} \langle E_n|B|E_m\rangle\langle E_m|A|E_n\rangle * \\ * e^{-\beta E_n}(e^{\beta E} - \eta)\delta\left(E - (E_n - E_m)\right) \tag{B.86}$$

The arguments of the delta functions contain the possible excitation energies! Let us compare this with (B.57), the intensity formula for r-particle spectroscopies and their simple relationships (B.61) and (B.62) with the respective spectral densities.

We now try to express the Green's functions in terms of spectral density. This is possible with the following representation of the step function:

$$\Theta(t - t') = \frac{i}{2\pi} \int_{-\infty}^{+\infty} dx \, \frac{e^{-ix(t-t')}}{x + i0^+} \tag{B.87}$$

This is proved using the residue theorem (Problem B.4). Using this we first rewrite the retarded function

$$\begin{aligned}
G^{ret}_{AB}(E) &= \int_{-\infty}^{+\infty} d(t - t') e^{\frac{i}{\hbar}E(t-t')} \left(-i\Theta(t - t')\right) \left(2\pi S_{AB}(t - t')\right) \\
&= \int_{-\infty}^{+\infty} d(t - t') e^{\frac{i}{\hbar}E(t-t')} \left(\frac{1}{2\pi} \int_{-\infty}^{+\infty} dx \, \frac{e^{-ix(t-t')}}{x + i0^+}\right) * \\
&\qquad * \left(\frac{1}{\hbar} \int_{-\infty}^{+\infty} dE' S_{AB}(E') e^{-\frac{i}{\hbar}E'(t-t')}\right) \\
&= \int_{-\infty}^{+\infty} dE' S_{AB}(E') \int_{-\infty}^{+\infty} dx \frac{1}{x + i0^+} * \\
&\qquad * \frac{1}{2\pi\hbar} \int_{-\infty}^{+\infty} d(t - t') e^{\frac{i}{\hbar}(E - E' - \hbar x)(t-t')} \\
&= \int_{-\infty}^{+\infty} dE' S_{AB}(E') \int_{-\infty}^{+\infty} dx \frac{1}{x + i0^+} \delta(E - E' - \hbar x) \\
&= \int_{-\infty}^{+\infty} dE' \frac{1}{\hbar} \frac{S_{AB}(E')}{\frac{1}{\hbar}(E - E') + i0^+}
\end{aligned}$$

This gives the "spectral representation of the retarded Green's function":

$$G^{ret}_{AB}(E) = \int_{-\infty}^{+\infty} dE' \frac{S_{AB}(E')}{E - E' + i0^+} \tag{B.88}$$

The advanced Green's function can be treated completely analogously:

$$
\begin{aligned}
G_{AB}^{ad}(E) &= \int_{-\infty}^{+\infty} d(t-t') e^{\frac{i}{\hbar}E(t-t')} \left(+i\Theta(t'-t)\right) \left(2\pi S_{AB}(t-t')\right) \\
&= \int_{-\infty}^{+\infty} d(t-t') e^{\frac{i}{\hbar}E(t-t')} \left(\frac{-1}{2\pi} \int_{-\infty}^{+\infty} dx \, \frac{e^{-ix(t'-t)}}{x+i0^+}\right) * \\
&\quad * \left(\frac{1}{\hbar} \int_{-\infty}^{+\infty} dE' S_{AB}(E') e^{-\frac{i}{\hbar}E'(t-t')}\right) \\
&= \int_{-\infty}^{+\infty} dE' S_{AB}(E') \int_{-\infty}^{+\infty} dx \frac{-1}{x+i0^+} * \\
&\quad * \frac{1}{2\pi\hbar} \int_{-\infty}^{+\infty} d(t-t') e^{\frac{i}{\hbar}(E-E'+\hbar x)(t-t')} \\
&= \int_{-\infty}^{+\infty} dE' S_{AB}(E') \int_{-\infty}^{+\infty} dx \frac{-1}{x+i0^+} \delta(E-E'+\hbar x) \\
&= \int_{-\infty}^{+\infty} dE' \frac{1}{\hbar} \frac{S_{AB}(E')}{\frac{1}{\hbar}(E-E')-i0^+}
\end{aligned}
$$

This gives the "spectral representation of the advanced Green's function":

$$
G_{AB}^{ad}(E) = \int_{-\infty}^{+\infty} dE' \frac{S_{AB}(E')}{E-E'-i0^+} \qquad \text{(B.89)}
$$

The different sign of $i0^+$ in the denominator of the integrand is the only but important difference between the retarded and the advanced functions. The retarded and advanced functions have poles, respectively, in the lower and upper half-planes. This results in different analytical behaviours of the two functions.

$G_{AB}^{ret}(E)$ in the upper and $G_{AB}^{av}(E)$ in the lower complex half-plane can be analytically continued!

Finally, we can substitute (B.86) in (B.88) and (B.89):

$$
\begin{aligned}
G_{AB}^{\substack{ret\\ad}}(E) &= \frac{\hbar}{\Xi} \sum_{n,m} \langle E_n|B|E_m\rangle\langle E_m|A|E_n\rangle * \\
&\quad * e^{-\beta E_n} \frac{e^{\beta(E_n-E_m)}-\eta}{E-(E_n-E_m)\pm i0^+}
\end{aligned} \qquad \text{(B.90)}
$$

We see that we have meromorphic functions with simple poles at the *exact (!)* excitation energies of the interacting system. With a suitable choice of the operators A

and B special poles can be *extracted*, i.e. $E = E_n - E_m$ appears as a pole only if $\langle E_n|B|E_m\rangle \neq 0$ and $\langle E_m|A|E_n\rangle \neq 0$.

Because of the identical predictive power and the same equation of motion the retarded and advanced functions are considered to be the two branches of a unified Green's function in the complex plane:

$$G_{AB}(E) = \int_{-\infty}^{+\infty} dE' \frac{S_{AB}(E')}{E - E'} = \begin{cases} G_{AB}^{ret}(E)\,, \text{ if } Im E > 0 \\ G_{AB}^{ad}(E)\,, \text{ if } Im E < 0 \end{cases} \tag{B.91}$$

In the following, we want to call this the *combined* Green's function. Its poles lie on the real axis.

With the "Dirac identity",

$$\frac{1}{x - x_0 \pm i0^+} = \mathcal{P}\frac{1}{x - x_0} \mp i\pi\delta(x - x_0) \tag{B.92}$$

where $\mathcal{P}$ represents the Cauchy's principal value, one finds the following relation between the spectral density and the Green's functions:

$$S_{AB}(E) = \frac{i}{2\pi}\left(G_{AB}^{ret}(E) - G_{AB}^{av}(E)\right) \tag{B.93}$$

If $S_{AB}(E)$ is real, which it is in many of the important cases (e.g. $B = A^\dagger$), this relation becomes even simpler:

$$S_{AB}(E) = \mp\frac{1}{\pi}\, Im\, G_{AB}^{\substack{ret\\ av}}(E) \tag{B.94}$$

Equations (B.88) and (B.89) show that the Green's functions are completely fixed by the spectral density. On the other hand according to (B.94) only the imaginary part of the Green's function determines the spectral density. This means there must be relations between the real and the imaginary parts of the Green's function; they are not independent of each other. These relations are called the *Kramers–Kronig relations* which we will explicitly derive in Sect. B.3.5.

B.3.3 Spectral Theorem

The discussion of the last section has shown that the Green's functions and the spectral density, in addition to their importance in the context of the response functions (Sect. B.1) and intensities of certain spectroscopies, also provide valuable microscopic information. Their singularities correspond to the exact excitation energies of the system. We now want to demonstrate that the whole macroscopic thermodynamics can be determined using suitably defined Green's functions and spectral densities.

The starting point is the spectral representation of the correlation functions introduced in Sect. B.3.2.

$$\langle B(t')\,A(t)\rangle = \frac{1}{\Xi}\sum_{n,m} e^{-\beta E_n}\langle E_n|B|E_m\rangle\langle E_m|A|E_n\rangle e^{-\frac{i}{\hbar}(E_n-E_m)(t-t')}$$

Comparing this expression with that for the spectral representation of the spectral density (B.86), we get the fundamental *spectral theorem*:

$$\langle B(t')\,A(t)\rangle = \frac{1}{\hbar}\int_{-\infty}^{+\infty} dE\,\frac{S_{AB}^{(\eta)}(E)}{e^{\beta E}-\eta}\,e^{-\frac{i}{\hbar}E(t-t')} + \frac{1}{2}(1+\eta)\,D \tag{B.95}$$

Except for the second summand, this result follows directly from the comparison with (B.86). The second term comes into play only for the commutator ($\eta = +1$) spectral density. It disappears for the anti-commutator function ($\eta = -1$). The reason for this is clear from (B.86). For $E = 0$, i.e. $E_n = E_m$, the commutator spectral densities do not contribute the corresponding term because $(e^{\beta E} - 1) = 0$, even though they may give a contribution $\frac{1}{\hbar}D$ unequal zero in the correlation function, where

$$D = \frac{\hbar}{\Xi}\sum_{n,m}^{E_n=E_m} e^{-\beta E_n}\langle E_n|B|E_m\rangle\langle E_m|A|E_n\rangle \tag{B.96}$$

Experience, however, shows that in fact in many cases $D = 0$. But the fact that this is not always necessary is made easily clear from the following example: The operator pairs A, B and $\widetilde{A} = A - \langle A\rangle \mathbb{1}$, $\widetilde{B} = B - \langle B\rangle \mathbb{1}$ form identical spectral densities:

$$S_{AB}^{(+)}(t-t') \equiv S_{\widetilde{A}\widetilde{B}}^{(+)}(t-t') \iff S_{AB}^{(+)}(E) \equiv S_{\widetilde{A}\widetilde{B}}^{(+)}(E)$$

On the other hand holds

$$\langle \widetilde{B}(t')\,\widetilde{A}(t)\rangle = \langle B(t')\,A(t)\rangle - \langle B(t')\rangle\langle A(t)\rangle$$

which would lead to a contradiction without the second term in (B.95) in case $\langle A(t)\rangle \neq 0$ and $\langle B(t')\rangle \neq 0$. Thus without the extra term D, the spectral theorem would be incomplete for the commutator spectral density.

How does one determine D? It is possible with the *combined* Green's function (B.91) whose spectral representation

$$G_{AB}^{(\eta)}(E) = \frac{\hbar}{\Xi}\sum_{n,m}\langle E_n|B|E_m\rangle\langle E_m|A|E_n\rangle e^{-\beta E_n}\,\frac{e^{\beta(E_n-E_m)}-\eta}{E-(E_n-E_m)} \tag{B.97}$$

leads to the following relation:

$$\lim_{E\to 0} E \cdot G^{(\eta)}_{AB}(E) = (1-\eta)\, D \tag{B.98}$$

The limit must be taken in complex plane. One recognizes the following important consequences:

- Even though D is needed for the *commutator* spectral density, the determination succeeds only using the *anti-commutator* Green's function.
- The commutator Green's function $G^{(+)}_{AB}(E)$ is always regular at $E = 0$, i.e. it has no pole there.
- The anti-commutator Green's function $G^{(-)}_{AB}(E)$, for $D \neq 0$, has a pole of first order with the residue $2D$.

B.3.4 Spectral Moments

Green's functions and spectral densities for realistic problems are in general not exactly solvable. Therefore, one must tolerate approximations. These approximations can be checked for correctness by using *exactly solvable limiting cases, symmetry relations, sum rules, etc.*

In this sense, the moments of spectral density are found to be extraordinarily useful. Let n and p be non-negative integers:

$$n = 0, 1, 2, \cdots \qquad ; \qquad 0 \le p \le n$$

Then for the time-dependent spectral density holds

$$\begin{aligned}
&\left(i\hbar\frac{\partial}{\partial t}\right)^{n-p}\left(-i\hbar\frac{\partial}{\partial t'}\right)^{p}\left(2\pi\, S_{AB}(t-t')\right) = \\
&= \left(i\hbar\frac{\partial}{\partial t}\right)^{n-p}\left(-i\hbar\frac{\partial}{\partial t'}\right)^{p}\frac{1}{\hbar}\int_{-\infty}^{+\infty} dE\, S_{AB}(E)\, e^{-\frac{i}{\hbar}E(t-t')} \\
&= \frac{1}{\hbar}\int_{-\infty}^{+\infty} dE\, E^n\, S_{AB}(E)\, e^{-\frac{i}{\hbar}E(t-t')} \\
&\overset{!}{=} \left(i\hbar\frac{\partial}{\partial t}\right)^{n-p}\left(-i\hbar\frac{\partial}{\partial t'}\right)^{p}\left\langle\left[A(t),\, B(t')\right]_{-\eta}\right\rangle
\end{aligned}$$

For $t = t'$ from the first part of the system of equations one gets the *spectral moments*

$$M^{(n)}_{AB} = \frac{1}{\hbar}\int_{-\infty}^{+\infty} dE\, E^n\, S_{AB}(E) \tag{B.99}$$

An alternative representation results from the last part by using the equation of motion for the time-dependent Heisenberg operators

$$M_{AB}^{(n)} =$$
$$= \langle \Big[\underbrace{[\cdots[[A,\,\mathcal{H}]_-,\,\mathcal{H}]_- \cdots,\,\mathcal{H}]_-}_{(n-p)-\text{fold}},\,\underbrace{[\mathcal{H},\cdots,\,[\mathcal{H},\,B]_- \cdots]_-}_{p-\text{fold}} \Big]_{-\eta} \rangle \tag{B.100}$$

With this last relation it is possible *in principle* to calculate exactly all the moments *independently* of the respective spectral densities if the Hamiltonian is known. Then one has the possibility to control certain approximate procedures for the spectral density using (B.99).

With the spectral moments, it is possible to formulate a very often useful "high-energy expansion". For the "combined Green's function" (B.91) holds

$$\begin{aligned} G_{AB}(E) &= \int_{-\infty}^{+\infty} dE' \, \frac{S_{AB}(E')}{E - E'} \\ &= \frac{1}{E} \int_{-\infty}^{+\infty} dE' \, \frac{S_{AB}(E')}{1 - \frac{E'}{E}} \\ &= \frac{1}{E} \sum_{n=0}^{\infty} \int_{-\infty}^{+\infty} dE' \, S_{AB}(E') \left(\frac{E}{E'}\right)^n \end{aligned}$$

A comparison with (B.99) gives

$$G_{AB}(E) = \hbar \sum_{n=0}^{\infty} \frac{M_{AB}^{(n)}}{E^{n+1}} \tag{B.101}$$

For the extreme high-energy behaviour ($E \to \infty$) this means

$$G_{AB}(E) \approx \frac{\hbar}{E} M_{AB}^{(0)} = \frac{\hbar}{E} \langle [A,\,B]_{-\eta} \rangle \tag{B.102}$$

The right-hand side is in general easy to calculate and therefore, e.g. the high-energy behaviour of the response functions in Sect. B.1.1 is known.

B.3.5 Kramer's–Kronig Relations

We have already seen that the Green's function $G_{AB}^{ret,ad}(E)$ is completely determined by the spectral density $S_{AB}(E)$ (B.88) and (B.89). In case it is real, then it can be determined by the imaginary part of the Green's function alone (B.94). Therefore the real and imaginary parts of the Green's function cannot be independent of each other. We will now derive the relationship between them.

One calculates the following integral for real E:

$$I_C(E) = \oint_C d\hat{E} \frac{G^{ret}_{AB}(\hat{E})}{E - \hat{E} - i0^+} \tag{B.103}$$

Integration is performed over the path C which follows the real axis and is closed in the upper half of the complex plane (Fig. B.3). The integrand has a pole at $\hat{E} = E - i0^+$, i.e. in the lower half-plane. The retarded Green's function also has a pole only in the lower half-plane so that there is no pole in the region enclosed by C. Therefore it holds

$$I_C(E) = 0 \tag{B.104}$$

The semicircle is extended to infinity. The high-energy expansion (B.101) shows that the integrand in (B.103) then goes to zero at least as $1/\hat{E}^2$. As a result, the semicircle does not contribute to the integral (B.103). It remains when the Dirac identity (B.92) is used:

$$\begin{aligned} 0 &= \int_{-\infty}^{+\infty} d\hat{E} \frac{G^{ret}_{AB}(\hat{E})}{E - \hat{E} - i0^+} \\ &= \mathcal{P} \int_{-\infty}^{+\infty} d\hat{E} \frac{G^{ret}_{AB}(\hat{E})}{E - \hat{E}} + i\pi G^{ret}_{AB}(E) \end{aligned}$$

That means

$$G^{ret}_{AB}(E) = \frac{i}{\pi} \mathcal{P} \int_{-\infty}^{+\infty} d\hat{E} \frac{G^{ret}_{AB}(\hat{E})}{E - \hat{E}} \tag{B.105}$$

Completely analogous considerations hold for the advanced function $G^{ad}_{AB}(E)$. Now the semicircle will be in the lower half-plane and in the denominator of the integrand (B.103) $-i0^+$ is replaced by $+i0^+$. Then we have

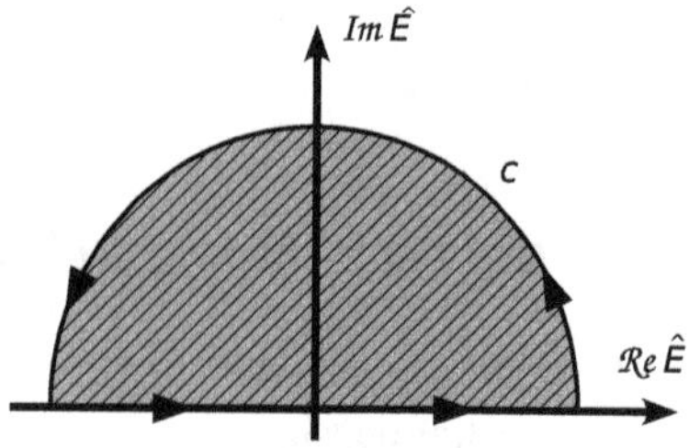

Fig. B.3 Integration path C in the complex E-plane for the integral (B.103)

$$G^{ad}_{AB}(E) = -\frac{i}{\pi}\mathcal{P}\int_{-\infty}^{+\infty} d\hat{E}\,\frac{G^{ad}_{AB}(\hat{E})}{E-\hat{E}} \tag{B.106}$$

Thus it is not at all necessary to know the full Green's function. The real and imaginary parts account for each other ("Kramers–Kronig relations")

$$Re\overset{ret}{G^{ad}_{AB}}(E) = \mp\frac{i}{\pi}\mathcal{P}\int_{-\infty}^{+\infty} d\hat{E}\,\frac{Im\overset{ret}{G^{ad}_{AB}}(\hat{E})}{E-\hat{E}} \tag{B.107}$$

$$Im\overset{ret}{G^{ad}_{AB}}(E) = \pm\frac{i}{\pi}\mathcal{P}\int_{-\infty}^{+\infty} d\hat{E}\,\frac{Re\overset{ret}{G^{ad}_{AB}}(\hat{E})}{E-\hat{E}} \tag{B.108}$$

In case the respective spectral density is real the additional relation (B.94) holds. That leads to

$$ImG^{ret}_{AB}(E) = -ImG^{ad}_{AB}(E) = -\pi S_{AB}(E) \tag{B.109}$$

$$ReG^{ret}_{AB}(E) = ReG^{ad}_{AB}(E) = \mathcal{P}\int_{-\infty}^{+\infty} d\hat{E}\,\frac{S_{AB}(\hat{E})}{E-\hat{E}} \tag{B.110}$$

The retarded and advanced Green's functions are therefore very closely related.

B.3.6 Simple Applications

We want to apply the above introduced Green's function formalism to two simple systems.

B.3.6.1 Free Band Electrons

As the first example, we discuss a system of electrons in a solid which do not interact with each other, interacting only with a periodic lattice potential. This is described by the following one-particle Hamiltonian:

$$\mathcal{H}_0 = H_0 - \mu\widehat{N} \quad , \quad H_0 = \sum_{\mathbf{k}\sigma}\varepsilon(\mathbf{k})\,c^{\dagger}_{\mathbf{k}\sigma}c_{\mathbf{k}\sigma} \tag{B.111}$$

$$\widehat{N} = \sum_{\mathbf{k}\sigma}c^{\dagger}_{\mathbf{k}\sigma}c_{\mathbf{k}\sigma} \tag{B.112}$$

All the interesting properties of the electron system can be calculated from the so-called one-electron Green's function:

$$G_{\mathbf{k}\sigma}^{ret,ad}(E) = \langle\langle c_{\mathbf{k}\sigma}; c_{\mathbf{k}\sigma}^{\dagger} \rangle\rangle_{E}^{ret,ad} \tag{B.113}$$

Since we are dealing with a pure Fermi system, the choice of the anti-commutator Green's function is natural, however, not mandatory.

The first step is setting up the equation of motion (B.83):

$$E\, G_{\mathbf{k}\sigma}^{ret,ad}(E) = \hbar \langle [c_{\mathbf{k}\sigma}, c_{\mathbf{k}\sigma}^{\dagger}]_{+} \rangle + \langle\langle [c_{\mathbf{k}\sigma}, \mathcal{H}_0]_{-}; c_{\mathbf{k}\sigma}^{\dagger} \rangle\rangle^{ret,ad} \tag{B.114}$$

With the help of the fundamental commutation relations for Fermions, one easily gets

$$[c_{\mathbf{k}\sigma}, \mathcal{H}_0]_{-} = (\varepsilon(\mathbf{k}) - \mu)\, c_{\mathbf{k}\sigma} \tag{B.115}$$

On substituting this leads to a simple equation of motion

$$E\, G_{\mathbf{k}\sigma}^{ret,ad}(E) = \hbar + (\varepsilon(\mathbf{k}) - \mu)\, G_{\mathbf{k}\sigma}^{ret,ad}(E)$$

Solving this and introducing $\pm i0^{+}$ in order to satisfy the boundary conditions one gets

$$G_{\mathbf{k}\sigma}^{ret,ad}(E) = \frac{\hbar}{E - (\varepsilon(\mathbf{k}) - \mu) \pm i0^{+}} \tag{B.116}$$

The Green's function is singular at the energy which is required to add an electron of wavevector **k** to the non-interacting electron system. That means the singularities of the Green's function (B.113) correspond to the one-particle excitations of the system. The *combined* Green's function is naturally directly given from (B.116) by removing the infinitesimal imaginary term:

$$G_{\mathbf{k}\sigma}(E) = \frac{\hbar}{E - (\varepsilon(\mathbf{k}) - \mu)} \tag{B.117}$$

The energy E is thought to be complex here. It is interesting to confirm the result (B.117) by an exact evaluation of high-energy expansion (B.101) (Problem B.7).

Finally the *one-electron spectral density* is another important quantity, for which with the Dirac identity (B.92) along with (B.94) directly from (B.116) follows:

$$S_{\mathbf{k}\sigma}(E) = \hbar\, \delta\,(E - \varepsilon(\mathbf{k}) + \mu) \tag{B.118}$$

Using the spectral theorem (B.95) one can easily calculate the average occupation number of the $(\mathbf{k}, \sigma)$-level:

$$\langle n_{\mathbf{k}\sigma} \rangle = \langle c_{\mathbf{k}\sigma}^{\dagger} c_{\mathbf{k}\sigma} \rangle = \frac{1}{\hbar} \int_{-\infty}^{+\infty} dE\, \frac{S_{\mathbf{k}\sigma}(E)}{e^{\beta E} + 1} = \frac{1}{e^{\beta(\varepsilon(\mathbf{k}) - \mu)} + 1} \tag{B.119}$$

This is of course a well-known result of quantum statistics. The average occupation number is given by the Fermi function $f_{-}(E) = (e^{\beta(E-\mu)} + 1)^{-1}$ at $E = \varepsilon(\mathbf{k})$.

From $\langle n_{\mathbf{k}\sigma}\rangle$ by summing over all the wavevectors $\mathbf{k}$ and the two spin projections σ we can fix the total number of electrons N_e:

$$
\begin{aligned}
N_e = \sum_{\mathbf{k}\sigma} \langle n_{\mathbf{k}\sigma}\rangle &= \sum_{\mathbf{k}\sigma} \frac{1}{\hbar} \int_{-\infty}^{+\infty} dE \, \frac{S_{\mathbf{k}\sigma}(E)}{e^{\beta E} + 1} \\
&= \sum_{\mathbf{k}\sigma} \frac{1}{\hbar} \int_{-\infty}^{+\infty} dE \, f_{-}(E) S_{\mathbf{k}\sigma}(E - \mu)
\end{aligned}
$$

We denote by $\rho_\sigma(E)$ the density of states per spin, where self-evidently for the free Fermion system $\rho_\sigma(E) \equiv \rho_{-\sigma}(E)$ holds, so that we can write N_e as

$$
N_e = N \sum_{\sigma} \int_{-\infty}^{+\infty} dE f_{-}(E) \rho_\sigma(E)
$$

Here N is the number of lattice sites with which $\rho_\sigma(E)$ is normalized to 1 since the number of energy band states per each spin direction should be equal to the number of lattice sites. The comparison of the last two equations leads to the important definition of the "quasiparticle density of states":

$$
\rho_\sigma(E) = \frac{1}{N\hbar} \sum_{\mathbf{k}} S_{\mathbf{k}\sigma}(E - \mu) \qquad \text{(B.120)}
$$

In the case of the non-interacting electrons considered here with (B.118)

$$
\rho_\sigma(E) = \rho_{-\sigma}(E) \equiv \rho_0(E) = \frac{1}{N} \sum_{\mathbf{k}} \delta\left(E - \varepsilon(\mathbf{k})\right) \qquad \text{(B.121)}
$$

follows. Without the lattice potential, for the one-particle energies we get the well-known parabolic dispersion $\varepsilon(\mathbf{k}) = \frac{\hbar^2 k^2}{2m}$. One can then easily show that the density of states has a $\sqrt{E}$-dependence.

The considerations for the electron number are correct not only for free electron system but also valid in general. That is why (B.120) will be accepted as the general (!) definition of the quasiparticle density of states for any interacting electron system.

The "internal energy" U as the thermodynamic expectation value of the Hamiltonian is fixed in a simple manner by $\langle n_{\mathbf{k}\sigma}\rangle$:

$$
\begin{aligned}
U = \langle H_0 \rangle &= \sum_{\mathbf{k}\sigma} \varepsilon(\mathbf{k}) \langle n_{\mathbf{k}\sigma} \rangle \\
&= \frac{1}{\hbar} \sum_{\mathbf{k}\sigma} \varepsilon(\mathbf{k}) \int_{-\infty}^{+\infty} dE f_-(E) S_{\mathbf{k}\sigma}(E-\mu) \\
&= \frac{1}{2\hbar} \sum_{\mathbf{k}\sigma} \int_{-\infty}^{+\infty} dE (E+\varepsilon(\mathbf{k})) f_-(E) S_{\mathbf{k}\sigma}(E-\mu) \qquad \text{(B.122)}
\end{aligned}
$$

The bit more complicated representation of the last line will turn out to be the definition of U which is valid in general for an interacting electron system (see B.4.5).

Finally the time-dependent functions are interesting. For the spectral density (B.118), it is trivial to perform the Fourier transformation:

$$
S_{\mathbf{k}\sigma}(t-t') = \frac{1}{2\pi} \exp\left(-\frac{i}{\hbar}(\varepsilon(\mathbf{k})-\mu)(t-t')\right) \qquad \text{(B.123)}
$$

This represents an undamped oscillations with a frequency that corresponds to an exact excitation energy of the system (Fig. B.4). This is typical for non-interacting particle systems. Exactly similarly we find the time-dependent Green's functions:

$$
G^{ret}_{\mathbf{k}\sigma}(t,t') = -i\Theta(t-t') \exp\left(-\frac{i}{\hbar}(\varepsilon(\mathbf{k})-\mu)(t-t')\right) \qquad \text{(B.124)}
$$

$$
G^{av}_{\mathbf{k}\sigma}(t,t') = +i\Theta(t'-t) \exp\left(-\frac{i}{\hbar}(\varepsilon(\mathbf{k})-\mu)(t-t')\right) \qquad \text{(B.125)}
$$

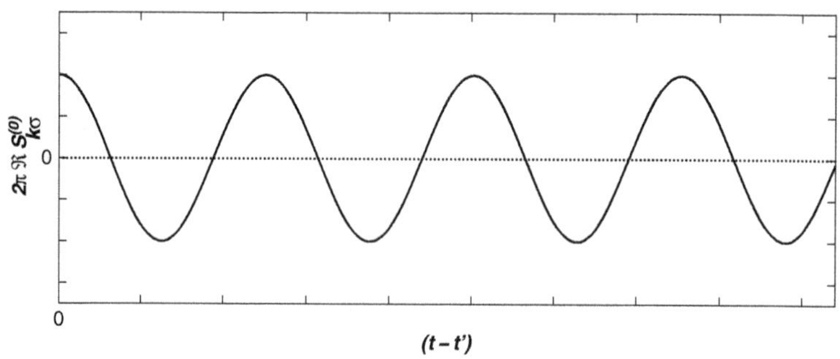

Fig. B.4 Time dependence of the real part of the single-particle spectral density of non-interacting Bloch electrons

B.3.6.2 Spin Waves

As another example we consider the (linear) spin waves (magnons) of a ferromagnet. At low temperatures the Heisenberg Hamiltonian (7.1) can be simplified in the Holstein–Primakoff approximation (7.1.1) as follows:

$$H_{SW} = E_0 + \sum_{\mathbf{q}} \hbar\omega(\mathbf{q})\, a^{\dagger}_{\mathbf{q}} a_{\mathbf{q}} \tag{B.126}$$

$a^{\dagger}_{\mathbf{q}}(a_{\mathbf{q}})$ are the creation (annihilation) operators for magnons. According to this, the ferromagnet is modelled as a *gas* of non-interacting magnons with one-particle energies

$$\hbar\omega(\mathbf{q}) = 2S\hbar^2\,(J_0 - J(\mathbf{q})) + g_J \mu_B B_0 \tag{B.127}$$

$J(\mathbf{q})$ are the exchange integrals with $J_0 = J(\mathbf{q} = 0)$. The second summand describes the influence of an external magnetic field on the one-particle energies. A precondition for the concept of spin waves is that at $T = 0$ the system is ferromagnetic (E_0 : ground state energy (7.246)). A symmetry breaking field

$$B_0 \geq 0^+ \tag{B.128}$$

has to be present necessarily.

Particle number conservation does not hold for magnons. At a given temperature T, it gives exactly the magnon number for which the free energy is minimum:

$$\left(\frac{\partial F}{\partial N}\right)_{T,V} \stackrel{!}{=} 0 \tag{B.129}$$

The differential fraction on the left-hand side is just the definition of the chemical potential μ. So for magnons holds

$$\mu = 0 \tag{B.130}$$

This means that we can substitute $\mathcal{H} = H - \mu\widehat{N} = H$ in the equation of motion for the Green's function. We need the commutator

$$\begin{aligned}
\left[a_{\mathbf{q}},\, H_{SW}\right]_- &= \sum_{\mathbf{q}'} \hbar\omega(\mathbf{q}')\left[a_{\mathbf{q}},\, a^{\dagger}_{\mathbf{q}'} a_{\mathbf{q}'}\right]_- \\
&= \sum_{\mathbf{q}'} \hbar\omega(\mathbf{q}')\left[a_{\mathbf{q}},\, a^{\dagger}_{\mathbf{q}'}\right]_- a_{\mathbf{q}'} \\
&= \hbar\omega(\mathbf{q})\, a_{\mathbf{q}}
\end{aligned}$$

Then the equation of motion becomes very simple:

$$E\, G^{ret,av}_{\mathbf{q}}(E) = \hbar + \hbar\omega(\mathbf{q})\, G^{ret,av}_{\mathbf{q}}(E)$$

Solving it along with the boundary conditions then gives

$$G_{\mathbf{q}}^{ret,av}(E) = \frac{\hbar}{E - \hbar\omega(\mathbf{q}) \pm i0^+} \tag{B.131}$$

Here also the pole represents the excitation energy which indicates either a creation or annihilation of a **q**-magnon. That is exactly $\hbar\omega(\mathbf{q})$ in the absence of interaction. With (B.92) and (B.94), the important "one-magnon spectral density" follows:

$$S_{\mathbf{q}}(E) = \hbar\,\delta\,(E - \hbar\omega(\mathbf{q})) \tag{B.132}$$

One can quickly calculate the time-dependent function and see that it represents as in the case of free Bloch electrons, an undamped oscillation:

$$S_{\mathbf{q}}(t - t') = \frac{1}{2\pi} \exp\left(-i\omega(\mathbf{q})(t - t')\right) \tag{B.133}$$

The frequency of oscillation again corresponds to an exact excitation energy of the system.

With the help of the spectral theorem (B.95) and the spectral density (B.132) we obtain the "magnon occupation density":

$$m_{\mathbf{q}} = \langle a_{\mathbf{q}}^{\dagger} a_{\mathbf{q}} \rangle = \frac{1}{\exp\left(\beta\hbar\omega(\mathbf{q})\right) - 1} + D_{\mathbf{q}} \tag{B.134}$$

As we started with the commutator Green's function, we must determine the constant $D_{\mathbf{q}}$ using the respective anti-commutator Green's function. The fundamental commutation relation for Bosons give for the inhomogeneity in the equation of motion

$$\langle \left[a_{\mathbf{q}}, a_{\mathbf{q}}^{\dagger}\right]_{+} \rangle = 1 + 2m_{\mathbf{q}}$$

Except for this the anti-commutator Green's function satisfies the same equation of motion as the commutator Green's function. One obtains

$$G_{\mathbf{q}}^{(-)}(E) = \frac{\hbar(1 + 2m_{\mathbf{q}})}{E - \hbar\omega(\mathbf{q})} \tag{B.135}$$

In the presence of at least an infinitesimal symmetry breaking external field ($B_0 \geq 0$), the magnon energies are always unequal zero and are positive. That according to (B.98) means

$$2\hbar D_{\mathbf{q}} = \lim_{E \to 0} E G_{\mathbf{q}}^{(-)}(E) = 0$$

So that for the occupation density we get

$$m_{\mathbf{q}} = \frac{1}{\exp\left(\beta\hbar\omega(\mathbf{q})\right) - 1} \tag{B.136}$$

This is the Bose–Einstein distribution function which is a well-known result of elementary quantum statistics for the ideal Bose gas.

B.4 The Quasiparticle Concept

The really interesting many-body problems unfortunately can not be solved exactly. Therefore one must tolerate approximations. For describing the interacting many-particle systems, the concept of "quasiparticles" has proved to be very successful and will be discussed in this section. The basis for this is the following idea:

complex interacting systems of "real" particles
$\Longrightarrow$ non- (or weakly-) interacting system of quasiparticles

This replacement is valid only if one assigns certain special properties to the quasiparticles which will be discussed in the following:

- energy renormalization;
- damping, finite lifetimes;
- effective masses;
- spectral weights, etc.

B.4.1 Interacting Electrons

In order to be concrete, we here want to concentrate on a system of electrons in a non-degenerate energy band interacting via Coulomb interaction. For such a system holds in *Bloch representation*:

$$\mathcal{H} = H - \mu\widehat{N} \qquad ; \qquad \widehat{N} = \sum_{\mathbf{k}\sigma} c^{\dagger}_{\mathbf{k}\sigma}\, c_{\mathbf{k}\sigma} \tag{B.137}$$

$$H = \sum_{\mathbf{k}\sigma} \varepsilon(\mathbf{k}) c^{\dagger}_{\mathbf{k}\sigma}\, c_{\mathbf{k}\sigma} + \frac{1}{2} \sum_{\mathbf{kpq}\sigma\sigma'} v_{\mathbf{kp}}(\mathbf{q}) c^{\dagger}_{\mathbf{k+q}\sigma} c^{\dagger}_{\mathbf{p-q}\sigma'} c_{\mathbf{p}\sigma'} c_{\mathbf{k}\sigma} \tag{B.138}$$

Let the matrix elements be built with Bloch functions $\psi_{\mathbf{k}}(\mathbf{r})$:

$$\varepsilon(\mathbf{k}) = \int d^3r\, \psi^*_{\mathbf{k}}(\mathbf{r}) \left[-\frac{\hbar^2}{2m}\Delta + V(\mathbf{r}) \right] \psi_{\mathbf{k}}(\mathbf{r}) \tag{B.139}$$

$$v_{\mathbf{kp}}(\mathbf{q}) = \frac{1}{4\pi\varepsilon_0} \int\!\!\int d^3r_1\, d^3r_2 \frac{\psi^*_{\mathbf{k+q}}(\mathbf{r}_1)\psi^*_{\mathbf{p-q}}(\mathbf{r}_2)\psi_{\mathbf{p}}(\mathbf{r}_2)\psi_{\mathbf{k}}(\mathbf{r}_1)}{|\mathbf{r}_1 - \mathbf{r}_2|} \tag{B.140}$$

$$v_{\mathbf{kp}}(\mathbf{q}) = v_{\mathbf{pk}}(-\mathbf{q}) \tag{B.141}$$

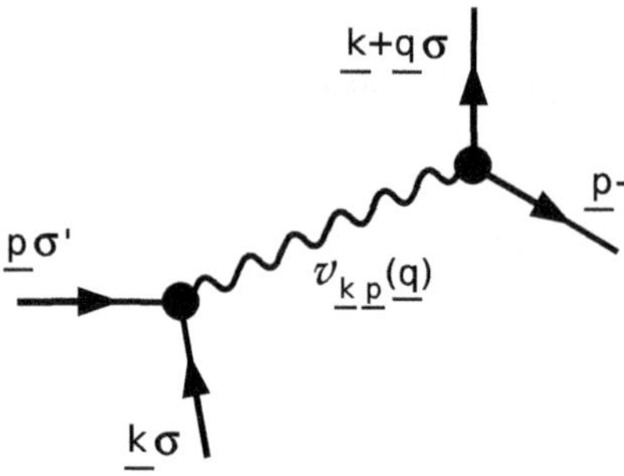

Fig. B.5 Diagram of the Coulomb interaction between electrons of a non-degenerate energy band

All the wavevectors **k**, **p**, **q** stem from the first Brillouin zone. Actually the detailed structure and interpretation of the matrix elements is not important for the following considerations. What is decisive is that the interaction (Fig. B.5) makes the problem definitely unsolvable. In spite of that, is it possible to make a few basic statements?

We will see in the following that also for interacting electrons the *one-electron Green's function* can provide a large part of the interesting information:

$$G_{\mathbf{k}\sigma}^{ret,ad}(E) = \langle\langle c_{\mathbf{k}\sigma};\, c_{\mathbf{k}\sigma}^{\dagger}\rangle\rangle_E^{ret,ad} \tag{B.142}$$

Completely equivalent to this is the *one-electron spectral density*:

$$S_{\mathbf{k}\sigma}(E) = \frac{1}{2\pi}\int_{-\infty}^{+\infty} d(t-t')\, e^{-\frac{i}{\hbar}E(t-t')}\langle\Big[c_{\mathbf{k}\sigma}(t),\, c_{\mathbf{k}\sigma}^{\dagger}(t')\Big]_+\rangle \tag{B.143}$$

We want to determine the spectral density using the equation of motion of the Green's function, where from now onwards, we consider only the retarded Green's function and for convenience, drop the index *ret*. Starting point is the following commutator whose derivation is suggested as an exercise (Problem B.6):

$$[c_{\mathbf{k}\sigma},\, \mathcal{H}]_- = (\varepsilon(\mathbf{k})-\mu)\, c_{\mathbf{k}\sigma} + \sum_{\mathbf{pq}\sigma'} v_{\mathbf{pk+q}}(\mathbf{q}) c_{\mathbf{p+q}\sigma'}^{\dagger} c_{\mathbf{p}\sigma'} c_{\mathbf{k+q}\sigma} \tag{B.144}$$

With the *higher* Green's function

$$\Gamma_{\mathbf{pk;q}}^{\sigma'\sigma}(E) = \langle\langle c_{\mathbf{p+q}\sigma'}^{\dagger} c_{\mathbf{p}\sigma'} c_{\mathbf{k+q}\sigma};\, c_{\mathbf{k}\sigma}^{\dagger}\rangle\rangle_E \tag{B.145}$$

the equation of motion reads as

$$(E-\varepsilon(\mathbf{k})+\mu)\; G_{\mathbf{k}\sigma}(E) = \hbar + \sum_{\mathbf{pq}\sigma'} v_{\mathbf{pk+q}}(\mathbf{q})\Gamma_{\mathbf{pk;q}}^{\sigma'\sigma}(E) \tag{B.146}$$

The function Γ prevents the exact solution. A *formal* solution, however, is possible with the following separation

$$\langle\langle [c_{\mathbf{k}\sigma}, \mathcal{H} - \mathcal{H}_0]_- \,;\, c_{\mathbf{k}\sigma}^\dagger \rangle\rangle_E \stackrel{!}{\equiv} \Sigma_{\mathbf{k}\sigma}(E) \cdot G_{\mathbf{k}\sigma}(E) \tag{B.147}$$

This equation defines the fundamental "self-energy" which, in general, is a complex function:

$$\Sigma_{\mathbf{k}\sigma}(E) = R_{\mathbf{k}\sigma}(E) + i \cdot I_{\mathbf{k}\sigma}(E) \tag{B.148}$$

With this we can now write the Green's function as follows:

$$G_{\mathbf{k}\sigma}(E) = \frac{\hbar}{E - (\varepsilon(\mathbf{k}) - \mu + \Sigma_{\mathbf{k}\sigma}(E))} \tag{B.149}$$

One should note that in case the self-energy is real, one has to include the term $+i0^+$ in the denominator. Let it be mentioned as a side remark that because of the Kramers–Kronig relations (B.109) and (B.110), $\left(G_{\mathbf{k}\sigma}^{av}(E)\right)^* = G_{\mathbf{k}\sigma}^{ret}(E)$ follows. Then it must also hold:

$$\left(\Sigma_{\mathbf{k}\sigma}^{av}(E)\right)^* = \Sigma_{\mathbf{k}\sigma}^{ret}(E) \tag{B.150}$$

We can therefore restrict the discussion to only the retarded self-energy. For simplicity we from now on drop the index *ret*.

If we switch off the interaction, then we get the Green's function for the free system (B.116) which we want to denote by $G_{\mathbf{k}\sigma}^{(0)}(E)$:

$$G_{\mathbf{k}\sigma}^{(0)}(E) = \frac{\hbar}{E - (\varepsilon(\mathbf{k}) - \mu) + i0^+} \tag{B.151}$$

So the total influence of the interaction is contained in the self-energy, whose knowledge solves the many-body problem. If (B.150) is substituted in (B.149) then we have

$$\hbar\left(G_{\mathbf{k}\sigma}^{(0)}(E)\right)^{-1} G_{\mathbf{k}\sigma}(E) = \hbar + \Sigma_{\mathbf{k}\sigma}(E) G_{\mathbf{k}\sigma}(E)$$

This leads to the "Dyson equation":

$$G_{\mathbf{k}\sigma}(E) = G_{\mathbf{k}\sigma}^{(0)}(E) + G_{\mathbf{k}\sigma}^{(0)}(E)\,\frac{1}{\hbar}\,\Sigma_{\mathbf{k}\sigma}(E) G_{\mathbf{k}\sigma}(E) \tag{B.152}$$

This equation can be solved by iteration up to any given accuracy by knowing (approximately) the self-energy:

$$G_{\mathbf{k}\sigma}(E) = G_{\mathbf{k}\sigma}^{(0)}(E) + G_{\mathbf{k}\sigma}^{(0)}(E) \sum_{n=1}^{\infty} \left(\frac{1}{\hbar}\,\Sigma_{\mathbf{k}\sigma}(E) G_{\mathbf{k}\sigma}^{(0)}(E)\right)^n \tag{B.153}$$

B.4.2 Electronic Self-energy

We want to try to get a picture of the general structure of the fundamental many-body terms such as self-energy, Green's function and spectral density, without being able to calculate explicitly the influence of the interaction. In doing this we want to restrict to the retarded functions whose relationship with other types is simple because of the Kramers–Kronig relations (see (B.150)). First we rewrite a bit the formal solution for the Green's function using (B.149):

$$G_{\mathbf{k}\sigma}(E) = \hbar \frac{[E - \varepsilon(\mathbf{k}) + \mu - R_{\mathbf{k}\sigma}(E)] + i\, I_{\mathbf{k}\sigma}(E)}{[E - \varepsilon(\mathbf{k}) + \mu - R_{\mathbf{k}\sigma}(E)]^2 + I^2_{\mathbf{k}\sigma}(E)}$$

$S_{\mathbf{k}\sigma}(E)$ is real and non-negative as one can easily see from the spectral representation (B.86) for the case $A = c_{\mathbf{k}\sigma}$ und $B = c^{\dagger}_{\mathbf{k}\sigma}$. Then holds (B.94)

$$S_{\mathbf{k}\sigma}(E) = -\frac{\hbar}{\pi} \frac{I_{\mathbf{k}\sigma}(E)}{[E - \varepsilon(\mathbf{k}) + \mu - R_{\mathbf{k}\sigma}(E)]^2 + I^2_{\mathbf{k}\sigma}(E)} \tag{B.154}$$

Obviously for the imaginary part of the self-energy must hold

$$I_{\mathbf{k}\sigma}(E) \leq 0 \tag{B.155}$$

In the following we want to analyse spectral density for the argument $E - \mu$ and we expect prominent maxima at points of *resonances* given by

$$E_{n\sigma}(\mathbf{k}) - \varepsilon(\mathbf{k}) - R_{\mathbf{k}\sigma}(E_{n\sigma}(\mathbf{k}) - \mu) \stackrel{!}{=} 0 \qquad n = 1, 2, \cdots \tag{B.156}$$

We must distinguish two cases:

(a) In the immediate E-neighbourhood of a resonance let

$$I_{\mathbf{k}\sigma}(E - \mu) \equiv 0 \tag{B.157}$$

Then with $I_{\mathbf{k}\sigma} \to -0^+$ and the well-known expression

$$\delta(E - E_0) = \frac{1}{\pi} \lim_{x \to 0} \frac{x}{(E - E_0)^2 + x^2} \tag{B.158}$$

for the delta function, the following representation holds for the spectral density:

$$S_{\mathbf{k}\sigma}(E - \mu) = \hbar\, \delta\, (E - \varepsilon(\mathbf{k}) - R_{\mathbf{k}\sigma}(E - \mu)) \tag{B.159}$$

For the case where there lies more than one resonance in the region under consideration (B.157), we use another well-known property of the delta function:

$$\delta\left(f(x)\right)=\sum_i \frac{1}{|f'(x_i)|}\delta(x-x_i) \quad ; \quad f(x_i)=0 \tag{B.160}$$

x_i are the zeros of the function $f(x)$. With this holds

$$S_{\mathbf{k}\sigma}(E-\mu)=\hbar\sum_{n=1}^{n_0}\alpha_{n\sigma}(\mathbf{k})\,\delta\left(E-E_{n\sigma}(\mathbf{k})\right) \tag{B.161}$$

where the coefficients $\alpha_{n\sigma}(\mathbf{k})$ are known as *spectral weights.*

$$\alpha_{n\sigma}(\mathbf{k})=\left|1-\frac{\partial}{\partial E}R_{\mathbf{k}\sigma}(E-\mu)\right|^{-1}_{E=E_{n\sigma}(\mathbf{k})} \tag{B.162}$$

Summation is over all resonances that lie in the region (B.157). Thus the spectral density appears as a linear combination of positively weighted delta functions in whose arguments the resonance energies appear.

(b) Now let it hold

$$I_{\mathbf{k}\sigma}(E-\mu)\not\equiv 0 \tag{B.163}$$

It may still be assumed that, however, in the immediate neighbourhood of a resonance holds:

$$|I_{\mathbf{k}\sigma}(E-\mu)|\ll|\varepsilon(\mathbf{k})+R_{\mathbf{k}\sigma}(E-\mu)| \tag{B.164}$$

Then a more or less prominent maximum is to be expected at $E=E_{n\sigma}(\mathbf{k})$. We therefore expand the bracket in the denominator of (B.154):

$$\begin{aligned}&E_{n\sigma}(\mathbf{k})-\varepsilon(\mathbf{k})-R_{\mathbf{k}\sigma}(E_{n\sigma}(\mathbf{k})-\mu)\\&=0+(E-E_{n\sigma}(\mathbf{k}))\left(1-\frac{\partial}{\partial E}R_{\mathbf{k}\sigma}(E-\mu)\right)_{E=E_{n\sigma}(\mathbf{k})}\\&+\mathcal{O}\left((E-E_{n\sigma}(\mathbf{k}))^2\right)\end{aligned}$$

We substitute this in (B.154) and also further assume that the imaginary part of the self-energy is a *well-behaved* function of energy, so that we can further simplify in the immediate neighbourhood of the resonance:

$$I_{\mathbf{k}\sigma}(E-\mu)\approx I_{\mathbf{k}\sigma}(E_{n\sigma}(\mathbf{k})-\mu)\equiv I_{n\mathbf{k}\sigma} \tag{B.165}$$

This gives in the surroundings of resonance a Lorentzian structure of the spectral density:

$$S^{(n)}_{\mathbf{k}\sigma}(E-\mu)\approx-\frac{\hbar}{\pi}\frac{\alpha^2_{n\sigma}I_{n\mathbf{k}\sigma}}{(E-E_{n\sigma}(\mathbf{k}))^2+(\alpha_{n\sigma}I_{n\mathbf{k}\sigma})^2} \tag{B.166}$$

The results (B.161) and (B.166) lead finally to the classical quasiparticle picture. Under the assumptions (B.164) and (B.165) the spectral density is made up of one or more Lorentz curves or delta functions, whose widths and positions are to a large extent determined by the imaginary and the real parts of the self-energy, respectively. However, the assumptions made are not verifiable directly but only after the complete solution of the many-body problem.

We want to physically interpret the general structure of the spectral density as depicted in Fig. B.6. For this, a Fourier transformation into the time domain is very useful:

$$S_{\mathbf{k}\sigma}(t-t') = \frac{1}{2\pi\hbar}\int_{-\infty}^{+\infty} dE\, e^{-\frac{i}{\hbar}(E-\mu)(t-t')} S_{\mathbf{k}\sigma}(E-\mu) \tag{B.167}$$

This acquires a particularly simple structure for the case (A). With (B.161) follows directly

$$S_{\mathbf{k}\sigma}(t-t') = \frac{1}{2\pi}\sum_{n=1}^{n_0} \alpha_{n\sigma}(\mathbf{k}) \exp\left(-\frac{i}{\hbar}(E_{n\sigma}(\mathbf{k})-\mu)(t-t')\right) \tag{B.168}$$

Thus it is a sum of *undamped* oscillations with frequencies which correspond to the *resonance energies*. This is similar to the result (B.123) depicted in Fig. B.4 for the non-interacting electron system.

The transformation of the function of type B is a little more complicated. For simplicity, we will assume that the Lorentz structure (B.166) is valid over the entire energy range for the spectral density $S^{(n)}_{\mathbf{k}\sigma}(E-\mu)$:

$$\begin{aligned} S^{(n)}_{\mathbf{k}\sigma}(t-t') \approx{}& \frac{1}{4\pi^2 i}\int_{-\infty}^{+\infty} dE\, e^{-\frac{i}{\hbar}(E-\mu)(t-t')}\alpha_{n\sigma}(\mathbf{k}) \,* \\ &* \left[\frac{1}{E-(E_{n\sigma}(\mathbf{k})-i\alpha_{n\sigma}(\mathbf{k})I_{n\mathbf{k}\sigma})}\right. \\ &\left. -\frac{1}{E-(E_{n\sigma}(\mathbf{k})+i\alpha_{n\sigma}(\mathbf{k})I_{n\mathbf{k}\sigma})}\right] \end{aligned} \tag{B.169}$$

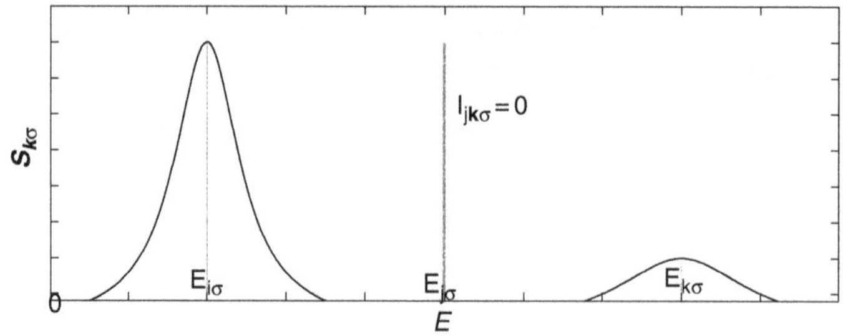

Fig. B.6 Classical quasiparticle picture of the single-electron spectral density

The evaluation is done using the residue theorem noting that $\alpha_{n\sigma}(\mathbf{k})I_{n\mathbf{k}\sigma} \le 0$. Therefore the first summand has a pole in the upper and the second summand in the lower half-plane. For $(t-t') > 0$ the integral in (B.169) is replaced by a contour integral where the path of integration runs along the real axis and closes with an infinite semicircle in the lower half-plane. Only the second summand in (B.169) then has a pole in the region enclosed by the integration path. The semicircle has no contribution because of the exponential function. For $(t-t') < 0$, for the same reason, the semicircle must be in the upper half-plane. In this region only the first summand in (B.169) has a pole. Then finally the residue theorem gives

$$\begin{aligned}S^{(n)}_{\mathbf{k}\sigma}(t-t') \approx{}& \frac{2\pi i}{4\pi^2 i}\alpha_{n\sigma}(\mathbf{k})e^{-\frac{i}{\hbar}(E_{n\sigma}(\mathbf{k})-\mu)(t-t')} * \\ &* \left[\Theta\left(t-t'\right)e^{\frac{1}{\hbar}\alpha_{n\sigma}(\mathbf{k})I_{n\mathbf{k}\sigma}(t-t')}\right. \\ &\left.+\Theta\left(t'-t\right)e^{-\frac{1}{\hbar}\alpha_{n\sigma}(\mathbf{k})I_{n\mathbf{k}\sigma}(t-t')}\right]\end{aligned}$$

This can obviously be summarized to

$$\begin{aligned}S^{(n)}_{\mathbf{k}\sigma}(t-t') \approx{}& \frac{1}{2\pi}\alpha_{n\sigma}(\mathbf{k})\exp\left(-\frac{i}{\hbar}(E_{n\sigma}(\mathbf{k})-\mu)(t-t')\right) * \\ &*\exp\left(-\frac{1}{\hbar}\left|\alpha_{n\sigma}(\mathbf{k})I_{n\mathbf{k}\sigma}(t-t')\right|\right)\end{aligned} \tag{B.170}$$

It now represents a "damped" oscillation. The frequency corresponds again to a *resonance energy*. The amount of damping is determined by the imaginary part $I_{n\mathbf{k}\sigma}$ of the self-energy. $I_{n\mathbf{k}\sigma} \to 0$ reproduces the result (B.108).

The time-dependent spectral density $S_{\mathbf{k}\sigma}(t-t')$ for the interacting electron system consists of an overlap of damped and undamped oscillations with frequencies which correspond to the *resonances* $E_{n\sigma}(\mathbf{k})$. The resulting total time dependence can be naturally quite complicated. In the next section we want to make it clear what these (un)damped oscillations have to do with quasiparticles.

B.4.3 Quasiparticles

What does one understand by quasiparticles in many-body theory? It certainly has something to do with the *resonance peaks* discussed in the last section and shall at least be qualitatively explained here. For that, for simplicity, we consider the special case:

$$T = 0 \;;\; |\mathbf{k}| > k_F \;;\; t > t' \tag{B.171}$$

That means, we assume that there is something like a Fermi edge. In the case of non-interacting electrons it is just the Fermi wavevector, which is the radius of the Fermi sphere. $|\mathbf{k}| > k_F$ here means only that at $T = 0$ the one-particle state with $|\mathbf{k}|$ is unoccupied. Let the system be in the normalized ground state $|E_0\rangle$. At time t' a $(\mathbf{k}, \sigma)$-electron is introduced into the N-particle system. The resulting state

$$|\varphi_0(t')\rangle = c^{\dagger}_{\mathbf{k}\sigma}(t')|E_0\rangle \tag{B.172}$$

is not necessarily an eigenstate of the Hamiltonian H. What happens to it in course of time? Because $|\mathbf{k}| > k_F$, we have $c_{\mathbf{k}\sigma}(t)|E_0\rangle = 0$. Therefore the spectral density simplifies to

$$2\pi\ S_{\mathbf{k}\sigma}(t-t') = \langle E_0| \left[c_{\mathbf{k}\sigma}(t),\ c^{\dagger}_{\mathbf{k}\sigma}(t') \right]_+ |E_0\rangle = \langle \varphi_0(t)|\ \varphi_0(t')\rangle \tag{B.173}$$

The spectral density is therefore the probability amplitude that the state that resulted by introducing of $(\mathbf{k}, \sigma)$-electron at time t' still exists at time $t > t'$. That means it describes the "propagation" of an "extra electron" in an N-electron system. Similarly, for $|\mathbf{k}| < k_F$, the spectral density would describe the propagation of a *hole*. As typical limiting cases one recognizes

- $\left|\langle \varphi_0(t)|\ \varphi_0(t')\rangle\right|^2 = const. \leftrightarrow$ "stationary state"
- $\left|\langle \varphi_0(t)|\ \varphi_0(t')\rangle\right|^2_{(t-t'\to\infty)} = 0 \leftrightarrow$ "state with finite lifetime"

First let us consider once again the case of non-interacting (band-)electrons of Sect. B.3.6.1 with the Hamiltonian (B.111) for the special case (B.171). Using the commutator (B.115) we show that $c^{\dagger}_{\mathbf{k}\sigma}|E_0\rangle$ is an eigenstate of $\mathcal{H}_0$

$$\begin{aligned}
\mathcal{H}_0 \left(c^{\dagger}_{\mathbf{k}\sigma}|E_0\rangle \right) &= c^{\dagger}_{\mathbf{k}\sigma}\mathcal{H}_0|E_0\rangle + \left[\mathcal{H}_0,\ c^{\dagger}_{\mathbf{k}\sigma} \right]_- |E_0\rangle \\
&= (E_0 + \varepsilon(\mathbf{k}) - \mu) \left(c^{\dagger}_{\mathbf{k}\sigma}|E_0\rangle \right)
\end{aligned}$$

With this we calculate

$$\begin{aligned}
|\varphi_0(t)\rangle &= \exp(\frac{i}{\hbar}\mathcal{H}_0 t)\, c^{\dagger}_{\mathbf{k}\sigma}\ \exp(-\frac{i}{\hbar}\mathcal{H}_0 t)|E_0\rangle \\
&= \exp(-\frac{i}{\hbar}E_0 t)\ \exp(\frac{i}{\hbar}\mathcal{H}_0 t) \left(c^{\dagger}_{\mathbf{k}\sigma}\ |E_0\rangle \right) \\
&= \exp(-\frac{i}{\hbar}E_0 t)\ \exp(\frac{i}{\hbar}(E_0 + \varepsilon(\mathbf{k}) - \mu)\, t) \left(c^{\dagger}_{\mathbf{k}\sigma}\ |E_0\rangle \right)
\end{aligned}$$

Then it holds

$$|\varphi_0(t)\rangle = \exp(\frac{i}{\hbar}(\varepsilon(\mathbf{k}) - \mu)\, t)\, |\varphi_0(t=0)\rangle \tag{B.174}$$

Let us consider further the normalization of the state:

$$\begin{aligned}\langle \varphi_0(t=0)|\,\varphi_0(t=0)\rangle &= \langle E_0|\, c_{\mathbf{k}\sigma}\, c^{\dagger}_{\mathbf{k}\sigma}\,|E_0\rangle \\ &= \langle E_0|\, E_0\rangle - \langle E_0|\, c^{\dagger}_{\mathbf{k}\sigma}\, c_{\mathbf{k}\sigma}\,|E_0\rangle \\ &= \langle E_0|\, E_0\rangle = 1\end{aligned}$$

We finally get for the probability amplitude

$$\langle \varphi_0(t)|\,\varphi_0(t')\rangle = \exp(-\frac{i}{\hbar}(\varepsilon(\mathbf{k})-\mu)\,(t-t')) = 2\pi\; S^{(0)}_{\mathbf{k}\sigma}(t-t') \tag{B.175}$$

This naturally agrees exactly with the result (B.123) which we have found in Sect. B.3.6.1 using the equation of motion method for the free band electrons. It gives the undamped harmonic oscillations shown in Fig. B.4. In particular we have

$$\left|\langle \varphi_0(t)|\,\varphi_0(t')\rangle\right|^2 = 1 \tag{B.176}$$

Thus, for the case of free electrons it is a stationary state. This is not surprising because $|\varphi_0\rangle = c^{\dagger}_{\mathbf{k}\sigma}|E_0\rangle$ turns out to be an eigenstate of $\mathcal{H}_0$. It is, however, no more the case for *interacting (band-) electrons*

This one recognizes from the spectral representation of the spectral density. If one carries out the average over the ground state $|E_0\rangle$ as is required by definition for $T=0$, and then goes through exactly the same steps as in deriving (B.86), then one gets the spectral density

$$2\pi\; S_{\mathbf{k}\sigma}(t-t') = \sum_n \left|\langle E_n|\, c^{\dagger}_{\mathbf{k}\sigma}\,|E_0\rangle\right|^2 \exp(-\frac{i}{\hbar}(E_n-E_0)(t-t')) \tag{B.177}$$

In the free system, $c^{\dagger}_{\mathbf{k}\sigma}|E_0\rangle$ is an eigenstate of $\mathcal{H}_0$, so that due to the orthogonality of the eigenstates, only one term in the sum contributes. That does not hold any more for the interacting system. $c^{\dagger}_{\mathbf{k}\sigma}|E_0\rangle$ is no more an eigenstate, but can be expanded in terms of the eigenstates:

$$|\varphi_0\rangle = c^{\dagger}_{\mathbf{k}\sigma}|E_0\rangle = \sum_n \gamma_n|E_n\rangle$$

where an arbitrary number, but at least two coefficients γ_n are unequal zero. Every summand in (B.177) represents a harmonic oscillation but with different frequencies. The overlap sees to it that the sum is maximum for $t = t'$. For increasing $t-t'$ the phase factors $\exp(-\frac{i}{\hbar}(E_n-E_0)(t-t'))$ distribute themselves over the entire unit circle in the complex plane and see to it that, because of the destructive interference, possibly a very complicated and certainly no more harmonic time-dependence results as depicted in Fig. B.7. The state created at t' is not stationary

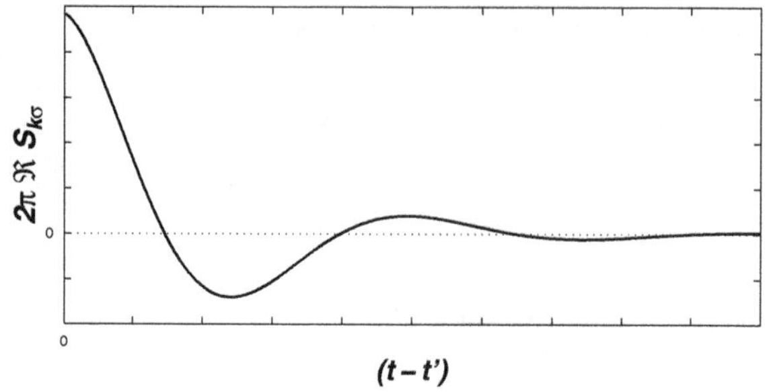

Fig. B.7 Schematic plot of the time-dependent spectral density of an interacting electron system

but has to some extent a *finite lifetime*. One could, however, imagine that the complicated time dependence could be simulated by a few weighted damped oscillations with well-defined frequencies:

$$2\pi\ S_{\mathbf{k}\sigma}(t-t') = \sum_n \alpha_{n\sigma}(\mathbf{k}) \exp(-\frac{i}{\hbar}(\eta_{n\sigma}(\mathbf{k}) - \mu)(t-t')) \tag{B.178}$$

This expression formally has the same structure as for the free system (B.175) with, however, in general complex one-particle energies:

$$\eta_{n\sigma}(\mathbf{k}) = Re\ \eta_{n\sigma}(\mathbf{k}) + i\ Im\ \eta_{n\sigma}(\mathbf{k}) \tag{B.179}$$

In order to realize damping, one must have

$$Im\ \eta_{n\sigma}(\mathbf{k}) \leq 0 \qquad (t - t' > 0) \tag{B.180}$$

The ansatz (B.178) gives the impression as if the extra $(\mathbf{k}, \sigma)$-electron *decays* into one or more *quasiparticles* with the following properties:

- *Quasiparticle energy* ⇔ $Re\ \eta_{n\sigma}(\mathbf{k})$
- *Quasiparticle lifetime* ⇔ $\hbar \cdot |Im\ \eta_{n\sigma}(\mathbf{k})|^{-1}$
- *Quasiparticle weight ("spectral")* ⇔ $\alpha_{n\sigma}(\mathbf{k})$

The *lifetime* is defined here as the time that is required for the respective summand to decrease from its initial value by a fraction e. Because of the particle conservation, for the spectral weights of the quasiparticles

$$\sum_n \alpha_{n\sigma}(\mathbf{k}) = 1 \tag{B.181}$$

must still hold. Formally this follows from

$$2\pi\ S_{\mathbf{k}\sigma}(0) = \sum_n \alpha_{n\sigma}(\mathbf{k}) = \langle \left[c_{\mathbf{k}\sigma}, c^{\dagger}_{\mathbf{k}\sigma}\right]_+ \rangle = 1$$

We will carry these interpretations now on to the results (B.108) and (B.110) of the last section

- *Quasiparticle energy* $\Leftrightarrow$ $Re\, E_{n\sigma}(\mathbf{k})$

$$E_{n\sigma}(\mathbf{k}) \stackrel{!}{=} \varepsilon(\mathbf{k}) + R_{\mathbf{k}\sigma}(E_{n\sigma}(\mathbf{k}) - \mu) \tag{B.182}$$

- *Quasiparticle lifetime* $\Leftrightarrow$ $\tau_{n\sigma}(\mathbf{k})$

$$\tau_{n\sigma}(\mathbf{k}) = \frac{\hbar}{|\alpha_{n\sigma}(\mathbf{k})\, I_{\mathbf{k}\sigma}\, (E_{n\sigma}(\mathbf{k}) - \mu)|} \tag{B.183}$$

- *Quasiparticle weight ("spectral")* $\Leftrightarrow$ $\alpha_{n\sigma}(\mathbf{k})$

$$\alpha_{n\sigma}(\mathbf{k}) = \left|1 - \frac{\partial}{\partial E} R_{\mathbf{k}\sigma}(E - \mu)\right|^{-1}_{E=E_{n\sigma}} \tag{B.184}$$

The *Lorentz-type* peaks in the spectral weight are to be assigned to the quasiparticles, whose energies are given by the positions and their lifetimes by the widths of the peaks. The quasiparticle energies are given only by the real part and the lifetimes are given mainly by the imaginary part of the self-energy. Because of $\alpha_{n\sigma}(\mathbf{k})$, in a limited way, the lifetime is of course influenced by the real part too. $I_{\mathbf{k}\sigma} = 0$ always means an infinite lifetime. Delta functions in the spectral weight indicate stable, i.e. infinitely long-living quasiparticles.

To conclude, it should be mentioned that these considerations for the classical quasiparticle picture are based on the preconditions (B.164) and (B.165) whose validity can be verified only after the complete solution of the many-body problem.

B.4.4 Quasiparticle Density of States

While discussing the free electrons as an application of the abstract Green's function formalism in 3.6., we have learned about the important concept of quasiparticle density of states. Now we want to introduce this quantity for interacting electron system and understand its relation to the one-particle Green's function and one-particle spectral density.

The starting point is the average occupation number $\langle n_{\mathbf{k}\sigma}\rangle$ of the $(\mathbf{k}, \sigma)$-level, which with the help of the spectral theorem can be expressed by the one-particle spectral density:

$$\langle n_{\mathbf{k}\sigma}\rangle = \langle c^{\dagger}_{\mathbf{k}\sigma}\, c_{\mathbf{k}\sigma}\rangle = \frac{1}{\hbar}\int_{-\infty}^{+\infty} dE\, f_{-}(E) S_{\mathbf{k}\sigma}(E - \mu) \tag{B.185}$$

Here $f_{-}(E)$ is again the Fermi function. A summation over all the wavevectors and both the spin directions gives the total number of electrons N_e

$$N_e = \sum_{\mathbf{k}\sigma} \langle n_{\mathbf{k}\sigma} \rangle \tag{B.186}$$

Alternatively, the electron number can be expressed in terms of a density of states. $\hat{\rho}_\sigma(E)dE$ is the number of σ-states in the energy interval $[E, E+dE]$ that can be occupied. Then, $f_-(E)\hat{\rho}_\sigma(E)dE$ is the density of the occupied states. Therefore

$$N_e = \sum_{\sigma} \int_{-\infty}^{+\infty} f_-(E)\hat{\rho}_\sigma(E)dE \tag{B.187}$$

must hold. A non-degenerate energy band (s-band) contains $2N$ states, where N is the number of lattice sites. The factor 2 comes because of the two spin directions. For the completely occupied band ($f_-(E) \equiv 1$) therefore

$$\int_{-\infty}^{+\infty} dE\, f_-(E)\hat{\rho}_\sigma(E) = N$$

holds. However, the density of states is usually normalized to one: $\rho_\sigma(E) \equiv \frac{1}{N}\hat{\rho}_\sigma(E)$. A comparison of the two expressions for N_e then gives the "Quasiparticle density of states"

$$\rho_\sigma(E) = \frac{1}{N\hbar} \sum_{\mathbf{k}} S_{\mathbf{k}\sigma}(E-\mu) \tag{B.188}$$

All the properties of the spectral density transfer to the quasiparticle density of states. It is in general dependent on temperature and particle number and naturally also depends on lattice structure. We have seen in Sect. B.2 that the one-electron spectral density has a direct relation to *angle resolved* photoemission. In contrast, the quasiparticle density of states is seen directly in *angle averaged* photoemission.

We want to investigate the quasiparticle density of states for an illustrative special case. For that we consider a real, **k**-independent self-energy:

$$R_{\mathbf{k}\sigma}(E) \equiv R_\sigma(E) \quad ; \quad I_{\mathbf{k}\sigma}(E) \equiv 0 \tag{B.189}$$

This corresponds to the case A of Sect. B.4.2. Therefore (B.159) holds:

$$S_{\mathbf{k}\sigma}(E-\mu) = \hbar\,\delta\,(E - R_\sigma(E-\mu) - \varepsilon(\mathbf{k})) \tag{B.190}$$

For the quasiparticle density of states this means

$$\rho_\sigma(E) = \frac{1}{N} \sum_{\mathbf{k}} \delta\,(E - R_\sigma(E-\mu) - \varepsilon(\mathbf{k})) \tag{B.191}$$

Comparing with the *Bloch density of states* for the non-interacting band electrons (B.121), we get in this special case

$$\rho_\sigma(E) = \rho_0(E - R_\sigma(E - \mu)) \tag{B.192}$$

$\rho_\sigma(E)$ is unequal zero for such energies for which $E - R_\sigma(E - \mu)$ lies between the lower and the upper edge of the free Bloch band. If $R_\sigma(E)$ is only a slowly varying smooth function of E, then $\rho_\sigma(E)$ will only be slightly deformed from $\rho_0(E)$ (Fig. B.8). The influence of the particle interaction can possibly be taken into account by a renormalization of certain parameters. On the contrary, new kind of phenomena appear if $E - R_\sigma(E - \mu)$ is strongly structured, if as in Fig. B.9 the self-energy, e.g. shows a singularity in the interesting region. The result can be a band splitting which cannot be understood from the one-electron picture. At an appropriate band filling, it can happen that in one-electron picture (Bloch picture) the system is metallic, whereas in reality electronic correlations can make it an insulator. Such a system is called a "Mott–Hubbard insulator".

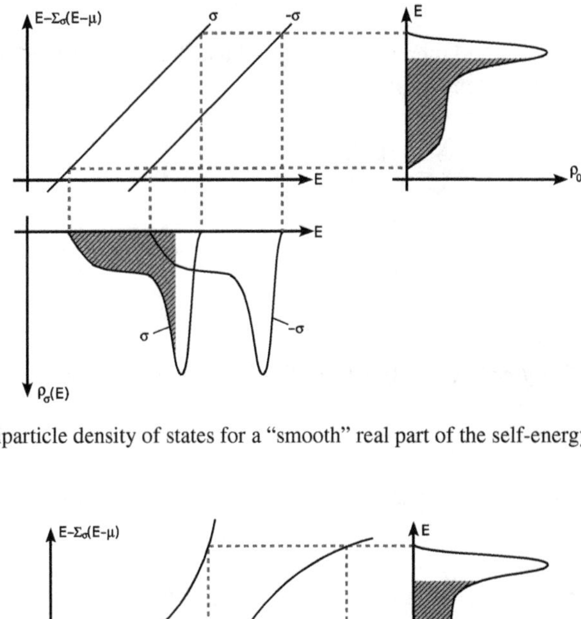

Fig. B.8 Quasiparticle density of states for a "smooth" real part of the self-energy

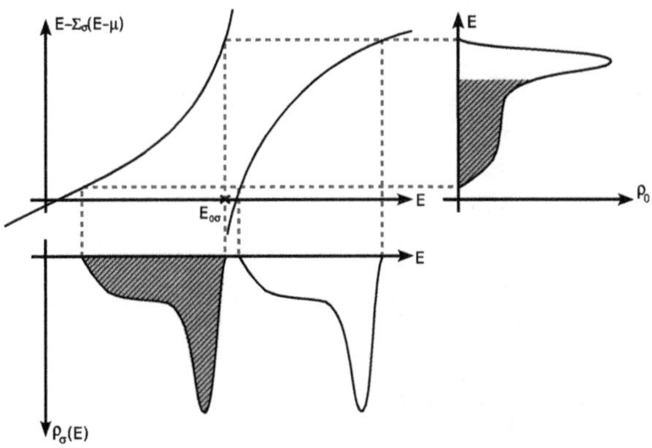

Fig. B.9 Quasiparticle density of states of a self-energy with a singularity

B.4.5 Thermodynamics

To conclude we want to show that the one-electron Green's function (B.142) or the corresponding spectral density (B.143) can provide complete information regarding the macroscopic thermodynamics of the interacting electron system. We start with the "internal energy" which is defined as the expectation value of the Hamiltonian (B.138):

$$\begin{aligned} U = \langle H \rangle = &\sum_{\mathbf{k}\sigma} \varepsilon(\mathbf{k}) \left\langle c^{\dagger}_{\mathbf{k}\sigma} c_{\mathbf{k}\sigma} \right\rangle + \\ &+ \frac{1}{2} \sum_{\mathbf{k},\mathbf{p},\mathbf{q},\sigma,\sigma'} v_{\mathbf{k},\mathbf{p}}(\mathbf{q}) \left\langle c^{\dagger}_{\mathbf{k}+\mathbf{q}\sigma}\, c^{\dagger}_{\mathbf{p}-\mathbf{q}\sigma'} c_{\mathbf{p}\sigma'} c_{\mathbf{k}\sigma} \right\rangle \end{aligned} \tag{B.193}$$

We substitute $\mathbf{q} \to -\mathbf{q}$ and then $\mathbf{k} \to \mathbf{k}+\mathbf{q}$ and use the symmetry relation (B.141). Then U can also be written as follows:

$$\begin{aligned} U = \langle H \rangle = &\sum_{\mathbf{k}\sigma} \varepsilon(\mathbf{k}) \left\langle c^{\dagger}_{\mathbf{k}\sigma} c_{\mathbf{k}\sigma} \right\rangle + \\ &+ \frac{1}{2} \sum_{\mathbf{k},\mathbf{p},\mathbf{q},\sigma,\sigma'} v_{\mathbf{p},\mathbf{k}+\mathbf{q}}(\mathbf{q}) \left\langle c^{\dagger}_{\mathbf{k}\sigma}\, c^{\dagger}_{\mathbf{p}+\mathbf{q}\sigma'} c_{\mathbf{p}\sigma'} c_{\mathbf{k}+\mathbf{q}\sigma} \right\rangle \end{aligned} \tag{B.194}$$

With the help of *higher* Green's function (B.145) and the spectral theorem (B.95) it follows that

$$\begin{aligned} U = &\frac{1}{\hbar} \int_{-\infty}^{+\infty} \frac{dE}{\exp(\beta E)+1} \left[\sum_{\mathbf{k}\sigma} \varepsilon(\mathbf{k}) \left(-\frac{1}{\pi}\, Im G_{\mathbf{k}\sigma}(E) \right) \right. \\ &\left. + \frac{1}{2} \sum_{\mathbf{k},\mathbf{p},\mathbf{q},\sigma,\sigma'} v_{\mathbf{p},\mathbf{k}+\mathbf{q}}(\mathbf{q}) \left(-\frac{1}{\pi}\, Im \Gamma^{\sigma'\sigma}_{\mathbf{pk};\mathbf{q}}(E) \right) \right] \end{aligned} \tag{B.195}$$

From the equation of motion we read off

$$\begin{aligned} &\frac{1}{2} \sum_{\mathbf{k},\mathbf{p},\mathbf{q},\sigma,\sigma'} v_{\mathbf{p},\mathbf{k}+\mathbf{q}}(\mathbf{q}) \left(-\frac{1}{\pi}\, Im \Gamma^{\sigma'\sigma}_{\mathbf{pk};\mathbf{q}}(E) \right) = \\ &= \frac{1}{2} \sum_{\mathbf{k}\sigma} \left(-\frac{1}{\pi} Im \left((E - \varepsilon(\mathbf{k}) + \mu) G_{\mathbf{k}\sigma}(E) - \hbar \right) \right) \\ &= \frac{1}{2} \sum_{\mathbf{k}\sigma} (E - \varepsilon(\mathbf{k}) + \mu) \left(-\frac{1}{\pi} Im G_{\mathbf{k}\sigma}(E)) \right) \end{aligned}$$

This we substitute in (B.195)

$$U = \frac{1}{2\hbar}\sum_{\mathbf{k}\sigma}\int_{-\infty}^{+\infty}\frac{dE}{\exp(\beta E)+1}(E+\mu+\varepsilon(\mathbf{k}))\left(-\frac{1}{\pi}\,ImG_{\mathbf{k}\sigma}(E)\right) \qquad \text{(B.196)}$$

Once again substituting $E \to E-\mu$, we finally obtain

$$U = \frac{1}{2\hbar}\sum_{\mathbf{k}\sigma}\int_{-\infty}^{+\infty} dE\, f_{-}(E)(E+\varepsilon(\mathbf{k}))\; S_{\mathbf{k}\sigma}(E-\mu) \qquad \text{(B.197)}$$

This is a very remarkable result because the contribution of a two-particle Coulomb interaction could be expressed in terms of one-particle spectral density. The result (B.197) was already formally obtained for the case of non-interacting band electrons (B.122).

From U the "free energy" follows from the generally valid relation

$$F(T) = U(0) - T\int_0^T dT'\,\frac{U(T')-U(0)}{T'^2} \qquad \text{(B.198)}$$

which we prove as problem (B.10). Therefore the whole of macroscopic thermodynamics is determined by the one-particle spectral density itself. In this context, for the various energy integrals, particularly the prominent quasiparticle peaks (see Fig. B.6) are important.

B.5 Problems

Problem B.1 Let $A(t)$ be an arbitrary operator in the Heisenberg picture and ρ be the statistical operator:

$$\rho = \frac{\exp(-\beta H)}{Tr(\exp(-\beta H))}$$

Prove the Kubo identity:

$$\frac{i}{\hbar}\,[A(t),\,\rho]_{-} = \rho\int_0^\beta d\lambda\,\dot{A}(t-i\lambda\hbar)$$

Assume that the Hamiltonian H is not explicitly time dependent!

Problem B.2 With the help of the Kubo identity (Problem B.1) show that the retarded commutator Green's function can be written as follows:

$$\langle\langle A(t);\,B(t')\rangle\rangle^{ret} = -\hbar\,\Theta(t-t')\int_0^\beta d\lambda\,\langle\dot{B}(t'-i\lambda\hbar)A(t)\rangle$$

Problem B.3 For the time-dependent correlation functions, show that

$$\langle B(0)\, A(t + i\hbar\beta)\rangle = \langle A(t)\, B(0)\rangle$$

holds provided the Hamiltonian does not explicitly depend on time.

Problem B.4 Prove the following representation of the step function:

$$\Theta(t - t') = \frac{i}{2\pi} \int_{-\infty}^{+\infty} dx\, \frac{\exp(-ix(t - t'))}{x + i0^+}$$

Problem B.5 Show that a complex function $F(E)$ has an analytical continuation in the upper (lower) half-plane, if its Fourier transform $f(t)$ for $t < 0$ $(t > 0)$ vanishes.

Problem B.6 For an interacting electron system

$$H = \sum_{\mathbf{k}\sigma} \varepsilon(\mathbf{k}) c^{\dagger}_{\mathbf{k}\sigma} c_{\mathbf{k}\sigma} + \frac{1}{2} \sum_{\mathbf{k},\mathbf{p},\mathbf{q}} \sum_{\sigma,\sigma'} v_{\mathbf{kp}}(\mathbf{q}) c^{\dagger}_{\mathbf{k}+\mathbf{q}\sigma} c^{\dagger}_{\mathbf{p}-\mathbf{q}\sigma'} c_{\mathbf{p}\sigma'} c_{\mathbf{k}\sigma}$$

derive the equation of motion for the retarded one-particle Green's function.

Problem B.7 For a system of non-interacting electrons

$$H = \sum_{\mathbf{k}\sigma} (\varepsilon(\mathbf{k}) - \mu)\, c^{\dagger}_{\mathbf{k}\sigma} c_{\mathbf{k}\sigma}$$

calculate all the spectral moments $M^{(n)}_{\mathbf{k}\sigma}$ and from there the exact spectral density.

Problem B.8 The BCS theory of superconductivity can be carried out with the simplified Hamiltonian

$$H = \sum_{\mathbf{k}\sigma} (\varepsilon(\mathbf{k}) - \mu)\, c^{\dagger}_{\mathbf{k}\sigma} c_{\mathbf{k}\sigma} - \Delta \sum_{\mathbf{k}} \left(b_{\mathbf{k}} + b^{\dagger}_{\mathbf{k}} \right) + \frac{\Delta^2}{V}$$

Here

$$b^{\dagger}_{\mathbf{k}} = c^{\dagger}_{\mathbf{k}\uparrow} c^{\dagger}_{-\mathbf{k}\downarrow}$$

is the "Cooper pair creation operator" and

$$\Delta = \Delta^* = V \sum_{\mathbf{k}} \langle b_{\mathbf{k}} \rangle = V \sum_{\mathbf{k}} \langle b^{\dagger}_{\mathbf{k}} \rangle$$

1. Calculate the commutation relations of the operators $b_{\mathbf{k}}, b_{\mathbf{k}}^{\dagger}$. Are the Cooper pairs Bosons?
2. Using the one-electron Green's function (B.142) calculate the excitation spectrum of the superconductor. Show that it has a "gap" Δ.
3. Determine the equation satisfied by Δ!

Problem B.9 1. For the superconductivity model in Problem B.8, calculate all the spectral moments of the one-electron spectral density.
2. Choose a two-pole ansatz for the spectral density

$$S_{\mathbf{k}\sigma}(E) = \hbar \sum_{i=1}^{2} \alpha_{i\sigma}(\mathbf{k})\, \delta\left(E - E_{i\sigma}(\mathbf{k})\right)$$

and determine the spectral weights $\alpha_{i\sigma}(\mathbf{k})$ and the quasiparticle energies $E_{i\sigma}(\mathbf{k})$! By inspecting the spectral moments show that the above ansatz is exact.

Problem B.10 Prove the following relation between the internal and the free energies:

$$F(T, V) = U(0, V) - T \int_0^T dT' \, \frac{U(T', V) - U(0, V)}{T'^2}$$

Problem B.11 Let $|E_0\rangle$ be the ground state of the interaction free electron system (Fermi sphere). Calculate the time dependence of the state

$$|\psi_0\rangle = c_{\mathbf{k}\sigma}^{\dagger} c_{\mathbf{k}'\sigma'} |E_0\rangle$$

Is it a stationary state?

Problem B.12 For the one-electron Green's function of an interacting electron system

$$G_{\mathbf{k}\sigma}^{ret}(E - \mu) = \frac{\hbar}{E - 2\varepsilon(\mathbf{k}) + \frac{E^2}{\varepsilon(\mathbf{k})} + i\gamma|E|} \qquad (\gamma > 0)$$

may hold.

1. Determine the electronic self-energy $\Sigma_{\mathbf{k}\sigma}(E)$.
2. Calculate the energies and lifetimes of the quasiparticles.
3. Under what conditions is the classical quasiparticle concept applicable?
4. Calculate the effective masses of the quasiparticles.

Problem B.13 For an interacting electron system, let the self-energy

$$\Sigma_{\sigma}(E) = \frac{a_{\sigma}(E + \mu - b_{\sigma})}{E + \mu - c_{\sigma}} \qquad (a_{\sigma}, b_{\sigma}, c_{\sigma} \text{ positive, real}\ ;\ c_{\sigma} > b_{\sigma})$$

be calculated. For the density of states of the interaction free system holds

$$\rho_0(E) = \begin{cases} \frac{1}{W} & \text{for } 0 \leq E \leq W \\ 0 & \text{otherwise} \end{cases}$$

Calculate the quasiparticle density of states. Is there a band splitting?

Appendix C
Solutions to Problems

Problem 1.1
Let us take $\mathbf{R}_i = 0$
With $\nabla \times (\varphi \mathbf{a}) = \varphi\, \nabla \times \mathbf{a} - \mathbf{a} \times \nabla\varphi$ follows:

$$\mathbf{j}_m^{(i)} = \nabla \times (\mathbf{m}_i f(r)) = f\, \nabla \times \mathbf{m}_i - \mathbf{m}_i \times \nabla f$$
$$\nabla \times\ \mathbf{m}_i = 0,\ \textit{since}\ \mathbf{m}_i\ :\ \textit{particle property}$$

Substitute:

$$\begin{aligned}
\mathbf{m}_i &= \frac{1}{2}\int d^3r\ \mathbf{r} \times \mathbf{j}_m^{(i)} \\
&= \frac{1}{2}\int d^3r\ \ [\mathbf{r} \times (\nabla f \times \mathbf{m}_i)] \\
&= \frac{1}{2}\int d^3r\ \ [\nabla f(\mathbf{r}\cdot\mathbf{m}_i) - \mathbf{m}_i(\mathbf{r}\cdot\nabla f)] \\
&= \frac{1}{2}\int d^3r\ \nabla f(\mathbf{r}\cdot\mathbf{m}_i) - \frac{1}{2}\int d^3r\ \mathbf{m}_i(div(f\mathbf{r}) - f\ div\,\mathbf{r}) \\
&= \frac{1}{2}\int d^3r\ \nabla f(\mathbf{r}\cdot\mathbf{m}_i) - \frac{1}{2}\mathbf{m}_i\int d^3r\ div(f\mathbf{r}) \\
&\quad + \frac{3}{2}\mathbf{m}_i\underbrace{\int d^3r\ f}_{\equiv 1,\ cond.2.} \\
&= \frac{1}{2}\int d^3r\ \nabla f(\mathbf{r}\cdot\mathbf{m}_i) + \frac{3}{2}\mathbf{m}_i - \frac{1}{2}\mathbf{m}_i\underbrace{\oint_{\partial T} d\mathbf{S}\cdot(f\mathbf{r})}_{\equiv 0,\ cond.1.}
\end{aligned}$$

Intermediate result:

$$\begin{aligned}
\mathbf{m}_i &= \frac{1}{2}\int d^3r\ \nabla f(\mathbf{r}\cdot\mathbf{m}_i) + \frac{3}{2}\mathbf{m}_i \\
\Leftrightarrow\ -\frac{1}{2}\mathbf{m}_i &= \frac{1}{2}\int d^3r\ [\nabla(f(\mathbf{r}\cdot\mathbf{m}_i)) - f\nabla(\mathbf{r}\cdot\mathbf{m}_i)] \\
&= \frac{1}{2}\int d^3r\ \nabla(f(\mathbf{r}\cdot\mathbf{m}_i)) - \frac{1}{2}\mathbf{m}_i \underbrace{\int d^3r\ f}_{\equiv 1,\ cond.2.} \\
\Leftrightarrow\ 0 &= \frac{1}{2}\int d^3r\ \nabla(f(\mathbf{r}\cdot\mathbf{m}_i)) \\
&= \frac{1}{2}\oint_{\partial T} d\mathbf{S}\ f(\mathbf{r}\cdot\mathbf{m}_i) = 0 \quad q.e.d.
\end{aligned}$$

Problem 1.2
From the definition of the canonical partition function

$$Z = Tr\left(e^{-\beta\widehat{H}}\right)$$

The average magnetic moment is given by

$$\begin{aligned}
\langle\widehat{m}\rangle &= \frac{1}{Z}Tr\left(-\frac{d}{dB_0}\widehat{H}e^{-\beta\widehat{H}}\right) \\
&= \frac{1}{\beta}\frac{1}{Z}\frac{\partial}{\partial B_0}Tr\left(e^{-\beta\widehat{H}}\right) \\
&= \frac{1}{\beta}\frac{1}{Z}\frac{\partial Z}{\partial B_0}
\end{aligned}$$

From this follows the susceptibility:

$$\begin{aligned}
\chi_T &= \frac{\mu_0}{V}\left(\frac{\partial}{\partial B_0}\langle\widehat{m}\rangle\right)_T \\
&= \frac{\mu_0}{\beta V}\left(-\frac{1}{Z^2}\left(\frac{\partial Z}{\partial B_0}\right)^2 + \frac{1}{Z}\frac{\partial^2 Z}{\partial B_0^2}\right)
\end{aligned}$$

The first term is clear:

$$\frac{1}{Z^2}\left(\frac{\partial Z}{\partial B_0}\right)^2 = \beta^2\langle\widehat{m}\rangle^2 \tag{1.1}$$

The second term is somewhat more complicated:

$$
\begin{aligned}
\frac{1}{Z}\frac{\partial^2 Z}{\partial B_0^2} &= \frac{1}{Z}\frac{\partial}{\partial B_0}\left(-\beta Tr\left(\frac{\partial \widehat{H}}{\partial B_0}e^{-\beta\widehat{H}}\right)\right) \\
&= \frac{-\beta}{Z}Tr\left(\frac{\partial^2 \widehat{H}}{\partial B_0^2}e^{-\beta\widehat{H}} - \beta\left(\frac{\partial \widehat{H}}{\partial B_0}\right)^2 e^{-\beta\widehat{H}}\right) \\
&= \frac{\beta^2}{Z}Tr\left(\left(\frac{\partial \widehat{H}}{\partial B_0}\right)^2 e^{-\beta\widehat{H}}\right) \\
&= \beta^2\langle\widehat{m}^2\rangle
\end{aligned}
$$

In the third step we have exploited the condition that we are dealing with a permanent magnetic moment. Therefore for the susceptibility we have

$$
\chi_T = \frac{\mu_0}{V}\beta\left(\langle\widehat{m}^2\rangle - \langle\widehat{m}\rangle^2\right) = \frac{1}{k_B T}\frac{\mu_0}{V}\langle(\widehat{m} - \langle\widehat{m}\rangle)^2\rangle \tag{1.2}
$$

Problem 1.3
In a magnetic field $\mathbf{B}$, the magnetic dipole $\mathbf{m}$ has the potential energy

$$
V = -\mathbf{m}\cdot\mathbf{B}
$$

It will therefore try to orient itself parallel to the field. If the field is only a homogeneous external field

$$
\mathbf{B}_0 = B_0\mathbf{e}_x = \begin{pmatrix} B_0 \\ 0 \\ 0 \end{pmatrix}
$$

then $\mathbf{m}$ will be oriented parallel to x-axis. According to elementary electrodynamics, the current creates an additional azimuthal field of the form

$$
\mathbf{B}_I = \mu_0\frac{I}{2\pi\rho}\mathbf{e}_\varphi
$$

It is convenient to use cylindrical coordinates:

$$
x = \rho\cos\varphi\ ,\ \ y = \rho\sin\varphi\ ,\ \ z = z
$$

$$
\rho = \sqrt{x^2 + y^2}
$$

$$
\mathbf{e}_\varphi = \begin{pmatrix} -\sin\varphi \\ \cos\varphi \\ 0 \end{pmatrix} = \frac{1}{\rho}\begin{pmatrix} -y \\ x \\ 0 \end{pmatrix}
$$

Total field

$$\mathbf{B} = \mathbf{B}_0 + \mathbf{B}_I = B_0 \begin{pmatrix} 1 \\ 0 \\ 0 \end{pmatrix} + \mu_0 \frac{I}{2\pi (x^2 + y^2)} \begin{pmatrix} -y \\ x \\ 0 \end{pmatrix}$$

Field at the position $\mathbf{r}_0 = (x_0, 0, 0)$:

$$\mathbf{B}(\mathbf{r}_0) = B_0 \mathbf{e}_x + \frac{\mu_0 I}{2\pi x_0} \mathbf{e}_y$$

Dipole is oriented parallel to $\mathbf{B}$, i.e. it makes an angle α given by

$$\tan \alpha = \mu_0 \frac{I}{2\pi x_0 B_0}$$

which for small angles gives

$$\alpha \approx \mu_0 \frac{I}{2\pi x_0 B_0}$$

with the x-axis.

Problem 1.4

For a magnetic system, we use the version (1.80) of the first law. Then the work done is given by

$$\delta W = V B_0 dM = \mu_0 V H dM$$

M: magnetization; V: constant volume (not a real thermodynamic variable!). From Curie law follows:

$$(dM)_T = \frac{C}{T} dH$$

$$\Leftrightarrow \qquad (\delta W)_T = \mu_0 \frac{CV}{T} H dH$$

$$\Leftrightarrow \qquad (\Delta W)_{12} = \int_{H_1}^{H_2} (\delta W)_T = \frac{\mu_0 CV}{2T} \left(H_2^2 - H_1^2\right) = \mu_0 \frac{VT}{2C} \left(M_2^2 - M_1^2\right)$$

Problem 1.5

1.

$$c_M = \left(\frac{\partial U}{\partial T}\right)_M$$

follows directly from the first law (1.80). Further holds with $U = U(T, M)$ and $N = const.$:

$$TdS = dU - VB_0 dM$$

$$= \left(\frac{\partial U}{\partial T}\right)_M dT + \left(\frac{\partial U}{\partial M}\right)_T dM - VB_0 dM$$

$$\curvearrowright c_H - c_M = \left[\left(\frac{\partial U}{\partial M}\right)_T - VB_0\right]\left(\frac{\partial M}{\partial T}\right)_H$$

Curie law:

$$\left(\frac{\partial M}{\partial T}\right)_H = -\frac{C}{T^2}H = -\frac{M^2}{CH}$$

Substitute

$$c_H = \left(\frac{\partial U}{\partial T}\right)_M + \left(\frac{\partial U}{\partial M}\right)_T \left(\frac{\partial M}{\partial T}\right)_H - VB_0 \left(\frac{\partial M}{\partial T}\right)_H$$

$$= \left(\frac{\partial U}{\partial T}\right)_M + \left(\frac{\partial U}{\partial M}\right)_T \left(\frac{\partial M}{\partial T}\right)_H + \mu_0 \frac{V}{C} M^2$$

Since $U = U(T, M)$ further holds

$$dU = \left(\frac{\partial U}{\partial T}\right)_M dT + \left(\frac{\partial U}{\partial M}\right)_T dM$$

$$\curvearrowright \left(\frac{\partial U}{\partial T}\right)_H = \left(\frac{\partial U}{\partial T}\right)_M + \left(\frac{\partial U}{\partial M}\right)_T \left(\frac{\partial M}{\partial T}\right)_H$$

$$c_H = \left(\frac{\partial U}{\partial T}\right)_H + \mu_0 \frac{V}{C} M^2$$

2. It holds

$$\left(\frac{\partial M}{\partial H}\right)_S = \left(\frac{\partial M}{\partial T}\right)_S \left(\frac{\partial T}{\partial H}\right)_S$$

The two factors are determined separately.

(a) First law (1.80):

$$dU = TdS + V\mu_0 H dM$$

adiabatic means $dS = 0$. Therefore

$$dU = \left(\frac{\partial U}{\partial T}\right)_M dT + \left(\frac{\partial U}{\partial M}\right)_T dM = V\mu_0 H dM$$

$$\curvearrowright \left(\left(\frac{\partial U}{\partial T}\right)_M dT\right)_S = \left(\left(V\mu_0 H - \left(\frac{\partial U}{\partial M}\right)_T\right) dM\right)_S$$

$$\curvearrowright \left(\frac{\partial M}{\partial T}\right)_S = \frac{c_M}{V\mu_0 H - \left(\frac{\partial U}{\partial M}\right)_T}$$

(b) Once again the first law for adiabatic changes of state

$$0 = dU - V\mu_0 H dM$$

With $U = U(T, H)$ and $M = M(T, H)$ follows:

$$0 = \left[\left(\frac{\partial U}{\partial T}\right)_H - V\mu_0 H \left(\frac{\partial M}{\partial T}\right)_H\right] dT +$$
$$+ \left[\left(\frac{\partial U}{\partial H}\right)_T - V\mu_0 H \left(\frac{\partial M}{\partial H}\right)_T\right] dH$$

According to part 1 the first parenthesis is equal to c_H. Therefore what remains is

$$c_H \, dT = \left[V\mu_0 H \left(\frac{\partial M}{\partial H}\right)_T - \left(\frac{\partial U}{\partial H}\right)_T\right] dH$$
$$= \left[V\mu_0 M - \left(\frac{\partial U}{\partial H}\right)_T\right] dH$$

That means

$$\left(\frac{\partial T}{\partial H}\right)_S = \frac{V\mu_0 M - \left(\frac{\partial U}{\partial H}\right)_T}{c_H}$$

This equation is combined with the final result of part 1:

$$\left(\frac{\partial M}{\partial H}\right)_S = \frac{c_M}{c_H} \cdot \frac{V\mu_o M - \left(\frac{\partial U}{\partial H}\right)_T}{V\mu_0 H - \left(\frac{\partial U}{\partial M}\right)_T}$$

That was the original proposition!

Problem 1.6

$$S = S(T, H) \Leftrightarrow dS = \left(\frac{\partial S}{\partial T}\right)_H dT + \left(\frac{\partial S}{\partial H}\right)_T dH \stackrel{!}{=} 0$$

This means

$$\left(\frac{\partial T}{\partial H}\right)_S = -\frac{\left(\frac{\partial S}{\partial H}\right)_T}{\left(\frac{\partial S}{\partial T}\right)_H} = -\frac{T}{c_H}\left(\frac{\partial S}{\partial H}\right)_T$$

Free enthalpy ($N = const.$):

$$dG = -SdT - MV\mu_0 dH$$

The corresponding Maxwell's equation gives

$$\left(\frac{\partial S}{\partial H}\right)_T = V\mu_0\left(\frac{\partial M}{\partial T}\right)_H = -V\mu_0\frac{C}{T^2}H$$

With this it finally follows:

$$\left(\frac{\partial T}{\partial H}\right)_S = \mu_0 V\frac{C}{c_H T}H$$

Problem 2.1

- $L > S$

$$\begin{aligned}
&\sum_{J=|L-S|}^{L+S}(2J+1) \\
&= \sum_{J=L-S}^{L+S}(2J+1) \\
&= \frac{1}{2}(2S+1)\left[2(L+S)+1+2(L-S)+1\right] \\
&= (2S+1)(2L+1)
\end{aligned}$$

- $L < S$

$$\begin{aligned}
&\sum_{J=|L-S|}^{L+S}(2J+1) \\
&= \sum_{J=S-L}^{L+S}(2J+1) \\
&= \frac{1}{2}(2L+1)\left[2(S+L)+1+2(S-L)+1\right] \\
&= (2S+1)(2L+1)
\end{aligned}$$

Problem 2.2

1. We obviously have

$$\sigma_x^2 = \begin{pmatrix} 0 & 1 \\ 1 & 0 \end{pmatrix} \cdot \begin{pmatrix} 0 & 1 \\ 1 & 0 \end{pmatrix} = \begin{pmatrix} 1 & 0 \\ 0 & 1 \end{pmatrix} = \mathbb{1}_2$$
$$\sigma_y^2 = \begin{pmatrix} 0 & -i \\ i & 0 \end{pmatrix} \cdot \begin{pmatrix} 0 & -i \\ i & 0 \end{pmatrix} = \begin{pmatrix} 1 & 0 \\ 0 & 1 \end{pmatrix} = \mathbb{1}_2$$
$$\sigma_z^2 = \begin{pmatrix} 1 & 0 \\ 0 & -1 \end{pmatrix} \cdot \begin{pmatrix} 1 & 0 \\ 0 & -1 \end{pmatrix} = \begin{pmatrix} 1 & 0 \\ 0 & 1 \end{pmatrix} = \mathbb{1}_2$$

2. Similarly one can easily get

$$\begin{aligned} &[\sigma_x, \sigma_y]_+ \\ &= \begin{pmatrix} 0 & 1 \\ 1 & 0 \end{pmatrix} \cdot \begin{pmatrix} 0 & +i \\ i & 0 \end{pmatrix} - \begin{pmatrix} 0 & -i \\ i & 0 \end{pmatrix} \cdot \begin{pmatrix} 0 & 1 \\ 1 & 0 \end{pmatrix} \\ &= \begin{pmatrix} i & 0 \\ 0 & -i \end{pmatrix} + \begin{pmatrix} -i & 0 \\ 0 & i \end{pmatrix} = 0 \end{aligned}$$

$$\begin{aligned} &[\sigma_y, \sigma_z]_+ \\ &= \begin{pmatrix} 0 & -i \\ i & 0 \end{pmatrix} \cdot \begin{pmatrix} 1 & 0 \\ 0 & -1 \end{pmatrix} + \begin{pmatrix} 1 & 0 \\ 0 & -1 \end{pmatrix} \cdot \begin{pmatrix} 0 & -i \\ i & 0 \end{pmatrix} \\ &= \begin{pmatrix} 0 & i \\ i & 0 \end{pmatrix} + \begin{pmatrix} 0 & -i \\ -i & 0 \end{pmatrix} = 0 \end{aligned}$$

$$\begin{aligned} &[\sigma_z, \sigma_x]_+ \\ &= \begin{pmatrix} 1 & 0 \\ 0 & -1 \end{pmatrix} \cdot \begin{pmatrix} 0 & 1 \\ 1 & 0 \end{pmatrix} + \begin{pmatrix} 0 & 1 \\ 1 & 0 \end{pmatrix} \cdot \begin{pmatrix} 1 & 0 \\ 0 & -1 \end{pmatrix} \\ &= \begin{pmatrix} 0 & 1 \\ -1 & 0 \end{pmatrix} + \begin{pmatrix} 0 & -1 \\ 1 & 0 \end{pmatrix} = 0 \end{aligned}$$

3.

$$\begin{aligned} &[\sigma_x, \sigma_y]_- \\ &= \begin{pmatrix} 0 & 1 \\ 1 & 0 \end{pmatrix} \cdot \begin{pmatrix} 0 & -i \\ i & 0 \end{pmatrix} - \begin{pmatrix} 0 & -i \\ i & 0 \end{pmatrix} \begin{pmatrix} 0 & 1 \\ 1 & 0 \end{pmatrix} \\ &= \begin{pmatrix} i & 0 \\ 0 & -i \end{pmatrix} - \begin{pmatrix} -i & 0 \\ 0 & i \end{pmatrix} \\ &= 2i \begin{pmatrix} 1 & 0 \\ 0 & -1 \end{pmatrix} = 2i\sigma_z \end{aligned}$$

The other two components are obtained analogously.

4.

$$\sigma_x \sigma_y \sigma_z = \begin{pmatrix} 0 & 1 \\ 1 & 0 \end{pmatrix} \cdot \begin{pmatrix} 0 & -i \\ i & 0 \end{pmatrix} \cdot \begin{pmatrix} 1 & 0 \\ 0 & -1 \end{pmatrix}$$

$$= \begin{pmatrix} 0 & 1 \\ 1 & 0 \end{pmatrix} \cdot \begin{pmatrix} 0 & i \\ i & 0 \end{pmatrix} = \begin{pmatrix} i & 0 \\ 0 & i \end{pmatrix}$$

$$= i\,\mathbb{1}_2$$

Problem 2.3
One immediately recognizes $(i, j = x, y, z)$

$$\widehat{\alpha}_i\,\widehat{\alpha}_j = \begin{pmatrix} \sigma_i \sigma_j & 0 \\ 0 & \sigma_i \sigma_j \end{pmatrix}$$

With the commutation relations for the Pauli spin matrices (see Problem 2.1) follows

$$\left[\widehat{\alpha}_i, \widehat{\alpha}_j\right]_+ = \begin{pmatrix} \left[\sigma_i, \sigma_j\right]_+ & 0 \\ 0 & \left[\sigma_i, \sigma_j\right]_+ \end{pmatrix}$$

$$= 2\delta_{ij} \begin{pmatrix} \mathbb{1}_2 & 0 \\ 0 & \mathbb{1}_2 \end{pmatrix} = 2\delta_{ij}\,\mathbb{1}_4$$

One can further calculate

$$\widehat{\alpha}_i\,\widehat{\beta} = \begin{pmatrix} 0 & \sigma_i \\ \sigma_i & 0 \end{pmatrix} \begin{pmatrix} \mathbb{1}_2 & 0 \\ 0 & -\mathbb{1}_2 \end{pmatrix} = \begin{pmatrix} 0 & -\sigma_i \\ \sigma_i & 0 \end{pmatrix}$$

$$\widehat{\beta}\,\widehat{\alpha}_i = \begin{pmatrix} \mathbb{1}_2 & 0 \\ 0 & -\mathbb{1}_2 \end{pmatrix} \begin{pmatrix} 0 & \sigma_i \\ \sigma_i & 0 \end{pmatrix} = \begin{pmatrix} 0 & \sigma_i \\ -\sigma_i & 0 \end{pmatrix}$$

$$\Rightarrow \left[\widehat{\alpha}_i, \widehat{\beta}\right]_+ = 0$$

Finally, it remains to be verified

$$\widehat{\beta}^2 = \begin{pmatrix} \mathbb{1}_2 & 0 \\ 0 & -\mathbb{1}_2 \end{pmatrix} \begin{pmatrix} \mathbb{1}_2 & 0 \\ 0 & -\mathbb{1}_2 \end{pmatrix} = \begin{pmatrix} \mathbb{1}_2 & 0 \\ 0 & \mathbb{1}_2 \end{pmatrix} = \mathbb{1}_4$$

So that (2.23), (2.24) and (2.25) are proved.

Problem 2.4
In solving the problem one uses the commutation relations for the Pauli spin matrices proved in Problem 2.2:

$$\begin{aligned}
\left[\widehat{s}_i, \widehat{s}_j\right]_- &= \frac{\hbar^2}{4}\begin{pmatrix} [\sigma_i, \sigma_j]_- & 0 \\ 0 & [\sigma_i, \sigma_j]_- \end{pmatrix} \\
&= \frac{\hbar^2}{4}\begin{pmatrix} 2i\sum_k \varepsilon_{ijk}\sigma_k & 0 \\ 0 & 2i\sum_k \varepsilon_{ijk}\sigma_k \end{pmatrix} \\
&= i\frac{\hbar^2}{2}\sum_k \varepsilon_{ijk}\begin{pmatrix} \sigma_k & 0 \\ 0 & \sigma_k \end{pmatrix} \\
&= i\hbar\sum_k \varepsilon_{ijk}\widehat{s}_k
\end{aligned}$$

Problem 2.5

$H_D^{(0)}$ is defined in (2.38). $\widehat{s}$ commutes with the momentum operator $\mathbf{p}$. We now have to calculate

$$\begin{aligned}
\widehat{s}_i\,\widehat{\alpha}_j &= \frac{\hbar}{2}\begin{pmatrix} \sigma_i & 0 \\ 0 & \sigma_i \end{pmatrix}\begin{pmatrix} 0 & \sigma_j \\ \sigma_j & 0 \end{pmatrix} = \frac{\hbar}{2}\begin{pmatrix} 0 & \sigma_i\sigma_j \\ \sigma_i\sigma_j & 0 \end{pmatrix} \\
\widehat{\alpha}_j\,\widehat{s}_i &= \frac{\hbar}{2}\begin{pmatrix} 0 & \sigma_j \\ \sigma_j & 0 \end{pmatrix}\begin{pmatrix} \sigma_i & 0 \\ 0 & \sigma_i \end{pmatrix} = \frac{\hbar}{2}\begin{pmatrix} 0 & \sigma_j\sigma_i \\ \sigma_j\sigma_i & 0 \end{pmatrix}
\end{aligned}$$

From the result of Problem 2.3 this means

$$\begin{aligned}
\left[\widehat{s}_i\,\widehat{\alpha}_j\right]_- &= \frac{\hbar}{2}\begin{pmatrix} 0 & [\sigma_i\sigma_j]_- \\ [\sigma_i\sigma_j]_- & 0 \end{pmatrix} \\
&= \frac{\hbar}{2}\begin{pmatrix} 0 & 2i\sum_k \varepsilon_{ijk}\sigma_k \\ 2i\sum_k \varepsilon_{ijk}\sigma_k & 0 \end{pmatrix} \\
&= i\hbar\sum_k \varepsilon_{ijk}\begin{pmatrix} 0 & \sigma_k \\ \sigma_k & 0 \end{pmatrix} = i\hbar\sum_k \varepsilon_{ijk}\widehat{\alpha}_k
\end{aligned}$$

So that we have

$$\begin{aligned}
[\widehat{s}_i, \widehat{\alpha}\cdot\mathbf{p}]_- &= i\hbar\sum_{jk}\varepsilon_{ijk}p_j\widehat{\alpha}_k = i\hbar(\mathbf{p}\times\widehat{\mathbf{a}})_i \\
[\widehat{\mathbf{s}}, \widehat{\alpha}\cdot\mathbf{p}]_- &= i\hbar(\mathbf{p}\times\widehat{\mathbf{a}})
\end{aligned}$$

To this one finds

$$\widehat{s}_i\cdot\widehat{\beta} = \frac{\hbar}{2}\begin{pmatrix} \sigma_i & 0 \\ 0 & \sigma_i \end{pmatrix}\begin{pmatrix} \mathbb{1}_2 & 0 \\ o & -\mathbb{1}_2 \end{pmatrix} = \frac{\hbar}{2}\begin{pmatrix} \sigma_i & 0 \\ 0 & -\sigma_i \end{pmatrix} = \widehat{\beta}\cdot\widehat{s}_i$$

So that

$$\left[\widehat{\mathbf{s}},\, H_D^{(0)}\right]_- = i\hbar c\,(\mathbf{p}\times\widehat{\alpha})$$

For the orbital angular momentum we use the well-known commutation relations with the linear momentum

$$\left[l_i,\, p_j\right]_- = i\hbar\sum_k \varepsilon_{ijk}\,p_k$$

So that it follows:

$$[l_i,\, \widehat{\alpha}\cdot\mathbf{p}]_- = i\hbar\sum_k \varepsilon_{ijk}\widehat{\alpha}_j\,p_k = i\hbar(\widehat{\alpha}\times\mathbf{p})_i$$

With $[l_i,\, \widehat{\beta}]_- = 0$ it eventually follows:

$$\left[\widehat{\mathbf{l}},\, H_D^{(0)}\right]_- = i\hbar c\,(\widehat{\alpha}\times\mathbf{p})$$

Problem 2.6
First we use the commutation relations from Problem 2.2:

$$\left[\sigma_i,\, \sigma_j\right]_- = 2i\sum_k \varepsilon_{ijk}\sigma_k$$
$$\left[\sigma_i,\, \sigma_j\right]_+ = 2\delta_{ij}\,\mathbb{1}_2$$

From this we get the important relation:

$$\sigma_i\sigma_j = \delta_{ij}\,\mathbb{1}_2 + i\sum_k \varepsilon_{ijk}\sigma_k$$

Due to the commutability it holds

$$\begin{aligned}(\sigma\cdot\mathbf{a})(\sigma\cdot\mathbf{b}) &= \sum_{i,j} a_i b_j\sigma_i\sigma_j = \sum_{i,j} a_i b_j\left(\delta_{ij} + i\sum_k \varepsilon_{ijk}\sigma_k\right)\\ &= \left(\sum_i a_i b_i\right) + i\sum_{ijk}\varepsilon_{ijk}a_i b_j\sigma_k\\ &= (\mathbf{a}\cdot\mathbf{b})\mathbb{1}_2 + i\mathbf{a}\cdot(\mathbf{b}\times\sigma)\end{aligned}$$

Since σ commutes with **a** and **b**, we can cyclically permute the operators in second summand (triple product!).

Problem 2.7

$$H_D = c\widehat{\boldsymbol{\alpha}}\cdot(\mathbf{p}+e\mathbf{A}) + \widehat{\beta}m_e c^2 - e\varphi$$

Heisenberg's equations of motion:

$$i\hbar\frac{\mathrm{d}}{\mathrm{d}t}\mathbf{r} =[\mathbf{r}, H_D]_- = c[\mathbf{r}, \mathbf{p}]_- \cdot \widehat{\boldsymbol{\alpha}} = i\hbar\widehat{\boldsymbol{\alpha}}c$$
$$\curvearrowright \quad \dot{\mathbf{r}}(t) = c\widehat{\boldsymbol{\alpha}}$$

$$\begin{aligned}
&i\hbar\frac{\mathrm{d}}{\mathrm{d}t}(\mathbf{p}+e\mathbf{A})\\
&= [(\mathbf{p}+e\mathbf{A}, H_D]_- + i\hbar\frac{\partial}{\partial t}(\mathbf{p}+e\mathbf{A}) =\\
&= ce[\mathbf{p}, \widehat{\boldsymbol{\alpha}}\cdot\mathbf{A}]_- - e[\mathbf{p}, \varphi]_- + ec[\mathbf{A}, \widehat{\boldsymbol{\alpha}}\cdot\mathbf{p}]_- + i\hbar e\frac{\partial\mathbf{A}}{\partial t} =\\
&= ec\frac{\hbar}{i}[-\widehat{\boldsymbol{\alpha}}\cdot\nabla\mathbf{A} + \nabla(\widehat{\boldsymbol{\alpha}}\cdot\mathbf{A})] - e\frac{\hbar}{i}\nabla\varphi + i\hbar e\frac{\partial\mathbf{A}}{\partial t} =\\
&= ec\frac{\hbar}{i}(\widehat{\boldsymbol{\alpha}}\times(\nabla\times\mathbf{A})) - e\frac{\hbar}{i}\nabla\varphi + i\hbar e\frac{\partial\mathbf{A}}{\partial t}
\end{aligned}$$

$$\curvearrowright \quad \frac{\mathrm{d}}{\mathrm{d}t}(\mathbf{p}+e\mathbf{A}) = -ec(\widehat{\boldsymbol{\alpha}}\times\mathbf{B}) + e\left(\nabla\varphi + \frac{\partial\mathbf{A}}{\partial t}\right)$$

With

$$\mathbf{E} = -\nabla\varphi - \frac{\partial\mathbf{A}}{\partial t}$$

finally follows:

$$\frac{\mathrm{d}}{\mathrm{d}t}(\mathbf{p}+e\mathbf{A}) = -e(\dot{\mathbf{r}}\times\mathbf{B}+\mathbf{E}).$$

On the right-hand side is the Lorentz force.

Problem 2.8

1.

$$\begin{aligned}
[H_{\mathrm{SB}}, L_i]_- &= \sum_{j=1}^{3}\lambda\left[L_j S_j, L_i\right]_- = \sum_{j=1}^{3}\lambda\left[L_j, L_i\right]_- S_j =\\
&= \sum_{j=1}^{3}\lambda\sum_k \varepsilon_{jik}L_k S_j i\hbar = i\hbar\lambda\sum_{jk}\varepsilon_{kji}L_k S_j =\\
&= i\hbar\lambda(\mathbf{L}\times\mathbf{S})_i
\end{aligned}$$
$$\Longrightarrow [H_{\mathrm{SB}}, \mathbf{L}]_- = i\hbar\lambda(\mathbf{L}\times\mathbf{S})$$

2.

$$
\begin{aligned}
[H_{\text{SB}}, S_i]_- &= \sum_{j=1}^{3} \lambda \left[L_j S_j, S_i\right]_- = \lambda \sum_{j=1}^{3} L_j \left[S_j, S_i\right]_- = \\
&= \lambda \sum_{j=1}^{3} L_j i\hbar \sum_k \varepsilon_{jik} S_k = i\hbar\lambda \sum_{jk} \varepsilon_{kji} S_k L_j = \\
&= i\hbar\lambda(\mathbf{S} \times \mathbf{L})_i \\
\Longrightarrow [H_{\text{SB}}, \mathbf{S}]_- &= i\hbar\lambda(\mathbf{S} \times \mathbf{L})
\end{aligned}
$$

3.

$$
\left[H_{\text{SB}}, \mathbf{L}^2\right]_- = \sum_{i=1}^{3} \lambda \left[L_i S_j, \mathbf{L}^2\right]_- = \sum_{i=1}^{3} \lambda \left[L_i, \mathbf{L}^2\right]_- S_i = 0
$$

4.

$$
\left[H_{\text{SB}}, \mathbf{S}^2\right]_- = \lambda \sum_{i=1}^{3} L_i \left[S_i, \mathbf{S}^2\right]_- = 0
$$

5. From 1 and 2 follows:

$$
\begin{aligned}
&[H_{\text{SB}}, J_i]_- = 0 \quad \text{for } i = x, y, z \\
\Longrightarrow &\left[H_{\text{SB}}, \mathbf{J}^2\right]_- = \sum_i \left[H_{\text{SB}}, J_i^2\right]_- = 0
\end{aligned}
$$

Problem 2.9

1. According to (2.162) it must hold

$$
R_z(\varepsilon) T_q^{(k)} R_z^{-1}(\varepsilon) = \sum_{q'=-k}^{+k} (R_z(\varepsilon))_{q'q}^{(k)} T_{q'}^{(k)}
$$

Here we have according to (2.145)

$$
R_z(\varepsilon) = 1 - \frac{i}{\hbar} \varepsilon J_z
$$

and according to (2.158)

$$
\begin{aligned}
(R_z(\varepsilon))^{(k)}_{q'q} &= \langle kq' \,|R_z(\varepsilon)|\, kq\rangle \\
&= \left\langle kq' \left|(1 - \frac{i}{\hbar}\varepsilon J_z)\right| kq\right\rangle \\
&= \delta_{qq'} - \frac{i}{\hbar}\varepsilon\hbar q\delta_{qq'} \\
&= (1 - i\varepsilon q)\delta_{qq'}
\end{aligned}
$$

$$
\curvearrowright \quad \sum_{q'=-k}^{k} (R_z(\varepsilon))^{(k)}_{q'q} = (1 - i\varepsilon q)T^{(k)}_q
$$

On the other hand:

$$
\begin{aligned}
R_z(\varepsilon)T^{(k)}_q R_z^{-1}(\varepsilon) &= (1 - \frac{i}{\hbar}\varepsilon J_z)T^{(k)}_q(1 + \frac{i}{\hbar}\varepsilon J_z) \\
&= T^{(k)}_q - \frac{i}{\hbar}\varepsilon\left[J_z, T^{(k)}_q\right]_- + O(\varepsilon^2)
\end{aligned}
$$

Compare:

$$
\left[J_z, T^{(k)}_q\right]_- = \hbar q T^{(k)}_q
$$

2. According to (2.160) holds

$$
\begin{aligned}
(R_x(\varepsilon))^{(k)}_{q'q} = \delta_{q'q} &- \frac{1}{2}i\varepsilon\sqrt{k(k+1) - q(q+1)}\delta_{q',q+1} \\
&- \frac{1}{2}i\varepsilon\sqrt{k(k+1) - q(q-1)}\delta_{q',q-1}
\end{aligned}
$$

So that we have

$$
\begin{aligned}
\sum_{q'=-k}^{+k} (R_x(\varepsilon))^{(k)}_{q'q} T^{(k)}_{q'} = T^{(k)}_q &- \frac{1}{2}i\varepsilon\sqrt{k(k+1) - q(q+1)}T^{(k)}_{q+1} \\
&- \frac{1}{2}i\varepsilon\sqrt{k(k+1) - q(q-1)}T^{(k)}_{q-1}
\end{aligned}
$$

On the other hand we have as in 1.:

$$
R_x(\varepsilon)T^{(k)}_q R_x^{-1}(\varepsilon) = T^{(k)}_q - \frac{i}{\hbar}\varepsilon\left[J_x, T^{(k)}_q\right]_- + O(\varepsilon^2)
$$

Then a comparison gives

$$\left[J_x, T_q^{(k)}\right]_- = \frac{\hbar}{2}\sqrt{k(k+1) - q(q+1)}T_{q+1}^{(k)} + \frac{\hbar}{2}\sqrt{k(k+1) - q(q-1)}T_{q-1}^{(k)}$$

Completely analogously one finds

$$\left[J_y, T_q^{(k)}\right]_- = -i\frac{\hbar}{2}\sqrt{k(k+1) - q(q+1)}T_{q+1}^{(k)} + i\frac{\hbar}{2}\sqrt{k(k+1) - q(q-1)}T_{q-1}^{(k)}$$

With $J_\pm = J_x \pm i J_y$ follows the final result:

$$\left[J_\pm, T_q^{(k)}\right]_- = \hbar\sqrt{k(k+1) - q(q \pm 1)}T_{q\pm 1}^{(k)}$$

Problem 2.10
We use (2.150)

$$[(\mathbf{n} \cdot \mathbf{J}), (\overline{\mathbf{n}} \cdot \mathbf{K})]_- = i\hbar(\mathbf{n} \times \overline{\mathbf{n}}) \cdot \mathbf{K} \quad \mathbf{n}, \overline{\mathbf{n}} : \text{unit vectors}$$

1. $\mathbf{n} = \mathbf{e}_z$
   If further $\overline{\mathbf{n}} = \mathbf{e}_z$ is valid, we have

$$\mathbf{n} \times \overline{\mathbf{n}} = 0 \curvearrowright [J_z, K_z]_- = 0 = \hbar \cdot 0 \cdot K_0^{(1)}$$

   For $\overline{\mathbf{n}} = \mathbf{e}_x$ we have

$$\mathbf{n} \times \overline{\mathbf{n}} = \mathbf{e}_y \curvearrowright [J_z, K_x]_- = i\hbar K_y$$

   For $\overline{\mathbf{n}} = \mathbf{e}_y$ we have

$$\mathbf{n} \times \overline{\mathbf{n}} = -\mathbf{e}_x \curvearrowright \left[J_z, K_y\right]_- = -i\hbar K_x$$

$$K_{\pm 1}^{(1)} = \mp\frac{1}{\sqrt{2}}(K_x \pm i K_y) \curvearrowright$$

$$\left[J_z, K_{\pm 1}^{(1)}\right]_- = \mp\frac{1}{\sqrt{2}}i\hbar(K_y \mp i K_x) = \hbar\frac{1}{\sqrt{2}}(-K_x \mp i K_y) = \hbar \cdot (\pm 1) \cdot K_{\pm 1}^{(1)}$$

Therefore altogether we have

$$\left[J_z, K_q^{(1)}\right]_- = \hbar q K_q^{(1)}$$

This is the first part of (2.163).

2. $\sqrt{k(k+1) - q(q \pm 1)} = \sqrt{2 - q(q \pm 1)}$

(a) $q = 0$:

$$\begin{aligned}
K_0^{(1)} &= K_z \\
\left[J_x, K_0^{(1)}\right]_- &= [J_x, K_z]_- = i\hbar(\mathbf{e}_x \times \mathbf{e}_z) \cdot \mathbf{K} = -i\hbar K_y \\
\left[J_y, K_0^{(1)}\right]_- &= i\hbar(\mathbf{e}_y \times \mathbf{e}_z) \cdot \mathbf{K} = i\hbar K_x \\
\curvearrowright \left[J_\pm, K_0^{(1)}\right]_- &= -i\hbar K_y \pm i(i\hbar K_x) \\
&= \mp\hbar(K_x \pm i K_y) \\
&= \hbar\sqrt{2} K_{\pm 1}^{(1)}
\end{aligned}$$

For $q = 0$ we therefore have

$$\left[J_\pm, K_q^{(1)}\right]_- = \hbar\sqrt{2 - q(q \pm 1)} K_{q \pm 1}^{(1)}$$

(b) $q = +1$:

$$\begin{aligned}
K_{+1}^{(1)} &= -\frac{1}{\sqrt{2}} K_+ \\
\left[J_x, K_{+1}^{(1)}\right]_- &= -\frac{1}{\sqrt{2}} \left[J_x, K_x + i K_y\right]_- \\
&= -\frac{i}{\sqrt{2}} \left[J_x, K_y\right] \\
&= \frac{\hbar}{\sqrt{2}} (\mathbf{e}_x \times \mathbf{e}_y) \cdot \mathbf{K} = \frac{\hbar}{\sqrt{2}} K_z
\end{aligned}$$

analogously

$$\begin{aligned}\left[J_y, K_{+1}^{(1)}\right]_- &= -\frac{1}{\sqrt{2}}\left[J_y, K_x + iK_y\right]_- \\ &= -\frac{1}{\sqrt{2}}\left[J_y, K_x\right]_- \\ &= -\frac{i\hbar}{\sqrt{2}}(\mathbf{e}_y \times \mathbf{e}_x)\cdot\mathbf{K} = i\frac{\hbar}{\sqrt{2}}K_z \\ \curvearrowright \left[J_+, K_{+1}^{(1)}\right]_- &= 0 \\ \left[J_-, K_{+1}^{(1)}\right]_- &= \hbar\sqrt{2}K_z = \sqrt{2}\hbar K_0^{(1)}\end{aligned}$$

For $q = +1$ then holds

$$\left[J_\pm, K_q^{(1)}\right]_- = \hbar\sqrt{2 - q(q \pm 1)}K_{q\pm1}^{(1)}$$

(c) $q = -1$:

$$\begin{aligned}K_{-1}^{(1)} &= \frac{1}{\sqrt{2}}K_- \\ \left[J_x, K_{-1}^{(1)}\right]_- &= \frac{1}{\sqrt{2}}(-i)\left[J_x, K_y\right]_- \\ &= \frac{\hbar}{\sqrt{2}}(\mathbf{e}_x \times \mathbf{e}_y)\cdot\mathbf{K} = \frac{\hbar}{\sqrt{2}}K_z \\ \left[J_y, K_{-1}^{(1)}\right]_- &= \frac{1}{\sqrt{2}}\left[J_y, K_x\right]_- \\ &= i\frac{\hbar}{\sqrt{2}}(\mathbf{e}_y \times \mathbf{e}_x)\cdot\mathbf{K} = -i\frac{\hbar}{\sqrt{2}}K_z \\ \curvearrowright \left[J_\pm, K_{-1}^{(1)}\right]_- &= \frac{\hbar}{\sqrt{2}}(K_z \pm K_z)\end{aligned}$$

For $q = -1$ we can write the following:

$$\left[J_\pm, K_q^{(1)}\right]_- = \hbar\sqrt{2 - q(q \pm 1)}K_{q\pm1}^{(1)}$$

2a, 2b and 2c together give the second part of (2.163):

$$\left[J_\pm, K_q^{(1)}\right]_- = \hbar\sqrt{2 - q(q \pm 1)}K_{q\pm1}^{(1)}$$

Problem 2.11

1. $\left[I_z, T_q^{(2)}\right]_- \overset{!}{=} \hbar q T_q^{(2)}$
   One immediately recognizes

$$\left[I_z, T_0^{(2)}\right]_- = \left[I_z, I^2 - 3I_z^2\right]_- = 0$$

On the other hand with a little more effort we have

$$\begin{aligned}\left[I_z, T_{\pm 1}^{(2)}\right]_- &= \pm\frac{1}{2}\sqrt{6}\left\{\left[I_z, I_z I_\pm\right]_- + \left[I_z, I_\pm I_z\right]_-\right\}\\ &= \pm\frac{1}{2}\sqrt{6}\left\{I_z\left[I_z, I_\pm\right]_- + \left[I_z, I_\pm\right]_- I_z\right\}\\ &= \frac{1}{2}\hbar\sqrt{6}\left\{I_z I_\pm + I_\pm I_z\right\}\\ &= \pm\hbar T_{\pm 1}^{(2)}\end{aligned}$$

$$\begin{aligned}\left[I_z, T_{\pm 2}^{(2)}\right]_- &= -\frac{1}{2}\sqrt{6}\left[I_z, (I_\pm)^2\right]_-\\ &= -\frac{1}{2}\sqrt{6}\left\{I_\pm\left[I_z, I_\pm\right]_- + \left[I_z, I_\pm\right]_- I_\pm\right\}\\ &= -(\pm\hbar)\sqrt{6}(I_\pm)^2\\ &= \pm 2\hbar T_{\pm 2}^{(2)}\end{aligned}$$

Thus the first commutation relation is obviously fulfilled

2. $\left[I_\pm, T_q^{(2)}\right]_- \stackrel{!}{=} \sqrt{6 - q(q \pm 1)}\hbar q T_{q\pm 1}^{(2)}$

$$\begin{aligned}\left[I_\pm, T_0^{(2)}\right]_- &= \left[I_\pm, I^2\right]_- - 3\left[I_\pm, I_z^2\right]_-\\ &= 0 - 3I_z\left[I_\pm, I_z\right]_- - 3\left[I_\pm, I_z\right]_- I_z\\ &= \pm 3\hbar(I_z I_\pm + I_\pm I_z)\\ &= \hbar\sqrt{6}\left(\pm\frac{1}{2}\sqrt{6}(I_z I_\pm + I_\pm I_z)\right)\\ &= \hbar\sqrt{6}T_{\pm 1}^{(2)}\end{aligned}$$

$$\begin{aligned}\left[I_+, T_{+1}^{(2)}\right]_- &= \frac{1}{2}\sqrt{6}\left(\left[I_+, I_z I_+\right]_- + \left[I_+, I_+ I_z\right]_-\right)\\ &= \frac{1}{2}\sqrt{6}\left(\left[I_+, I_z\right]_- I_+ + I_+\left[I_+, I_z\right]_-\right)\\ &= \frac{1}{2}\sqrt{6}(-2\hbar(I_+)^2)\\ &= 2\hbar T_{+2}^{(2)}\end{aligned}$$

$$\left[I_+, T_{-1}^{(2)}\right]_- = -\frac{1}{2}\sqrt{6}\left([I_+, I_zI_-]_- + [I_+, I_-I_z]_-\right)$$
$$= -\frac{1}{2}\sqrt{6}\left(I_z2\hbar I_z - \hbar I_+I_- - \hbar I_-I_+ + 2\hbar I_z^2\right)$$
$$= -\frac{1}{2}\sqrt{6}\hbar(4I_z^2 - 2I_x^2 - 2I_y^2)$$
$$= -\hbar\sqrt{6}(3I_z^2 - I^2)$$
$$= \hbar\sqrt{6}T_0^{(2)}$$

$$\left[I_-, T_{+1}^{(2)}\right]_- = -\frac{1}{2}\sqrt{6}\left([I_-, I_zI_+]_- + [I_-, I_+I_z]_-\right)$$
$$= \frac{1}{2}\sqrt{6}\left(I_z(-2\hbar I_z) + \hbar I_-I_+ + \hbar I_+I_- + (-2\hbar I_z)I_z\right)$$
$$= \frac{1}{2}\sqrt{6}\hbar(-4I_z^2 + 2I_x^2 + 2I_y^2)$$
$$= \hbar\sqrt{6}(I^2 - 3I_z^2)$$
$$= \hbar\sqrt{6}T_0^{(2)}$$

$$\left[I_-, T_{-1}^{(2)}\right]_- = -\frac{1}{2}\sqrt{6}\left([I_-, I_zI_-]_- + [I_-, I_-I_z]_-\right)$$
$$= -\frac{1}{2}\sqrt{6}(\hbar(I_-)^2 + I_-(+\hbar I_-))$$
$$= 2\hbar T_{-2}^{(2)}$$

$$\left[I_+, T_{+2}^{(2)}\right]_- = -\frac{1}{2}\sqrt{6}\left[I_+, (I_+)^2\right]_- = 0$$

$$\left[I_+, T_{-2}^{(2)}\right]_- = -\frac{1}{2}\sqrt{6}\left[I_+, (I_-)^2\right]_-$$
$$= -\frac{1}{2}\sqrt{6}(I_-2\hbar I_z + 2\hbar I_zI_-)$$
$$= 2\hbar T_{-1}^{(2)}$$

$$\left[I_-, T_{+2}^{(2)}\right]_- = -\frac{1}{2}\sqrt{6}\left[I_-, (I_+)^2\right]_-$$
$$= -\frac{1}{2}\sqrt{6}(-2\hbar I_+ I_z - 2\hbar I_z I_+)$$
$$= 2\hbar T_{+1}^{(2)}$$

$$\left[I_-, T_{-2}^{(2)}\right]_- = -\frac{1}{2}\sqrt{6}\left[I_-, (I_-)^2\right]_- = 0$$

Problem 2.12

$$A = \sum_{q=-2}^{+2} t_q^{(2)} \cdot \left(T_q^{(2)}\right)^\dagger = \sum_{i=-2}^{+2} A_i$$

$\underline{A_0 = \left(J^2 - 3J_z^2\right)\left(I^2 - 3I_z^2\right)}$

$$\curvearrowright \quad A_0 = J^2 I^2 - 3J_z^2 I^2 - 3I_z^2 J^2 + 9J_z^2 I_z^2$$

$\underline{A_{\pm 1} = \frac{3}{2}\left(J_z J_\pm + J_\pm J_z\right)\left(I_z I_\mp + I_\mp I_z\right)}$

$$\begin{aligned}
&A_{+1} + A_{-1} \\
&= \frac{3}{2}(J_z J_+ I_z I_- + J_z J_+ I_- I_z + J_+ J_z I_z I_- + J_+ J_z I_- I_z + \\
&\quad + J_z J_- I_z I_+ + J_z J_- I_+ I_z + J_- J_z I_z I_+ + J_- J_z I_+ I_z) \\
&= \frac{3}{2}\left(J_z I_z \left(J_+ I_- + J_- I_+\right) + \left(J_+ I_- + J_- I_+\right) J_z I_z\right) + \\
&\quad + \frac{3}{2} J_z \left(J_+ I_- + J_- I_+\right) I_z + \frac{3}{2}\left(J_+ \left(J_z I_z\right) I_- + J_- \left(J_z I_z\right) I_+\right) \\
&= 3\left(J_z I_z \left(J_+ I_- + J_- I_+\right) + \left(J_+ I_- + J_- I_+\right) J_z I_z\right) + \\
&\quad + \frac{3}{2}\hbar\left(J_+ I_- I_z - J_+ I_z I_-\right) + \frac{3}{2}\hbar\left(-J_- I_+ I_z + J_- I_z I_+\right) \\
&= 6\left(J_z I_z \left(J_x I_x + J_y I_y\right) + \left(J_x I_x + J_y I_y\right) J_z I_z\right) + \\
&\quad + \frac{3}{2}\hbar J_+ \left[I_-, I_z\right]_- + \frac{3}{2}\hbar J_- \left[I_z, I_+\right]_- \\
&= 6\left(J_x I_x + J_y I_y + J_z I_z\right)^2 - 6J_z^2 I_z^2 - 6\left(J_x I_x + J_y I_y\right)^2 + \\
&\quad \frac{3}{2}\hbar^2\left(J_+ I_- + J_- I_+\right)
\end{aligned}$$

↷

$$A_{+1} + A_{-1} = 6\,(\mathbf{J}\cdot\mathbf{I})^2 - 6J_z^2 I_z^2 - \frac{3}{2}\,(J_+I_- + J_-I_+)^2 \\ + 3\hbar^2\,(\mathbf{J}\cdot\mathbf{I}) - 3\hbar^2 J_z I_z$$

$A_{\pm 2} = \frac{3}{2} J_\pm^2 I_\mp^2$

↷

$$A_{+2} + A_{-2} = \frac{3}{2}\left(J_+^2 I_-^2 + J_-^2 I_+^2\right)$$

We now sum the terms

$$\begin{aligned} A &= J^2I^2 - 3J_z^2I^2 - 3I_z^2J^2 + 9J_z^2I_z^2 + 6\,(\mathbf{J}\cdot\mathbf{I})^2 \\ &\quad - 6J_z^2I_z^2 - \frac{3}{2}\,(J_+I_- + J_-I_+)^2 + \\ &\quad + 3\hbar^2\,(\mathbf{J}\cdot\mathbf{I}) - 3\hbar^2 J_zI_z + \frac{3}{2}\left(J_+^2I_-^2 + J_-^2I_+^2\right) \\ &= 6\,(\mathbf{J}\cdot\mathbf{I})^2 + 3\hbar^2\,(\mathbf{J}\cdot\mathbf{I}) - 2J^2I^2 + D \end{aligned}$$

$$\begin{aligned} D &= 3J^2I^2 - 3J_z^2I^2 - 3I_z^2J^2 + 3J_z^2I_z^2 - \\ &\quad - 3\hbar^2 J_zI_z\frac{3}{2}\,(J_+J_-I_-I_+ + J_-J_+I_+I_-) - \\ &= 3\left(J_x^2 + J_y^2\right)I^2 - 3\left(J_x^2 + J_y^2\right)I_z^2 - 3Re\,(J_+J_-I_-I_+) \\ &\quad - 3\hbar^2 J_zI_z \\ &= 3\left(J_x^2 + J_y^2\right)\left(I_x^2 + I_y^2\right) - \\ &\quad - 3Re\left(\left(J_x^2 + J_y^2 + i\left[J_y, J_x\right]_-\right)\left(I_x^2 + I_y^2 - i\left[I_y, I_x\right]_-\right)\right) \\ &\quad - 3\hbar^2 J_zI_z \\ &= -3Re\,((-i\hbar J_z)(-i\hbar I_z)) - 3\hbar^2 J_zI_z \\ &= 0 \end{aligned}$$

Then we have

$$A = 6\,(\mathbf{J}\cdot\mathbf{I})^2 + 3\hbar^2\,(\mathbf{J}\cdot\mathbf{I}) - 2J^2I^2$$

This corresponds to the quadruple term (2.245).

Problem 2.13

1. Build

$$\begin{aligned}\mathrm{div}(gf\mathbf{j}) &= (gf)\mathrm{div}\mathbf{j} + \mathbf{j}\cdot\mathrm{grad}(gf)\\ &= \mathbf{j}\cdot\mathrm{grad}(fg)\\ &= f\mathbf{j}\cdot\nabla g + g\mathbf{j}\cdot\nabla f\end{aligned}$$

That is exactly the integrand of D. With Gauss theorem we have

$$D = \int d^3r\ \mathrm{div}(gf\mathbf{j}) = \int\limits_{S\to\infty} d\mathbf{f}\cdot(gf\mathbf{j}) = 0$$

since $\mathbf{j}$ vanishes at infinity.

2. Set $f \equiv 1,\ g = x, y, z$:

$$\begin{aligned}&\curvearrowright 0 = D = \int d^3r\ \mathbf{j}\cdot\mathbf{e}_{x,y,z} = \int d^3r\ j_{x,y,z}\\ &\curvearrowright \int d^3r\ \mathbf{j}(\mathbf{r}) = 0\end{aligned}$$

3. Set $f = x_i,\quad g = x_j$ with $x_{i,j} \in \{x, y, z\}$: Then with 1. we have

$$\begin{aligned}0 &= \int d^3r(x_i j_j + x_j j_i)\\ \curvearrowright \int d^3r\ x_j j_i &= -\int d^3r x_i j_j\end{aligned}$$

Now let $\mathbf{a}$ be an arbitrary vector:

$$\begin{aligned}\mathbf{a}\cdot\int d^3r\ \mathbf{r}\ j_i(\mathbf{r}) &= \sum_j a_j \int d^3r\ x_j j_i(\mathbf{r})\\ &= \frac{1}{2}\sum_j a_j \int d^3r\ (x_j j_i(\mathbf{r}) - x_i j_j(\mathbf{r}))\\ &= \frac{1}{2}\sum_j a_j\varepsilon_{jik}\int d^3r\ (\mathbf{r}\times\mathbf{j})_k\\ &= -\frac{1}{2}\sum_j \varepsilon_{ijk}a_j\int d^3r\ (\mathbf{r}\times\mathbf{j})_k\\ &= -\frac{1}{2}\left(\mathbf{a}\times\int d^3r(\mathbf{r}\times\mathbf{j})\right)_i\end{aligned}$$

This is valid for all the components i:

$$\mathbf{a} \cdot \int d^3r \, \mathbf{r} \, \mathbf{j}(\mathbf{r}) = -\frac{1}{2} \left(\mathbf{a} \times \int d^3r (\mathbf{r} \times \mathbf{j}) \right)$$

Problem 3.1

Because of the circular motion, every electron has an angular momentum and therefore a magnetic orbital momentum $\mathbf{m}^{(i)}$. In the absence of an external field, the orientations of the electron moments are statistically distributed and hence compensate each other. After the application of a field, the angular momenta precess about the direction of the field while the motion of the electrons in the orbital planes remains unchanged.

Because of the precession there is an extra current:

$$\Delta I = \frac{-e}{\tau} = \frac{-e\omega_L}{2\pi}$$

which according to classical electrodynamics induces an additional magnetic moment:

$$\Delta \mathbf{m}^{(i)} = \Delta I \, \mathbf{F}_i = \frac{-e\omega_L}{2\pi} \pi r_{i\perp}^2 \mathbf{e}_z = -\frac{e^2}{4m} \left(x_i^2 + y_i^2 \right) B_0 \mathbf{e}_z$$

$\mathbf{F}_i$ is the vector area of the circle along which the i th electron moves. Finally the (average) magnetization of the diamagnet (N, the number of atoms; V, the volume, N_e, the number of electrons per atom) is given by

$$\Delta \mathbf{M} = \mathbf{M} = \frac{N}{V} \sum_{i=1}^{N_e} \langle \Delta \mathbf{m}^{(i)} \rangle = -\frac{Ne^2}{6mV} B_0 \sum_{i=1}^{N_e} \langle r_i^2 \rangle$$

In calculating this we have used

$$\sum_{i=1}^{N_e} \langle x_i^2 \rangle = \sum_{i=1}^{N_e} \langle y_i^2 \rangle = \sum_{i=1}^{N_e} \langle z_i^2 \rangle = \frac{1}{3} \sum_{i=1}^{N_e} \langle r_i^2 \rangle$$

Then the diamagnetic susceptibility is given by

$$\chi^{dia} = \mu_0 \left(\frac{\partial M}{\partial B_0} \right)_T = -\frac{Ne^2 \mu_0}{6mV} \sum_{i=1}^{N_e} \langle r_i^2 \rangle$$

This expression agrees with the quantum mechanically correct expression (3.21). Since the calculation is *not* strictly classical, there is no contradiction with Bohr–van Leeuwen theorem. The assumption of stationary electron orbits is classically untenable!

Problem 3.2

$T = 0$:

$$\bar{\varepsilon} = \frac{1}{N}\left(\underbrace{2}_{\text{Spin}}\sum_{\mathbf{k}}^{k\le k_F}\varepsilon(\mathbf{k})\right) = \frac{2}{N}\frac{1}{\Delta k}\int_{k\le k_F}\frac{\hbar^2k^2}{2m}$$

$$\Delta k = \frac{(2\pi)^3}{V} \qquad \text{grid volume : Volume per } \mathbf{k}\text{-state in } \mathbf{k}\text{-space.}$$

$$\curvearrowright \quad \bar{\varepsilon} = 2\frac{V}{N}\frac{4\pi}{8\pi^3}\frac{\hbar^2}{2m}\int_0^{k_F} dk\, k^4 = \frac{V}{N}\frac{\hbar^2}{2m\pi^2}\frac{1}{5}k_F^5 = \varepsilon_F\frac{V}{N}\frac{1}{5\pi^2}k_F^3$$

With $k_F^3 = 3\pi^2\frac{N}{V}$ follows:

$$\bar{\varepsilon} = \frac{3}{5}\varepsilon_F$$

Problem 3.3

$$\hat{H} = \sum_{\mathbf{k}\sigma}\varepsilon(\mathbf{k})a_{\mathbf{k}\sigma}^+a_{\mathbf{k}\sigma} = \sum_{\mathbf{k}\sigma}\varepsilon(\mathbf{k})\hat{n}_{\mathbf{k}\sigma}$$

1. Statistical operator of the grand canonical ensemble

$$\rho = \exp(-\beta(\hat{H} - \mu\hat{N})) \qquad \text{(unnormalized)}$$

$$\hat{N} = \sum_{\mathbf{k}\sigma}a_{\mathbf{k}\sigma}^+a_{\mathbf{k}\sigma} = \sum_{\mathbf{k}\sigma}\hat{n}_{\mathbf{k}\sigma}$$

$$\curvearrowright \quad \left[\hat{H}, \hat{N}\right]_- = 0 \quad \curvearrowright \quad \text{combined eigenstates (Fock states)}$$

So that

$$\begin{aligned}
\operatorname{Tr}\rho &= \sum_{N=0}^{\infty}\sum_{\{n_{\mathbf{k}\sigma}\}}^{(\sum n_{\mathbf{k}\sigma}=N)}\exp(-\beta\sum_{\mathbf{k}\sigma}(\varepsilon(\mathbf{k})-\mu)n_{\mathbf{k}\sigma}) \\
&= \sum_{N=0}^{\infty}\sum_{\{n_{\mathbf{k}\sigma}\}}^{(\sum n_{\mathbf{k}\sigma}=N)}\prod_{\mathbf{k}\sigma}\exp(-\beta(\varepsilon(\mathbf{k})-\mu)n_{\mathbf{k}\sigma}), \qquad (n_{\mathbf{k}\sigma} = 0, 1) \\
&= \sum_{\{n_{\mathbf{k}_1\sigma_1}\}}\sum_{\{n_{\mathbf{k}_2\sigma_2}\}}\cdots\prod_{\mathbf{k}\sigma}\exp(-\beta(\varepsilon(\mathbf{k})-\mu)n_{\mathbf{k}\sigma})
\end{aligned}$$

$$= \left[\sum_{n_{\mathbf{k}_1\sigma_1}=0,1} \exp(-\beta(\varepsilon(\mathbf{k}_1) - \mu) n_{\mathbf{k}_1\sigma_1}) \right]^2 \times$$

$$\times \left[\sum_{n_{\mathbf{k}_2\sigma_2}=0,1} \exp(-\beta(\varepsilon(\mathbf{k}_2) - \mu) n_{\mathbf{k}_2\sigma_2}) \right]^2 \cdots$$

$$= \left[1 + \exp(-\beta(\varepsilon(\mathbf{k}_1) - \mu))\right]^2 \left[1 + \exp(-\beta(\varepsilon(\mathbf{k}_2) - \mu))\right]^2 \cdots$$

$$\curvearrowright \quad \Xi_\mu(T, V) = \prod_{\mathbf{k}_\nu} \left[1 + \exp(-\beta(\varepsilon(\mathbf{k}_\nu) - \mu))\right]^2$$

2. Average occupation number:

$$\begin{aligned}
\langle \hat{n}_{\mathbf{k}\sigma} \rangle &= \frac{1}{\Xi_\mu} \mathrm{Sp}\,(\rho \hat{n}_{\mathbf{k}\sigma}) \\
&= \frac{1}{2} \sum_\sigma \langle \hat{n}_{\mathbf{k}\sigma} \rangle \\
&= -\frac{1}{2} \frac{1}{\beta} \frac{\partial}{\partial \varepsilon(\mathbf{k})} \underbrace{\ln \Xi_\mu(T, V)}_{2\sum_{k\nu} \ln\left[1+\exp(-\beta(\varepsilon(\mathbf{k}_\nu)-\mu))\right]} \\
&= \frac{\exp(-\beta(\varepsilon(\mathbf{k}) - \mu))}{1 + \exp(-\beta(\varepsilon(\mathbf{k}) - \mu))} \\
&= \frac{1}{\exp(\beta(\varepsilon(\mathbf{k}) - \mu)) + 1} = f_-(\varepsilon(\mathbf{k}))
\end{aligned}$$

3. Entropy:

$$\begin{aligned}
S &= k_B \frac{\partial}{\partial T} (T \ln \Xi_\mu) \\
&= k_B \sum_{\mathbf{k}\sigma} \ln\left[1 + \exp(-\beta(\varepsilon(\mathbf{k}) - \mu))\right] + \\
&+ k_B T \frac{1}{k_B T^2} \sum_{\mathbf{k}\sigma} \frac{\exp(-\beta(\varepsilon(\mathbf{k}) - \mu))(\varepsilon(\mathbf{k}) - \mu)}{1 + \exp(-\beta(\varepsilon(\mathbf{k}) - \mu))}
\end{aligned}$$

$$\frac{\partial \mu}{\partial T} \approx 0 \quad \curvearrowright$$

$$S = \sum_{\mathbf{k}\sigma} \left\{ k_B \ln \frac{1}{1- < \hat{n}_{\mathbf{k}\sigma} >} + k_B \beta \left(\varepsilon(\mathbf{k}) - \mu\right) < \hat{n}_{\mathbf{k}\sigma} > \right\}$$

$$\begin{aligned}
&- \beta \left(\varepsilon(\mathbf{k}) - \mu\right) \\
&= \ln \exp(-\beta \left(\varepsilon(\mathbf{k}) - \mu\right)) \\
&= \ln \left\{ \frac{\exp(-\beta \left(\varepsilon(\mathbf{k}) - \mu\right))}{1 + \exp(-\beta \left(\varepsilon(\mathbf{k}) - \mu\right))} (1 + \exp(-\beta \left(\varepsilon(\mathbf{k}) - \mu\right))) \right\} \\
&= \ln < \hat{n}_{\mathbf{k}\sigma} > - \ln \frac{1}{1 + \exp(-\beta \left(\varepsilon(\mathbf{k}) - \mu\right))} \\
&= \ln < \hat{n}_{\mathbf{k}\sigma} > - \ln \{1- < \hat{n}_{\mathbf{k}\sigma} >\}
\end{aligned}$$

$$\curvearrowright \quad S = -k_B \sum_{\mathbf{k}\sigma} \{\ln(1- < \hat{n}_{\mathbf{k}\sigma} >) + < \hat{n}_{\mathbf{k}\sigma} > \ln < \hat{n}_{\mathbf{k}\sigma} > - \\ - < \hat{n}_{\mathbf{k}\sigma} > \ln(1- < \hat{n}_{\mathbf{k}\sigma} >)\}$$

$$\curvearrowright \quad S = -k_B \sum_{\mathbf{k}\sigma} \left\{ \underbrace{(1- < \hat{n}_{\mathbf{k}\sigma} >)}_{\text{holes}} \ln(1- < \hat{n}_{\mathbf{k}\sigma} >) + \\ < \hat{n}_{\mathbf{k}\sigma} > \ln \underbrace{< \hat{n}_{\mathbf{k}\sigma} >}_{\text{electrons}} \right\}$$

$S = 0$ for filled band ($< \hat{n}_{\mathbf{k}\sigma} > \equiv 1$)

3rd law: (Behaviour for $T \to 0$)

$\varepsilon(\mathbf{k}) > \mu$: $< \hat{n}_{\mathbf{k}\sigma} > \overset{T\to 0}{\longrightarrow} 0$: $\ln(1- < \hat{n}_{\mathbf{k}\sigma} >) \overset{T\to 0}{\longrightarrow} 0 \quad \curvearrowright \quad S \overset{T\to 0}{\longrightarrow} 0$

$\varepsilon(\mathbf{k}) < \mu$: $< \hat{n}_{\mathbf{k}\sigma} > \overset{T\to 0}{\longrightarrow} 1$: $\ln < \hat{n}_{\mathbf{k}\sigma} > \overset{T\to 0}{\longrightarrow} 0$

$\curvearrowright \quad S \overset{T\to 0}{\longrightarrow} 0$

$\curvearrowright$ 3rd law is satisfied!

Problem 3.4

Periodic boundary conditions $\curvearrowright$ grid volume

$$\Delta k^{(1)} = \frac{2\pi}{L}; \qquad \Delta k^{(2)} = \frac{(2\pi)^2}{L_x L_y}$$

1. $d = 1$ Density of states:

$$\rho_0^{(1)}(E) = \frac{2}{\Delta k^{(1)}} \frac{d}{dE} \varphi_1(E)$$

Phase volume:

$$\varphi_1(E) = \int\limits_{\varepsilon(k)\le E} dk = \int\limits_{-\frac{1}{\hbar}\sqrt{2mE}}^{+\frac{1}{\hbar}\sqrt{2mE}} dk = \frac{2}{\hbar}\sqrt{2mE}$$

$$\curvearrowright \quad \frac{d}{dE}\varphi_1(E) = \frac{1}{\hbar}\frac{1}{\sqrt{2mE}}2m = \sqrt{\frac{2m}{E\hbar^2}}$$

$$\boxed{\rho_0^{(1)}(E) = \begin{cases} d_1 \cdot \frac{1}{\sqrt{E}}, & \text{if } E > 0 \\ 0, & \text{otherwise} \end{cases}} \qquad ; d_1 = \frac{\sqrt{2m}}{\pi\hbar}L$$

2. $d = 2$

$$\varphi_2(E) = \int\limits_{\varepsilon(k)\le E} d^2k = \pi k^2|_{\varepsilon(k)\le E} = \pi\frac{2mE}{\hbar^2}$$

$$\curvearrowright \quad \frac{d}{dE}\varphi_2(E) = \frac{2m\pi}{\hbar^2}$$

$$\boxed{\rho_0^{(2)}(E) = \begin{cases} d_2 > 0, & \text{if } E > 0 \\ 0, & \text{otherwise} \end{cases}} \qquad ; d_2 = \frac{L_x L_y}{\pi}\cdot\frac{m}{\hbar^2}$$

3. *d arbitrary*:

$$\rho_0(E)\,dE = \frac{2}{\Delta^{(d)}k}\int\limits_{E\le\varepsilon(k)\le E+dE} d^dk$$

$\varepsilon(k) = \frac{\hbar^2k^2}{2m}$; cell volume: $\Delta^{(d)}(k) = \frac{(2\pi)^d}{Vd}$
Phase volume:

$$\varphi_d(E) = \Omega_d\int\limits_0^{k_0} dk\,k^{d-1} = \Omega_d\frac{{k_0}^d}{d} \overset{k_0=\left(\frac{2m}{\hbar^2}\right)^{\frac{1}{2}}E^{\frac{1}{2}}}{\widehat{=}} \frac{\Omega_d}{d}\left(\frac{2m}{\hbar^2}\right)^{\frac{d}{2}}E^{\frac{d}{2}}$$

Ω_d: Surface of the d-dimensional unit sphere

$\curvearrowright$ *density of states*

$$\rho_0(E) = 2\frac{V_d}{(2\pi)^d}\frac{d}{dE}\varphi_d(E)\Theta(E)$$
$$= V_d\frac{\Omega_d}{(2\pi)^d}\left(\frac{2m}{\hbar^2}\right)^{\frac{d}{2}} E^{\frac{d}{2}-1}\Theta(E)$$

The surface Ω_d is still to be determined! Gauss integral in the d-dimensional space:

$$g_d = \int d^d r e^{-r^2} = \prod_{i=1}^{d}\int_{-\infty}^{+\infty} dx_i\, e^{-x_i^2} = (\sqrt{\pi})^d = \pi^{\frac{d}{2}}$$

Spherical coordinates:

$$g_d = \Omega_d \int_0^\infty dx\, x^{d-1} e^{-x^2}$$
$$= \Omega_d \frac{1}{2}\int_0^\infty dy\, y^{\frac{d}{2}-1} e^{-y}$$
$$= \frac{1}{2}\Omega_d \Gamma\left(\frac{d}{2}\right)$$

$(y = x^2 \curvearrowright dy = 2xdy \curvearrowright dx = \frac{1}{2}\frac{dy}{\sqrt{y}})$
by comparing:

$$\Omega_d = \frac{2\pi^{\frac{d}{2}}}{\Gamma(\frac{d}{2})}$$

$\curvearrowright$ *Density of states*

$$\boxed{\rho_0^{(d)}(E) = 2\frac{V_d\pi^{\frac{d}{2}}}{(2\pi)^d}\cdot\frac{1}{\Gamma(\frac{d}{2})}\left(\frac{2m}{\hbar^2}\right)^{\frac{d}{2}} E^{\frac{d}{2}-1}\Theta(E)}$$

Check:

1. $d = 1$:

$$\rho_0^{(1)}(E) = d_1\frac{1}{\sqrt{E}}\Theta(E); \qquad d_1 = \frac{L}{\pi}\sqrt{\frac{2m}{\hbar^2}}$$

2. $d = 2$:

$$\rho_0^{(2)}(E) = d_2\Theta(E); \qquad d_2 = \frac{L_x L_y}{\pi}\frac{m}{\hbar^2}$$

3. $d = 3$:

$$\rho_0^{(3)}(E) = d_3\sqrt{E}\Theta(E)$$
$$d_3 = \frac{V\pi^{\frac{3}{2}}}{4\pi^3}\frac{2}{\sqrt{\pi}}\left(\frac{2m}{\hbar^2}\right)^{\frac{3}{2}} = \frac{V}{2\pi^2}\left(\frac{2m}{\hbar^2}\right)^{\frac{3}{2}}$$

Problem 3.5

$$\varepsilon(\mathbf{k}) = \varepsilon(k) = c\hbar k$$

1. *Density of states*:

$$\rho(E) = \frac{2V}{(2\pi)^3}\frac{d}{dE}\varphi(E)$$
$$\varphi(E) = \int\limits_{\varepsilon(k)\le E} d^3k = \frac{4\pi}{3}k^3|_{E=c\hbar k} = \frac{4\pi}{3c^3\hbar^3}E^3$$
$$\curvearrowright \quad \boxed{\rho(E) = \alpha E^2\Theta(E)} \qquad ;\alpha = \frac{V}{\pi^2\hbar^3c^3}$$

$\curvearrowright$ Fermi energy:

$$\varepsilon_F = c\hbar k_F = c\hbar\left(3\pi^2\frac{N}{V}\right)^{\frac{1}{3}}$$

where k_F is derived from the particle number:

$$N = \frac{2V}{(2\pi)^3}\int\limits_{k\le k_F} d^3k = \frac{V}{4\pi^3}\cdot\frac{4\pi}{3}k_F^3$$
$$\curvearrowright \quad k_F = \left(3\pi^2\frac{N}{V}\right)^{\frac{1}{3}}$$

2. *Chemical potential*:
With the help of the particle number:

$$N = \underbrace{\int\limits_{-\infty}^{\varepsilon_F} dE\,\rho(E)}_{N(T=0)} = \underbrace{\int\limits_{-\infty}^{+\infty} dE\,f_-(E)\rho(E)}_{N(T)}$$
$$\curvearrowright \quad \frac{\alpha}{3}\varepsilon_F^3 = \int\limits_{-\infty}^{\mu} dE\,\rho(E) + \frac{\pi^2}{6}(k_BT)^2\rho'(\mu) + \ldots$$

(Sommerfeld expansion)

$$\curvearrowright \quad \frac{1}{3}\varepsilon_F^3 = \frac{1}{3}\mu^3 + \frac{\pi^2}{3}(k_BT)^2\mu + \dots$$
$$\approx \frac{1}{3}\mu^3\left\{1+\pi^2\left(\frac{k_BT}{\mu}\right)^2\right\}$$
$$\approx \frac{1}{3}\mu^3\left\{1+\pi^2\left(\frac{k_BT}{\varepsilon_F}\right)^2\right\}$$

$$\curvearrowright \quad \mu \approx \varepsilon_F\left\{1+\pi^2\underbrace{\left(\frac{k_BT}{\varepsilon_F}\right)^2}_{\approx 10^{-4}}\right\}^{-\frac{1}{3}}$$
$$\approx \varepsilon_F\left\{1-\frac{\pi^2}{3}\left(\frac{k_BT}{\varepsilon_F}\right)^2\right\}$$

The same structure as in the non-relativistic case, only the numerical factor in front of the correction term is changed from $\frac{\pi^2}{12}$ to $\frac{\pi^2}{3}$

3. *Internal energy and heat capacity*:

$$U(T=0) = \int_{-\infty}^{\varepsilon_F} dE\; E\cdot\rho(E) = \frac{\alpha}{4}\varepsilon_F^4$$

$$U(T) = \int_{-\infty}^{\mu} dE\; E\cdot\rho(E) + \frac{\pi^2}{6}(k_BT)^2(3\alpha\mu^2) + \dots$$
$$= \frac{\alpha}{4}\mu^4 + \frac{\pi^2}{6}(k_BT)^2(3\alpha\mu^2) + \dots$$
$$= \frac{\alpha}{4}\varepsilon_F^4\left\{\left(\frac{\mu}{\varepsilon_F}\right)^4 + 2\pi^2\left(\frac{k_BT}{\varepsilon_F}\right)^2\left(\frac{\mu}{\varepsilon_F}\right)^2\right\} + \dots$$
$$\stackrel{2.)}{=} U(0)\left\{1-\frac{4}{3}\pi^2\left(\frac{k_BT}{\varepsilon_F}\right)^2 + 2\pi^2\left(\frac{k_BT}{\varepsilon_F}\right)^2\right\} + \dots$$

$$\curvearrowright \quad \boxed{U(T) \approx U(0)\left\{1+\frac{2\pi^2}{3}\left(\frac{k_BT}{\varepsilon_F}\right)^2\right\}}$$

$N = \frac{\alpha}{3}\varepsilon_F^3 \curvearrowright \alpha = \frac{3N}{\varepsilon_F^3} \curvearrowright \underline{U(0) = \frac{3}{4}N\varepsilon_F}$

Heat capacity

$$c_V = \hat{\gamma} T; \qquad \hat{\gamma} = U(0) \frac{4\pi^2}{3} \frac{k_B^2}{\varepsilon_F^2} = N\pi^2 \frac{k_B^2}{\varepsilon_F}$$

Non-relativistic:

$$\gamma = \frac{1}{2} N\pi^2 \frac{k_B^2}{\varepsilon_F}$$

which means

$$\frac{\hat{\gamma}}{\gamma} = 2\frac{\varepsilon_F^{nr}}{\varepsilon_F^{r}} = 2\frac{\hbar^2 k_F^2}{2mc\hbar k_F} = \frac{\hbar^2}{m} \cdot \frac{k_F}{c\hbar} = \frac{\hbar}{mc} \left(3\pi^2 \frac{N}{V}\right)^{\frac{1}{3}} << 1$$

Problem 3.6

1. Degree of degeneracy of a Landau level (3.117):

$$2g_y(B_0) = 2\frac{eL_x L_y}{2\pi\hbar} B_0$$

   Factor 2 because of the spin degeneracy. $B_0^{(0)}$ is determined from

$$N_e = 2g_y\left(B_0^{(0)}\right)$$

   Therefore

$$B_0^{(0)} = N_e \frac{\pi\hbar}{eL_x L_y}$$

2. The degeneracy of the Landau level is independent of the quantum number n. The first n_0 levels are then exactly fully occupied and the levels $n > n_0$ are completely empty if $B_0 = B_0^{(n_0-1)}$ so that

$$\frac{N_e}{n_0} = 2g_y\left(B_0^{(n_0-1)}\right)$$

$$\curvearrowright \; B_0^{(n_0-1)} = \frac{1}{n_0} B_0^{(0)}$$

3. The degree of degeneracy at an arbitrary field B_0

$$2g_y(B_0) = 2g_y\left(B_0^{(0)}\right) \cdot \frac{B_0}{B_0^{(0)}} = N_e \frac{B_0}{B_0^{(0)}}$$

Now let

$$B_0^{(n_0-1)} \geq B_0 \geq B_0^{(n_0)}$$

$\curvearrowright$ n_0 levels are fully occupied and the (n_0+1)th level is partially occupied. The number of electrons in the fully occupied levels:

$$N^* = n_0 \cdot 2g_y(B_0) = n_0 N_e \frac{B_0}{B_0^{(0)}}$$

The number of electrons in the uppermost level:

$$N_{n_0} = N_e - N^* = N_e \left(1 - n_0 \frac{B_0}{B_0^{(0)}}\right)$$

$\curvearrowright$ Energy contribution of the highest level:

$$\begin{aligned} E_{n_0} &= \hbar\omega_c^* \left(n_0 + \frac{1}{2}\right) \cdot N_{n_0} \\ &= N_e \hbar\omega_c^* \left(n_0 + \frac{1}{2} - n_0 \left(n_0 + \frac{1}{2}\right) \frac{B_0}{B_0^{(0)}}\right) \\ &= N_e \mu_B^* B_0 \left(2n_0 + 1 - n_0(2n_0+1) \frac{B_0}{B_0^{(0)}}\right) \end{aligned}$$

Energy contribution of the fully occupied levels:

$$\begin{aligned} E^* &= \sum_{n=0}^{n_0-1} \hbar\omega_c^* \left(n + \frac{1}{2}\right) N_e \frac{B_0}{B_0^{(0)}} \\ &= \hbar\omega_c^* N_e \frac{B_0}{B_0^{(0)}} \left(n_0 - \frac{1}{2} + \frac{1}{2}\right) \frac{1}{2} n_0 \\ &= N_e \mu_B^* B_0 \frac{B_0}{B_0^{(0)}} n_0^2 \end{aligned}$$

$\curvearrowright$ Total energy:

$$E(B_0) = E_{n_0} + E^* = N_e \mu_B^* B_0 \left(2n_0 + 1 - n_0(n_0+1) \frac{B_0}{B_0^{(0)}}\right)$$

This gives the curve of Fig. 3.5.

4. $B_0 = B_0^{(n_0)}$
With (2) we have

$$\frac{B_0^{(n_0)}}{B_0^{(0)}} = \frac{1}{n_0 + 1}$$

$$\curvearrowright \quad E\left(B_0^{(n_0)}\right) = N_e \mu_B^* B_0^{(n_0)} \left(2n_0 + 1 - n_0(n_0+1)\frac{B_0^{(n_0)}}{B_0^{(0)}}\right)$$

$$= N_e \mu_B^* B_0^{(0)}$$

that is, independent of n_0, that means for all critical fields it is the same!

Problem 3.7

1. Energy levels (3.112):

$$E_n(k_z) = 2\mu_B B_0 \left(n + \frac{1}{2}\right) + \frac{\hbar^2 k_z^2}{2m}$$

Degeneracy (3.117):

$$g_y(B_0) = \frac{eL_x L_y}{2\pi\hbar} B_0$$

Partition function:

$$Z_1 = \frac{1}{2\pi/L_z} \int_{-\infty}^{+\infty} dk_z \sum_{n=0}^{\infty} g_y(B_0) \exp[-\beta E_n(k_z)]$$

$$= \frac{eVB_0}{(2\pi\hbar)^2} \left[\int_{-\infty}^{+\infty} dp_z \exp\left(-\beta \frac{p_z^2}{2m}\right)\right] e^{-\beta\mu_B B_0} \sum_{n=0}^{\infty} e^{-\beta 2\mu_B B_0 n}$$

$$= \frac{eVB_0}{(2\pi\hbar)^2} \sqrt{\frac{2\pi m}{\beta}} \frac{e^{-\beta\mu_B B_0}}{1 - e^{-2\beta\mu_B B_0}}$$

$$\curvearrowright$$

$$Z_1 = V \left(\frac{m}{2\pi\hbar^2\beta}\right)^{3/2} \frac{\beta\mu_B B_0}{\sinh(\beta\mu_B B_0)} \quad \left(\mu_B = \frac{e\hbar}{2m}\right)$$

2. Free energy:

$$dF = -S\,dT - m\,dB_0$$

(to magnetization's work see Sect. 1.5)

$$\curvearrowright \quad m = -\frac{\partial F}{\partial B_0} = k_B T \frac{\partial}{\partial B_0} \ln Z_N = N k_B T \frac{\partial}{\partial B_0} \ln Z_1 =$$
$$= N k_B T \frac{\partial}{\partial B_0} \ln \frac{\beta \mu_B B_0}{\sinh(\beta \mu_B B_0)}$$
$$= -N \mu_B \left(\frac{d}{dx} \ln \frac{\sinh x}{x} \right)_{x = \beta \mu_B B_0}$$

In the bracket is the classical Langevin function (see Problem 4.6):

$$L(x) = \coth x - \frac{1}{x}$$

$$\curvearrowright \quad m = -N \mu_B L \left(\frac{\mu_B B_0}{k_B T} \right)$$

negative sign

$\curvearrowright$

induced magnetic moment is oriented opposite to the field

$\curvearrowright$ Diamagnetism

Problem 3.8
For $f(x)$ one can write

$$f(x) = \delta\left(x - \frac{1}{2}\right) + \delta\left(x + \frac{1}{2}\right) \quad \text{if} \quad -1 \le x \le +1$$

$$\text{with} \quad f(x) = f(x+2)$$

$f(x)$ is thus periodic with the period 2 and is also symmetric

$$f(-x) = f(x)$$

Then an ansatz for the Fourier series is possible

$$f(x) = f_0 + \sum_{m=1}^{\infty} \left[a_m \cos(m\pi x) + b_m \sin(m\pi x) \right]$$

$$f_0 = \frac{1}{2} \int_{-1}^{+1} f(x)\, dx = 1$$

$$a_m = \int_{-1}^{+1} f(x) \cos(m\pi x)\, dx$$
$$= \begin{cases} 0 & \text{for } m = 2p+1 \\ 2(-1)^p & \text{for } m = 2p \end{cases}$$

$$b_m \equiv 0 \;, \quad \text{since } f(x) \text{ symmetric}$$

Then it follows:

$$
\begin{aligned}
f(x) &= 1+\sum_{p=1}^{+\infty} 2(-1)^p \cos(2p\pi x) \\
&= 1+\sum_{p=1}^{+\infty}(-1)^p \left(e^{i2p\pi x}+e^{-i2p\pi x}\right) \\
&= \sum_{p=-\infty}^{+\infty} (-1)^p e^{i2p\pi x}
\end{aligned}
$$

Problem 4.1

$\mathbf{q} = \mathbf{q}'$

$$
\frac{1}{V}\int_V d^3r\, e^{i(\mathbf{q}-\mathbf{q}')\cdot\mathbf{r}} = \frac{1}{V}\int_V d^3r = 1
$$

$\mathbf{q} \neq \mathbf{q}'$

$$
\begin{aligned}
&\frac{1}{V}\int_V d^3r\, e^{i(\mathbf{q}-\mathbf{q}')\cdot\mathbf{r}} = \\
&= \frac{1}{V}\int_0^{L_x} dx \int_0^{L_y} dy \int_0^{L_z} dz \exp\left[2\pi i\left(\frac{n_x-n_x'}{L_x}x+\right.\right. \\
&\qquad\qquad \left.\left.\frac{n_y-n_y'}{L_y}y+\frac{n_z-n_z'}{L_z}z\right)\right]
\end{aligned}
$$

Here we must have $(n_x, n_y, n_z) \neq (n_x', n_y', n_z')$. For example, let $n_x \neq n_x'$. Then the integral over x gives

$$
\begin{aligned}
&\int_0^{L_x} dx \exp\left[2\pi i\left(\frac{n_x-n_x'}{L_x}x\right)\right] \\
&= \frac{L_x}{2\pi i\left(n_x-n_x'\right)} \exp\left[2\pi i\left(\frac{n_x-n_x'}{L_x}x\right)\right]\Bigg|_0^{L_x} \\
&= 0 \;\text{ since }\; n_x-n_x' \in \mathbb{Z}
\end{aligned}
$$

So that the proposition is proved.

Problem 4.2

1. The solid is a three-dimensional periodic array of primitive unit cells V_{UC} $(\mathbf{a}_1\cdot(\mathbf{a}_2\times\mathbf{a}_3))$ with the total volume

$$
V = N_1N_2N_3\left(\mathbf{a}_1\cdot(\mathbf{a}_2\times\mathbf{a}_3)\right)
$$

Periodic boundary conditions for Bloch functions:

$$\psi_{\mathbf{k}}(\mathbf{r}) \stackrel{!}{=} \psi_{\mathbf{k}}(\mathbf{r}+N_1\mathbf{a}_1) \stackrel{!}{=} \psi_{\mathbf{k}}(\mathbf{r}+N_2\mathbf{a}_2) \stackrel{!}{=} \psi_{\mathbf{k}}(\mathbf{r}+N_3\mathbf{a}_3)$$

The Bloch functions are

$$\psi_{\mathbf{k}}(\mathbf{r}) = e^{i\mathbf{k}\cdot\mathbf{r}}\, u_{\mathbf{k}}(\mathbf{r})$$

where the amplitude functions have the periodicity of the lattice:

$$u_{\mathbf{k}}(\mathbf{r}) = u_{\mathbf{k}}(\mathbf{r}+\mathbf{R}^{\mathbf{n}})$$

Periodic boundary conditions therefore demand

$$e^{i\mathbf{k}\cdot(N_1\mathbf{a}_1)} \stackrel{!}{=} e^{i\mathbf{k}\cdot(N_2\mathbf{a}_2)} \stackrel{!}{=} e^{i\mathbf{k}\cdot(N_3\mathbf{a}_3)} \stackrel{!}{=} 1$$

This means

$$\mathbf{k}\cdot(N_i\mathbf{a}_i) = 2\pi z_i \quad \text{with} \quad z_i \in \mathbb{Z}$$

Then for the allowed wavevectors holds:

$$\mathbf{k} = \frac{z_1}{N_1}\mathbf{b}_1 + \frac{z_1}{N_2}\mathbf{b}_2 + \frac{z_1}{N_3}\mathbf{b}_3$$

Here $\mathbf{b}_i$ are the primitive translation vectors of the reciprocal lattice, defined by

$$\mathbf{a}_i \cdot \mathbf{b}_j = 2\pi\,\delta_{ij} \iff \mathbf{b}_1 = 2\pi\,\frac{\mathbf{a}_2\times\mathbf{a}_3}{\mathbf{a}_1\cdot(\mathbf{a}_2\times\mathbf{a}_3)} \quad \cdots$$

The first Brillouin zone ≡ Wigner–Seitz cell of the reciprocal lattice. Therefore for the wavevectors of the first Brillouin zone holds

$$-\frac{1}{2}N_i < z_i \le +\frac{1}{2}N_i$$

2. The proposition is valid for $\mathbf{k} = \mathbf{k}'$, since

$$\frac{1}{N}\sum_{\mathbf{n}} 1 = 1$$

$\mathbf{k} \neq \mathbf{k}'$:
According to 1. holds

$$(\mathbf{k}-\mathbf{k}')\cdot\mathbf{R}^{\mathbf{n}} = 2\pi\sum_{j=1}^{3}\frac{n_j}{N_j}(z_j - z_j')$$

So that we calculate

$$\sum_{\mathbf{n}} e^{i(\mathbf{k}-\mathbf{k}')\cdot\mathbf{R}^{\mathbf{n}}} = \sum_{n_1 n_2 n_3} \exp\left(2\pi i \sum_{j=1}^{3} \frac{n_j}{N_j}(z_j - z_j')\right) =$$

$$= \prod_{j=1}^{3} \sum_{n_j=0}^{N_j-1} \exp\left(2\pi i \frac{n_j}{N_j}(z_j - z_j')\right)$$

$z_j \neq z_j'$ at least for one j. Then we have

$$\sum_{n_j=0}^{N_j-1} \left(\exp\left(2\pi i \frac{z_j - z_j'}{N_j}\right)\right)^{n_j} = \frac{1 - a^{N_j}}{1 - a} = 0$$

This holds because

$$a = \exp\left(2\pi i \frac{z_j - z_j'}{N_j}\right) \neq 1$$

since $z_j \neq z_j'$ and in addition $-N_j < z_j - z_j' < +N_j$. On the other hand

$$a^{N_j} = \exp\left(2\pi i (z_j - z_j')\right) = 1$$

since $z_j - z_j'$ is an integer. Thus the proposition is proved.

3. The proposition is trivial for $\mathbf{R}^{\mathbf{n}} = \mathbf{R}^{\mathbf{m}}$. Therefore let $\mathbf{R}^{\mathbf{n}} \neq \mathbf{R}^{\mathbf{m}}$:

With

$$\mathbf{k} \cdot \left(\mathbf{R}^{\mathbf{n}} - \mathbf{R}^{\mathbf{m}}\right) = 2\pi \sum_{j=1}^{3} \frac{z_j}{N_j}(n_j - m_j)$$

now holds

$$\sum_{\mathbf{k}}^{1.BZ} e^{i\mathbf{k}\cdot(\mathbf{R}^{\mathbf{n}}-\mathbf{R}^{\mathbf{m}})} = \sum_{z_1 z_2 z_3} \exp\left(2\pi i \sum_{j=1}^{3} \frac{z_j}{N_j}(n_j - m_j)\right) =$$

$$= \prod_{j=1}^{3} \sum_{z_j=-1/2N_j+1}^{1/2N_j} \exp\left(2\pi i z_j \frac{n_j - m_j}{N_j}\right)$$

$n_j \neq m_j$ at least for one j. Then we have

$$\sum_{z_j=-1/2N_j+1}^{1/2N_j} \exp\left(2\pi i z_j \frac{n_j - m_j}{N_j}\right)$$
$$= \sum_{p_j=0}^{N_j-1} \exp\left(2\pi i (p_j - \frac{1}{2}N_j + 1)\frac{n_j - m_j}{N_j}\right)$$
$$\propto \sum_{p_j=0}^{N_j-1} \left(\exp\left(2\pi i \frac{n_j - m_j}{N_j}\right)\right)^{p_j}$$
$$= \frac{1 - b^{N_j}}{1 - b} = 0$$

since

$$b = \exp\left(2\pi i \frac{n_j - m_j}{N_j}\right) \neq 1$$

because $n_j \neq m_j$ and $-N_j + 1 \leq n_j - m_j \leq N_j - 1$. On the other hand it holds

$$b^{N_j} = \exp\left(2\pi i (n_j - m_j)\right) = 1$$

because $n_j - m_j$ is an integer. The proposition is thus proved.

Problem 4.3

For both the integrals, it is meaningful (because of $V \to \infty$) to introduce relative and centre of mass coordinates:

$$\mathbf{x} = \mathbf{r} - \mathbf{r}' \;\; ; \;\; \mathbf{R} = \frac{1}{2}(\mathbf{r} + \mathbf{r}')$$
$$\mathbf{r} = \frac{1}{2}\mathbf{x} + \mathbf{R} \;\; ; \;\; \mathbf{r}' = -\frac{1}{2}\mathbf{x} + \mathbf{R}$$

With the help of the Jacobi determinant one can show

$$d^3r\, d^3r' = d^3R\, d^3x$$

1.

$$I_1 = \int_V d^3r \int_V d^3r' \, \frac{e^{-\alpha|\mathbf{r}-\mathbf{r}'|}}{|\mathbf{r}-\mathbf{r}'|}$$
$$= \int d^3R \int d^3x \frac{e^{-\alpha x}}{x} = V \cdot 4\pi \int_0^\infty dx\, x\, e^{-\alpha x}$$
$$= V \cdot 4\pi \left(-\frac{d}{d\alpha}\right) \int_0^\infty dx\, e^{-\alpha x}$$
$$= V \cdot 4\pi \left(-\frac{d}{d\alpha}\right) \frac{1}{\alpha}$$

So that we have

$$I_1 = \frac{4\pi V}{\alpha^2}$$

2. From the result of Problem 4.1 it follows directly:

$$\begin{aligned}
I_2 &= \int_V d^3r \int_V d^3r' \, \frac{\exp(i(\mathbf{q}\cdot\mathbf{r}+\mathbf{q}'\cdot\mathbf{r}'))}{|\mathbf{r}-\mathbf{r}'|} \\
&= \int\int d^3R\, d^3x\, \frac{1}{x} \exp\left(\frac{i}{2}(\mathbf{q}-\mathbf{q}')\cdot\mathbf{x}\right) * \\
&\qquad\qquad * \exp\left(i(\mathbf{q}+\mathbf{q}')\cdot\mathbf{R}\right) \\
&= V\,\delta_{\mathbf{q},-\mathbf{q}'} \cdot \widehat{I}
\end{aligned}$$

In order to calculate $\widehat{I}$ it is advisable to introduce a *factor that ensures convergence*:

$$\begin{aligned}
\widehat{I} &= \lim_{\alpha\to 0^+} \int d^3x \, \frac{1}{x}\, e^{i\mathbf{q}\cdot\mathbf{x}}\, e^{-\alpha x} \\
&= \lim_{\alpha\to 0^+} 2\pi \int_0^\infty dx\, x \int_{-1}^{+1} d\cos\vartheta\, e^{iqx\cos\vartheta}\, e^{-\alpha x} \\
&= \lim_{\alpha\to 0^+} 2\pi \int_0^\infty dx\, \frac{x}{iqx}\, e^{-\alpha x}\left(e^{iqx} - e^{-iqx}\right) \\
&= \lim_{\alpha\to 0^+} \frac{2\pi}{iq} \int_0^\infty dx \left(e^{iqx-\alpha x} - e^{-iqx-\alpha x}\right) \\
&= \lim_{\alpha\to 0^+} \frac{2\pi}{iq} \left\{ \frac{1}{iq-\alpha}\, e^{iqx-\alpha x}\Big|_0^\infty - \frac{1}{-iq-\alpha}\, e^{-iqx-\alpha x}\Big|_0^\infty \right\} \\
&= \lim_{\alpha\to 0^+} \frac{2\pi}{iq} \left\{ \frac{-1}{iq-\alpha} - \frac{1}{iq+\alpha} \right\} \\
&= \frac{2\pi}{iq}\left(\frac{-2}{iq}\right) = \frac{4\pi}{q^2}
\end{aligned}$$

Thus we find

$$I_2 = \frac{4\pi V}{q^2}\,\delta_{\mathbf{q},-\mathbf{q}'}$$

Problem 4.4
First we calculate

$$
\begin{aligned}
I_1(\mathbf{r}) &\equiv \int_{FK} d^3k\, e^{i\mathbf{k}\cdot\mathbf{r}} \\
&= 2\pi \int_{-1}^{+1} dx \int_0^{k_F} dk\, k^2 e^{ikrx} \\
&= \frac{2\pi}{ir} \int_0^{k_F} dk\, k\, 2i \sin(kr) \\
&= \frac{4\pi}{r} \left[\frac{1}{r^2} \sin(kr) - \frac{k}{r} \cos(kr) \right]_0^{k_F} \\
&= -4\pi \frac{k_F r \cos(kr) - \sin(k_F r)}{r^3}
\end{aligned}
$$

Then we have the intermediate result

$$
\begin{aligned}
I &= \int d^3r \frac{I_1^2(\mathbf{r})}{r} \\
&= (4\pi)^3\, k_F^4 \int_0^{\infty} dx \frac{1}{x^5} (x \cos x - \sin x)^2
\end{aligned}
$$

To calculate the integral we use integration by parts a number of times.

$$
\begin{aligned}
I_2 &= \int_0^{\infty} dx \frac{1}{x^5} (x \cos x - \sin x)^2 \\
&= - \frac{1}{4x^4} (x \cos x - \sin x)^2 \Big|_0^{\infty} + \\
&+ \frac{1}{4} \int_0^{\infty} \frac{dx}{x^4} 2(x \cos x - \sin x)(-x \sin x) \\
&= - \frac{1}{2} \int_0^{\infty} \frac{dx}{x^3} (x \cos x - \sin x) \sin x \\
&= \frac{1}{4x^2} (x \cos x - \sin x) \sin x \Big|_0^{\infty} + \\
&+ \frac{1}{4} \int_0^{\infty} \frac{dx}{x^2} (x(\sin^2 x - \cos^2 x) + \sin x \cos x) \\
&= \frac{1}{4} \int_0^{\infty} \frac{dx}{x^2} (-x \cos 2x + \frac{1}{2} \sin 2x) \\
&= \frac{1}{4} \int_0^{\infty} \frac{dy}{y^2} (\sin y - y \cos y)
\end{aligned}
$$

$$
\begin{aligned}
&= -\frac{1}{4y}(\sin y - y\cos y)\Big|_0^\infty + \frac{1}{4}\int_0^\infty \frac{dy}{y}\, y \sin y \\
&= -\frac{1}{4y}\sin y\Big|_0^\infty = \frac{1}{4}\lim_{y\to 0}\frac{\sin y}{y} \\
&= \frac{1}{4}
\end{aligned}
$$

Then we finally have the integral

$$I = 16\pi^3\, k_F^4$$

Problem 4.5

- S integer
  To show

$$\sum_{n=-S}^{+S} n^2 = 2\sum_{n=1}^{+S} n^2 = \frac{1}{3}S(S+1)(2S+1)$$

  Proof by full induction: The statement is certainly true for $S = 1$

$$2\sum_{n=1}^{+1} n^2 = 2$$

  We assume that it is valid for S and substitute $S \curvearrowright S+1$

$$
\begin{aligned}
2\sum_{n=1}^{S+1} n^2 &= 2\sum_{n=1}^{+S} n^2 + 2(S+1)^2 \\
&= \frac{1}{3}S(S+1)(2S+1) + 2(S+1)^2 \\
&= \frac{1}{3}(S+1)(S(2S+1) + 6(S+1)) \\
&= \frac{1}{3}(S+1)(2S^2 + 7S + 6) \\
&= \frac{1}{3}(S+1)(S+2)(2S+3))
\end{aligned}
$$

  Thus the proposition is proved!
- S half-integer

$$\sum_{n=-S}^{+S} n^2 = 2\sum_{n=1}^{S+\frac{1}{2}} \left(n - \frac{1}{2}\right)^2$$
$$= 2\sum_{n=1}^{S+\frac{1}{2}} n^2 - 2\sum_{n=1}^{S+\frac{1}{2}} n + \frac{1}{2}\sum_{n=1}^{S+\frac{1}{2}} 1$$
$$= \frac{1}{3}\left(S+\frac{1}{2}\right)\left(S+\frac{3}{2}\right)(2S+2)$$
$$-2\frac{1}{2}\left(S+\frac{1}{2}\right)\left(S+\frac{3}{2}\right)+\frac{1}{2}\left(S+\frac{1}{2}\right)$$
$$= \left(S+\frac{1}{2}\right)\left[\frac{1}{3}\left(S+\frac{3}{2}\right)(2S+2)-(S+1)\right]$$
$$= \frac{1}{3}(2S+1)(S+1)\left[\left(S+\frac{3}{2}\right)-\frac{3}{2}\right]$$
$$= \frac{1}{3}S(S+1)(2S+1)$$

This is the proposition!

Problem 4.6

1. Energy of a magnetic dipole in a magnetic field

$$E = -\mu \cdot \mathbf{B}$$
$$\Leftrightarrow\ H_1 = -\mu\, B \sum_{i=1}^{N} \cos\vartheta_i$$

ϑ_i is the angle between the magnetic moment of the ith atom μ_i and the field $\mathbf{B}$.

2. Canonical partition function

$$Z_N(T,B) = \frac{1}{h^{3N}N!}\int\int d^{3N}q\, d^{3N}p\, e^{-\beta H(\mathbf{q},\mathbf{p})}$$
$$= \frac{Z_N(T,0)}{(4\pi)^N}(2\pi)^N \int_{-1}^{+1} d\cos\vartheta_1 \cdots$$
$$\cdots \int_{-1}^{+1} d\cos\vartheta_N\, e^{\beta\mu B\sum_{i=1}^{N}\cos\vartheta_i}$$
$$= Z_N(T,0)\left(\frac{1}{2}\int_{-1}^{+1} dx\, e^{\beta\mu B x}\right)^N$$

$$= Z_N(T,0)\left(\frac{e^{\beta\mu B} - e^{-\beta\mu B}}{2\beta\mu B}\right)^N$$

$$= Z_N(T,0)\left(\frac{\sinh\beta\mu B}{\beta\mu B}\right)^N$$

3.

$$\mu_i = \mu\,(\sin\vartheta_i\ \cos\varphi_i, \sin\vartheta_i\ \sin\varphi_i, \cos\vartheta_i)$$

Average value:

$$\mathbf{m} = \frac{\int\int d^{3N}q\, d^{3N}p\, \left(\sum_i \mu_i\right)\, e^{-\beta H(\mathbf{q},\mathbf{p})}}{\int\int\, d^{3N}q\, d^{3N}p\, e^{-\beta H(\mathbf{q},\mathbf{p})}}$$

The φ_i-integrations give $m_x = m_y = 0$ and therefore

$$\mathbf{m} = m\,\mathbf{e}_z \qquad m = \frac{d}{d(\beta B)}\ln Z_N(T,B)$$

Partition function from 2.

$$m = N\left(\frac{d}{dx}\ln\sinh\mu x - \ln\mu x\right)(x = \beta B)$$

$$= N\mu\left(\frac{\cosh\mu x}{\sinh\mu x} - \frac{1}{\mu x}\right)(x = \beta B)$$

$$\curvearrowright \qquad \mathbf{m} = N\mu\, L(\beta\mu B)\mathbf{e}_z$$

$$L(x) = \coth x - \frac{1}{x}$$

$L(x)$ is the Langevin function. This is the *classical Langevin paramagnetism*.

4. Low temperatures, strong fields: $\beta\mu B \gg 1$
That means

$$\coth\beta\mu B\ \to\ 1\ ;\quad \frac{1}{\beta\mu B}\ \to\ 0$$

The system is in *saturation*:

$$\mathbf{m} \approx N\mu\mathbf{e}_z$$

High temperatures, weak fields: $\beta\mu B \gg 1$

$$\begin{aligned} \coth x &= \frac{1}{x} + \frac{x}{3} + \mathcal{O}(x^3) \\ \curvearrowright \qquad L(x) &\approx \frac{x}{3} \end{aligned}$$

This result is the *Curie law*:

$$\mathbf{m} \approx N\mu \frac{\mu B}{3k_B T}\mathbf{e}_z$$

Problem 4.7

1. Free energy ($N = const.$):

$$F = U - T\,S$$

$$dF = dU - TdS - SdT = -SdT + V B_0 dM$$

$$\curvearrowright \qquad \left(\frac{\partial F}{\partial M}\right)_T = \mu_0 V H$$

$$\curvearrowright \qquad \left(\frac{\partial^2 F}{\partial M^2}\right)_T = \mu_0 V \left(\frac{\partial H}{\partial M}\right)_T = \frac{\mu_0 V}{\left(\frac{\partial M}{\partial H}\right)_T} = \frac{\mu_0 V}{\chi_T}$$

From the last equation follows:

$$\left(\frac{\partial F}{\partial M}\right)_T = \mu_0 V \frac{M}{\chi_T} + f(T)$$

since χ_T is only a function of temperature. The comparison with the above expression requires $f(T) \equiv 0$. Then we have

$$F(T, M) = F(T, 0) + \frac{1}{2}\mu_0 V \frac{M^2}{\chi_T}$$

2. Entropy

$$S(T, M) = -\left(\frac{\partial F}{\partial T}\right)_M = S(T, 0) - \frac{1}{2}\mu_0 V M^2 \left(\frac{d}{dT}\chi_T^{-1}\right)$$

Internal energy

$$\begin{aligned} U(T, M) &= F(T, M) + TS(T, M) \\ &= U(T, 0) + \frac{1}{2}\mu_0 V M^2 \left(\chi_T^{-1} - T \frac{d}{dT} \chi_T^{-1} \right) \\ U(T, 0) &= F(T, 0) + TS(T, 0) \end{aligned}$$

We assume that the Curie law is valid:

$$\chi_T = \frac{C}{T} \Leftrightarrow \chi_T^{-1} = \frac{T}{C} \Leftrightarrow T \frac{d}{dT} \chi_T^{-1} = \frac{T}{C}$$

Then we have

$$\begin{aligned} U(T, M) &= U(T, 0) \\ S(T, M) &= S(T, 0) - \frac{1}{2}\mu_0 V M^2 \frac{1}{C} \\ F(T, M) &= F(T, 0) + \frac{1}{2}\mu_0 M^2 \frac{T}{C} \end{aligned}$$

3. Third law:
   This requires that for $T \to 0$ entropy vanishes, independent of the value of the second variable. We therefore must assume in particular

$$\lim_{T \to 0} S(T, 0) = 0$$

   If we take $M(T) \neq 0$ if $H \neq 0$, then according to the entropy obtained in 2. it must be concluded that if the Curie law is valid, then the third law is violated. On the contrary, it must further hold

$$\lim_{T \to 0} \frac{d}{dT} \chi_T^{-1}(T) = \lim_{T \to 0} \frac{1}{\chi_T^2} \frac{d\chi_T}{dT} \stackrel{!}{=} 0$$

$$\curvearrowright \qquad \chi_T^{-1}(T \to 0) = c + \mathcal{O}(T^2)$$

   with a constant $c \neq 0$. $\chi_T(T)$ therefore should not diverge, specially for $T \to 0$.

Problem 4.8

1. Internal energy $U = U(T, M)$ and the equation of state in the form $M = f(T, H)$ are given. First law (1.80)

$$dU = \delta Q + \mu_0 V H \, dM$$

$$\curvearrowright \qquad c_M = \left(\frac{\partial U}{\partial T} \right)_M$$

   So that it holds

$$\delta Q = c_M dT + \left[\left(\frac{\partial U}{\partial M}\right)_T - \mu_0 V H\right] dM$$

$$\curvearrowright$$

$$c_H = \left(\frac{\delta Q}{dT}\right)_H = c_M + \left[\left(\frac{\partial U}{\partial M}\right)_T - \mu_0 V H\right]\left(\frac{\partial M}{\partial T}\right)_H$$

Therefore

$$c_H - c_M = \left[\left(\frac{\partial U}{\partial M}\right)_T - \mu_0 V H\right]\left(\frac{\partial M}{\partial T}\right)_H$$

2. Ideal paramagnet:

$$M = \frac{C}{T} H \ ; \ \left(\frac{\partial U}{\partial M}\right)_T = 0 \qquad C : Curie\ constant$$

$$\curvearrowright \qquad \left(\frac{\partial M}{\partial T}\right)_H = -\frac{C}{T^2} H$$

$$\curvearrowright \qquad c_H - c_M = \frac{\mu_0 V}{C} M^2 \geq 0$$

3. (a) Maxwell relation obtained from the free energy is

$$dF = -SdT + \mu_0 V H\, dM$$

$$\curvearrowright \qquad \left(\frac{\partial S}{\partial M}\right)_T = -\mu_0 V \left(\frac{\partial H}{\partial T}\right)_M$$

(b) Maxwell relation of the free enthalpy is

$$dG = -SdT - \mu_0 V M\, dH$$

$$\curvearrowright \qquad \left(\frac{\partial S}{\partial H}\right)_T = \mu_0 V \left(\frac{\partial M}{\partial T}\right)_H$$

(c) According to 1. from the first law follows:

$$\delta Q = TdS = c_M dT + \left[\left(\frac{\partial U}{\partial M}\right)_T - \mu_0 V H\right] dM$$

$$\curvearrowright T\left(\frac{\partial S}{\partial M}\right)_T = \left[\left(\frac{\partial U}{\partial M}\right)_T - \mu_0 V H\right]$$

This is the proposition!

4.

$$
\begin{aligned}
c_H - c_M &\overset{1.)}{=} \left[\left(\frac{\partial U}{\partial M}\right)_T - \mu_0 V H\right]\left(\frac{\partial M}{\partial T}\right)_H \\
&\overset{3.c)}{=} T\left(\frac{\partial S}{\partial M}\right)_T\left(\frac{\partial M}{\partial T}\right)_H \\
&\overset{3.a)}{=} -\mu_0 V T\left(\frac{\partial H}{\partial T}\right)_M\left(\frac{\partial M}{\partial T}\right)_H
\end{aligned}
$$

5.

$$
\left(\frac{\partial H}{\partial T}\right)_M = \frac{M}{C} \curvearrowright c_H - c_M = -\mu_0 V T\frac{M}{C}\left(\frac{\partial M}{\partial T}\right)_H
$$

Equation of state:

$$
dH = \frac{M}{C}dT + \frac{1}{C}(T - T_C)dM + 3bM^2 dM
$$

$$
\curvearrowright \qquad \left(\frac{\partial M}{\partial T}\right)_H\left[\frac{1}{C}(T - T_C) + 3bM^2\right] = -\frac{M}{C}
$$

So that it follows:

$$
c_H - c_M = \mu_0 V \frac{T M^2}{C(T - T_C) + 3bC^2 M^2}
$$

6.

$$
\begin{aligned}
\left(\frac{\partial c_M}{\partial M}\right)_T &= \left(\frac{\partial}{\partial M}\left(T\left(\frac{\partial S}{\partial T}\right)_M\right)\right)_T \\
&= T\left(\frac{\partial}{\partial T}\left(\frac{\partial S}{\partial M}\right)_T\right)_M \\
&\overset{3a.)}{=} T(-\mu_0 V)\left(\frac{\partial^2 H}{\partial T^2}\right)_M \\
&= 0
\end{aligned}
$$

7. Because of 6. it holds:

$$\left(\frac{\partial U}{\partial T}\right)_M = c_M(T)$$

According to 3. it also holds

$$\begin{aligned}\left(\frac{\partial U}{\partial T}\right)_M &= T\left(\frac{\partial S}{\partial M}\right)_T + \mu_0 V H \\ &= -\mu_0 V T\left(\frac{\partial H}{\partial T}\right)_M + \mu_0 V H \\ &= -\mu_0 V T\frac{M}{C} + \mu_0 V\frac{1}{C}(T - T_C)M \\ &+ \mu_0 V b M^3 \\ &= \mu_0 V\left(bM^3 - \frac{T_C}{C}M\right)\end{aligned}$$

Integration:

$$U(T, M) = \mu_0 V\left(\frac{1}{4}bM^4 - \frac{T_C}{2C}M^2\right) + f(T)$$

Then it must be

$$c_M(T) = f'(T)$$

So that we have

$$\begin{aligned}U(T, M) &= \mu_0 V\left(\frac{1}{4}bM^4 - \frac{T_C}{2C}M^2\right) \\ &+ \int^T c_M(T')dT' + U_0\end{aligned}$$

One gets entropy analogously:

$$\left(\frac{\partial S}{\partial T}\right)_M = \frac{1}{T}c_M(T)$$

$$\left(\frac{\partial S}{\partial M}\right)_T \overset{3a.)}{=} -\mu_0 V\left(\frac{\partial H}{\partial T}\right)_M = -\mu_0 V\frac{M}{C}$$

$$\curvearrowright \quad S(T, M) = -\mu_0 V\frac{M^2}{2C} + \int^T \frac{c_M(T')}{T'}dT' + S_0$$

That means for the free energy

$$\begin{aligned} F &= U - TS \\ &= F_0 + \frac{1}{4}\mu_0 V b M^4 + \mu_0 V \frac{M^2}{2C}(T - T_C) \\ &+ \int^T \frac{c_M(T')}{T'} dT' \\ F_0 &= U_0 - T S_0 \end{aligned}$$

8.

$$H = \frac{1}{C}(T - T_C)M + bM^3$$

For $H \to 0$ we get the trivial solution $M = 0$, but also

$$M_S = \pm\sqrt{\frac{1}{bC}(T_C - T)}$$

These solutions are real for $T \le T_C$. For the free energy we take the result from 7.:

$$F(T, M) = f(T) + \frac{\mu_0 V}{2C}(T - T_C)M^2 + \frac{1}{4}\mu_0 V b M^4$$

We substitute in this expression both the mathematical solutions:

$$\begin{aligned} F(T, M = 0) &= f(T) \\ F(T; M = \pm M_S) &= f(T) + \frac{\mu_0 V}{2C}(T - T_C)\frac{1}{bC}(T_C - T) \\ &+ \frac{1}{4}\mu_0 V b \frac{1}{b^2 C^2}(T_C - T)^2 \\ &= f(T) - \frac{1}{4}\frac{\mu_0 V}{bC^2}(T_C - T)^2 \end{aligned}$$

It obviously holds

$$F(T, M = \pm M_S) \le F(T, M = 0)$$

The equality sign holds for $T = T_C$. Therefore the *ferromagnetic* solution is stable. This exists as a real solution only for $T \le T_C$.

9. Magnetic susceptibility

$$\chi_T = \left(\frac{\partial M}{\partial H}\right)_T = \frac{1}{\left(\frac{\partial H}{\partial M}\right)_T} \overset{5.)}{=} \frac{1}{\frac{1}{C}(T - T_C) + 3bM^2}$$

$$\curvearrowright \quad \lim_{H\to 0} \chi_T = \frac{1}{\frac{1}{C}(T - T_C) + 3bM_S^2} = \frac{C}{2(T_C - T)}$$

χ_T diverges for $T \overset{<}{\to} T_C$, as required by the general theory of phase transitions. Finally we again use the result from 5.:

$$\begin{aligned}\lim_{H\to 0}(c_H - c_M) &= \mu_0 V \frac{T M_S^2}{C(T - T_C) + 3bC^2 M_S^2} \\ &= \mu_0 V \frac{T \frac{1}{bC}(T_C - T)}{C(T - T_C) + 3C(T_C - T)} \\ &= \frac{\mu_0 V}{bC^2} T\end{aligned}$$

For $T \to 0$ the two heat capacities are therefore equal to zero as demanded by the third law.

Problem 4.9

1. According to Problem 4.7 we have

$$U(T, M) = U(T, 0)$$

Since

$$c_M = \left(\frac{\partial U}{\partial T}\right)_{M=0} = \gamma T$$

follows:

$$U(T, M) = U_0 + \frac{1}{2}\gamma T^2$$

Isothermal change of field from 0 to $H \neq 0$:

$$\Delta U = 0$$

That means

$$\begin{aligned}\Delta Q &= -\Delta W = -V\mu_0 \int_0^H H' dH' \\ dM' &= \frac{C}{T_1} dH' \\ \curvearrowright \Delta Q &= -\frac{V\mu_0 C}{T_1} \int_0^H H' dH' = -\frac{\mu_0 V C}{2T_1} H^2 < 0\end{aligned}$$

2. *Adiabatic and reversible* means: $dS = 0$, therefore

$$S(T_1, H) \stackrel{!}{=} S(T_f, 0)$$

From Problem 4.7 we have

$$\begin{aligned} S(T, M) &= S(T, 0) - \frac{1}{2}\mu_0 V \frac{1}{C} M^2 \\ S(T, 0) - S_0 &= \int_0^T \frac{c_{M=0}(T')}{T'} dT' = \gamma T \\ \curvearrowright S(T, M) &= S_0 + \gamma T - \frac{1}{2}\mu_0 V \frac{1}{C} M^2 \\ \curvearrowright S(T, H) &= S_0 + \gamma T - \frac{1}{2}\mu_0 C V \frac{H^2}{T^2} \end{aligned}$$

Switching off the field

$$\begin{aligned} S(T_1, H) &= S_0 + \gamma T_1 - \frac{1}{2}\mu_0 C V \frac{H^2}{T_1^2} \\ &\stackrel{!}{=} S(T_f, 0) = S_0 + \gamma T_f \end{aligned}$$

$\curvearrowright$ final temperature:

$$T_f = T_1 - \frac{1}{2}\mu_0 C V \frac{H^2}{\gamma T_1^2} < T_1$$

$\curvearrowright$ Cooling effect.

Problem 5.1
Commutation relations:

1.

$$\begin{aligned} \left[S_i^+, S_i^-\right]_- &= \hbar^2 \left[c_{i\uparrow}^+ c_{i\downarrow}, c_{i\downarrow}^+ c_{i\uparrow}\right]_- \\ &= \hbar^2 \left(c_{i\uparrow}^+ (1 - n_{i\downarrow}) c_{i\uparrow} - c_{i\downarrow}^+ (1 - n_{i\uparrow}) c_{i\downarrow}\right) \\ &= \hbar^2 \left(n_{i\uparrow}(1 - n_{i\downarrow}) - n_{i\downarrow}(1 - n_{i\uparrow})\right) \\ &= \hbar^2 \left(n_{i\uparrow} - n_{i\downarrow}\right) \\ &= 2\hbar S_i^z \end{aligned}$$

$$
\begin{aligned}
\left[S_i^z, S_i^+\right]_- &= \frac{1}{2}\hbar^2\left(\left[n_{i\uparrow}, c_{i\uparrow}^+ c_{i\downarrow}\right]_- - \left[n_{i\downarrow}, c_{i\uparrow}^+ c_{i\downarrow}\right]_-\right)\\
&= \frac{1}{2}\hbar^2\left((n_{i\uparrow} - n_{i\downarrow})c_{i\uparrow}^+ c_{i\downarrow} - c_{i\uparrow}^+ c_{i\downarrow}(n_{i\uparrow} - n_{i\downarrow})\right)\\
&= \frac{1}{2}\hbar^2\left(\underbrace{n_{i\uparrow}c_{i\uparrow}^+}_{c_{i\uparrow}^+} c_{i\downarrow} - c_{i\uparrow}^+ \underbrace{n_{i\downarrow}c_{i\downarrow}}_{\equiv 0} - \underbrace{c_{i\uparrow}^+ n_{i\uparrow}}_{\equiv 0} c_{i\downarrow} + c_{i\uparrow}^+ \underbrace{c_{i\downarrow}n_{i\downarrow}}_{c_{i\downarrow}}\right)\\
&= \hbar^2 c_{i\uparrow}^+ c_{i\downarrow}\\
&= \hbar S_i^+
\end{aligned}
$$

$$
\begin{aligned}
\left[S_i^z, S_i^-\right]_- &= \frac{1}{2}\hbar^2\left((n_{i\uparrow} - n_{i\downarrow})c_{i\downarrow}^+ c_{i\uparrow} - c_{i\downarrow}^+ c_{i\uparrow}(n_{i\uparrow} - n_{i\downarrow})\right)\\
&= \frac{1}{2}\hbar^2\left(-\underbrace{n_{i\downarrow}c_{i\downarrow}^+}_{c_{i\downarrow}^+} c_{i\uparrow} - c_{i\downarrow}^+ \underbrace{c_{i\uparrow}n_{i\uparrow}}_{c_{i\uparrow}}\right)\\
&= -\hbar^2 c_{i\downarrow}^+ c_{i\uparrow}\\
&= -\hbar S_i^-
\end{aligned}
$$

$$
\begin{aligned}
\mathbf{S}_i^2 &= S_i^{x\,2} + S_i^{y\,2} + S_i^{z\,2}\\
&= \frac{1}{2}\left(S_i^+ S_i^- + S_i^- S_i^+\right) + S_i^{z\,2}\\
&= \frac{\hbar^2}{2}\left(\underbrace{c_{i\uparrow}^+ c_{i\downarrow} c_{i\downarrow}^+}_{n_{i\uparrow}(1-n_{i\downarrow})} c_{i\uparrow} + \underbrace{c_{i\downarrow}^+ c_{i\uparrow} c_{i\uparrow}^+ c_{i\downarrow}}_{n_{i\downarrow}(1-n_{i\uparrow})}\right) + \frac{\hbar^2}{4}(\underbrace{n_{i\uparrow}^2}_{n_{i\uparrow}} + \underbrace{n_{i\downarrow}^2}_{n_{i\downarrow}} - 2n_{i\uparrow}n_{i\downarrow})\\
&= \hbar^2\left(\frac{3}{4}(n_{i\uparrow} + n_{i\downarrow}) - \frac{3}{2}n_{i\uparrow}n_{i\downarrow}\right)\\
&= \frac{3}{4}\hbar^2(n_{i\uparrow} + n_{i\downarrow} - 2n_{i\uparrow}n_{i\downarrow})\\
&= \frac{3}{4}\hbar^2(n_{i\uparrow} - n_{i\downarrow})^2\\
&\overset{!}{=} \hbar^2 S(S+1) \qquad (S = \frac{1}{2})
\end{aligned}
$$

only if there is exactly one electron per lattice site

2.

$$\left[c_{\mathbf{q}\sigma},\, c^+_{\mathbf{k}\sigma'}\right]_+ = \frac{1}{N}\sum_{i,j} e^{-i\mathbf{q}\mathbf{R}_i + i\mathbf{k}\mathbf{R}_j} \underbrace{\left[c_{i\sigma},\, c^+_{j\sigma'}\right]_+}_{\delta_{ij}\delta_{\sigma\sigma'}}$$
$$= \delta_{\sigma\sigma'}\frac{1}{N}\sum_i e^{-i(\mathbf{q}-\mathbf{k})\mathbf{R}_i}$$
$$= \delta_{\mathbf{q},\mathbf{k}}\delta_{\sigma\sigma'}$$

$$\left[c_{\mathbf{q}\sigma},\, c_{\mathbf{k}\sigma'}\right]_+ = \left[c^+_{\mathbf{q}\sigma},\, c^+_{\mathbf{k}\sigma'}\right]_+ = 0$$

Problem 5.2
Free energy:

$$F = U - TS\,; \qquad dF = -SdT + \mu_0 V H dM$$

Integrability condition:

$$\left(\frac{\partial S}{\partial M}\right)_T = -\mu_0 V\left(\frac{\partial H}{\partial T}\right)_M$$

Curie–Weiss law:

$$M = \frac{C}{T - T_C}H \qquad (C:\ \text{Curie constant})$$

Heat capacity:

$$c_M(T,M) = T\left(\frac{\partial S}{\partial T}\right)_M$$
$$\left(\frac{\partial c_M}{\partial M}\right)_T = T\left(\frac{\partial^2 S}{\partial M\partial T}\right) = T\left[\frac{\partial}{\partial T}\frac{\partial S}{\partial M}\right]$$
$$= -\mu_0 VT\left[\frac{\partial}{\partial T}\left(\frac{\partial H}{\partial T}\right)_M\right]$$
$$=0$$
$$\curvearrowright\ c_M(T,M) \equiv c_M(T)$$

Internal energy:
Equation (1.80): $dU = TdS + \mu_0 V H dM$

$$
\begin{aligned}
\curvearrowright \left(\frac{\partial U}{\partial T}\right)_M &= T\left(\frac{\partial S}{\partial T}\right)_M = c_M(T) \\
\left(\frac{\partial U}{\partial M}\right)_T &= T\left(\frac{\partial S}{\partial M}\right)_T + \mu_0 V H \\
&\overset{\text{s.o.}}{=} -\mu_0 V T\left(\frac{\partial H}{\partial T}\right)_M + \mu_0 V H \\
&= -\mu_0 V T\frac{M}{C} + \mu_0 V\frac{T - T_C}{C}M \\
&= -\mu_0 V\frac{M}{C}T_C
\end{aligned}
$$

This means

$$
U(T, M) = -\mu_0 V T_C \frac{M^2}{2C} + f(T)
$$

Because of

$$
\left(\frac{\partial U}{\partial T}\right)_M = f'(T) = c_M(T)
$$

altogether we have

$$
U(T, M) = \int_0^T c_M(T')dT' - \mu_0 V T_C \frac{M^2}{2C} + U_0
$$

Entropy:

$$
\begin{aligned}
S(T, M) &= \int_0^T \frac{c_M(T')}{T'}dT' + \sigma(M) \\
\left(\frac{\partial S}{\partial M}\right)_T &= \sigma'(M) = -\mu_0 V\left(\frac{\partial H}{\partial T}\right)_M = -\mu_0 V\frac{M}{C} \\
\curvearrowright\ \ \sigma(M) &= -\mu_0 V\frac{M^2}{2C} + \sigma_0 \\
S(T, M) &= \sigma_0 + \int_0^T \frac{c_M(T')}{T'}dT' - \mu_0 V\frac{M^2}{2C}
\end{aligned}
$$

Free energy:

$$
\begin{aligned}
F(T, M) =& U(T, M) - T S(T, M) \\
=& F_0(T) + \int_0^T c_M(T') \left(1 - \frac{T}{T'}\right) dT' + \frac{\mu_0 V}{2C} M^2 (T - T_C)
\end{aligned}
$$

$$F_0(T) = U_0 - T\sigma_0$$

Check:

$$S(T, M) = -\left(\frac{\partial F(T, M)}{\partial T}\right)_M$$

Free enthalpy:

$$
\begin{aligned}
& G(T, H) \\
=& F - \mu_0 V M H \\
=& F - \frac{\mu_0 V}{C} (T - T_C) M^2 \\
=& F_0(T) + \int_0^T c_M(T') \left(1 - \frac{T}{T'}\right) dT' - \frac{\mu_0 V}{2C} M^2 (T - T_C) \\
& \curvearrowright \\
& G(T, H) \\
=& F_0(T) + \int_0^T c_M(T') \left(1 - \frac{T}{T'}\right) dT' - \frac{1}{2} \mu_0 V C \frac{1}{T - T_C} H^2
\end{aligned}
$$

Problem 5.3

1.

$$c_M(T, M = 0) = T \left(\frac{\partial S}{\partial T}\right)_{M=0}$$

$$\curvearrowright \; S(T, 0) = \int_0^T \frac{c_M(T', M = 0)}{T'} dT' = \gamma T$$

With the result of Problem 5.2

$$S(T, M) = \gamma T - \mu_0 V \frac{M^2}{2C} + \sigma_0$$

Because

$$M = \frac{C}{T - T_C} H$$

directly follows:

$$S(T, H) = \gamma T - \frac{1}{2}\mu_0 C V \frac{H^2}{(T - T_c)^2} + \sigma_0$$

For the free energy F we use

$$\left(\frac{\partial F}{\partial T}\right)_M = -S(T, M)$$

$$F(T, M) = F_0(M) - \frac{1}{2}\gamma T^2 + \mu_0 V \frac{M^2}{2C} T - \sigma_0 T$$

According to the considerations of Problem 5.2 we must have

$$F(T, M) = F_0(T) + \frac{1}{2}\gamma T^2 - \gamma T^2 + \frac{\mu_0 V}{2C} M^2 (T - T_C)$$

$$\curvearrowright \ F_0(M) = U_0 - \frac{\mu_0 V}{2C} M^2 T_C$$

What remains is only

$$F(T, M) = U_0 - \frac{1}{2}\gamma T^2 + \frac{\mu_0 V}{2C} M^2 (T - T_C)$$

Internal energy again from Problem 5.2:

$$U(T, M) = U_0 + \frac{1}{2}\gamma T^2 - \mu_0 V T_C \frac{M^2}{2C}$$

Check:

$$\begin{aligned} F(T, M) &= U(T, M) - T S(T, M) \\ &= U_0 - \frac{1}{2}\gamma T^2 + \frac{\mu_0 V}{2C} M^2 (T - T_C) - \sigma_0 T \end{aligned}$$

$$\curvearrowright \ \sigma_0 = 0$$

2. Heat capacities

$$c_M(T, M) = T \left(\frac{\partial S}{\partial T}\right)_M = \gamma T = c_M(T, M = 0)$$

$$c_H(T, H) = T \left(\frac{\partial S}{\partial T}\right)_H = \gamma T + \mu_0 C V \frac{T H^2}{(T - T_C)^3}$$

Since $T > T_C$, we have $c_H \geq c_M$.
From elementary thermodynamics

$$\chi_S(T, H) = \chi_T \frac{c_M}{c_H} = \frac{C}{T - T_C + \frac{\mu_0 C V}{\gamma} \frac{H^2}{(T-T_C)^2}}$$

Problem 5.4

1. Spontaneous sub-lattice magnetization ($B_0 = 0$)

$$\begin{aligned} M_{1S}(T) &= -M_{2S}(T) \\ \curvearrowright \quad B_A^{(i)} &= \mu_0(\lambda - \rho) M_{iS}(T) \\ \curvearrowright \quad M_{iS}(T) &= M_0^* B_J(\beta g_J J \mu_0 \mu_B (\lambda - \rho) M_{iS}(T)) \end{aligned}$$

$T \to T_c \quad \curvearrowright \quad M_{iS}$ will be very small

$$\curvearrowright \quad M_{iS} = \frac{J+1}{3J}(n^* g_J J \mu_B) \beta g_J J \mu_0 \mu_B (\lambda - \rho) M_{iS}$$

Curie constant: $C = n \frac{\mu_0 \mu_B^2}{3k_B} g_J^2 J(J+1); \quad n^* = \frac{N}{2V} = \frac{1}{2} n$

$$\curvearrowright \quad \boxed{M_{iS} \cong \frac{1}{2}(\lambda - \rho) C \frac{M_{iS}}{T}}$$

Condition for intersection at finite M_{is}:

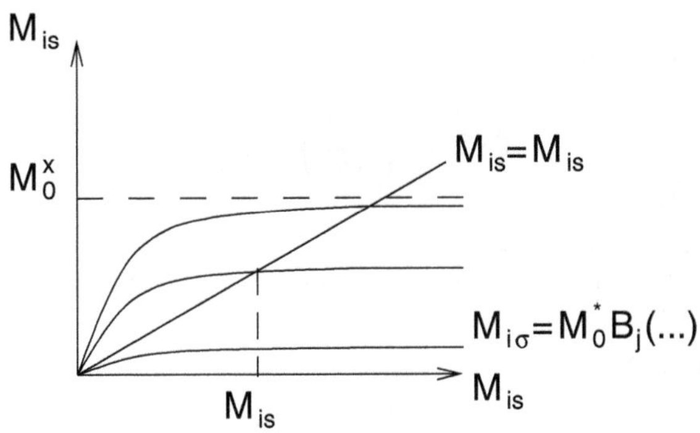

$$\left. \frac{d}{dM_{is}}(M_0 B_j(\ldots)) \right|_{M_{is}=0} \geq 1$$

$$\curvearrowright \quad \frac{1}{2}(\lambda - \rho)\frac{C}{T} \geq 1$$

$$\Longrightarrow \quad \boxed{T_N = \tfrac{1}{2}(\lambda - \rho)C}$$

2. High temperature behaviour:
   No spontaneous sub-lattice magnetization

$$\beta \mu_B B_0 << 1$$

$$\curvearrowright \quad M_i(T, B_0) \cong \frac{J+1}{3J}(n^* g_J J \mu_B)\beta g_J J \mu_B (B_0 + B_A^{(i)})$$
$$= \frac{1}{2}\frac{C}{T}(B_0 + B_A^{(i)})\frac{1}{\mu_0}$$

Total magnetization:

$$M(T, B_0) = M_1(T, B_0) + M_2(T, B_0)$$
$$= \mu_0 \frac{C}{T} B_0 + \frac{1}{2}\frac{C}{T}\underbrace{(B_A^{(1)} + B_A^{(2)})}_{\mu_0(\lambda+\rho)(M_1+M_2)}\frac{1}{\mu_0}$$

$$\curvearrowright \quad M(T, B_0)\left(1 - \frac{1}{2}\frac{C}{T}(\lambda + \rho)\right) = \frac{C}{\mu_0 T} B_0$$

$$\curvearrowright \quad M(T, B_0) = \frac{C}{T - \frac{1}{2}C(\lambda + \rho)}\frac{1}{\mu_0} B_0$$

$\curvearrowright$ *Curie–Weiss law*

$$\chi(T) = \mu_0 \left(\frac{\partial M}{\partial B_0}\right)_T = \frac{C}{T - \Theta}$$

$\boxed{\Theta = \tfrac{1}{2}C(\lambda + \rho)}$ *paramagnetic Curie temperature*

3. necessary: $\rho < 0$

$$-\frac{\Theta}{T_N} = \frac{\lambda + \rho}{\rho - \lambda} = \frac{|\rho| - \lambda}{|\rho| + \lambda}$$

(α) $\lambda > 0$: ferromagnetic coupling within the sub-lattice

$$-\frac{\Theta}{T_N} < 1 \qquad \text{i.a.: } \Theta < 0 \quad (\lambda < |\rho|)$$

Example: EuTe: $\Theta = -4.0K$; $T_N = 9.6K$

(β) $\lambda < 0$: antiferromagnetic coupling in the sub-lattice in order to have $T_N > 0$: $|\lambda| < |\rho|$

$$-\frac{\Theta}{T_N} = \frac{|\rho| + |\lambda|}{|\rho| - |\lambda|} \qquad \text{always: } \Theta < 0$$

$$-\frac{\Theta}{T_N} > 1$$

Examples:

	MnO	NiO	MnF_2
$-\frac{\Theta}{T_N}$	5.3	5.7	1.7

Typically: $\Theta < 0$

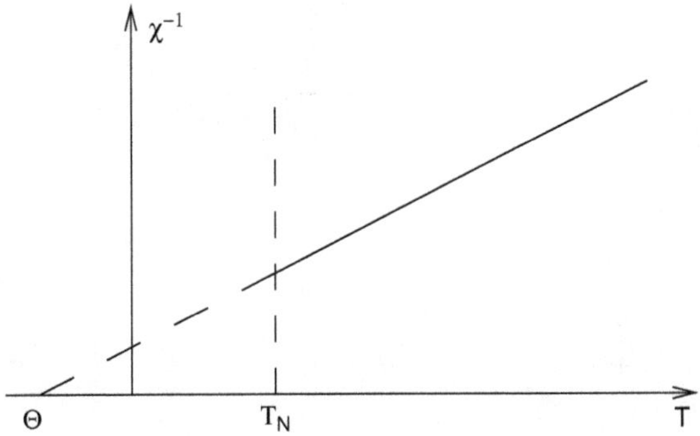

Problem 5.5

According to Problem 4.6 the equation of state of Weiss ferromagnet reads

$$M = M_0 L\left(m\frac{\mu_0 H + \lambda\mu_0 M}{k_B T}\right)$$

1.

$$\frac{m\lambda\mu_0 M}{k_B T} = \frac{M}{M_0}\frac{\frac{N}{V}m^2\lambda\mu_0}{k_B T} = \hat{M}\frac{3k_B C\lambda}{k_B T}$$

Classical Curie constant (Problem 4.6)

$$C = \mu_0\frac{N}{V}\frac{m^2}{3k_B}$$

Further using (5.13): $T_C = \lambda C$ and then we have

$$\hat{M} = L\left(b + 3\frac{\hat{M}}{\varepsilon + 1}\right)$$

2. Series expansion of the Langevin function

$$L(x) = \frac{1}{3}x - \frac{1}{45}x^3 + O(x^5)$$
$$B_0 = \mu_0 H = 0 \curvearrowright b = 0$$
$$T \overset{<}{\to} T_c \curvearrowright \hat{M} \text{ very small}$$

Then we can write approximately

$$\hat{M} \approx \frac{\hat{M}}{\varepsilon + 1} - \frac{3}{5}\frac{\hat{M}^3}{(\varepsilon + 1)^3}$$

For $\hat{M} \neq 0$ that means

$$\frac{\varepsilon}{\varepsilon + 1} \approx -\frac{3}{5}\frac{\hat{M}^2}{(\varepsilon + 1)^3} \curvearrowright \hat{M}^2 \approx -\frac{5}{3}\varepsilon(\varepsilon + 1)^2$$

Since $(\varepsilon + 1)^2 \to 1$ for $T \to T_C$, it follows that

$$\hat{M} \sim \sqrt{\frac{5}{3}}(-\varepsilon)^{\frac{1}{2}}$$

The critical exponent β of the order parameter then has the classical value:

$$\beta = \frac{1}{2}$$

3. Critical isotherm: $T = T_c$; $B_0 \to 0$

$$\curvearrowright \varepsilon = 0 \text{ and } \hat{M} \text{ as well as } b \text{ very small.}$$

This means

$$\begin{aligned}
&\hat{M} \approx \frac{1}{3}b + \hat{M} - \frac{1}{45}(b+3\hat{M})^3 \\
\curvearrowright\quad & 15b \approx (b+3\hat{M})^3 \quad \leftrightarrow \quad b+3\hat{M} \approx (15b)^{\frac{1}{3}} \\
\leftrightarrow\quad & 3\hat{M} \approx (15b)^{\frac{1}{3}} - b \approx (15b)^{\frac{1}{3}}, \text{ since } b \to 0 \\
\curvearrowright\quad & b \sim \frac{9}{5}\hat{M}^3
\end{aligned}$$

From this we read off the critical exponent:

$$\delta = 3$$

That is also well-known value for classical theories.

4. Susceptibility:

$$\begin{aligned}
\chi_T = \left(\frac{\partial M}{\partial H}\right)_T &= \frac{M_0 \mu_0 m}{k_B T}\left(\frac{\partial \hat{M}}{\partial b}\right)_{T,b=0} \\
&= \frac{\frac{N}{V} m^2 \mu_0}{k_B(\varepsilon+1)T_c}\left(\frac{\partial \hat{M}}{\partial b}\right)_{T,b=0} \\
&= \frac{3}{\lambda(\varepsilon+1)}\left(\frac{\partial \hat{M}}{\partial b}\right)_{T,b=0}
\end{aligned}$$

In the critical region $\hat{M}$ is very small. Therefore we can expand

$$\begin{aligned}
&\left.\frac{\partial L}{\partial b}\right|_{b=0} = \left.\frac{\partial x}{\partial b}\left(\frac{1}{3} - \frac{1}{15}x^2\right)\right|_{b=0} \\
&\frac{\partial x}{\partial b} = 1 + \frac{3}{\varepsilon+1}\frac{\partial \hat{M}}{\partial b} \\
\curvearrowright\quad & \left.\frac{\partial \hat{M}}{\partial b}\right|_{b=0} = \left(1 + \frac{3}{\varepsilon+1}\left.\frac{\partial \hat{M}}{\partial b}\right|_{b=0}\right)\left(\frac{1}{3} - \frac{1}{15}\frac{9\hat{M}^2}{(\varepsilon+1)^2}\right) \\
\curvearrowright\quad & \left.\frac{\partial \hat{M}}{\partial b}\right|_{b=0}\left\{1 - \frac{1}{\varepsilon+1} + \frac{9}{5}\frac{\hat{M}^2}{(\varepsilon+1)^3}\right\} = \frac{1}{3}\left(1 - \frac{9}{5}\frac{\hat{M}^2}{(\varepsilon+1)^2}\right) \\
\curvearrowright\quad & \left.\frac{\partial \hat{M}}{\partial b}\right|_{b=0} = \frac{1}{3}\frac{1 - \frac{9}{5}\frac{\hat{M}^2}{(\varepsilon+1)^2}}{\frac{\varepsilon}{\varepsilon+1} + \frac{9}{5}\frac{\hat{M}^2}{(\varepsilon+1)^3}}
\end{aligned}$$

$T \to T_c$ means $\hat{M} \to 0$:

$$\left.\frac{\partial \hat{M}}{\partial b}\right|_{b=0} \approx \frac{1}{3}\left[\left(\frac{\varepsilon}{\varepsilon+1}+\frac{9}{5}\frac{\hat{M}^2}{(\varepsilon+1)^3}\right)\left(1+\frac{9}{5}\frac{\hat{M}^2}{(\varepsilon+1)^2}\right)\right]^{-1}$$
$$\approx \frac{1}{3}\left[\frac{\varepsilon}{\varepsilon+1}+\frac{9}{5}\frac{\hat{M}^2}{(\varepsilon+1)^2}\right]^{-1}$$

(a)$T \overset{>}{\to} T_c$:
then $\hat{M} \equiv 0$ and $\varepsilon + 1 \overset{T\to T_c}{\longrightarrow} 1$:

$$\left.\frac{\partial \hat{M}}{\partial b}\right|_{b=0} \approx \frac{1}{3}\varepsilon^{-1}$$
$$\curvearrowright \quad \chi_T \sim \frac{1}{\lambda}\varepsilon^{-1}$$

$\curvearrowright$ critical exponent:

$$\gamma = 1$$

(b)$T \overset{<}{\to} T_c$:
then according to part 2 we have

$$\hat{M} \sim \sqrt{\frac{5}{3}}(-\varepsilon)^{\frac{1}{2}}$$

This means

$$\left.\frac{\partial \hat{M}}{\partial b}\right|_{b=0} \approx \frac{1}{3}\left[\frac{\varepsilon}{\varepsilon+1}+3\frac{-\varepsilon}{(\varepsilon+1)^2}\right]^{-1}$$
$$\overset{T\to T_c}{\longrightarrow} \frac{1}{3}[-2\varepsilon]^{-1}$$
$$\curvearrowright \quad \chi_T \sim \frac{1}{2\lambda}(-\varepsilon)^{-1} \quad \Rightarrow \quad \gamma' = 1$$

Critical amplitudes:

$$C \sim \frac{1}{\lambda}; \quad C' \sim \frac{1}{2\lambda} \quad \curvearrowright \quad \frac{C}{C'} = 2$$

The results obtained for γ, γ' and $\frac{C}{C'}$ are typical for the classical theories of phase transitions. The concluding remark is as follows.

The sign " $\sim$" in the above formulae need not necessarily mean proportionality but should be understood as, "for $T \to T_c$ behaves like".

Problem 5.6

1.

$$
\begin{aligned}
&\left(\sigma^{(1)} \cdot \sigma^{(2)}\right)^2 \\
&= \left(\sigma_x^{(1)}\sigma_x^{(2)} + \sigma_y^{(1)}\sigma_y^{(2)} + \sigma_z^{(1)}\sigma_z^{(2)}\right)^2 = \\
&= \left(\sigma_x^{(1)}\right)^2 \left(\sigma_x^{(2)}\right)^2 + \left(\sigma_y^{(1)}\right)^2 \left(\sigma_y^{(2)}\right)^2 + \left(\sigma_z^{(1)}\right)^2 \left(\sigma_z^{(2)}\right)^2 + \\
&+ \sigma_x^{(1)}\,\sigma_x^{(2)}\,\sigma_y^{(1)}\,\sigma_y^{(2)} + \sigma_x^{(1)}\,\sigma_x^{(2)}\,\sigma_z^{(1)}\,\sigma_z^{(2)} + \\
&+ \sigma_y^{(1)}\,\sigma_y^{(2)}\,\sigma_x^{(1)}\,\sigma_x^{(2)} + \sigma_y^{(1)}\,\sigma_y^{(2)}\,\sigma_z^{(1)}\,\sigma_z^{(2)} + \\
&+ \sigma_z^{(1)}\,\sigma_z^{(2)}\,\sigma_x^{(1)}\,\sigma_x^{(2)} + \sigma_z^{(1)}\,\sigma_z^{(2)}\,\sigma_y^{(1)}\,\sigma_y^{(2)} = \\
&= 3\,\mathbb{1}^2 - \sigma_z^{(1)}\,\sigma_z^{(2)} - \sigma_y^{(1)}\,\sigma_y^{(2)} - \sigma_z^{(1)}\,\sigma_z^{(2)} - \\
&- \sigma_x^{(1)}\,\sigma_x^{(2)} - \sigma_y^{(1)}\,\sigma_y^{(2)} - \sigma_x^{(1)}\,\sigma_x^{(2)} = \\
&= 3\,\mathbb{1} - 2\sigma^{(1)} \cdot \sigma^{(2)}
\end{aligned}
$$

Here the properties (5.81) and (5.82) of the Pauli spin matrices are used many times.

2. Representatively, we calculate the x-component

$$
\begin{aligned}
&Q_{12}\,\sigma_x^{(1)} \\
&= \frac{1}{2}\left(\mathbb{1} + \sigma_x^{(1)}\,\sigma_x^{(2)} + \sigma_y^{(1)}\,\sigma_y^{(2)} + \sigma_z^{(1)}\,\sigma_z^{(2)}\right)\,\sigma_x^{(1)} = \\
&= \frac{1}{2}\left(\sigma_x^{(1)} + \left(\sigma_x^{(1)}\right)^2 \sigma_x^{(2)} + \sigma_y^{(1)}\sigma_x^{(1)}\sigma_y^{(2)} + \sigma_z^{(1)}\sigma_x^{(1)}\sigma_z^{(2)}\right) \\
&= \frac{1}{2}\left(\sigma_x^{(1)} + \sigma_x^{(2)} - i\,\sigma_z^{(1)}\sigma_y^{(2)} + i\,\sigma_y^{(1)}\sigma_z^{(2)}\right)
\end{aligned}
$$

$$
\begin{aligned}
&\sigma_x^{(2)}\,Q_{12} \\
&= \frac{1}{2}\,\sigma_x^{(2)}\left(\mathbb{1} + \sigma_x^{(1)}\,\sigma_x^{(2)} + \sigma_y^{(1)}\,\sigma_y^{(2)} + \sigma_z^{(1)}\,\sigma_z^{(2)}\right) = \\
&= \frac{1}{2}\left(\sigma_x^{(2)} + \sigma_x^{(1)}\left(\sigma_x^{(2)}\right)^2 + \sigma_y^{(1)}\sigma_x^{(2)}\sigma_y^{(2)} + \sigma_z^{(1)}\sigma_x^{(2)}\sigma_z^{(2)}\right) \\
&= \frac{1}{2}\left(\sigma_x^{(2)} + \sigma_x^{(1)} + i\,\sigma_y^{(1)}\sigma_z^{(2)} - i\,\sigma_z^{(1)}\sigma_y^{(2)}\right)
\end{aligned}
$$

That means

$$
Q_{12}\,\sigma_x^{(1)} = \sigma_x^{(2)}\,Q_{12} \quad ,
$$

so that the proposition for the x-component is proved. Analogously one can prove for the other components.

3. We use

$$\sigma_x \begin{pmatrix} 1 \\ 0 \end{pmatrix} = \begin{pmatrix} 0 & 1 \\ 1 & 0 \end{pmatrix} \begin{pmatrix} 1 \\ 0 \end{pmatrix} = \begin{pmatrix} 0 \\ 1 \end{pmatrix}$$

$$\sigma_x \begin{pmatrix} 0 \\ 1 \end{pmatrix} = \begin{pmatrix} 0 & 1 \\ 1 & 0 \end{pmatrix} \begin{pmatrix} 0 \\ 1 \end{pmatrix} = \begin{pmatrix} 1 \\ 0 \end{pmatrix}$$

$$\sigma_y \begin{pmatrix} 1 \\ 0 \end{pmatrix} = \begin{pmatrix} 0 & -i \\ i & 0 \end{pmatrix} \begin{pmatrix} 1 \\ 0 \end{pmatrix} = i \begin{pmatrix} 0 \\ 1 \end{pmatrix}$$

$$\sigma_y \begin{pmatrix} 0 \\ 1 \end{pmatrix} = \begin{pmatrix} 0 & -i \\ i & 0 \end{pmatrix} \begin{pmatrix} 0 \\ 1 \end{pmatrix} = -i \begin{pmatrix} 1 \\ 0 \end{pmatrix}$$

$$\sigma_z \begin{pmatrix} 1 \\ 0 \end{pmatrix} = \begin{pmatrix} 1 \\ 0 \end{pmatrix} \; ; \; \sigma_z \begin{pmatrix} 0 \\ 1 \end{pmatrix} = -\begin{pmatrix} 0 \\ 1 \end{pmatrix}$$

So that one immediately recognizes

$$\begin{aligned} \sigma^{(1)} \cdot \sigma^{(2)} \, |\uparrow\uparrow\rangle &= |\downarrow\downarrow\rangle + i^2 \, |\downarrow\downarrow\rangle + |\uparrow\uparrow\rangle && = |\uparrow\uparrow\rangle \\ \sigma^{(1)} \cdot \sigma^{(2)} \, |\downarrow\downarrow\rangle &= |\uparrow\uparrow\rangle + (-i)^2 \, |\uparrow\uparrow\rangle + |\downarrow\downarrow\rangle = |\downarrow\downarrow\rangle \end{aligned}$$

$$\begin{aligned} &\sigma^{(1)} \cdot \sigma^{(2)} (|\uparrow\downarrow\rangle + |\downarrow\uparrow\rangle) \\ &= \left(|\downarrow\uparrow\rangle + |\uparrow\downarrow\rangle - i^2 |\downarrow\uparrow\rangle - i^2 \, |\uparrow\downarrow\rangle - |\uparrow\downarrow\rangle - |\downarrow\uparrow\rangle\right) \\ &= (|\uparrow\downarrow\rangle + |\downarrow\uparrow\rangle) \end{aligned}$$

$$\begin{aligned} &\sigma^{(1)} \cdot \sigma^{(2)} (|\uparrow\downarrow\rangle - |\downarrow\uparrow\rangle) \\ &= \left(|\downarrow\uparrow\rangle - |\uparrow\downarrow\rangle - i^2 |\downarrow\uparrow\rangle + i^2 \, |\uparrow\downarrow\rangle - |\uparrow\downarrow\rangle + |\downarrow\uparrow\rangle\right) \\ &= 3(|\uparrow\downarrow\rangle - |\downarrow\uparrow\rangle) \end{aligned}$$

The spin states are therefore eigenstates of the operator $\sigma^{(1)} \cdot \sigma^{(2)}$ with the eigenvalues $(1, 1, 1, -3)$. Therefore they are also the eigenstates of Q_{12} with the eigenvalues $(1, 1, 1, -1)$.

Problem 5.7

$$R = \int_0^{\infty} dx \, x \, \frac{\sin x}{x^2 - y^2} = \frac{1}{4i} \int_{-\infty}^{+\infty} dx \, x \, \frac{\exp(ix) - \exp(-ix)}{x^2 - y^2}$$

We use the residue theorem and for that we choose the integration path parallel to the real axis from $-\infty + i0^+$ to $+\infty + i0^+$, close it with a semicircle at infinity in the upper half-plane for the first exponential function and in the lower half-plane for the second exponential function. Then the exponentials see to it that the semicircles do not contribute to the integral. The poles at $x = \pm y$ lie inside the (mathematically negatively running) integration path C only for the second term. Then from the residue theorem we get

$$\begin{aligned} R &= -\frac{1}{8i}\int_C dx\, e^{-ix}\left(\frac{1}{x+y}+\frac{1}{x-y}\right) \\ &= \frac{2\pi i}{8i}\left(e^{+iy}+e^{-iy}\right) \\ &= \frac{\pi}{2}\cos y \end{aligned}$$

Problem 5.8

1. The diagonal terms $\mu = \nu$ in $\widehat{H}_1$ directly give H_U:

$$\widehat{H}_1(\mu = \nu) = H_U$$

That is why from now on we will consider only the non-diagonal terms $\mu \neq \nu$. For these we have to calculate

$$\begin{aligned} &\widehat{H}_1(\mu \neq \nu) \\ &= \frac{1}{2}\sum_{i\sigma\sigma'}\sum_{\mu\nu}^{\mu\neq\nu}\left[U_{\mu\nu} n_{i\mu\sigma} n_{i\nu\sigma'} + J_{\mu\nu}\, c^{\dagger}_{i\mu\sigma} c^{\dagger}_{i\nu\sigma'} c_{i\mu\sigma'} c_{i\nu\sigma}\right] \end{aligned}$$

The first summand leads to

$$\widehat{H}_{1a}(\mu \neq \nu) = \frac{1}{2}\sum_{i\mu\nu}^{\mu\neq\nu} U_{\mu\nu} n_{i\mu} n_{i\nu}$$

The second summand we somewhat reformulate

$$\begin{aligned} &\widehat{H}_{1b}(\mu \neq \nu) \\ &= \frac{1}{2}\sum_{i\sigma\sigma'}\sum_{\mu\nu}^{\mu\neq\nu} J_{\mu\nu}\, c^{\dagger}_{i\mu\sigma} c^{\dagger}_{i\nu\sigma'} c_{i\mu\sigma'} c_{i\nu\sigma} = \\ &= \frac{1}{2}\sum_{i\mu\nu}^{\mu\neq\nu} J_{\mu\nu}\left[c^{\dagger}_{i\mu\uparrow} c^{\dagger}_{i\nu\uparrow} c_{i\mu\uparrow} c_{i\nu\uparrow} + c^{\dagger}_{i\mu\uparrow} c^{\dagger}_{i\nu\downarrow} c_{i\mu\downarrow} c_{i\nu\uparrow} + \right. \\ &\left. + c^{\dagger}_{i\mu\downarrow} c^{\dagger}_{i\nu\uparrow} c_{i\mu\uparrow} c_{i\nu\downarrow} + c^{\dagger}_{i\mu\downarrow} c^{\dagger}_{i\nu\downarrow} c_{i\mu\downarrow} c_{i\nu\downarrow}\right] \\ &= \frac{1}{2}\sum_{i\mu\nu}^{\mu\neq\nu} J_{\mu\nu}\left[-\sigma^{+}_{i\mu}\sigma^{-}_{i\nu} - \sigma^{-}_{i\mu}\sigma^{+}_{i\nu} - n_{i\mu\uparrow} n_{i\nu\uparrow} - n_{i\mu\downarrow} n_{i\nu\downarrow}\right] \\ &= \frac{1}{2}\sum_{i\mu\nu}^{\mu\neq\nu} J_{\mu\nu}\left[-2\sigma^{x}_{i\mu}\sigma^{x}_{i\nu} - 2\sigma^{y}_{i\mu}\sigma^{y}_{i\nu} - 2\sigma^{z}_{i\mu}\sigma^{z}_{i\nu} - \right. \end{aligned}$$

$$- \frac{1}{2}(n_{i\mu\uparrow} + n_{i\mu\downarrow})(n_{i\nu\uparrow} + n_{i\nu\downarrow})\Big]$$

$$= -\frac{1}{4}\sum_{i\mu\nu}^{\mu\neq\nu} J_{\mu\nu}\, n_{i\mu}n_{i\nu} - \frac{4}{\hbar^2}\sum_{i\mu\nu}^{\mu\neq\nu} J_{\mu\nu}\mathbf{s}_{i\mu}\cdot\mathbf{s}_{i\nu}$$

We recognize

$$\widehat{H}_1(\mu\neq\nu) = \widehat{H}_{1a}(\mu\neq\nu) + \widehat{H}_{1b}(\mu\neq\nu) = H_d + H_{ex}$$

Thus the proposition is proved.

2. As an example we consider EuO. This has 5d-conduction bands which are empty and seven half-filled 4f-bands (levels). For the empty (!) conduction bands H_U does not play any role but for sufficiently large $U_{\mu\mu}$, it splits each of the f-bands into two sub-bands out of which the lower one is occupied and the upper one is empty. This leads to the localized magnetic 4f-moment.
H_d provides only an unimportant energy shift of the f-levels. H_{ex} becomes important if μ is an index of the conduction band and ν is that of an f-band or the converse. Let us assume that the coupling between the conduction band and the f-level is same for all the conduction bands,

$$J_\mu = \frac{4}{\hbar}J_{\mu\nu} \quad \forall\nu \ ,$$

and define as localized spin:

$$\mathbf{S}_{if} = \sum_{\nu}\mathbf{s}_{i\nu}$$

where the ν-summation runs exclusively over the f-levels, then, H_{ex} becomes the interaction operator of the multi-band Kondo lattice model:

$$H_{df} = -\sum_{i\mu} J_\mu\,\mathbf{S}_{if}\cdot\sigma_{i\mu}$$

The μ-summation runs only over the conduction bands.

Problem 5.9

1. We use the matrix representation with the basis

$$|\,\mathbf{k}\sigma\rangle_\alpha = c_{\mathbf{k}\sigma\alpha}\,|\,0\rangle \qquad \alpha = A, B$$

Then the "free" matrix reads as

$$H_{\mathbf{k}\sigma}^{(0)} \equiv \sum_{\alpha\beta}\varepsilon_{\alpha\beta}(\mathbf{k})c_{\mathbf{k}\sigma\alpha}^{\dagger}c_{\mathbf{k}\sigma\beta}$$

with the elements

$$\left(H_{\mathbf{k}\sigma}^{(0)}\right)^{\gamma\delta} = \langle 0 \mid c_{\mathbf{k}\sigma\gamma} \sum_{\alpha\beta} \varepsilon_{\alpha\beta}(\mathbf{k}) c_{\mathbf{k}\sigma\alpha}^{\dagger} c_{\mathbf{k}\sigma\beta} c_{\mathbf{k}\sigma\delta}^{\dagger} \mid 0\rangle$$

$$= \sum_{\alpha\beta} \varepsilon_{\alpha\beta}(\mathbf{k}) \delta_{\gamma\alpha} \delta_{\beta\delta} \langle 0 \mid 0\rangle = \varepsilon_{\gamma\delta}(\mathbf{k})$$

Hamilton Matrix:

$$H_{\mathbf{k}\sigma}^{(0)} = \begin{pmatrix} \varepsilon(\mathbf{k}) & t(\mathbf{k}) \\ t^*(\mathbf{k}) & \varepsilon(\mathbf{k}) \end{pmatrix}$$

Eigenenergies:

$$\det\left(E_{0\mathbf{k}} - H_{\mathbf{k}\sigma}^{(0)}\right) \stackrel{!}{=} 0$$

$$\curvearrowright \ (\varepsilon(\mathbf{k}) - E_{0\mathbf{k}})^2 - |t(\mathbf{k})|^2 \stackrel{!}{=} 0$$

$$\curvearrowright \ E_{0\mathbf{k}}^{(\pm)} = \varepsilon(\mathbf{k}) \pm |t(\mathbf{k})|$$

Eigenstates:

$$\begin{pmatrix} \mp|t(\mathbf{k})| & t(\mathbf{k}) \\ t^*(\mathbf{k}) & \mp|t(\mathbf{k})| \end{pmatrix} \begin{pmatrix} c_A \\ c_B \end{pmatrix} = 0$$

$$\curvearrowright \quad c_A^{(\pm)} = \pm\gamma c_B^{(\pm)} \ ; \ \gamma = \frac{t(\mathbf{k})}{|t(\mathbf{k})|}$$

Normalization $\curvearrowright$

$$|E_{0\mathbf{k}\sigma}^{(\pm)}\rangle = \frac{1}{\sqrt{2}}\left(\gamma c_{\mathbf{k}\sigma A}^{\dagger} \pm c_{\mathbf{k}\sigma B}^{\dagger}\right)|0\rangle$$

2. Schrödinger's first-order perturbation theory:

$$\langle E_{0\mathbf{k}\sigma}^{(\pm)}|H_1|E_{0\mathbf{k}\sigma}^{(\pm)}\rangle$$
$$= \frac{1}{2}\left(-\frac{1}{2}Jz_\sigma\right)\sum_{\alpha}\langle S_\alpha^z\rangle\langle 0|\left(\gamma^* c_{\mathbf{k}\sigma A} \pm c_{\mathbf{k}\sigma B}\right) c_{\mathbf{k}\sigma\alpha}^{\dagger} c_{\mathbf{k}\sigma\alpha} \, *$$
$$* \ \left(\gamma c_{\mathbf{k}\sigma A}^{\dagger} \pm c_{\mathbf{k}\sigma B}^{\dagger}\right)|0\rangle$$
$$= -\frac{1}{4}Jz_\sigma\langle S^z\rangle\left(|\gamma|^2\langle 0|c_{\mathbf{k}\sigma A}c_{\mathbf{k}\sigma A}^{\dagger}c_{\mathbf{k}\sigma A}c_{\mathbf{k}\sigma A}^{\dagger}|0\rangle -\right.$$
$$\left. - \ \langle 0|c_{\mathbf{k}\sigma B}c_{\mathbf{k}\sigma B}^{\dagger}c_{\mathbf{k}\sigma B}c_{\mathbf{k}\sigma B}^{\dagger}|0\rangle\right)$$
$$= -\frac{1}{4}Jz_\sigma\langle S^z\rangle(|\gamma|^2 - 1)\langle 0|0\rangle$$

The first order therefore does not contribute

$$E^{(\pm)}_{1\mathbf{k}\sigma} \equiv 0$$

The energy correction in second order:
We need the following matrix element:

$$\begin{aligned}
&\langle E^{(+)}_{0\mathbf{k}\sigma}|H_1|E^{(-)}_{0\mathbf{k}\sigma}\rangle \\
&= \frac{1}{2}\left(-\frac{1}{2}Jz_\sigma\right)\sum_\alpha \langle S^z_\alpha\rangle\langle 0|\left(\gamma^* c_{\mathbf{k}\sigma A} + c_{\mathbf{k}\sigma B}\right) c^\dagger_{\mathbf{k}\sigma\alpha}c_{\mathbf{k}\sigma\alpha} \, * \\
&* \left(\gamma c^\dagger_{\mathbf{k}\sigma A} - c^\dagger_{\mathbf{k}\sigma B}\right)|0\rangle \\
&= -\frac{1}{4}Jz_\sigma\left(\langle S^z_A\rangle|\gamma|^2\langle 0|c_{\mathbf{k}\sigma A}c^\dagger_{\mathbf{k}\sigma A}c_{\mathbf{k}\sigma A}c^\dagger_{\mathbf{k}\sigma A}|0\rangle + \right. \\
&\left. + \langle S^z_B\rangle(-1)\langle 0|c_{\mathbf{k}\sigma B}c^\dagger_{\mathbf{k}\sigma B}c_{\mathbf{k}\sigma B}c^\dagger_{\mathbf{k}\sigma B}|0\rangle\right) \\
&= -\frac{1}{4}Jz_\sigma\langle S^z\rangle(|\gamma|^2+1)\langle 0|0\rangle \\
&= -\frac{1}{2}Jz_\sigma\langle S^z\rangle
\end{aligned}$$

This gives as the energy correction in the second order:

$$E^{(\pm)}_{2\mathbf{k}\sigma} = \frac{\left|\langle E^{(\pm)}_{0\mathbf{k}\sigma}|H_1|E^{(\mp)}_{0\mathbf{k}\sigma}\rangle\right|^2}{E^{(\pm)}_{0\mathbf{k}} - E^{(\mp)}_{0\mathbf{k}}} = \frac{\frac{1}{4}J^2\langle S^z\rangle^2}{\pm 2|t(\mathbf{k})|}$$

Thus the Schrödinger's second-order perturbation theory leads to the following spin-independent expression:

$$E^{(\pm)}_{\mathbf{k}} = \varepsilon(\mathbf{k}) \pm |t(\mathbf{k})| \pm \frac{1}{8}J^2\frac{\langle S^z\rangle^2}{|t(\mathbf{k})|}$$

3. Brillouin–Wigner perturbation theory:
The energy correction in the first order is identical to the Schrödinger's perturbation theory, i.e. it vanishes in this case also. In the second order, one has to calculate

$$E^{(\pm)}_{2\mathbf{k}\sigma} = \frac{\left|\langle E^{(\pm)}_{0\mathbf{k}\sigma}|H_1|E^{(\mp)}_{0\mathbf{k}\sigma}\rangle\right|^2}{E^{(\pm)}_{\mathbf{k}} - E^{(\mp)}_{0\mathbf{k}}}$$

It should be noted that in the denominator the "full" eigenenergy $E^{(\pm)}_{\mathbf{k}}$ appears. We have already used the matrix element in part 3 We then have the following determining equation:

$$E_{\mathbf{k}}^{(\pm)} = \varepsilon(\mathbf{k}) \pm |t(\mathbf{k})| + \frac{\frac{1}{4}J^2\langle S^z\rangle^2}{E_{\mathbf{k}}^{(\pm)} - \varepsilon(\mathbf{k}) \pm |t(\mathbf{k})|}$$

This gives a quadratic equation for the energy,

$$\left(E_{\mathbf{k}}^{(\pm)} - \varepsilon(\mathbf{k})\right)^2 - |t(\mathbf{k})|^2 = \frac{1}{4}J^2\langle S^z\rangle^2$$

with the solution:

$$E_{\mathbf{k}}^{(\pm)} = \varepsilon(\mathbf{k}) \pm \sqrt{|t(\mathbf{k})|^2 + \frac{1}{4}J^2\langle S^z\rangle^2}$$

4. The problem can be easily exactly solved:

$$H = \sum_{\mathbf{k}\sigma\alpha\beta}\left(\varepsilon_{\alpha\beta}(\mathbf{k}) - \frac{1}{2}Jz_\sigma\langle S^z_\alpha\rangle\delta_{\alpha\beta}\right)c^\dagger_{\mathbf{k}\sigma\alpha}c_{\mathbf{k}\sigma\beta} \equiv \sum_{\mathbf{k}\sigma} H_{\mathbf{k}\sigma}$$

Matrix representation:

$$H_{\mathbf{k}\sigma} = \begin{pmatrix} \varepsilon(\mathbf{k}) - \frac{1}{2}Jz_\sigma\langle S^z\rangle & t(\mathbf{k}) \\ t(\mathbf{k})^* & \varepsilon(\mathbf{k}) + \frac{1}{2}Jz_\sigma\langle S^z\rangle \end{pmatrix}$$

The secular determinant

$$\det(E - H_{\mathbf{k}\sigma}) \overset{!}{=} 0$$

is solved by

$$(E - \varepsilon(\mathbf{k}))^2 - \frac{1}{4}J^2\langle S^z\rangle^2 = |t(\mathbf{k})|^2$$

That means

$$E_\pm(\mathbf{k}) = \varepsilon(\mathbf{k}) \pm \sqrt{|t(\mathbf{k})|^2 + \frac{1}{4}J^2\langle S^z\rangle^2}$$

The Brillouin–Wigner perturbation theory gives the exact result already in the second order. If the root is expanded for small J, then the coefficient of the first term in the expansion is the result of Schrödinger' perturbation theory in the second order.

Problem 5.10

1. The total spin $\widehat{\mathbf{S}} = \mathbf{S}_1 + \mathbf{S}_2$ has the quantum numbers $\widehat{S} = 0, 1, 2, \cdots, 2S$ and the z-component $\widehat{S}^z = S_1^z + S_2^z$ the quantum numbers $M = -\widehat{S}, -\widehat{S}+1, \cdots, +\widehat{S}$. Additionally holds

$$(\mathbf{S}_1+\mathbf{S}_2)^2 = \mathbf{S}_1^2+\mathbf{S}_2^2+2\mathbf{S}_1\cdot\mathbf{S}_2 \equiv \hbar^2\widehat{S}(\widehat{S}+1)\mathbb{1}$$

$$\curvearrowright\ \mathbf{S}_1\cdot\mathbf{S}_2 = \hbar^2\left(\frac{1}{2}\widehat{S}(\widehat{S}+1)-S(S+1)\right)\mathbb{1}$$

Possible energy levels:

$$E_{\widehat{S}M} = -2J\hbar^2\left(\frac{1}{2}\widehat{S}(\widehat{S}+1)-S(S+1)\right)-Mg\mu_B\widehat{B}$$

With the abbreviation,

$$b \equiv g\mu_B\widehat{B}$$

we have for the canonical partition function:

$$\begin{aligned} Z_S &= Sp\left(e^{-\beta H}\right) \\ &= e^{-\beta\hbar^2 JS(S+1)}\sum_{\widehat{S}=0}^{2S}\sum_{M=-\widehat{S}}^{+\widehat{S}} e^{\beta\frac{1}{2}\hbar^2 J\widehat{S}(\widehat{S}+1)}\,e^{\beta bM} \end{aligned}$$

M-summation as in (4.94):

$$\sum_{M=-\widehat{S}}^{+\widehat{S}} e^{\beta bM} = \frac{\sinh(\beta b(\widehat{S}+\frac{1}{2}))}{\sinh(\frac{1}{2}\beta b)}$$

Partition function:

$$Z_S = \frac{e^{-\beta\hbar^2 JS(S+1)}}{\sinh(\frac{1}{2}\beta b)}\sum_{\widehat{S}=0}^{2S} e^{\beta\frac{1}{2}\hbar^2 J\widehat{S}(\widehat{S}+1)}\sinh(\beta b(\widehat{S}+\frac{1}{2}))$$

2. Special case $S_1 = S_2 = \frac{1}{2}$:

$$Z_{1/2} = \frac{e^{-\frac{3}{4}\beta\hbar^2 J}}{\sinh(\frac{1}{2}\beta b)}\left(\sinh(\frac{1}{2}\beta b)+e^{\beta\hbar^2 J}\sinh(\frac{3}{2}\beta b)\right)$$

With

$$\frac{\sinh(\frac{3}{2}x)}{\sinh(\frac{1}{2}x)} = 1+2\cosh(x)$$

we finally get

$$Z_{1/2} = e^{\frac{1}{4}\beta\hbar^2 J}\left(e^{-\beta\hbar^2 J} + 1 + 2\cosh(\beta g\mu_B \widehat{B})\right)$$

The magnetization satisfies the following implicit equation:

$$\begin{aligned} M &= nk_B T \frac{\partial}{\partial B_0} \ln Z_{1/2} \\ &= ng\mu_B \frac{2\sinh(\beta g\mu_B \widehat{B})}{e^{-\beta\hbar^2 J} + 1 + 2\cosh(\beta g\mu_B \widehat{B})} \end{aligned}$$

For $J = 0$, and with $\tanh x/2 = \frac{\sinh x}{1+\cosh x}$ we get the result of the Weiss theory (5.8) for $S = 1/2$.

Problem 6.1
With the abbreviations

$$j = \beta J \quad ; \quad b = \beta\mu_B B_0 \quad ; \quad \beta = \frac{1}{k_B T}$$

holds

$$\widehat{T} - E\mathbb{1} = \begin{pmatrix} e^{j+b} - E & e^{-j} \\ e^{-j} & e^{j-b} - E \end{pmatrix}$$

The eigenvalues follow from

$$\begin{aligned} 0 \stackrel{!}{=} \det(\widehat{T} - E\mathbb{1}) &= \left(e^{j+b} - E\right)\left(e^{j-b} - E\right) - e^{-2j} \\ &= E^2 - E\left(e^{j+b} + e^{j-b}\right) + e^{2j} - e^{-2j} \\ &= E^2 - E\, e^j\, 2\cosh b + 2\sinh 2j \\ &= \left(E - e^j\, \cosh b\right)^2 - e^{2j}\, \cosh^2 b + 2\sinh 2j \end{aligned}$$

This gives the eigenvalues

$$E_\pm = e^{\beta J}\left(\cosh \beta\mu_B B_0 \pm \sqrt{\cosh^2 \beta\mu_B B_0 - 2e^{-2\beta J}\sinh 2\beta J}\right)$$

Problem 6.2
Partition function (6.35) in thermodynamic limit:

$$Z_N(T, B_0) = E_+^N\left(1 + \left(\frac{E_-}{E_+}\right)^N\right) \longrightarrow E_+^N \quad \text{f'ur } N \to \infty$$

E_+ as in Problem 6.1.

Free energy per spin:

$$\begin{aligned} f(T, B_0) &= -k_B T \lim_{N\to\infty} \frac{1}{N} \ln T_N(T, B_0) = -k_B T \ln E_+ = \\ &= -J - k_B T \ln(\cosh \beta\mu_B B_0 \pm sqrt) \end{aligned}$$

where

$$sqrt = \sqrt{\cosh^2 \beta\mu_B B_0 - 2e^{-2\beta J} \sinh 2\beta J}$$

Magnetization:

$$\begin{aligned} m &= -\left(\frac{\partial}{\partial B_0} f(T, B_0)\right)_T \\ &= \mu_B \frac{\sinh b + \frac{\cosh b\ \sinh b}{sqrt}}{\cosh b + sqrt} \\ &= \mu_B \sinh b \frac{sqrt + \cosh b}{(\cosh b + sqrt)\, sqrt} \\ &= \mu_B \frac{\sinh b}{\sqrt{\cosh^2 b - 2e^{-2\beta J} \sinh 2\beta J}} \end{aligned}$$

That gives

$$m(T, B_0) = \mu_B \frac{\sinh(\beta\mu_B B_0)}{\sqrt{\cosh^2 \beta\mu_B B_0 - 2e^{-2\beta J} \sinh 2\beta J}}$$

Susceptibility:

$$\begin{aligned} \chi_T(T, B_0) &= \mu_0 \left(\frac{\partial m}{\partial B_0}\right)_T \\ &= \beta\mu_B^2 \left\{ \frac{\cosh b}{\sqrt{\cosh^2 b - 2e^{-2\beta J} \sinh 2\beta J}} \right. \\ &\qquad \left. - \frac{\sinh^2 b \cosh b}{\left(\sqrt{\cosh^2 b - 2e^{-2\beta J} \sinh 2\beta J}\right)^3} \right\} \end{aligned}$$

$$= \beta\mu_B^2 \frac{\cosh b}{\sqrt{\cosh^2 b - 2e^{-2\beta J}\sinh 2\beta J}} *$$

$$* \left\{ 1 - \frac{\sinh^2 b}{\cosh^2 b - 2e^{-2\beta J}\sinh 2\beta J} \right\}$$

$$= \beta\mu_B^2 \frac{\cosh(\beta\mu_B B_0)\left(1 - 2e^{-2\beta J}\sinh 2\beta J\right)}{\left(\cosh^2(\beta\mu_B B_0) - 2e^{-2\beta J}\sinh 2\beta J\right)^{3/2}}$$

In the limit $B_0 \to 0$ this expression simplifies to

$$\chi_T(T, B_0 \to 0) = \beta\,\mu_B^2\, e^{2\beta J}$$

Problem 6.3
In the classical Ising Model the magnetic moment is given by

$$m = \mu \sum_i S_i$$

where μ is a positive constant. With the Hamiltonian function

$$H = -J \sum_{ij} S_i S_j - m B_0$$

the canonical partition function is given by

$$Z(T, m) = \sum_{\{S_i\}} \exp\left(-\beta\left(-J \sum_{ij} S_i S_j - m B_0\right)\right)$$

Here the summation is over *all* conceivable spin configurations, where the individual spins can have the values ± 1. Therefore the substitution $S_i \to -S_i\ \forall i$ cannot affect the partition function. Then in the exponents the first term does not change the sign but for the second term however, $m \to -m$. That means

$$Z(T, m) = Z(T, -m)$$

so that

$$F(T, m) = -k_B T \ln Z(T, m) = -k_B T \ln Z(T, -m) = F(T, -m)$$

Problem 6.4

1.

$$
\begin{aligned}
& Z_N(T, B_0) = Tr\,(\exp(-\beta H)) \\
&= Tr(\mathbb{1}) - \beta Tr(H) + \frac{1}{2}\beta^2 Tr(H^2) - \frac{1}{3!}\beta^3 Sp(H^3) + \cdots \\
&= \sum_{l=0}^{\infty} \frac{1}{l!} (-\beta)^l \, Tr(H^l) \\
&= Tr(\mathbb{1}) \left[1 + \sum_{l=1}^{\infty} \frac{(-\beta)^l}{l!} m_l \right]
\end{aligned}
$$

Each spin has two possible orientations $S_i = \pm 1$. That gives a total 2^N spin configurations. Therefore

$$
Sp(\mathbb{1}) = 2^N
$$

2.

$$
\begin{aligned}
c_{B_0} &= -T \left(\frac{\partial^2 F_N(T, B_0)}{\partial T^2} \right)_{B_0} \\
&= -T \left(\frac{\partial^2}{\partial T^2} \left(-k_B T \ln Z_N(T, B_0) \right) \right)_{B_0} \\
&= k_B \beta^2 \left(\frac{\partial^2}{\partial \beta^2} \ln Z_N(T, B_0) \right)_{B_0} \\
&= k_B \beta^2 \left(\frac{\partial}{\partial \beta} \frac{\frac{\partial}{\partial \beta} Z_N(T, B_0)}{Z_N(T, B_0)} \right)_{B_0} \\
&= k_B \beta^2 \left(\frac{1}{Z_N} \frac{\partial^2 Z_N}{\partial \beta^2} - \frac{1}{Z_N^2} \left(\frac{\partial Z_N}{\partial \beta} \right)^2 \right) \\
&= k_B \beta^2 \left[\frac{Tr(\mathbb{1})}{Z_N} \sum_{l=1}^{\infty} \frac{l(l-1)}{l!} (-\beta)^{l-2} m_l \right. \\
&\qquad \left. - \left(\frac{Tr(\mathbb{1})}{Z_N} \sum_{l=1}^{\infty} \frac{l}{l!} (-\beta)^{l-1} m_l \right)^2 \right] \\
&= k_B \beta^2 \left(\frac{2}{2!} m_2 - m_1^2 + \mathcal{O}(\beta) \right) \\
&= \frac{1}{k_B T^2} (m_2 - m_1^2) + \cdots
\end{aligned}
$$

In the last step we have restricted ourselves to the lowest term in $1/T$. One should notice that the moments are temperature independent. This result for the high-temperature behaviour of the heat capacity is of course valid not only for the Ising spins but also for all magnetic systems (!).

Problem 6.5

1. According to Problem 1.2:

$$\chi_T = \frac{1}{k_B T}\frac{\mu_0}{V}\langle(\widehat{m} - \langle\widehat{m}\rangle)^2\rangle$$

This means

$$\chi_T = \frac{1}{k_B T}\frac{\mu_0}{V}g^2\mu_B^2\sum_{ij}\langle(S_i - \langle S_i\rangle)\left(S_j - \langle S_j\rangle\right)\rangle$$

So that we directly get

$$\chi_T = \frac{1}{k_B T}\frac{\mu_0}{V}\left(g^2\mu_B^2\sum_{ij}\langle S_i S_j\rangle - \langle\widehat{m}\rangle^2\right)$$

2. The spin chain shows no spontaneous magnetization. When an external field is switched off, then, $\langle\widehat{m}\rangle \equiv 0$. According to (6.19) for the spin correlation of a one-dimensional chain we have

$$\langle S_i S_j\rangle = v^{|i-j|}$$

One can easily see that, in the double summation, N terms with $|i-j| = 0$ give the contribution $v^0 = 1$; $2(N-1)$ terms with $|i-j| = 1$ give the contribution v^1; $2(N-2)$ terms with $|i-j| = 2$ give the contribution v^2; ....; and finally two terms with $|i-j| = N-1$ give the contribution v^{N-1}. Then all of them together give

$$\chi_T(T, B_0) = \frac{1}{k_B T}\frac{\mu_0}{V}g^2\mu_B^2\left(N + 2\sum_{k=1}^{N-1}(N-k)v^k\right)$$

One calculates

$$2\sum_{k=1}^{N-1} Nv^k = 2N\frac{1-v^N}{1-v} - 2N$$

$$-2\sum_{k=1}^{N-1} kv^k = -2\sum_{k=0}^{N-1} kv^k = 2v\frac{d}{dv}\frac{1-v^N}{1-v}$$

$$= 2v\,\frac{(1-v)(-Nv^{N-1}) + (1-v^N)}{(1-v)^2}$$

$$\curvearrowright \quad 2\sum_{k=1}^{N-1}(N-k)v^k = 2N\frac{v}{1-v} - 2v\frac{1-v^N}{(1-v)^2}$$

From this it follows:

$$\chi_T(T, B_0 = 0) = \frac{1}{k_B T}\frac{\mu_0}{V} g^2 \mu_B^2 \left(N\left(1 + \frac{2v}{1-v}\right) - 2v\frac{1-v^N}{(1-v)^2}\right)$$

3. For $N \to \infty$ the expression for susceptibility can be simplified:

$$\frac{1}{N}\chi_T(T, B_0 = 0)$$
$$= \frac{1}{k_B T}\frac{\mu_0}{V} g^2 \mu_B^2 \,\frac{1+v}{1-v}$$
$$= \frac{1}{k_B T}\frac{\mu_0}{V} g^2 \mu_B^2 \,\frac{1+\tanh(\beta J)}{1-\tanh(\beta J)}$$
$$= \frac{1}{k_B T}\frac{\mu_0}{V} g^2 \mu_B^2 \,\frac{e^{\beta J} + e^{-\beta J} + e^{\beta J} - e^{-\beta J}}{e^{\beta J} + e^{-\beta J} - e^{\beta J} + e^{-\beta J}}$$
$$= \frac{1}{k_B T}\frac{\mu_0}{V} g^2 \mu_B^2 \, e^{2\beta J}$$

This agrees with the result (6.46) for the Ising ring in the thermodynamic limit.

Problem 6.6

1. The partition function

$$Z_N(T) = \sum_{S_1}\sum_{S_2}\cdots\sum_{S_N} \exp\left(\sum_{i=1}^{N-1} j_i S_i S_{i+1}\right)$$

where $j_i = \beta J_i$ is valid, we have already calculated with (6.14):

$$Z_N(T) = 2^N \prod_{i=1}^{N-1} \cosh j_i$$

Four-spin correlation function $i \neq j$:

$$\begin{aligned}
&\langle S_i S_{i+1} S_j S_{j+1} \rangle \\
&= \frac{1}{Z_N} \sum_{S_1} \sum_{S_2} \cdots \sum_{S_N} S_i S_{i+1} S_j S_{j+1} \exp\left(\sum_{i=1}^{N-1} j_i S_i S_{i+1} \right) \\
&= \frac{1}{Z_N} \frac{\partial^2 Z_N}{\partial j_i \, \partial j_j} \\
&= \frac{\cosh j_1 \cdots \sinh j_i \cdots \sinh j_j \cdots \cosh j_{N-1}}{\cosh j_1 \cdots \cosh_{N-1}} \\
&= \tanh j_i \tanh j_j
\end{aligned}$$

For $i = j$ the four-spin correlation function is equal to 1. Let us now set $j_i = j \ \forall i$, then we have

$$\langle S_i S_{i+1} S_j S_{j+1} \rangle = \begin{cases} 1 & \text{falls } i = j \\ \tanh^2 j & \text{falls } i \neq j \end{cases}$$

2. From Problem 6.4 we have

$$\begin{aligned}
c_{B_0} &= k_B \beta^2 \left(\frac{1}{Z_N} \frac{\partial^2 Z_N}{\partial \beta^2} - \frac{1}{Z_N^2} \left(\frac{\partial Z_N}{\partial \beta} \right)^2 \right) \\
&= k_B \beta^2 \left(\langle H^2 \rangle - \langle H \rangle^2 \right)
\end{aligned}$$

With

$$\langle H^2 \rangle = J^2 \sum_{ij} \langle S_i S_{i+1} S_j S_{j+1} \rangle$$

follows:

$$
\begin{aligned}
c_{B_0=0} &= k_B\beta^2 J^2 \sum_{i,j=1}^{N-1} \big(\langle S_i S_{i+1} S_j S_{j+1}\rangle \\
&\qquad\qquad\qquad - \langle S_i S_{i+1}\rangle\langle S_j S_{j+1}\rangle\big) \\
&= k_B\beta^2 J^2 \sum_{i,j=1}^{N-1} \left(\delta_{ij} + (1-\delta_{ij})\tanh^2 j - \tanh^2 j\right) \\
&= k_B\beta^2 J^2 \sum_{i,j=1}^{N-1} \delta_{ij}(1-\tanh^2 j) \\
c_{B_0=0} &= (N-1)k_B\beta^2 J^2 \frac{1}{\cosh^2 \beta J}
\end{aligned}
$$

Compare this result with (6.45).

Problem 7.1

1.

$$
\left[S_i^x,\, S_j^y\right]_- = i\hbar S_i^z \delta_{ij} \text{ and cyclic}
$$

$$
\begin{aligned}
\left[S_i^z,\, S_j^{\pm}\right]_- &= \left[S_i^z,\, S_j^x\right]_- \pm i\left[S_i^z,\, S_j^y\right]_- \\
&= i\hbar S_i^y \delta_{ij} \mp \hbar\left(-S_i^x\delta_{ij}\right) \\
&= \pm\hbar\left(S_i^x \pm i S_i^y\right)\delta_{ij} \\
&= \pm\hbar S_i^{\pm}\delta_{ij}
\end{aligned}
$$

2.

$$
\begin{aligned}
\left[S_i^+,\, S_j^-\right]_- &= \left[S_i^x,\, S_j^x\right]_- + \left[S_i^y,\, S_j^y\right]_- + i\left[S_i^y,\, S_j^x\right]_- - i\left[S_i^x,\, S_j^y\right]_- \\
&= i(-i\hbar S_i^z)\delta_{ij} - i(i\hbar S_i^z)\delta_{ij} \\
&= 2\hbar S_i^z \delta_{ij}
\end{aligned}
$$

3.

$$
\begin{aligned}
S_i^{\pm} S_i^{\mp} &= (S_i^x \pm i S_i^y)(S_i^x \mp i S_i^y) \\
&= (S_i^x)^2 + (S_i^y)^2 \pm i\left[S_i^y,\, S_i^x\right]_- \\
&= \mathbf{S}_i^2 - (S_i^z)^2 \pm \hbar S_i^z \\
&\text{(where } \mathbf{S}_i^2 = \hbar^2 S(S+1)\mathbb{1} \text{ in the space of the spin states)}
\end{aligned}
$$

4.

$$\frac{1}{2}\left(S_i^+S_j^- + S_i^-S_j^+\right) = \frac{1}{2}\left(S_i^xS_j^x + S_i^yS_j^y + i\,S_i^yS_j^x - i\,S_i^xS_j^y + \right.$$
$$\left. + S_i^xS_j^x + S_i^yS_j^y - i\,S_i^yS_j^x + i\,S_i^xS_j^y\right)$$
$$= S_i^xS_j^x + S_i^yS_j^y$$
$$= \mathbf{S}_i\cdot\mathbf{S}_j - S_i^zS_j^z$$
$$\curvearrowright\quad \mathbf{S}_i\cdot\mathbf{S}_j = \frac{1}{2}\left(S_i^+S_j^- + S_i^-S_j^+\right) + S_i^zS_j^z$$

5.

$$H = -\sum_{i,j} J_{ij}\mathbf{S}_i\cdot\mathbf{S}_j$$
$$\overset{4.)}{=} -\frac{1}{2}\sum_{i,j}(S_i^+S_j^- + S_i^-S_j^+)J_{ij} - \sum_{i,j} J_{ij}S_i^zS_j^z$$
$$= -\frac{1}{2}\sum_{i,j} J_{ij}S_i^+S_j^- - \frac{1}{2}\underbrace{\sum_{i,j}^{i\neq j} J_{ij}\,S_j^+S_i^-}_{i\leftrightarrow j\ (J_{ij}=J_{ji})} - \sum_{i,j} J_{ij}S_i^zS_j^z$$
$$= -\sum_{i,j} J_{ij}\left(S_i^+S_j^- + S_i^zS_j^z\right)$$

Problem 7.2

1. $\left[(S_i^-)^n, S_i^z\right]_- = n\hbar(S_i^-)^n;\ n = 1, 2, \ldots$
   Complete induction:
   $n = 1$:

$$\left[S_i^-, S_i^z\right]_- = \hbar S_i^-$$

   $n \to n+1$:

$$\left[(S_i^-)^{n+1}, S_i^z\right]_- = \left[(S_i^-)^n, S_i^z\right]_- S_i^- + (S_i^-)^n\left[S_i^-, S_i^z\right]_-$$
$$= n\hbar(S_i^-)^nS_i^- + (S_i^-)^n\hbar S_i^-$$
$$= (n+1)\hbar(S_i^-)^{n+1} \qquad \text{q.e.d.}$$

2. $\left[(S_i^-)^n, (S_i^z)^2\right]_- = n^2\hbar^2(S_i^-)^n + 2n\hbar S_i^z(S_i^-)^n;\ n = 1, 2, \ldots$

$$\left[(S_i^-)^n, (S_i^z)^2\right]_- = S_i^z\left[(S_i^-)^n, S_i^z\right]_- + \left[(S_i^-)^n, S_i^z\right]_- S_i^z$$
$$\overset{1.)}{=} S_i^z n\hbar(S_i^-)^n + n\hbar(S_i^-)^nS_i^z$$
$$\overset{1.)}{=} 2n\hbar S_i^z(S_i^-)^n + n^2\hbar^2(S_i^-)^n \qquad \text{q.e.d.}$$

3. $\left[S_i^+, (S_i^-)^n\right]_- = (2n\hbar S_i^z + \hbar^2 n(n-1))(S_i^-)^{n-1};\ n = 1, 2, \dots$
Complete induction:
$n = 1$:

$$\left[S_i^+, S_i^-\right]_- = 2\hbar S_i^z$$

$n \to n+1$:

$$\begin{aligned}
&\left[S_i^+, (S_i^-)^{n+1}\right]_- \\
&= \left[S_i^+, (S_i^-)^n\right]_- S_i^- + (S_i^-)^n \left[S_i^+, S_i^-\right]_- \\
&= (2n\hbar S_i^z + \hbar^2 n(n-1))(S_i^-)^{n-1} S_i^- + (S_i^-)^n 2\hbar S_i^z \\
&\overset{1.)}{=} (2n\hbar S_i^z + \hbar^2 n(n-1))(S_i^-)^n + 2n\hbar^2 (S_i^-)^n + \\
&\qquad\qquad +2\hbar S_i^z (S_i^-)^n \\
&= (2\hbar(n+1) S_i^z + \hbar^2 n(n+1))(S_i^-)^n \qquad \text{q.e.d.}
\end{aligned}$$

Problem 7.3

Starting point is the identity (7.485):

$$\prod_{m_S=-S}^{+S} \left(S_i^z - \hbar m_s\right) = 0$$

$S = \frac{1}{2}$

$$\begin{aligned}
0 &= \left(S_i^z + \frac{\hbar}{2}\right)\left(S_i^z - \frac{\hbar}{2}\right) = \left(S_i^z\right)^2 - \frac{\hbar^2}{4} \\
&\curvearrowright \langle \left(S_i^z\right)^2 \rangle = \frac{\hbar^2}{4} \\
&\curvearrowright \alpha_0\left(\frac{1}{2}\right) = \frac{\hbar^2}{4}, \quad \alpha_1\left(\frac{1}{2}\right) = 0
\end{aligned}$$

$S = 1$

$$\begin{aligned}
0 &= \left(S_i^z + \hbar\right) S_i^z \left(S_i^z - \hbar\right) \\
&\curvearrowright \langle \left(S_i^z\right)^3 \rangle = \hbar^2 \langle S_i^z \rangle \\
&\curvearrowright \alpha_0(1) = 0, \quad \alpha_1(1) = \hbar^2, \quad \alpha_2(1) = 0
\end{aligned}$$

$S = \frac{3}{2}$

$$
\begin{aligned}
0 &= \left(S_i^z + \frac{3}{2}\hbar\right)\left(S_i^z + \frac{1}{2}\hbar\right)\left(S_i^z - \frac{1}{2}\hbar\right)\left(S_i^z - \frac{3}{2}\hbar\right) \\
&= \left(\left(S_i^z\right)^2 - \frac{9}{4}\hbar^2\right)\left(\left(S_i^z\right)^2 - \frac{1}{4}\hbar^2\right) \\
\curvearrowright\ & \langle\left(S_i^z\right)^4\rangle = \frac{5}{2}\hbar^2\langle\left(S_i^z\right)^2\rangle - \frac{9}{16}\hbar^4 \\
\curvearrowright\ & \alpha_0\left(\frac{3}{2}\right) = -\frac{9}{16}\hbar^4, \quad \alpha_1\left(\frac{3}{2}\right) = 0, \quad \alpha_2\left(\frac{3}{2}\right) = \frac{5}{2}\hbar^2, \\
& \alpha_3\left(\frac{3}{2}\right) = 0
\end{aligned}
$$

$S = 2$

$$
\begin{aligned}
0 &= \left(S_i^z + 2\hbar\right)\left(S_i^z + \hbar\right) S_i^z \left(S_i^z - \hbar\right)\left(S_i^z - 2\hbar\right) \\
&= \left(\left(S_i^z\right)^2 - 4\hbar^2\right)\left(\left(S_i^z\right)^2 - \hbar^2\right) S_i^z \\
\curvearrowright\ & \langle\left(S_i^z\right)^5\rangle = 5\hbar^2\langle\left(S_i^z\right)^3\rangle - 4\hbar^4\langle S_i^z\rangle \\
\curvearrowright\ & \alpha_0(2) = 0, \quad \alpha_1(2) = -4\hbar^4, \quad \alpha_2(2) = 0, \\
& \alpha_3(2) = 5\hbar^2, \quad \alpha_4(2) = 0
\end{aligned}
$$

$S = \frac{7}{2}$

$$
\begin{aligned}
0 &= \left(S_i^z + \frac{7}{2}\hbar\right)\left(S_i^z + \frac{5}{2}\hbar\right)\left(S_i^z + \frac{3}{2}\hbar\right)\left(S_i^z + \frac{1}{2}\hbar\right) \cdot \\
&\quad \cdot \left(S_i^z - \frac{1}{2}\hbar\right)\left(S_i^z - \frac{3}{2}\hbar\right)\left(S_i^z - \frac{5}{2}\hbar\right)\left(S_i^z - \frac{7}{2}\hbar\right) \\
&= \left(\left(S_i^z\right)^2 - \frac{49}{4}\hbar^2\right)\left(\left(S_i^z\right)^2 - \frac{25}{4}\hbar^2\right)\left(\left(S_i^z\right)^2 - \frac{9}{4}\hbar^2\right) * \\
&\quad * \left(\left(S_i^z\right)^2 - \frac{1}{4}\hbar^2\right) \\
\curvearrowright\ & \langle\left(S_i^z\right)^8\rangle = 21\hbar^2\langle\left(S_i^z\right)^6\rangle - \frac{987}{8}\hbar^4\langle\left(S_i^z\right)^4\rangle + \\
& + \frac{3229}{16}\hbar^6\langle\left(S_i^z\right)^2\rangle - \frac{617}{8}\hbar^8 \\
\curvearrowright\ & \alpha_0\left(\frac{7}{2}\right) = -\frac{617}{8}\hbar^8, \quad \alpha_1\left(\frac{7}{2}\right) = 0, \quad \alpha_2\left(\frac{7}{2}\right) = \frac{3229}{16}\hbar^6, \\
& \alpha_3\left(\frac{7}{2}\right) = 0, \alpha_4\left(\frac{7}{2}\right) = -\frac{987}{8}\hbar^4, \quad \alpha_5\left(\frac{7}{2}\right) = 0, \\
& \alpha_6\left(\frac{7}{2}\right) = 21\hbar^2, \quad \alpha_7\left(\frac{7}{2}\right) = 0
\end{aligned}
$$

Problem 7.4

Eu^{2+} on f.c.c.-sites

↷ each atom of a (111)-plane has six nearest neighbours in the same (111)-plane and three each in the two neighbouring planes and six next nearest neighbours, three each in the two neighbouring (111)-planes.

EuSe:

$$2.8\text{K} \le T \le 4.6\text{K} \quad \text{NNSS-Antiferromagnet}$$

↷ 12 nearest neighbours, 9 in the same and 3 in the other sub-lattice.

$$\curvearrowright \quad k_B T_N = \frac{2}{3}\hbar^2 S(S+1)\left\{\sum_j^{\in 1} J_{1j} - \sum_j^{\in 2} J_{1j}\right\}$$

$$= \frac{2}{3}\hbar^2 S(S+1)\{9J_1 + 3J_2 - 3J_1 - 3J_2\}$$

$$= \frac{2}{3}\hbar^2 S(S+1)(6J_1)$$

$$k_B\Theta = \frac{2}{3}\hbar^2 S(S+1)\left\{\sum_j^{\in 1} J_{1j} + \sum_j^{\in 2} J_{1j}\right\}$$

$$= \frac{2}{3}\hbar^2 S(S+1)\{9J_1 + 3J_2 + 3J_1 + 3J_2\}$$

$$= \frac{2}{3}\hbar^2 S(S+1)\{12J_1 + 6J_2\}$$

$$\curvearrowright \quad \boxed{J_1 = \frac{k_B}{4\hbar^2 S(S+1)} T_N}$$

$$k_B\Theta - 2k_B T_N = 4\hbar^2 S(S+1) J_2$$

$$\curvearrowright \quad \boxed{J_2 = \frac{k_B}{4\hbar^2 S(S+1)}(\Theta - 2T_N)}$$

Problem 7.5

1. It is convenient to first split the matrix as follows:

$$A = bA' + a\mathbb{1}$$

Here the reduced matrix A' is given by

$$A' = \begin{pmatrix} 0\,1\,0\,0\,0 \cdots 0\,0 \\ 1\,0\,1\,0\,0 \cdots 0\,0 \\ 0\,1\,0\,1\,0 \cdots 0\,0 \\ \cdots\cdots\cdots \\ \cdots\cdots\cdots \\ 0\,0\,0\,0\,0 \cdots 1\,0 \end{pmatrix}$$

Eigenvalues of A:

$$\lambda = a + b\lambda' \;\; ; \;\; \lambda' \; : \; \text{Eigenvalue of } A'$$

Now we have to solve

$$0 \stackrel{!}{=} \det(A' - \lambda' \mathbb{1}) \equiv D_d(\lambda')$$

One recognizes

$$D_1(\lambda') = -\lambda' \;\; ; \;\; D_2(\lambda') = \lambda'^2 - 1$$

In general one obtains by expanding after the first row:

$$\begin{aligned} D_d(\lambda') &= -\lambda' \, D_{d-1}(\lambda') - 1 \, * \\ &* \begin{vmatrix} 1 & 1 & 0 & 0\,0 \cdots 0 & 0 \\ 0 & -\lambda' & 1 & 0\,0 \cdots 0 & 0 \\ 0 & 1 & -\lambda' & 1\,0 \cdots 0 & 0 \\ & & \cdots\cdots\cdots & & \\ & & \cdots\cdots\cdots & & \\ 0 & 0 & 0 & 0\,0 \cdots 1 & -\lambda' \end{vmatrix} \\ &= -\lambda' \, D_{d-1}(\lambda') - \; D_{d-2}(\lambda') \end{aligned}$$

In the last step, the remaining determinant is expanded after the first column. With the ansatz for solution

$$D_d(\lambda') = e^{\pm i d \alpha}$$

follows:

$$\begin{aligned} e^{\pm i d \alpha} &= \lambda' e^{\pm i(d-1)\alpha} - e^{\pm i(d-2)\alpha} \\ \curvearrowright \; e^{\pm i \alpha} &= -\lambda' - e^{\mp i \alpha} \; \curvearrowright \; \lambda' = \\ &= -2\cos\alpha \; \curvearrowright \; \alpha = \arccos(-\frac{\lambda'}{2}) \end{aligned}$$

General solution

$$D_d(\lambda') = c_1 \cos(d\alpha) + c_2 \sin(d\alpha)$$

c_1,c_2 from the initial conditions:

$$\begin{aligned} D_1(\lambda') &= -\lambda' = c_1 \cos(\alpha) + c_2 \sin(\alpha) = \\ &= 2\cos(\alpha) \\ D_2(\lambda') &= \lambda'^2 - 1 = c_1 \cos(2\alpha) + c_2 \sin(2\alpha) = \\ &= 4\cos^2(\alpha) - 1 \end{aligned}$$

$$c_1 = 2 - c_2 \tan\alpha$$

$$\begin{aligned} \curvearrowright\ 4\cos^2\alpha - 1 &= c_1(2\cos^2\alpha - 1) + c_2\, 2\sin\alpha\cos\alpha \\ &= 4\cos^2\alpha - 2 - 2c_2 \sin\alpha\cos\alpha + \\ &\quad + c_2\tan\alpha + 2c_2 \sin\alpha\cos\alpha \end{aligned}$$

$$\curvearrowright\ c_2 = \cot\alpha \quad \curvearrowright\ c_1 = 1$$

Intermediate result:

$$\begin{aligned} D_d(\lambda') &= \cos(d\alpha) + \frac{\cos\alpha}{\sin\alpha}\sin(d\alpha) \\ &= \frac{1}{\sin\alpha}(\sin\alpha\cos(d\alpha) + \cos\alpha\sin(d\alpha)) \\ &= \frac{\sin((d+1)\alpha)}{\sin\alpha} \end{aligned}$$

Requirement:

$$D_d(\lambda') \stackrel{!}{=} 0 \quad \curvearrowright\ \alpha = \frac{r\pi}{d+1}\ ;\ r = 1, \ldots, d$$

$$\curvearrowright\ \lambda'_r = -2\cos\frac{r\pi}{d+1}$$

With this we have the *Eigenvalues of the tridiagonal matrix A*:

$$\lambda_r = a - 2b\cos\frac{r\pi}{d+1}\quad ,\ r = 1, \ldots, d$$

2. Heisenberg model for films:
Hamiltonian in molecular field approximation (7.124):

$$H_{MFA} = -2\sum_{i,j} J_{ij}\langle S_j^z\rangle S_i^z$$

That means we have an *effective paramagnet* in the molecular field $2\sum_{ij} J_{ij}\langle S_j^z\rangle/g_J\mu_B$. Due to the film structure, translational symmetry is applicable only in the plane of the film and not in three dimensions. Let B_S be the Brillouin function in the following. Then according to (7.134) holds

$$\langle S_i^z\rangle = \hbar\, S\, B_S\left(2\hbar\beta S\sum_j J_{ij}\langle S_j^z\rangle\right)$$

Exchange only between the nearest neighbours:

$$\sum_j J_{ij}\langle S_j^z\rangle = qJ\langle S_\alpha^z\rangle + pJ\langle S_{\alpha+1}^z\rangle + pJ\langle S_{\alpha-1}^z\rangle$$

$\langle S_\alpha^z\rangle$: layer magnetization, where α numbers the layers; $q(p)$: coordination number (the number of nearest neighbours) within the layer (between the layers). $q+2p=z$: volume coordination number. Examples:

$$sc\ (100):\ q=4,\ p=1$$
$$sc\ (110):\ q=2,\ p=2$$
$$sc\ (111):\ q=0,\ p=3$$

For $T\to T_C$ linearization of Brillouin function: $B_S(x)\approx\frac{S+1}{3S}x$:

$$\langle S_\alpha^z\rangle = \frac{2}{3}\hbar^2(S+1)S\beta\left(qJ\langle S_\alpha^z\rangle + pJ\langle S_{\alpha+1}^z\rangle + pJ\langle S_{\alpha-1}^z\rangle\right)$$

This gives a system of homogeneous equations:

$$0=\sum_\gamma M_{\alpha\gamma}\langle S_\gamma^z\rangle$$

Here the matrix $\widehat{M}$ is given by

$$\widehat{M}\equiv\begin{pmatrix} qJ-x & pJ & 0 & 0 & 0 & \cdots & 0\\ pJ & qJ-x & pJ & 0 & 0 & \cdots & 0\\ 0 & pJ & qJ-x & pJ & 0 & \cdots & 0\\ & \cdots & \cdots & \cdots & \cdots & \cdots & \\ 0 & \cdots & & & \cdots & pJ & qJ-x \end{pmatrix}$$

$$x=\frac{1}{\frac{2}{3}\hbar^2 S(S+1)\beta}$$

We thus have a tridiagonal matrix. There exist non-trivial solutions ($\langle S_\gamma^z \rangle \overset{>}{\to} 0$) only for

$$\det \widehat{M} \overset{!}{=} 0 = \prod_{r=1}^{d} \lambda_r$$

λ_r is the rth eigenvalue of $\widehat{M}$. According to part 1, for the eigenvalues we have

$$\lambda_r = qJ - x - 2pJ\cos\frac{r\pi}{d+1}$$

At least one of these eigenvalues must be equal to zero. This gives the following equation for T_C:

$$k_B T_C^{(r)} = \frac{2}{3}\hbar^2 S(S+1)J\left\{q - 2p\cos\frac{r\pi}{d+1}\right\}$$

The *physical* solution must satisfy the well-known limiting cases

$$k_B T_C(d=1) = \frac{2}{3}\hbar^2 S(S+1)Jq$$
$$k_B T_C(d=\infty) = \frac{2}{3}\hbar^2 S(S+1)J(q+2p)$$

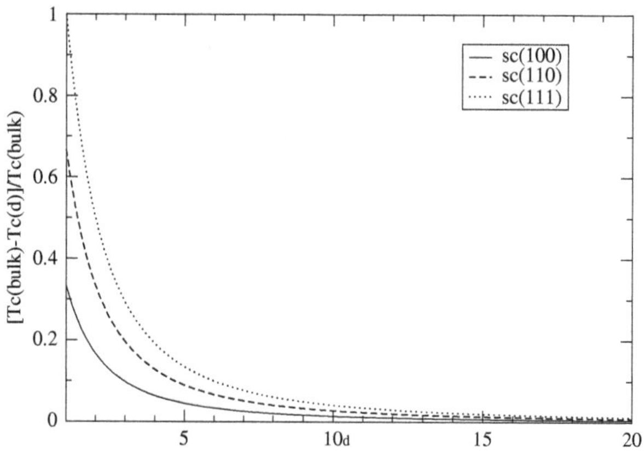

Fig. C.1 Relative change of the Curie temperature of a film with its thickness. d: number of monolayers

That is possible only for $r = d$ (Fig. C.1):

$$k_B T_C = \frac{2}{3}\hbar^2 S(S+1)J\left\{q - 2p\cos\frac{d\pi}{d+1}\right\}$$

It is instructive to compare the critical temperature of the film with that of the bulk material:

$$\frac{T_C(\infty) - T_C(d)}{T_C(\infty)} = 2\frac{z - z_S}{z}\left(1 + \cos\frac{d\pi}{d+1}\right)$$

Here $z = q + 2p$ is the coordination number of the bulk and $z_S = q + p$ that of the surface.

Problem 7.6

We use the Heisenberg Hamiltonian in the wavenumber representation (7.105):

$$H = -\frac{1}{N}\sum_{\mathbf{k}} J(\mathbf{k})\left\{S^+(\mathbf{k})S^-(-\mathbf{k}) + S^z(\mathbf{k})S^z(-\mathbf{k})\right\} - \\ - \frac{1}{\hbar}g_J\mu_B B_0 S^z(0)$$

For the spin operators (7.106) and (7.107) hold

$$S^z(\mathbf{k})\,|0\rangle = \hbar N S\,|0\rangle\,\delta_{\mathbf{k},0} \qquad ; \quad S^+(\mathbf{k})\,|0\rangle = 0$$

With this we calculate

$$\begin{aligned} -\frac{1}{N}\sum_{\mathbf{k}} J(\mathbf{k})S^+(\mathbf{k})S^-(-\mathbf{k})\,|0\rangle &= \\ &= -\frac{1}{N}\sum_{\mathbf{k}} J(\mathbf{k})\left\{2\hbar S^z(0) + S^-(-\mathbf{k})S^+(\mathbf{k})\right\}|0\rangle \\ &= (2\hbar^2 N S\,|0\rangle)\left(-\frac{1}{N}\sum_{\mathbf{k}} J(\mathbf{k})\right) \\ &= -(2\hbar^2 N S\,|0\rangle)J_{ii} \\ &\overset{(7.12)}{=} 0 \end{aligned}$$

$$\begin{aligned} -\frac{1}{N}\sum_{\mathbf{k}} J(\mathbf{k})S^z(\mathbf{k})S^z(-\mathbf{k})\,|0\rangle &= -\hbar S\sum_{\mathbf{k}} J(\mathbf{k})S^z(\mathbf{k})\delta_{-\mathbf{k},0}\,|0\rangle \\ &= -\hbar S J(0) S^z(0)\,|0\rangle \\ &= -N\hbar^2 S^2 J_0\,|0\rangle \qquad (J(0) \equiv J_0) \end{aligned}$$

$$-\frac{1}{\hbar}g_J\mu_B B_0 S^z(0)\,|0\rangle = -N g_J S \mu_B B_0\,|0\rangle$$

With this it is shown that $|0\rangle$ is the eigenstate with the eigenvalue

$$E_0(B_0) = -N\hbar^2 S^2 J_0 - N g_J \mu_B B_0 S$$

Problem 7.7

$$\begin{aligned}
\langle \mathbf{k}| S_i^z |\mathbf{k}\rangle &= \frac{1}{2SN\hbar^2} \langle 0| S^+(-\mathbf{k}) S_i^z S^-(\mathbf{k}) |0\rangle \\
&= \frac{1}{2SN^2\hbar^2} \sum_{\mathbf{q}} e^{i\mathbf{q}\cdot\mathbf{R}_i} \langle 0| S^+(-\mathbf{k}) S^z(\mathbf{q}) S^-(\mathbf{k}) |0\rangle \\
&= \frac{1}{2SN^2\hbar^2} \sum_{\mathbf{q}} e^{i\mathbf{q}\cdot\mathbf{R}_i} \langle 0| S^+(-\mathbf{k}) \big(S^-(\mathbf{k}) S^z(\mathbf{q}) - \\
&\quad -\hbar S^-(\mathbf{k}+\mathbf{q}) \big) |0\rangle \\
&= \frac{1}{2N\hbar} \langle 0| (2\hbar S^z(0) + S^-(\mathbf{k}) S^+(-\mathbf{k})) |0\rangle - \\
&\quad - \frac{1}{2SN^2\hbar} \sum_{\mathbf{q}} e^{i\mathbf{q}\cdot\mathbf{R}_i} \langle 0| (2\hbar S^z(\mathbf{q}) \\
&\quad + S^-(\mathbf{k}+\mathbf{q}) S^+(-\mathbf{k})) |0\rangle \\
&= \frac{1}{2N\hbar} 2\hbar \cdot N\hbar S - \frac{1}{2SN^2\hbar} 2\hbar \cdot N\hbar S \\
&= \hbar S - \frac{1}{N}\hbar \\
&= \hbar \left(S - \frac{1}{N} \right) \qquad \text{q.e.d.}
\end{aligned}$$

Problem 7.8
Holstein–Primakoff transformation:

$$\begin{aligned}
S_i^+ &= \hbar\sqrt{2S}\varphi(n_i) a_i \qquad & ;\varphi(n_i) = \sqrt{1 - \frac{n_i}{2S}} \\
S_i^- &= \hbar\sqrt{2S} a_i^+ \varphi(n_i) & \\
S_i^z &= \hbar(S - n_i) & ;n_i = a_i^+ a_i
\end{aligned}$$

a_i, a_i^+: Bose operators

$$\left[a_i, a_j^+\right]_- = \delta_{ij}; \qquad [a_i, a_j]_- = \left[a_i^+, a_j^+\right]_- = 0$$

$$\curvearrowright [n_i, a_j]_- = -a_i \delta_{ij}; \qquad \left[n_i, a_j^+\right]_- = a_i^+ \delta_{ij}$$

1. *Commutation relations*:

$$
\begin{aligned}
&\left[S_i^+, S_j^-\right]_- = \\
&= 2S\hbar^2\delta_{ij}\left[\varphi(n_i)a_i,\ a_i^+\varphi(n_i)\right]_- \\
&= 2S\hbar^2\delta_{ij}\left(\varphi(n_i)(1+n_i)\varphi(n_i) - a_i^+\left(1-\frac{n_i}{2S}\right)a_i\right) \\
&= 2S\hbar^2\delta_{ij}\left((1+n_i)\left(1-\frac{n_i}{2S}\right) - n_i + \frac{1}{2S}a_i^+ n_i a_i\right) \\
&= 2S\hbar^2\delta_{ij}\left(1-\frac{n_i}{2S}-\frac{n_i^2}{2S}-\frac{1}{2S}n_i+\frac{1}{2S}n_i^2\right) \\
&= 2\hbar\delta_{ij}\hbar(S-n_i) \\
&= 2\hbar\delta_{ij}S_i^z
\end{aligned}
$$

$$
\begin{aligned}
\left[S_i^z, S_j^+\right]_- &= \left[\hbar(S-n_i),\ \hbar\sqrt{2S}\varphi(n_j)a_j\right]_- \\
&= -\hbar^2\sqrt{2S}\left[n_i,\ \varphi(n_i)a_i\right]_-\delta_{ij} \\
&= -\hbar^2\sqrt{2S}\varphi(n_i)\underbrace{\left[n_i,\ a_i\right]_-}_{-a_i}\delta_{ij} \\
&= \hbar^2\sqrt{2S}\varphi(n_i)a_i\delta_{ij} \\
&= +\hbar\delta_{ij}S_i^+
\end{aligned}
$$

$$
\begin{aligned}
\left[S_i^z, S_j^-\right]_- &= -\hbar^2\sqrt{2S}\left[n_i,\ a_i^+\varphi(n_i)\right]_-\delta_{ij} \\
&= -\hbar^2\sqrt{2S}\underbrace{\left[n_i,\ a_i^+\right]_-}_{a_i^+}\varphi(n_i)\delta_{ij} \\
&= -\hbar\left(\hbar\sqrt{2S}a_i^+\varphi(n_i)\right)\delta_{ij} \\
&= -\hbar\delta_{ij}S_i^-
\end{aligned}
$$

2. S_i^2:

$$
\begin{aligned}
\mathbf{S}_i^2 &= \frac{1}{2}\left(S_i^+S_i^- + S_i^-S_i^+\right) + (S_i^z)^2 \\
&= \frac{1}{2}2S\hbar^2\left(\varphi(n_i)a_ia_i^+\varphi(n_i) + \right. \\
&\quad \left. + a_i^+\left(1-\frac{n_i}{2S}\right)a_i\right) + \hbar^2(S-n_i)^2
\end{aligned}
$$

$$
\begin{aligned}
&= \hbar^2 S\left(\left(1-\frac{n_i}{2S}\right)+n_i\left(1-\frac{n_i}{2S}\right)+n_i-\frac{1}{2S}a_i^+\underbrace{n_i a_i}_{-a_i+a_i n_i}\right)+\\
&\quad +\hbar^2(S^2-2Sn_i+n_i^2)\\
&= \hbar^2(S+2n_iS-n_i^2+S^2-2Sn_i+n_i^2)\\
&= \hbar^2 S(S+1)
\end{aligned}
$$

Problem 7.9
Dyson–Maléev transformation:

$$
\begin{aligned}
S_i^+ &= \hbar\sqrt{2S}\alpha_i\\
S_i^- &= \hbar\sqrt{2S}\alpha_i^+\left(1-\frac{n_i}{2S}\right)\\
S_i^z &= \hbar(S-n_i) \qquad\qquad ; n_i=\alpha_i^+\alpha_i
\end{aligned}
$$

α_i, α_i^+: Bose operators

1. *Commutation relations*:

$$
\begin{aligned}
\left[S_i^+, S_j^-\right]_- &= 2S\hbar^2\left[\alpha_i,\ \alpha_j^+\left(1-\frac{n_j}{2S}\right)\right]_-\\
&= 2S\hbar^2\delta_{ij}-\hbar^2\delta_{ij}\left[\alpha_i,\ \alpha_i^+ n_i\right]_-\\
&= 2S\hbar^2\delta_{ij}-\hbar^2\delta_{ij}n_i-\hbar^2\delta_{ij}\alpha_i^+\underbrace{[\alpha_i,\ n_i]_-}_{\alpha_i}\\
&= 2\hbar^2\delta_{ij}(S-n_i)\\
&= 2\hbar\delta_{ij}S_i^z
\end{aligned}
$$

$$
\begin{aligned}
\left[S_i^z, S_j^+\right]_- &= \hbar^2\sqrt{2S}\left[S-n_i,\ \alpha_j\right]_-\\
&= -\hbar^2\sqrt{2S}\delta_{ij}\underbrace{[n_i,\ \alpha_i]_-}_{-\alpha_i}\\
&= \hbar^2\sqrt{2S}\alpha_i\delta_{ij}\\
&= \hbar\delta_{ij}S_i^+
\end{aligned}
$$

$$
\begin{aligned}
\left[S_i^z, S_j^-\right]_- &= \hbar^2\sqrt{2S}\left[S-n_i,\ \alpha_j^+\left(1-\frac{n_j}{2S}\right)\right]_-\\
&= -\hbar^2\sqrt{2S}\delta_{ij}\left[n_i,\ \alpha_i^+\left(1-\frac{n_i}{2S}\right)\right]_-\\
&= -\hbar^2\sqrt{2S}\delta_{ij}\underbrace{\left[n_i,\ \alpha_i^+\right]_-}_{\alpha_i^+}\left(1-\frac{n_i}{2S}\right)
\end{aligned}
$$

$$= -\hbar^2\sqrt{2S}\delta_{ij}\alpha_i^+\left(1-\frac{n_i}{2S}\right)$$
$$= -\hbar\delta_{ij}S_i^-$$

2. S_i^2:

$$\begin{aligned}
\mathbf{S}_i^2 &= \frac{1}{2}\left(S_i^+S_i^- + S_i^-S_i^+\right) + (S_i^z)^2 \\
&= S\hbar^2\left(\alpha_i\alpha_i^+\left(1-\frac{n_i}{2S}\right) + \alpha_i^+\left(1-\frac{n_i}{2S}\right)\alpha_i\right) + \\
&\quad + \hbar^2(S^2 - 2Sn_i + n_i^2) \\
&= \hbar^2\left(S(1+n_i) - \frac{1}{2}(n_i+n_i^2) + Sn_i - \right. \\
&\qquad \left. -\frac{1}{2}\alpha_i^+n_i\alpha_i + S^2 - 2Sn_i + n_i^2\right) \\
&= \hbar^2\left(S(S+1) - \frac{1}{2}n_i + \frac{1}{2}n_i^2 + \frac{1}{2}n_i - \frac{1}{2}n_i^2\right) \\
&= \hbar^2S(S+1)
\end{aligned}$$

Problem 7.10

$$S_i^+ \approx \hbar\sqrt{2S}a_i\ ; \qquad S_i^- \approx \hbar\sqrt{2S}a_i^+\ ; \qquad S_i^z = \hbar(S-n_i)$$

Fourier transformation:

$$\begin{aligned}
S^+(\mathbf{k}) &= \sum_i e^{-i\mathbf{k}\cdot\mathbf{R_i}}S_i^+ \\
&\approx \hbar\sqrt{2S}\underbrace{\sum_i e^{-i\mathbf{k}\cdot\mathbf{R_i}}a_i}_{\sqrt{N}a_{\mathbf{k}}} \\
&= \hbar\sqrt{2SN}a_{\mathbf{k}}
\end{aligned}$$

$$\begin{aligned}
S^-(\mathbf{k}) &= \sum_i e^{-i\mathbf{k}\cdot\mathbf{R_i}}S_i^- \\
&\approx \hbar\sqrt{2S}\underbrace{\sum_i e^{-i\mathbf{k}\cdot\mathbf{R_i}}a_i^+}_{\sqrt{N}a_{-\mathbf{k}}} \\
&= \hbar\sqrt{2SN}a_{-\mathbf{k}}^+
\end{aligned}$$

$$
\begin{aligned}
S^z(\mathbf{k}) &= \sum_i e^{-i\mathbf{k}\cdot\mathbf{R_i}}(\hbar S - \hbar a_i^+ a_i) \\
&= \hbar S N \delta_{\mathbf{k},0} - \hbar \sum_i e^{-i\mathbf{k}\cdot\mathbf{R}_i} \frac{1}{N} \sum_{\mathbf{k}',\mathbf{q}} a_{\mathbf{q}}^+ a_{\mathbf{k}'} e^{i(\mathbf{k}'-\mathbf{q})\mathbf{R}_i} \\
&= \hbar S N \delta_{\mathbf{k},0} - \hbar \sum_{\mathbf{k}',\mathbf{q}} a_{\mathbf{q}}^+ a_{\mathbf{k}'} \delta_{\mathbf{k}',\mathbf{k}+\mathbf{q}} \\
&= \hbar S N \delta_{\mathbf{k},0} - \hbar \sum_{\mathbf{q}} a_{\mathbf{q}}^+ a_{\mathbf{k}+\mathbf{q}}
\end{aligned}
$$

Verification of commutation relations:

1.

$$
\begin{aligned}
\left[S^+(\mathbf{k}),\ S^-(\mathbf{q})\right]_- &= \hbar^2 2SN \left[a_{\mathbf{k}},\ a_{-\mathbf{q}}^+\right]_- \\
&= 2SN\hbar^2 \delta_{\mathbf{k}+\mathbf{q}} \\
&\approx 2\hbar S^z(\mathbf{k}+\mathbf{q})
\end{aligned}
$$

2.

$$
\begin{aligned}
\left[S^z(\mathbf{k}),\ S^+(\mathbf{q})\right]_- &= -\hbar \sum_{\mathbf{q}'} \left[a_{\mathbf{q}'}^+ a_{\mathbf{k}+\mathbf{q}'},\ a_{\mathbf{q}}\right]_- \hbar\sqrt{2SN} \\
&= -\hbar^2\sqrt{2SN} \sum_{\mathbf{q}'} \underbrace{\left[a_{\mathbf{q}'}^+,\ a_{\mathbf{q}}\right]_-}_{-\delta_{\mathbf{q}\mathbf{q}'}} a_{\mathbf{k}+\mathbf{q}'} \\
&= \hbar^2\sqrt{2SN} a_{\mathbf{k}+\mathbf{q}} \\
&= \hbar S^+(\mathbf{k}+\mathbf{q})
\end{aligned}
$$

3.

$$
\begin{aligned}
\left[S^z(\mathbf{k}),\ S^-(\mathbf{q})\right]_- &= -\hbar^2\sqrt{2SN} \sum_{\mathbf{q}'} \left[a_{\mathbf{q}'}^+ a_{\mathbf{k}+\mathbf{q}'},\ a_{-\mathbf{q}}^+\right]_- \\
&= -\hbar^2\sqrt{2SN} \sum_{\mathbf{q}'} a_{\mathbf{q}'}^+ \underbrace{\left[a_{\mathbf{k}+\mathbf{q}'},\ a_{-\mathbf{q}}^+\right]_-}_{-\delta_{\mathbf{q}',-\mathbf{q}-\mathbf{k}}} \\
&= -\hbar^2\sqrt{2SN} a_{-\mathbf{q}-\mathbf{k}}^+ \\
&= -\hbar S^-(\mathbf{k}+\mathbf{q})
\end{aligned}
$$

Problem 7.11

$$\alpha_{\mathbf{q}} = \frac{1}{\sqrt{N}} \sum_i e^{-i\mathbf{q}\cdot\mathbf{R}_i} \alpha_i$$

$$\alpha_i = \frac{1}{\sqrt{N}} \sum_{\mathbf{q}} e^{i\mathbf{q}\cdot\mathbf{R}_i} \alpha_{\mathbf{q}}$$

$$J(\mathbf{q}) = \frac{1}{N} \sum_{i,j} J_{ij} e^{i\mathbf{q}(\mathbf{R}_i - \mathbf{R}_j)}$$

Dyson–Maléev:

$$S_i^+ = \hbar\sqrt{2S}\alpha_i$$

$$S_i^- = \hbar\sqrt{2S}\alpha_i^+ \left(1 - \frac{\hat{n}_i}{2S}\right)$$

$$S_i^z = \hbar(S - \hat{n}_i)$$

Heisenberg model:

$$H = E_0 + H_2 + H_4$$

$$\boxed{H_2 = 2S\hbar^2 J_0 \textstyle\sum_i \hat{n}_i - 2S\hbar^2 \sum_{i,j} J_{ij}\alpha_i^+\alpha_j}$$

$$\begin{aligned}
\sum_i \hat{n}_i &= \frac{1}{N} \sum_{\mathbf{q},\mathbf{q}',i} e^{-i(\mathbf{q}-\mathbf{q}')\mathbf{R}_i} \alpha_{\mathbf{q}}^+ \alpha_{\mathbf{q}'} \\
&= \sum_{\mathbf{q},\mathbf{q}'} \alpha_{\mathbf{q}}^+ \alpha_{\mathbf{q}'} \delta_{\mathbf{q}\mathbf{q}'} \\
&= \sum_{\mathbf{q}} \alpha_{\mathbf{q}}^+ \alpha_{\mathbf{q}}
\end{aligned}$$

$$\begin{aligned}
&\sum_{i,j} J_{ij}\alpha_i^+\alpha_j = \\
&= \frac{1}{N^2} \sum_{i,j} \sum_{\mathbf{q},\mathbf{q}',\mathbf{q}''} J(\mathbf{q}) \alpha_{\mathbf{q}'}^+ \alpha_{\mathbf{q}''} e^{i\mathbf{q}(\mathbf{R}_i - \mathbf{R}_j)} e^{-i\mathbf{q}'\mathbf{R}_i} e^{i\mathbf{q}''\mathbf{R}_j} \\
&= \sum_{\mathbf{q},\mathbf{q}',\mathbf{q}''} J(\mathbf{q}) \alpha_{\mathbf{q}'}^+ \alpha_{\mathbf{q}''} \delta_{\mathbf{q},\mathbf{q}'} \delta_{\mathbf{q},\mathbf{q}''} \\
&= \sum_{\mathbf{q}} J(\mathbf{q}) \alpha_{\mathbf{q}}^+ \alpha_{\mathbf{q}}
\end{aligned}$$

$$\hbar\omega(\mathbf{q}) = 2S\hbar^2(J_0 - J(\mathbf{q}))$$

$$\curvearrowright \quad H_2 = \sum_{\mathbf{q}} \hbar\omega(\mathbf{q})\alpha^+_{\mathbf{q}}\alpha_{\mathbf{q}}$$

$$\boxed{H_4 = -\hbar^2 \sum_{i,j} J_{ij}\hat{n}_i\hat{n}_j + \hbar^2 \sum_{i,j} J_{ij}\alpha^+_i\hat{n}_i\alpha_j}$$

$$\begin{aligned}
&\sum_{i,j} J_{ij}\hat{n}_i\hat{n}_j = \\
&= \frac{1}{N^3}\sum_{i,j}\sum_{\mathbf{q}_1\cdots\mathbf{q}_4,\mathbf{q}} \alpha^+_{\mathbf{q}_1}\alpha_{\mathbf{q}_3}\alpha^+_{\mathbf{q}_2}\alpha_{\mathbf{q}_4} J(\mathbf{q}) * \\
&\qquad\qquad * e^{-\mathbf{q}(\mathbf{R}_i-\mathbf{R}_j)} e^{i(\mathbf{q}_1\mathbf{R}_i-\mathbf{q}_3\mathbf{R}_i+\mathbf{q}_2\mathbf{R}_j-\mathbf{q}_4\mathbf{R}_j)} \\
&= \frac{1}{N}\sum_{\mathbf{q}_1\cdots\mathbf{q}_4,\mathbf{q}} J(\mathbf{q})\alpha^+_{\mathbf{q}_1}\alpha_{\mathbf{q}_3}\alpha^+_{\mathbf{q}_2}\alpha_{\mathbf{q}_4}\delta_{\mathbf{q},\mathbf{q}_1-\mathbf{q}_3}\delta_{\mathbf{q},\mathbf{q}_4-\mathbf{q}_2} \\
&= \frac{1}{N}\sum_{\mathbf{q}_1\cdots\mathbf{q}_4} J(\mathbf{q}_4-\mathbf{q}_2)\delta_{\mathbf{q}_1+\mathbf{q}_2,\mathbf{q}_3+\mathbf{q}_4}\alpha^+_{\mathbf{q}_1}\alpha_{\mathbf{q}_3}\alpha^+_{\mathbf{q}_2}\alpha_{\mathbf{q}_4} \\
&= \frac{1}{N}\sum_{\mathbf{q}_1\mathbf{q}_2\mathbf{q}_4} J(\mathbf{q}_4-\mathbf{q}_2)\delta_{\mathbf{q}_1+\mathbf{q}_2,\mathbf{q}_2+\mathbf{q}_4}\alpha^+_{\mathbf{q}_1}\alpha_{\mathbf{q}_4} + \\
&\quad + \frac{1}{N}\sum_{\mathbf{q}_1\ldots\mathbf{q}_4} J(\mathbf{q}_4-\mathbf{q}_2)\delta_{\mathbf{q}_1+\mathbf{q}_2,\mathbf{q}_3+\mathbf{q}_4}\alpha^+_{\mathbf{q}_1}\alpha^+_{\mathbf{q}_2}\alpha_{\mathbf{q}_3}\alpha_{\mathbf{q}_4} \\
&= \frac{1}{N}\sum_{\mathbf{q}_1,\mathbf{q}_2} J(\mathbf{q}_1-\mathbf{q}_2)\alpha^+_{\mathbf{q}_1}\alpha_{\mathbf{q}_1} + \\
&\quad + \frac{1}{N}\sum_{\mathbf{q}_1\ldots\mathbf{q}_4} J(\mathbf{q}_4-\mathbf{q}_2)\delta_{\mathbf{q}_1+\mathbf{q}_2,\mathbf{q}_3+\mathbf{q}_4}\alpha^+_{\mathbf{q}_1}\alpha^+_{\mathbf{q}_2}\alpha_{\mathbf{q}_3}\alpha_{\mathbf{q}_4}
\end{aligned}$$

$$\begin{aligned}
&\sum_{i,j} J_{ij}\alpha^+_i\hat{n}_i\alpha_j = \\
&= \frac{1}{N^3}\sum_{i,j}\sum_{\mathbf{q},\mathbf{q}_1\ldots\mathbf{q}_4} J(\mathbf{q})\alpha^+_{\mathbf{q}_1}\alpha^+_{\mathbf{q}_2}\alpha_{\mathbf{q}_3}\alpha_{\mathbf{q}_4} e^{-i\mathbf{q}(\mathbf{R}_i-\mathbf{R}_j)} * \\
&\qquad\qquad * e^{i(\mathbf{q}_1\mathbf{R}_i+\mathbf{q}_2\mathbf{R}_i-\mathbf{q}_3\mathbf{R}_i-\mathbf{q}_4\mathbf{R}_j)} \\
&= \frac{1}{N}\sum_{\mathbf{q}_1\ldots\mathbf{q}_4,\mathbf{q}} J(\mathbf{q})\alpha^+_{\mathbf{q}_1}\alpha^+_{\mathbf{q}_2}\alpha_{\mathbf{q}_3}\alpha_{\mathbf{q}_4}\delta_{\mathbf{q},\mathbf{q}_4}\delta_{\mathbf{q},\mathbf{q}_1+\mathbf{q}_2-\mathbf{q}_3} \\
&= \frac{1}{N}\sum_{\mathbf{q}_1\ldots\mathbf{q}_4} J(\mathbf{q}_4)\delta_{\mathbf{q}_1+\mathbf{q}_2,\mathbf{q}_3+\mathbf{q}_4}\alpha^+_{\mathbf{q}_1}\alpha^+_{\mathbf{q}_2}\alpha_{\mathbf{q}_3}\alpha_{\mathbf{q}_4}
\end{aligned}$$

In the first term $\mathbf{q}_1$ and $\mathbf{q}_2$ commute:

$$\curvearrowright \quad H_4 = \frac{\hbar^2}{N} \sum_{\mathbf{q}_1 \ldots \mathbf{q}_4} (J(\mathbf{q}_4) - J(\mathbf{q}_4 - \mathbf{q}_1)) * \\ * \delta_{\mathbf{q}_1+\mathbf{q}_2, \mathbf{q}_3+\mathbf{q}_4} \alpha^+_{\mathbf{q}_1} \alpha^+_{\mathbf{q}_2} \alpha_{\mathbf{q}_3} \alpha_{\mathbf{q}_4}$$

Here we exploit

$$\begin{aligned} \frac{1}{N} \sum_{\mathbf{q}_2} J(\mathbf{q}_1 - \mathbf{q}_2) &= \frac{1}{N^2} \sum_{\mathbf{q}_2} \sum_{i,j} J_{ij} e^{i(\mathbf{q}_1 - \mathbf{q}_2)(\mathbf{R}_i - \mathbf{R}_j)} \\ &= \frac{1}{N} \sum_{i,j} J_{ij} e^{i\mathbf{q}_1(\mathbf{R}_i - \mathbf{R}_j)} \delta_{ij} \\ &= \frac{1}{N} \sum_i J_{ii} = 0 \end{aligned}$$

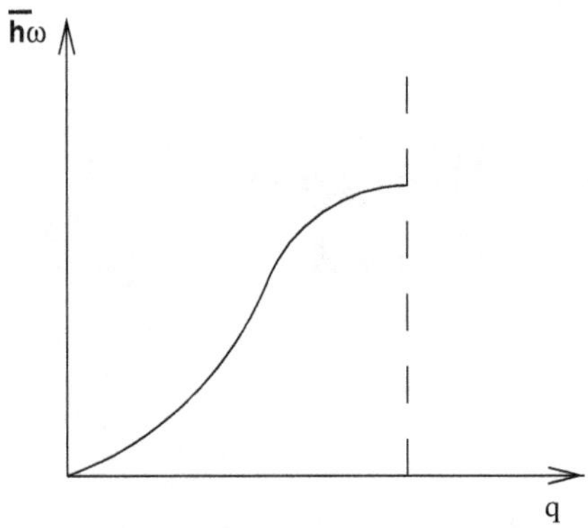

Problem 7.12

$$E_0^0 = E_0(B_0 = 0) = -N J_0 \hbar^2 S^2$$

$$U = < H_{SW} > = E_0^0 + \sum_{\mathbf{q}} \hbar\omega(\mathbf{q}) < \hat{n}_{\mathbf{q}} >$$

$$\hbar\omega(\mathbf{q}) = 2S\hbar^2 (J_0 - J(\mathbf{q})) \approx D q^2 \quad \text{for small } |\mathbf{q}|$$

Low temperatures → only a few magnons are excited:

$$
\begin{aligned}
U &\approx E_0^0 + D \sum_{\mathbf{q}} \frac{q^2}{e^{\beta D q^2} - 1} \\
&= E_0^0 + D \frac{V}{(2\pi)^3} \int_{\mathrm{BZ}} d^3 q \, q^2 e^{-\beta D q^2} \sum_{n=0}^{\infty} e^{-n\beta D q^2} \\
&\approx E_0^0 + \frac{DV}{2\pi^2} \sum_{n=1}^{\infty} \int_0^{\infty} dq \, q^4 e^{-n\beta D q^2}
\end{aligned}
$$

Substitution

$$
t = n\beta D q^2 \qquad dt = 2n\beta D q \, dq = 2\sqrt{t} \frac{n\beta D}{\sqrt{n\beta D}} dq
$$

$$
dq = \frac{1}{2\sqrt{n\beta D}} \frac{dt}{\sqrt{t}}
$$

$$
\begin{aligned}
\curvearrowright \quad U &\approx E_0^0 + \\
&+ \frac{DV}{4\pi^2} \sum_{n=1}^{\infty} (n\beta D)^{-\frac{5}{2}} \underbrace{\int_0^{\infty} dt \, t^{\frac{3}{2}} e^{-t}}_{\Gamma(\frac{5}{2}) = \frac{3}{2}\Gamma(\frac{3}{2}) = \frac{3}{4}\Gamma(\frac{1}{2}) = \frac{3}{4}\sqrt{\pi}} \\
U &\approx E_0^0 + \frac{3DV}{16\pi^{\frac{3}{2}}} \zeta\left(\frac{5}{2}\right) \left(\frac{k_B}{D}\right)^{\frac{5}{2}} T^{\frac{5}{2}} = \\
&= E_0^0 + \eta T^{\frac{5}{2}}
\end{aligned}
$$

$$
\curvearrowright \quad C_{B_0=0} = \left(\frac{\partial U}{\partial T}\right)_{B_0=0} = \frac{5}{2} \eta T^{\frac{3}{2}}
$$

is experimentally uniquely confirmed.
Measurement of $C_{B_0=0} \to \eta \to D \to J_0 = z_1 J_1$.

Problem 7.13

1. Proof by complete induction:
   $p = 0$; trivial.
   $p = 1$

$$
\begin{aligned}
\left[\hat{n}_{\mathbf{q}}, a_{\mathbf{q}}^{\dagger}\right]_- &= a_{\mathbf{q}}^{\dagger} \left[a_{\mathbf{q}}, a_{\mathbf{q}}^{\dagger}\right]_- + \left[a_{\mathbf{q}}^{\dagger}, a_{\mathbf{q}}^{\dagger}\right]_- a_{\mathbf{q}} \\
&= a_{\mathbf{q}}^{\dagger}
\end{aligned}
$$

   $p \curvearrowright p + 1$

$$\begin{aligned}\left[\hat{n}_{\mathbf{q}},(a_{\mathbf{q}}^{\dagger})^{p+1}\right]_- &= \\ &= (a_{\mathbf{q}}^{\dagger})^p\left[\hat{n}_{\mathbf{q}},a_{\mathbf{q}}^{\dagger}\right]_- + \left[\hat{n}_{\mathbf{q}},(a_{\mathbf{q}}^{\dagger})^p\right]_- a_{\mathbf{q}}^{\dagger} \\ &= (a_{\mathbf{q}}^{\dagger})^p a_{\mathbf{q}}^{\dagger} + p(a_{\mathbf{q}}^{\dagger})^p a_{\mathbf{q}}^{\dagger} \\ &= (p+1)(a_{\mathbf{q}}^{\dagger})^{p+1} \qquad \text{q.e.d.}\end{aligned}$$

2.

$$\begin{aligned}\hat{n}_{\mathbf{k}}\left(\prod_{\mathbf{q}}(a_{\mathbf{q}}^{\dagger})^{n_{\mathbf{q}}}\right)|0\rangle &= \\ &= \hat{n}_{\mathbf{k}}(a_{\mathbf{k}}^{\dagger})^{n_{\mathbf{k}}}\prod_{\mathbf{q}}^{\neq\mathbf{k}}(a_{\mathbf{q}}^{\dagger})^{n_{\mathbf{q}}}|0\rangle \\ &= \prod_{\mathbf{q}}^{\neq\mathbf{k}}(a_{\mathbf{q}}^{\dagger})^{n_{\mathbf{q}}}\left((a_{\mathbf{k}}^{\dagger})^{n_{\mathbf{k}}}\hat{n}_{\mathbf{k}} + n_{\mathbf{k}}(a_{\mathbf{k}}^{\dagger})^{n_{\mathbf{k}}}\right)|0\rangle \\ &= 0 + n_{\mathbf{k}}\prod_{\mathbf{q}}^{\neq\mathbf{k}}(a_{\mathbf{q}}^{\dagger})^{n_{\mathbf{q}}}(a_{\mathbf{k}}^{\dagger})^{n_{\mathbf{k}}}|0\rangle \\ &= n_{\mathbf{k}}\prod_{\mathbf{q}}(a_{\mathbf{q}}^{\dagger})^{n_{\mathbf{q}}}|0\rangle \\ &= n_{\mathbf{k}}|\psi\rangle\end{aligned}$$

$|\psi\rangle$ is therefore an eigenstate of $\hat{n}_{\mathbf{k}}$ with the eigenvalue $n_{\mathbf{k}}$. Thus $|\psi\rangle$ is also eigenstate of H:

$$H|\psi\rangle = \left(E_0(B_0) + \sum_{\mathbf{k}}\hbar\omega(\mathbf{k})n_{\mathbf{k}}\right)|\psi\rangle$$

Problem 7.14

$$U = < H_{SW} > = \hat{E}_a + \sum_{\mathbf{q}}\left\{E_\alpha(\mathbf{q}) < \alpha_{\mathbf{q}}^{+}\alpha_{\mathbf{q}} > + E_\beta(\mathbf{q}) < \beta_{\mathbf{q}}^{+}\beta_{\mathbf{q}} >\right\}$$

$B_0 = 0$; $B_A = 0$:

$$E_\alpha(\mathbf{q}) = E_\beta(q) \approx D\cdot q = \varepsilon(q)$$

$$\curvearrowright\quad U = \hat{E}_a + 2D\sum_{\mathbf{q}}\frac{q}{e^{\beta\varepsilon(q)}-1}$$

$$U \approx \hat{E}_a + \frac{DV}{\pi^2} \sum_{n=1}^{\infty} \int_0^{\infty} dq\, q^3 e^{-\beta D q n}$$

$$= \hat{E}_a + \frac{DV}{\pi^2} \frac{1}{(\beta D)^4} \sum_{n=1}^{\infty} \frac{1}{n^4} \underbrace{\int_0^{\infty} dy\, y^3 e^{-y}}_{\Gamma(4)=3!=6}$$

Abbreviation:

$$C_4 = \frac{6DV}{\pi^2 (k_B D)^4} \zeta(4)$$

$$\curvearrowright \quad \boxed{U \approx \hat{E}_a + C_4 \cdot T^4}$$

Heat capacity:

$$C_{B_0=0} = \left(\frac{\partial U}{\partial T}\right)_{B_0=0} = 4C_4 \cdot T^3$$

Problem 7.15
Sub-lattice magnetization of an antiferromagnet (Eq. (7.304)):

$$M_A(T) = \frac{1}{V} g_J \mu_B \left\{ \frac{N}{2} S - \sum_{\mathbf{q}} \sinh^2 \eta_q - \sum_{\mathbf{q}} \left(\frac{\cosh^2 \eta_q}{e^{\beta E_\alpha(q)} - 1} + \right.\right.$$
$$\left.\left. + \frac{\sinh^2 \eta_q}{e^{\beta E_\beta(q)} - 1} \right) \right\}$$

$B_0 = 0$; $B_A = 0$:

$$(7.356) \curvearrowright \tanh 2\eta_q \approx -\frac{J(q)}{J(0)} = -\gamma_q \qquad \text{(T independent)}$$

$$\boxed{M_A(0) = \tfrac{1}{V} g_J \mu_B \left\{ \tfrac{N}{2} - \textstyle\sum_{\mathbf{q}} \sinh^2 q \right\}}$$

$$E_\alpha(\mathbf{q}) \equiv E_\beta(\mathbf{q}) = 2S\hbar^2 \sqrt{(J_0 + J(\mathbf{q}))(J_0 - J(\mathbf{q}))}$$
$$\overset{J_0 - J(\mathbf{q}) \approx d J_0 q^2}{\approx} \underbrace{(2S\hbar^2 |J_0| \sqrt{2d})}_{D} \cdot q \equiv \varepsilon(\mathbf{q})$$

$$\curvearrowright M_A(T) - M_A(0) = -\frac{1}{V} g_J \mu_B \sum_q \frac{1 + 2\sinh^2 \eta_q}{e^{\beta\varepsilon(\mathbf{q})} - 1}$$

<u>$T \to 0; \beta \to \infty$:</u>

$$\begin{aligned}
&\sum_{\mathbf{q}} \frac{1}{e^{\beta\varepsilon(q)} - 1} = \\
&= \frac{V}{(2\pi)^3} \int_{BZ} d^3q \, \frac{e^{-\beta\varepsilon(q)}}{1 - e^{-\beta\varepsilon(q)}} \qquad (\underline{\text{without}} \text{ correction term}) \\
&= \frac{V}{(2\pi)^3} \int_{BZ} d^3q \sum_{n=1}^{\infty} e^{-n\beta\varepsilon(q)} \\
&\approx \frac{V}{2\pi^2} \sum_{n=1}^{\infty} \int_0^{\infty} dq \; q^2 e^{-n\beta D q} \qquad (\text{justification as for ferromagnets!}) \\
&= \frac{V}{2\pi^2} \frac{1}{(\beta D)^3} \sum_{n=1}^{\infty} \frac{1}{n^3} \underbrace{\int_0^{\infty} dy \; y^2 e^{-y}}_{\Gamma(3)=2!=2} \\
&= \left(\frac{V}{\pi^2} \frac{1}{(k_B^{-1} D)^3} \zeta(3) \right) T^3
\end{aligned}$$

Abbreviation: $C_3 = \frac{g_J \mu_B}{\pi^2 (k_B^{-1} D)^3} \zeta(3)$

$$\curvearrowright \quad \boxed{M_A(T) - M(0) = C_3 \cdot T^3} \quad \text{if } \textit{without} \text{ correction term}$$

With correction term:

$$1 + 2\sinh^2 \eta_q = \frac{1}{\sqrt{1 - \tanh^2 2\eta_q}} = \frac{1}{\sqrt{1 - \gamma_q^2}}$$

Consider

$$\gamma_{\mathbf{q}} \approx 1 - dq^2 \quad \curvearrowright$$

$$\frac{1}{\sqrt{1-\gamma_q^2}} = \frac{1}{\sqrt{2dq^2 - d^2q^4}} =$$

$$= \frac{1}{\sqrt{2d}} \cdot \frac{1}{q} \cdot \frac{1}{\sqrt{1-\frac{1}{2}dq^2}}$$

$$\curvearrowright \qquad 1-\gamma_q^2 \approx 2dq^2$$

(only small q play a role in the spin wave approximation)

$$\curvearrowright \quad \sum_{\mathbf{q}} \frac{1+2\sinh^2\eta_q}{e^{\beta\varepsilon(q)}-1} \approx \frac{1}{\sqrt{2d}} \sum_{\mathbf{q}} \frac{1}{q} \frac{1}{e^{\beta\varepsilon(q)}-1}$$

$$= \frac{1}{\sqrt{2d}} \frac{V}{2\pi^2} \frac{1}{(\beta D)^2} \sum_{n=1}^{\infty} \frac{1}{n^2} \underbrace{\int_0^\infty dy\, ye^{-y}}_{=1}$$

Abbreviation: $C_2 = \frac{gJ\mu_B}{\sqrt{2d}2\pi^2(k_B D)^2}\zeta(2)$

$$\curvearrowright \quad \boxed{M_A(T) - M(0) = C_2 \cdot T^2}$$

Problem 7.16

$$H = E_a + b_A \sum_{\mathbf{q}} a_{\mathbf{q}}^+ a_{\mathbf{q}} + b_B \sum_{\mathbf{q}} b_{\mathbf{q}}^+ b_{\mathbf{q}} + \sum_{\mathbf{q}} c(\mathbf{q}) \left\{a_{\mathbf{q}} b_{\mathbf{q}} + a_{\mathbf{q}}^+ b_{\mathbf{q}}^+\right\}$$

$\curvearrowright$ in the new operators:

$$\begin{aligned} H =& E_a + b_A \sum_q \left(\cosh\eta \cdot \alpha_q^+ + \sinh\eta \cdot \beta_q\right) * \\ & \qquad * \left(\cosh\eta \cdot \alpha_q + \sinh\eta \cdot \beta_q^+\right) + \\ &+ b_B \sum_{\mathbf{q}} \left(\sinh\eta \cdot \alpha_q + \cosh\eta \cdot \beta_q^+\right) * \\ & \qquad * \left(\sinh\eta \cdot \alpha_q^+ + \cosh\eta \cdot \beta q\right) + \\ &+ \sum_{\mathbf{q}} c(\mathbf{q}) \left\{(\cosh\eta \cdot \alpha_q + \sinh\eta \cdot \beta_q^+) * \right. \\ & \qquad * (\sinh\eta \cdot \alpha_q^+ + \cosh\eta \cdot \beta_q) + \end{aligned}$$

$$
\begin{aligned}
&+ \left(\cosh\eta \cdot \alpha_q^+ + \sinh\eta \cdot \beta_q\right) * \\
&\qquad * \left(\sinh\eta \cdot \alpha_q + \cosh\eta \cdot \beta_q^+\right)\} \\
=& E_a + \sum_{\mathbf{q}} \left[b_A \left(\cosh^2\eta \cdot \alpha_q^+\alpha_q + \sinh^2\eta \cdot \beta_q\beta_q^+ \right.\right.\\
&\qquad + \sinh\eta \cdot \cosh\eta \left(\alpha_q^+\beta_q^+ + \beta_q\alpha_q\right)) \\
&\qquad . + b_B \left(\sinh^2\eta \cdot \alpha_q\alpha_q^+ + \cosh^2\eta \cdot \beta_q^+\beta_q + \right.\\
&\qquad + \sinh\eta \cdot \cosh\eta(\alpha_q\beta_q + \beta_q^+\alpha_q^+)) + \\
&\qquad + c(q)\left(\cosh^2\eta(\alpha_q\beta_q + \alpha_q^+\beta_q^+) + \right.\\
&\qquad + \sinh^2\eta(\beta_q^+\alpha_q^+ + \beta_q\alpha_q) + \\
&\qquad + \cosh\eta \cdot \sinh\eta \left(\alpha_q\alpha_q^+ + \beta_q^+\beta_q + \right.\\
&\qquad + \alpha_q^+\alpha_q + \beta_q\beta_q^+)))]
\end{aligned}
$$

Use

$$
\begin{aligned}
\sinh\eta \cdot \cosh\eta &= \frac{1}{2}\sinh 2\eta \\
\cosh^2\eta &= \frac{1}{2}(\cosh 2\eta + 1) \\
\sinh^2\eta &= \frac{1}{2}(\cosh 2\eta - 1)
\end{aligned}
$$

$$
\begin{aligned}
\curvearrowright H =& E_a + \frac{1}{2}\sum_q \left[b_A(\alpha_q^+\alpha_q - \beta_q\beta_q^+) + b_B(-\alpha_q\alpha_q^+ + \beta_q^+\beta_q) + \right.\\
&\qquad \left. + c(q)\left(\alpha_q\beta_q + \alpha_q^+\beta_q^+ - \beta_q^+\alpha_q^+ - \beta_q\alpha_q\right)\right] \\
&+ \frac{1}{2}\sum_q \sinh(2\eta)\left[b_A(\alpha_q^+\beta_q^+ + \beta_q\alpha_q) + b_B(\alpha_q\beta_q + \beta_q^+\alpha_q^+) + \right.\\
&\qquad \left. + c(q)\left(\alpha_q\alpha_q^+ + \beta_q^+\beta_q + \alpha_q^+\alpha_q + \beta_q\beta_q^+\right)\right] \\
&+ \frac{1}{2}\sum_q \cosh(2\eta)\left[b_A(\alpha_q^+\alpha_q + \beta_q\beta_q^+) + b_B(\alpha_q\alpha_q^+ + \beta_q^+\beta_q) + \right.\\
&\qquad \left. + c(q)\left(2\alpha_q\beta_q + 2\alpha_q^+\beta_q^+\right)\right] \\
=& E_a + \frac{1}{2}\sum_q \left((b_A - b_B)\alpha_q^+\alpha_q + (b_B - b_A)\beta_q^+\beta_q - (b_A + b_B)\right) \\
&+ \frac{1}{2}\sum_q \sinh(2\eta)\left((b_a + b_B)(\alpha_q\beta_q + \alpha_q^+\beta_q^+) + \right.\\
&\qquad \left. + c(q)\left(2\alpha_q^+\alpha_q + 2\beta_q^+\beta_q + 2\right)\right)
\end{aligned}
$$

$$+\frac{1}{2}\sum_q \cosh(2\eta)\Big[(b_a+b_B)(\alpha_q^+\alpha_q+\beta_q^+\beta_q)+(b_A+b_B)+$$

$$\underbrace{2c(q)}_{-(b_A+b_B)\tanh(2\eta)}\left(\alpha_q\beta_q+\alpha_q^+\beta_q^+\right)\Big]$$

$$=E_a-\frac{N}{4}(b_A+b_B)+\frac{1}{2}\sum_q((b_A+b_B)\cosh 2\eta+$$

$$+\,2c(q)\sinh 2\eta)+$$

$$+\sum_q \alpha_q^+\alpha_q\left(\frac{1}{2}(b_A-b_B)+c(q)\sinh 2\eta+\right.$$

$$\left.+\frac{1}{2}(b_A+b_B)\cosh 2\eta\right)+$$

$$+\sum_q \beta_q^+\beta_q\left(\frac{1}{2}(b_B-b_A)+c(q)\sinh 2\eta+\right.$$

$$\left.+\frac{1}{2}(b_A+b_B)\cosh 2\eta\right)$$

$$c(q)\sinh 2\eta+\frac{1}{2}(b_A+b_B)\cosh 2\eta=$$

$$=c(q)\frac{\tanh 2\eta}{\sqrt{1-\tanh^2 2\eta}}+\frac{1}{2}(b_A+b_B)\frac{1}{\sqrt{1-\tanh^2 2\eta}}$$

$$=c(q)\frac{\frac{-2c(q)}{b_A+b_B}}{\sqrt{1-\frac{4c^2}{(b_A+b_B)^2}}}+\frac{1}{2}(b_A+b_B)\frac{1}{\sqrt{1-\frac{4c^2}{(b_A+b_B)^2}}}$$

$$=\frac{-4c^2(q)}{2\sqrt{(b_A+b_B)^2-4c^2}}+\frac{1}{2}\frac{(b_A+b_B)^2}{\sqrt{(b_A+b_B)^2-4c^2}}$$

$$=\frac{1}{2}\sqrt{(b_A+b_B)^2-4c^2}$$

$$=E_\alpha(q)-\frac{1}{2}(b_A-b_B)$$

$$=E_\beta(q)+\frac{1}{2}(b_A-b_B)$$

$$\curvearrowright\quad \boxed{\hat{E}_a=E_a-\frac{N}{4}(b_A+b_B)+\frac{1}{2}\sum_{\mathbf{q}}\sqrt{(b_A+b_B)^2-4c^2(\mathbf{q})}}$$

$$\boxed{H=\hat{E}_a+\sum_{\mathbf{q}}\left(E_\alpha(\mathbf{q})\alpha_{\mathbf{q}}^+\alpha_{\mathbf{q}}+E_\beta(\mathbf{q})\beta_{\mathbf{q}}^+\beta_{\mathbf{q}}\right)}$$

Problem 7.17

$$\gamma_{\mathbf{q}} = \frac{1}{z_1} \sum_{\Delta_1} e^{i\mathbf{q}\cdot\mathbf{R}_{\Delta_1}}$$

z_1: number of nearest neighbours
$\mathbf{R}_{\Delta_1}$: lattice vector from the origin to a nearest neighbouring site

$$\begin{aligned}
\sum_{\mathbf{q}_1} \gamma_{\mathbf{q}-\mathbf{q}_1} \langle \hat{n}_{\mathbf{q}_1} \rangle &= \\
&= \frac{1}{z_1} \sum_{\mathbf{q}_1} \sum_{\Delta_1} e^{i(\mathbf{q}-\mathbf{q}_1)\cdot\mathbf{R}_{\Delta_1}} \frac{1}{N} \sum_{i,j} \langle a_i^\dagger a_j \rangle e^{i\mathbf{q}_1\cdot(\mathbf{R}_i - \mathbf{R}_j)} \\
&= \frac{1}{z_1} \sum_{i,j} \sum_{\Delta_1} \langle a_i^\dagger a_j \rangle e^{i\mathbf{q}\cdot\mathbf{R}_{\Delta_1}} \delta_{i-j,\Delta_1} \\
&= \frac{1}{z_1} \sum_{\Delta_1} e^{i\mathbf{q}\cdot\mathbf{R}_{\Delta_1}} \sum_i \langle a_i^\dagger a_{i-\Delta_1} \rangle
\end{aligned}$$

Due to translational symmetry, the expectation value is the same for all nearest neighbours. That is, independent of any particular Δ_1:

$$\curvearrowright$$

$$\begin{aligned}
&\sum_{\mathbf{q}_1} \gamma_{\mathbf{q}-\mathbf{q}_1} \langle \hat{n}_{\mathbf{q}_1} \rangle \\
&= \left(\frac{1}{z_1} \sum_{i,\Delta_1} \langle a_i^\dagger a_{i-\Delta_1} \rangle \right) \frac{1}{z_1} \sum_{\Delta_1} e^{i\mathbf{q}\cdot\mathbf{R}_{\Delta_1}} \\
&= \gamma_{\mathbf{q}} \frac{1}{z_1} \sum_{i,\Delta_1} \langle a_i^\dagger a_{i-\Delta_1} \rangle \\
&= \gamma_{\mathbf{q}} \frac{1}{z_1} \sum_{i,\Delta_1} \frac{1}{N} \sum_{\mathbf{q}_1,\mathbf{q}_2} e^{-i\mathbf{q}_1\cdot\mathbf{R}_i} e^{i\mathbf{q}_2(\mathbf{R}_i - \mathbf{R}_{\Delta_1})} \langle a_{\mathbf{q}_1}^\dagger a_{\mathbf{q}_2} \rangle \\
&= \gamma_{\mathbf{q}} \frac{1}{z_1} \sum_{\Delta_1} \sum_{\mathbf{q}_1,\mathbf{q}_2} e^{-i\mathbf{q}_2\cdot\mathbf{R}_{\Delta_1}} \delta_{\mathbf{q}_1,\mathbf{q}_2} \langle a_{\mathbf{q}_1}^\dagger a_{\mathbf{q}_2} \rangle \\
&= \gamma_{\mathbf{q}} \sum_{\mathbf{q}_1} \left(\frac{1}{z_1} \sum_{\Delta_1} e^{-i\mathbf{q}_1\cdot\mathbf{R}_{\Delta_1}} \right) \langle a_{\mathbf{q}_1}^\dagger a_{\mathbf{q}_1} \rangle \\
&= \gamma_{\mathbf{q}} \sum_{\mathbf{q}_1} \gamma_{\mathbf{q}_1} \langle \hat{n}_{\mathbf{q}_1} \rangle
\end{aligned}$$

Problem 7.18

$$\gamma_{\mathbf{q}} = \frac{1}{z_1} \sum_{\Delta_1} e^{i\mathbf{q}\cdot\mathbf{R}_{\Delta_1}}$$

$$\mathbf{R}_{\Delta_1} \in \{a(\pm 1, 0, 0), a(0, \pm 1, 0), a(0, 0, \pm 1)\}$$

$c_1 = \frac{1}{N} \sum_{\mathbf{q}} (1 - \gamma_{\mathbf{q}})$:

$$\frac{1}{N} \sum_{\mathbf{q}} 1 = 1$$

$$\frac{1}{N} \sum_{\mathbf{q}} \gamma_{\mathbf{q}} = \frac{1}{N} \sum_{\mathbf{q}} \frac{1}{z_1} \sum_{\Delta_1} e^{i\mathbf{q}\cdot\mathbf{R}_{\Delta_1}}$$

$$= \frac{1}{z_1} \sum_{\Delta_1} \delta_{\Delta_1, 0} = 0$$

since $\mathbf{R}_0 = (0, 0, 0)$ is not a nearest neighbour.

$$\curvearrowright \underline{c_1 = 1}$$

$c_2 = \frac{1}{N} \sum_{\mathbf{q}} (1 - \gamma_{\mathbf{q}})^2$:

$$c_2 = \frac{1}{N} \sum_{\mathbf{q}} \left(1 - 2\gamma_{\mathbf{q}} + \gamma_{\mathbf{q}}^2\right)$$

$$\frac{1}{N} \sum_{\mathbf{q}} \gamma_{\mathbf{q}}^2 = \frac{1}{N} \sum_{\mathbf{q}} \frac{1}{z_1^2} \sum_{\Delta_1} \sum_{\Delta_1'} e^{i\mathbf{q}\cdot\left(\mathbf{R}_{\Delta_1} + \mathbf{R}_{\Delta_1'}\right)}$$

$$= \frac{1}{z_1^2} \sum_{\Delta_1} \sum_{\Delta_1'} \delta_{\Delta_1, -\Delta_1'}$$

$$= \frac{1}{z_1^2} \sum_{\Delta_1} 1$$

$$= \frac{1}{z_1}$$

$$\curvearrowright c_2 = 1 + \frac{1}{z_1}$$

$c_3 = \frac{1}{N} \sum_{\mathbf{q}} (1 - \gamma_{\mathbf{q}})^3$:

$$c_3 = \frac{1}{N}\sum_{\mathbf{q}}\left(1 - 3\gamma_{\mathbf{q}} + 3\gamma_{\mathbf{q}}^2 - \gamma_{\mathbf{q}}^3\right)$$

$$\begin{aligned}\frac{1}{N}\sum_{\mathbf{q}}\gamma_{\mathbf{q}}^3 &= \frac{1}{N}\sum_{\mathbf{q}}\frac{1}{z_1^3}\sum_{\Delta_1}\sum_{\Delta_1'}\sum_{\Delta_1''} e^{i\mathbf{q}\cdot\left(\mathbf{R}_{\Delta_1}+\mathbf{R}_{\Delta_1'}+\mathbf{R}_{\Delta_1''}\right)} \\ &= \frac{1}{z_1^3}\sum_{\Delta_1}\sum_{\Delta_1'}\sum_{\Delta_1''}\delta_{\Delta_1'',-\Delta_1-\Delta_1'} \\ &= 0\end{aligned}$$

since $(-\mathbf{R}_{\Delta_1} - \mathbf{R}_{\Delta_1'})$ for s.c. lattice cannot be a nearest neighbour.

Problem 7.19

Calculation of the anisotropy contribution of the dipole interaction

$$H_a = -3\sum_{i,j} D_{ij}\left(\mathbf{S}_i \cdot \mathbf{e}_{ij}\right)\left(\mathbf{S}_j \cdot \mathbf{e}_{ij}\right)$$

in spin wave approximation:

$$\begin{aligned} S_i^+ &= \hbar\sqrt{2S}a_i \\ S_i^- &= \hbar\sqrt{2S}a_i^\dagger \\ S_i^z &= \hbar\left(S - a_i^\dagger a_i\right)\end{aligned}$$

This gives

$$\begin{aligned} H_a &= -3\sum_{i,j} D_{ij}\left(\mathbf{S}_i \cdot \mathbf{e}_{ij}\right)\left(\mathbf{S}_j \cdot \mathbf{e}_{ij}\right) \\ &= -3\sum_{i,j} D_{ij}\left\{\left(S_i^x x_{ij} + S_i^y y_{ij} + S_i^z z_{ij}\right) * \right. \\ &\qquad\qquad \left. * \left(S_j^x x_{ij} + S_j^y y_{ij} + S_j^z z_{ij}\right)\right\} \\ &= -3\sum_{i,j} D_{ij}\left\{x_{ij}^2 S_i^x S_j^x + y_{ij}^2 S_i^y S_j^y + z_{ij}^2 S_i^z S_j^z + \right. \\ &\quad + x_{ij}y_{ij}\left(S_i^x S_j^y + S_i^y S_j^x\right) + x_{ij}z_{ij}\left(S_i^x S_j^z + S_i^z S_j^x\right) + \\ &\quad \left. + y_{ij}z_{ij}\left(S_i^y S_j^z + S_i^z S_j^y\right)\right\} \\ &= -3\hbar^2\sum_{i,j} D_{ij}\left\{x_{ij}^2\frac{2S}{4}\left(a_i + a_i^\dagger\right)\left(a_j + a_j^\dagger\right) + \right.\end{aligned}$$

$$
\begin{aligned}
&+ y_{ij}^2 \frac{-2S}{4}\left(a_i - a_i^\dagger\right)\left(a_j - a_j^\dagger\right) + \\
&+ z_{ij}^2\left(S^2 - S\left(n_i + n_j\right) + n_i n_j\right) + \\
&+ x_{ij} y_{ij} \frac{2S}{4i}\left(\left(a_i + a_i^\dagger\right)\left(a_j - a_j^\dagger\right) + \right. \\
&\qquad\qquad \left. + \left(a_i - a_i^\dagger\right)\left(a_j + a_j^\dagger\right)\right) + \\
&+ x_{ij} z_{ij} \frac{\sqrt{2S}}{2}\left(\left(a_i + a_i^\dagger\right)\left(S - n_j\right) + \right. \\
&\qquad\qquad \left. + \left(S - n_i\right)\left(a_j + a_j^\dagger\right)\right) + \\
&+ y_{ij} z_{ij} \frac{\sqrt{2S}}{2i}\left(\left(a_i - a_i^\dagger\right)\left(S - n_j\right) + \right. \\
&\qquad\qquad \left. + \left(S - n_i\right)\left(a_j - a_j^\dagger\right)\right)\Big\} \\
= &-3\hbar^2 \sum_{i,j} D_{ij}\left\{\frac{S}{2} x_{ij}^2\left(a_i a_j + a_i a_j^\dagger + a_i^\dagger a_j + a_i^\dagger a_j^\dagger\right) - \right. \\
&- \frac{S}{2} y_{ij}^2\left(a_i a_j + a_i^\dagger a_j^\dagger - a_i a_j^\dagger - a_i^\dagger a_j\right) + \\
&+ z_{ij}^2 S\left(S - n_i - n_j\right) + \\
&+ \frac{S}{2i} x_{ij} y_{ij}\left(a_i a_j - a_i a_j^\dagger + a_i^\dagger a_j - a_i^\dagger a_j^\dagger + \right. \\
&\qquad\qquad \left. + a_i a_j + a_i a_j^\dagger - a_i^\dagger a_j - a_i^\dagger a_j^\dagger\right) + \\
&+ S\frac{\sqrt{2S}}{2} x_{ij} z_{ij}\left(a_i + a_i^\dagger + a_j + a_j^\dagger\right) + \\
&\left. + S\frac{\sqrt{2S}}{2i} y_{ij} z_{ij}\left(a_i - a_i^\dagger + a_j - a_j^\dagger\right)\right\} \\
= &-3\hbar^2 S \sum_{i,j} D_{ij} z_{ij}^2\left(S - 2n_i\right) - \\
&-3\hbar^2 S \sum_{i,j} D_{ij}\left\{a_i^\dagger a_j^\dagger\left(\frac{1}{2}x_{ij}^2 - \frac{1}{2}y_{ij}^2 + i x_{ij} y_{ij}\right) + \right. \\
&\qquad + a_i a_j\left(\frac{1}{2}x_{ij}^2 - \frac{1}{2}y_{ij}^2 - i x_{ij} y_{ij}\right) + \\
&\qquad + \left(a_i a_j^\dagger + a_i^\dagger a_j\right)\left(\frac{1}{2}x_{ij}^2 + \frac{1}{2}y_{ij}^2\right) + \\
&\qquad \left. S\sqrt{2S} x_{ij} z_{ij}\left(a_i + a_i^\dagger\right) - i S\sqrt{2S} y_{ij} z_{ij}\left(a_i - a_i^\dagger\right)\right\}
\end{aligned}
$$

The mixed terms vanish. In order to see that, hold z_{ij} fixed. All $\mathbf{R}_i$ and $\mathbf{R}_j$ with $z_{ij} = const$ define an x, y-plane. In this for each $\mathbf{R}_j$ there is a $\mathbf{R}_j'$ with $x_{ij} = -x_{ij}'$

and $y_{ij} = -y'_{ij}$ and $|\mathbf{R}_{ij}| = |\mathbf{R}'_{ij}|$, i.e. $D_{ij} = D'_{ij}$. That means $\sum_j D_{ij}x_{ij}z_{ij} = 0$ in the plane and therefore in the entire space. Analogously: $\sum_i D_{ij}y_{ij}z_{ij} = 0$. Therefore what remains is (Fig. C.2)

$$\begin{aligned} H_a = & -3\hbar^2 S \sum_{i,j} D_{ij} z_{ij}^2 (S - 2n_i) - \\ & -3\hbar^2 S \sum_{i,j} D_{ij} \left\{ a_i^\dagger a_j^\dagger \frac{1}{2} (x_{ij} + i y_{ij})^2 + a_i a_j \frac{1}{2} (x_{ij} - i y_{ij})^2 + \right. \\ & \left. + a_i^\dagger a_j \left(x_{ij}^2 + y_{ij}^2 \right) \right\} \end{aligned}$$

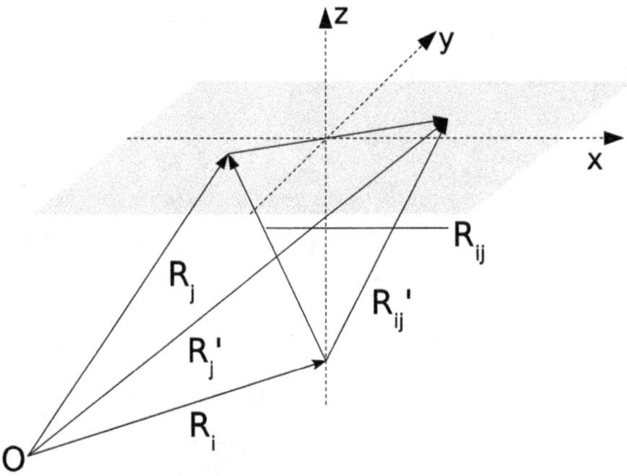

Fig. C.2 Graphical illustration for the evaluation of the "mixed" terms in H_a (see text)

Problem 7.20
First derivative

$$\begin{aligned} \frac{d\omega}{da} &= \hbar x \omega - \frac{e^{\hbar a x} \hbar (1+\varphi) e^{\hbar a}}{\left((1+\varphi)e^{\hbar a} - \varphi\right)^2} \\ &= \hbar \omega (x - \alpha) \end{aligned}$$

with

$$\alpha = \frac{(1+\varphi)e^{\hbar a}}{(1+\varphi)e^{\hbar a} - \varphi} = 1 + \frac{\varphi}{(1+\varphi)e^{\hbar a} - \varphi}$$

Second derivative:

$$\begin{aligned}\frac{\mathrm{d}^2\omega}{\mathrm{d}a^2} &= \hbar^2\omega(x-\alpha)^2 + \hbar\omega\frac{\hbar\varphi(1+\varphi)e^{\hbar a}}{\left((1+\varphi)e^{\hbar a}-\varphi\right)^2}\\ &= \hbar^2\omega(x-\alpha)^2 + \hbar^2\omega\alpha\frac{\varphi}{(1+\varphi)e^{\hbar a}-\varphi}\\ &= \hbar^2\omega\left((x-\alpha)^2+\alpha(\alpha-1)\right)\end{aligned}$$

Coefficient of the second term in the differential equation:

$$\frac{(1+\varphi)+\varphi e^{-a\hbar}}{(1+\varphi)-\varphi e^{-a\hbar}} = \alpha + \frac{\varphi}{(1+\varphi)e^{a\hbar}-\varphi} = \alpha+\alpha-1$$

Thus what remains is

$$\begin{aligned}&\hbar^2\omega\left((x-a)^2+\alpha(\alpha-1)\right)+\\ &+\hbar^2\omega(x-\alpha)(2\alpha-1)-\hbar^2 S(S+1)\omega = 0\\ \curvearrowright\; &x^2-2ax+\alpha^2+\alpha^2-\alpha+2x\alpha-\\ &-x-2\alpha^2+\alpha-S(S+1)=0\\ \curvearrowright\; &x^2-x-S(S+1)=0\\ \curvearrowright\; &x_1=-S;\; x_2=S+1\end{aligned}$$

general solution:

$$\Omega(a) = c_1\omega(-S,a)+c_2\omega(S+1,a)$$

Problem 8.1
For the Hamiltonian

$$H_0 = \sum_{ij\sigma}\sum_{\mu\nu} T_{ij}^{\mu\nu} c_{i\mu\sigma}^{+} c_{j\nu\sigma}$$

holds after substituting the Fourier integrals:

$$
\begin{aligned}
H_0 &= \sum_{ij\sigma}\sum_{\mu,\nu}\frac{1}{N_i}\sum_{\mathbf{k}} T_{\mathbf{k}}^{\mu\nu} e^{i\mathbf{k}\cdot(\mathbf{R}_i-\mathbf{R}_j)} * \\
&* \frac{1}{N_i}\sum_{\mathbf{q},\mathbf{p},m,m'} e^{-i\mathbf{q}\cdot\mathbf{R}_i-\mathbf{p}\cdot\mathbf{R}_j} c_{\mathbf{q}m\sigma}^{+} c_{\mathbf{p}m'\sigma} U_{\mathbf{q}\sigma}^{m\mu}\left(U_{\mathbf{p}\sigma}^{m'\nu}\right)^{*} \\
&= \sum_{\mathbf{k},\mathbf{q},\mathbf{p}}\sum_{m,m'}\sum_{\sigma,\nu,\mu} T_{\mathbf{k}}^{\mu\nu} U_{\mathbf{q}\sigma}^{m\mu}\left(U_{\mathbf{p}\sigma}^{m'\nu}\right)^{*}\delta_{\mathbf{k},\mathbf{q}}\delta_{\mathbf{k},\mathbf{p}} c_{\mathbf{q}m\sigma}^{+} c_{\mathbf{p}m'\sigma} \\
&= \sum_{\mathbf{k}}\sum_{m,m'}\sum_{\sigma,\nu,\mu} T_{\mathbf{k}}^{\mu\nu}\left(U_{\mathbf{k}\sigma}^{m'\nu}\right)^{*} U_{\mathbf{k}\sigma}^{m\mu} c_{\mathbf{k}m\sigma}^{+} c_{\mathbf{k}m'\sigma} \\
&= \sum_{\mathbf{k}}\sum_{m,m'}\sum_{\sigma,\mu} \varepsilon_{m'}(\mathbf{k})\left(U_{\mathbf{k}\sigma}^{m'\mu}\right)^{*} U_{\mathbf{k}\sigma}^{m\mu} c_{\mathbf{k}m\sigma}^{+} c_{\mathbf{k}m'\sigma} \\
&= \sum_{\mathbf{k},m,m',\sigma} \varepsilon_{m'}(\mathbf{k})\delta_{m',m} c_{\mathbf{k}m\sigma}^{+} c_{\mathbf{k}m'\sigma} \\
&= \sum_{\mathbf{k},m,\sigma} \varepsilon_{m}(\mathbf{k}) c_{\mathbf{k}m\sigma}^{+} c_{\mathbf{k}m\sigma}
\end{aligned}
$$

Problem 8.2
Wannier representation:

$$
H = \sum_{ij\sigma} T_{ij} c_{i\sigma}^{\dagger} c_{j\sigma} + \frac{1}{2}U\sum_{i,\sigma} n_{i\sigma} n_{i-\sigma}
$$

Hopping integrals:

$$
T_{ij} = \frac{1}{N}\sum_{\mathbf{k}} \varepsilon(\mathbf{k}) e^{i\mathbf{k}\cdot(\mathbf{R}_i-\mathbf{R}_j)}
$$

Construction operators:

$$
c_{i\sigma} = \frac{1}{\sqrt{N}}\sum_{\mathbf{k}} c_{\mathbf{k}\sigma} e^{i\mathbf{k}\cdot\mathbf{R}_i}
$$

One-particle part:

$$
\begin{aligned}
H_0 &= \sum_{ij\sigma} T_{ij} c_{i\sigma}^{\dagger} c_{j\sigma} \\
&= \frac{1}{N^2} \sum_{\mathbf{k},\mathbf{p},\mathbf{q},\sigma} \varepsilon(\mathbf{k}) c_{\mathbf{p}\sigma}^{\dagger} c_{\mathbf{q}\sigma} \sum_{i,j} e^{i\mathbf{k}\cdot(\mathbf{R}_i - \mathbf{R}_j)} e^{-i\mathbf{p}\cdot\mathbf{R}_i} e^{i\mathbf{q}\cdot\mathbf{R}_j} \\
&= \sum_{\mathbf{k},\mathbf{p},\mathbf{q},\sigma} \varepsilon(\mathbf{k}) c_{\mathbf{p}\sigma}^{\dagger} c_{\mathbf{q}\sigma} \delta_{\mathbf{k},\mathbf{p}} \delta_{\mathbf{k},\mathbf{q}} \\
&= \sum_{\mathbf{k}\sigma} \varepsilon(\mathbf{k}) c_{\mathbf{k}\sigma}^{\dagger} c_{\mathbf{k}\sigma}
\end{aligned}
$$

Interaction part:

$$
\begin{aligned}
H_1 &= \frac{1}{2} U \sum_{i\sigma} n_{i\sigma} n_{i-\sigma} \\
&= \frac{1}{2} U \frac{1}{N^2} \sum_{\mathbf{k}_1 \ldots \mathbf{k}_4, \sigma} c_{\mathbf{k}_1\sigma}^{\dagger} c_{\mathbf{k}_2\sigma} c_{\mathbf{k}_3-\sigma}^{\dagger} c_{\mathbf{k}_4-\sigma} * \\
&\quad * \sum_{i} e^{i(-\mathbf{k}_1+\mathbf{k}_2)\mathbf{R}_i} e^{i(-\mathbf{k}_3+\mathbf{k}_4)\mathbf{R}_i} \\
&= \frac{1}{2} U \sum_{\mathbf{k}_1 \ldots \mathbf{k}_4, \sigma} \delta_{\mathbf{k}_1+\mathbf{k}_3, \mathbf{k}_2+\mathbf{k}_4} c_{\mathbf{k}_1\sigma}^{\dagger} c_{\mathbf{k}_2\sigma} c_{\mathbf{k}_3-\sigma}^{\dagger} c_{\mathbf{k}_4-\sigma} \\
&\quad \mathbf{k}_1 \to \mathbf{k}+\mathbf{q}, \quad \mathbf{k}_3 \to \mathbf{p}-\mathbf{q}, \quad \mathbf{k}_4 \to \mathbf{p}, \quad \mathbf{k}_2 \to \mathbf{k} \\
\Leftrightarrow \quad H_1 &= \frac{1}{2} U \sum_{\mathbf{k},\mathbf{p},\mathbf{q},\sigma} c_{\mathbf{k}+\mathbf{q}\sigma}^{\dagger} c_{\mathbf{p}-\mathbf{q}-\sigma}^{\dagger} c_{\mathbf{p}-\sigma} c_{\mathbf{k}\sigma}
\end{aligned}
$$

Total:

$$
H = \sum_{\mathbf{k}\sigma} \varepsilon(\mathbf{k}) c_{\mathbf{k}\sigma}^{\dagger} c_{\mathbf{k}\sigma} + \frac{1}{2} U \sum_{\mathbf{k},\mathbf{p},\mathbf{q},\sigma} c_{\mathbf{k}+\mathbf{q}\sigma}^{\dagger} c_{\mathbf{p}-\mathbf{q}\sigma}^{\dagger} c_{\mathbf{p}-\sigma} c_{\mathbf{k}\sigma}
$$

Problem 8.3
Band limit:

$$
G_{\mathbf{k}\sigma}^{U\to 0}(E) = \frac{\hbar}{E - \varepsilon(\mathbf{k}) + \mu}
$$

Zero bandwidth limit ($W = 0$):

$$
G_{\sigma}^{W=0} = \frac{\hbar(1 - n_{-\sigma})}{E - T_0 + \mu} + \frac{\hbar n_{-\sigma}}{E - T_0 - U + \mu}
$$

Stoner approximation:

$$G_{\mathbf{k}\sigma}^{(Stoner)}(E) = \frac{\hbar}{E - \varepsilon(\mathbf{k}) - Un_{-\sigma} + \mu}$$

obviously satisfies the band limit but not the limit of infinitely narrow band.

Problem 8.4

$$H = \sum_{\mathbf{k}\sigma,\alpha,\beta} \varepsilon_\sigma^{\alpha\beta}(\mathbf{k}) c_{\mathbf{k}\alpha\sigma}^\dagger c_{\mathbf{k}\beta\sigma}, \qquad \alpha, \beta \in \{A, B\}$$

$$\varepsilon_\sigma^{AA}(\mathbf{k}) = \varepsilon(\mathbf{k}) + \frac{1}{2}Un - \frac{1}{2}z_\sigma Um - \mu$$
$$\varepsilon_\sigma^{BB}(\mathbf{k}) = \varepsilon(\mathbf{k}) + \frac{1}{2}Un + \frac{1}{2}z_\sigma Um - \mu$$
$$\varepsilon_\sigma^{AB}(\mathbf{k}) = t(\mathbf{k}) = \varepsilon_\sigma^{BA*}(\mathbf{k})$$

Here we have used

$$< n_\sigma^{AA} >= - < n_\sigma^{BB} >= n_\sigma$$
$$m = n_\uparrow - n_\downarrow; \quad m_A = -m_B = m$$

Green's functions:

$$G_{\mathbf{k}\sigma}^{\alpha,\beta}(E) =<< c_{\mathbf{k}\alpha\sigma}; c_{\mathbf{k}\beta\sigma}^\dagger >>_E$$

1. Quasiparticle energies:

$$[c_{\mathbf{k}A\sigma}, H]_- = \sum_\beta \varepsilon_\sigma^{A\beta}(\mathbf{k}) c_{\mathbf{k}\beta\sigma}$$
$$= \varepsilon_\sigma^{AA}(\mathbf{k}) c_{\mathbf{k}A\sigma} + \varepsilon_\sigma^{AB}(\mathbf{k}) c_{\mathbf{k}B\sigma}$$

Equation of motion:

$$(E - \varepsilon_\sigma^{AA}(\mathbf{k})) G_{\mathbf{k}\sigma}^{AA}(E) = \hbar + \varepsilon_\sigma^{AB}(\mathbf{k}) G_{\mathbf{k}\sigma}^{BA}(E)$$

With

$$[c_{\mathbf{k}B\sigma}, H]_- = \varepsilon_\sigma^{BA}(\mathbf{k}) c_{\mathbf{k}A\sigma} + \varepsilon_\sigma^{BB}(\mathbf{k}) c_{\mathbf{k}B\sigma}$$

also follows:

$$
\begin{aligned}
(E-\varepsilon_\sigma^{BB}(\mathbf{k}))G_{\mathbf{k}\sigma}^{BA}(E) &= \varepsilon_\sigma^{BA}(\mathbf{k})G_{\mathbf{k}\sigma}^{AA}(E)\\
\curvearrowright G_{\mathbf{k}\sigma}^{BA}(E) &= \frac{\varepsilon_\sigma^{BA}(\mathbf{k})}{E-\varepsilon_\sigma^{BB}(\mathbf{k})}G_{\mathbf{k}\sigma}^{AA}(E)
\end{aligned}
$$

Substituting in the equation of motion for $G_{\mathbf{k}\sigma}^{AA}(E)$:

$$
\begin{aligned}
(E-\varepsilon_\sigma^{AA}(\mathbf{k}))G_{\mathbf{k}\sigma}^{AA}(E) &= \hbar + \frac{|\varepsilon_\sigma^{AB}(\mathbf{k})|^2}{E-\varepsilon_\sigma^{BB}(\mathbf{k})}G_{\mathbf{k}\sigma}^{AA}(E)\\
\curvearrowright\; G_{\mathbf{k}\sigma}^{AA}(E) &= \hbar\frac{E-\varepsilon_\sigma^{BB}(\mathbf{k})}{(E-\varepsilon_\sigma^{AA}(\mathbf{k}))(E-\varepsilon_\sigma^{BB}(\mathbf{k}))-|\varepsilon_\sigma^{AB}(\mathbf{k})|^2}
\end{aligned}
$$

Poles:

$$
(E_\pm-\varepsilon_\sigma^{AA}(\mathbf{k}))(E-\varepsilon_\sigma^{BB}(\mathbf{k}))-|\varepsilon_\sigma^{AB}(\mathbf{k})|^2 \overset{!}{=} 0
$$

$$
\begin{aligned}
\curvearrowright\quad E_\pm(\mathbf{k}) = &\frac{1}{2}(\varepsilon_\sigma^{AA}(\mathbf{k})+\varepsilon_\sigma^{BB}(\mathbf{k}))\\
&\pm\sqrt{\frac{1}{4}\left(\varepsilon_\sigma^{AA}(\mathbf{k})-\varepsilon_\sigma^{BB}(\mathbf{k})\right)^2+|\varepsilon_\sigma^{AB}(\mathbf{k})|^2}
\end{aligned}
$$

That means the spin-*in*dependent quasiparticle energies:

$$
E_\pm(\mathbf{k}) = \varepsilon(\mathbf{k})+\frac{1}{2}Un\pm\sqrt{\frac{1}{4}U^2m^2+|t(\mathbf{k})|^2}-\mu
$$

2. Spectral weights:

$$
\begin{aligned}
G_{\mathbf{k}\sigma}^{AA}(E) &= \hbar\frac{E-\varepsilon_\sigma^{BB}(\mathbf{k})}{(E-E_+(\mathbf{k}))(E-E_-(\mathbf{k}))}\\
\hbar\alpha_\sigma^{(\pm)}(\mathbf{k}) &= \lim_{E\to E_\pm(\mathbf{k})} G_{\mathbf{k}\sigma}^{AA}(E)(E-E_\pm(\mathbf{k}))\\
\alpha_\sigma^{(\pm)}(\mathbf{k}) &= \frac{E_\pm(\mathbf{k})-\varepsilon_\sigma^{BB}(\mathbf{k})}{E_\pm(\mathbf{k})-E_\mp(\mathbf{k})}\\
\Leftrightarrow\quad \alpha_\sigma^{(\pm)}(\mathbf{k}) &= \frac{1}{2}\left(1\mp z_\sigma\frac{Um}{\sqrt{U^2m^2+4|t(\mathbf{k})|^2}}\right)
\end{aligned}
$$

Spectral weights are obviously spin dependent
Spectral density:

$$
S_{\mathbf{k}\sigma}^{(\pm)}(E) = \hbar\alpha_\sigma^{(+)}(\mathbf{k})\delta(E-E_+(\mathbf{k}))+\hbar\alpha_\sigma^{(-)}(\mathbf{k})\delta(E-E_-(\mathbf{k}))
$$

Quasiparticle density of states:

$$\rho_\sigma^{(\pm)}(E) = \frac{1}{\hbar}\frac{1}{N}\sum_{\mathbf{k}}\left\{\alpha_\sigma^{(+)}(\mathbf{k})\delta(E-\mu-E_+(\mathbf{k}))\right.$$
$$\left.+\alpha_\sigma^{(-)}(\mathbf{k})\delta(E-\mu-E_-(\mathbf{k}))\right\}$$

k: wavevector of the first Brillouin zone of the sub-lattice.

Problem 8.5
Let $x \neq 0$:

$$\frac{1}{2}\lim_{\beta\to\infty}\frac{\beta}{1+\cosh(\beta x)} = \lim_{\beta\to\infty}\beta e^{-\beta|x|} = 0$$

The expression diverges for $x = 0$. In addition it holds

$$\int_{-\infty}^{+\infty} dx\,\frac{1}{2}\lim_{\beta\to\infty}\frac{\beta}{1+\cosh(\beta x)} = \lim_{\beta\to\infty}\int_0^{\infty} dx\,\frac{\beta}{1+\cosh(\beta x)},$$

$$\int_0^{\infty} dx\,\frac{\beta}{1+\cosh(\beta x)} = \int_0^{\infty} dy\,\frac{1}{1+\cosh(y)} = \int_0^{\infty} dy\,\frac{1}{2\cosh^2\frac{y}{2}}$$
$$= \int_0^{\infty} dz\,\frac{1}{\cosh^2 z} = \tanh z\Big|_0^{\infty}$$
$$= 1 - 0 = 1.$$

Thus we have the defining properties of the δ-function satisfied!

Problem 8.6
According to (8.97) the Hubbard Hamiltonian can be written as follows:

$$H = \sum_{\mathbf{k}\sigma}\varepsilon(\mathbf{k})c_{\mathbf{k}\sigma}^{+}c_{\mathbf{k}\sigma} - \frac{2U}{3N}\sum_{\mathbf{k}}\sigma(\mathbf{k})\cdot\sigma(-\mathbf{k}) + \frac{1}{2}U\hat{N} - 2\mu_B B_0\sigma^z(\mathbf{0})$$

1. We calculate the commutator termwise:

$$
\begin{aligned}
&\left[\sigma^+(\mathbf{k}), \sum_{m,n,\sigma} T_{mn} c^+_{m\sigma} c_{n\sigma}\right]_- \\
&= \sum_i e^{-i\mathbf{k}\cdot\mathbf{R}_i} \sum_{m,n,\sigma} T_{mn} \left[c^+_{i\uparrow} c_{i\downarrow}, c^+_{m\sigma} c_{n\sigma}\right]_- \\
&= \sum_i e^{-i\mathbf{k}\cdot\mathbf{R}_i} \sum_{m,n,\sigma} T_{mn} \left(\delta_{im}\delta_{\sigma\downarrow} c^+_{i\uparrow} c_{n\sigma} - \delta_{in}\delta_{\sigma\uparrow} c^+_{m\sigma} c_{i\downarrow}\right) \\
&= \sum_{m,n} T_{mn} \left(c^+_{m\uparrow} c_{n\downarrow} e^{-i\mathbf{k}\cdot\mathbf{R}_m} - c^+_{m\uparrow} c_{n\downarrow} e^{-i\mathbf{k}\cdot\mathbf{R}_n}\right) \\
&= \sum_{m,n} T_{mn} \left(e^{-i\mathbf{k}\cdot\mathbf{R}_m} - e^{-i\mathbf{k}\cdot\mathbf{R}_n}\right) c^+_{m\uparrow} c_{n\downarrow}
\end{aligned}
$$

We further calculate

$$
\begin{aligned}
&\left[\sigma^+(\mathbf{k}), \sum_{\mathbf{p}} \sigma(\mathbf{p})\sigma(-\mathbf{p})\right]_- \\
&= \sum_{\mathbf{p}} \left[\sigma^+(\mathbf{k}), \sigma^z(\mathbf{p})\sigma^z(-\mathbf{p} + \frac{1}{2}\sigma^+(\mathbf{p})\sigma^-(-\mathbf{p})\right. \\
&\qquad\qquad \left. + \frac{1}{2}\sigma^-(\mathbf{p})\sigma^+(-\mathbf{p})\right]_- \\
&= \sum_{\mathbf{p}} \left\{\sigma^z(\mathbf{p})\left[\sigma^+(\mathbf{k}), \sigma^z(-\mathbf{p})\right]_- + \left[\sigma^+(\mathbf{k}), \sigma^z(\mathbf{p})\right]_- *\right. \\
&\qquad *\sigma^z(-\mathbf{p}) + \frac{1}{2}\sigma^+(\mathbf{p})\left[\sigma^+(\mathbf{k}), \sigma^-(-\mathbf{p})\right]_- \\
&\qquad \left. + \frac{1}{2}\left[\sigma^+(\mathbf{k}), \sigma^-(\mathbf{p})\right]_- \sigma^+(-\mathbf{p})\right\} \\
&= \sum_{\mathbf{p}} \left\{-\sigma^z(\mathbf{p})\sigma^+(\mathbf{k}-\mathbf{p}) - \sigma^+(\mathbf{k}+\mathbf{p})\sigma^z(-\mathbf{p}) + \right. \\
&\qquad \left. + \sigma^+(\mathbf{p})\sigma^z(\mathbf{k}-\mathbf{p}) + \sigma^z(\mathbf{k}+\mathbf{p})\sigma^+(-\mathbf{p})\right\} \\
&= 0
\end{aligned}
$$

One recognizes this when one replaces $\mathbf{p}$ by $\mathbf{p}+\mathbf{k}$ in the term before the last and $\mathbf{p}$ by $\mathbf{p}-\mathbf{k}$ in the last term:

$$\begin{aligned}\left[\sigma^+(\mathbf{k}), \hat{N}\right]_- &= \sum_{i,m,\sigma} e^{-i\mathbf{k}\cdot\mathbf{R}_i}\left[c^+_{i\uparrow}c_{i\downarrow}, c^+_{m\sigma}c_{m\sigma}\right]_- \\ &= \sum_{i,m,\sigma} e^{-i\mathbf{k}\cdot\mathbf{R}_i}\left\{\delta_{\sigma\downarrow}\delta_{im}c^+_{i\uparrow}c_{m\sigma} - \delta_{im}\delta_{\sigma\uparrow}c^+_{m\sigma}c_{i\downarrow}\right\} \\ &= \sum_i e^{-i\mathbf{k}\cdot\mathbf{R}_i}\left\{c^+_{i\uparrow}c_{i\downarrow} - c^+_{i\uparrow}c_{i\downarrow}\right\} \\ &= 0 \\ \left[\sigma^+(\mathbf{k}), \sigma^z(\mathbf{0})\right]_- &= -\sigma^+(\mathbf{k})\end{aligned}$$

Therefore what remains is

$$\begin{aligned}\left[\sigma^+(\mathbf{k}), H\right]_- &= 2\mu_B B_0 \sigma^+(\mathbf{k}) \\ &\quad + \sum_{m,n} T_{mn}\left(e^{-i\mathbf{k}\cdot\mathbf{R}_m} - e^{-i\mathbf{k}\cdot\mathbf{R}_n}\right)c^+_{m\uparrow}c_{n\downarrow}\end{aligned}$$

This corresponds to (8.106).

2. For the double commutator we need the results of 1.:

$$\begin{aligned}&\sum_{m,n} T_{mn}\left(e^{-i\mathbf{k}\cdot\mathbf{R}_m} - e^{-i\mathbf{k}\cdot\mathbf{R}_n}\right)\sum_i e^{i\mathbf{k}\cdot\mathbf{R}_i}\left[c^+_{m\uparrow}c_{n\downarrow}, c^+_{i\downarrow}c_{i\uparrow}\right]_- \\ &= \sum_{m,n} T_{mn}\left(e^{-i\mathbf{k}\cdot\mathbf{R}_m} - e^{-i\mathbf{k}\cdot\mathbf{R}_n}\right) * \\ &\qquad * \sum_i e^{i\mathbf{k}\cdot\mathbf{R}_i}\left\{\delta_{ni}c^+_{m\uparrow}c_{i\uparrow} - \delta_{mi}c^+_{i\downarrow}c_{n\downarrow}\right\} \\ &= \sum_{m,n} T_{mn}\left\{\left(e^{i\mathbf{k}(\mathbf{R}_n-\mathbf{R}_m)} - 1\right)c^+_{m\uparrow}c_{n\uparrow} - \right. \\ &\qquad \left. - \left(1 - e^{i\mathbf{k}(\mathbf{R}_m-\mathbf{R}_n)}\right)\delta_{mi}c^+_{m\downarrow}c_{n\downarrow}\right\}\end{aligned}$$

So that it follows:

$$\begin{aligned}&\left[\left[\sigma^+(\mathbf{k}), H\right]_-, \sigma^-(-\mathbf{k})\right]_- \\ &= 4\mu_B B_0 \sigma^z(\mathbf{0}) + \sum_{mn\sigma} T_{mn}\left(e^{iz_\sigma \mathbf{k}\cdot(\mathbf{R}_n-\mathbf{R}_m)} - 1\right)c^+_{m\sigma}c_{n\sigma}\end{aligned}$$

This is exactly (8.107).

Problem 8.7
We use (8.137) and (8.138) and then have the following determining equation for the chemical potential μ:

$$n = n_\sigma + n_{-\sigma} = \frac{2f_-(T_0)}{1 + f_-(T_0) - f_-(T_0 + U)} \stackrel{!}{=} 1$$

That is the same thing as

$$\begin{aligned}
& f_-(T_0) = 1 - f_-(T_0 + U) \\
\curvearrowright\ & \frac{1}{e^{\beta(T_0-\mu)} + 1} = 1 - \frac{1}{e^{\beta(T_0+U-\mu)} + 1} = \frac{e^{\beta(T_0+U-\mu)}}{e^{\beta(T_0+U-\mu)} + 1} \\
\curvearrowright\ & e^{\beta(T_0-\mu)} + 1 = 1 + e^{-\beta(T_0+U-\mu)} \\
\curvearrowright\ & T_0 - \mu = -(T_0 + U - \mu) \\
\curvearrowright\ & 2T_0 + U = 2\mu \\
\curvearrowright\ & \mu = T_0 + \frac{U}{2}
\end{aligned}$$

Problem 8.8
Substituting the spectral density (8.139) for

$$n_\sigma = n_{-\sigma} = \frac{1}{2}$$

in (8.77), it directly follows:

$$Z_{W=0} = -\frac{1}{2}\left(f_-{}'(T_0) + f_-{}'(T_0 + U)\right)$$

On the other hand substituting in (8.78) gives

$$N_{W=0} = -f_-(T_0) + f_-(T_0 + U)$$

For the susceptibility holds (8.79)

$$\bar{\chi}_{W=0} = 2\mu_B \frac{Z_{W=0}}{1 + N_{W=0}}$$

The chemical potential for half-filling is known from Problem 8.7:

$$\mu(n = 1) = T_0 + \frac{U}{2}$$

So that we calculate

$$
\begin{aligned}
(f_-(T_0+U) - f_-(T_0)) &= \left(\frac{1}{e^{\beta(T_0+U-\mu)}+1} - \frac{1}{e^{\beta(T_0-\mu)}+1}\right) \\
&= \left(\frac{1}{e^{\beta\frac{U}{2}}+1} - \frac{1}{e^{-\beta\frac{U}{2}}+1}\right) \\
&= \frac{e^{-\beta\frac{U}{4}} - e^{\beta\frac{U}{4}}}{e^{\beta\frac{U}{4}} + e^{-\beta\frac{U}{4}}} \\
\curvearrowright \quad N_{W=0} &= -\tanh(\beta\frac{W}{4})
\end{aligned}
$$

For $Z_{W=0}$ we need the derivative of the Fermi function:

$$
\begin{aligned}
f_-{}'(E) &= -\beta\frac{e^{\beta(E-\mu)}}{(e^{\beta(E-\mu)}+1)^2} \\
&= -\beta\frac{1}{\left(e^{\frac{1}{2}\beta(E-\mu)} + e^{-\frac{1}{2}\beta(E-\mu)}\right)^2}
\end{aligned}
$$

Then it follows:

$$
\begin{aligned}
-\frac{1}{2}\left(f_-{}'(T_0) + f_-{}'(T_0+U)\right) &= \frac{\beta}{2}\left(\frac{1}{\left(e^{-\beta\frac{U}{4}} + e^{+\beta\frac{U}{4}}\right)^2} + \right. \\
&\qquad \left. + \frac{1}{\left(e^{+\beta\frac{U}{4}} + e^{-\beta\frac{U}{4}}\right)^2}\right) \\
&= \frac{\beta}{2}\cdot 2\cdot\frac{1}{4}\cdot\frac{1}{\cosh^2(\beta\frac{U}{4})} \\
\curvearrowright \quad Z_{W=0} &= \frac{1}{4}\beta\frac{1}{\cosh^2(\beta\frac{U}{4})}
\end{aligned}
$$

Susceptibility:

$$
\begin{aligned}
\bar{\chi} &= \frac{1}{2}\beta\mu_B\frac{\frac{1}{\cosh^2(\beta\frac{U}{4})}}{1-\tanh(\beta\frac{U}{4})} \\
&= \frac{1}{2}\beta\mu_B\frac{1+\tanh(\beta\frac{U}{4})}{\cosh^2(\beta\frac{U}{4})(1-\tanh^2(\beta\frac{U}{4}))} \\
&= \frac{1}{2}\beta\mu_B\frac{1+\tanh(\beta\frac{U}{4})}{\cosh^2(\beta\frac{U}{4}) - \sinh^2(\beta\frac{U}{4})}
\end{aligned}
$$

Then we finally have

$$\bar{\chi}_{W=0} = \frac{1}{2}\beta\mu_B\left(1+\tanh\left(\frac{\beta U}{4}\right)\right)$$

Problem 8.9
Complete basis:

$$\left|E_1^{(1)}\right\rangle = c_{1\sigma}^+ \,|0\rangle$$
$$\left|E_2^{(1)}\right\rangle = c_{2\sigma}^+ \,|0\rangle$$

Hamiltonian matrix:

$$H^{(1)} = \begin{pmatrix} 0 & t \\ t & 0 \end{pmatrix}$$

Diagonalization:

$$Det\left(H^{(1)} - E\right) \stackrel{!}{=} 0$$
$$Det\begin{vmatrix} -E & t \\ t & -E \end{vmatrix} = E^2 - t^2$$
$$\Leftrightarrow \quad E_1^{(1)} = -t \quad ; \qquad E_2^{(1)} = +t$$

For $E_1^{(1)}$:

$$\begin{pmatrix} t & t \\ t & t \end{pmatrix}\begin{pmatrix} \alpha_1 \\ \alpha_2 \end{pmatrix} \stackrel{!}{=} 0$$
$$\Leftrightarrow \quad t(\alpha_1+\alpha_2) = 0 \quad \Leftrightarrow \quad \alpha_2 = -\alpha_1$$

Normalization:

$$\left|E_1^{(1)}\right\rangle = \frac{1}{\sqrt{2}}\left(c_{1\sigma}^+\,|0\rangle - c_{2\sigma}^+\,|0\rangle\right)$$

For $E_2^{(1)}$:

$$\begin{pmatrix} -t & t \\ t & -t \end{pmatrix}\begin{pmatrix} \beta_1 \\ \beta_2 \end{pmatrix} \stackrel{!}{=} 0$$
$$\Leftrightarrow \qquad t(-\beta_1+\beta_2) = 0 \quad \Leftrightarrow \quad \beta_1 = \beta_2$$

Normalization:

$$\left|E_2^{(1)}\right\rangle = \frac{1}{\sqrt{2}}\left(c_{1\sigma}^+\,|0\rangle + c_{2\sigma}^+\,|0\rangle\right)$$

Problem 8.10

1. We use

$$\begin{aligned}
\left[c_{i\sigma}, c_{j\sigma'}^{+}\right]_{+} &= \delta_{ij}\delta_{\sigma\sigma'} \\
\left[c_{i\sigma}, c_{j\sigma'}\right]_{+} &= \left[c_{i\sigma}^{+}, c_{j\sigma'}^{+}\right]_{+} = 0 \\
c_{i\sigma}\left|0\right\rangle = 0 \ ; &\qquad \left\langle 0\right| c_{j\sigma}^{+} = 0 \\
\left\langle \varepsilon_1^{(2)} \mid \varepsilon_1^{(2)}\right\rangle &= \left\langle 0\right| c_{2-\sigma} c_{1\sigma} c_{1\sigma}^{+} c_{2-\sigma}^{+} \left|0\right\rangle \\
&= \left\langle 0\right| c_{2-\sigma} c_{2-\sigma}^{+} \left|0\right\rangle = \left\langle 0 \mid 0\right\rangle = 1
\end{aligned}$$

analogously:

$$\begin{aligned}
\left\langle \varepsilon_i^{(2)} \middle| \varepsilon_i^{(2)}\right\rangle &= 1 \qquad i = 2, \cdots, 4 \\
\left\langle \varepsilon_1^{(2)} \middle| \varepsilon_2^{(2)}\right\rangle &= \left\langle 0\right| c_{2-\sigma} c_{1\sigma} c_{2\sigma}^{\dagger} c_{1-\sigma}^{\dagger} \left|0\right\rangle \\
&= \left\langle 0 \middle| c_{2-\sigma} c_{2\sigma}^{\dagger} c_{1-\sigma}^{\dagger} c_{1\sigma} \middle| 0\right\rangle \\
&= 0 \\
\left\langle \varepsilon_1^{(2)} \middle| \varepsilon_3^{(2)}\right\rangle &= \left\langle 0\right| c_{2-\sigma} c_{1\sigma} c_{1\sigma}^{\dagger} c_{1-\sigma}^{\dagger} \left|0\right\rangle = 0 \\
\left\langle \varepsilon_1^{(2)} \middle| \varepsilon_4^{(2)}\right\rangle &= \left\langle 0\right| c_{2-\sigma} c_{1\sigma} c_{2\sigma}^{\dagger} c_{2-\sigma}^{\dagger} \left|0\right\rangle = 0 \\
\left\langle \varepsilon_2^{(2)} \middle| \varepsilon_3^{(2)}\right\rangle &= \left\langle 0\right| c_{1-\sigma} c_{2\sigma} c_{1\sigma}^{\dagger} c_{1-\sigma}^{\dagger} \left|0\right\rangle = 0 \\
\left\langle \varepsilon_2^{(2)} \middle| \varepsilon_4^{(2)}\right\rangle &= \left\langle 0\right| c_{1-\sigma} c_{2\sigma} c_{2\sigma}^{\dagger} c_{2-\sigma}^{\dagger} \left|0\right\rangle = 0 \\
\left\langle \varepsilon_3^{(2)} \middle| \varepsilon_4^{(2)}\right\rangle &= \left\langle 0\right| c_{1-\sigma} c_{1\sigma} c_{2\sigma}^{\dagger} c_{2-\sigma}^{\dagger} \left|0\right\rangle = 0
\end{aligned}$$

Therefore it holds

$$\left\langle \varepsilon_i^{(2)} \middle| \varepsilon_j^{(2)}\right\rangle = \delta_{ij}; \qquad i, j = 1 \ldots 4$$

2.

$$
\begin{aligned}
H\left|\varepsilon_1^{(2)}\right\rangle &= t\sum_{\sigma'}\left(c_{1\sigma'}^{\dagger}c_{2\sigma'}+c_{2\sigma'}^{\dagger}c_{1\sigma'}\right)c_{1\sigma}^{\dagger}c_{2-\sigma}^{\dagger}\left|0\right\rangle + \\
&+\frac{1}{2}U\sum_{i\sigma'}n_{i\sigma'}n_{i-\sigma'}c_{1\sigma}^{\dagger}c_{2-\sigma}^{\dagger}\left|0\right\rangle \\
&= -tc_{1-\sigma}^{\dagger}c_{1\sigma}^{\dagger}\left|0\right\rangle + tc_{2\sigma}^{\dagger}c_{2-\sigma}^{\dagger}\left|0\right\rangle + \\
&+\frac{1}{2}U\sum_{\sigma'}n_{1\sigma'}n_{1-\sigma'}c_{1\sigma}^{\dagger}c_{2-\sigma}^{\dagger}\left|0\right\rangle + \\
&+\frac{1}{2}U\sum_{\sigma'}n_{2\sigma'}n_{2-\sigma'}c_{1\sigma}^{\dagger}c_{2-\sigma}^{\dagger}\left|0\right\rangle \\
&= t\left(\left|\varepsilon_3^{(2)}\right\rangle+\left|\varepsilon_4^{(2)}\right\rangle\right)+\frac{1}{2}U\left(n_{1-\sigma}c_{1\sigma}^{\dagger}c_{2-\sigma}^{\dagger}\left|0\right\rangle + \right.\\
&\quad +c_{1\sigma}^{\dagger}n_{1-\sigma}c_{2-\sigma}^{\dagger}\left|0\right\rangle + c_{1\sigma}^{\dagger}n_{2\sigma}c_{2-\sigma}^{\dagger}\left|0\right\rangle + \\
&\qquad\qquad \left. +c_{1\sigma}^{\dagger}c_{2-\sigma}^{\dagger}n_{2\sigma}\left|0\right\rangle\right) \\
&= t\left(\left|\varepsilon_3^{(2)}\right\rangle+\left|\varepsilon_4^{(2)}\right\rangle\right)
\end{aligned}
$$

$$
\begin{aligned}
H\left|\varepsilon_2^{(2)}\right\rangle &= t\sum_{\sigma'}\left(c_{1\sigma'}^{\dagger}c_{2\sigma'}+c_{2\sigma'}^{\dagger}c_{1\sigma'}\right)c_{2\sigma}^{\dagger}c_{1-\sigma}^{\dagger}\left|0\right\rangle + \\
&+\frac{1}{2}U\sum_{i\sigma'}n_{i\sigma'}n_{i-\sigma'}c_{2\sigma}^{\dagger}c_{1-\sigma}^{\dagger}\left|0\right\rangle \\
&= tc_{1\sigma}^{\dagger}c_{1-\sigma}^{\dagger}\left|0\right\rangle - tc_{2-\sigma}^{\dagger}c_{2\sigma}^{\dagger}\left|0\right\rangle \\
&= t\left(\left|\varepsilon_3^{(2)}\right\rangle+\left|\varepsilon_4^{(2)}\right\rangle\right)
\end{aligned}
$$

$$
\begin{aligned}
H\left|\varepsilon_3^{(2)}\right\rangle &= t\sum_{\sigma'}\left(c_{1\sigma'}^{\dagger}c_{2\sigma'}+c_{2\sigma'}^{\dagger}c_{1\sigma'}\right)c_{1\sigma}^{\dagger}c_{1-\sigma}^{\dagger}\left|0\right\rangle + \\
&+\frac{1}{2}U\sum_{\sigma'}n_{1\sigma'}n_{1-\sigma'}c_{1\sigma}^{\dagger}c_{1-\sigma}^{\dagger}\left|0\right\rangle \\
&= t\sum_{\sigma'}c_{1\sigma'}^{\dagger}c_{2\sigma'}c_{1\sigma}^{\dagger}c_{1-\sigma}^{\dagger}\left|0\right\rangle + Un_{1\sigma}n_{1-\sigma}c_{1\sigma}^{\dagger}c_{1-\sigma}^{\dagger}\left|0\right\rangle \\
&= \left(tc_{2\sigma}^{\dagger}c_{1-\sigma}^{\dagger}-tc_{2-\sigma}^{+}c_{1\sigma}^{\dagger}\right)\left|0\right\rangle + Uc_{1\sigma}^{\dagger}c_{1-\sigma}^{\dagger}\left|0\right\rangle \\
&= t\left(\left|\varepsilon_1^{(2)}\right\rangle+\left|\varepsilon_2^{(2)}\right\rangle\right)+U\left|\varepsilon_3^{(2)}\right\rangle
\end{aligned}
$$

$$
\begin{aligned}
H\left|\varepsilon_4^{(2)}\right\rangle &= t\sum_{\sigma'}\left(c_{1\sigma'}^{\dagger}c_{2\sigma'} + c_{2\sigma'}^{\dagger}c_{1\sigma'}\right)c_{2\sigma}^{\dagger}c_{2-\sigma}^{\dagger}\left|0\right\rangle + \\
&+ \frac{1}{2}U\sum_{i\sigma'} n_{i\sigma'}n_{i-\sigma'}c_{2\sigma}^{\dagger}c_{2-\sigma}^{\dagger}\left|0\right\rangle \\
&= t\left(c_{1\sigma}^{\dagger}c_{2-\sigma}^{\dagger}\left|0\right\rangle - c_{1-\sigma}^{\dagger}c_{2\sigma}^{\dagger}\left|0\right\rangle\right) + \\
&+ U n_{2\sigma}n_{2-\sigma}c_{2\sigma}^{\dagger}c_{2-\sigma}^{\dagger}\left|0\right\rangle \\
&= t\left(c_{1\sigma}^{\dagger}c_{2-\sigma}^{\dagger}\left|0\right\rangle + c_{2\sigma}^{\dagger}c_{1-\sigma}^{\dagger}\left|0\right\rangle\right) + U c_{2\sigma}^{\dagger}c_{2-\sigma}^{\dagger}\left|0\right\rangle \\
&= t\left(\left|\varepsilon_1^{(2)}\right\rangle + \left|\varepsilon_2^{(2)}\right\rangle\right) + U\left|\varepsilon_4^{(2)}\right\rangle
\end{aligned}
$$

So that we get from the matrix elements

$$
\left\langle\varepsilon_i^{(2)}\right| H \left|\varepsilon_j^{(2)}\right\rangle; \qquad i,j = 1\ldots 4
$$

the Hamiltonian matrix:

$$
H^{(2)} = \begin{pmatrix} 0 & 0 & t & t \\ 0 & 0 & t & t \\ t & t & U & 0 \\ t & t & 0 & U \end{pmatrix}
$$

3. The eigenvalues are determined from the secular determinant:

$$
det\left(H^{(2)} - E\,\mathbb{1}\right) \stackrel{!}{=} 0
$$

$$
\begin{aligned}
& det\begin{pmatrix} -E & 0 & t & t \\ 0 & -E & t & t \\ t & t & U-E & 0 \\ t & t & 0 & U-E \end{pmatrix} \\
&= -E\,det\begin{pmatrix} -E & t & t \\ t & U-E & 0 \\ t & 0 & U-E \end{pmatrix} + \\
&+ t\,det\begin{pmatrix} 0 & -E & t \\ t & t & 0 \\ t & t & U-E \end{pmatrix} - t\,det\begin{pmatrix} 0 & -E & t \\ t & t & U-E \\ t & t & 0 \end{pmatrix} \\
&= (-E)\left((-E)(U-E)^2 - 2t^2(U-E)\right) + \\
&+ t(t^3 - t^3 + E(U-E)t) - t(-E(U-E)t + t^3 - t^3) \\
&= E(U-E)(E(U-E) + 4t^2)
\end{aligned}
$$

It must therefore hold

$$0 = E(U - E)(E(U - E) + 4t^2)$$

Eigenvalues

$$\begin{aligned} E_1^{(2)} &= E_- \\ E_2^{(2)} &= 0 \\ E_3^{(2)} &= U \\ E_4^{(2)} &= E_+ \end{aligned}$$

with

$$E_\pm = \frac{1}{2}U \pm \sqrt{\frac{1}{4}U^2 + 4t^2}$$

Eigenstates:
$E_1^{(2)} = E_-$

$$\begin{pmatrix} -E_- & 0 & t & t \\ 0 & -E_- & t & t \\ t & t & U - E_- & 0 \\ t & t & 0 & U - E_- \end{pmatrix} \begin{pmatrix} x_1 \\ x_2 \\ x_3 \\ x_4 \end{pmatrix} \overset{!}{=} 0$$

$$\begin{aligned} \curvearrowright \quad -E_- x_1 + t(x_3 + x_4) &= 0 \\ -E_- x_2 + t(x_3 + x_4) &= 0 \\ t(x_1 + x_2) + (U - E_-)x_3 &= 0 \\ t(x_1 + x_2) + (U - E_-)x_4 &= 0 \end{aligned}$$

$$\curvearrowright \quad x_1 = x_2; \quad x_3 = x_4; \quad x_3 = \frac{E_-}{2t}\, x_2 \, ;$$

We define (8.168)

$$\gamma_\pm = \frac{E_\pm}{2t}$$

Normalization:

$$\begin{aligned} x_1^2 + x_2^2 + x_3^2 + x_4^2 &= 1 \\ 2x_1^2 + 2x_3^2 &= 1 \\ 2x_1^2 + 2\gamma_-^2 x^2 &= 1 \\ x_1^2 &= \frac{1}{2(1+\gamma_-^2)} \end{aligned}$$

$$\curvearrowright \quad x_1 = \frac{1}{\sqrt{2(1+\gamma_-^2)}} = x_2$$

$$x_3 = \frac{\gamma_-}{\sqrt{2(1+\gamma_-^2)}} = x_4$$

$$\curvearrowright \quad \left|E_1^{(2)}\right\rangle = \frac{1}{\sqrt{2(1+\gamma_-^2)}} \left(\left|\varepsilon_1^{(2)}\right\rangle + \left|\varepsilon_2^{(2)}\right\rangle + \gamma_- \left(\left|\varepsilon_3^{(2)}\right\rangle + \left|\varepsilon_4^{(2)}\right\rangle \right) \right); \qquad (8.164)$$

$E_2^{(2)} = 0$

$$\begin{pmatrix} 0 & 0 & t & t \\ 0 & 0 & t & t \\ t & t & U & 0 \\ t & t & 0 & U \end{pmatrix} \begin{pmatrix} y_1 \\ y_2 \\ y_3 \\ y_4 \end{pmatrix} \overset{!}{=} 0$$

$$\begin{aligned} \curvearrowright \quad t(y_3 + y_4) &= 0 \\ t(y_3 + y_4) &= 0 \\ t(y_1 + y_2) + U y_3 &= 0 \\ t(y_1 + y_2) + U y_4 &= 0 \end{aligned}$$

$$\curvearrowright \quad y_3 = -y_4; \quad y_3 = +y_4; \quad y_3 = y_4 = 0; \quad y_1 = -y_2;$$

Normalization:

$$y_1 = \frac{1}{\sqrt{2}} = -y_2$$

$$\curvearrowright \quad \left|E_2^{(2)}\right\rangle = \frac{1}{\sqrt{2}} \left(\left|\varepsilon_1^{(2)}\right\rangle - \left|\varepsilon_2^{(2)}\right\rangle \right); \qquad (8.165)$$

$E_3^{(2)} = U$

$$\begin{pmatrix} -U & 0 & t & t \\ 0 & -U & t & t \\ t & t & 0 & 0 \\ t & t & 0 & 0 \end{pmatrix} \begin{pmatrix} z_1 \\ z_2 \\ z_3 \\ z_4 \end{pmatrix} \stackrel{!}{=} 0$$

$$\curvearrowright \quad -Uz_1 + t(z_3 + z_4) = 0$$

$$-Uz_2 + t(z_3 + z_4) = 0$$

$$t(z_1 + z_2) = 0$$

$$t(z_1 + z_2) = 0$$

$$\curvearrowright \quad z_1 = z_2; \quad z_1 = -z_2; \quad z_1 = z_2 = 0; \quad z_3 = -z_4;$$

Normalization:

$$z_3 = \frac{1}{\sqrt{2}} = -z_4$$

$$\curvearrowright \quad \left| E_3^{(2)} \right\rangle = \frac{1}{\sqrt{2}} \left(\left| \varepsilon_3^{(2)} \right\rangle - \left| \varepsilon_4^{(2)} \right\rangle \right); \qquad (8.166)$$

$E_4^{(2)} = E_+$

$$\begin{pmatrix} -E_+ & 0 & t & t \\ 0 & -E_+ & t & t \\ t & t & U - E_+ & 0 \\ t & t & 0 & U - E_+ \end{pmatrix} \begin{pmatrix} w_1 \\ w_2 \\ w_3 \\ w_4 \end{pmatrix} \stackrel{!}{=} 0$$

Formally as for $E_1^{(2)} = E_-$, only γ_- is replaced by γ_+:

$$\curvearrowright \quad \left| E_4^{(2)} \right\rangle = \frac{1}{\sqrt{2(1 + \gamma_+^2)}} \left(\left| \varepsilon_1^{(2)} \right\rangle + \left| \varepsilon_2^{(2)} \right\rangle + \right.$$

$$\left. + \gamma_+ \left(\left| \varepsilon_3^{(2)} \right\rangle + \left| \varepsilon_4^{(2)} \right\rangle \right) \right); \qquad (8.167)$$

4.
(i)

$$
\begin{aligned}
\left\langle E_1^{(2)}\right| c_{1\sigma}^\dagger \left|E_1^{(1)}\right\rangle &= \frac{1}{\sqrt{2}}\left\langle E_1^{(2)}\right| c_{1\sigma}^\dagger (c_{1-\sigma}^\dagger - c_{2-\sigma}^\dagger)\left|0\right\rangle \\
&= \frac{1}{\sqrt{2}}\left\langle E_1^{(2)} \middle| \varepsilon_3^{(2)}\right\rangle - \frac{1}{\sqrt{2}}\left\langle E_1^{(2)} \middle| \varepsilon_1^{(2)}\right\rangle \\
&= \frac{1}{\sqrt{2(1+\gamma_-^2)}}(\gamma_- - 1)\frac{1}{\sqrt{2}} \\
&= \frac{1}{2}\frac{\gamma_- - 1}{\sqrt{1+\gamma_-^2}}\ ; \\
E_1^{(2)} - E_1^{(1)} &= E_- + t
\end{aligned}
$$

(ii)

$$
\begin{aligned}
\left\langle E_2^{(2)}\right| c_{1\sigma}^\dagger \left|E_1^{(1)}\right\rangle &= \frac{1}{\sqrt{2}}\left(\left\langle E_2^{(2)} \middle| \varepsilon_3^{(2)}\right\rangle - \left\langle E_2^{(2)} \middle| \varepsilon_1^{(2)}\right\rangle\right) \\
&= \frac{1}{2}\ ; \\
E_2^{(2)} - E_1^{(1)} &= +t
\end{aligned}
$$

(iii)

$$
\begin{aligned}
\left\langle E_3^{(2)}\right| c_{1\sigma}^\dagger \left|E_1^{(1)}\right\rangle &= \frac{1}{\sqrt{2}}\left(\left\langle E_3^{(2)} \middle| \varepsilon_3^{(2)}\right\rangle - \left\langle E_3^{(2)} \middle| \varepsilon_1^{(2)}\right\rangle\right) \\
&= \frac{1}{2}\ ; \\
E_3^{(2)} - E_1^{(1)} &= U + t
\end{aligned}
$$

(iv)

$$
\begin{aligned}
\left\langle E_4^{(2)}\right| c_{1\sigma}^\dagger \left|E_1^{(1)}\right\rangle &= \frac{1}{\sqrt{2}}\left(\left\langle E_4^{(2)} \middle| \varepsilon_3^{(2)}\right\rangle - \left\langle E_4^{(2)} \middle| \varepsilon_1^{(2)}\right\rangle\right) \\
&= \frac{1}{2}\frac{1}{\sqrt{1+\gamma_+^2}}(\gamma_+ - 1)\ ; \\
E_4^{(2)} - E_1^{(1)} &= E_+ + t
\end{aligned}
$$

(v)

$$\left\langle E_1^{(2)}\right| c_{1\sigma}^\dagger \left|E_2^{(1)}\right\rangle = \frac{1}{\sqrt{2}}\left(\left\langle E_1^{(2)}\right|\left.\varepsilon_3^{(2)}\right\rangle + \left\langle E_1^{(2)}\right|\left.\varepsilon_1^{(2)}\right\rangle\right)$$
$$= \frac{1}{2}\frac{1}{\sqrt{1+\gamma_-^2}}(1+\gamma_-)\,;$$
$$E_1^{(2)} - E_2^{(1)} = E_- + t$$

(vi)

$$\left\langle E_2^{(2)}\right| c_{1\sigma}^\dagger \left|E_2^{(1)}\right\rangle = \frac{1}{\sqrt{2}}\left(\left\langle E_2^{(2)}\right|\left.\varepsilon_3^{(2)}\right\rangle + \left\langle E_2^{(2)}\right|\left.\varepsilon_1^{(2)}\right\rangle\right)$$
$$= \frac{1}{2}\,;$$
$$E_2^{(2)} - E_2^{(1)} = -t$$

(vii)

$$\left\langle E_3^{(2)}\right| c_{1\sigma}^\dagger \left|E_2^{(1)}\right\rangle = \frac{1}{\sqrt{2}}\left(\left\langle E_3^{(2)}\right|\left.\varepsilon_3^{(2)}\right\rangle + \left\langle E_3^{(2)}\right|\left.\varepsilon_1^{(2)}\right\rangle\right)$$
$$= \frac{1}{2}\,;$$
$$E_3^{(2)} - E_2^{(1)} = U - t$$

(viii)

$$\left\langle E_4^{(2)}\right| c_{1\sigma}^\dagger \left|E_2^{(1)}\right\rangle = \frac{1}{\sqrt{2}}\left(\left\langle E_4^{(2)}\right|\left.\varepsilon_3^{(2)}\right\rangle + \left\langle E_4^{(2)}\right|\left.\varepsilon_1^{(2)}\right\rangle\right)$$
$$= \frac{1}{2}\frac{1}{\sqrt{1+\gamma_+^2}}(\gamma_+ + 1)\,;$$
$$E_4^{(2)} - E_2^{(1)} = E_+ - t$$

For the partition function of the one-particle system holds

$$Z = e^{+\beta t} + e^{-\beta t};$$

and therefore

$$\frac{e^{-\beta E_1^{(1)}}}{Z} = \frac{e^{\beta t}}{e^{\beta t} + e^{-\beta t}} = \frac{1}{1+e^{-2\beta t}}$$
$$\frac{e^{-\beta E_2^{(1)}}}{Z} = \frac{e^{-\beta t}}{e^{\beta t} + e^{-\beta t}} = \frac{e^{-2\beta t}}{1+e^{-2\beta t}}$$

Then with (8.147) the density of states is given by

$$\rho_{\sigma}^{(-\sigma)}(E) = \rho_{t\sigma}^{(-\sigma)}(E) + \rho_{U\sigma}^{(-\sigma)}(E)$$

$$\rho_{t\sigma}(E) = \frac{1}{4(1+e^{-2\beta t})} *$$
$$* \left\{ \frac{(1-\gamma_-)^2}{1+\gamma_-^2} \delta(E-(E_- + t)) + \delta(E-t) + \right.$$
$$\left. + \frac{(1+\gamma_-)^2}{1+\gamma_-^2} e^{-2\beta t} \delta(E-(E_- + t)) + e^{-2\beta t} \delta(E+t) \right\}$$

$$\rho_{U\sigma}(E) = \frac{1}{4(1+e^{-2\beta t})} *$$
$$* \left\{ \frac{(1-\gamma_+)^2}{1+\gamma_+^2} \delta(E-(E_+ + t)) + \delta(E-(U+t)) + \right.$$
$$+ \frac{(1+\gamma_+)^2}{1+\gamma_+^2} e^{-2\beta t} \delta(E-(E_+ - t))$$
$$\left. + e^{-2\beta t} \delta(E-(U-t)) \right\}$$

Problem 8.11

1. We demonstrate the correctness of the eigenstates by substituting of $\left|E_{1,2}^{(3)}\right\rangle$ in the corresponding eigenvalue equation:

$$H_0 \left|E_{1,2}^{(3)}\right\rangle = \frac{1}{\sqrt{2}} t \sum_{\sigma'} \left(c_{1\sigma'}^{\dagger} c_{2\sigma'} + c_{2\sigma'}^{\dagger} c_{1\sigma'} \right) *$$
$$* \left(c_{1-\sigma}^{\dagger} \mp c_{2-\sigma}^{\dagger} \right) c_{1\sigma}^{\dagger} c_{2\sigma}^{\dagger} |0\rangle$$
$$= \frac{t}{\sqrt{2}} \left(c_{1-\sigma}^{\dagger} c_{2-\sigma} + c_{2-\sigma}^{\dagger} c_{1-\sigma} \right) *$$
$$* \left(c_{1-\sigma}^{\dagger} \mp c_{2-\sigma}^{\dagger} \right) c_{1\sigma}^{\dagger} c_{2\sigma}^{\dagger} |0\rangle$$
$$= \frac{t}{\sqrt{2}} \left(\mp c_{1-\sigma}^{\dagger} (1 - c_{2-\sigma}^{\dagger} c_{2-\sigma}) + \right.$$
$$\left. + c_{2-\sigma}^{\dagger} (1 - c_{1-\sigma}^{\dagger} c_{1-\sigma}) \right) c_{1\sigma}^{\dagger} c_{2\sigma}^{\dagger} |0\rangle$$
$$= \frac{t}{\sqrt{2}} \left(c_{2-\sigma}^{\dagger} \mp c_{1-\sigma}^{\dagger} \right) c_{1\sigma}^{\dagger} c_{2\sigma}^{\dagger} |0\rangle$$
$$= \mp t \left|E_{1,2}^{(3)}\right\rangle$$

$$
\begin{aligned}
H_1 \left| E_{1,2}^{(3)} \right\rangle &= \frac{1}{\sqrt{2}} \cdot \frac{1}{2} U \sum_{\sigma'} (n_{1\sigma'} n_{1-\sigma'} + n_{2\sigma'} n_{2-\sigma'}) * \\
&\quad * \left(c_{1-\sigma}^{\dagger} \mp c_{2-\sigma}^{\dagger} \right) c_{1\sigma}^{\dagger} c_{2\sigma}^{\dagger} |0\rangle \\
&= \frac{U}{\sqrt{2}} (n_{1\sigma} n_{1-\sigma} + n_{2\sigma} n_{2-\sigma}) * \\
&\quad * \left(c_{1-\sigma}^{\dagger} \mp c_{2-\sigma}^{\dagger} \right) c_{1\sigma}^{\dagger} c_{2\sigma}^{\dagger} |0\rangle \\
&= \frac{U}{\sqrt{2}} \left(n_{1\sigma} n_{1-\sigma} c_{1-\sigma}^{\dagger} \mp n_{2\sigma} n_{2-\sigma} c_{2-\sigma}^{\dagger} \right) c_{1\sigma}^{\dagger} c_{2\sigma}^{\dagger} |0\rangle \\
&= \frac{U}{\sqrt{2}} \left(n_{1\sigma} c_{1-\sigma}^{\dagger} \mp n_{2\sigma} c_{2-\sigma}^{\dagger} \right) c_{1\sigma}^{\dagger} c_{2\sigma}^{\dagger} |0\rangle \\
&= \frac{U}{\sqrt{2}} \left(-c_{1\sigma}^{\dagger} c_{1-\sigma}^{\dagger} c_{2\sigma}^{\dagger} \mp c_{2\sigma}^{\dagger} c_{2-\sigma}^{\dagger} c_{1\sigma}^{\dagger} \right) |0\rangle \\
&= \frac{U}{\sqrt{2}} \left(c_{1-\sigma}^{\dagger} \mp c_{2-\sigma}^{\dagger} \right) c_{1\sigma}^{\dagger} c_{2\sigma}^{\dagger} |0\rangle \\
&= U \left| E_{1,2}^{(3)} \right\rangle
\end{aligned}
$$

Therefore they are indeed eigenstates with eigenenergies:

$$
E_1^{(3)} = U - t; \qquad E_2^{(3)} = U + t
$$

2. Density of states

$$
\begin{aligned}
\rho_{\sigma}^{(\sigma,-\sigma)}(E) &= \sum_n \left| \left\langle E_n^{(3)} \left| c_{1\sigma}^{\dagger} \right| E_1^{(2)} \right\rangle \right|^2 \delta(E - (E_n^{(3)} - E_-)) + \\
&\quad + \sum_m \left| \left\langle E_m^{(1)} \left| c_{1\sigma} \right| E_1^{(2)} \right\rangle \right|^2 \delta(E - (E_- - E_m^{(1)}))
\end{aligned}
$$

At $T = 0$, for the averaging in $\rho_{\sigma}^{(\sigma,-\sigma)}(E)$, the ground state $\left| E_1^{(2)} \right\rangle$ of the two-electron system can be used (see Problem 8.10, Eq. (8.164)) with the ground state energy $E_1^{(2)} = E_-$ (8.163).
Matrix elements:

$$
\begin{aligned}
&\left\langle E_{1,2}^{(3)} \middle| c_{1\sigma}^\dagger \middle| E_1^{(2)} \right\rangle \\
&= \frac{1}{\sqrt{2(1+\gamma_-^2)}} \left\langle E_{1,2}^{(3)} \right| c_{1\sigma}^\dagger \Big(c_{1\sigma}^\dagger c_{2-\sigma}^\dagger + \\
&\qquad + c_{2\sigma}^\dagger c_{1-\sigma}^\dagger + \gamma_- \sum_{i=1}^{2} c_{i\sigma}^\dagger c_{i-\sigma}^\dagger \Big) \Big| 0 \Big\rangle \\
&= \frac{1}{\sqrt{2(1+\gamma_-^2)}} \left\langle E_{1,2}^{(3)} \right| (c_{1\sigma}^\dagger c_{2\sigma}^\dagger c_{1-\sigma}^\dagger + \\
&\qquad + \gamma_- c_{1\sigma}^\dagger c_{2\sigma}^\dagger c_{2-\sigma}^\dagger) \Big| 0 \Big\rangle \\
&= \frac{1}{2} \frac{1}{\sqrt{(1+\gamma_-^2)}} \left\langle 0 \right| c_{2\sigma} c_{1\sigma} c_{1-\sigma} c_{1-\sigma}^\dagger c_{1\sigma}^\dagger c_{2\sigma}^\dagger) \Big| 0 \Big\rangle \\
&\mp \frac{1}{2} \frac{1}{\sqrt{(1+\gamma_-^2)}} \gamma_- \left\langle 0 \right| c_{2\sigma} c_{1\sigma} c_{2-\sigma} c_{1\sigma}^\dagger c_{2\sigma}^\dagger c_{2-\sigma}^\dagger) \Big| 0 \Big\rangle \\
&= \frac{1}{2} \frac{1}{\sqrt{(1+\gamma_-^2)}} \langle 0 \mid 0 \rangle \mp \frac{1}{2} \frac{1}{\sqrt{(1+\gamma_-^2)}} \gamma_- \langle 0 \mid 0 \rangle \\
&= \frac{1}{2} \frac{1 \mp \gamma_-}{\sqrt{(1+\gamma_-^2)}}
\end{aligned}
$$

With (8.160) and (8.161) we calculate

$$
\begin{aligned}
\left\langle E_{1,2}^{(1)} \right| c_{1\sigma} \left| E_1^{(2)} \right\rangle &= \frac{1}{\sqrt{2}} \left\langle 0 \right| (c_{1-\sigma} \mp c_{2-\sigma}) c_{1\sigma} \left| E_1^{(2)} \right\rangle \\
&= \frac{1}{2} \frac{1}{\sqrt{1+\gamma_-^2}} \Big\{ \left\langle 0 \middle| c_{1-\sigma} c_{1\sigma} \gamma_- c_{1\sigma}^\dagger c_{1-\sigma}^\dagger \middle| 0 \right\rangle \\
&\qquad \mp \left\langle 0 \middle| c_{2-\sigma} c_{1\sigma} c_{1\sigma}^\dagger c_{2-\sigma}^\dagger \middle| 0 \right\rangle \Big\} \\
&= \frac{1}{2} \frac{\gamma_- \mp 1}{\sqrt{1+\gamma_-^2}}
\end{aligned}
$$

Density of states:

$$\rho_\sigma^{(\sigma,-\sigma)}(E) = \frac{(1-\gamma_-)^2}{4(1+\gamma_-^2)}\{\delta(E-(U-t-E_-))+$$
$$+\delta(E-(E_-+t))\}+$$
$$\frac{(1+\gamma_-)^2}{4(1+\gamma_-^2)}\{\delta(E-(U+t-E_-))+$$
$$+\delta(E-(E_--t))\}$$

Problem 8.12
The expression

$$\frac{1}{N}\sum_{i,j}^{i\neq j} T_{ij}\left\langle c_{i-\sigma}^{\dagger}c_{j-\sigma}(2n_{i\sigma}-1)\right\rangle$$

should be expressed in terms of the one-electron spectral density!

(a) Spectral theorem:

$$\frac{1}{N}\sum_{i,j}^{i\neq j} T_{ij}\left\langle c_{i-\sigma}^{\dagger}c_{j-\sigma}\right\rangle$$
$$=\frac{1}{N}\sum_{i,j} T_{ij}\left\langle c_{i-\sigma}^{\dagger}c_{j-\sigma}\right\rangle - T_0\left\langle c_{i-\sigma}^{\dagger}c_{i-\sigma}\right\rangle$$
$$=\frac{1}{N}\sum_{i,j} T_{ij}\int\limits_{-\infty}^{+\infty} dE\left(-\frac{1}{\pi\hbar}ImG_{ji-\sigma}(E-\mu)\right)f_-(E)-$$
$$-T_0\int\limits_{-\infty}^{+\infty} dE\left(-\frac{1}{\pi\hbar}ImG_{ii-\sigma}(E-\mu\right)f_-(E)$$
$$=\frac{1}{N}\sum_{i,j}\frac{1}{N}\sum_{\mathbf{k}_1} e^{i\mathbf{k}_1(\mathbf{R}_i-\mathbf{R}_j)}\varepsilon(\mathbf{k}_1)\int\limits_{-\infty}^{+\infty} dE\; f_-(E)\,\cdot$$
$$\cdot\frac{1}{N}\sum_{\mathbf{k}_2} e^{i\mathbf{k}_2(\mathbf{R}_j-\mathbf{R}_i)}\left(-\frac{1}{\pi\hbar}ImG_{\mathbf{k}_2-\sigma}(E-\mu\right)-$$

$$- T_0 \int_{-\infty}^{+\infty} dE\ f_-(E) \frac{1}{N} \sum_{\mathbf{k}} \left(-\frac{1}{\pi\hbar} \right) Im G_{\mathbf{k}-\sigma}(E-\mu)$$

$$= \frac{1}{N\hbar} \sum_{\mathbf{k}_1,\mathbf{k}_2} \int_{-\infty}^{+\infty} dE\ f_-(E) \delta_{\mathbf{k}_1,\mathbf{k}_2} \delta_{\mathbf{k}_1,\mathbf{k}_2} \varepsilon(\mathbf{k}_1) S_{\mathbf{k}_2-\sigma}(E-\mu) -$$

$$- T_0 \frac{1}{N\hbar} \sum_{\mathbf{k}} \int_{-\infty}^{+\infty} dE\ f_-(E) S_{\mathbf{k}-\sigma}(E-\mu)$$

$$= \frac{1}{N\hbar} \sum_{\mathbf{k}} (\varepsilon(\mathbf{k} - T_0) \int_{-\infty}^{+\infty} dE\ f_-(E) S_{\mathbf{k}-\sigma}(E-\mu)$$

(b) Real expectation value:

$$\left\langle c^{\dagger}_{i-\sigma} c_{j-\sigma} n_{i\sigma} \right\rangle \stackrel{!}{=} \left\langle n_{i\sigma} c^{\dagger}_{j-\sigma} c_{i-\sigma} \right\rangle = \left\langle c^{\dagger}_{j-\sigma} n_{i\sigma} c_{i-\sigma} \right\rangle$$

$$\Leftrightarrow \quad \frac{1}{N} \sum_{i,j}^{i\neq j} T_{ij} \left\langle c^{\dagger}_{i-\sigma} c_{j-\sigma} n_{i\sigma} \right\rangle$$

$$= \frac{1}{N} \sum_{i,j} T_{ij} \left\langle c^{\dagger}_{j-\sigma} n_{i\sigma} c_{i-\sigma} \right\rangle - T_0 \left\langle c^{\dagger}_{i-\sigma} n_{i\sigma} c_{i-\sigma} \right\rangle$$

$$= \frac{1}{N} \sum_{i,j} T_{ij} \left(-\frac{1}{\pi\hbar} \right) \int_{-\infty}^{+\infty} dE\ f_-(E) Im \Gamma_{iii;j-\sigma}(E-\mu) -$$

$$- T_0 \int_{-\infty}^{+\infty} dE\ f_-(E) \left(-\frac{1}{\pi\hbar} \right) Im \Gamma_{iii;i-\sigma}(E-\mu)$$

The "higher" Green's function is defined in (8.124). From (8.125) one reads off

$$Im \Gamma_{iii;j-\sigma}(E-\mu) = \frac{1}{U} \sum_m (E\delta_{im} - T_{im}) Im G_{mi-\sigma}(E-\mu)$$

Therewith follows:

$$\frac{1}{N}\sum_{i,j}^{i\neq j} T_{ij} < c_{i-\sigma}^{\dagger} c_{j-\sigma} n_{i\sigma} >$$

$$= \frac{1}{N}\sum_{i,j}(T_{ij} - T_0\delta_{ij}) \int\limits_{-\infty}^{+\infty} dE\ f_{-}(E) *$$

$$* \frac{1}{U\hbar}\sum_{m}(E\delta_{mi} - T_{im})S_{mj-\sigma}(E-\mu)$$

$$= \frac{1}{N}\sum_{i,j}\frac{1}{N^3}\sum_{\mathbf{k}_1,\mathbf{k}_2,\mathbf{k}_3} e^{i\mathbf{k}_1(\mathbf{R}_i-\mathbf{R}_j)}(\varepsilon(\mathbf{k}_1 - T_0) *$$

$$* \int_{-\infty}^{+\infty} dE\ f_{-}(E)\frac{1}{U\hbar}\sum_{m} e^{i\mathbf{k}_2(\mathbf{R}_i-\mathbf{R}_m)} e^{i\mathbf{k}_3(\mathbf{R}_m-\mathbf{R}_j)} *$$

$$*(E - \varepsilon(\mathbf{k}_2))S_{\mathbf{k}_3-\sigma}(E-\mu)$$

$$= \frac{1}{N\hbar}\sum_{\mathbf{k}_1,\mathbf{k}_2,\mathbf{k}_3}(\varepsilon(\mathbf{k}_1) - T_0) \int\limits_{-\infty}^{+\infty} dE\ f_{-}(E)\frac{1}{U}(E - \varepsilon(\mathbf{k}_2)) *$$

$$*S_{\mathbf{k}_3-\sigma}(E-\mu)\delta_{\mathbf{k}_1,-\mathbf{k}_2}\delta_{\mathbf{k}_2,\mathbf{k}_3}\delta_{\mathbf{k}_1,-\mathbf{k}_3}$$

$$= \frac{1}{N\hbar}\sum_{\mathbf{k}}(\varepsilon(\mathbf{k}) - T_0) \int\limits_{-\infty}^{+\infty} dE\ f_{-}(E)\frac{1}{U}(E - \varepsilon(\mathbf{k})) *$$

$$*S_{\mathbf{k}-\sigma}(E-\mu)$$

Together with the result of (a) we have found

$$n_{-\sigma}(1 - n_{-\sigma})B_{-\sigma}$$

$$= \frac{1}{N}\sum_{i,j}^{i\neq j} T_{ij}\left\langle c_{i-\sigma}^{\dagger} c_{j-\sigma}(2n_{i\sigma} - 1)\right\rangle$$

$$= \frac{1}{N\hbar}\sum_{\mathbf{k}}(\varepsilon(\mathbf{k}) - T_0) \int\limits_{-\infty}^{+\infty} dE\ f_{-}(E)\left(\frac{2}{U}(E - \varepsilon(\mathbf{k})) - 1\right) *$$

$$*S_{\mathbf{k}-\sigma}(E-\mu)$$

This is exactly (8.217)

Problem 8.13

0th spectral moment:

$$M_{ij\sigma}^{(0)} = \left\langle\left[c_{i\sigma}, c_{j\sigma}^{\dagger}\right]_{-}\right\rangle = \delta_{ij} \quad \Leftrightarrow \quad M_{\mathbf{k}\sigma}^{(0)} = 1 \qquad \text{(a)}$$

1st spectral moment:

$$M^{(1)}_{ij\sigma} = \left\langle \left[[c_{i\sigma}, H]_-, c^\dagger_{j\sigma} \right]_+ \right\rangle$$

$$\begin{aligned}
[c_{i\sigma}, H]_- &= \sum_{m,n,\sigma'} (T_{mn} - \mu\delta_{mn}) \left[c_{i\sigma}, c^\dagger_{m\sigma'} c_{n\sigma'} \right]_- + \\
&+ \frac{1}{2} U \sum_{m\sigma'} \left[c_{i\sigma}, n_{m\sigma'} n_{m-\sigma'} \right]_- \\
&= \sum_{m,n,\sigma'} (T_{mn} - \mu\delta_{mn}) \delta_{im} \delta_{\sigma\sigma'} c_{n\sigma'} + \\
&+ \frac{1}{2} U \sum_{m\sigma'} \delta_{im} (\delta_{\sigma\sigma'} c_{m\sigma'} n_{m-\sigma'} + \delta_{\sigma-\sigma'} n_{m\sigma'} c_{m-\sigma'}) \\
&= \sum_n (T_{in} - \mu\delta_{in}) c_{n\sigma} + U n_{i-\sigma} c_{i\sigma} \qquad \text{(b)}
\end{aligned}$$

$$\curvearrowright \left[[c_{i\sigma}, H]_-, c^\dagger_{j\sigma} \right]_+ = (T_{ij} - \mu\delta_{ij}) + U \delta_{ij} n_{i-\sigma} \qquad \text{(c)}$$

$$\curvearrowright M^{(1)}_{ij\sigma} = (T_{ij} - \mu\delta_{ij}) + U n_{-\sigma} \delta_{ij} \qquad \text{(d)}$$

$$\curvearrowright M^{(1)}_{\mathbf{k}\sigma} = \varepsilon(\mathbf{k}) - \mu + U n_{-\sigma} \qquad \text{(e)}$$

2nd spectral moment:

$$M^{(2)}_{ij\sigma} = \left\langle \left[\left[[c_{i\sigma}, H]_-, H \right]_-, c^\dagger_{j\sigma} \right]_+ \right\rangle \qquad \text{(f)}$$

$$abbreviation: \qquad t_{ij} = T_{ij} - \mu\delta_{ij}$$

Then with (b) holds

$$\left[[c_{i\sigma}, H]_-, H \right]_- = \sum_n t_{in} [c_{n\sigma}, H]_- + U \left[n_{i-\sigma} c_{i\sigma}, H \right]_-$$

$$
\begin{aligned}
&\left[n_{i-\sigma}c_{i\sigma}, H\right]_- \\
&= \sum_{m,n,\sigma'} t_{mn}\left[n_{i-\sigma}c_{i\sigma}, c^\dagger_{m\sigma'}c_{n\sigma'}\right]_- + \\
&+\frac{1}{2}U\sum_{m,\sigma'}\left[n_{i-\sigma}c_{i\sigma}, n_{m\sigma'}n_{m-\sigma'}\right]_- \\
&= \sum_{m,n,\sigma'} t_{mn}(\delta_{im}\delta_{\sigma\sigma'}n_{i-\sigma}c_{n\sigma'} + \delta_{im}\delta_{\sigma-\sigma'}c^\dagger_{i-\sigma}c_{n\sigma'}c_{i\sigma} - \\
&- \delta_{in}\delta_{\sigma-\sigma'}c^\dagger_{m\sigma'}c_{i-\sigma}c_{i\sigma}) + \frac{1}{2}U\sum_{m\sigma'}(\delta_{im}\delta_{\sigma\sigma'}n_{i-\sigma}c_{m\sigma'}n_{m-\sigma'} + \\
&+\delta_{im}\delta_{\sigma-\sigma'}n_{i-\sigma}n_{m\sigma'}c_{m-\sigma'}) \\
&= \sum_n t_{in}n_{i-\sigma}c_{n\sigma} + \sum_n t_{in}c^\dagger_{i-\sigma}c_{n-\sigma}c_{i\sigma} - \\
&- \sum_m t_{mi}c^\dagger_{m-\sigma}c_{i-\sigma}c_{i\sigma} + \frac{1}{2}U(n_{i-\sigma}c_{i\sigma}n_{i-\sigma} + n_{i-\sigma}n_{i-\sigma}c_{i\sigma}) \\
&= \sum_m t_{im}(n_{i-\sigma}c_{m\sigma} + c^\dagger_{i-\sigma}c_{m-\sigma}c_{i\sigma} - c^\dagger_{m-\sigma}c_{i-\sigma}c_{i\sigma}) + \\
&\qquad + Un_{i-\sigma}c_{i\sigma}
\end{aligned} \tag{g}
$$

Here we have used $n^2_{i-\sigma} = n_{i-\sigma}$ which is an identity valid for Fermions. Then what remains is

$$
\begin{aligned}
&\left[[c_{i\sigma}, H]_-, H\right]_- \\
&= \sum_n t_{in}\,[c_{n\sigma}, H]_- + \\
&+ U\sum_m t_{im}(n_{i-\sigma}c_{m\sigma} + c^\dagger_{i-\sigma}c_{m-\sigma}c_{i\sigma} - c^\dagger_{m-\sigma}c_{i-\sigma}c_{i\sigma}) + \\
&+ U^2 n_{i-\sigma}c_{i\sigma}
\end{aligned} \tag{h}
$$

With this follows:

$$
\begin{aligned}
M^{(2)}_{ij\sigma} &= \sum_n t_{in}M^{(1)}_{nj\sigma} + Ut_{ij}n_{-\sigma} + \delta_{ij}U\sum_m t_{im}\left\langle c^\dagger_{i-\sigma}c_{m-\sigma}\right\rangle - \\
&- \delta_{ij}U\sum_m t_{im}\left\langle c^\dagger_{m-\sigma}c_{i-\sigma}\right\rangle + U^2\delta_{ij}n_{-\sigma}
\end{aligned}
$$

Because of (8.268), the third and the fourth terms cancel each other. Then it remains with (d):

$$M_{ij\sigma}^{(2)} = \sum_n t_{in} t_{nj} + 2U t_{ij} n_{-\sigma} + U^2 n_{-\sigma} \delta_{ij} \tag{i}$$

After Fourier transformation follows:

$$M_{\mathbf{k}\sigma}^{(2)} = (\varepsilon(\mathbf{k}) - \mu)^2 + 2(\varepsilon(\mathbf{k}) - \mu) U n_{-\sigma} + U^2 n_{-\sigma}$$

3rd spectral moment:

$$M_{ij\sigma}^{(3)} = \left\langle \left[\left[\left[[c_{i\sigma}, H]_- , H \right]_- , H \right]_- , c_{j\sigma}^\dagger \right]_+ \right\rangle$$

For the triple commutator holds with (h):

$$\begin{aligned}
&\left[\left[[c_{i\sigma}, H]_- , H \right]_- , H \right]_- \\
&= \sum_n t_{in} \left[[c_{n\sigma}, H]_- , H \right]_- + \\
&+ U \sum_m t_{im} \left[(n_{i-\sigma} c_{m\sigma} + c_{i-\sigma}^\dagger c_{m-\sigma} c_{i\sigma} - c_{m-\sigma}^\dagger c_{i-\sigma} c_{i\sigma}), H \right]_- + \\
&+ U^2 \left[n_{i-\sigma} c_{i\sigma}, H \right]_- \\
&= \sum_n t_{in} \left[[c_{n\sigma}, H]_- , H \right]_- + U \left[[c_{i\sigma}, H]_- , H \right]_- - \\
&- U \sum_n t_{in} [c_{n\sigma}, H]_- + U \sum_m t_{im} \left\{ \left[n_{i-\sigma} c_{m\sigma}, H \right]_- + \right. \\
&\quad \left. + \left[c_{i-\sigma}^\dagger c_{m-\sigma} c_{i\sigma}, H \right]_- - \left[c_{m-\sigma}^\dagger c_{i-\sigma} c_{i\sigma}, H \right]_- \right\}
\end{aligned} \tag{j}$$

Three commutators remain to be calculated:

$$
\begin{aligned}
(I) &= U \sum_m t_{im} \left[n_{i-\sigma} c_{m\sigma}, H\right]_- \\
&= U \sum_m t_{im} \sum_{s,t,\sigma'} t_{st} \left[n_{i-\sigma} c_{m\sigma}, c^\dagger_{s\sigma'} c_{t\sigma'}\right]_- + \\
&+ \frac{1}{2} U^2 \sum_m t_{im} \sum_{s,\sigma'} \left[n_{i-\sigma} c_{m\sigma}, n_{s\sigma'} n_{s-\sigma'}\right]_- \\
&= U \sum_{m,s,t,\sigma'} t_{im} t_{st} \left\{\delta_{ms} \delta_{\sigma\sigma'} n_{i-\sigma} c_{t\sigma'} + \right. \\
&\qquad \left. + \delta_{is} \delta_{\sigma-\sigma'} c^\dagger_{i-\sigma} c_{t\sigma'} c_{m\sigma} - \delta_{ti} \delta_{\sigma-\sigma'} c^\dagger_{s\sigma'} c_{i-\sigma} c_{m\sigma}\right\} + \\
&+ \frac{1}{2} U^2 \sum_{m,s,\sigma'} t_{im} \left\{\delta_{ms} \delta_{\sigma\sigma'} n_{i-\sigma} c_{s\sigma'} n_{s-\sigma'} + \right. \\
&\qquad \left. + \delta_{ms} \delta_{\sigma-\sigma'} n_{i-\sigma} n_{s\sigma'} c_{s-\sigma'}\right\} \\
&= U \sum_{m,t} t_{im} t_{mt} n_{i-\sigma} c_{t\sigma} + U \sum_{m,t} t_{im} t_{it} c^\dagger_{i-\sigma} c_{t-\sigma} c_{m\sigma} - \\
&- U \sum_{m,s} t_{im} t_{si} c^\dagger_{s-\sigma} c_{i-\sigma} c_{m\sigma} + \frac{1}{2} U^2 \sum_m t_{im} n_{i-\sigma} c_{m\sigma} n_{m-\sigma} + \\
&+ \frac{1}{2} U^2 \sum_m t_{im} n_{i-\sigma} n_{m-\sigma} c_{m\sigma} \\
&= U \sum_{m,t} t_{im} t_{mt} n_{i-\sigma} c_{t\sigma} + U \sum_{m,t} t_{im} t_{it} (c^\dagger_{i-\sigma} c_{t-\sigma} c_{m\sigma} - \\
&\qquad - c^\dagger_{t-\sigma} c_{i-\sigma} c_{m\sigma}) + + U^2 \sum_m t_{im} n_{i-\sigma} n_{m-\sigma} c_{m\sigma}
\end{aligned}
$$

$$
\begin{aligned}
\curvearrowright \quad \left[(I), c^\dagger_{j\sigma}\right]_+ &= U \sum_m t_{im} t_{mj} n_{i-\sigma} + \\
&+ U \sum_t t_{it} t_{ij} (c^\dagger_{i-\sigma} c_{t-\sigma} - c^\dagger_{t-\sigma} c_{i-\sigma}) + \\
&+ U^2 t_{ij} n_{i-\sigma} n_{j-\sigma}
\end{aligned}
$$

We again use (8.268):

$$
\left\langle \left[(I), c^\dagger_{j\sigma}\right]_+ \right\rangle = U n_{-\sigma} \sum_m t_{im} t_{mj} + U^2 t_{ij} \left\langle n_{i-\sigma} n_{j-\sigma} \right\rangle \tag{k}
$$

In the same manner we calculate

$$
\begin{aligned}
(II) &= U\sum_m t_{im}\left[c^\dagger_{i-\sigma}c_{m-\sigma}c_{i\sigma}, H\right]_- \\
&= U\sum_m t_{im}\sum_{s,t,\sigma'} t_{st}\left[c^\dagger_{i-\sigma}c_{m-\sigma}c_{i\sigma}, c^\dagger_{s\sigma'}c_{t\sigma'}\right]_- + \\
&+ \frac{1}{2}U^2\sum_m t_{im}\sum_{s,\sigma'}\left[c^\dagger_{i-\sigma}c_{m-\sigma}c_{i\sigma}, n_{s\sigma'}n_{s-\sigma'}\right]_- \\
&= U\sum_{m,s,t,\sigma'} t_{im}t_{st}\left\{\delta_{\sigma\sigma'}\delta_{is}c^\dagger_{i-\sigma}c_{m-\sigma}c_{t\sigma'} + \right. \\
&\qquad \left. +\delta_{\sigma-\sigma'}\delta_{ms}c^\dagger_{i-\sigma}c_{t\sigma'}c_{i\sigma} - \delta_{\sigma-\sigma'}\delta_{it}c^\dagger_{s\sigma'}c_{m-\sigma}c_{i\sigma}\right\} + \\
&+ \frac{1}{2}U^2\sum_{m,s,\sigma'} t_{im}\left\{\delta_{is}\delta_{\sigma\sigma'}c^\dagger_{i-\sigma}c_{m-\sigma}c_{s\sigma'}n_{s-\sigma'} + \right. \\
&\qquad +\delta_{is}\delta_{\sigma-\sigma'}c^\dagger_{i-\sigma}c_{m-\sigma}n_{s\sigma'}c_{s-\sigma'} + \\
&\qquad +\delta_{ms}\delta_{\sigma'-\sigma'}c^\dagger_{i-\sigma}c_{s\sigma'}n_{s-\sigma'}c_{i\sigma} + \\
&\qquad +\delta_{ms}\delta_{\sigma\sigma'}c^\dagger_{i-\sigma}n_{s\sigma'}c_{s-\sigma'}c_{i\sigma} - \\
&\qquad -\delta_{is}\delta_{\sigma-\sigma'}c^\dagger_{s\sigma'}n_{s-\sigma'}c_{m-\sigma}c_{i\sigma} - \\
&\qquad \left. -\,\delta_{is}\delta_{\sigma\sigma'}n_{s\sigma'}c^\dagger_{s-\sigma'}c_{m-\sigma}c_{i\sigma}\right\} \\
&= U\sum_{m,t} t_{im}t_{it}c^\dagger_{i-\sigma}c_{m-\sigma}c_{t\sigma} + U\sum_{m,t} t_{im}t_{mt}c^\dagger_{i-\sigma}c_{t-\sigma}c_{i\sigma} - \\
&\qquad -U\sum_{m,s} t_{im}t_{si}c^\dagger_{s-\sigma}c_{m-\sigma}c_{i\sigma} + \\
&+ \frac{1}{2}U^2\sum_m t_{im}\left\{c^\dagger_{i-\sigma}c_{m-\sigma}c_{i\sigma}n_{i-\sigma} + c^\dagger_{i-\sigma}c_{m-\sigma}n_{i-\sigma}c_{i\sigma} + \right. \\
&\qquad +c^\dagger_{i-\sigma}c_{m-\sigma}n_{m\sigma}c_{i\sigma} + c^\dagger_{i-\sigma}n_{m\sigma}c_{m-\sigma}c_{i\sigma} - \\
&\qquad \left. -\,c^\dagger_{i-\sigma}n_{i\sigma}c_{m-\sigma}c_{i\sigma} - n_{i\sigma}c^\dagger_{i-\sigma}c_{m-\sigma}c_{i\sigma}\right\}
\end{aligned}
$$

$$
\begin{aligned}
\curvearrowright\ \left[(II), c^\dagger_{j\sigma}\right]_+ &= U\sum_m t_{im}t_{ij}c^\dagger_{i-\sigma}c_{m-\sigma} + \\
&+ U\delta_{ij}\sum_{m,t} t_{im}(t_{mt}c^\dagger_{i-\sigma}c_{t-\sigma} - t_{it}c^\dagger_{t-\sigma}c_{m-\sigma}) + \\
&+ U^2\sum_m t_{im}\left\{\delta_{ij}c^\dagger_{i-\sigma}c_{m-\sigma}n_{i-\sigma} + \right. \\
&+ \delta_{ij}c^\dagger_{i-\sigma}c_{m-\sigma}n_{m\sigma} - \delta_{mj}c^\dagger_{i-\sigma}c_{m-\sigma}c^\dagger_{m\sigma}c_{i\sigma} - \\
&\quad \left. -\delta_{ij}c^\dagger_{i-\sigma}n_{i\sigma}c_{m-\sigma} - \delta_{ij}c^\dagger_{i-\sigma}c^\dagger_{i\sigma}c_{m-\sigma}c_{i\sigma}\right\} \\
&= Ut_{ij}\sum_m t_{im}c^\dagger_{i-\sigma}c_{m-\sigma} +
\end{aligned}
$$

$$+ U\delta_{ij}\sum_{m,t} t_{im}(t_{mt}c^\dagger_{i-\sigma}c_{t-\sigma} - t_{it}c^\dagger_{t-\sigma}c_{m-\sigma}) -$$

$$- U^2 t_{ij}c^\dagger_{i-\sigma}c^\dagger_{j\sigma}c_{i\sigma}c_{j-\sigma} +$$

$$+ U^2\delta_{ij}\sum_m t_{im}\left\{c^\dagger_{i-\sigma}c_{m-\sigma}n_{i-\sigma} + c^\dagger_{i-\sigma}c_{m-\sigma}n_{m\sigma}\right\}$$

$$(III) = -U\sum_m t_{im}\left[c^\dagger_{m-\sigma}c_{i-\sigma}c_{i\sigma}, H\right]_-$$

$$= -U\sum_m t_{im}\sum_{s,t,\sigma'} t_{st}\left[c^\dagger_{m-\sigma}c_{i-\sigma}c_{i\sigma}, c^\dagger_{s\sigma'}c_{t\sigma'}\right]_- -$$

$$- \frac{1}{2}U^2\sum_m t_{im}\sum_{s,\sigma'}\left[c^\dagger_{m-\sigma}c_{i-\sigma}c_{i\sigma}, n_{s\sigma'}n_{s-\sigma'}\right]_-$$

$$= -U\sum_m t_{im}\sum_{s,t,\sigma'} t_{st}\left\{\delta_{is}\delta_{\sigma\sigma'}c^\dagger_{m-\sigma}c_{i-\sigma}c_{t\sigma'} + \right.$$

$$+\delta_{is}\delta_{\sigma-\sigma'}c^\dagger_{m-\sigma}c_{t\sigma'}c_{i\sigma} -$$

$$\left. -\delta_{mt}\delta_{\sigma-\sigma'}c^\dagger_{s\sigma'}c_{i-\sigma}c_{i\sigma}\right\} -$$

$$- \frac{1}{2}U^2\sum_{m,s,\sigma'} t_{im}\left\{\delta_{is}\delta_{\sigma\sigma'}c^\dagger_{m-\sigma}c_{i-\sigma}c_{s\sigma'}n_{s-\sigma'} + \right.$$

$$+\delta_{is}\delta_{\sigma-\sigma'}c^\dagger_{m-\sigma}c_{i-\sigma}n_{s\sigma'}c_{s-\sigma'} +$$

$$+\delta_{is}\delta_{\sigma-\sigma'}c^\dagger_{m-\sigma}c_{s\sigma'}n_{s-\sigma'}c_{i\sigma} +$$

$$+\delta_{is}\delta_{\sigma\sigma'}c^\dagger_{m-\sigma}n_{s\sigma'}c_{s-\sigma'}c_{i\sigma} -$$

$$-\delta_{ms}\delta_{\sigma-\sigma'}c^\dagger_{s\sigma'}n_{s-\sigma'}c_{i-\sigma}c_{i\sigma} -$$

$$\left. -\delta_{ms}\delta_{\sigma\sigma'}n_{s\sigma'}c^\dagger_{s-\sigma'}c_{i-\sigma}c_{i\sigma}\right\}$$

$$= -U\sum_{m,t} t_{im}t_{it}c^\dagger_{m-\sigma}c_{i-\sigma}c_{t\sigma} -$$

$$- U\sum_{m,t} t_{im}t_{it}c^\dagger_{m-\sigma}c_{t-\sigma}c_{i\sigma} +$$

$$+ U\sum_{m,s} t_{im}t_{sm}c^\dagger_{s-\sigma}c_{i-\sigma}c_{i\sigma} -$$

$$- \frac{1}{2}U^2\sum_m t_{im}\left\{c^\dagger_{m-\sigma}c_{i-\sigma}c_{i\sigma}n_{i-\sigma} + \right.$$

$$+c^\dagger_{m-\sigma}c_{i-\sigma}n_{i-\sigma}c_{i\sigma} -$$

$$-c^\dagger_{m-\sigma}n_{m\sigma}c_{i-\sigma}c_{i\sigma} -$$

$$
\begin{aligned}
& -n_{m\sigma} c^{\dagger}_{m-\sigma} c_{i-\sigma} c_{i\sigma} \Big\} \\
&= -U \sum_{m,t} t_{im} t_{it} (c^{\dagger}_{m-\sigma} c_{i-\sigma} c_{t\sigma} + c^{\dagger}_{m-\sigma} c_{t-\sigma} c_{i\sigma}) + \\
&+ U \sum_{m,s} t_{im} t_{sm} c^{\dagger}_{s-\sigma} c_{i-\sigma} c_{i\sigma} - \\
&- U^2 \sum_{m} t_{im} \left\{ c^{\dagger}_{m-\sigma} c_{i-\sigma} n_{i-\sigma} c_{i\sigma} - c^{\dagger}_{m-\sigma} c_{i-\sigma} n_{m\sigma} c_{i\sigma} \right\}
\end{aligned}
$$

$$
\begin{aligned}
\curvearrowright \left[(III), c^{\dagger}_{j\sigma}\right]_+ &= -U \sum_{m} t_{im} t_{ij} c^{\dagger}_{m-\sigma} c_{i-\sigma} + \\
&+ U \delta_{ij} \sum_{m,t} t_{im} t_{it} c^{\dagger}_{m-\sigma} c_{t-\sigma} - \\
&- U \delta_{ij} \sum_{m,t} t_{im} t_{tm} c^{\dagger}_{t-\sigma} c_{i-\sigma} - \\
&- U^2 \delta_{ij} \sum_{m} t_{im} c^{\dagger}_{m-\sigma} c_{i-\sigma} n_{i-\sigma} + \\
&+ U^2 \delta_{ij} \sum_{m} t_{im} c^{\dagger}_{m-\sigma} c_{i-\sigma} n_{m\sigma} - \\
&- U^2 t_{ij} c^{\dagger}_{j-\sigma} c_{i-\sigma} c^{\dagger}_{j\sigma} c_{i\sigma}
\end{aligned}
$$

With this follows:

$$
\begin{aligned}
& \left\langle \left[(II) + (III), c^{\dagger}_{j\sigma}\right]_+ \right\rangle \\
&= U t_{ij} \sum_{m} t_{im} \left(\left\langle c^{\dagger}_{i-\sigma} c_{m-\sigma} \right\rangle - \left\langle c^{\dagger}_{m-\sigma} c_{i-\sigma} \right\rangle \right) + \\
&+ U \delta_{ij} \sum_{m,t} t_{im} t_{mt} \left(\left\langle c^{\dagger}_{i-\sigma} c_{t-\sigma} \right\rangle - \left\langle c^{\dagger}_{t-\sigma} c_{i-\sigma} \right\rangle \right) - \\
&- U \delta_{ij} \sum_{m,t} t_{im} t_{it} \left(\left\langle c^{\dagger}_{t-\sigma} c_{m-\sigma} \right\rangle - \left\langle c^{\dagger}_{m-\sigma} c_{t-\sigma} \right\rangle \right) + \\
&+ U^2 \delta_{ij} \sum_{m} t_{im} \left(\left\langle c^{\dagger}_{i-\sigma} c_{m-\sigma} n_{i-\sigma} \right\rangle - \left\langle c^{\dagger}_{m-\sigma} c_{i-\sigma} n_{i-\sigma} \right\rangle \right) \\
&+ U^2 \delta_{ij} \sum_{m} t_{im} \left(\left\langle c^{\dagger}_{i-\sigma} c_{m-\sigma} n_{m\sigma} \right\rangle + \left\langle c^{\dagger}_{m-\sigma} c_{i-\sigma} n_{m\sigma} \right\rangle \right) - \\
&- U^2 t_{ij} \left\langle c^{\dagger}_{i-\sigma} c^{\dagger}_{j\sigma} c_{i\sigma} c_{j-\sigma} \right\rangle + U^2 t_{ij} \left\langle c^{\dagger}_{j-\sigma} c^{\dagger}_{j\sigma} c_{i-\sigma} c_{i\sigma} \right\rangle
\end{aligned}
$$

Translational symmetry:

$$\sum_m t_{im}\left(\left\langle c^\dagger_{i-\sigma}c_{m-\sigma}\right\rangle - \left\langle c^\dagger_{m-\sigma}c_{i-\sigma}\right\rangle\right)$$
$$= \frac{1}{N}\sum_{i,m} t_{im}\left(\left\langle c^\dagger_{i-\sigma}c_{m-\sigma}\right\rangle - \left\langle c^\dagger_{m-\sigma}c_{i-\sigma}\right\rangle\right)$$
$$= \frac{1}{N}\sum_{i,m}(t_{im}-t_{mi})\left\langle c^\dagger_{i-\sigma}c_{m-\sigma}\right\rangle = 0$$
$$\sum_{m,t} t_{im}t_{mt}\left(\left\langle c^\dagger_{i-\sigma}c_{t-\sigma}\right\rangle - \left\langle c^\dagger_{t-\sigma}c_{i-\sigma}\right\rangle\right)$$
$$= \frac{1}{N}\sum_{m,t,i}(t_{im}t_{mt}-t_{tm}t_{mi})\left\langle c^\dagger_{i-\sigma}c_{t-\sigma}\right\rangle = 0$$
$$\sum_{m,t} t_{im}t_{it}\left(\left\langle c^\dagger_{t-\sigma}c_{m-\sigma}\right\rangle - \left\langle c^\dagger_{m-\sigma}c_{t-\sigma}\right\rangle\right)$$
$$= \sum_{m,t}(t_{im}t_{it}-t_{it}t_{im})\left\langle c^\dagger_{t-\sigma}c_{m-\sigma}\right\rangle = 0$$
$$\sum_m t_{im}\left(\left\langle c^\dagger_{i-\sigma}c_{m-\sigma}n_{i-\sigma}\right\rangle - \left\langle c^\dagger_{m-\sigma}c_{i-\sigma}n_{i-\sigma}\right\rangle\right)$$
$$= \sum_m t_{im}\left(\delta_{im}n_{-\sigma} - \left\langle c^\dagger_{m-\sigma}c_{i-\sigma}\right\rangle\right)$$
$$= (T_0-\mu)n_{-\sigma} - \sum_m t_{im}\left\langle c^\dagger_{m-\sigma}c_{i-\sigma}\right\rangle$$

Real expectation values:

$$\left\langle c^\dagger_{i-\sigma}c_{m-\sigma}n_{m\sigma}\right\rangle \overset{!}{=} \left\langle n_{m\sigma}c^\dagger_{m-\sigma}c_{i-\sigma}\right\rangle$$

Then we finally have

$$\begin{aligned}&\left\langle\left[(I)+(II)+(III), c^\dagger_{j\sigma}\right]_+\right\rangle\\
&= U^2\delta_{ij}\left((T_0-\mu)n_{-\sigma} - \sum_m t_{im}\left\langle c^\dagger_{m-\sigma}c_{i-\sigma}\right\rangle\right) +\\
&+ Un_{-\sigma}\sum_m t_{im}t_{mj} +\\
&+ U^2\delta_{ij}\sum_m t_{im}\left(2\left\langle c^\dagger_{m-\sigma}c_{i-\sigma}n_{m\sigma}\right\rangle\right) +\\
&+ U^2 t_{ij}\left\{\left\langle n_{i-\sigma}n_{j-\sigma}\right\rangle + \left\langle c^\dagger_{j-\sigma}c^\dagger_{j\sigma}c_{i-\sigma}c_{i\sigma}\right\rangle +\right.\\
&\qquad\left.\left\langle c^\dagger_{j\sigma}c^\dagger_{i-\sigma}c_{i\sigma}c_{j-\sigma}\right\rangle\right\}\end{aligned}$$

Fourier transformation:

$$\frac{1}{N}\sum_{i,j} e^{-i\mathbf{k}(\mathbf{R}_i-\mathbf{R}_j)} t_{ij}\left\{\left\langle n_{i-\sigma}n_{j-\sigma}\right\rangle + \left\langle c^{\dagger}_{j-\sigma}c^{\dagger}_{j\sigma}c_{i-\sigma}c_{i\sigma}\right\rangle + \right.$$
$$\left. + \left\langle c^{\dagger}_{j\sigma}c^{\dagger}_{i-\sigma}c_{i\sigma}c_{j-\sigma}\right\rangle\right\} =$$
$$= t_0\left\{n_{-\sigma} - 2\left\langle n_{i-\sigma}n_{i\sigma}\right\rangle\right\} + \frac{1}{N}\sum_{i,j}^{i\neq j} e^{-i\mathbf{k}(\mathbf{R}_i-\mathbf{R}_j)} t_{ij}\left\{\cdots\right\}$$
$$= (T_0-\mu)\left(n_{-\sigma} - 2\left\langle n_{i\sigma}n_{i-\sigma}\right\rangle\right) + n^2_{-\sigma}(\varepsilon(\mathbf{k}) - T_0) +$$
$$+ n_{-\sigma}(1-n_{-\sigma})F_{\mathbf{k}-\sigma}$$

The second summand is exactly the bandwidth correction of (8.213):

$$\frac{1}{N}\sum_{i,j} e^{-i\mathbf{k}(\mathbf{R}_i-\mathbf{R}_j)}\delta_{ij}\sum_m t_{im}\left(2\left\langle c^{\dagger}_{m-\sigma}c_{i-\sigma}n_{m\sigma}\right\rangle\right.$$
$$\left. - \left\langle c^{\dagger}_{m-\sigma}c_{i-\sigma}\right\rangle\right) =$$
$$= \frac{1}{N}\sum_{i,m} t_{im}\left(\cdots\right)$$
$$= \frac{1}{N}\sum_{i,m}^{i\neq m} t_{im}\left(\cdots\right) + t_0\left(2\left\langle n_{i-\sigma}n_{i\sigma}\right\rangle - n_{-\sigma}\right)$$
$$= n_{-\sigma}(1-n_{-\sigma})B_{-\sigma} + (T_0-\mu)\left(2\left\langle n_{i-\sigma}n_{i\sigma}\right\rangle - n_{-\sigma}\right)$$

We have used here the definition (8.212) of spin-dependent band shift. Then finally what remains is

$$\frac{1}{N}\sum_{i,j} e^{-i\mathbf{k}(\mathbf{R}_i-\mathbf{R}_j)}\left\langle\left[(I)+(II)+(III), c^{\dagger}_{j\sigma}\right]_+\right\rangle =$$
$$= U n_{-\sigma}(\varepsilon(\mathbf{k})-\mu)^2 + U^2 n_{-\sigma}(T_0-\mu) +$$
$$+ U^2 n^2_{-\sigma}(\varepsilon(\mathbf{k}) - T_0) + U^2 n_{-\sigma}(1-n_{-\sigma})B_{\mathbf{k}-\sigma} \qquad (1)$$

We substitute (l) in (j) and then have

$$
\begin{aligned}
M_{\mathbf{k}\sigma}^{(3)} &= (\varepsilon(\mathbf{k})-\mu)M_{\mathbf{k}\sigma}^{(2)} + UM_{\mathbf{k}\sigma}^{(2)} - U(\varepsilon(\mathbf{k})-\mu)M_{\mathbf{k}\sigma}^{(1)} + \\
&+ Un_{-\sigma}(\varepsilon(\mathbf{k})-\mu)^2 + \\
&+ U^2 n_{-\sigma}(T_0-\mu) + U^2 n_{-\sigma}^2(\varepsilon(\mathbf{k})-T_0) + \\
&+ U^2 n_{-\sigma}(1-n_{-\sigma})B_{\mathbf{k}-\sigma} \\
&= (\varepsilon(\mathbf{k})-\mu)^3 + 2(\varepsilon(\mathbf{k})-\mu)^2 Un_{-\sigma} + \\
&+ 2U^2(\varepsilon(\mathbf{k})-\mu)n_{-\sigma} + U^3 n_{-\sigma} + \\
&+ Un_{-\sigma}(\varepsilon(\mathbf{k})-\mu)^2 + U^2 n_{-\sigma}(T_0-\mu) + \\
&+ U^2 n_{-\sigma}^2(\varepsilon(\mathbf{k})-T_0) + U^2 n_{-\sigma}(1-n_{-\sigma})B_{\mathbf{k}-\sigma} \\
&= (\varepsilon(\mathbf{k})-\mu)^3 + 3Un_{-\sigma}(\varepsilon(\mathbf{k})-\mu)^2 + \\
&+ U^2 n_{-\sigma}(\varepsilon(\mathbf{k})-\mu)(2+n_{-\sigma}) + U^3 n_{-\sigma} + \\
&+ U^2 n_{-\sigma}(1-n_{-\sigma})(B_{\mathbf{k}-\sigma} + T_0 - \mu)
\end{aligned}
$$

This is the 3rd spectral moment (8.224).

Problem 8.14

Hamiltonian of the Stoner model (8.34):

$$
H_S = \sum_{\mathbf{k}\sigma} (\varepsilon(\mathbf{k}) + Un_{-\sigma} - \mu) c_{\mathbf{k}\sigma}^{\dagger} c_{\mathbf{k}\sigma}
$$

That means

$$
\begin{aligned}
[c_{\mathbf{k}\sigma}, H]_- &= (\varepsilon(\mathbf{k}) + Un_{-\sigma} - \mu)c_{\mathbf{k}\sigma} \\
\underbrace{[\cdots[c_{\mathbf{k}\sigma}, H]_- \cdots, H]_-}_{n-fold} &= (\varepsilon(\mathbf{k}) + Un_{-\sigma} - \mu)^n c_{\mathbf{k}\sigma}
\end{aligned}
$$

$\curvearrowright$ Spectral moments:

$$
M_{\mathbf{k}\sigma}^{(n)} = (\varepsilon(\mathbf{k}) + Un_{-\sigma} - \mu)^n
$$

Then it holds

$$
\begin{aligned}
\Delta_{\mathbf{k}\sigma}^{(0)} &\equiv M_{\mathbf{k}\sigma}^{(0)} = 1 \\
\Delta_{\mathbf{k}\sigma}^{(1)} &= \begin{pmatrix} M_{\mathbf{k}\sigma}^{(0)} & M_{\mathbf{k}\sigma}^{(1)} \\ M_{\mathbf{k}\sigma}^{(1)} & M_{\mathbf{k}\sigma}^{(2)} \end{pmatrix} \\
&= M_{\mathbf{k}\sigma}^{(0)} M_{\mathbf{k}\sigma}^{(2)} - (M_{\mathbf{k}\sigma}^{(1)})^2 \\
&= 0
\end{aligned}
$$

$\curvearrowright$ Therefore the spectral density is a one-pole function!

Problem 8.15

1. $$H = T_0 \sum_{i,\sigma} n_{i\sigma} + \frac{1}{2} U \sum_{i,\sigma} n_{i\sigma} n_{i-\sigma}$$

It holds

$$[c_{i\sigma}, n_{i\sigma'}]_- = \delta_{\sigma\sigma'} c_{i\sigma}$$

With this it directly follows:

$$\begin{aligned}
[c_{i\sigma}, \mathcal{H}]_- &= (T_0 - \mu) c_{i\sigma} + U c_{i\sigma} n_{i-\sigma} \\
\left[c_{i\sigma} n_{i-\sigma}, \mathcal{H}\right]_- &= [c_{i\sigma}, \mathcal{H}]_- n_{i-\sigma} \\
&= (T_0 - \mu + U) c_{i\sigma} n_{i-\sigma}
\end{aligned}$$

Here we have used once more $n_{i-\sigma}^2 = n_{i-\sigma}$

$$\begin{aligned}
M_{ii\sigma}^{(0)} &= 1 \\
M_{ii\sigma}^{(1)} &= (T_0 - \mu) + U n_{-\sigma} \\
&= (T_0 - \mu)^1 + \left[(T_0 + U - \mu)^1 - (T_0 - \mu)^1\right] n_{-\sigma}
\end{aligned}$$

Complete induction:
Let the proposition be true for n. That means

$$\begin{aligned}
&\underbrace{\left[\cdots\left[[c_{i\sigma}, \mathcal{H}]_-, \mathcal{H}\right]_- \cdots, \mathcal{H}\right]_-}_{n-fold} = \\
&= (T_0 - \mu)^n c_{i\sigma} + \left[(T_0 + U - \mu)^n - (T_0 - \mu)^n\right] c_{i\sigma} n_{i-\sigma} \\
&\curvearrowright \underbrace{\left[\cdots\left[[c_{i\sigma}, \mathcal{H}]_-, \mathcal{H}\right]_- \cdots, \mathcal{H}\right]_-}_{(n+1)-fold} = \\
&= (T_0 - \mu)^n \, [c_{i\sigma}, \mathcal{H}]_- - \\
&- \left[(T_0 + U - \mu)^n - (T_0 - \mu)^n\right] \left[c_{i\sigma} n_{i-\sigma}, \mathcal{H}\right]_- \\
&= (T_0 - \mu)^n ((T_0 - \mu) c_{i-\sigma} + U n_{i-\sigma} c_{i\sigma}) + \\
&+ \left[(T_0 - \mu + U)^n - (T_0 - \mu)^n\right] (T_0 - \mu + U) c_{i\sigma} n_{i-\sigma} \\
&= (T_0 - \mu)^{n+1} c_{i\sigma} + U (T_0 - \mu)^n n_{i-\sigma} c_{i\sigma} + \\
&+ (T_0 - \mu + U)^{n+1} c_{i\sigma} n_{i-\sigma} - \\
&- (T_0 - \mu)^n (T_0 - \mu + U) c_{i\sigma} n_{i-\sigma} \\
&= (T_0 - \mu)^{n+1} c_{i\sigma} + \\
&+ \left[(T_0 - \mu + U)^{n+1} - (T_0 - \mu)^{n+1}\right] c_{i\sigma} n_{i-\sigma}
\end{aligned}$$

So that it holds

$$\left[\cdots\left[[c_{i\sigma},\mathcal{H}]_-,\mathcal{H}\right]_-\cdots,\mathcal{H}\right]_- =$$
$$= (T_0-\mu)^n c_{i\sigma} + \left[(T_0-\mu+U)^n - (T_0-\mu)^n\right] c_{i\sigma} n_{i-\sigma}$$

Then the spectral moments are

$$M_{ii\sigma}^{(n)} = (T_0-\mu)^n + \left[(T_0-\mu+U)^n - (T_0-\mu)^n\right] n_{-\sigma}$$

2. Lonke theorem [24]

$$\begin{aligned}\Delta_{ii\sigma}^{(1)} &= \begin{pmatrix} M_{ii\sigma}^{(0)} & M_{ii\sigma}^{(1)} \\ M_{ii\sigma}^{(1)} & M_{ii\sigma}^{(2)} \end{pmatrix} \\ &= M_{ii\sigma}^{(0)} M_{ii\sigma}^{(2)} - (M_{ii\sigma}^{(0)})^2 \\ &= (T_0-\mu)^2 + \left[(T_0-\mu+U)^2 - (T_0-\mu)^2\right] n_{-\sigma} - \\ &- (T_0-\mu+Un_{-\sigma})^2 \\ &= -2Un_{-\sigma}(T_0-\mu) - U^2 n_{-\sigma}^2 + 2U(T_0-\mu)n_{-\sigma} + \\ &+ U^2 n_{-\sigma} \\ &= U^2 n_{-\sigma}(1-n_{-\sigma}) \neq 0\,, \quad If\ n_{-\sigma} \neq 0, 1\end{aligned}$$

For empty bands ($n = 0$), fully occupied bands ($n = 2$) and fully polarized and half-filled bands ($n_\sigma = 1, n_{-\sigma} = 0$) the spectral density consists of only one (!) δ-function. In all other cases

$$\Delta_{ii\sigma}^{(1)} > 0$$

We now calculate

$$\begin{aligned}\Delta_{ii\sigma}^{(2)} &= \begin{pmatrix} M_{ii\sigma}^{(0)} & M_{ii\sigma}^{(1)} & M_{ii\sigma}^{(2)} \\ M_{ii\sigma}^{(1)} & M_{ii\sigma}^{(2)} & M_{ii\sigma}^{(3)} \\ M_{ii\sigma}^{(2)} & M_{ii\sigma}^{(3)} & M_{ii\sigma}^{(4)} \end{pmatrix} \\ &= M_{ii\sigma}^{(0)} M_{ii\sigma}^{(2)} M_{ii\sigma}^{(4)} + 2M_{ii\sigma}^{(1)} M_{ii\sigma}^{(3)} M_{ii\sigma}^{(2)} - \\ &- (M_{ii\sigma}^{(2)})^3 - M_{ii\sigma}^{(0)} (M_{ii\sigma}^{(3)})^2 - (M_{ii\sigma}^{(1)})^2 M_{ii\sigma}^{(4)} \\ &= \left\{M_{ii\sigma}^{(0)} M_{ii\sigma}^{(2)} - (M_{ii\sigma}^{(1)})^2\right\} M_{ii\sigma}^{(4)} + \\ &+ \left\{M_{ii\sigma}^{(1)} M_{ii\sigma}^{(3)} - (M_{ii\sigma}^{(2)})^2\right\} M_{ii\sigma}^{(2)} + \\ &+ \left\{M_{ii\sigma}^{(1)} M_{ii\sigma}^{(2)} - M_{ii\sigma}^{(0)} M_{ii\sigma}^{(3)}\right\} M_{ii\sigma}^{(3)}\end{aligned}$$

We calculate the individual terms with the abbreviation:

$$t_0 = T_0 - \mu$$

$$\begin{aligned}
&\left\{M^{(0)}_{ii\sigma}M^{(2)}_{ii\sigma} - (M^{(1)}_{ii\sigma})^2\right\} = \\
&= t_0^2 + ((t_0+U)^2 - t_0^2)n_{-\sigma} - (t_0 + (t_0 + U - t)n_{-\sigma})^2 \\
&= t_0^2 - t_0^2 + n_{-\sigma}\left\{2t_0U + U^2 - 2t_0U\right\} - n_{-\sigma}^2U^2 \\
&= U^2n_{-\sigma}(1 - n_{-\sigma}) \\
&\left\{M^{(1)}_{ii\sigma}M^{(3)}_{ii\sigma} - (M^{(2)}_{ii\sigma})^2\right\} = \\
&= (t_0 + Un_{-\sigma})(t_0^3 + ((t_0+U)^3 - t_0^3)n_{-\sigma} - \\
&- (t_0^2 + ((t_0+U)^2 - t_0^2)n_{-\sigma})^2 \\
&= (t_0 + Un_{-\sigma})(t_0^3 + n_{-\sigma}(3t_0^2U + 3t_0U^2 + U^3)) - \\
&- (t_0^2 + n_{-\sigma}(2t_0U + U^2))^2 \\
&= t_0^4 + n_{-\sigma}(3t_0^3U + 3t_0^2U^2 + t_0U^3) + Un_{-\sigma}t_0^3 + \\
&+ Un_{-\sigma}^2(3t_0^2U + 3t_0U^2 + U^3) - t_0^4 - \\
&- n_{-\sigma}^2(2t_0U + U^2)^2 - 2n_{-\sigma}t_0^2(2t_0U + U^2) \\
&= n_{-\sigma}\left(3t_0^3U + 3t_0^2U^2 + t_0U^3 + Ut_0^3 - 4t_0^3U - \right. \\
&- \left. 2t_0^2U^2\right) + n_{-\sigma}^2\left(3t_0^2U^2 + 3t_0U^3 + U^4 - 4t_0^2U^2 - \right. \\
&- \left. U^4 - 4t_0U^3\right) \\
&= n_{-\sigma}(t_0^2U^2 + t_0U^3) + n_{-\sigma}^2(-t_0^2U^2 - t_0U^3) \\
&= U^2n_{-\sigma}(t_0^2 + t_0U) - U^2n_{-\sigma}^2(t_0^2 + t_0U) \\
&= U^2n_{-\sigma}(1 - n_{-\sigma})t_0(t_0 + U)
\end{aligned}$$

$$\begin{aligned}
&\left\{M^{(1)}_{ii\sigma}M^{(2)}_{ii\sigma} - M^{(0)}_{ii\sigma}M^{(3)}_{ii\sigma}\right\} \\
&= (t_0 + Un_{-\sigma})(t_0^2 + n_{-\sigma}\left((t_0+U)^2 - t_0^2\right)) - \\
&- \left(t_0^3 + n_{-\sigma}\left((t_0+U)^3 - t_0^3\right)\right) \\
&= t_0^3 + n_{-\sigma}\left(2t_0^2U + t_0U^2\right) + Un_{-\sigma}\left(t_0^2 + \right. \\
&+ \left. n_{-\sigma}(2t_0U + U^2)\right) - \\
&- (t_0^3 + n_{-\sigma}(3t_0^2U + 3t_0U^2 + U^3)) \\
&= n_{-\sigma}\left(2t_0^2U + t_0U^2 + Ut_0^2 - \right. \\
&- \left. 3t_0^2U - 3t_0U^2 - U^3\right) + n_{-\sigma}^2\left(2t_0U^2 + U^3\right) \\
&= n_{-\sigma}(-2t_0U^2 - U^3) + n_{-\sigma}^2U^2(2t_0 + U) \\
&= -U^2n_{-\sigma}(1 - n_{-\sigma})(2t_0 + U)
\end{aligned}$$

Intermediate result:

$$\frac{\Delta_{ii\sigma}^{(2)}}{U^2 n_{-\sigma}(1-n_{-\sigma})} = M_{ii\sigma}^{(4)} + t_0(t_0+U)M_{ii\sigma}^{(2)} - \\ - (2t_0+U)M_{ii\sigma}^{(3)}$$

We now finally calculate the right-hand side:

$$\begin{aligned}
&\frac{\Delta_{ii\sigma}^{(2)}}{U^2 n_{-\sigma}(1-n_{-\sigma})} = \\
&= t_0^4 + ((t_0+U)^4 - t_0^4)n_{-\sigma} + \\
&+ t_0(t_0+U)\left[t_0^2 + \left((t_0+U)^2 - t_0^2\right)n_{-\sigma}\right] - \\
&- (2t_0+U)\left[t_0^3 + \left((t_0+U)^3 - t_0^3\right)n_{-\sigma}\right] \\
&= \left(t_0^4 + t_0^3(t_0+U) - 2t_0^4 - Ut_0^3\right) + \\
&+ n_{-\sigma}\left\{\left((t_0+U)^2 - t_0^2\right)(t_0(t_0+U) + \right. \\
&+ \left. (t_0+U)^2 + t_0^2) - (2t_0+U)((t_0+U)^3 - t_0^3)\right\} \\
&= n_{-\sigma}\left\{(2t_0U + U^2)(3t_0^2 + t_0U + 2t_0U + U^2) - \right. \\
&- \left. (2t_0+U)(3t_0^2U + 3t_0U^2 + U^3)\right\} \\
&= n_{-\sigma}\left\{6t_0^2U^2 + 3U^2t_0^2 - 6t_0^2U^2 - 3t_0^2U^2 + \right. \\
&+ \left. 2t_0U^3 + 3t_0U^3 - 2t_0U^3 - 3t_0U^3\right\} \\
&= 0
\end{aligned}$$

With this it is proved that the one-electron spectral density in the limit of infinitely narrow band is a two-pole function:

$$S_\sigma(E) = \alpha_{1\sigma}\delta(E - E_{1\sigma}) + \alpha_{2\sigma}\delta(E - E_{2\sigma})$$

3. Spectral moments:

$$M_{ii\sigma}^{(n)}(E) = (T_0 - \mu)^n(1 - n_{-\sigma}) + (T_0 + U - \mu)^n n_{-\sigma}$$

On the other hand it follows from part 2

$$M_{ii\sigma}^{(n)}(E) = \alpha_{1\sigma}E_{1\sigma}^n + \alpha_{2\sigma}E_{2\sigma}^n$$

Compare:

$$\begin{aligned}
E_{1\sigma} &= T_0 - \mu; \quad \alpha_{1\sigma} = 1 - n_{-\sigma} \\
E_{2\sigma} &= T_0 + U - \mu\,; \quad \alpha_{2\sigma} = n_{-\sigma}
\end{aligned}$$

This agrees with (8.129), (8.130) and (8.131)!

Problem 8.16

One can easily calculate

$$[c_{d\sigma}, H]_- = (\varepsilon_d - \mu)c_{d\sigma} + \\ + \sum_{\mathbf{k}} V_{\mathbf{k}d} c_{\mathbf{k}\sigma} + \left[c_{d\sigma}, \frac{1}{2}U \sum_{\sigma'} n_{d\sigma'} n_{d-\sigma'}\right]_-$$

With (8.334) follows then the equation of motion:

$$(E + \mu - \varepsilon_d - \Sigma_{d\sigma}(E))G_{d\sigma}(E) = \hbar + \sum_{\mathbf{k}} V_{\mathbf{k}d} \left\langle\left\langle c_{\mathbf{k}\sigma}; c_{d\sigma}^\dagger \right\rangle\right\rangle_E$$

We calculate the "mixed" Green's function

$$[c_{\mathbf{k}\sigma}, H]_- = (\varepsilon(\mathbf{k}) - \mu)c_{\mathbf{k}\sigma} + V_{\mathbf{k}d} c_{d\sigma}$$

$$\curvearrowright \quad (E + \mu - \varepsilon(\mathbf{k})) \left\langle\left\langle c_{\mathbf{k}\sigma}; c_{d\sigma}^\dagger \right\rangle\right\rangle_E = V_{\mathbf{k}d} \left\langle\left\langle c_{d\sigma}; c_{d\sigma}^\dagger \right\rangle\right\rangle$$

So that it follows:

$$\left\langle\left\langle c_{\mathbf{k}\sigma}; c_{d\sigma}^\dagger \right\rangle\right\rangle_E = \frac{V_{\mathbf{k}d}}{E + \mu - \varepsilon(\mathbf{k})} G_{d\sigma}(E)$$

With the definition of (8.335) of the "hybridization function", what remains is:

$$(E + \mu - \varepsilon_d - \Sigma_{d\sigma}(E))G_{d\sigma}(E) = \hbar + \Delta(E)G_{d\sigma}(E)$$

This proves the proposition:

$$G_{d\sigma}(E) = \frac{\hbar}{E + \mu - \varepsilon_d - \Sigma_{d\sigma}(E) - \Delta(E)}$$

Problem 8.17

$$n_{d\sigma}(1 - n_{d\sigma})(B_{d\sigma} - \varepsilon_d) = \sum_{\mathbf{k}} V_{\mathbf{k}d} \left\langle c_{\mathbf{k}\sigma}^\dagger c_{d\sigma}(2n_{d-\sigma} - 1)\right\rangle$$

We begin with

$$\sum_{\mathbf{k}} V_{\mathbf{k}d} \left\langle c_{\mathbf{k}\sigma}^\dagger c_{d\sigma}\right\rangle = \\ = -\frac{1}{\pi\hbar} Im \int_{-\infty}^{+\infty} dE\; f_-(E) \sum_{\mathbf{k}} V_{\mathbf{k}d} \left\langle\left\langle c_{d\sigma}; c_{\mathbf{k}\sigma}^\dagger \right\rangle\right\rangle_{E-\mu} \quad (1)$$

We have calculated the mixed Green's function in Problem 8.16:

$$\langle\langle c_{\mathbf{k}\sigma}; c_{d\sigma}^{\dagger} \rangle\rangle = \frac{V_{\mathbf{k}d}}{E + \mu - \varepsilon(\mathbf{k})} G_{d\sigma}(E) = \langle\langle c_{d\sigma}; c_{\mathbf{k}\sigma}^{\dagger} \rangle\rangle \quad (2)$$

Because of the assumption that $V_{\mathbf{k}d}$ is real, the last step can be easily proved:

$$\sum_{\mathbf{k}} V_{\mathbf{k}d} \left\langle c_{\mathbf{k}\sigma}^{\dagger} c_{d\sigma} \right\rangle =$$

$$= -\frac{1}{\pi\hbar} Im \int_{-\infty}^{+\infty} dE\ f_{-}(E)\Delta(E-\mu)G_{d\sigma}(E-\mu) \quad (3)$$

Δ: "hybridization function" (8.335):

$$\Delta(E) = \sum_{\mathbf{k}} \frac{V_{\mathbf{k}d}^2}{E + \mu - \varepsilon(\mathbf{k})}$$

It holds

$$[c_{d\sigma}, \mathcal{H}]_{-} = (\varepsilon_d - \mu)c_{d\sigma} + U c_{d\sigma} n_{d-\sigma} + \sum_{\mathbf{p}} V_{\mathbf{p}d} c_{\mathbf{p}\sigma}$$

So that

$$\left\langle c_{\mathbf{k}\sigma}^{\dagger} c_{d\sigma} n_{d-\sigma} \right\rangle = -\frac{1}{U}(\varepsilon_d - \mu) \left\langle c_{\mathbf{k}\sigma}^{\dagger} c_{d\sigma} \right\rangle - \\ -\frac{1}{U} \sum_{\mathbf{p}} V_{\mathbf{p}d} \left\langle c_{\mathbf{k}\sigma}^{\dagger} c_{\mathbf{p}\sigma} \right\rangle + \frac{1}{U} \left\langle c_{\mathbf{k}\sigma}^{\dagger} [c_{d\sigma}, \mathcal{H}]_{-} \right\rangle$$

Now for the band shift we still have to calculate

$$n_{d\sigma}(1 - n_{d\sigma})(B_{d\sigma} - \varepsilon_d) = \left(-2\frac{\varepsilon_d - \mu}{U} - 1\right) \sum_{\mathbf{k}} V_{\mathbf{k}d} \left\langle c_{\mathbf{k}\sigma}^{\dagger} c_{d\sigma} \right\rangle - \\ -\frac{2}{U} \sum_{\mathbf{k},\mathbf{p}} V_{\mathbf{k}d} V_{\mathbf{p}d} \left\langle c_{\mathbf{k}\sigma}^{\dagger} c_{\mathbf{p}\sigma} \right\rangle + \\ +\frac{2}{U} \sum_{\mathbf{k}} V_{\mathbf{k}d} \left\langle c_{\mathbf{k}\sigma}^{\dagger} [c_{\mathbf{d}\sigma}, \mathcal{H}]_{-} \right\rangle \quad (4)$$

The first summand is known from (3). For the second summand we need $< c_{\mathbf{k}\sigma}^{\dagger} c_{\mathbf{p}\sigma} >$:

$$(E+\mu-\varepsilon(\mathbf{p}))\left\langle\left\langle c_{\mathbf{p}\sigma};c^{\dagger}_{\mathbf{k}\sigma}\right\rangle\right\rangle_E = \hbar\delta_{\mathbf{pk}} + V_{\mathbf{p}d}\left\langle\left\langle c_{d\sigma};c^{\dagger}_{\mathbf{k}\sigma}\right\rangle\right\rangle$$
$$\stackrel{(2)}{=} \hbar\delta_{\mathbf{pk}} + \frac{V_{\mathbf{k}d}V_{\mathbf{p}d}}{E+\mu-\varepsilon(\mathbf{k})}G_{d\sigma}(E)$$

That means

$$\sum_{\mathbf{k},\mathbf{p}} V_{\mathbf{k}d}V_{\mathbf{p}d}\left\langle c^{\dagger}_{\mathbf{k}\sigma}c_{\mathbf{p}\sigma}\right\rangle =$$
$$= -\frac{1}{\pi\hbar}Im\int_{-\infty}^{+\infty} dE\; f_-(E)\sum_{\mathbf{k},\mathbf{p}} V_{\mathbf{k}d}V_{\mathbf{p}d}\left\langle\left\langle c_{\mathbf{p}\sigma};c^{\dagger}_{\mathbf{k}\sigma}\right\rangle\right\rangle_{E-\mu}$$
$$= -\frac{1}{\pi\hbar}Im\int_{-\infty}^{+\infty} dE\; f_-(E)\left\{\hbar\sum_{\mathbf{k}}\frac{V^2_{\mathbf{k}d}}{E-\varepsilon(\mathbf{k})}+\right.$$
$$\left.+\sum_{\mathbf{k},\mathbf{p}}\frac{V^2_{\mathbf{p}d}V^2_{\mathbf{k}d}}{(E-\varepsilon(\mathbf{k}))(E-\varepsilon(\mathbf{p}))}G_{d\sigma}(E-\mu)\right\}$$
$$= -\frac{1}{\pi\hbar}Im\int_{-\infty}^{+\infty} dE\; f_-(E)\Delta(E-\mu) *$$
$$* \{\hbar + \Delta(E-\mu)G_{d\sigma}(E-\mu)\} \quad (5)$$

Finally it still holds

$$\left\langle\left\langle [c_{d\sigma},\mathcal{H}]_-;c^{\dagger}_{\mathbf{k}\sigma}\right\rangle\right\rangle_E = E\left\langle\left\langle c_{d\sigma};c^{\dagger}_{\mathbf{k}\sigma}\right\rangle\right\rangle_E - \hbar\left\langle\left[c_{d\sigma},c^{\dagger}_{\mathbf{k}\sigma}\right]_+\right\rangle$$
$$\stackrel{(2)}{=} E\frac{V_{\mathbf{k}d}}{E+\mu-\varepsilon(\mathbf{k})}G_{d\sigma}(E) - \hbar\left\langle\left[c_{d\sigma},c^{\dagger}_{\mathbf{k}\sigma}\right]_+\right\rangle$$

Spectral theorem:

$$\sum_{\mathbf{k}} V_{\mathbf{k}d}\left\langle c^{\dagger}_{\mathbf{k}\sigma}\,[c_{d\sigma},\mathcal{H}]_-\right\rangle =$$
$$= -\frac{1}{\pi\hbar}Im\int_{-\infty}^{+\infty} dE\; f_-(E)(E-\mu)\Delta(E-\mu)G_{d\sigma}(E-\mu) \quad (6)$$

In (4) we need (3), (5) and (6):

$$
\begin{aligned}
I &= (-2\frac{\varepsilon_d-\mu}{U}-1)\Delta(E-\mu)G_{d\sigma}(E-\mu)- \\
&-\frac{2}{U}\Delta(E-\mu)\{\hbar+\Delta(E-\mu)G_{d\sigma}(E-\mu)\}+ \\
&+\frac{2}{U}\Delta(E-\mu)(E-\mu)G_{d\sigma}(E-\mu) \\
&= -\Delta(E-\mu)\frac{2}{U}\hbar+\Delta(E-\mu)G_{d\sigma}(E-\mu)* \\
&* \left\{-1+\frac{2}{U}(-\varepsilon_d+\mu-\Delta(E-\mu)+E-\mu)\right\}
\end{aligned}
$$

Equation of motion:

$$
\begin{aligned}
&(E+\mu-\varepsilon_d-\Sigma_{d\sigma}(E)-\Delta(E))G_{d\sigma}(E)=\hbar \\
\Leftrightarrow\quad I &= -\Delta(E-\mu)\frac{2}{U}\hbar-\Delta(E-\mu)G_{d\sigma}(E-\mu)+ \\
&+\frac{2}{U}\Delta(E-\mu)(\hbar+\Sigma_{d\sigma}(E-\mu)G_{d\sigma}(E-\mu)) \\
&= \Delta(E-\mu)G_{d\sigma}(E-\mu)\left(\frac{2}{U}\Sigma_{d\sigma}(E-\mu)-1\right)
\end{aligned}
$$

Then it finally follows:

$$
\begin{aligned}
&n_{d\sigma}(1-n_{d\sigma})(B_{d\sigma}-\varepsilon_d)= \\
&= -\frac{1}{\pi\hbar}\,Im\int_{-\infty}^{+\infty}dE\;f_-(E)\Delta(E-\mu)G_{d\sigma}(E-\mu)* \\
&\qquad\qquad * \left(\frac{2}{U}\Sigma_{d\sigma}(E-\mu)-1\right)
\end{aligned}
$$

This is exactly the proposition (8.352).

Problem A.1

With (A.24) and (A.30) we first have

$$
\begin{aligned}
c_\beta\left(c_\gamma^\dagger|\varphi_{\alpha_1}\cdots\varphi_{\alpha_N}\rangle^{(\varepsilon)}\right) &= \sqrt{N+1}\,a_\beta|\varphi_\gamma\varphi_{\alpha_1}\cdots\varphi_{\alpha_N}\rangle^{(\varepsilon)} \\
&= \left\{\delta(\varphi_\beta-\varphi_\gamma)|\varphi_{\alpha_1}\cdots\varphi_{\alpha_N}\rangle^{(\varepsilon)}+\right. \\
&+\varepsilon^1\delta(\varphi_\beta-\varphi_{\alpha_1})|\varphi_\gamma\varphi_{\alpha_2}\cdots\varphi_{\alpha_N}\rangle^{(\varepsilon)}+ \\
&+\cdots+ \\
&\left.+\varepsilon^N\delta(\varphi_\beta-\varphi_{\alpha_N})|\varphi_\gamma\varphi_{\alpha_1}\cdots\varphi_{\alpha_{N-1}}\rangle^{(\varepsilon)}\right\}
\end{aligned}
$$

On the other hand it is also valid that

$$c_\gamma^\dagger \left(c_\beta \,|\varphi_{\alpha_1} \cdots \varphi_{\alpha_N}\rangle^{(\varepsilon)}\right) = \delta(\varphi_\beta - \varphi_{\alpha_1})|\varphi_\gamma \varphi_{\alpha_2} \cdots \varphi_{\alpha_N}\rangle^{(\varepsilon)} \\ + \cdots + \\ + \varepsilon^{N-1}\delta(\varphi_\beta - \varphi_{\alpha_N})|\varphi_\gamma \varphi_{\alpha_1} \cdots \varphi_{\alpha_{N-1}}\rangle^{(\varepsilon)}$$

One multiplies the last equation by ε and then subtracts one equation from the other to get

$$\left(c_\beta c_\gamma^\dagger - \varepsilon c_\gamma^\dagger c_\beta\right) |\varphi_{\alpha_1} \cdots \varphi_{\alpha_N}\rangle^{(\varepsilon)} = \delta(\varphi_\beta - \varphi_\gamma)|\varphi_{\alpha_1} \cdots \varphi_{\alpha_N}\rangle^{(\varepsilon)}$$

Problem A.2

Bosons : $\left|\cdots n_{\alpha_r} \cdots n_{\alpha_s} \cdots\right\rangle^{(+)}$: arbitrary Fock state.

$r \neq s$:

$$c_{\alpha_r}^\dagger c_{\alpha_s}^\dagger \left|\cdots n_{\alpha_r} \cdots n_{\alpha_s} \cdots\right\rangle^{(+)} \\ = \sqrt{n_{\alpha_r}+1}\sqrt{n_{\alpha_s}+1} \left|\cdots n_{\alpha_r}+1 \cdots n_{\alpha_s}+1 \cdots\right\rangle^{(+)} \\ = c_{\alpha_s}^\dagger c_{\alpha_r}^\dagger \left|\cdots n_{\alpha_r} \cdots n_{\alpha_s} \cdots\right\rangle^{(+)} \\ \Longrightarrow \left[c_{\alpha_r}^\dagger, c_{\alpha_s}^\dagger\right]_- = 0.$$

For $r = s$ this relation is trivially valid.

Since

$$\left[c_{\alpha_r}, c_{\alpha_s}\right]_- = \left(\left[c_{\alpha_s}^\dagger, c_{\alpha_r}^\dagger\right]_-\right)^\dagger$$

directly follows:

$$\left[c_{\alpha_r}, c_{\alpha_s}\right]_- = 0.$$

$r \neq s$:

$$c_{\alpha_r} c_{\alpha_s}^\dagger \left|\cdots n_{\alpha_r} \cdots n_{\alpha_s} \cdots\right\rangle^{(+)} \\ = \sqrt{n_{\alpha_r}}\sqrt{n_{\alpha_s}+1} \left|\cdots n_{\alpha_r}-1 \cdots n_{\alpha_s}+1 \cdots\right\rangle^{(+)} \\ = c_{\alpha_s}^\dagger c_{\alpha_r} \left|\cdots n_{\alpha_r} \cdots n_{\alpha_s} \cdots\right\rangle^{(+)}.$$

$r = s$:

$$c_{\alpha_r} c^\dagger_{\alpha_r} \left|\cdots n_{\alpha_r} \cdots\right)^{(+)} = c_{\alpha_r} \sqrt{n_{\alpha_r}+1} \left|\cdots n_{\alpha_r}+1 \cdots\right)^{(+)}$$
$$= (n_{\alpha_r}+1) \left|\cdots n_{\alpha_r} \cdots\right)^{(+)},$$
$$c^\dagger_{\alpha_r} c_{\alpha_r} \left|\cdots n_{\alpha_r} \cdots\right)^{(+)} = \sqrt{n_{\alpha_r}} c^\dagger_{\alpha_r} \left|\cdots n_{\alpha_r}-1 \cdots\right)^{(+)}$$
$$= n_{\alpha_r} \left|\cdots n_{\alpha_r} \cdots\right)^{(+)}$$
$$\Longrightarrow \left[c_{\alpha_r}, c^\dagger_{\alpha_s}\right]_- = \delta_{r,s}.$$

Fermions:

$$(c^\dagger_{\alpha_r})^2 \left|\cdots n_{\alpha_r} \cdots\right)^{(-)} = 0 \quad \text{(Pauli principle: Problem A.4).}$$

$r < s$:

$$c^\dagger_{\alpha_r} c^\dagger_{\alpha_s} \left|\cdots n_{\alpha_r} \cdots n_{\alpha_s} \cdots\right)^{(-)}$$
$$= c^\dagger_{\alpha_r} (-1)^{N_s} \delta_{n_{\alpha_s},0} \left|\cdots n_{\alpha_r} \cdots n_{\alpha_s}+1 \cdots\right)^{(-)}$$
$$= (-1)^{N_r} (-1)^{N_s} \delta_{n_{\alpha_s},0} \delta_{n_{\alpha_r},0} \left|\cdots n_{\alpha_r}+1 \cdots n_{\alpha_s}+1 \cdots\right)^{(-)},$$
$$c^\dagger_{\alpha_s} c^\dagger_{\alpha_r} \left|\cdots n_{\alpha_r} \cdots n_{\alpha_s} \cdots\right)^{(-)}$$
$$= (-1)^{N_r} \delta_{n_{\alpha_r},0} c^\dagger_{\alpha_s} \left|\cdots n_{\alpha_r}+1 \cdots n_{\alpha_s} \cdots\right)^{(-)}$$
$$= (-1)^{N_r} (-1)^{N'_s} \delta_{n_{\alpha_r},0} \delta_{n_{\alpha_s},0} \left|\cdots n_{\alpha_r}+1 \cdots n_{\alpha_s}+1 \cdots\right)^{(-)},$$

$$N'_s = N_s + 1$$
$$\Longrightarrow (c^\dagger_{\alpha_r} c^\dagger_{\alpha_s} + c^\dagger_{\alpha_s} c^\dagger_{\alpha_r}) \left|\cdots n_{\alpha_r} \cdots n_{\alpha_s} \cdots\right)^{(-)} = 0$$
$$\Longrightarrow \left[c^\dagger_{\alpha_r}, c^\dagger_{\alpha_s}\right]_+ = 0.$$

Since

$$\left[c_{\alpha_r}, c_{\alpha_s}\right]_+ = (\left[c^\dagger_{\alpha_s}, c^\dagger_{\alpha_r}\right]_+)^\dagger$$

again the second anti-commutator relation follows directly:

$$\left[c_{\alpha_r}, c_{\alpha_s}\right]_+ = 0$$

$r = s$:

$$c_{\alpha_r} c^\dagger_{\alpha_r} \left|\cdots n_{\alpha_r} \cdots\right)^{(-)} = c_{\alpha_r} (-1)^{N_r} \delta_{n_{\alpha_r},0} \left|\cdots n_{\alpha_r}+1 \cdots\right)^{(-)} =$$
$$= (-1)^{2N_r} \delta_{n_{\alpha_r},0} \left|\cdots n_{\alpha_r} \cdots\right)^{(-)} =$$
$$= \delta_{n_{\alpha_r},0} \left|\cdots n_{\alpha_r} \cdots\right)^{(-)},$$
$$c^\dagger_{\alpha_r} c_{\alpha_r} \left|\cdots n_{\alpha_r} \cdots\right)^{(-)} = \delta_{n_{\alpha_r},1} \left|\cdots n_{\alpha_r} \cdots\right)^{(-)}.$$

Since in every case $n_{\alpha_r} = 0$ or 1, we have

$$\left(c_{\alpha_r} c^{\dagger}_{\alpha_r} + c^{\dagger}_{\alpha_r} c_{\alpha r}\right) \left|\cdots n_{\alpha_r} \cdots\right\rangle^{(-)} = \left|\cdots n_{\alpha_r} \cdots\right\rangle^{(-)}.$$

$r < s$:

$$\begin{aligned}
&c_{\alpha_r} c^{\dagger}_{\alpha_s} \left|\cdots n_{\alpha_r} \cdots n_{\alpha_s} \cdots\right\rangle^{(-)} \\
&= c_{\alpha_r} (-1)^{N_s} \delta_{n_{\alpha_s},0} \left|\cdots n_{\alpha_r} \cdots n_{\alpha_s} + 1 \cdots\right\rangle^{(-)} \\
&= (-1)^{N_r + N_s} \delta_{n_{\alpha r},1} \delta_{n_{\alpha s},0} \left|\cdots n_{\alpha_r} - 1 \cdots n_{\alpha_s} + 1 \cdots\right\rangle^{(-)}, \\
&c^{\dagger}_{\alpha_s} c_{\alpha_r} \left|\cdots n_{\alpha_r} \cdots n_{\alpha_s} \cdots\right\rangle^{(-)} \\
&= c^{\dagger}_{\alpha_s} (-1)^{N_r} \delta_{n_{\alpha r},1} \left|\cdots n_{\alpha_r} - 1 \cdots n_{\alpha_s} \cdots\right\rangle^{(-)} \\
&= (-1)^{N_r + N''_s} \delta_{n_{\alpha r},1} \delta_{n_{\alpha s},0} \left|\cdots n_{\alpha_r} - 1 \cdots n_{\alpha_s} + 1 \cdots\right\rangle^{(-)},
\end{aligned}$$

$$\begin{aligned}
&N''_s = N_s - 1 \\
&\Longrightarrow \left(c_{\alpha_r} c^{\dagger}_{\alpha_s} + c^{\dagger}_{\alpha_s} c_{\alpha_r}\right) \left|\cdots n_{\alpha_r} \cdots n_{\alpha_s} \cdots\right\rangle^{(-)} = 0.
\end{aligned}$$

So altogether we have

$$\left[c_{\alpha_r}, c^{\dagger}_{\alpha_s}\right]_+ = \delta_{r,s}.$$

Problem A.3

1. Bosons:

$$\begin{aligned}
\widehat{n}_{\alpha}\, c^{\dagger}_{\beta} &= c^{\dagger}_{\alpha} c_{\alpha} c^{\dagger}_{\beta} \\
&= c^{\dagger}_{\alpha} c^{\dagger}_{\beta} c_{\alpha} + \delta_{\alpha\beta}\, c^{\dagger}_{\alpha} \\
&= c^{\dagger}_{\beta} c^{\dagger}_{\alpha} c_{\alpha} + \delta_{\alpha\beta}\, c^{\dagger}_{\alpha} \\
&= c^{\dagger}_{\beta}\, \widehat{n}_{\alpha} + \delta_{\alpha\beta}\, c^{\dagger}_{\alpha}
\end{aligned}$$

So that we have

$$\left[\widehat{n}_{\alpha}, c^{\dagger}_{\beta}\right]_- = \delta_{\alpha\beta}\, c^{\dagger}_{\alpha}$$

Fermions:

$$
\begin{aligned}
\widehat{n}_\alpha\, c_\beta^\dagger &= c_\alpha^\dagger c_\alpha c_\beta^\dagger \\
&= -c_\alpha^\dagger c_\beta^\dagger c_\alpha + \delta_{\alpha\beta}\, c_\alpha^\dagger \\
&= c_\beta^\dagger c_\alpha^\dagger c_\alpha + \delta_{\alpha\beta}\, c_\alpha^\dagger \\
&= c_\beta^\dagger\, \widehat{n}_\alpha + \delta_{\alpha\beta}\, c_\alpha^\dagger
\end{aligned}
$$

So that just as in the case of Bosons we get

$$
\left[\widehat{n}_\alpha,\, c_\beta^\dagger\right]_- = \delta_{\alpha\beta}\, c_\alpha^\dagger
$$

2. Bosons :

$$
\begin{aligned}
\widehat{n}_\alpha\, c_\beta &= c_\alpha^\dagger c_\alpha c_\beta \\
&= c_\alpha^\dagger c_\beta c_\alpha = c_\beta c_\alpha^\dagger c_\alpha - \delta_{\alpha\beta}\, c_\alpha \\
&= c_\beta\, \widehat{n}_\alpha - \delta_{\alpha\beta}\, c_\alpha
\end{aligned}
$$

With this follows:

$$
\left[\widehat{n}_\alpha,\, c_\beta\right]_- = -\delta_{\alpha\beta}\, c_\alpha
$$

Fermions:

$$
\begin{aligned}
\widehat{n}_\alpha\, c_\beta &= c_\alpha^\dagger c_\alpha c_\beta \\
&= -c_\alpha^\dagger c_\beta c_\alpha = c_\beta c_\alpha^\dagger c_\alpha - \delta_{\alpha\beta}\, c_\alpha \\
&= c_\beta\, \widehat{n}_\alpha - \delta_{\alpha\beta}\, c_\alpha
\end{aligned}
$$

With this, as in the case of Bosons we get

$$
\left[\widehat{n}_\alpha,\, c_\beta\right]_- = -\delta_{\alpha\beta}\, c_\alpha
$$

3. For Bosons as well as for Fermions, with part 1

$$
\left[\widehat{N},\, c_\alpha^\dagger\right]_- = \sum_\gamma \left[\widehat{n}_\gamma,\, c_\alpha^\dagger\right]_- = \sum_\gamma \delta_{\alpha\gamma} c_\alpha^\dagger = c_\alpha^\dagger
$$

is valid.

4. For Bosons as well as Fermions with part 2

$$
\left[\widehat{N},\, c_\alpha\right]_- = \sum_\gamma \left[\widehat{n}_\gamma,\, c_\alpha\right]_- = \sum_\gamma (-\delta_{\alpha\gamma} c_\alpha) = -c_\alpha
$$

is valid.

Problem A.4

1.

$$\left[c_\alpha,\, c_\beta\right]_+ = 0 \;\curvearrowright\; [c_\alpha,\, c_\alpha]_+ = 2c_\alpha^2 = 0 \;\curvearrowright\; c_\alpha^2 = 0$$

$$\left[c_\alpha^\dagger,\, c_\beta^\dagger\right]_+ = 0 \;\curvearrowright\; \left[c_\alpha^\dagger,\, c_\alpha^\dagger\right]_+ = 2\left(c_\alpha^\dagger\right)^2 = 0$$
$$\curvearrowright\; \left(c_\alpha^\dagger\right)^2 = 0 \quad \text{(Pauli principle)}$$

2.

$$\begin{aligned}\widehat{n}_\alpha^2 &= c_\alpha^\dagger c_\alpha c_\alpha^\dagger c_\alpha = c_\alpha^\dagger\left(1 - c_\alpha^\dagger c_\alpha\right)c_\alpha \\ &= c_\alpha^\dagger c_\alpha - \left(c_\alpha^\dagger\right)^2 (c_\alpha)^2 \overset{1.)}{=} \widehat{n}_\alpha \qquad \text{(Pauli principle)}\end{aligned}$$

3.

$$\begin{aligned}c_\alpha \widehat{n}_\alpha &= c_\alpha c_\alpha^\dagger c_\alpha = \left(1 - c_\alpha^\dagger c_\alpha\right)c_\alpha \overset{1.)}{=} c_\alpha \\ c_\alpha^\dagger \widehat{n}_\alpha &= c_\alpha^\dagger c_\alpha^\dagger c_\alpha \overset{1.)}{=} 0\end{aligned}$$

4.

$$\begin{aligned}\widehat{n}_\alpha\, c_\alpha &= c_\alpha^\dagger c_\alpha c_\alpha \overset{1.)}{=} 0 \\ \widehat{n}_\alpha\, c_\alpha^\dagger &= c_\alpha^\dagger c_\alpha c_\alpha^\dagger = c_\alpha^\dagger\left(1 - c_\alpha^\dagger c_\alpha\right) \overset{1.)}{=} c_\alpha^\dagger\end{aligned}$$

Problem A.5

Proof by complete induction

$N = 1:$

$$\begin{aligned}\langle 0|\, c_{\beta_1} c_{\alpha_1}^\dagger\, |0\rangle &= \langle 0|\left[\delta(\beta_1, \alpha_1) \pm c_{\alpha_1}^\dagger c_{\beta_1}\right]|0\rangle \\ &= \delta(\beta_1, \alpha_1)\,\langle 0|0\rangle \pm \langle 0|\, c_{\alpha_1}^\dagger c_{\beta_1}\, |0\rangle = \delta(\beta_1, \alpha_1)\end{aligned}$$

because $c_{\beta_1} 10 >= 0$.

$N - 1 \longrightarrow N:$

$$
\begin{aligned}
&\langle 0|c_{\beta_N}\cdots c_{\beta_1}c^{\dagger}_{\alpha_1}\cdots c^{\dagger}_{\alpha_N}|0\rangle \\
&\swarrow \text{taking } c_{\beta_1} \text{ to the right} \\
&=\delta(\beta_1,\alpha_1)\,\langle 0|\,c_{\beta_N}\cdots c_{\beta_2}c^{\dagger}_{\alpha_2}\cdots c^{\dagger}_{\alpha_N}\,|0\rangle + \\
&\quad + (\pm)^1\delta(\beta_1,\alpha_2)\,\langle 0|\,c_{\beta_N}\cdots c_{\beta_2}c^{\dagger}_{\alpha_1}c^{\dagger}_{\alpha_3}\cdots c^{\dagger}_{\alpha_N}\,|0\rangle + \\
&\quad + \cdots + \\
&\quad + (\pm)^{N-1}\delta(\beta_1,\alpha_N)\,\langle 0|\,c_{\beta_N}\cdots c_{\beta_2}c^{\dagger}_{\alpha_1}c^{\dagger}_{\alpha_2}\cdots c^{\dagger}_{\alpha_{N-1}}\,|0\rangle = \\
&\swarrow \text{condition for induction} \\
&=\delta(\beta_1,\alpha_1)\sum_{\mathcal{P}_\alpha}(\pm)^{p_\alpha}\mathcal{P}_\alpha\left[\delta(\beta_2,\alpha_2)\cdots\delta(\beta_N,\alpha_N)\right] + \\
&\quad (\pm)^1\delta(\beta_1,\alpha_2)\sum_{\mathcal{P}_\alpha}(\pm)^{p_\alpha}\mathcal{P}_\alpha\left[\delta(\beta_2,\alpha_1)\delta(\beta_3,\alpha_3)\cdots\delta(\beta_N,\alpha_N)\right] + \\
&\quad + \cdots + \\
&\quad + (\pm)^{N-1}\delta(\beta_1,\alpha_N)\sum_{\mathcal{P}_\alpha}(\pm)^{p_\alpha}\mathcal{P}_\alpha\big[\delta(\beta_2,\alpha_1)\delta(\beta_3,\alpha_2)\cdots \\
&\qquad\qquad\qquad \cdots\delta(\beta_N,\alpha_{N-1})\big] \\
&=\sum_{\mathcal{P}_\alpha}(\pm)^{p_\alpha}\mathcal{P}_\alpha\left[\delta(\beta_1,\alpha_1)\delta(\beta_2,\alpha_2)\cdots\delta(\beta_N,\alpha_N)\right] \quad \text{q.e.d.}
\end{aligned}
$$

Problem A.6

1.

$$
\begin{aligned}
\left[\widehat{n}_\alpha,\, c^{\dagger}_\beta\right]_- &= c^{\dagger}_\alpha c_\alpha c^{\dagger}_\beta - c^{\dagger}_\beta c^{\dagger}_\alpha c_\alpha \\
&= \delta(\alpha-\beta)\,c^{\dagger}_\alpha \pm c^{\dagger}_\alpha c^{\dagger}_\beta c_\alpha - c^{\dagger}_\beta c^{\dagger}_\alpha c_\alpha \\
&= \delta(\alpha-\beta)\,c^{\dagger}_\alpha + c^{\dagger}_\beta c^{\dagger}_\alpha c_\alpha - c^{\dagger}_\beta c^{\dagger}_\alpha c_\alpha \\
&= \delta(\alpha-\beta)\,c^{\dagger}_\alpha
\end{aligned}
$$

2.

$$
\begin{aligned}
\left[\widehat{n}_\alpha,\, c_\beta\right]_- &= c^{\dagger}_\alpha c_\alpha c_\beta - c_\beta c^{\dagger}_\alpha c_\alpha \\
&= c^{\dagger}_\alpha c_\alpha c_\beta - \delta(\alpha-\beta)\,c_\alpha \mp c^{\dagger}_\alpha c_\beta c_\alpha \\
&= c^{\dagger}_\alpha c_\alpha c_\beta - \delta(\alpha-\beta)\,c_\alpha - c^{\dagger}_\alpha c^{\dagger}_\alpha c_\beta \\
&= -\delta(\alpha-\beta)\,c_\alpha
\end{aligned}
$$

These relations are equally valid for both Bosons and Fermions.

Problem A.7

$$
\hat{N} = \int d\alpha\; \hat{n}_\alpha .
$$

We first calculate the following commutators:

$$\left[\hat{N}, c_\beta^\dagger\right]_- = \int d\alpha \left[\hat{n}_\alpha, c_\beta^\dagger\right]_- \overset{\text{Problem A.6}}{=} \int d\alpha\, c_\alpha^\dagger \delta(\alpha - \beta) = c_\beta^\dagger,$$
$$\left[\hat{N}, c_\beta\right]_- = \int d\alpha \left[\hat{n}_\alpha, c_\beta\right]_- \overset{\text{Problem A.6}}{=} \int d\alpha \left[-\delta(\alpha - \beta) c_\alpha\right] = -c_\alpha.$$

We therefore have

$$\hat{N} c_\beta^\dagger = c_\beta^\dagger (\hat{N} + \mathbb{1});$$
$$\hat{N} c_\beta = c_\beta (\hat{N} - \mathbb{1}).$$

1.

$$\hat{N}\left(c_\beta^\dagger \left|\varphi_{\alpha_1} \cdots\right\rangle^{(\pm)}\right) = c_\beta^\dagger (\hat{N} + \mathbb{1}) \left|\varphi_{\alpha_1} \cdots\right\rangle^{(\pm)}$$
$$= (N + 1)\left(c_\beta^\dagger \left|\varphi_{\alpha_1} \cdots\right\rangle^{(\pm)}\right)$$

As proposed, it is an eigenstate. The eigenvalue is $N + 1$. The name *creator* for $c_\beta^\dagger$ is therefore appropriate.

2.

$$\hat{N}\left(c_\beta \left|\varphi_{\alpha_1} \cdots\right\rangle^{(\pm)}\right) = c_\beta (\hat{N} - \mathbb{1}) \left|\varphi_{\alpha_1} \cdots\right\rangle^{(\pm)}$$
$$= (N - 1)\left(c_\beta \left|\varphi_{\alpha_1} \cdots\right\rangle^{(\pm)}\right)$$

$c_\beta \left|\varphi_{\alpha_1} \cdots\right\rangle^{(\pm)}$ is also an eigenstate of the particle number operator $\hat{N}$ with the eigenvalue $N - 1$. The name *annihilator* for c_β is therefore appropriate.

Problem A.8
Plane waves:

$$\varphi_{\mathbf{k}}(\mathbf{r}) = \frac{1}{\sqrt{V}} e^{i\mathbf{k}\cdot\mathbf{r}} = \langle \mathbf{r} | \mathbf{k} \rangle$$

Kinetic energy:
One-particle basis: $|\mathbf{k}\sigma\rangle = |\mathbf{k}\rangle |\sigma\rangle$

$$\left\langle \mathbf{k}\sigma \left| \frac{\mathbf{p}^2}{2m} \right| \mathbf{k}'\sigma' \right\rangle = \frac{\hbar^2 k'^2}{2m} \langle \mathbf{k}\sigma \mid \mathbf{k}'\sigma' \rangle = \frac{\hbar^2 k'^2}{2m} \delta_{\mathbf{k}\mathbf{k}'} \delta_{\sigma\sigma'}$$

$$\curvearrowright \sum_{i=1}^{N} \frac{\mathbf{p}_i^2}{2m} = \sum_{\mathbf{k}\mathbf{k}'\sigma\sigma'} \left\langle \mathbf{k}\sigma \left| \frac{\mathbf{p}^2}{2m} \right| \mathbf{k}'\sigma' \right\rangle c^{\dagger}_{\mathbf{k}\sigma} c_{\mathbf{k}'\sigma'}$$

$$= \sum_{\mathbf{k}\sigma} \frac{\hbar^2 k^2}{2m} c^{\dagger}_{\mathbf{k}\sigma} c_{\mathbf{k}\sigma}$$

Interaction:

$$\left\langle \mathbf{k}_1\sigma_1,\, \mathbf{k}_2\sigma_2 \left| \frac{1}{\left|\widehat{r}^{(1)} - \widehat{r}^{(2)}\right|} \right| \mathbf{k}_3\sigma_3,\, \mathbf{k}_4\sigma_4 \right\rangle$$

$$= \delta_{\sigma_1\sigma_3} \delta_{\sigma_2\sigma_4} \left\langle \mathbf{k}_1\, \mathbf{k}_2 \left| \frac{1}{\left|\widehat{r}^{(1)} - \widehat{r}^{(2)}\right|} \right| \mathbf{k}_3\, \mathbf{k}_4 \right\rangle$$

The interaction is spin independent. Therefore the spin parts of the eigenstates can be evaluated directly and they yield the two Kronecker deltas. The two-particle states used are not symmetrized:

$$\left\langle \mathbf{k}_1\, \mathbf{k}_2 \left| \frac{1}{\left|\widehat{r}^{(1)} - \widehat{r}^{(2)}\right|} \right| \mathbf{k}_3\, \mathbf{k}_4 \right\rangle$$

$$= \int\!\!\int d^3r_1 d^3r_2 \left\langle \mathbf{k}_1\, \mathbf{k}_2 \left| \frac{1}{\left|\widehat{r}^{(1)} - \widehat{r}^{(2)}\right|} \right| \mathbf{r}_1\, \mathbf{r}_2 \right\rangle \langle \mathbf{r}_1\, \mathbf{r}_2 \mid \mathbf{k}_3\, \mathbf{k}_4 \rangle$$

$$= \int\!\!\int d^3r_1 d^3r_2 \frac{1}{|r_1 - r_2|} \left\langle \mathbf{k}_1^{(1)} \mid \mathbf{r}_1^{(1)} \right\rangle \left\langle \mathbf{k}_2^{(2)} \mid \mathbf{r}_2^{(2)} \right\rangle *$$

$$* \left\langle \mathbf{r}_1^{(1)} \mid \mathbf{k}_3^{(1)} \right\rangle \left\langle \mathbf{r}_2^{(2)} \mid \mathbf{k}_4^{(2)} \right\rangle$$

$$= \frac{1}{V^2} \int\!\!\int d^3r_1 d^3r_2 \frac{1}{|r_1 - r_2|} e^{i(\mathbf{k}_3 - \mathbf{k}_1)\cdot\mathbf{r}_1} e^{i(\mathbf{k}_4 - \mathbf{k}_2)\cdot\mathbf{r}_2}$$

$$= \delta_{\mathbf{k}_1+\mathbf{k}_2,\mathbf{k}_3+\mathbf{k}_4} \frac{1}{V} \int d^3r\, \frac{1}{r}\, e^{i(\mathbf{k}_3 - \mathbf{k}_1)\cdot\mathbf{r}}$$

The last step is obtained by introducing the centre of mass and relative coordinates. We get the Kronecker delta when the centre of mass part is integrated out. The remaining integral is calculated using a *convergence ensuring factor* α:

$$\begin{aligned}\lim_{\alpha\to 0}\int d^3r\,\frac{1}{r}e^{i\mathbf{q}\cdot\mathbf{r}}\,e^{\alpha r} &= \lim_{\alpha\to 0}\,2\pi\int_{-1}^{+1}dx\int_0^{\infty}dr\,r\,e^{iqrx}\,e^{\alpha r}\\ &= \lim_{\alpha\to 0}\frac{2\pi}{iq}\int_0^{\infty}dr\left(e^{iqr}-e^{-iqr}\right)e^{-\alpha r}\\ &= \lim_{\alpha\to 0}\frac{4\pi}{q^2+\alpha^2}\\ &= \frac{4\pi}{q^2}\end{aligned}$$

Then the interaction matrix element reads as

$$\left\langle \mathbf{k}_1\sigma_1,\,\mathbf{k}_2\sigma_2\left|\frac{1}{\left|\widehat{r}^{(1)}-\widehat{r}^{(2)}\right|}\right|\mathbf{k}_3\sigma_3,\,\mathbf{k}_4\sigma_4\right\rangle = \delta_{\sigma_1\sigma_3}\delta_{\sigma_2\sigma_4}\delta_{\mathbf{k}_1+\mathbf{k}_2,\mathbf{k}_3+\mathbf{k}_4}\frac{4\pi}{V\,\left|\mathbf{k}_3-\mathbf{k}_1\right|^2}$$

Interaction operator in second quantization:

$$\begin{aligned}&\frac{1}{2}\sum_{ij}^{i\neq j}\frac{1}{\left|\widehat{\mathbf{r}}_i-\widehat{\mathbf{r}}_j\right|}=\\ &=\frac{1}{2}\sum_{\mathbf{k}_1\sigma_1,\mathbf{k}_2\sigma_2,\mathbf{k}_3\sigma_3,\mathbf{k}_4\sigma_4}\left\langle \mathbf{k}_1\sigma_1,\,\mathbf{k}_2\sigma_2\left|\frac{1}{\left|\widehat{r}^{(1)}-\widehat{r}^{(2)}\right|}\right|\mathbf{k}_3\sigma_3,\,\mathbf{k}_4\sigma_4\right\rangle *\\ &\qquad * c^{\dagger}_{\mathbf{k}_1\sigma_1}c^{\dagger}_{\mathbf{k}_2\sigma_2}c_{\mathbf{k}_4\sigma_4}c_{\mathbf{k}_3\sigma_3}\\ &=\frac{1}{2}\sum_{\mathbf{k}_1\sigma_1,\mathbf{k}_2\sigma_2,\mathbf{k}_3}\frac{4\pi}{V\,\left|\mathbf{k}_3-\mathbf{k}_1\right|^2}c^{\dagger}_{\mathbf{k}_1\sigma_1}c^{\dagger}_{\mathbf{k}_2\sigma_2}c_{\mathbf{k}_1+\mathbf{k}_2-\mathbf{k}_3\sigma_2}c_{\mathbf{k}_3\sigma_1}\end{aligned}$$

We further set

$$\mathbf{k}_1\to\mathbf{k}+\mathbf{q}\ ;\ \mathbf{k}_2\to\mathbf{p}-\mathbf{q}\ ;\ \mathbf{k}_3\to\mathbf{k}\ ;\ \sigma_1\to\sigma\ ;\ \sigma_2\to\sigma'$$

and have the Hamiltonian of the N-electron system in second quantization:

$$H_N=\sum_{\mathbf{k}\sigma}\varepsilon_0(\mathbf{k})c^{\dagger}_{\mathbf{k}\sigma}c_{\mathbf{k}\sigma}+\frac{1}{2}\sum_{\mathbf{kpq}\sigma\sigma'}v_0(\mathbf{q})c^{\dagger}_{\mathbf{k}+\mathbf{q}\sigma}c^{\dagger}_{\mathbf{p}-\mathbf{q}\sigma'}c_{\mathbf{p}\sigma'}c_{\mathbf{k}\sigma}$$

$$\varepsilon_0(\mathbf{k})=\frac{\hbar^2k^2}{2m}\quad ;\quad v_0(\mathbf{q})=\frac{e^2}{\varepsilon_0 V q^2}$$

Problem A.9

$$\hat{n}_{\mathbf{k}\sigma} = c^{\dagger}_{\mathbf{k}\sigma} c_{\mathbf{k}\sigma} \implies [\hat{n}_{\mathbf{k}\sigma}, \hat{n}_{\mathbf{k}'\sigma'}]_{-} = 0$$

Therefore the kinetic energy commutes in any case with $\hat{N}$. Therefore we only have to calculate the commutator with the interaction:

$$\frac{1}{2}\sum_{\substack{\mathbf{kpq}\\ \sigma\sigma'\\ \mathbf{k}'\sigma''}} v_0(\mathbf{q}) \left[c^{\dagger}_{\mathbf{k}+\mathbf{q}\sigma} c^{\dagger}_{\mathbf{p}-\mathbf{q}\sigma'} c_{\mathbf{p}\sigma'} c_{\mathbf{k}\sigma}, c^{\dagger}_{\mathbf{k}'\sigma''} c_{\mathbf{k}'\sigma''} \right]_{-}$$

$$= \frac{1}{2}\sum_{\substack{\mathbf{k},\mathbf{p},\mathbf{q}\\ \mathbf{k}',\sigma,\sigma',\sigma''}} v_0(\mathbf{q}) \left\{ \delta_{\mathbf{k}\mathbf{k}'}\delta_{\sigma\sigma''} c^{\dagger}_{\mathbf{k}+\mathbf{q}\sigma} c^{\dagger}_{\mathbf{p}-\mathbf{q}\sigma'} c_{\mathbf{p}\sigma'} c_{\mathbf{k}'\sigma''} - \right.$$

$$- \delta_{\mathbf{p},\mathbf{k}'}\delta_{\sigma'\sigma''} c^{\dagger}_{\mathbf{k}+\mathbf{q}\sigma} c^{\dagger}_{\mathbf{p}-\mathbf{q}\sigma'} c_{\mathbf{k}\sigma} c_{\mathbf{k}'\sigma''} +$$

$$+ \delta_{\mathbf{p}-\mathbf{q}\mathbf{k}'}\delta_{\sigma'\sigma''} c^{\dagger}_{\mathbf{k}'\sigma''} c^{\dagger}_{\mathbf{k}+\mathbf{q}\sigma} c_{\mathbf{p}\sigma'} c_{\mathbf{k}\sigma} -$$

$$\left. \delta_{\mathbf{k}+\mathbf{q}\mathbf{k}'}\delta_{\sigma\sigma''} c^{\dagger}_{\mathbf{k}'\sigma''} c^{\dagger}_{\mathbf{p}-\mathbf{q}\sigma'} c_{\mathbf{p}\sigma'} c_{\mathbf{k}\sigma} \right\}$$

$$= \frac{1}{2}\sum_{\substack{\mathbf{k},\mathbf{p},\mathbf{q}\\ \sigma,\sigma'}} v_0(\mathbf{q}) \left\{ c^{\dagger}_{\mathbf{k}+\mathbf{q}\sigma} c^{\dagger}_{\mathbf{p}-\mathbf{q}\sigma'} c_{\mathbf{p}\sigma'} c_{\mathbf{k}\sigma} - c^{\dagger}_{\mathbf{k}+\mathbf{q}\sigma} c^{\dagger}_{\mathbf{p}-\mathbf{q}\sigma'} c_{\mathbf{k}\sigma} c_{\mathbf{p}\sigma'} + \right.$$

$$\left. + c^{\dagger}_{\mathbf{p}-\mathbf{q}\sigma'} c^{\dagger}_{\mathbf{k}+\mathbf{q}\sigma} c_{\mathbf{p}\sigma'} c_{\mathbf{k}\sigma} - c^{\dagger}_{\mathbf{k}+\mathbf{q}\sigma} c^{\dagger}_{\mathbf{p}-\mathbf{q}\sigma'} c_{\mathbf{p}\sigma'} c_{\mathbf{k}\sigma} \right\} = 0$$

$$\implies \left[H_N, \hat{N} \right]_{-} = 0$$

H_N and $\hat{N}$ have common eigenstates. The particle number is a conserved quantity.

Problem A.10

1. Hamiltonian of the two-particle system:

$$H = H_1 + H_2 = -\frac{\hbar^2}{2m}(\Delta_1 + \Delta_2) + V(x_1) + V(x_2)$$

Unsymmetrized eigenstate:

$$|\varphi_{\alpha_1}\varphi_{\alpha_2}\rangle = |\varphi^{(1)}_{\alpha_1}\rangle|\varphi^{(2)}_{\alpha_2}\rangle$$

Position space representation:

$$\langle x_1 x_2 | \varphi_{\alpha_1}\varphi_{\alpha_2}\rangle = \varphi_n(x_1)\varphi_m(x_2)\chi_S(m^{(1)}_S)\chi_S(m^{(2)}_{S'})$$

χ_S: spin function (identical particles have the same spin S)

$$\alpha_1 = (n, m_S); \qquad \alpha_2 = (m, m_{S'})$$

2. Solution of the one-particle problem:

$$\left(-\frac{\hbar^2}{2m}\Delta + V(x)\right)\varphi(x) = E\varphi(x)$$

We first have

$$\varphi(x) \equiv 0 \text{ for } x < 0 \text{ and } x > a$$

For $0 \le x \le a$ we have to solve

$$-\frac{\hbar^2}{2m}\Delta\varphi(x) = E\varphi(x)$$

Ansatz for solution:

$$\varphi(x) = c\sin(\gamma_1 x + \gamma_2)$$

Boundary conditions:

$$\begin{aligned}
\varphi(0) = 0 &\implies \gamma_2 = 0, \\
\varphi(a) = 0 &\implies \gamma_1 = n\frac{\pi}{a}; \qquad n = 1, 2, 3, \ldots
\end{aligned}$$

Energy eigenvalues:

$$E = \frac{\hbar^2}{2m}\gamma_1^2 \implies E_n = \frac{\hbar^2\pi^2}{2ma^2}n^2; \qquad n = 1, 2, \ldots$$

Eigen functions:

$$\begin{aligned}
\varphi_n(x) &= c\sin\left(n\frac{\pi}{a}x\right), \\
1 &\stackrel{!}{=} c^2\int_0^a \sin^2\left(n\frac{\pi}{a}x\right)\mathrm{d}x \implies c = \sqrt{\frac{2}{a}}, \\
\varphi_n(x) &= \begin{cases} \sqrt{\frac{2}{a}}\sin\left(n\frac{\pi}{a}x\right) & \text{for } 0 \le x \le a, \\ 0 & \text{otherwise} \end{cases}
\end{aligned}$$

3. Two-particle problem:

$$\begin{aligned}
|\varphi_{\alpha_1}\varphi_{\alpha_2}\rangle^{(\pm)} \longrightarrow\ & \frac{1}{\sqrt{2}}\{\varphi_n(x_1)\varphi_m(x_2)\chi_S(m_S^{(1)})\chi_S(m_{S'}^{(2)}) \pm \\
& \pm \varphi_n(x_2)\varphi_m(x_1)\chi_S(m_S^{(2)})\chi_S(m_{S'}^{(1)})\}
\end{aligned}$$

(+): Bosons,

(−): Fermions: $(n, m_S) \neq (m, m_{S'})$ because of Pauli's principle.

4. Ground state energy of the N-particle system:

Bosons:

All particles in the $n = 1$-state:

$$E_0 = N \frac{\hbar^2 \pi^2}{2ma^2}.$$

Fermions:

$$E_0 = 2 \sum_{n=1}^{\frac{N}{2}} \frac{\hbar^2 \pi^2}{2ma^2} n^2 \approx \frac{\hbar^2 \pi^2}{2ma^2} \frac{N^3}{24}$$

with

$$\sum_{n=1}^{\frac{N}{2}} n^2 \overset{N \gg 1}{\approx} \int_1^{\frac{N}{2}} n^2 \mathrm{d}n = \frac{1}{3}\left(\frac{N^3}{8} - 1\right) \approx \frac{N^3}{24}$$

Problem A.11

1. Non-interacting, identical Bosons or Fermions:

$$H = \sum_{i=1}^{N} H_1^{(i)}$$

Eigenvalue equation:

$$H_1^{(i)} |\varphi_r^{(i)}\rangle = \epsilon_r |\varphi_r^{(i)}\rangle, \quad \langle \varphi_r^{(i)} | \varphi_s^{(i)} \rangle = \delta_{rs}$$

One-particle operator in second quantization:

$$H = \sum_{r,s} \langle \varphi_r | H_1 | \varphi_s \rangle a_r^\dagger a_s = \sum_{r,s} \epsilon_s \delta_{rs} a_r^\dagger a_s$$

$$\Longrightarrow H = \sum_r \epsilon_r a_r^\dagger a_r = \sum_r \epsilon_r \hat{n}_r .$$

2. Unnormalized density matrix of the grand canonical ensemble:

$$\rho = \exp[-\beta(H - \mu \hat{N})],$$
$$\hat{N} = \sum_r \hat{n}_r$$

The normalized Fock states

$$|N; n_1 n_2 \dots n_i \dots\rangle^{(\epsilon)}$$

are the eigenstates of $\hat{n}_r$ and therefore also of $\hat{N}$ und H:

$$H|N; n_1 \dots\rangle^{(\epsilon)} = \left(\sum_r \epsilon_r n_r\right) |N; n_1 \dots\rangle^{(\epsilon)},$$
$$\hat{N}|N; n_1 \dots\rangle^{(\epsilon)} = N|N; n_1 \dots\rangle^{(\epsilon)}$$

That is why it is convenient to build the trace with these Fock states:

$$^{(\epsilon)}\left\langle N; n_1 n_2 \dots \left| \exp[-\beta(H - \mu\hat{N})] \right| N; n_1 n_2 \dots \right\rangle^{(\epsilon)}$$
$$= \exp\left[-\beta \sum_r (\epsilon_r - \mu) n_r\right] \quad \text{with } \sum_r n_r = N$$

From this follows:

$$\begin{aligned}
\mathrm{Tr}\rho &= \sum_{N=0}^{\infty} \sum_{\substack{\{n_r\} \\ (\sum n_r = N)}} \exp\left[-\beta \sum_r (\epsilon_r - \mu) n_r\right] \\
&= \sum_{N=0}^{\infty} \sum_{\substack{\{n_r\} \\ (\sum n_r = N)}} \prod_r e^{-\beta(\epsilon_r - \mu) n_r} \\
&= \sum_{n_1} \sum_{n_2} \cdots \sum_{n_r} \cdots \prod_r e^{-\beta(\epsilon_r - \mu) n_r} \\
&= \left(\sum_{n_1} e^{-\beta n_1(\epsilon_1 - \mu)}\right) \left(\sum_{n_2} e^{-\beta n_2(\epsilon_2 - \mu)}\right) \cdots
\end{aligned}$$

Grand canonical partition function:

$$\Xi(T, V, \mu) = \mathrm{Tr}\rho = \prod_r \left(\sum_{n_r} e^{-\beta n_r(\epsilon_r - \mu)}\right)$$

Bosons ($n_r = 0, 1, 2, \dots$):

$$\Xi_B(T, V, \mu) = \prod_r \frac{1}{1 - e^{-\beta(\epsilon_r - \mu)}}$$

Fermions ($n_r = 0, 1$):

$$\Xi_F(T, V, \mu) = \prod_r \left(1 + e^{-\beta(\epsilon_r - \mu)}\right)$$

3. Expectation value of the particle number:

$$\langle \hat{N} \rangle = \frac{1}{\Xi} \mathrm{Sp}(\rho \hat{N})$$

To build the trace Fock states are preferred because they are the eigenstates of $\hat{N}$:

$$\begin{aligned}
\langle \hat{N} \rangle &= \frac{1}{\Xi} \sum_{N=0}^{\infty} \sum_{\substack{\{n_r\} \\ (\sum n_r = N)}} \left\{ N \exp\left[-\beta \sum_r (\epsilon_r - \mu) n_r \right] \right\} \\
&= \frac{1}{\beta} \frac{\partial}{\partial \mu} \ln \Xi
\end{aligned}$$

With part 2

$$\begin{aligned}
\frac{\partial}{\partial \mu} \ln \Xi_B &= \frac{\partial}{\partial \mu} \left\{ -\sum_r \ln\left[1 - e^{-\beta(\epsilon_r - \mu)}\right] \right\} \\
&= -\sum_r \frac{-\beta e^{-\beta(\epsilon_r - \mu)}}{1 - e^{-\beta(\epsilon_r - \mu)}} \\
&= \beta \sum_r \frac{1}{e^{\beta(\epsilon_r - \mu)} - 1}, \\
\frac{\partial}{\partial \mu} \ln \Xi_F &= \frac{\partial}{\partial \mu} \left\{ \sum_r \ln\left[1 + e^{-\beta(\epsilon_r - \mu)}\right] \right\} \\
&= \beta \sum_r \frac{e^{-\beta(\epsilon_r - \mu)}}{1 + e^{-\beta(\epsilon_r - \mu)}} \\
&= \beta \sum_r \frac{1}{e^{\beta(\epsilon_r - \mu)} + 1}
\end{aligned}$$

This means

$$\langle \hat{N} \rangle = \begin{cases} \sum_r \frac{1}{e^{\beta(\epsilon_r - \mu)} - 1} & \text{Bosons} \\ \sum_r \frac{1}{e^{\beta(\epsilon_r - \mu)} + 1} & \text{Fermions} \end{cases}$$

4. Internal energy:

$$U = \langle H \rangle = \frac{1}{\Xi} \mathrm{Tr}(\rho H)$$

Fock states are the eigenstates of H and therefore appropriate for building the trace required here:

$$
\begin{aligned}
U &= \frac{1}{\Xi}\sum_{N=0}^{\infty}\sum_{\substack{\{n_r\}\\(\sum n_r=N)}}\left[\left(\sum_i \epsilon_i n_i\right)e^{-\beta\sum_r(\epsilon_r-\mu)n_r}\right] \\
&= -\frac{\partial}{\partial\beta}\ln\Xi + \mu\langle\hat{N}\rangle \\
-\frac{\partial}{\partial\beta}\ln\Xi_B &= \sum_r \frac{(\epsilon_r-\mu)e^{-\beta(\epsilon_r-\mu)}}{1-e^{-\beta(\epsilon_r-\mu)}} \\
&= -\mu\langle\hat{N}\rangle + \sum_r \frac{\epsilon_r}{e^{\beta(\epsilon_r-\mu)}-1}, \\
-\frac{\partial}{\partial\beta}\ln\Xi_F &= -\sum_r \frac{-(\epsilon_r-\mu)e^{-\beta(\epsilon_r-\mu)}}{1+e^{-\beta(\epsilon_r-\mu)}} \\
&= -\mu\langle\hat{N}\rangle + \sum_r \frac{\epsilon_r}{e^{\beta(\epsilon_r-\mu)}+1}
\end{aligned}
$$

We finally get

$$
U = \begin{cases} \sum_r \frac{\epsilon_r}{e^{\beta(\epsilon_r-\mu}-1} & \text{Bosons} \\ \sum_r \frac{\epsilon_r}{e^{\beta(\epsilon_r-\mu)}+1} & \text{Fermions.} \end{cases}
$$

5. Fock states are also eigenstates of the occupation number operator:

$$
\begin{aligned}
\langle\hat{n}_i\rangle &= \frac{1}{\Xi}\mathrm{Tr}(\rho\hat{n}_i) \\
&= \frac{1}{\Xi}\sum_{N=0}^{\infty}\sum_{\substack{\{n_r\}\\(\sum n_r=N)}}\left[n_i e^{-\beta\sum_r(\epsilon_r-\mu)n_r}\right] \\
&= -\frac{1}{\beta}\frac{\partial}{\partial\epsilon_i}\ln\Xi, \\
-\frac{1}{\beta}\frac{\partial}{\partial\epsilon_i}\ln\Xi_B &= +\frac{1}{\beta}\sum_r \frac{+\beta e^{-\beta(\epsilon_r-\mu)}}{1-e^{-\beta(\epsilon_r-\mu)}}\frac{\partial\epsilon_r}{\partial\epsilon_i} \\
&= \frac{1}{e^{\beta(\epsilon_i-\mu)}-1} \qquad \text{(Bose function)}, \\
-\frac{1}{\beta}\frac{\partial}{\partial\epsilon_i}\ln\Xi_F &= -\frac{1}{\beta}\sum_r \frac{-\beta e^{-\beta(\epsilon_r-\mu)}}{1+e^{-\beta(\epsilon_r-\mu)}}\frac{\partial\epsilon_r}{\partial\epsilon_i} \\
&= \frac{1}{e^{\beta(\epsilon_i-\mu)}+1} \qquad \text{(Fermi function)}.
\end{aligned}
$$

It follows:

$$\langle \hat{n}_i \rangle = \begin{cases} \{\exp[\beta(\epsilon_i - \mu) - 1\}^{-1} & Bosons, \\ \{\exp[\beta(\epsilon_i - \mu)] + 1\}^{-1} & Fermions. \end{cases}$$

One immediately recognizes by comparison with the earlier problems:

$$\langle \hat{N} \rangle = \sum_r \langle \hat{n}_r \rangle; \qquad U = \sum_r \epsilon_r \langle \hat{n}_r \rangle$$

Problem B.1

$$\begin{aligned}
\rho \int_0^\beta d\lambda \dot{A}(t - i\lambda\hbar) &= \rho \int_0^\beta d\lambda \frac{i}{\hbar} \frac{d}{d\lambda} A(t - i\lambda\hbar) = \\
&= \frac{i}{\hbar} \rho \left[A(t - i\hbar\beta) - A(t) \right] = \\
&= \frac{i}{\hbar} \rho \left[e^{\frac{i}{\hbar}(-i\hbar\beta)\mathcal{H}} A(t) e^{-\frac{i}{\hbar}(-i\hbar\beta)\mathcal{H}} - A(t) \right] = \\
&= \frac{i}{\hbar} \rho \left(e^{\beta\mathcal{H}} A(t) e^{-\beta\mathcal{H}} - A(t) \right) = \\
&= \frac{i}{\hbar} \left[\frac{e^{-\beta\mathcal{H}} e^{\beta\mathcal{H}} A(t) e^{-\beta\mathcal{H}}}{\mathrm{Sp}(e^{-\beta\mathcal{H}})} - \rho A(t) \right]_- = \\
&= \frac{i}{\hbar} (A(t)\rho - \rho A(t)) = \frac{i}{\hbar} \left[A(t), \rho \right]_- \quad \text{q.e.d.}
\end{aligned}$$

Problem B.2

$$\begin{aligned}
\left\langle \left[A(t), B(t') \right]_- \right\rangle &= \mathrm{Sp} \left\{ \rho \left[A(t), B(t') \right]_- \right\} \\
&= \mathrm{Sp} \left\{ \rho A(t) B(t') - \rho B(t') A(t) \right\} \\
&= \mathrm{Sp} \left\{ B(t') \rho A(t) - \rho B(t') A(t) \right\} \\
&= \mathrm{Sp} \left\{ \left[B(t'), \rho \right]_- A(t) \right\}
\end{aligned}$$

(cyclic invariance of trace).

Substitute Kubo identity:

$$\begin{aligned}
\langle\langle A(t); B(t') \rangle\rangle^{\text{ret}} &= -i\Theta(t - t') \left\langle \left[A(t), B(t') \right]_- \right\rangle = \\
&= -\hbar\Theta(t - t') \int_0^\beta d\lambda \mathrm{Sp} \left\{ \rho \dot{B}(t' - i\lambda\hbar) A(t) \right\} = \\
&= -\hbar\Theta(t - t') \int_0^\beta d\lambda \left\langle \dot{B}(t' - i\lambda\hbar) A(t) \right\rangle \quad \text{q.e.d.}
\end{aligned}$$

Problem B.3

$$\begin{aligned}\langle B(0)A(t+i\beta)\rangle &= \frac{1}{\Xi}\mathrm{Sp}\left\{e^{-\beta\mathcal{H}}Be^{\frac{i}{\hbar}\mathcal{H}(t+i\hbar\beta)}Ae^{-\frac{i}{\hbar}\mathcal{H}(t+i\hbar\beta)}\right\}\\ &= \frac{1}{\Xi}\mathrm{Sp}\left\{e^{\beta\mathcal{H}}e^{-\beta\mathcal{H}}Be^{\frac{i}{\hbar}\mathcal{H}t}e^{-\beta\mathcal{H}}Ae^{-\frac{i}{\hbar}\mathcal{H}t}\right\}\\ &= \frac{1}{\Xi}\mathrm{Sp}\left\{e^{-\beta\mathcal{H}}e^{\frac{i}{\hbar}\mathcal{H}t}Ae^{-\frac{i}{\hbar}\mathcal{H}t}B\right\}\\ &= \langle A(t)B(0)\rangle\end{aligned}$$

Here the cyclic invariance of trace has been used several times.

Problem B.4

1. $t-t'>0$:
   The integrand has a pole at $x=x_0=-i0^+$. Residue:

$$c_{-1}=\lim_{x\to x_0}(x-x_0)\frac{e^{-ix(t-t')}}{x+i0^+}=\lim_{x\to x_0}e^{-ix(t-t')}=1$$

   Since $t-t'>0$, the semicircle closes in the lower half-plane; then the exponential function sees to it that the contribution from the semicircle vanishes. The contour runs mathematically negatively. Therefore it follows that

$$\Theta(t-t')=\frac{i}{2\pi}(-2\pi i)1=1$$

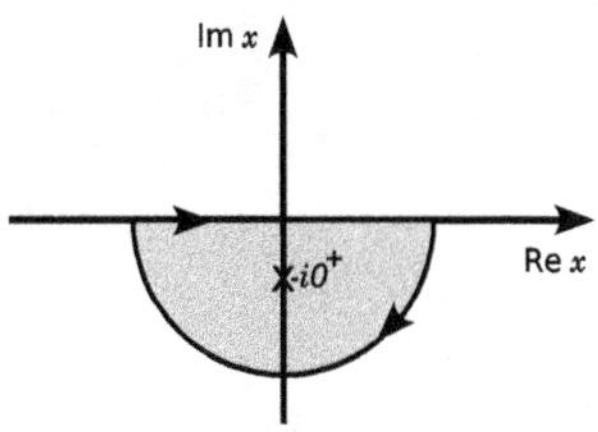

2. $t-t'<0$:
   In order that no contribution from the semicircle appears, now it closes in the upper half-plane. Then it follows that

$$\Theta(t-t')=0$$

   as there is no pole in the region of integration.

Problem B.5

$$f(\omega)=\int_{-\infty}^{+\infty}dt\ \bar{f}(t)e^{i\omega t}$$

Let the integral exist for real ω. Set

$$\omega = \omega_1 + i\omega_2$$

$$\Longrightarrow \; f(\omega) = \int_{-\infty}^{+\infty} dt\; \bar{f}(t) e^{i\omega_1 t} e^{-\omega_2 t}$$

1. $\bar{f}(t) = 0$ for $t < 0$:

$$\Longrightarrow \; f(\omega) = \int_{0}^{\infty} dt\; \bar{f}(t) e^{i\omega_1 t} e^{-\omega_2 t}$$

   Converges for all $\omega_2 > 0$, therefore it is possible to analytically continue in the upper half-plane.
2. $\bar{f}(t) = 0$ for $t > 0$:

$$\Longrightarrow \; f(\omega) = \int_{-\infty}^{0} dt\; \bar{f}(t) e^{i\omega_1 t} e^{-\omega_2 t}$$

   Converges for all $\omega_2 < 0$, therefore it is possible to analytically continue in the lower half-plane.

Problem B.6
With

$$\mathcal{H}_0 = \sum_{\mathbf{k}\sigma} (\varepsilon(\mathbf{k}) - \mu) a^{\dagger}_{\mathbf{k}\sigma} a_{\mathbf{k}\sigma}$$

we first calculate

$$\begin{aligned}
[a_{\mathbf{k}\sigma}, \mathcal{H}_0]_- &= \sum_{\mathbf{k}'\sigma'} (\varepsilon(\mathbf{k}') - \mu) \left[a_{\mathbf{k}\sigma}, a^{\dagger}_{\mathbf{k}'\sigma'} a_{\mathbf{k}'\sigma'} \right]_- = \\
&= \sum_{\mathbf{k}'\sigma'} (\varepsilon(\mathbf{k}') - \mu) \delta_{\mathbf{k}\mathbf{k}'} \delta_{\sigma\sigma'} a_{\mathbf{k}'\sigma'} = (\varepsilon(\mathbf{k}) - \mu) a_{\mathbf{k}\sigma} .
\end{aligned}$$

The interaction term requires more effort:

$$
\begin{aligned}
&[a_{\mathbf{k}\sigma}, \mathcal{H} - \mathcal{H}_0]_- \\
&= \frac{1}{2} \sum_{\substack{\mathbf{k}'\mathbf{p}\mathbf{q} \\ \sigma''\sigma'}} v_{\mathbf{k}'\mathbf{p}}(\mathbf{q}) \left[a_{\mathbf{k}\sigma}, a^{\dagger}_{\mathbf{k}'+\mathbf{q}\sigma''} a^{\dagger}_{\mathbf{p}-\mathbf{q}\sigma'} a_{\mathbf{p}\sigma'} a_{\mathbf{k}'\sigma''} \right]_- \\
&= \frac{1}{2} \sum_{\substack{\mathbf{k}'\mathbf{p}\mathbf{q} \\ \sigma''\sigma'}} v_{\mathbf{k}'\mathbf{p}}(\mathbf{q}) \Big(\delta_{\sigma\sigma''} \delta_{\mathbf{k},\mathbf{k}'+\mathbf{q}} a^{\dagger}_{\mathbf{p}-\mathbf{q}\sigma'} a_{\mathbf{p}\sigma'} a_{\mathbf{k}'\sigma''} \\
&\qquad\qquad - \delta_{\sigma\sigma'} \delta_{\mathbf{k}\mathbf{p}-\mathbf{q}} a^{\dagger}_{\mathbf{k}'+\mathbf{q}\sigma''} a_{\mathbf{p}\sigma'} a_{\mathbf{k}'\sigma''} \Big) \\
&= \frac{1}{2} \sum_{\mathbf{p}\mathbf{q}\sigma'} v_{\mathbf{k}-\mathbf{q}\mathbf{p}}(\mathbf{q}) a^{\dagger}_{\mathbf{p}-\mathbf{q}\sigma'} a_{\mathbf{p}\sigma'} a_{\mathbf{k}-\mathbf{q}\sigma} \\
&\quad - \frac{1}{2} \sum_{\mathbf{k}'\mathbf{q}\sigma''} v_{\mathbf{k}'\mathbf{k}+\mathbf{q}}(\mathbf{q}) a^{\dagger}_{\mathbf{k}'+\mathbf{q}\sigma''} a_{\mathbf{k}+\mathbf{q}\sigma} a_{\mathbf{k}'\sigma''}
\end{aligned}
$$

In the first summand:

$$
\mathbf{q} \rightarrow -\mathbf{q}; \quad v_{\mathbf{k}+\mathbf{q},\mathbf{p}}(-\mathbf{q}) = v_{\mathbf{p},\mathbf{k}+\mathbf{q}}(\mathbf{q})
$$

In the second summand:

$$
\mathbf{k}' \rightarrow \mathbf{p}; \quad \sigma'' \rightarrow \sigma'
$$

Then the two summands can be combined:

$$
[a_{\mathbf{k}\sigma}, \mathcal{H} - \mathcal{H}_0]_- = \sum_{\mathbf{p}\mathbf{q}\sigma'} v_{\mathbf{p},\mathbf{k}+\mathbf{q}}(\mathbf{q}) a^{\dagger}_{\mathbf{p}+\mathbf{q}\sigma'} a_{\mathbf{p}\sigma'} a_{\mathbf{k}+\mathbf{q}\sigma}
$$

Equation of motion:

$$
\begin{aligned}
&(E - \varepsilon(\mathbf{k}) + \mu) G^{\mathrm{ret}}_{\mathbf{k}\sigma}(E) \\
&= \hbar + \sum_{\mathbf{p}\mathbf{q}\sigma'} v_{\mathbf{p},\mathbf{k}+\mathbf{q}}(\mathbf{q}) \langle\langle a^{\dagger}_{\mathbf{p}+\mathbf{q}\sigma'} a_{\mathbf{p}\sigma'} a_{\mathbf{k}+\mathbf{q}\sigma} ; a^{\dagger}_{\mathbf{k}\sigma} \rangle\rangle^{\mathrm{ret}}_E
\end{aligned}
$$

Problem B.7

$$
\mathcal{H} = \sum_{\mathbf{k}\sigma} \epsilon(\mathbf{k}) a^{\dagger}_{\mathbf{k}\sigma} a_{\mathbf{k}\sigma} - \mu \hat{N} = \sum_{\mathbf{k}\sigma} (\epsilon(\mathbf{k}) - \mu) a^{\dagger}_{\mathbf{k}\sigma} a_{\mathbf{k}\sigma}
$$

One can easily calculate

$$
\begin{aligned}
[a_{\mathbf{k}\sigma}, \mathcal{H}]_- &= \sum_{\mathbf{k}'\sigma'} (\epsilon(\mathbf{k}') - \mu)[a_{\mathbf{k}\sigma}, a^{\dagger}_{\mathbf{k}'\sigma'} a_{\mathbf{k}'\sigma'}]_- \\
&= \sum_{\mathbf{k}'\sigma'} (\epsilon(\mathbf{k}') - \mu)\delta_{\mathbf{k}\mathbf{k}'}\delta_{\sigma\sigma'} a_{\mathbf{k}'\sigma'} \\
&= (\epsilon(\mathbf{k}) - \mu) a_{\mathbf{k}\sigma}
\end{aligned}
$$

From this it further follows that

$$
[[a_{\mathbf{k}\sigma}, \mathcal{H}]_-, \mathcal{H}]_- = (\epsilon(\mathbf{k}) - \mu)[a_{\mathbf{k}\sigma}, \mathcal{H}]_- = (\epsilon(\mathbf{k}) - \mu)^2 a_{\mathbf{k}\sigma}
$$

For the spectral moments this means

$$
\begin{aligned}
M^{(0)}_{\mathbf{k}\sigma} &= \langle [a_{\mathbf{k}\sigma}, a^{\dagger}_{\mathbf{k}\sigma}]_+ \rangle \\
&= 1 \\
M^{(1)}_{\mathbf{k}\sigma} &= \left\langle [[a_{\mathbf{k}\sigma}, \mathcal{H}]_-, a^{\dagger}_{\mathbf{k}\sigma}]_+ \right\rangle \\
&= (\epsilon(\mathbf{k}) - \mu) \langle [a_{\mathbf{k}\sigma}, a^{\dagger}_{\mathbf{k}\sigma}]_+ \rangle \\
&= (\epsilon(\mathbf{k}) - \mu) \\
M^{(2)}_{\mathbf{k}\sigma} &= \left\langle \Big[[[a_{\mathbf{k}\sigma}, \mathcal{H}]_-, \mathcal{H}]_-, a^{\dagger}_{\mathbf{k}\sigma} \Big]_+ \right\rangle \\
&= (\epsilon(\mathbf{k}) - \mu)^2 \langle [a_{\mathbf{k}\sigma}, a^{\dagger}_{\mathbf{k}\sigma}]_+ \rangle \\
&= (\epsilon(\mathbf{k}) - \mu)^2 \\
&\vdots
\end{aligned}
$$

Then by complete induction one gets immediately

$$
M^{(n)}_{\mathbf{k}\sigma} = (\epsilon(\mathbf{k}) - \mu)^n; \quad n = 0, 1, 2, \ldots
$$

The relationship (B.99) with the spectral density,

$$
M^{(n)}_{\mathbf{k}\sigma} = \frac{1}{\hbar} \int_{-\infty}^{+\infty} dE \; E^n \, S_{\mathbf{k}\sigma}(E)
$$

then leads to the solution:

$$
S_{\mathbf{k}\sigma}(E) = \hbar\delta(E - \epsilon(\mathbf{k}) + \mu)
$$

Problem B.8

1. Creation and annihilation operators for Cooper pairs:

$$
b^{\dagger}_{\mathbf{k}} = a^{\dagger}_{\mathbf{k}\uparrow} a^{\dagger}_{-\mathbf{k}\downarrow} \quad ; \quad b_{\mathbf{k}} = a_{-\mathbf{k}\downarrow} a_{\mathbf{k}\uparrow}
$$

Fundamental commutation relations
(a)

$$[b_{\mathbf{k}}, b_{\mathbf{k}'}]_- = \left[b_{\mathbf{k}}^\dagger, b_{\mathbf{k}'}^\dagger\right]_- = 0$$

because the creation and annihilation operators of fermions anticommute among themselves. Therefore products of even number of Fermion construction operators then commute.
(b)

$$\begin{aligned}\left[b_{\mathbf{k}}, b_{\mathbf{k}'}^\dagger\right]_- &= \left[a_{-\mathbf{k}\downarrow}a_{\mathbf{k}\uparrow}, a_{\mathbf{k}'\uparrow}^\dagger a_{-\mathbf{k}'\downarrow}^\dagger\right]_- \\ &= \delta_{\mathbf{k}\mathbf{k}'}a_{-\mathbf{k}\downarrow}a_{-\mathbf{k}'\downarrow}^\dagger - \delta_{-\mathbf{k}-\mathbf{k}'}a_{\mathbf{k}'\uparrow}^\dagger a_{\mathbf{k}\uparrow} \\ &= \delta_{\mathbf{k}\mathbf{k}'}(1 - \hat{n}_{-\mathbf{k}\downarrow} - \hat{n}_{\mathbf{k}\uparrow})\end{aligned}$$

Therefore the Cooper pairs inspite of their total spin being zero are not *real* Bosons because only two of the three basic commutation relations are satisfied.
(c) Since

$$[b_{\mathbf{k}}, b_{\mathbf{k}'}]_+ = 2b_{\mathbf{k}}b_{\mathbf{k}'} \neq 0 \quad \text{for} \quad \mathbf{k} \neq \mathbf{k}'$$

they are naturally also not real Fermions, either, even though

$$\left(b_{\mathbf{k}}^\dagger\right)^2 = b_{\mathbf{k}}^2 = 0$$

is valid for them.

2. Equation of motion:

$$\begin{aligned}\left[a_{\mathbf{k}\sigma}, H^*\right]_- &= \sum_{\mathbf{p}\sigma'} t(\mathbf{p})\left[a_{\mathbf{k}\sigma}, a_{\mathbf{p}\sigma'}^\dagger a_{\mathbf{p}\sigma'}\right]_- \\ &\quad - \Delta\sum_{\mathbf{p}}\left[a_{\mathbf{k}\sigma}, a_{-\mathbf{p}\downarrow}a_{\mathbf{p}\uparrow} + a_{\mathbf{p}\uparrow}^\dagger a_{-\mathbf{p}\downarrow}^\dagger\right]_- \\ &= \sum_{\mathbf{p}\sigma'} t(\mathbf{p})\delta_{\sigma\sigma'}\delta_{\mathbf{k}\mathbf{p}}a_{\mathbf{p}\sigma'} \\ &\quad - \Delta\sum_{\mathbf{p}}\left(\delta_{\mathbf{k}\mathbf{p}}\delta_{\sigma\uparrow}a_{-\mathbf{p}\downarrow}^\dagger - \delta_{\mathbf{k}-\mathbf{p}}\delta_{\sigma\downarrow}a_{\mathbf{p}\uparrow}^\dagger\right) \\ &= t(\mathbf{k})a_{\mathbf{k}\sigma} - \Delta(\delta_{\sigma\uparrow} - \delta_{\sigma\downarrow})a_{-\mathbf{k}-\sigma}^\dagger, \\ z_\sigma &= \begin{cases}+1, & \text{for } \sigma = \uparrow, \\ -1, & \text{for } \sigma = \downarrow\end{cases}\end{aligned}$$

Then the equation of motion reads as

$$(E - t(\mathbf{k}))G_{\mathbf{k}\sigma}(E) = \hbar - \Delta z_\sigma \langle\langle a^\dagger_{-\mathbf{k}-\sigma}; a^\dagger_{\mathbf{k}\sigma}\rangle\rangle.$$

The Green's function on the right-hand side prevents a direct solution. Therefore we set up the corresponding equation of motion for this:

$$\begin{aligned}
&\left[a^\dagger_{-\mathbf{k}-\sigma}, H^*\right]_- \\
&= -t(-\mathbf{k})a^\dagger_{-\mathbf{k}-\sigma} - \Delta \sum_{\mathbf{p}} \left[a^\dagger_{-\mathbf{k}-\sigma}, a_{-\mathbf{p}\downarrow}a_{\mathbf{p}\uparrow}\right]_- \\
&= -t(\mathbf{k})a^\dagger_{-\mathbf{k}-\sigma} - \Delta \sum_{\mathbf{p}} \left(\delta_{\mathbf{k}\mathbf{p}}\delta_{-\sigma\downarrow}a^\dagger_{\mathbf{p}\uparrow} - \delta_{-\mathbf{k}\mathbf{p}}\delta_{-\sigma\uparrow}a_{-\mathbf{p}\downarrow}\right) \\
&= -t(\mathbf{k})a^\dagger_{-\mathbf{k}-\sigma} - \Delta z_\sigma a_{\mathbf{k}\sigma}
\end{aligned}$$

This gives us the following equation of motion:

$$\begin{aligned}
(E + t(\mathbf{k}))\langle\langle a^\dagger_{-\mathbf{k}-\sigma}; a^\dagger_{\mathbf{k}\sigma}\rangle\rangle &= -\Delta z_\sigma G_{\mathbf{k}\sigma}(E) \\
\langle\langle a^\dagger_{-\mathbf{k}-\sigma}; a^\dagger_{\mathbf{k}\sigma}\rangle\rangle &= -\frac{z_\sigma \Delta}{E + t(\mathbf{k})} G_{\mathbf{k}\sigma}(E)
\end{aligned}$$

This is substituted in the equation of motion for $G^{\text{ret}}_{\mathbf{k}\sigma}(E)$:

$$\left(E - t(\mathbf{k}) - \frac{\Delta^2}{E + t(\mathbf{k})}\right) G_{\mathbf{k}\sigma}(E) = \hbar$$

Excitation energies:

$$E(\mathbf{k}) = +\sqrt{t^2(\mathbf{k}) + \Delta^2} \underset{t\to 0}{\longrightarrow} \Delta \quad \textit{(Energy gap)}.$$

Green's function:

$$G_{\mathbf{k}\sigma}(E) = \frac{\hbar(E + t(\mathbf{k}))}{E^2 - E^2(\mathbf{k})}$$

Imposing the boundary conditions:

$$G^{\text{ret}}_{\mathbf{k}\sigma}(E) = \frac{\hbar}{2E(\mathbf{k})}\left[\frac{t(\mathbf{k}) + E(\mathbf{k})}{E - E(\mathbf{k}) + i0^+} - \frac{t(\mathbf{k}) - E(\mathbf{k})}{E + E(\mathbf{k}) + i0^+}\right]$$

3. For Δ we need the expectation value:

$$\langle a^\dagger_{\mathbf{k}\uparrow} a^\dagger_{-\mathbf{k}\downarrow}\rangle$$

Its determination is via spectral theorem and the Green's function used in part 2.:

$$\langle\langle a^{\dagger}_{-\mathbf{k}\downarrow}; a^{\dagger}_{\mathbf{k}\uparrow}\rangle\rangle = \frac{-\Delta}{E + t(\mathbf{k})} G_{\mathbf{k}\uparrow}(E) = \frac{-\hbar\Delta}{E^2 - E^2(\mathbf{k})}$$

Taking into account the boundary conditions we obtain for the corresponding retarded function:

$$\langle\langle a^{\dagger}_{-\mathbf{k}\downarrow}; a^{\dagger}_{\mathbf{k}\uparrow}\rangle\rangle^{\text{ret}}_{E} = \frac{\hbar\Delta}{2E(\mathbf{k})}\left[\frac{1}{E + E(\mathbf{k}) + i0^+} - \frac{1}{E - E(\mathbf{k}) + i0^+}\right]$$

The spectral density corresponding to this

$$S_{-\mathbf{k}\downarrow;\mathbf{k}\uparrow}(E) = \frac{\hbar\Delta}{2E(\mathbf{k})}\left[\delta(E + E(\mathbf{k})) - \delta(E - E(\mathbf{k}))\right]$$

Spectral theorem:

$$\begin{aligned}
&\langle a^{\dagger}_{\mathbf{k}\uparrow} a^{\dagger}_{-\mathbf{k}\downarrow}\rangle \\
&= \frac{1}{\hbar}\int\limits_{-\infty}^{+\infty} dE \frac{S_{-\mathbf{k}\downarrow;\mathbf{k}\uparrow}(E)}{\exp(\beta E) + 1} \\
&= \frac{\Delta}{2E(\mathbf{k})}\left(\frac{1}{\exp(-\beta E(\mathbf{k})) + 1} - \frac{1}{\exp(\beta E(\mathbf{k})) + 1}\right) \\
&= \frac{\Delta}{2E(\mathbf{k})} \tanh\left(\frac{1}{2}\beta E(\mathbf{k})\right)
\end{aligned}$$

Then we finally get

$$\Delta = \frac{1}{2}\Delta V \sum_{\mathbf{k}} \frac{\tanh\left(\frac{1}{2}\beta\sqrt{t^2(\mathbf{k}) + \Delta^2}\right)}{\sqrt{t^2(\mathbf{k}) + \Delta^2}}$$

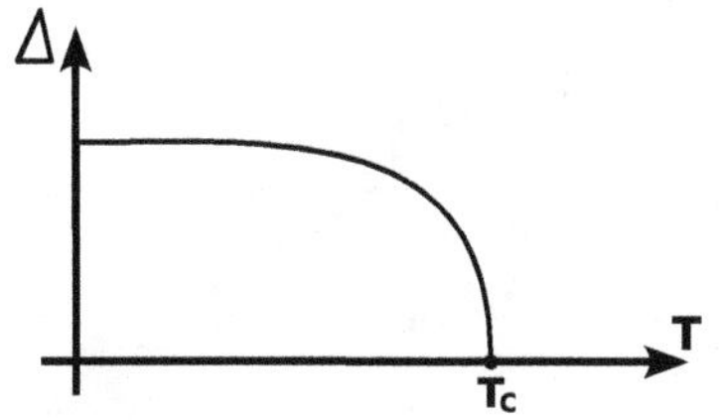

$\Delta = \Delta(T) \;\Rightarrow$ Energy gap is T dependent. Special case:

$$T \to 0 \;\Rightarrow\; \tanh\left(\frac{1}{2}\beta\sqrt{t^2(\mathbf{k}) + \Delta^2}\right) \to 1$$

Problem B.9

1. We first show that

$$\left[\cdots\left[[a_{\mathbf{k}\sigma}, H^*]_-, H^*\right]_- \cdots, H^*\right]_-$$
$$= \begin{cases} (t^2(\mathbf{k}) + \Delta^2)^n \,, & \text{if } p = 2n \\ (t^2(\mathbf{k}) + \Delta^2)^n (t(\mathbf{k})\, a_{\mathbf{k}\sigma} - z_\sigma \Delta a^\dagger_{-\mathbf{k}-\sigma})\,, & \text{if } p = 2n+1 \end{cases}$$

is valid.
Here $n = 0, 1, 2, \cdots$. We prove this by complete induction.
Induction's start $p = 1, 2$:

$$\left[a_{\mathbf{k}\sigma}, H^*\right]_- = t(\mathbf{k}) a_{\mathbf{k}\sigma} - z_\sigma \Delta a^\dagger_{-\mathbf{k}-\sigma} \qquad \text{(see Problem B.8)}$$

$$\begin{aligned}\left[\left[a_{\mathbf{k}\sigma}, H^*\right]_-, H^*\right]_- &= t(\mathbf{k})\left(t(\mathbf{k}) a_{\mathbf{k}\sigma} - z_\sigma \Delta a^\dagger_{-\mathbf{k}-\sigma}\right) \\ &\quad - z_\sigma \Delta \left(-t(\mathbf{k}) a^\dagger_{-\mathbf{k}-\sigma} - z_\sigma \Delta a_{\mathbf{k}\sigma}\right) \\ &= \left(t^2(\mathbf{k}) + \Delta^2\right) a_{\mathbf{k}\sigma}\,.\end{aligned}$$

Induction's end $p \longrightarrow p+1$:

(a) p even:

$$\begin{aligned}&\underbrace{\left[\cdots\left[\left[a_{\mathbf{k}\sigma}, H^*\right]_-, H^*\right]_-, \ldots, H^*\right]_-}_{(p+1)\text{-fold commutator}} \\ &= (t^2 + \Delta^2)^{\frac{p}{2}} \left[a_{\mathbf{k}\sigma}, H^*\right]_- \\ &= (t^2 + \Delta^2)^{\frac{p}{2}} \left(t a_{\mathbf{k}\sigma} - z_\sigma \Delta a^\dagger_{-\mathbf{k}-\sigma}\right)\end{aligned}$$

(b) p odd:

$$\begin{aligned}&\underbrace{\left[\cdots\left[\left[a_{\mathbf{k}\sigma}, H^*\right]_-, H^*\right]_-, \ldots, H^*\right]_-}_{(p+1)\text{-fold commutator}} \\ &= (t^2 + \Delta^2)^{\frac{1}{2}(p-1)} \left[t a_{\mathbf{k}\sigma} - z_\sigma \Delta a^\dagger_{-\mathbf{k}-\sigma}, H^*\right]_- \\ &= (t^2 + \Delta^2)^{\frac{1}{2}(p-1)} \Big[t(t a_{\mathbf{k}\sigma} - z_\sigma \Delta a^\dagger_{-\mathbf{k}-\sigma}) \\ &\qquad\qquad - z_\sigma \Delta(-t a^\dagger_{-\mathbf{k}-\sigma} - \Delta z_\sigma a_{\mathbf{k}\sigma})\Big] \\ &= (t^2 + \Delta^2)^{\frac{1}{2}(p+1)} a_{\mathbf{k}\sigma} \quad \text{q.e.d.}\end{aligned}$$

For the spectral moments of the one-electron spectral density we directly get $n = 0, 1, 2, \ldots$

$$M_{\mathbf{k}\sigma}^{(2n)} = \left(t^2(\mathbf{k}) + \Delta^2\right)^n ,$$
$$M_{\mathbf{k}\sigma}^{(2n+1)} = \left(t^2(\mathbf{k}) + \Delta^2\right)^n t(\mathbf{k}).$$

2. We use

$$M_{\mathbf{k}\sigma}^{(n)} = \frac{1}{\hbar}\int\limits_{-\infty}^{+\infty} dE\, E^n S_{\mathbf{k}\sigma}(E)$$

Determining equations from the first four spectral moments:

$$\begin{aligned}
\alpha_{1\sigma} + \alpha_{2\sigma} &= \hbar, \\
\alpha_{1\sigma} E_{1\sigma} + \alpha_{2\sigma} E_{2\sigma} &= \hbar t, \\
\alpha_{1\sigma} E_{1\sigma}^2 + \alpha_{2\sigma} E_{2\sigma}^2 &= \hbar(t^2 + \Delta^2), \\
\alpha_{1\sigma} E_{1\sigma}^3 + \alpha_{2\sigma} E_{2\sigma}^3 &= \hbar(t^2 + \Delta^2)t
\end{aligned}$$

Reformulating them:

$$\begin{aligned}
\alpha_{2\sigma}(E_{2\sigma} - E_{1\sigma}) &= \hbar(t - E_{1\sigma}), \\
\alpha_{2\sigma} E_{2\sigma}(E_{2\sigma} - E_{1\sigma}) &= \hbar\left[t^2 + \Delta^2 - t E_{1\sigma}\right], \\
\alpha_{2\sigma} E_{2\sigma}^2(E_{2\sigma} - E_{1\sigma}) &= \hbar\left[(t^2 + \Delta^2)(t - E_{1\sigma})\right]
\end{aligned}$$

After division follows:

$$E_{2\sigma}^2 = t^2 + \Delta^2 \quad \Longrightarrow \quad E_{2\sigma}(\mathbf{k}) = +\sqrt{t^2(\mathbf{k}) + \Delta^2} \equiv E(\mathbf{k})$$

This has the further consequence:

$$\begin{aligned}
E(\mathbf{k}) &= \frac{t^2 + \Delta^2 - t E_{1\sigma}}{t - E_{1\sigma}} = t + \frac{\Delta^2}{t - E_{1\sigma}} \\
\Longrightarrow \quad & (E(\mathbf{k}) - t(\mathbf{k}))^{-1}\Delta^2 = t(\mathbf{k}) - E_{1\sigma}(\mathbf{k}) \\
\Longrightarrow \quad & E_{1\sigma}(\mathbf{k}) = t(\mathbf{k}) - \frac{\Delta^2}{E(\mathbf{k}) - t(\mathbf{k})} = \frac{E(\mathbf{k})t(\mathbf{k}) - E^2(\mathbf{k})}{E(\mathbf{k}) - t(\mathbf{k})} \\
\Longrightarrow \quad & E_{1\sigma}(\mathbf{k}) = -E(\mathbf{k}) = -E_{2\sigma}(\mathbf{k})
\end{aligned}$$

Spectral weights:

$$\alpha_{2\sigma}(\mathbf{k})2E(\mathbf{k}) = \hbar(t(\mathbf{k}) + E(\mathbf{k}))$$
$$\Longrightarrow \quad \alpha_{2\sigma}(\mathbf{k}) = \hbar\frac{t(\mathbf{k}) + E(\mathbf{k})}{2E(\mathbf{k})},$$
$$\alpha_{1\sigma}(\mathbf{k}) = \hbar - \alpha_{2\sigma}(\mathbf{k}) = \hbar\frac{E(\mathbf{k}) - t(\mathbf{k})}{2E(\mathbf{k})},$$
$$\Longrightarrow \quad S_{\mathbf{k}\sigma}(E) = \hbar\left[\frac{E(\mathbf{k}) - t(\mathbf{k})}{2E(\mathbf{k})}\delta(E + E(\mathbf{k})) + \frac{E(\mathbf{k}) + t(\mathbf{k})}{2E(\mathbf{k})}\delta(E - E(\mathbf{k}))\right]$$

Problem B.10
Free energy:

$$F(T, V) = U(T, V) - TS(T, V) = U(T, V) + T\left(\frac{\partial F}{\partial T}\right)_V$$

So that for the internal energy we get

$$U(T, V) = -T^2\left\{\frac{\partial}{\partial T}\left(\frac{1}{T}F(T, V)\right)\right\}_V$$

$$F(0, V) \equiv U(0, V)$$

$$\curvearrowright \quad U(T, V) - U(T, 0) = -T^2\left\{\frac{\partial}{\partial T}(F(T, V) - F(0, V))\right\}$$

$$-\int_0^T dT'\,\frac{U(T', V) - U(0, V)}{T'^2} =$$
$$= \frac{1}{T}(F(T, V) - F(0, V)) - \lim_{T\to 0}\left\{\frac{1}{T}(F(T, V) - F(0, V))\right\}$$

Third law:

$$\lim_{T\to 0}\left\{\frac{1}{T}(F(T, V) - F(0, V))\right\} = \left(\frac{\partial F}{\partial T}\right)_V (T = 0)$$
$$= -S(T = 0, V) \stackrel{!}{=} 0$$

Then it follows that

$$F(T,V) = F(0,V) - T \int_0^T dT' \frac{U(T',V) - U(0,V)}{T'^2}$$

Problem B.11

$$\mathcal{H}_0 = \sum_{\mathbf{k}\sigma} (\epsilon(\mathbf{k}) - \mu) a^\dagger_{\mathbf{k}\sigma} a_{\mathbf{k}\sigma}$$

Then one can easily calculate

$$\begin{aligned}
[a_{\mathbf{k}\sigma}, \mathcal{H}_0]_- &= (\epsilon(\mathbf{k}) - \mu) a_{\mathbf{k}\sigma}, \\
[a^\dagger_{\mathbf{k}\sigma}, \mathcal{H}_0]_- &= -(\epsilon(\mathbf{k}) - \mu) a^\dagger_{\mathbf{k}\sigma}, \\
[a^\dagger_{\mathbf{k}\sigma} a_{\mathbf{k}'\sigma'}, \mathcal{H}_0]_- &= [a^\dagger_{\mathbf{k}\sigma}, \mathcal{H}_0]_- a_{\mathbf{k}'\sigma'} + a^\dagger_{\mathbf{k}\sigma} [a_{\mathbf{k}'\sigma'}, \mathcal{H}_0]_- \\
&= -(\epsilon(\mathbf{k}) - \mu) a^\dagger_{\mathbf{k}\sigma} a_{\mathbf{k}'\sigma'} + (\epsilon(\mathbf{k}') - \mu) a^\dagger_{\mathbf{k}\sigma} a_{\mathbf{k}'\sigma'} \\
&= (\epsilon(\mathbf{k}') - \epsilon(\mathbf{k})) a^\dagger_{\mathbf{k}\sigma} a_{\mathbf{k}'\sigma'}
\end{aligned}$$

$|\psi_0\rangle$ is eigenstate of $\mathcal{H}_0$, because

$$\begin{aligned}
\mathcal{H}_0 |\psi_0\rangle &= a^\dagger_{\mathbf{k}\sigma} a_{\mathbf{k}'\sigma'} \mathcal{H}_0 |E_0\rangle - [a^\dagger_{\mathbf{k}\sigma} a_{\mathbf{k}'\sigma'}, \mathcal{H}_0]_- |E_0\rangle \\
&= (E_0 - \epsilon(\mathbf{k}') + \epsilon(\mathbf{k})) |\psi_0\rangle
\end{aligned}$$

Time dependence:

$$\begin{aligned}
|\psi_0(t)\rangle &= a^\dagger_{\mathbf{k}\sigma}(t) a_{\mathbf{k}'\sigma'}(t) |E_0\rangle \\
&= e^{\frac{i}{\hbar}\mathcal{H}_0 t} a^\dagger_{\mathbf{k}\sigma} a_{\mathbf{k}'\sigma'} e^{-\frac{i}{\hbar}\mathcal{H}_0 t} |E_0\rangle \\
&= e^{-\frac{i}{\hbar}E_0 t} e^{\frac{i}{\hbar}\mathcal{H}_0 t} |\psi_0\rangle \\
&= e^{-\frac{i}{\hbar}E_0 t} e^{\frac{i}{\hbar}(E_0 + \epsilon(\mathbf{k}') - \epsilon(\mathbf{k}))t} |\psi_0\rangle \\
\Longrightarrow \quad |\psi_0(t)\rangle &= e^{-\frac{i}{\hbar}(\epsilon(\mathbf{k}') - \epsilon(\mathbf{k}))t} |\psi_0\rangle
\end{aligned}$$

Further with $\langle E_0 | E_0 \rangle = 1$ follows:

$$\begin{aligned}
\langle \psi_0 | \psi_0 \rangle &= \langle E_0 | a^\dagger_{\mathbf{k}'\sigma'} a_{\mathbf{k}\sigma} a^\dagger_{\mathbf{k}\sigma} a_{\mathbf{k}'\sigma'} | E_0 \rangle \\
&= \langle E_0 | a^\dagger_{\mathbf{k}'\sigma'} (1 - n_{\mathbf{k}\sigma}) a_{\mathbf{k}'\sigma'} | E_0 \rangle \\
&= \langle E_0 | a^\dagger_{\mathbf{k}'\sigma'} a_{\mathbf{k}'\sigma'} | E_0 \rangle \qquad (k > k_F) \\
&= \langle E_0 | (1 - a_{\mathbf{k}'\sigma'} a^\dagger_{\mathbf{k}'\sigma'}) | E_0 \rangle \\
&= \langle E_0 | E_0 \rangle \qquad (k' < k_F) \\
&= 1
\end{aligned}$$

Therewith we finally have

$$\langle\psi_0(t)|\psi_0(t')\rangle = \exp\left(-\frac{i}{\hbar}(\epsilon(\mathbf{k}') - \epsilon(\mathbf{k}))(t-t')\right)$$
$$\Longrightarrow |\langle\psi_0(t)|\psi_0(t')\rangle|^2 = 1: \quad \text{stationary state}$$

Problem B.12

$$G^{\text{ret}}_{\mathbf{k}\sigma} = \hbar(E - \epsilon(\mathbf{k}) + \mu - \Sigma_\sigma(\mathbf{k}, E))^{-1}$$

general representation

1. It must hold

$$E - \epsilon(\mathbf{k}) + \mu - \Sigma_\sigma(\mathbf{k}, E) \overset{!}{=} E - 2\epsilon(\mathbf{k}) + \frac{E^2}{\epsilon(\mathbf{k})} + i\gamma|E|$$

$$\begin{aligned}\Longrightarrow \Sigma_\sigma(\mathbf{k}, E) &= R_\sigma(\mathbf{k}, E) + i I_\sigma(\mathbf{k}, E)\\ &= \left(\epsilon(\mathbf{k}) + \mu - \frac{E^2}{\epsilon(\mathbf{k})}\right) - i\gamma|E|\\ \Longrightarrow R_\sigma(\mathbf{k}, E) &= \epsilon(\mathbf{k}) + \mu - \frac{E^2}{\epsilon(\mathbf{k})}, \quad I_\sigma(\mathbf{k}, E) = -\gamma|E|.\end{aligned}$$

2.

$$E_{i\sigma} \overset{!}{=} \epsilon(\mathbf{k}) - \mu + R_\sigma(\mathbf{k}, E_{i\sigma}(\mathbf{k})) = 2\epsilon(\mathbf{k}) - \frac{E^2_{i\sigma}(\mathbf{k})}{\epsilon(\mathbf{k})}$$
$$\Longrightarrow E^2_{i\sigma}(\mathbf{k}) + \epsilon(\mathbf{k})E_{i\sigma}(\mathbf{k}) = 2\epsilon^2(\mathbf{k}),$$
$$\left(E_{i\sigma}(\mathbf{k}) + \frac{1}{2}\epsilon(\mathbf{k})\right)^2 = \frac{9}{4}\epsilon^2(\mathbf{k})$$

Then we get two quasiparticle energies:

$$E_{1\sigma}(\mathbf{k}) = -2\epsilon(\mathbf{k}); \quad E_{2\sigma}(\mathbf{k}) = \epsilon(\mathbf{k}).$$

Spectral weights (B.162);

$$\alpha_{i\sigma}(\mathbf{k}) = \left|1 - \frac{\partial}{\partial E} R_\sigma(\mathbf{k}, E)\right|^{-1}_{E=E_{i\sigma}} = \left|1 + 2\frac{E_{i\sigma}(\mathbf{k})}{\epsilon(\mathbf{k})}\right|^{-1}$$

$$\Longrightarrow \alpha_{1\sigma}(\mathbf{k}) = \alpha_{2\sigma}(\mathbf{k}) = \frac{1}{3}$$

Lifetimes:

$$I_\sigma(\mathbf{k}, E_{1\sigma}(\mathbf{k})) = -2\gamma|\epsilon(\mathbf{k})| = I_{1\sigma}(\mathbf{k}),$$
$$I_\sigma(\mathbf{k}, E_{2\sigma}(\mathbf{k})) = -\gamma|\epsilon(\mathbf{k})| = I_{2\sigma}(\mathbf{k})$$
$$\Longrightarrow \tau_{1\sigma}(\mathbf{k}) = \frac{3\hbar}{2\gamma|\epsilon(\mathbf{k})|}; \quad \tau_{2\sigma}(\mathbf{k}) = \frac{3\hbar}{\gamma|\epsilon(\mathbf{k})}.$$

3. Quasi particle concept is applicable provided

$$|I_\sigma(\mathbf{k}, E)| \ll |\epsilon(\mathbf{k}) - \mu + R_\sigma(\mathbf{k}, E)|$$
$$\Longleftrightarrow |I_\sigma(\mathbf{k}, E_{i\sigma})| \ll |E_{i\sigma}(\mathbf{k})|$$
$$\Longleftrightarrow \gamma|E_{i\sigma}(\mathbf{k})| \ll |E_{i\sigma}(\mathbf{k})|$$
$$\Longleftrightarrow \gamma \ll 1$$

4.

$$\left(\frac{\partial R_\sigma(\mathbf{k}, E)}{\partial E}\right)_{\epsilon(\mathbf{k})} = -\frac{2E}{\epsilon(\mathbf{k})}$$
$$\left(\frac{\partial R_\sigma(\mathbf{k}, E)}{\partial \epsilon(\mathbf{k})}\right)_E = 1 + \frac{E^2}{\epsilon^2(\mathbf{k})}$$
$$\Longrightarrow m^*_{1\sigma}(\mathbf{k}) = m\frac{1-4}{1+5} = -\frac{1}{2}m,$$
$$m^*_{2\sigma}(\mathbf{k}) = m\frac{1+2}{1+2} = m$$

Problem B.13

The self-energy is real and **k** independent. Then with (B.192),

$$\rho_\sigma(E) = \rho_0(E - \Sigma_\sigma(E-\mu)) = \rho_0\left(E - a_\sigma\frac{E - b_\sigma}{E - c_\sigma}\right)$$

Lower band edges:

$$\begin{aligned}
&0 \overset{!}{=} E - a_\sigma \frac{E - b_\sigma}{E - c_\sigma}\\
\Longleftrightarrow\ &0 = E^2 - (a_\sigma + c_\sigma)E + a_\sigma b_\sigma\\
&= \left(E - \frac{1}{2}(a_\sigma + c_\sigma)\right)^2 + a_\sigma b_\sigma - \frac{1}{4}(a_\sigma + c_\sigma)^2\\
\Longrightarrow\ &E^{(l)}_{1,2\sigma} = \frac{1}{2}\left(a_\sigma + c_\sigma \mp \sqrt{(a_\sigma + c_\sigma)^2 - 4a_\sigma b_\sigma}\right)
\end{aligned}$$

Upper band edges:

$$\begin{aligned}
W &\overset{!}{=} E - a_\sigma \frac{E - b_\sigma}{E - c_\sigma}\\
\Longleftrightarrow\ -c_\sigma W &= E^2 - (a_\sigma + c_\sigma + W)E + a_\sigma b_\sigma\\
0 &= \left(E - \frac{1}{2}(a_\sigma + c_\sigma + W)\right)^2 + (a_\sigma b_\sigma + c_\sigma W)\\
&\qquad - \frac{1}{4}(a_\sigma + c_\sigma + W)^2\\
&\Longrightarrow
\end{aligned}$$

$$E^{(u)}_{1,2\sigma} = \frac{1}{2}\left(a_\sigma + c_\sigma + W \mp \sqrt{(a_\sigma + c_\sigma + W)^2 - 4(a_\sigma b_\sigma + c_\sigma W)}\right)$$

Quasi particle density of states:

$$\rho_\sigma(E) = \begin{cases} \frac{1}{W}, & \text{falls } E^{(u)}_{1\sigma} \leq E \leq E^{(o)}_{1\sigma}\\ \frac{1}{W}, & \text{falls } E^{(u)}_{2\sigma} \leq E \leq E^{(o)}_{2\sigma}\\ 0, & \text{otherwise} \end{cases}$$

Band splitting into two quasi particle sub-bands.

Index

T

GPSR Compliance
The European Union's (EU) General Product Safety Regulation (GPSR) is a set of rules that requires consumer products to be safe and our obligations to ensure this.

If you have any concerns about our products, you can contact us on

ProductSafety@springernature.com

In case Publisher is established outside the EU, the EU authorized representative is:

Springer Nature Customer Service Center GmbH
Europaplatz 3
69115 Heidelberg, Germany

www.ingramcontent.com/pod-product-compliance
Lightning Source LLC
Chambersburg PA
CBHW071700180826
50031CB00003B/74
9783540854159